Tables
From *Beginning Statistics*

TABLE B – Standard Normal Distribution

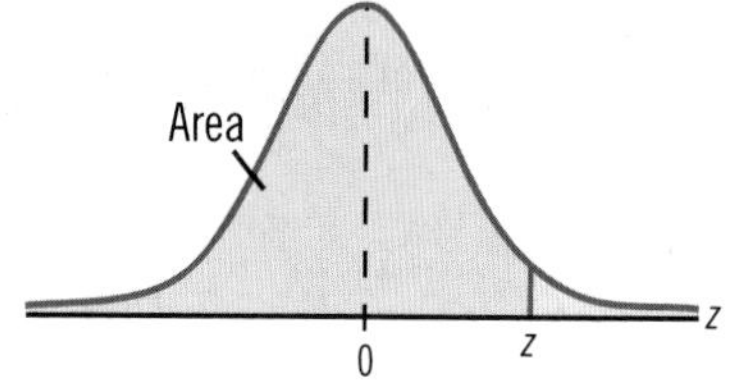

z	0.00	0.01	0.02	0.03	0.04	0.05	0.06	0.07	0.08	0.09
0.0	0.5000	0.5040	0.5080	0.5120	0.5160	0.5199	0.5239	0.5279	0.5319	0.5359
0.1	0.5398	0.5438	0.5478	0.5517	0.5557	0.5596	0.5636	0.5675	0.5714	0.5753
0.2	0.5793	0.5832	0.5871	0.5910	0.5948	0.5987	0.6026	0.6064	0.6103	0.6141
0.3	0.6179	0.6217	0.6255	0.6293	0.6331	0.6368	0.6406	0.6443	0.6480	0.6517
0.4	0.6554	0.6591	0.6628	0.6664	0.6700	0.6736	0.6772	0.6808	0.6844	0.6879
0.5	0.6915	0.6950	0.6985	0.7019	0.7054	0.7088	0.7123	0.7157	0.7190	0.7224
0.6	0.7257	0.7291	0.7324	0.7357	0.7389	0.7422	0.7454	0.7486	0.7517	0.7549
0.7	0.7580	0.7611	0.7642	0.7673	0.7704	0.7734	0.7764	0.7794	0.7823	0.7852
0.8	0.7881	0.7910	0.7939	0.7967	0.7995	0.8023	0.8051	0.8078	0.8106	0.8133
0.9	0.8159	0.8186	0.8212	0.8238	0.8264	0.8289	0.8315	0.8340	0.8365	0.8389
1.0	0.8413	0.8438	0.8461	0.8485	0.8508	0.8531	0.8554	0.8577	0.8599	0.8621
1.1	0.8643	0.8665	0.8686	0.8708	0.8729	0.8749	0.8770	0.8790	0.8810	0.8830
1.2	0.8849	0.8869	0.8888	0.8907	0.8925	0.8944	0.8962	0.8980	0.8997	0.9015
1.3	0.9032	0.9049	0.9066	0.9082	0.9099	0.9115	0.9131	0.9147	0.9162	0.9177
1.4	0.9192	0.9207	0.9222	0.9236	0.9251	0.9265	0.9279	0.9292	0.9306	0.9319
1.5	0.9332	0.9345	0.9357	0.9370	0.9382	0.9394	0.9406	0.9418	0.9429	0.9441
1.6	0.9452	0.9463	0.9474	0.9484	0.9495	0.9505	0.9515	0.9525	0.9535	0.9545
1.7	0.9554	0.9564	0.9573	0.9582	0.9591	0.9599	0.9608	0.9616	0.9625	0.9633
1.8	0.9641	0.9649	0.9656	0.9664	0.9671	0.9678	0.9686	0.9693	0.9699	0.9706
1.9	0.9713	0.9719	0.9726	0.9732	0.9738	0.9744	0.9750	0.9756	0.9761	0.9767
2.0	0.9772	0.9778	0.9783	0.9788	0.9793	0.9798	0.9803	0.9808	0.9812	0.9817
2.1	0.9821	0.9826	0.9830	0.9834	0.9838	0.9842	0.9846	0.9850	0.9854	0.9857
2.2	0.9861	0.9864	0.9868	0.9871	0.9875	0.9878	0.9881	0.9884	0.9887	0.9890
2.3	0.9893	0.9896	0.9898	0.9901	0.9904	0.9906	0.9909	0.9911	0.9913	0.9916
2.4	0.9918	0.9920	0.9922	0.9925	0.9927	0.9929	0.9931	0.9932	0.9934	0.9936
2.5	0.9938	0.9940	0.9941	0.9943	0.9945	0.9946	0.9948	0.9949	0.9951	0.9952
2.6	0.9953	0.9955	0.9956	0.9957	0.9959	0.9960	0.9961	0.9962	0.9963	0.9964
2.7	0.9965	0.9966	0.9967	0.9968	0.9969	0.9970	0.9971	0.9972	0.9973	0.9974
2.8	0.9974	0.9975	0.9976	0.9977	0.9977	0.9978	0.9979	0.9979	0.9980	0.9981
2.9	0.9981	0.9982	0.9982	0.9983	0.9984	0.9984	0.9985	0.9985	0.9986	0.9986
3.0	0.9987	0.9987	0.9987	0.9988	0.9988	0.9989	0.9989	0.9989	0.9990	0.9990
3.1	0.9990	0.9991	0.9991	0.9991	0.9992	0.9992	0.9992	0.9992	0.9993	0.9993
3.2	0.9993	0.9993	0.9994	0.9994	0.9994	0.9994	0.9994	0.9995	0.9995	0.9995
3.3	0.9995	0.9995	0.9995	0.9996	0.9996	0.9996	0.9996	0.9996	0.9996	0.9997
3.4	0.9997	0.9997	0.9997	0.9997	0.9997	0.9997	0.9997	0.9997	0.9997	0.9998

Critical Values

Level of Confidence	$z_{\alpha/2}$
0.80	1.28
0.85	1.44
0.90	1.645
0.95	1.96
0.98	2.33
0.99	2.575

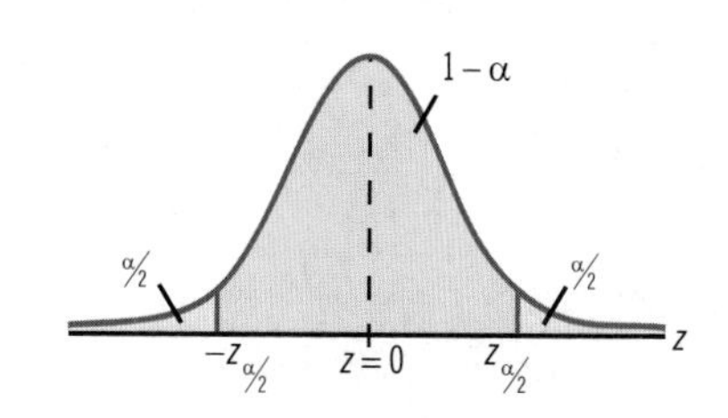

Beginning Statistics

Carolyn Warren
Kimberly Denley
Emily Atchley

Editor: Kimberly Scott
Project Director: Marcel Prevuznak
Production Editors: Harding Brumby, Phillip Bushkar, Cynthia Ellison, Amanda Glover, Bethany Loftis, Nina Miller
Editorial Assistants: Greg Hill, Susan Rackley, Kelly Epperson, Chris Yount, Brian Smith, Tristan Vogler
Art: Ayvin Samonte

1023 Wappoo Rd. Suite A-6/Charleston, SC 29407
www.hawkeslearning.com

Library of Congress Control Number: 2006920454

Printed in the United States of America

ISBN (Student): 978-1-932628-11-1

Software Bundle: 978-1-932628-12-8

CONTENTS

CHAPTER 5 ✦ Probability Distribution 200

CHAPTER 6 ✦ Continuous Random Variables 246

CHAPTER 7 ✦ Samples and Sampling Distributions 298

CHAPTER 8 ✦ Confidence Intervals 344

CHAPTER 9 ✦ Confidence Intervals for Two Samples 396

CHAPTER 10 ✦ Hypothesis Testing 466

CHAPTER 11 ✦ Hypothesis Testing (Two or More Populations) 572

Preface

Purpose and Style

Our goal for this text is to educate liberal arts students in the basics of statistics. At the same time, we aim to give students in other majors, such as business, education and psychology, a solid foundation for future statistics classes. To this end, we have tried to create a textbook that incorporates basic statistical calculations with heavy doses of the analytical reasoning students need to understand research in their field of study. As all teachers do, we want students to read the textbook; therefore, the writing style is relaxed and speaks directly to the reader, asking them to interact and verify various calculations as they read through the text. The writing style is informal, yet mathematically accurate without being difficult for the "non-math" people among us to understand. We have tried to include interesting, and at times humorous, examples and comments to encourage students to keep reading. We firmly believe that math textbooks do not have to be odious to read, and we hope that this one is not. In addition, we have included comments in the margin, called Memory Boosters, that point out key details and helpful hints for learning the material.

As we stated before, our main goal is for students to become comfortable interpreting the results of research and the statistical calculations used in that research. To further help, we encourage students to use available technology, specifically graphing calculators, to perform the statistical calculations so that they can focus on the meaning of their results. However, we have placed most of the technology discussion at the end of each chapter to allow instructors to choose between performing calculations manually or with the aid of technology. In addition, this textbook integrates nicely with the ***Hawkes Learning Systems: Statistics*** software package.

Lastly, we have incorporated a hands-on approach to statistics. We know from experience that the best way to learn and retain new information is to use that information. Thus hands-on projects are included at the end of each chapter to enable students to synthesize the material in each chapter. These projects require students to write-up their results in a formal manner, thus introducing students to communicating their own research findings. Students will then be better prepared, and more willing, to conduct research in their own field of study.

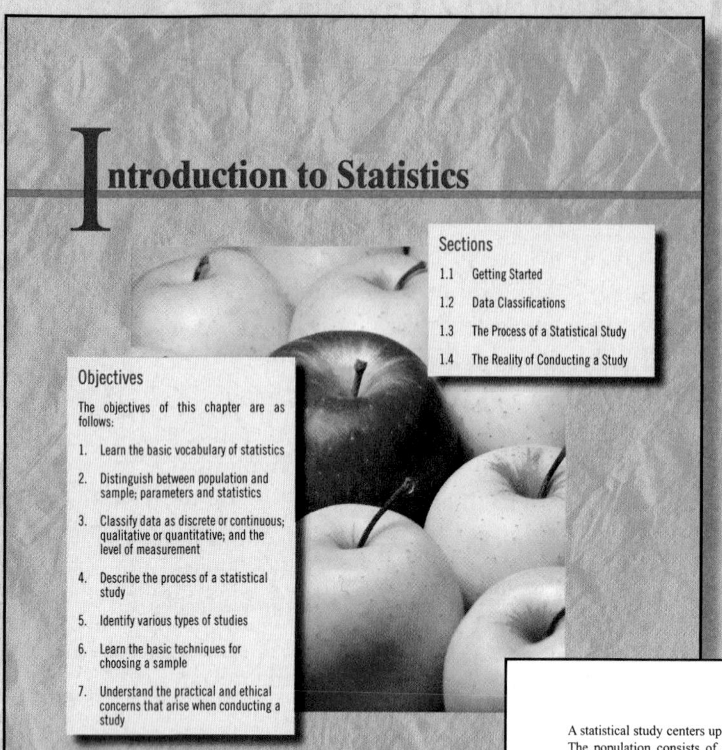

Introduction to Statistics

Sections

1.1 Getting Started
1.2 Data Classifications
1.3 The Process of a Statistical Study
1.4 The Reality of Conducting a Study

Objectives

The objectives of this chapter are as follows:

1. Learn the basic vocabulary of statistics
2. Distinguish between population and sample; parameters and statistics
3. Classify data as discrete or continuous; qualitative or quantitative; and the level of measurement
4. Describe the process of a statistical study
5. Identify various types of studies
6. Learn the basic techniques for choosing a sample
7. Understand the practical and ethical concerns that arise when conducting a study

Ch

Chapter Opener ✦

Each chapter begins with a list of sections and a list of objectives to prepare the students for the topics about to be covered.

Memory Booster ✦

Where applicable, memory boosters are placed in the margins to give students helpful hints to remember important topics.

Section 1.1 ✦ Getting Started

A statistical study centers upon a particular group of interest called a **population**. The population consists of all persons or things being studied. For example, a software company may wish to know the percentage of women that own a personal computer. In this case, the population is all women. Perhaps, instead, an instructor wants to know the number of statistics students with red hair. In this second example, the population to be studied is all statistics students.

When studying a population, the information gathered is called **data** (The singular form of data is datum.), and it may be in a variety of forms including counts, measurements, or observations. If the population is small enough, data may be obtained from every member. This is called a **census**. You can easily remember this term by thinking of the U.S. census that attempts to count each and every American.

Memory Booster!

Population
Parameter

Sample
Statistic

Notice that the first letters match!

The numerical description of a particular population characteristic is called a **parameter**. That is, the numbers that describe a population are called parameters. Suppose it is true that 65% of all Americans own a computer. Then 65% is a population parameter. An important note is that although the parameter is a fixed number, often it is impossible or impractical to determine it precisely. For example, suppose that the population for a study includes all Americans. Would it be possible to collect data from EVERY American? Probably not! Because of human limitations, usually the best we can do is to estimate a population's true parameter.

When the parameter cannot be determined, a subset of the population, called a **sample**, is chosen. Information is then collected only from the sample. Care must be taken in choosing your sample so that the population is represented well, and the results of the study are meaningful. (We will talk about how to choose a representative sample shortly.) A number that describes a characteristic of a sample is called a **sample statistic**. For example, suppose that a research group wants to know the percentage of college students that do laundry on a weekly basis. If 100 college students are chosen for the study, and 42% do laundry on a weekly basis, then the population would be all college students, and the 100 students chosen are the sample. The sample statistic would be 42% of students.

Definitions ✦

Data is information, in particular, information prepared for a study.
Population is the particular group of interest.
Parameter is a numerical description of a particular population characteristic.
Sample is a subset of the population from which data are collected.
Sample Statistics are the actual numerical descriptions of a particular sample characteristic.

5

Definitions & Properties ✦

Definitions and properties are presented in highly visible boxes for easy reference.

Section 3.1 ✦ Measures of Center

This formula is actually the **sample mean** because we are calculating the mean of a set of sample values. If we are calculating the mean of all the values in a population, we call it the **population mean**, μ. The procedure is the same: divide the sum of all the population values by the number of values in the population, N. The formula is written in a slightly different manner,

$$\mu = \frac{x_1 + x_2 + \cdots + x_N}{N} = \frac{\sum x_i}{N}$$

where x_i is the ith value in the population, and N is the number of values in the population.

Note that the mean might not be a value in the data set.

Math Symbols!

μ, population mean; Greek symbol mu

$\bar{x}$, sample mean; read as "x-bar"

Example 1

Calculate the mean of the following sample data.

5, 7, 8, 10, 4, 3, 12

Solution:

We have values from a sample, so we are calculating the sample mean.

$$\bar{x} = \frac{5+7+8+10+4+3+12}{7} = 7.0$$

Rounding Rule

When calculating the mean, round to one more decimal place than is given.

For certain data sets, the formula above for calculating a mean value is not sufficient. For example, suppose that Walter wants to calculate his overall average in his U.S. History course. If the course syllabus states that the final grade is determined by tests (40%), homework (20%), quizzes (10%) and final exam (30%), how should Walter calculate his grade?

When each data value in the set does not hold the same relative importance, a **weighted mean** must be used. To calculate a weighted mean for a sample, first multiply each value by its respective weight. Then divide the sum of these products by the sum of the weights to obtain the mean. The mathematical formula is given as follows:

Math Symbols

This feature, found in the margins, showcases specific symbols used in statistics.

Rounding Rules

Throughout the book rounding rules are given in the margins to help students learn the standardized rounding used in statistics.

Examples

Carefully explained examples appear throughout the text, giving clarification to the skills being presented and making use of tables, diagrams, and graphs.

Chapter 3 ✦ Numerical Descriptions of Data

Example 7

Determine which letter represents the mean, the median, and the mode in the graph below.

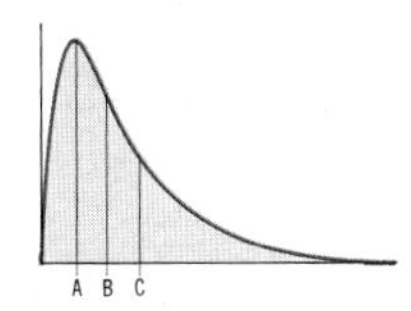

Solution:

Mean: This distribution is skewed to the right, so the mean will be the measure of center farthest to the right. Here it is represented by line C.

Median: The median is the value that divides the area of the distribution in half. Here it is represented by line B.

Mode: The mode is always located at the peak of a distribution, so it is line A.

✦ Exercises 3.1

Directions: Calculate the mean, median, and mode for the following data sets, and when applicable state whether the data set is unimodal, bimodal, or multimodal. Remember to round the mean to one decimal place more than is given in the data set.

1. 8, 9, 10, 12, 4, 3, 5, 7, 10
2. 5, 5, 5, 5, 5, 5, 5
3. 5, 6, 3, 5, 2, 7, 8, 35
4. 17, 19, 15, 16, 20, 19, 17, 3
5. 1, 2, 3, 4, 5, 7, 8, 11, 12, 15

110

Section Exercises

Each section concludes with a selection of exercises designed to allow the student to practice skills and master concepts.

Chapter Review

Section 1.1 - Getting Started	Pages
• **Definitions** **Data** – information, in particular, information prepared for a study **Population** – the particular group of interest **Parameter** – a numerical description of a particular population characteristic **Sample** – a subset of the population from which data are collected **Sample Statistics** – the actual numerical descriptions of a particular sample characteristic **Census** – data which are obtained from every member of the population	p. 5
• Population values are called parameters.	p. 5
• Sample values are called statistics.	p. 5
• **Branches of Statistics** **1. Descriptive Statistics** – gathers, sorts, summarizes, and displays the data **2. Inferential Statistics** – uses descriptive statistics to estimate population parameters	p. 8
Section 1.2 - Data Classification	**Pages**
• **Definitions** **Qualitative Data** – consists of labels or descriptions of traits **Quantitative Data** – consists of counts or measurements **Continous Data** – can take on any value in a given interval, are usually measurements **Discrete Data** – can take on only particular values, are usually counts	p. 12–13
• **Levels of Measurement** **1. Nominal** – consists of labels or names **2. Ordinal** – data can be arranged in a meaningful order, but calculations suc as addition and division do not make sense **3. Interval** – data can be arranged in a meaningful order and difference between entries are meaningful **4. Ratio** – data can be arranged and differences between entries are meaningfu and the zero point indicates the absence of something	p. 13–17
Section 1.3 - The Process of a Statistical Study	
• **Definitions** **Observational Study** – observes data that already exists **Representative Sample** – has the same relevant characteristics as the populatio and does not favor one group of the population over another	

Chapter Review ✦

These sections assist students in reviewing the central ideas, terms, and skills presented in each chapter.

Chapter Exercises

1. Eighty-nine percent of Americans polled are in favor of arts education being taught in public schools.
  a. Does the above numerical value describe a parameter or a statistic?
  b. After reading the above statement, you decide to campaign for the arts in education and use the statement: "A majority of Americans are in favor of art education which shows their love of arts. Americans support ballet teaching in public schools." What branch of statistics would you be applying?

2. The average life-span of Americans in 1990 was 75.37 years.
  a. Does the above numerical value describe a parameter or a statistic?
  b. Giving a complete breakdown of the above statement into categories of gender, sex and geographical regions would be an example of which branch of statistics?

3. Determine whether the following data is qualitative or quantitative and classify it according to its level of measurement.
  a. The street addresses of houses along North Lamar Avenue.
  b. The years in which Christmas falls on a Sunday.
  c. The monthly rates for digital cable service for five competing companies.

4. Jake wanted to know the most popular brand of cereal amongst teenagers. He stood in the center of the cereal aisle at the local grocery and surveyed every 10th teenager who approached him.
  a. How many biases can you name?
  b. In what ways could Jake better collect data to answer his question?

5. If boys are assigned odd-numbered rooms and an equal number of girls are assigned even-numbered rooms in a dorm, is choosing every 10th room a representative sample? What about every 5th?

6. If someone is looking at what proportion of students are left-handed, is a current statistics class a good sample? Is this class a good indicator of the male to female ratio on a campus?

7. A supermarket wants to decide whether or not to carry a new brand of salsa, and uses a free taste test stand in the store.
  a. What is the population being studied?
  b. What type of sampling technique is being used?
  c. Is the taste test representative of the population?

8. It is not always possible to obtain a sampling frame for a given population. Give one example of a population that has a sampling frame and one that does not.

Chapter Exercises

Chapter exercises are organized by sets which correlate to specific sections in the text. Actual, sourced data is used whenever possible and presented using easy-to-read graphs and tables. Problems make use of intriguing topics and are accented by colorful graphics.

Chapter Project

Suitable for individual work, group projects, or class discussion, the Chapter Project offers focused, in-depth questions regarding a single topic.

Chapter 1 ✦ Introduction to Statistics

Chapter Project: A – Analyzing a Given Study

Poker on the Rise with Young Adults

A mail survey conducted by the American Gaming Association in 2004, showed that 54.1 million American adults had made trips to a casino that year. They averaged 5.9 trips per person. The survey also showed that these people were more affluent than the average American adult with a household income of $55,500 versus the average $47,000. They also had a slightly higher education with 55% having some college versus 53% of average Americans. This has been the norm over the past few years. What has become a surprise is the amount of young adults who play poker. 18% of the young adults surveyed said they played poker in the past 12 months, up from 12% in 2003. Of those, 29% of young adults, ages 21-39, played poker which is an increase of 7% from the previous year. This was followed by 16% of adults age 40-49, 14% of adults age 50-65, and 12% of those 65 and over.

Source: The Detroit News Entertainment Insider, July 14, 2005

Analyze the above survey using the following questions as a guide.

1. Identify the population being studied.
2. Identify the sample and the approximate sample size. Is the sample representative of the population? Explain your answer.
3. Describe how the sample was chosen. Is there any potential bias in the sampling method? Explain your answer.
4. List the descriptive statistics given in the article.
5. What inferences does the article make from the descriptive statistics?
6. Who conducted the study? Is there any potential researcher bias? Explain your answer.

38

Technology

All chapters conclude with a special technology section which makes use of the TI-84 Plus calculator, Microsoft® Excel, and Minitab®. Precise instructions are given and screen shots are provided for each program used.

Chapter 1 ✦ Introduction to Statistics

Chapter Technology – Generating Random Numbers

TI-84 Plus

To generate a random number using a TI-84 Plus calculator, use the RANDINT function under the MATH - PRB (probability) menu. The function syntax is **randInt(*lower, upper, number of trials*)**. This will produce the specified number of random integers between the lower and upper values. If you only want 1 random number, the formula can be modified to **randInt(*lower, upper*)**.

For example, if we want to get 7 random numbers between 10 and 25, we would press the MATH button, then scroll over to PRB and then scroll down to option 5, 'randInt(' and press ENTER. Type in '10, 25, 7)'. Press ENTER. The calculator will then produce a string of 7 random whole numbers between 10 and 25.

Example T.1

Use your TI-84 Plus calculator to create a random sample of 4 integers between 2 and 9.

Solution:

Press the MATH button, then scroll over to PRB, choose option 5, 'randInt(', and press ENTER. Type in '2, 9, 4)'. Press ENTER and you will now have something similar to the screen on the right.

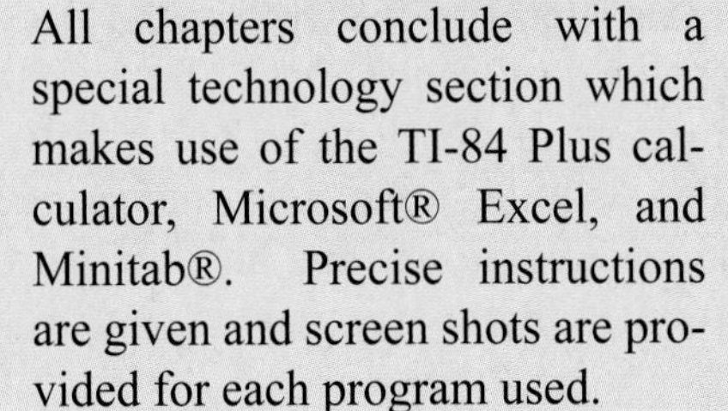

This screenshot shows a random sample of 7, 8, 3, and 7. Your random sample will obviously be different.

Duplicate numbers are a possibility, as shown in our example. If distinct numbers are required, you might need to increase the number of random digits you ask the calculator to generate.

MICROSOFT EXCEL

Creating a random number in Microsoft Excel is very easy. You can use the function command =RANDBETWEEN(*lower, upper*), which returns a random number between the numbers you specify. For example, if we want to generate a random number between zero and nine, we would type in the formula =RANDBETWEEN(0, 9) in one of the cells in the worksheet. We used this formula to create the random digits table in Appendix A.

NOTE: If RANDBETWEEN does not appear in your Excel menu, you can easily add this function. Under TOOLS, click on Add-Ins. Select Analysis Toolpack. When finished using this command, it might be a good idea to remove the Analysis Toolpack as it slows Excel down.

40

Content

This textbook assumes no previous knowledge of statistics. Furthermore, it includes enough topics ordered in such a way that it can be used in an introductory liberal arts course or a more rigorous business or psychology statistics course. We would like to make a few comments on specific content. First, the overall layout of the book follows the order of our definition of statistics – the science of gathering, organizing, and analyzing data. In Chapter 1 we discuss the details of how to gather data. In Chapters 2 – 3, we explore different ways of organizing data. The remainder of the book explores, in detail, numerous ways to analyze data, such as estimating population parameters, testing hypotheses, and describing relationships between data.

CHAPTER 1 – Introduction to Statistics

In this chapter we introduce students to basic statistical vocabulary and the process of a statistical study. We want students to view newspaper articles and research reports critically, and not take them at face value; therefore we specifically point out common problems and weaknesses with different sampling methods and types of studies, as well as emphasizing under what circumstances to use which techniques. We devote an entire section of this chapter to the practical and ethical concerns involved in conducting a study. Our goal here is to heighten students' awareness of possible problems in the research they read as well as to prepare them to think about these issues in their own research.

CHAPTER 2 – Graphical Descriptions of Data

We begin our discussion of organizing data with graphical representations of data, as this is probably more familiar to students than the numerical descriptions of data calculated in the next chapter. We specifically include a section on analyzing graphs, to encourage students to think critically about the information they receive.

CHAPTER 3 – Numerical Descriptions of Data

In this chapter, we discuss calculating numerical descriptions of center, dispersion, and position. We demonstrate calculations by hand, although we encourage using technology and reference the detailed directions for using various technologies presented at the end of the chapter. This allows for different approaches to this topic. One interesting point we ran into while researching this chapter is how to calculate percentiles and quartiles. We have tried to be as accurate as possible while pointing out that there is not always agreement on how to calculate these values. We have intentionally used the word "average" instead of the word "mean" in various places throughout the text. We want students to realize that the word "average" is the word used in everyday speech to indicate the mathematical concept of mean, but to also be aware of the fact that there are other mathematical calculations (median, mode) that can be expressed in laymen's terms as an average. We spend an entire subsection dealing with this topic. Our students are not statisticians, but nurses, teachers, journalists, social workers, etc. They will use and hear the word "average." We want them to equate that with the word mean, even when that is not what is explicitly said.

CHAPTER 4 – Probability, Randomness, and Uncertainty

Our discussion of probability begins with terminology and simple probability calculations. We also include a discussion of the various types of probability. We introduce counting techniques as a means of counting the number of outcomes in the sample space or event for more complicated probability calculations. We end the chapter with more advanced counting techniques. This last section could easily be omitted for those courses not needing this more advanced discussion.

CHAPTER 5 – Probability Distributions

This chapter continues our discussion of probability, but focuses on probability distributions. We discuss in detail the binomial, Poisson, and hypergeometric distributions. We chose to put these topics in a separate chapter, as we believe that the grouping together of these probability distributions into a single unit helps point out their similarities and differences and thus helps clarify the concepts of this chapter.

CHAPTER 6 – Continuous Random Variables

In this chapter, we discuss two important probability distributions – the normal distribution and the Student *t*-distribution. These two distributions form the backbone for the rest of the text, so we spend an entire chapter describing these distributions and working with finding probabilities using the normal distribution table and the Critical *t*-value table.

CHAPTER 7 – Sampling Distributions

In this chapter we continue our discussion of the normal distribution with a discussion of sampling distributions and the Central Limit Theorem. So much of statistics is based on this important theorem; we certainly feel that it deserves extensive treatment. In addition, word problems scare introductory students enough by themselves. They don't need word problems added into a chapter where they have to learn how to use confusing tables or software as well. We think it is pedagogically best to keep these two concepts separate. Furthermore, some textbooks have begun placing the discussion of population proportions before the discussion of population means. One of the reasons behind this change is that students see proportions more often than means. Though we agree that students are more comfortable with the concept of proportions, they are certainly not more comfortable with the more complicated formulas necessary to work with proportions. Therefore, we chose to present means first.

CHAPTER 8 – Confidence Intervals

We introduce confidence intervals for population means, proportions, and variance in this chapter. We emphasize to students that the confidence interval is not guaranteed to contain the population parameter. We introduce the χ^2 distribution in this chapter as well. One important note about this chapter is the way in which we decided to present the material in regards to the population standard deviation being known or unknown. We are of the opinion that if you are trying to estimate an unknown parameter such as the population mean, then it is very unlikely that the population standard deviation would be known either. We therefore chose to include only the methods and formulas for situations in which σ is unknown. We continued this same practice throughout the rest of the text where applicable.

CHAPTER 9 – Confidence Intervals for Two Samples

We continue our discussion of confidence intervals by extending our discussion to include comparing two population means, proportions, or variances. We split this chapter off from hypothesis testing for two populations so that the discussion of confidence intervals would not be interrupted by a discussion of hypothesis testing.

CHAPTER 10 – Hypothesis Testing

We begin this chapter with an introduction to the fundamental concepts behind performing a hypothesis test. We discuss setting up the hypotheses, the burden of proof, the benefit of the doubt, and drawing conclusions. We use the illustration of the U.S. legal system throughout the first section to tie this concept to something with which students are already familiar. We do not discuss types of errors or their relationship to the level of significance in the first section, as we do not want students thinking their answers could be wrong before they have even performed the first hypothesis test. In this chapter, we reversed the presentation order of small samples and large samples for population means. We discuss rejection regions using *t*-tests first so that students can visually see how we are drawing the conclusion. Then when moving on to large sample sizes, *z*-test statistics, we can use the more common method of comparing the *p*-value to the level of significance. We do include directions on using technology to calculate *p*-values for all situations at the end of the chapter.

CHAPTER 11 – Hypothesis Testing (Two or More Populations)

Just as we discussed confidence intervals for one population, then moved on to discuss two populations, we do the same for hypothesis testing. Similarly, we discuss at the end of the chapter all of the technology a student would need to perform these hypothesis tests more efficiently. We do, however, incorporate technology into the section on the ANOVA test, as calculating the necessary values for this test by hand would be tedious at best.

CHAPTER 12 – Regression, Inference, and Model Building

In this chapter, we discuss linear regression and regression models. We also emphasize when it is appropriate to use these models. At the end of the section, we tie linear regression to confidence intervals and hypothesis testing, using these techniques to discuss the appropriateness of the linear regression model for a given set of data. As the calculations in these sections would never be done by hand in today's technological society, we incorporate the necessary technology throughout this chapter.

Hawkes Learning Systems: Statistics

Overview

This multimedia courseware allows students to become better problem-solvers by creating a mastery level of learning in the classroom. The software includes an "Instruct," "Practice," "Tutor," and "Certify" mode in each lesson, allowing students to learn through step-by-step interactions with the software. This automated homework system's tutorial and assessment modes extend instructional influence beyond the classroom. Artificial intelligence is what makes the tutorials so unique. By offering intelligent tutoring and mastery level testing to measure what has been learned, the software extends the instructor's ability to influence students to solve problems. This courseware can be ordered either separately or bundled together with this text.

Minimum Requirements

In order to run *HLS: Statistics,* you will need:

Windows® 2000 or later
800 MHz or faster processor
128 MB RAM
300 MB hard drive space
16-bit color display (800 x 600 resolution)
Internet Explorer 6.0 or later
CD-ROM drive

Getting Started

Before you can run *HLS: Statistics*, you will need an access code. This 30 character code is your personal access code. To obtain an access code, go to http://www.hawkeslearning.com and follow the links to the access code request page (unless directed otherwise by your instructor.)

Installation

Insert the ***HLS: Statistics*** installation CD-ROM into the CD-ROM drive. Select the Start/Run command, type in the CD-ROM drive letter followed by \setup.exe. (For example, d:\setup.exe where d is the CD-ROM drive letter.)

After selecting the desired installation option, follow the on-screen instructions to complete your installation of ***HLS: Statistics***.

Starting the Courseware

After you install ***HLS: Statistics*** on your computer, to run the courseware select Start/Programs/Hawkes Learning Systems/Statistics.

You will be prompted to enter your access code with a message box similar to the following:

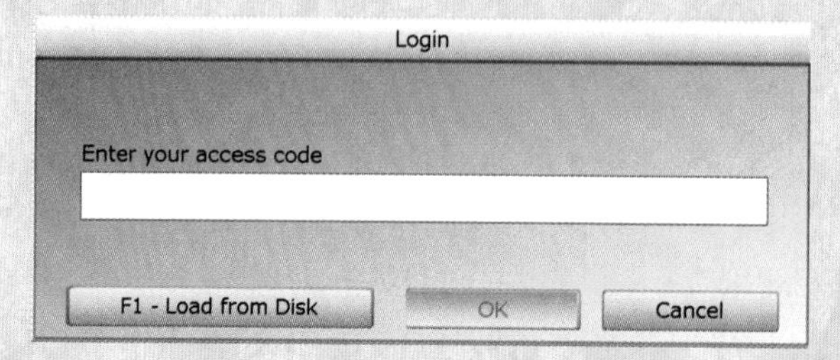

Type your access code into the box provided. When you are finished, press OK.

If you typed in your access code correctly, you will be prompted to save the code to disk. If you choose to save your code to disk, typing in the access code each time you run ***HLS: Statistics*** will not be necessary. Instead, select the F1 - Load from Disk button when prompted to enter your access code and choose the path to your saved access code.

Now that you have entered your access code and saved it to disk, you are ready to run a lesson. From the table of contents screen, choose the appropriate chapter and then choose the lesson you wish to run.

Features

Each lesson in *HLS: Statistics* has four modes: Instruct, Practice, Tutor, and Certify.

Instruct: Instruct provides a multimedia presentation of the material covered in the lesson. Instruct offers example problems, animation, and helpful tips to enhance the learning experience.

Practice: Practice allows you to hone your problem-solving skills. It provides an unlimited number of randomly generated problems. Practice also provides access to the Tutor mode by selecting the Tutor button located next to the Submit button.

Tutor: Tutor mode is broken up into several parts: Instruct, Explain Error, Hint, and Solution.

1. Instruct, which can also be selected directly from Practice mode, contains a multimedia lecture of the material covered in a lesson.
2. Explain Error is active whenever a problem is incorrectly answered. It will attempt to explain the error that caused you to incorrectly answer the problem.
3. Hint provides a hint to assist in solution of the problem.
4. Solution will provide you with a detailed "worked-out" solution to the problem.

Throughout the Tutor, you will see words or phrases colored green with a dashed underline. These are called Hot Words. Clicking on a Hot Word will provide you with more information on these word(s) or phrases.

Certify: Certify is the testing mode. You are given a finite number of problems and a certain number of strikes (problems you can get wrong). If you correctly answer the required number of questions, you will receive a certification code and a certificate. Write down your certification code and/or print out your certificate. The certification code will be used by your instructor to update your records. Note that the Tutor is not available in Certify.

Keypad in Practice and Certify.

A keypad is also available in the Practice and Certify modes to allow you to enter symbols, fractions, exponents, and mathematical characters that cannot be easily typed.

Support

If you have questions about ***HLS: Statistics*** or are having technical difficulties, we can be contacted as follows:

Phone: (843) 571-2825
Email: support@hawkeslearning.com
Web: www.hawkeslearning.com

Our support hours are 8:30 am to 5:30 pm, Eastern Time, Monday through Friday.

Acknowledgements

As a group, we would first like to thank Dr. Tristan Denley, Dr. Sunil Mathur, Dr. Hanxiang Peng, Dr. Xin Dang, Dr. Laura Sheppardson, Dr. Pam Smith, and the many others at the University of Mississippi who contributed to this project. Your knowledge and wise council have been invaluable to us throughout the entire text. We count ourselves blessed to work with such resourceful people.

Special thanks to our reviewers:

- Brenda Cates – Mount Olive College
- George Hilton – Pacific Union College
- Paul Holmes – University of Georgia
- Marcel Maupin – Oklahoma State University, Oklahoma City
- Lindsay Packer – Metropolitan State College of Denver
- George Perkins – Prince George's Community College
- Deanna Voehl – Indian River Community College, Main Campus

The text was much improved by your remarks and suggestions, and we are most grateful for your input.

Thank you to all of the staff at Hawkes Learning Systems. To our editors, Kim Scott and Marcel Prevuznak, as well as all those on staff who helped with proofing and editing the text. Thank you to Greg Hill for getting this project off the ground, and to Emily Omlor for helping us to choose the cover and working so diligently to market this text.

Carolyn would like to thank all her family and friends who continually supported and encouraged her throughout this long project, in particular the Warrens, Rutherfords, Willards, Carters, and Shirleys. To Mom, Dad, John, and Christen, thanks for always listening to me and helping however you could; especially thanks for all the pictures! You all mean the world to me. Lastly, thanks to Shelly and Chelsie, who sacrificed the most for me to get this project done, and tried really hard to not complain that I could not spend time with you – I love you both.

Kim would like to first thank her husband for continuously encouraging and supporting our efforts, and for all the long hours helping to get it 'just right'! Emma and Chloe, thank you for understanding that Mom started a project and had to finish – I love you! Thanks also to Mom, Tina, and Stephanie for listening even when you didn't quite understand. Your support meant more than you can imagine.

Emily would particularly like to thank her family for being a constant source of encouragement throughout this whole process. For Mom, Dad, Ashley, Bill, Amy, Randall, Lori, Blake, and Julie, I will always be grateful. Thank you to my children, Abby and Andrew, and especially my husband Peyton for the patience and support you have shown me these past two years. I love you all very much!

Beginning Statistics

Warren
Denley
Atchley

Introduction to Statistics

Sections

Objectives

The objectives of this chapter are as follows:

1. Learn the basic vocabulary of statistics
2. Distinguish between population and sample; parameters and statistics
3. Classify data as discrete or continuous; qualitative or quantitative; and the level of measurement
4. Describe the process of a statistical study
5. Identify various types of studies
6. Learn the basic techniques for choosing a sample
7. Understand the practical and ethical concerns that arise when conducting a study

Chapter 1

Introduction

In the 1950's, researchers first began looking into the health effects of smoking. The studies performed showed, in fact, that smokers were more likely to suffer from lung cancer, emphysema, and various other diseases related to tobacco use. The researchers were convinced that smoking was hazardous to one's health. If smoking posed a health risk, then the public needed to know. It was time to issue a health warning.

The tobacco industry, however, was not convinced. They claimed that factors other than smoking were the cause for the health problems observed throughout the study. Until a study could be performed that conclusively showed a connection between smoking and health risks, the tobacco industry would not acknowledge a problem. What alterations to the design of the study did the researchers have to make in order to convince even the tobacco industry? In this chapter, we begin the process of learning how a statistical study must be designed in order to produce significant results.

Getting Started 1.1

Whether a tobacco study, the federal budget report, or even batting averages from local little league games, statistics are everywhere in today's world. Sources, such as magazines, commercials, and the evening news, constantly give us information in the form of statistics. In order to be informed citizens and discerning consumers we must understand exactly what these statistics mean. The goal of the branch of mathematics called *statistics* is to provide information so that informed decisions can be made. Hopefully, this text will enable you to filter the statistics you encounter so that you can be better prepared for the decisions you make in your daily life.

We begin our study of statistics with some basic vocabulary. The word **statistics** itself may refer to either the science of gathering, describing, and analyzing data or the actual numerical descriptions of sample data. To illustrate the difference in meanings, note the following examples:

Section 1.1 is your first lesson in statistics (the science). Throughout this course, you will learn how to properly collect and analyze statistics (actual data).

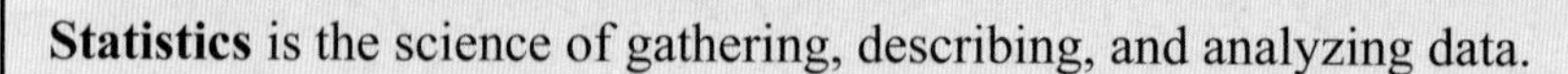

Definitions

Statistics is the science of gathering, describing, and analyzing data.

OR

Statistics are the actual numerical descriptions of sample data.

A statistical study centers upon a particular group of interest called a **population**. The population consists of all persons or things being studied. For example, a software company may wish to know the percentage of women that own a personal computer. In this case, the population is all women. Perhaps, instead, an instructor wants to know the number of statistics students with red hair. In this second example, the population to be studied is all statistics students.

When studying a population, the information gathered is called **data** (The singular form of data is datum), and it may be in a variety of forms including counts, measurements, or observations. If the population is small enough, data may be obtained from every member. This is called a **census**. You can easily remember this term by thinking of the U.S. census that attempts to count each and every American.

The numerical description of a particular population characteristic is called a **parameter**. That is, the numbers that describe a population are called parameters. Suppose it is true that 65% of all Americans own a computer. Then 65% is a population parameter. An important note is that although the parameter is a fixed number, often it is impossible or impractical to determine it precisely. For example, suppose that the population for a study includes all Americans. Would it be possible to collect data from EVERY American? Probably not! Because of human limitations, usually the best we can do is to estimate a population's true parameter.

When the parameter cannot be determined, a subset of the population, called a **sample**, is chosen. Information is then collected only from the sample. Care must be taken in choosing your sample so that the population is represented well, and the results of the study are meaningful. (We will talk about how to choose a representative sample shortly.) A number that describes a characteristic of a sample is called a **sample statistic**. For example, suppose that a research group wants to know the percentage of college students that do laundry on a weekly basis. If 100 college students are chosen for the study, and 42% do laundry on a weekly basis, then the population would be all college students, and the 100 students chosen are the sample. The sample statistic would be 42% of students.

Memory Booster!

Population
Parameter

Sample
Statistic

Notice that the first letters match!

Definitions ✦

Data is information, in particular, information prepared for a study.
Population is the particular group of interest.
Parameter is a numerical description of a particular population characteristic.
Sample is a subset of the population from which data are collected.
Sample Statistics are the actual numerical descriptions of a particular sample characteristic.

Throughout this text, it is essential that you are mindful of the relationship between population and sample. Figure 1 is a picture to help you visualize this relationship. The large circle represents the entire population, and the smaller circle represents the sample chosen from the population. Note that the sample is a subset of the population. That is, the sample is a group from within the population.

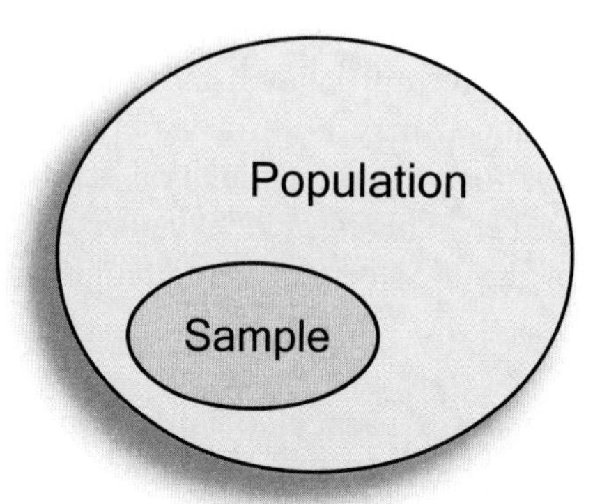

Figure 1 – Population versus Sample

Table 1 summarizes the differences between the population parameters and sample statistics.

Table 1 - Population versus Sample

Population	Sample
Whole group	**Part** of the group
Group I **want to know** about	Group I **do know** about
Characteristics are called **parameters**	Characteristics are called **statistics**
Parameters are generally **unknown**	Statistics are always **known**
Parameter is **fixed**	Statistics **change** with the sample

Example 1

Identify the population and the sample:

a. In a survey, 359 college students at the University of Jackson were asked if they had tried the October flavor of the month at the campus coffee shop. Eighty-three of the students surveyed said yes.

b. A survey of 1125 households in the U.S. found that 24% subscribe to satellite radio.

Solution:

a. Population: All college students at the University of Jackson.
Sample: The 359 college students who were surveyed.

b. Population: All households in the U.S.
Sample: The 1125 households in the U.S. that were surveyed.

Example 2

Is the numerical value given a population parameter or a sample statistic?

a. The average grade on Test 1 in Mrs. Shirley's statistics class was an 87%.

b. The average math score for all high school graduates who took the SAT was 519.

c. In a survey of college students, 52% said that they plan to attend graduate school.

Solution:

As a general rule, first consider whether each statement refers to a sample or a population. To be a population parameter, it must describe ALL members being studied and not a portion of them.

a. 87% is a population parameter because it is based on the grades of all students in Mrs. Shirley's statistics class.

b. 519 is a population parameter because it is based on the scores for all graduates who took the SAT, which is the population.

c. 52% is a sample statistic because it is a description of a sample of college students, since it would be impossible to survey every college student.

Finally, there are two main branches of statistics that we need to distinguish. The first, called **descriptive statistics**, gathers, sorts, summarizes, and presents data. Descriptive statistics involves the raw data, as well as the graphs, tables, numerical summaries, etc., used to describe these data. You may think of this branch of statistics as "just the facts". Descriptive statistics refer to the sample data without making any assumptions about the population from which the data was drawn. For example, if you wanted to know the average age of fast food employees, descriptive statistics would involve gathering the ages of a sample of employees, as well as calculating and reporting the average age of that sample.

On the other hand, the second branch of statistics, called **inferential statistics**, deals with the interpretation of the information collected. These interpretations are typically conclusions about some hypothesis. If in our previous example of fast food employees, we calculated the average age of our sample to be 20 years old, then we could infer from this information that the average age of fast food employees nationally is 20 years old. Of course, the accuracy of this inference is dependent upon how well our sample represents our population.

Definitions

The branch of **descriptive statistics**, as a science, gathers, sorts, summarizes, and displays the data.

The branch of **inferential statistics**, as a science, involves using descriptive statistics to estimate population parameters.

Usually both of these branches of statistics are used in conjunction within the same statistical study. A common scenario in statistics is as follows:

- A large population is to be studied.
- A sample is chosen that represents the population.
- Descriptive statistics is used to gather, sort, summarize, and present the information about the sample.
- Inferential statistics is used to estimate the population parameter from the sample statistics.

Example 3

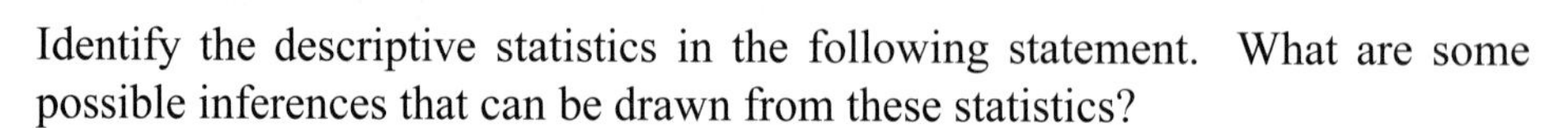

Identify the descriptive statistics in the following statement. What are some possible inferences that can be drawn from these statistics?

In a sample of weather services, the percent that did not correctly forecast the next day's weather was 75.8%.

Solution:

75.8% of weather services incorrectly predicting the weather is a descriptive statistic. Possible inferences that can be drawn are that the weather is difficult to forecast or that these meteorologists are particularly inaccurate.

Exercises

1.1

Population versus Sample

Directions: Determine whether the statement describes a population or a sample.

1. The diameter of every fifth soda can off an assembly line.

2. The ACT score of all students in Ms. Weigang's Algebra class.

3. The ages of all U.S. presidents.

4. The television shows watched by 1045 families from across the U.S. for the Nielson ratings.

5. The IQ score of all residents of a group home for drug rehabilitation.

6. The height of 15 out of 30 plants in a green house.

7. The annual salary of all professors at a large university.

8. The number of television shows watched each night by each of the children in Mr. Rutherford's fifth grade class.

9. The number of emergency calls received by five of the eight emergency response units in a tri-county area.

10. The number of times a group of 100 college freshmen order pizza in a month.

Population and Sample

Directions: For each scenario, identify the population being studied and the sample chosen.

11. A national magazine wishes to determine America's favorite celebrities. A ballot is included in the November issue of the magazine. Readers are encouraged to mail in their ballots.

12. An education professor wants to know the hometowns of students attending a particular Ivy League university. She obtains a list of registered students from the registrar's office and randomly chooses 300 students to survey.

13. A large discount store wants to determine the average income of its shoppers. A researcher chooses 100 shoppers at random between the hours of 1 and 5 p.m. on Thursday.

14. A hotel chain wants to build a new facility. The board of directors uses a list of the top 100 vacation spots in America and visits 20 of the cities on the list to determine their feasibility as the new hotel's location.

15. One Christmas season, you are interested in determining the average height of Christmas trees sold in your city. You visit two local Christmas tree farms and measure the height of 45 randomly selected trees.

Parameter versus Statistic

Directions: Identify if the numerical value in each statement describes a parameter or a statistic.

16. Thirty-five percent of all graduating seniors from a local university receive a degree in business.

17. The average number of hours per week a sample of eleven year olds spends watching TV is 20.6.

18. The average height of a sample of men entering the armed forces is 6.1 feet.

19. A sample of 1067 people reveals that 25% of them are overweight.

20. The students in Mrs. Allen's third period class have an average of 2.5 siblings.

21. The average number of pets per person reported in a recent survey is 1.8.

22. A survey reports that 78% of drivers talk on their cell phone while driving.

23. Eighty-seven percent of all clients at a mental health facility report having an alcohol or drug addiction.

24. The average price of all houses in a new subdivision is $135,000.

25. A poll of local voters reveals that the governor's approval rating is currently 36%.

Descriptive versus Inferential Statistics

Directions: Decide if the following statements are examples of descriptive or inferential statistics.

26. Eighty-two percent of the employees from a small local company attended the annual company picnic.

27. Based on information from a recent survey, social scientists believe that 21% of all married men would admit to having had an affair.

28. The average age of entering freshman at the University of Senatobia is 20.8 years old, based on information from the registrar's office.

29. Sixty-five percent of seniors at a local high school applying to college apply to at least one college out of state.

30. The average number of hours vacationers spend in national parks during the summer months is 4.5 hours.

Data Classifications

1.2

Just as animals can be classified into phylum and then further into species, data collected in a statistical study can be classified into different categories. The different categories group data based on the type of statistical analysis that can be performed on the data. Therefore, knowing the classification of a set of data is the first step in any statistical process.

Qualitative versus Quantitative Data

Memory Booster!

Qualitative data are descriptions. (qualities)

Quantitative data are counts and measurements. (quantities)

The first distinction we will make is between qualitative and quantitative data. **Qualitative data**, also known as categorical data, consist of labels or descriptions of traits of the sample. Examples of qualitative data include such information as favorite foods, hometown, eye color, or identification numbers. Generally, qualitative data will be in terms of words; however, sometimes numbers act like words. For example, numbers on sports jerseys serve only to distinguish different players during games. In this case, the numbers are merely labels and must be classified as qualitative.

In contrast, **quantitative data** consist of counts or measurements, and therefore are numerical. Test scores, average rainfall, and median weight are all examples of quantitative data. Notice that quantitative data can be manipulated in ways that qualitative data cannot. For example, you cannot add eye color, but you can add weights together.

Example 4

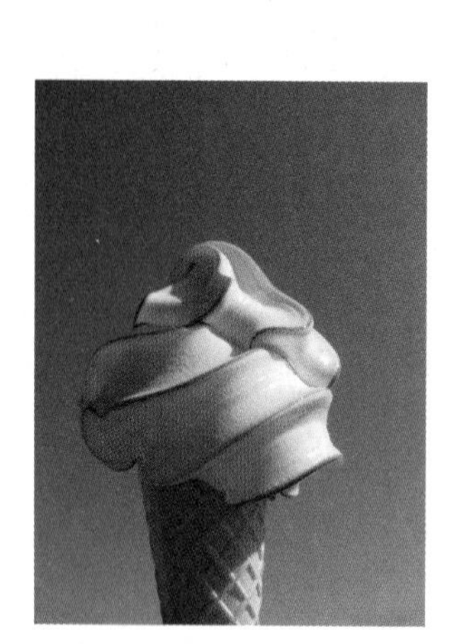

Classify the following data as either qualitative or quantitative.

a. The heights of members of a soccer team.
b. The flavors of ice cream at a local ice cream parlor.
c. The jersey numbers of a women's basketball team.

Solution:

a. Heights are measured, so these are quantitative data.
b. Flavors are descriptions, so these are qualitative data.
c. Jersey numbers are numeric but are not a measurement or count. They are considered labels, so these are qualitative data.

✦ Continuous versus Discrete

Quantitative data can be further classified as either continuous or discrete. **Continuous data** can take on any value in a given range of numbers. Continuous data are usually measurements, such as length and weight. If data are not continuous, then they are considered to be discrete. **Discrete data** can take on only particular values and cannot take on the values in between. Thus discrete data are usually counts. For example, the number of pets you have would be discrete because you can have either 2 pets or 3 pets, but not 2.75 pets.

Memory Booster!

Continuous data are usually measurements.

Discrete data are usually counts.

Example 5

Determine whether the following data are continuous or discrete.

a. The temperature in Fahrenheit of cities in South Carolina.
b. The number of houses in a neighborhood.
c. The number of peanuts in a jar, assuming all of the peanuts are whole.
d. The height of a door.

Solution:

a. Continuous - the temperature could be measured to any level of precision based on your thermometer.
b. Discrete - houses are counted in whole numbers. A house under construction is still a house.
c. Discrete - counts are generally considered to be discrete.
d. Continuous - again, depending on the ruler, measurements of any level of precision can determine the height.

✦ Levels of Measurement

Another way that data may be classified is according to its level of measurement. There are four levels of measurement: nominal, ordinal, interval, and ratio. The higher the level of measurement, the more mathematical calculations that can be performed on that data.

The first and lowest level of measurement is the nominal level. Data at the **nominal level** of measurement are qualitative, consisting of labels or names. Because we cannot add labels or names together, no calculations can be performed on data at the nominal level. You can think of the nominal level as labeling or categorizing a subject. For example, Susie belongs in the category of females, and jersey number 15 labels Bob.

Example 6

Suppose every student in a statistics class was asked what their favorite pizza topping is. Explain why this data is at the nominal level of measurement.

Solution:

This example is nominal because the data simply describe or label the different toppings of the pizza.

The second level of measurement is the **ordinal level**, and data at this level are usually qualitative. Data at the ordinal level can be arranged in a meaningful order, but calculations such as addition or division do not make sense. In short, ordinal data has everything nominal data does, but it also has a natural order to it. For example, the ranking of SEC sports teams would be at the ordinal level. (1. Ole Miss, 2. Auburn, etc.) However, note that there is no meaningful difference between any two data entries. For instance, in our SEC example, how much better is Ole Miss than Auburn? Clearly the ranking gives only an order, so this data is at the ordinal level. The survey responses of 'excellent', 'average', or 'poor', are another example of data at the ordinal level that definitely have a natural order.

Example 7

Consider the seat numbers on your opera tickets, such as A23. Is this data nominal or ordinal?

Solution:

Seat numbers are ordinal because there is a meaningful order to the data, namely, the position in the theater.

The third level of measurement is the **interval level**, and data at this level are quantitative. Like data at the ordinal level, data at the interval level can also be ordered, but the interval level is distinguished because differences between data entries are meaningful. For example, if we compared the average temperatures of various cities, the data collected could be ordered. Furthermore, the differences between temperatures could be calculated. Specifically, if Phoenix has an average temperature of 100° F, and Stockholm's average temperature is 50° F, we can say that Phoenix is hotter, and it is hotter by 50° F.

One important note regarding the interval level is that the zero point associated with these data is a position on a scale, but does not actually mean the absence of something. In the given example regarding temperatures, zero on the Fahrenheit scale does not mean the absence of heat, it is simply a really cold temperature. Zeros in the interval level are merely placeholders. Some common examples of interval data include calendar dates, temperature, and certain personality or intelligence tests. The exception to temperature being an interval level of data is the Kelvin scale of temperature. Absolute zero, 0 K, really does indicate the absence of heat.

Example 8

The birth years of your classmates are collected. What level of measurement is this data?

Solution:

Birth years can be ordered. It is also meaningful to subtract years to determine the difference in age. However, the year 0 A.D. does <u>not</u> mean the beginning of time. Therefore, birth years would be at the interval level of measurement.

Memory Booster!

Nominal ⇔ names

Ordinal ⇔ order

Interval ⇔ 0 is a placeholder

Ratio ⇔ 0 means the absence of something

The fourth and final level of measurement is the **ratio level** of measurement. In this highest level of measurement, data are quantitative, can be ordered with meaningful differences, and the zero point indicates the absence of something. At this level, not only can we add or subtract data values, we can also multiply or divide. For example, suppose that we compare the prices of two cars. If one car costs $10,000 and the other $20,000, then the second car costs $10,000 more than the first and is twice as expensive. Note, though, that in the earlier example, it does not make sense to say that the average temperature of Phoenix is twice as hot as the average temperature of Stockholm.

Example 9

The cost of a particular pair of jeans is collected from many different department stores. What level of measurement are these data?

Solution:

One pair of jeans may be more expensive than another, and we can determine how much more expensive. The key, however, is that we can also say that one pair is 50% more expensive than another pair. This is because $0.00 really means the absence of money, i.e. no cost. So these data are at the ratio level of measurement.

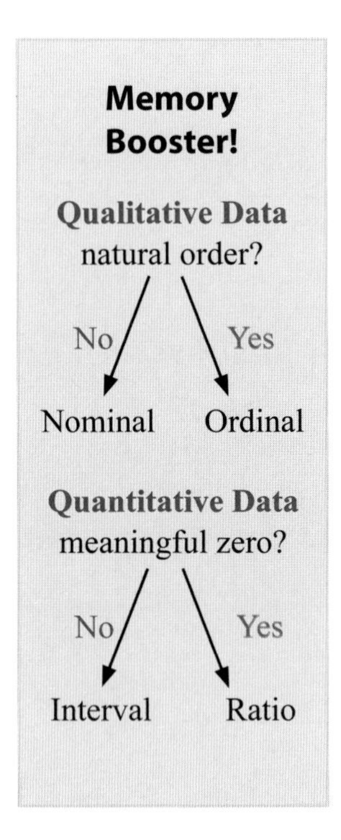

When categorizing data according to level of measurement, data should be associated with the highest level of measurement possible. The levels of measurement are like a staircase, and each level is the next stair-step. You should try to "walk up" as many stairs as possible for each data set. For example, letter grades on a test categorize a student's performance, thus satisfying the guidelines for nominal data, but we should not stop there. Letter grades can also be put in a meaningful order, so this data satisfies the guidelines for ordinal data as well. However, one cannot subtract a B from an A and have a meaningful calculation, so the data does not satisfy the guidelines for the interval level. Thus the highest level of measurement obtainable for letter grades is the ordinal level.

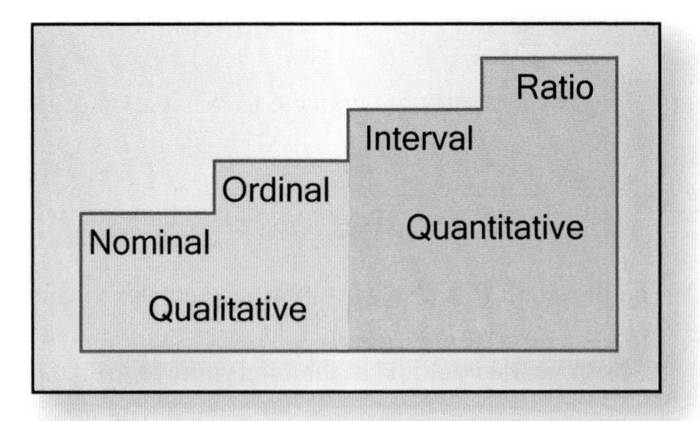

Figure 2 – Levels of Measurement

Example 10

Determine if the following data are qualitative or quantitative and then determine the level of measurement:

a. The finishing times for runners in the Labor Day 10K race.
b. The colors contained in a box of crayons.
c. The boiling point (on the Celsius scale) for various caramel candies.
d. The individual page numbers at the bottom of each page in this book.

Solution:

a. The amount of time it takes to run the race would be quantitative, since calculations on this data are meaningful. We could also say that Andrew finished the race in half of Peyton's time; thus, the data are at the ratio level of measurement.

b. Colors are labels, so the data is qualitative. The colors cannot be ordered, so the data are at the nominal level of measurement.

c. Calculations can be performed on boiling points because they are temperatures, making this data quantitative. The temperatures are measured using the Celsius scale, where 0 degrees does indicate an amount of heat. This puts the data at the interval level of measurement.

d. Individual page numbers cannot be meaningfully added or subtracted, placing this data as qualitative. Page numbers can be ordered, so they are at the ordinal level. This is different from the number of pages in a book, which would be quantitative and ratio.

Exercises

1.2

Directions: Classify each of the following variables as:

a. Qualitative or Quantitative.

b. If the data are quantitative, classify them as Discrete or Continuous.

c. Nominal, Ordinal, Interval, or Ratio.

1. The age in whole years of U.S. presidents.

2. The widths of the doors in a home.

3. The IQ scores, reported in whole numbers, of research scientists.

4. The names of television shows watched by families from across the U.S.

5. The yearly amount of snow fall in Cleveland over 10 years.

6. The height of orchids on a windowsill.

7. The amount of weight gained by each person from a group of college freshmen.

8. The number of antique cars collected by the Smith family.

9. The number of 6 foot wooden boards it takes to build a desk.

10. Bank account pin numbers.

11. The heights of men entering the armed forces.

12. The sizes of T-shirts on sale.

13. The temperature in kelvins of various sites on the planet Mars.

14. The number of siblings that students in Ms. Pitcock's third grade class have.

15. Jersey numbers on a lacrosse team.

16. The types of pets reported in a recent survey.

17. Your position in line at the checkout counter.

18. The letter grade on students' English essays.

19. Your birth order in the family.

20. The title that precedes your name (Dr., Mr., Ms., etc.).

21. The number of students taking college algebra at The Ohio State University.

22. The average temperature in degrees Celsius of the water in the Bahamas each month in 2004.

23. The birth-years of members of your immediate family.

The Process of a Statistical Study

1.3

Now that we have learned some of the vocabulary used in the field of statistics, let's turn to the process of a statistical study. Usually a statistical study seeks to determine the value of an unknown parameter. Thus mathematical methods, i.e. statistics, are used for estimating that parameter. Anytime you read a statistical study, it is up to you to decide on the validity of the researcher's conclusions. Unfortunately, many studies, even those with good intentions, use faulty data or draw inappropriate conclusions. Being aware of how the design of a statistics study helps in critiquing the results of a study.

Memory Booster!

Parameter is a numerical description of a population characteristic.

As shown in Figure 3, the process of a statistical study is somewhat cyclic. Once a researcher determines a question that needs answering, the cycle begins. This question determines the population and variables to be studied. From there, a method of data collection is chosen and data are collected, usually from a sample of the population. From the data, sample statistics can be calculated and population parameters can be estimated. The estimated population parameter can then help us answer the question that originated the study.

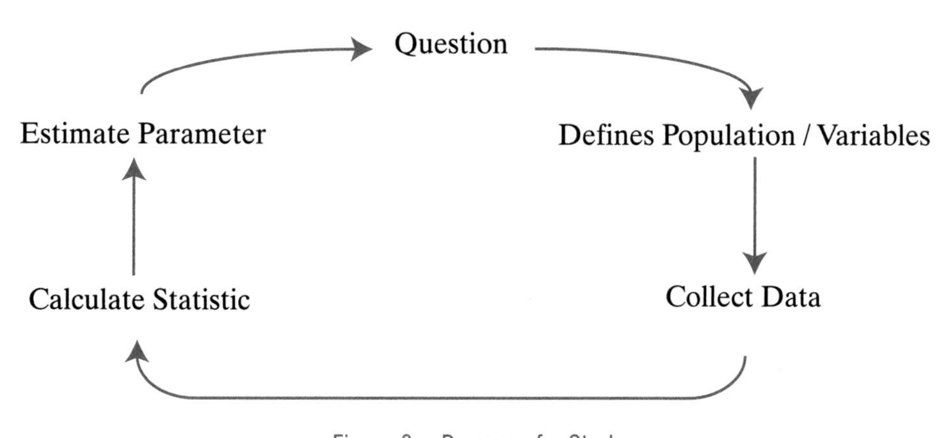

Figure 3 – Process of a Study

Let's examine each part of the process further. We'll begin with the question that prompts our statistical study. For example, take the question "What is the average amount of money a full-time college student spends eating out each week?" This question defines our population to be full-time college students. We also must determine the study's **variables**, which are values that can change amongst members of the population. Our statistical variable, which is the focus of our study, is the amount of money each student spends on eating out per week.

After our population and variables have been defined, data must be collected. Deciding the best method for collecting the data is an important step in the statistical process. The method will vary depending on the type of study you choose. Although there are many different types of studies, most studies can be classified as either an observational study or an experiment.

✦ Observational Studies

An **observational study** observes data that already exist, such as student SAT scores or favorite colors. No manipulation of the sample is performed by the researchers. One limitation of observational studies is that no cause and effect relationship can be determined between variables. For example, if we collect data on SAT scores and the GPA of freshmen college students, we might find a relationship between the variables, but it would be inappropriate to conclude a cause and effect relationship.

In an observational study, a researcher collects data from a sample of the population. When choosing a sample, the researcher should be careful to choose one that represents the population well. A **representative sample** is one that has the same relevant characteristics as the population and does not favor one group of the population over another. In choosing our sample, we need a list of all the members of the population from which the sample can be drawn. This list is called a **sampling frame**. If your population is small, then compiling the sampling frame can be an easy task. However, populations that are large might require more sophisticated methods. Ideally, the frame would include all members of the population; however, this is often not practical. If this list is not complete, the sample produced may not be a representative sample. Therefore, you should be conscious to choose the best possible sampling frame. For example, in the past, to obtain information about the citizens of a community, large sampling firms would use a telephone directory as their sampling frame. Now, with the prevalence of cellular phones, many people no longer have home phones. Therefore, a telephone directory is no longer the best choice for a sampling frame if the population you want to study is all people in a given geographical area.

There are several basic methods of choosing a sample from a population. Let's look at some of those techniques.

Sampling Techniques:

1. Census – If our goal is to collect information from a representative sample, the best one we could find would be the entire population. Recall that when you gather data from every member of the population, it is called a **census**.

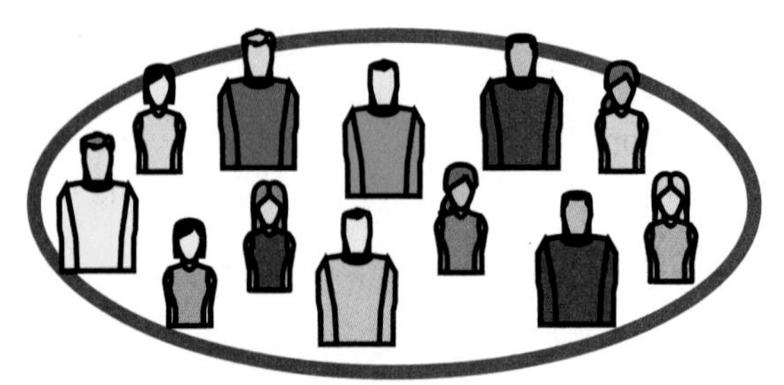

Everyone is included

Figure 4 – Census

However, a census is often impractical. It can be too time-consuming, too expensive, and/or the data collection could be destructive in nature. For example, consider the time and money that would be necessary to ask every single college student how much they spend eating out per week. It would be impractical. Furthermore, some tests are destructive in nature. Consider testing a batch of California wine. If we drank every glass of wine in the batch, there would be none left to sell! Because a census is often not an option, we usually choose other sampling techniques.

2. Random Sampling – A **random sample** is one in which every member of the population has an equal chance of being selected. An example of this kind of sampling occurs when drawing names from a hat. If every name in the hat has an equal chance of being chosen, such as when all the names in the hat are equal in size and well shuffled, then choosing a name from a hat is random sampling.

Figure 5 – Random Sampling

Random sampling can also be performed by assigning identification numbers to each member of the population and using a random number table to choose ID numbers. If you randomly choose campus ID numbers and survey those students about their spending on dining out, then you will have a random sample.

Often technology is used in place of a random number table. Some computer software programs can choose random samples from a list. Other software programs and calculators can generate random numbers that can be matched to a list of ID numbers. The technology section at the end of this chapter demonstrates how to use Microsoft Excel, Minitab, or a TI-84 Plus calculator to generate a series of random integers.

Why not let a human choose the random sample? In reality, it is against our human nature to choose members of the population truly at random. To understand this phenomenon, take a moment to choose 3 random numbers between 1 and 100. Try to make sure they are truly random. Now, consider these questions: Are the numbers spaced out or grouped closely together? Are they all even, odd, or some of both? Did you have a reason for choosing any of the numbers? Did you alter any of your original responses and if so why? Did you repeat any of the numbers? It is unlikely that you would have chosen consecutive numbers, such as 1, 2, and 3 when trying to create a random sample, though a random number generator could create that sequence at times.

3. Stratified Sampling – A **stratified sample** is one in which members of the population are divided into two or more subgroups, called **strata**, that share similar characteristics like age, gender, or ethnicity. A random sample from *each* stratum is then drawn. For instance, if we divide the population of college students into strata based on the number of years in school, then our strata would be freshmen, sophomores, juniors, and seniors. We would then select our sample by choosing a random sample of freshmen, a random sample of sophomores, and so on.

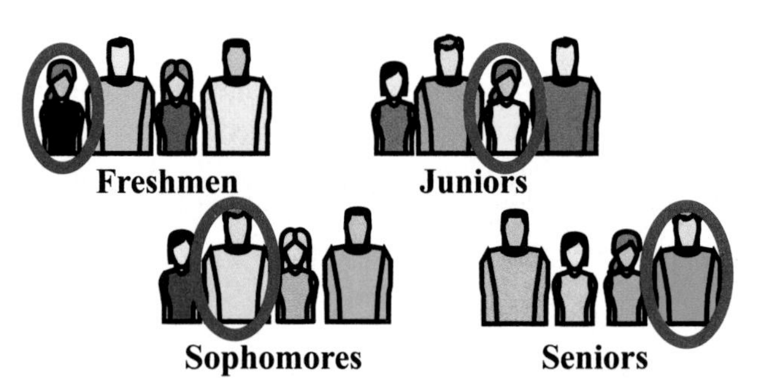

Figure 6 – Stratified Sampling

This technique is used when it is necessary to ensure that particular subsets of the population are represented in the sample. Since a random sample cannot guarantee that sophomores would be chosen, we would use a stratified sample if it were important that sophomores be included in our sample. Furthermore, by using stratified sampling you can preserve certain characteristics of the population. For example, if freshmen make up 40% of our *population*, then we can choose 40% of our *sample* from the freshmen stratum. Stratified sampling is one of the best ways to enforce "representativeness" on a sample.

4. Cluster Sampling – A **cluster sample** is one chosen by dividing the population into groups, called **clusters**, that are each similar to the entire population. Often populations lend themselves to clusters naturally, i.e., counties or voting precincts. The researcher then randomly selects some of the clusters. The sample consists of the data collected from *every* member of each cluster selected. As an example, if we have several already formed classes of students, and each class is similar to the student population, we could simply choose some classes at random and collect data from every member of those classes about their eating habits.

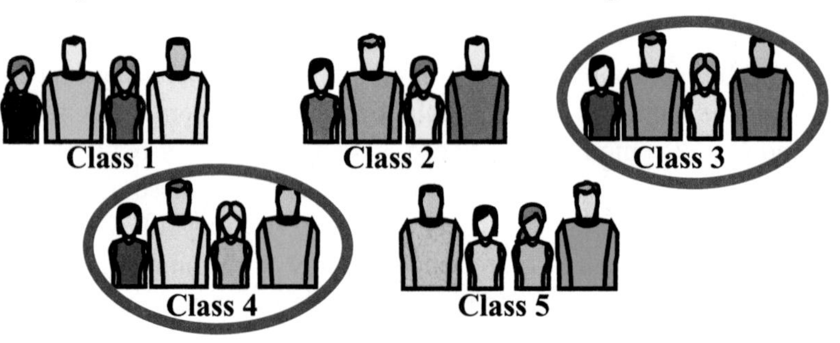

Figure 7 – Cluster Sampling

5. Systematic Sampling – A **systematic sample** is one chosen by selecting every n^{th} member of the population. This type of sampling is easy to do if you have an accurate sampling frame. We can choose one random number to decide where to start in the list and another to decide the increment. For example, if we have a master list of all full-time students at a certain institution, we might start with the 14th student and then choose every 8th student on the list until we have a sample of the size we desire. Systematic sampling should not be used if the sampling frame has a pattern, such as every other person on the list is female.

Figure 8 – Systematic Sampling

6. Convenience Sampling – A **convenience sample** is one in which the sample is "convenient" to select. It is so named because it is convenient for the *researcher*. Because this method often leads to members of the sample all having a similar characteristic, it is prone to creating non-representative samples. For example, a researcher standing in front of the campus cafeteria asking about how much students spend on eating out might get a different answer than if the researcher was standing at the door of a classroom on campus.

Figure 9 – Convenience Sampling

One example of convenience sampling is a **self-selected survey**. A self-selected survey is one in which the survey participants volunteer to be a part of the study, rather than having been chosen by the researcher. The problem with a self-selected survey is that usually only people with strong opinions will take the time to volunteer their time or information for the study. For instance, consider a magazine that wants to do research for a story regarding hospital care. The magazine lists a website in its June issue, inviting readers to log on and share stories about the care they received in the hospital. Now, do you think a person who stayed in the hospital an average length of time and received adequate care will take the time to log on and share their experience? Probably not. Most likely only readers who have had either an exceptionally good or an exceptionally bad hospital stay will "self-select" themselves for the study. While these types of stories make for better magazine articles, they rarely give a true picture of what is happening for the entire population. All in all, self-selected surveys are a <u>very</u> poor method of sampling.

Example 11

Identify the type of sampling used in each of the following cases:

a. A pollster surveys 50 people in each of a senator's 12 voting precincts.

b. The quality control department at a cereal manufacturer measures the weight of every 10th box off of the assembly line.

c. A female student walks down the hall in her sorority house asking how much money each person spends on CD's in a month.

d. An educator chooses 5 of the school districts in the Chicago area and asks each household in those districts how many school-age children are in the home.

e. To determine who will win a $100,000 shopping spree at the mall, the manager draws a name out of a box of entries.

f. The department secretary lists each professor and their salary in order to determine the average salary of professors in the English department at her university.

Solution:

a. Stratified sampling: the voting precincts make up the strata.

b. Systematic sampling: the system of selecting the sample is every 10th box.

c. Convenience sampling: this would be a very easy method of surveying; however, it is not representative of the population since everyone does not live in a sorority house.

d. Cluster sampling: every school district in the Chicago area makes up the clusters.

e. Random sampling: every name in the box has an equal chance of being choosen.

f. Census: it is practical to survey this population, as the information of every member of the population of English professors is being sampled.

In addition to collecting data from a sample, some studies collect data from previous studies or analyze a single member of the population. When a study compiles information from previous studies, it is referred to as a **meta-analysis**. This type of study seeks to identify patterns across many studies on a similar topic that would not be discernible when looking at a single study. A benefit of meta-analysis is that many smaller samples can be combined into a single larger sample. A drawback is that the combined study is only as good as its weakest link. In other words, if you use poorly constructed studies, your study will not be able to draw strong conclusions.

In contrast to a meta-analysis that looks at one variable over several studies, a **case study** looks at multiple variables that affect a single event. When you desire to look at a single case in depth and all of the possible variables associated with that case, then it would be most appropriate to perform a case study. If you are interested in studying how administrators respond to hazing incidents on college campuses, you could perform a meta-analysis of all of the research done on hazing. In contrast, if you are interested in studying a particular hazing incident at a certain institution, you would perform a case study of that hazing incident.

Example 12

Categorize the following studies as a case study or a meta-analysis:

a. Meteorologists study the Tsunami of December 2004 to try to identify warning signs.

b. Oceanographers study research from tsunamis dating from 1900 to 2000 to determine their effects on the ocean floor.

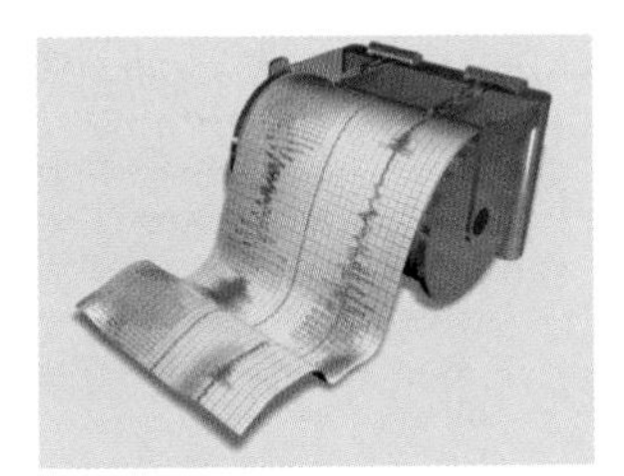

Solution:

a. Because the meteorologists are studying a single case, it is a case study.

b. Because oceanographers are looking at multiple studies relating to the single variable of tsunamis' effects on the ocean floor, this would be a meta-analysis.

✦ Experiment

An experiment is another method of data collection. As opposed to observational studies that measure data that already exists, **experiments** seek to create data to help identify cause and effect relationships. The people or things being studied in an experiment are called the **subjects**. If they are people, they can also be referred to as **participants**.

In an experiment, researchers apply a **treatment** to a group of subjects and measure the response. This group is referred to as the **treatment group**. In contrast, the **control group** consists of the subjects to which no treatment is applied. Ideally, each group should have approximately the same number of subjects, and the subjects should be assigned to the treatment or control group randomly. This helps to ensure that the only difference between the two groups is the treatment itself.

Memory Booster!

Control group does not have a treatment applied.

Treatment group does have a treatment applied.

To make sure that the researchers cannot influence the outcome of the experiment, the experiment may be **masked**, also known as blinded. A single-masked experiment is when the subjects do not know in which group they have been placed, although the people interacting with the subject of the experiment have that information. In a double-masked experiment, neither the subjects nor the people interacting with the subjects, such as doctors or nurses, know to which group the subjects belong. A double-masked design would be preferable when the researchers have to make subjective decisions regarding the participants, such as in clinical trials for psychotropic drugs.

Because people respond to suggestion, giving someone a drug for the common cold and telling them that it will cure them will often produce effects caused by the suggestion alone and not the drug itself. This response is known as the **placebo effect**. A **placebo** appears identical to the actual treatment, but contains no intrinsic beneficial elements. For example, if the treatment is a small orange pill that contains vitamin C, then the placebo would be a small orange pill that does not contain vitamin C. To counteract the placebo effect, a masked experiment is used so subjects do not know to which group they belong. This helps ensure that the only difference in the treatment and control groups is due to the active ingredients in the treatment.

Example 13

Which type of study would you conduct: observational or experimental? Name the sampling frame you would use, if appropriate.

a. You want to determine the average age of college students across the nation.
b. Researchers wish to determine if flu shots actually help prevent severe cases of the flu.

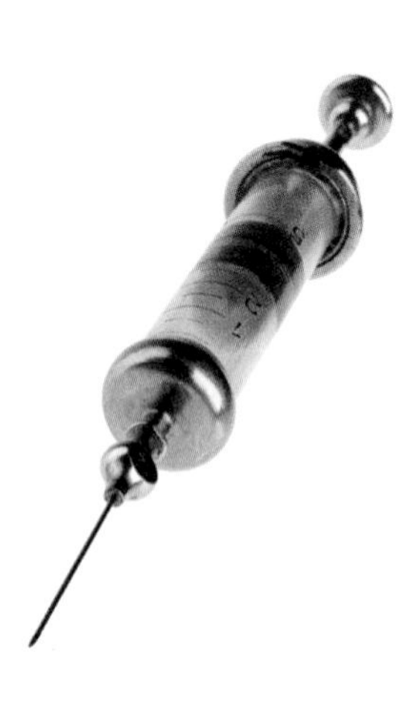

Solution:

a. Observational: no treatment is being applied to any group. A choice of sampling frame might be enrollment lists from colleges in the nation.
b. Experimental: this study would require a treatment group (flu shot) and a control group (given a placebo). Because this is an experiment, participants would not come from a sampling frame, but on a volunteer basis.

Exercises

1.3

Vocabulary

Directions: Determine if the following statements are **true** or **false**.

1. The population parameter is used to estimate the sample statistic.

2. Data collection must be complete before variables are chosen so that the researcher can be sure he has the data needed to answer the question.

3. A sample statistic is calculated from data collected in a study.

4. A statistical study will always precisely determine the population parameter once the study is complete.

5. The question in a statistical study dictates the population and variables.

6. Participants in an experiment should always be allowed to choose which group they are placed in so that they feel as comfortable as possible for the duration of the experiment.

Directions: Fill in the blank with the most appropriate word.

7. An experiment in which both the participant and the one administering the treatment are unaware of whether the participant is in the treatment group or the control group is referred to as a __________ experiment.

8. A member of the population that is being studied in an experiment is called a __________ .

9. The group in an experiment that receives a placebo is called the ______ group.

10. An experiment is a type of statistical study in which a _________ is applied to a group of the population.

11. A pill that looks like aspirin but has no active ingredient is called a ______.

12. An experiment in which only the participant is unaware of whether they are in the treatment group or the control group is referred to as a __________ experiment.

13. When a subject believes that he has recovered from an illness because he is taking a treatment drug, when in reality he is in the control group of an experiment, it is referred to as the __________.

14. A human subject in an experiment is referred to as a __________.

15. A researcher gives an active drug to the _______ group in an experiment.

Type of Study

Directions: Determine which type of study you would conduct: observational or experimental. Then name a possible sampling frame you would use.

16. A football coach wants to know the average weight of his offensive linemen.

17. A researcher wants to study the effectiveness of vitamin C in combating the common cold.

18. A city planner wants to know the average number of vehicles parked in downtown parking lots on any given business day.

19. A cell phone company wants to know the average length of time teenage girls spend on the phone.

20. A dentist wants to look at the effects of a new dental material used for fillings.

Sampling Techniques

Directions: Identify the sampling technique used for each of the following.

21. The FDA chooses 15 hospitals around the country at random. Every doctor in the chosen hospitals is required to participate in the study.

22. Every 4th dorm room is selected for a survey regarding study hours and campus security.

23. A state politician wants to gauge public opinion in his area before deciding to run for re-election. For the study, 200 registered voters are chosen at random from each county in his district.

24. A computer program is used to randomly generate a list of student ID numbers in order to gather a group to give feedback about the Greek system on campus.

25. In order to complete a psychology project, you pass out surveys to the first 25 people you find in the student union.

26. A student asks all the people living on the 1st, 5th, and 8th floors of his dorm to answer a survey about dorm life on your campus.

27. A superintendent wants to know the average reading level of all students in the third grade at a local elementary school.

28. A local church wanted to know the average age of its morning congregation, so they asked every 10th person leaving the service to put their age in a box.

29. For budget purposes, administrators need to know the average salary of tenured faculty at the University of Kalamazoo.

30. Ten students from each of the 15 sections of college algebra were asked about the quality of the textbook used in the course.

31. One thousand phone numbers were selected by a computer to be called for a telephone survey.

32. Local politicians ask 20 people in their neighborhood what they think about the new school board proposal.

Meta-Analysis versus Case Study
Directions: Classify each scenario as either a meta-analysis or case study.

33. A child prodigy's home life is examined in order to determine environmental factors that may have shaped him intellectually.

34. Studies performed on four different airlines are compared in order to determine which provides the best customer care.

35. In order to prepare a story for the January issue, a magazine reporter analyzes 8 different studies about dieting strategies in order to determine which is most effective long term.

36. For the purpose of studying sibling rivalry as affected by birth order, a typical American family is selected.

37. A medical researcher looks at multiple studies performed on a new drug in order to determine whether or not the drug is safe to put on the market.

The Reality of Conducting a Study

1.4

When you begin to study a research question, it is not always as easy as just going out and collecting data. There are both ethical concerns that must be resolved before the study begins, as well as practical concerns that must be monitored while performing the research, in order to ensure meaningful results.

✦ Observational Studies

Before beginning any research study, there are three issues that should be addressed. The first hurdle in performing a study is getting approval from an **Institutional Review Board (IRB)**. This is particularly necessary in medical or academic research. The IRB will require you to fill out documents describing your proposed study in detail, including how you will handle issues of informed consent, human or animal subjects, and confidentiality.

Informed consent involves completely disclosing to participants the goals and procedures involved in your study and obtaining their agreement to participate. There are grey areas, such as child participants or studies on the mentally handicapped. In these cases, questions arise about the meaning of informed consent and who should give it.

When dealing with human or animal subjects, IRBs will require extensive proof that there will be no physical or mental harm to the subjects in the study. Since animal subjects cannot give their informed consent to participate, IRB's require more extensive documentation on study procedures involving animals.

Research often involves a high level of trust between the researcher and the participants. Part of that trust is that any information acquired will be kept confidential. An IRB will require evidence that any data gathered from participants will be kept in a secure location and that only people who need to see the raw data will be allowed to do so. However, this does not mandate that the data must be gathered anonymously. For example, if an educator studies the number of class absences and their effect on students' final grades, then it is likely that the educator will know which student is associated with what data. So, although the data is kept confidential, it is not anonymous.

Something To Ponder...

Consider preparing a research proposal for the following question, "What percentage of 11 to 16 year olds smoke cigarettes?" When thinking about your proposal, be sure to consider the issues of informed consent, confidentiality, and working with human participants.

✦ Practical Concerns Resulting in Bias

Whether conducting or reading a study, you should look out for any problems in its design or execution. Even when unintentional, results sometimes produce erroneous conclusions. When a study tends to favor certain results, it is called biased. **Bias** can occur many ways, including but not limited to the size of the sample, the way the sample is chosen, or the conduct of the researcher.

When it is the researcher who influences the results of the study to favor a certain outcome, the bias is called **researcher bias**. Researcher bias may be intentional or unintentional. The researcher might intentionally choose a favorable sample, or unintentionally influence the sample's responses by their actions.

One form of researcher bias is in the wording of the survey questions. Researchers should be careful to word questions so as not to influence the answer. A question such as, "Do you oppose testing cosmetics on innocent animals, even if it could cause serious damage to those poor creatures?" could influence the responses of the sample. Often survey questions can be tested for neutrality by having a psychologist read through them. In some cases, it might even be useful to administer the survey to a small trial sample to determine the types of responses that will be received. This will point out any questions that the sample finds confusing, as well as any that might cause a certain response.

Even a properly worded question can lead to bias if it is asked in an improper manner. The researcher's facial expression, tone of voice, or physical proximity to the subject could all encourage a subject to respond with the answer they believe the researcher wants, instead of their true feelings. Researchers should also be mindful of asking about very sensitive information in a setting that would make the subjects feel uneasy.

Errors in a study resulting from the way that the sample is chosen are called **sampling errors**. When choosing a sample, a researcher should aim for a representative sample. If the characteristics of the sample differ significantly from the population, then the sample will be biased as well as the results. One key element in attempting to create a representative sample is choosing a large enough sample size, i.e., it would be silly to predict the winner of the next presidential election based on the results of one small town. One issue that affects the sample size in an experiment is a **dropout**, a participant who begins the experiment but then fails to complete it. It would be unfortunate for a researcher to begin with a representative sample, and then through numerous dropouts end up with a sample that no longer represents the population. Another sampling error occurs when the sampling frame is defective. When the sampling frame excludes certain members of the population, it is unlikely that you will be able to use it to obtain a representative sample.

A possible result of a sampling error is that of **participation bias**. Participation bias is created when there is a problem with either the participation—or lack thereof—of those chosen for the study. One type of study, mentioned in the previous section that is especially prone to participation bias is a self-selected survey. When the participants must make an extra effort to be a part of the sample, usually only those with strong opinions will be included. In this case, the lack of participation among those with a more neutral opinion results in a sample that does not adequately represent the population.

Non-sampling errors occur from sources other than the construction of the sample. First, a non-sampling error might be as simple as a typo in the data set. Errors of this form are called **processing errors**. Secondly, we realize that there is always the possibility that some participants in our sample might lie when answering a survey. However, we also need to be aware of the possibility that our sample could report false information unintentionally. For example, answer the following question as truthfully as possible: "How many carbonated beverages have you had in the last week?" How accurate do you think your response was? Unless carbonated beverages are not part of your regular diet, you were making an educated guess that could have been mis-remembered. A third type of non-sampling error can occur in an experiment when subjects do not follow the prescribed treatment. Subjects who stray from the directions they were given, but remain in the sample to the end are called **non-adheres**. Lastly, there are always factors that influence the results of a study that the researchers did not, or could not, account for. We refer to these as **confounding variables**. At an elementary level, when comparing the test results of two groups of students, confounding variables might include different teachers or even different class times.

Something To Ponder...

In the teenage smoking proposal, what types of sampling/non-sampling errors might occur? How would you watch out for researcher bias? What possible confounding variables might exist?

Exercises

1.4

Vocabulary
Directions: Fill in the blank with the most appropriate word.

1. An Institutional Review Board will require that a researcher disclose to participants the goals and procedures involved in her study and obtain their agreement to participate. This is called obtaining the participants' _______.

2. A _________ occurs when results from a study are tabulated incorrectly.

3. _________ occurs when the person administering the study influences the participants' response.

4. __________ occurs when the results of a study tend to favor one outcome over another.

5. A participant who begins a study but fails to complete it is referred to as a __________ .

6. A researcher must gain consent from the __________ in order to conduct a legal experiment on animals.

7. A participant that does not fully comply with the instructions for a study is called a __________.

8. Collecting data from a sample that misrepresents the population is called a __________ error.

9. Additional variables that are not measured by the study but affect the outcome would be called __________.

10. If participants are asked to answer survey questions in an uncomfortable setting, there is potential for __________ to occur.

Types of Bias

Group Discussion Questions: Identify as many potential sources of bias for the following studies as you can.

11. StarCrazy magazine wants to determine America's favorite celebrities. A survey is placed inside each subscription for readers to fill out and mail in to the company.

12. A struggling retailer wishes to improve sales. Store employees are given the task of polling local residents regarding their opinions about the store as well as asking them to make suggestions for improvement.

13. A major television network wants to know what TV shows people in one state are watching most often. For the study, televisions in 350 households in the four largest cities are monitored for one week.

14. American Star is a hit TV show in which Americans are asked to decide who has enough talent to be a star. Each week viewers call in and vote for their favorite performers. The contestant with the fewest votes is eliminated from the show.

15. A study regarding physical fitness is being conducted on campus. All Exercise Science majors are required to be a part of the study.

Chapter Review

Section 1.1 - Getting Started	Pages
• **Definitions** **Data** – information, in particular, information prepared for a study **Population** – the particular group of interest **Parameter** – a numerical description of a particular population characteristic **Sample** – a subset of the population from which data are collected **Sample Statistics** – the actual numerical descriptions of a particular sample characteristic **Census** – data which are obtained from every member of the population	p. 5
• Population values are called parameters.	p. 5
• Sample values are called statistics.	p. 5
• **Branches of Statistics** 1. **Descriptive Statistics** – gathers, sorts, summarizes, and displays the data 2. **Inferential Statistics** – uses descriptive statistics to estimate population parameters	p. 8

Section 1.2 - Data Classifications	Pages
• **Definitions** **Qualitative Data** – consists of labels or descriptions of traits **Quantitative Data** – consists of counts or measurements **Continuous Data** – can take on any value in a given interval, are usually measurements **Discrete Data** – can take on only particular values, are usually counts	p. 12 – 13
• **Levels of Measurement** 1. **Nominal** – consists of labels or names 2. **Ordinal** – data can be arranged in a meaningful order, but calculations such as addition and division do not make sense 3. **Interval** – data can be arranged in a meaningful order and differences between entries are meaningful 4. **Ratio** – data can be arranged and differences between entries are meaningful, and the zero point indicates the absence of something	p. 13 – 17

Section 1.3 - The Process of a Statistical Study	Pages
• **Definitions** **Observational Study** – observes data that already exists **Representative Sample** – has the same relevant characteristics as the population and does not favor one group of the population over another	p. 20 – 26

Section 1.3 - The Process of a Statistical Study (cont.)	Pages
• **Definitions (cont.)** **Sampling Frame** – a list of all the members of the population from which the sample can be drawn **Meta-Analysis** – study that compiles information from previous studies **Case Study** – looks at multiple variables from a single unit **Experiments** – seek to create data to help identify cause and effect relationships **Subjects** – people or things being studied in an experiment **Participants** – people in a study or experiment **Treatment Group** – a group in which researchers apply a treatment **Control Group** – a group in which no treatment is applied **Placebo** – a substance that appears identical to the actual treatment	p. 20 – 26
• **Sampling Techniques:** 1. **Census** – gathering data from every member of the population 2. **Random Sampling** – every member of the population has an equal chance of being selected 3. **Stratified Sampling** – dividing members of the population into subgroups, called **strata**, that share similar characteristics and selecting the same number of subjects from each 4. **Cluster Sampling** – dividing the population into groups, called **clusters**, that are each similar to the entire population and selecting whole clusters, at random, to sample 5. **Systematic Sampling** – selecting every n^{th} member of the population 6. **Convenience Sampling** – the sample is "convenient" for the researcher to select	p. 20 – 23

Section 1.4 - The Reality of Conducting a Study	Pages
• **Definitions** **Informed Consent** – completely disclosing to participants the goals and procedures involved in the study and obtaining their agreement to participate **Biased** – when a study tends to favor certain results **Researcher Bias** – if the researcher influences the results of a study **Sampling Errors** – errors resulting from how the sample is chosen **Dropout** – a participant who begins the study but fails to complete it **Non-Sampling Error** – errors resulting from sources other than the construction of the sample **Non-Adheres** – subjects who stray from the directions they were given, but remain in the sample **Confounding Variables** – factors that influence the results of a study that the researchers did not, or could not, account for	p. 30 – 32

Chapter Exercises

1. Eighty-nine percent of Americans polled are in favor of arts education being taught in public schools.
   a. Does the above numerical value describe a parameter or a statistic?
   b. After reading the above statement, you decide to campaign for the arts in education and use the statement: "A majority of Americans are in favor of art education which shows their love of arts. Americans support ballet teaching in public schools." What branch of statistics would you be applying?

2. The average life-span of Americans in 1990 was 75.37 years.
   a. Does the above numerical value describe a parameter or a statistic?
   b. Giving a complete breakdown of the above statement into categories of gender, sex and geographical regions would be an example of which branch of statistics?

3. Determine whether the following data is qualitative or quantitative and classify it according to its level of measurement.
   a. The street addresses of houses along North Lamar Avenue.
   b. The years in which Christmas falls on a Sunday.
   c. The monthly rates for digital cable service for five competing companies.

4. Jake wanted to know the most popular brand of cereal amongst teenagers. He stood in the center of the cereal aisle at the local grocery and surveyed every 10th teenager who approached him.
   a. How many biases can you name?
   b. In what ways could Jake better collect data to answer his question?

5. If boys are assigned odd-numbered rooms and an equal number of girls are assigned even-numbered rooms in a dorm, is choosing every 10th room a representative sample? What about every 5th?

6. If someone is looking at what proportion of students are left-handed, is a current statistics class a good sample? Is this class a good indicator of the male to female ratio on a campus?

7. A supermarket wants to decide whether or not to carry a new brand of salsa, and uses a free taste test stand in the store.
   a. What is the population being studied?
   b. What type of sampling technique is being used?
   c. Is the taste test representative of the population?

8. It is not always possible to obtain a sampling frame for a given population. Give one example of a population that has a sampling frame and one that does not.

9. How could you use cluster sampling if you want to know the mean price of gasoline at gas stations located within a mile of rental car return locations at airports?

10. A researcher wants to determine the average age of people attending the state fair.
  a. Could a sampling frame be found for this population?
  b. Give 3 different methods of sampling that could be used in this scenario. Which method is best and why?
  c. List any potential for bias.

11. Suppose that a researcher wants to determine the optimal number of hours of sleep that the average adult requires each night.
  a. Identify the following: the population and appropriate type of study to conduct.
  b. Give a brief discussion on how you would set up and conduct this type of study.

12. Convenience sampling is the sampling technique most prone to bias. However, it is possible to obtain a representative sample using the convenience method. Give one example of how the convenience method could produce a representative sample and one example of it producing a biased sample.

13. Suppose that the governor of Nebraska wants to determine his approval rating among his constituents. A team from the governor's staff is assigned to the task. Four of the state's largest cities are chosen, and 100 registered voters from each city are chosen to be included in the sample.
  a. Identify the sampling technique for this scenario.
  b. Identify all potential sources of bias.

Directions: Answer the following questions for exercises 14 and 15:
  a. What is the population being studied?
  b. What is the sample?
  c. What are the sample statistics?
  d. List any potential sources of bias.

14. A recent survey of approximately 1000 adults showed that Great Britain's population found the following events stressful:

Divorce/separation	78%
Moving	77%
Studying for an exam	70%
Christmas shopping	49%
Visiting the dentist	43%

Source: http://www.mori.com/polls/2004/argos.shtml

15. A travel agency wanted to know how many adults use the internet to book their vacations. Of the adults who purchased a vacation package in the past year, 724 were surveyed. It was found that one in five vacationers book their vacation package online.

Source: http://www.mori.com/polls/2004/abta.shtml

Chapter Project: A – Analyzing a Given Study

Poker on the Rise with Young Adults

A mail survey conducted by the American Gaming Association in 2004, showed that 54.1 million American adults had made trips to a casino that year. They averaged 5.9 trips per person. The survey also showed that these people were more affluent than the average American adult with a household income of $55,500 versus the average $47,000. They also had a slightly higher education with 55% having some college versus 53% of average Americans. This has been the norm over the past few years. What has become a surprise is the amount of young adults who play poker. 18% of the young adults surveyed said they played poker in the past 12 months, up from 12% in 2003. Of those, 29% of young adults, ages 21-39, played poker which is an increase of 7% from the previous year. This was followed by 16% of adults age 40-49, 14% of adults age 50-65, and 12% of those 65 and over.

Source: The Detroit News Entertainment Insider, July 14, 2005

Analyze the above survey using the following questions as a guide.

1. Identify the population being studied.
2. Identify the sample and the approximate sample size. Is the sample representative of the population? Explain your answer.
3. Describe how the sample was chosen. Is there any potential bias in the sampling method? Explain your answer.
4. List the descriptive statistics given in the article.
5. What inferences does the article make from the descriptive statistics?
6. Who conducted the study? Is there any potential researcher bias? Explain your answer.

Chapter Project: B – Analyzing a Study You Find

Find a study of interest on the Internet, in a newspaper or magazine, or in an educational journal. If you are reading studies in another course, an excellent choice is to use an assignment from that class. This would also be a good opportunity to begin reading journal articles from your major field of study. The periodical section of the library or library internet resources are wonderful starting places. Once you have found an article, make a copy to include with your analysis.

Note: Though it may be tempting to choose a very short article, if the article is too short, it might not contain all of the necessary elements.

Once you have found a study of interest, analyze it using the following questions as a guide. Write up a formal summary of your analysis.

1. Who conducted the study, when and where?
2. What question(s) does the study seek to answer?
3. Identify the population and variables being studied.
4. Identify the sample and the approximate sample size. Is the sample representative of the population? Explain your answer.
5. How do the researchers deal with the issues of confidentiality or informed consent?
6. Describe how the sample was chosen. Is there any potential bias in the sampling method? Explain your answer.
7. Are any non-sampling errors identified?
8. List the descriptive statistics given in the article.
9. Are there any confounding variables?
10. What inferences do the researchers draw from the descriptive statistics?
11. Is there any potential researcher bias?
12. Do you feel comfortable believing the results of the study based on your analysis? Explain your answer.

Chapter Technology – Generating Random Numbers

TI-84 Plus

To generate a random number using a TI-84 Plus calculator, use the RANDINT function under the MATH-PRB (probability) menu. The function syntax is **randInt(*lower, upper, number of trials*)**. This will produce the specified number of random integers between the lower and upper values. If you only want 1 random number, the formula can be modified to **randInt(*lower, upper*)**.

For example, if we want to get 7 random numbers between 10 and 25, we would press the MATH button, then scroll over to PRB and then scroll down to option 5, 'randInt(' and press ENTER. Type in '10, 25, 7)'. Press ENTER. The calculator will then produce a string of 7 random whole numbers between 10 and 25.

Example T.1

Use your TI-84 Plus calculator to create a random sample of 4 integers between 2 and 9.

Solution:

Press the MATH button, then scroll over to PRB, choose option 5, 'randInt(', and press ENTER. Type in '2, 9, 4)'. Press ENTER and you will now have something similar to the screen on the right.

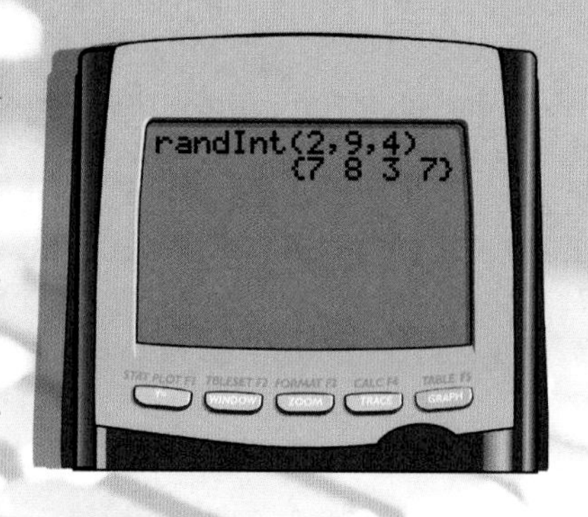

This screenshot shows a random sample of 7, 8, 3, and 7. Your random sample will obviously be different.

Duplicate numbers are a possibility, as shown in our example. If distinct numbers are required, you might need to increase the number of random digits you ask the calculator to generate.

MICROSOFT EXCEL

Creating a random number in Microsoft Excel is very easy. You can use the function command =RANDBETWEEN(*lower*, *upper*), which returns a random number between the numbers you specify. For example, if we want to generate a random number between zero and nine, we would type in the formula =RANDBETWEEN(0, 9) in one of the cells in the worksheet. We used this formula to create the random digits table in Appendix A.

NOTE: If RANDBETWEEN does not appear in your Excel menu, you can easily add this function. Under TOOLS, click on Add-Ins. Select Analysis Toolpack. When finished using this command, it might be a good idea to remove the Analysis Toolpack as it slows Excel down.

Example T.2

Use Microsoft Excel to create a list of 5 random numbers between 200 and 299.

Solution:

In cell A1, type in =RANDBETWEEN(200, 299). This should give you one random number. Copy that formula in 4 additional cells to get the remaining 4 random numbers. Your worksheet might look like the screen on the right.

	A	B
1	260	
2	280	
3	241	
4	262	
5	227	
6		
7		

For this screenshot, the random numbers would be 260, 280, 241, 262, and 227.

It is possible for Excel to duplicate random numbers here. If distinct random numbers are required, you will likely need to generate additional numbers.

MINITAB

Random numbers can also be created using Minitab.

Example T.3

Generate 20 random integers between 1 and 100 using Minitab.

Solution:

Go to CALC then RANDOM DATA and *Integer*. Enter the following parameters: **Generate** 20 rows of data, **Store in column(s):** c1, **Minimum value:** 1, **Maximum value:** 100; and select OK. A menu is displayed below. Your results will be produced in column c1.

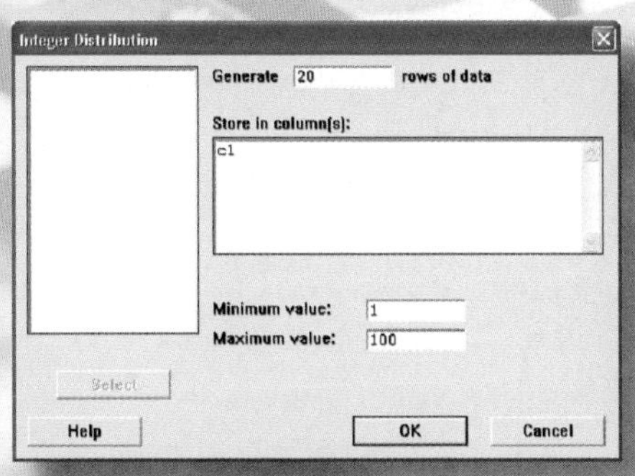

Graphical Descriptions of Data

Sections

Objectives

The objectives of this chapter are as follows:

1. Determine how to construct a frequency table
2. Create and interpret the basic types of graphs used to display data
3. Distinguish between the basic shapes of a distribution
4. Identify misleading characteristics of a graph

Chapter 2

Introduction

In our modern, fast-paced society, graphs are useful for quickly providing a wealth of information to a reader. However, you must always use a discriminating eye and beware of graphs that are poorly constructed and/or misleading. For instance, Figure 1 illustrates the long distance charges of various schools in the South.

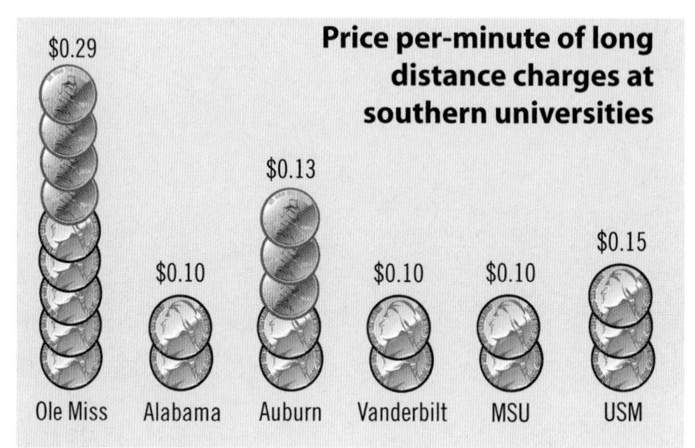

Figure 1 – Long Distance Prices at Southern Universities

Which school has the second-highest long distance rate? Take another look, because it is not Auburn. Although using coins for the bars was a clever idea, the overall picture is incorrect. USM, at $0.15 per minute, actually has a higher rate than Auburn's $0.13 per minute.

Unfortunately, you can find misleading graphs everywhere. Sometimes these poorly constructed graphs occur by accident. At other times, they are purposefully misleading. In either case, as an informed reader, you should be able to spot an incorrect graph. Don't be fooled by slick graphics.

Frequency Distributions

2.1

After collecting data as described in Chapter 1, you need to organize it so that inferences and conclusions can be drawn. Raw data is not very meaningful to an audience. Imagine looking up the prices of new plasma screen TVs on an internet search engine and getting the following dollar values.

Table 1 - Plasma Screen TV Prices (in Dollars)				
1999	1599	1999	1899	1899
1699	1699	1899	1799	1685
1888	1787	1984	1699	1799
1699	1885	1999	1595	1757

This list is not very useful if you want to know the lowest price of a plasma screen TV or the average price. If you create an **ordered array**, an ordered list of the data from largest to smallest or vice versa, then the lowest price and highest price are easy to see.

Table 2 - Plasma Screen TV Prices (in Dollars and in an ordered array)				
1595	1599	1685	1699	1699
1699	1699	1757	1787	1799
1799	1885	1888	1899	1899
1899	1984	1999	1999	1999

✦ Constructing a Frequency Distribution

Now we can see that the lowest price for the TV is $1595, and the highest price is $1999. But which price is most common? Is there a pattern to the data? If we look at a distribution of the data, we can answer these questions. A **distribution** displays data values that occur and how frequently each value occurs. One type of distribution is a frequency distribution. A **frequency distribution** is a table that divides the data into groups, called **classes**, and shows how many data values occur in each group. The **frequency**, f, of a group, or class, is the number of data values in that class.

There are two basic types of frequency distributions: grouped distributions and ungrouped distributions. If the data set is relatively small, or contains only a few possible values, then each class might represent a single value. For example, if you were looking at the distribution of letter grades, the first class could be the A's, the second class could be the B's, and so forth. This would produce a frequency distribution with 5 classes (one for each possible letter grade). A frequency distribution such as this one, where each class represents a single value, is called an **ungrouped frequency distribution**.

With large data sets containing many different values for the variable, an ungrouped frequency distribution might have hundreds of classes, and thus would be unreasonable and probably not depict a clear pattern to the data. Our example of plasma screen TV prices would result in 12 different classes if we used each value as a separate class. (Look again at the ordered data set above and see if you can verify this.) In cases such as this, the data is often grouped into ranges of values, thus allowing for fewer classes and a more manageable frequency distribution. A frequency distribution of this type, where the classes are ranges of possible values, is called a **grouped frequency distribution**. Grouped distributions are more common and take more skill to create, so we will focus the remainder of our discussion on grouped frequency distributions. The procedure for creating a grouped frequency distribution (from this point on, we will refer to this simply as a frequency distribution) is outlined on the following page.

Steps for Constructing a Frequency Distribution 

1. *Decide how many classes should be in the distribution.* There are typically between 5 and 20 classes in a frequency distribution. Several different methods can be used to determine the number of classes that will show the data most clearly, but for this textbook, the number of classes for a given data set will be suggested.

2. *Choose an appropriate class width.* To find an appropriate **class width**, begin by subtracting the lowest number in the data set from the highest number in the data set and dividing the difference by the number of classes. Rounding this number up gives a good starting point from which to choose the class width. You will want to choose a width so that the classes formed present a clear representation of the data, so make a sensible choice. Also note that the class width is the difference between lower limits of consecutive classes, which we will define in the next step.

3. *Find the class limits.* The **lower class limit** is the smallest number that can belong to a particular class, and the **upper class limit** is the largest number that can belong to a class. Using the minimum data entry, or a smaller number, as the lower limit of the first class is a good place to begin. However, judgment is required. You should choose the first lower limit so that reasonable classes will be produced, and it should have the same number of decimals as the data set. After choosing the lower limit of the first class, add the class width to it to find the lower limit of the second class. Continue this pattern until the classes are complete. The upper limit of each class is determined such that the classes do not overlap. If, after creating your classes, there is data that falls outside the class limits, you must adjust either the class width or the choice for the first lower class limit.

4. *Determine the frequency of each class.* Make a tally mark for each piece of data in the appropriate class. Count the marks to find the total frequency for each class.

Rounding Rule

Class limits should have the same number of decimal places as the data set.

Remember that the objective for creating a frequency distribution is to provide an overview of the data. There will be many ways to create this overview. By using common sense when choosing the classes, one can present the data clearly and concisely. Let's look at an example of creating a frequency distribution.

Example 1

Create a frequency distribution for the list of plasma screen TV prices given earlier using 5 classes.

Solution:

Begin by finding a starting point for the choice of class width. Subtract the lowest piece of data from the highest and divide by the number of classes, as shown below:

$$\frac{1999-1595}{5}=80.8\approx 81.$$

This would give us a class width of \$81. We will stop here and consider some options. Choosing a class width of \$81 does seem perfectly reasonable from a theoretical point of view. However, one should consider the impression created by having TV prices grouped in intervals of \$81. Can you imagine presenting this data to a client? Instead, it would be more reasonable to group TV prices by intervals of \$100. Therefore, we will choose our class width to be \$100.

Next, we need to choose a starting point for the classes, i.e., the first lower class limit. One should always first consider using the smallest data entry for the beginning point. In this case, if we choose the smallest TV price, we would be starting the first class at \$1595 with a width of \$100. However, given that we've chosen a class width of \$100, it is more natural to begin the first class at \$1500.

Now let's continue building the class limits. Adding the class width of \$100 to \$1500, we obtain a second lower class limit of \$1600. The next lower limit is found by adding \$100 to \$1600. We continue in this fashion until we have five lower class limits, one for each of our five classes.

Finally, we need to determine an appropriate cut-off for the upper class limits. Again, be reasonable. Remember, too, that the classes are not allowed to overlap. Because the data is in whole dollar amounts, it makes sense, then, to choose upper class limits that are one dollar less than the next lower limit. The classes we have produced are as follows:

Table 3 - Plasma Screen TV Prices

Class	Frequency
\$1500–\$1599	
\$1600–\$1699	
\$1700–\$1799	
\$1800–\$1899	
\$1900–\$1999	

cont'd on the next page

Note that the last upper class limit is also the maximum value in the data set. This will not necessarily occur in every frequency table. However, we have included all the data values in our range of classes, so no adjustments to the classes are necessary.

Tabulating the number of data values that occur in each class produces the following frequency table:

Table 4 - Plasma Screen TV Prices

Class	Frequency
$1500–$1599	2
$1600–$1699	5
$1700–$1799	4
$1800–$1899	5
$1900–$1999	4

Note that the sum of the frequency column should equal the number of data values in the set. Check for yourself that this is true.

✦ Characteristics of a Frequency Distribution

There are other characteristics of a frequency distribution that can be calculated once the basic frequency table has been constructed. Let's look at four of them.

The first calculation is that of **class boundaries**, which are similar to the class limits. The class boundaries split the difference in the gap between the upper limit of one class and the lower limit of the next class. To find a class boundary, add the upper limit of one class to the lower limit of the next class and divide by two. For example, if an upper class limit is 10, and the next lower class limit is 11, the class boundary would be

$$\frac{10+11}{2}=10.5.$$

Thus 10.5 is the boundary between those two classes. You can use one class boundary to find all of the other class boundaries by adding or subtracting the class width to this boundary. Class boundaries are useful in constructing frequency histograms, which we will cover in the next section on graphing distributions.

Example 2

Calculate the class boundaries for each class in Example 1.

Solution:

Look at the first and second classes. The upper limit of class one is 1599. The lower limit of class two is 1600. This gives you a class boundary of

$$\frac{1599+1600}{2}=1599.5.$$

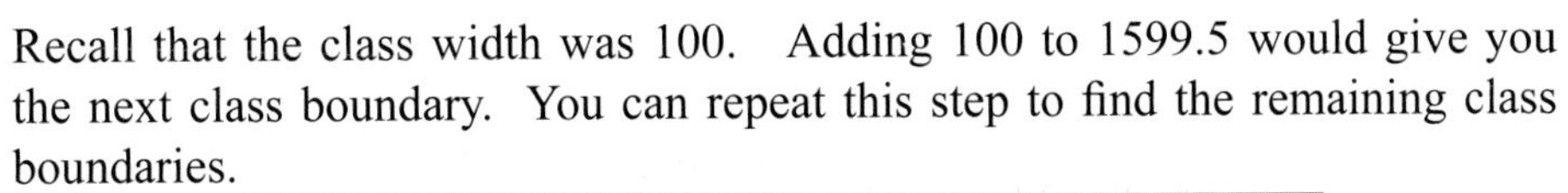

Recall that the class width was 100. Adding 100 to 1599.5 would give you the next class boundary. You can repeat this step to find the remaining class boundaries.

Table 5 - Plasma Screen TV Prices		
Class	**Frequency**	**Class Boundaries**
\$1500–\$1599	2	1499.5–1599.5
\$1600–\$1699	5	1599.5–1699.5
\$1700–\$1799	4	1699.5–1799.5
\$1800–\$1899	5	1799.5–1899.5
\$1900–\$1999	4	1899.5–1999.5

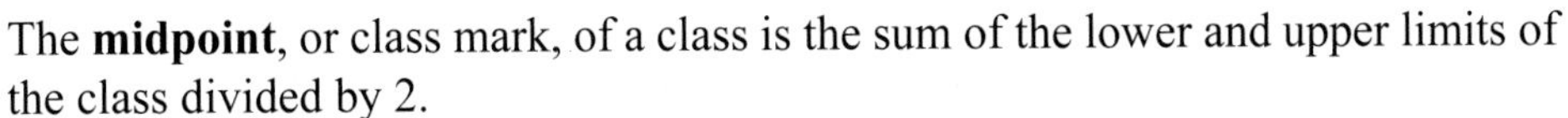

The **midpoint**, or class mark, of a class is the sum of the lower and upper limits of the class divided by 2.

$$\text{Midpoint} = \frac{\text{Upper Limit} + \text{Lower Limit}}{2}$$

The midpoints are often used for estimating the average value in each class.

Example 3

Calculate the midpoint for each class in Example 1.

Solution:

The midpoint is the sum of the class limits divided by two. For the first class, the midpoint is

$$\frac{1500+1599}{2}=1549.5.$$

We can use this same calculation to construct the midpoint of the remaining classes. Another method is to add 100 to the first midpoint, as we did with class boundaries.

cont'd on the next page

Table 6 - Plasma Screen TV Prices		
Class	**Frequency**	**Midpoint**
\$1500–\$1599	2	1549.5
\$1600–\$1699	5	1649.5
\$1700–\$1799	4	1749.5
\$1800–\$1899	5	1849.5
\$1900–\$1999	4	1949.5

The third calculation we will discuss, **relative frequency**, is the percentage of the data set that falls into a particular class. It is calculated by dividing the class frequency by the sample size. The sample size, *n*, for a frequency distribution can be found by adding all of the class frequencies together.

$$\text{Relative Frequency} = \frac{\text{Class Frequency}}{\text{Sample Size}} = \frac{f}{n}$$

Relative frequencies are useful because percentages make it easier to quickly analyze the data as a whole.

Example 4

Calculate the relative frequency for each class in Example 1.

Solution:

We first find the sample size by summing down the frequency column. Thus

$$n = 2 + 5 + 4 + 5 + 4 = 20.$$

Then divide each class frequency by 20.

Table 7 - Plasma Screen TV Prices		
Class	**Frequency**	**Relative Frequency**
\$1500–\$1599	2	$\frac{2}{20} = \frac{1}{10} = 10\%$
\$1600–\$1699	5	$\frac{5}{20} = \frac{1}{4} = 25\%$
\$1700–\$1799	4	$\frac{4}{20} = \frac{1}{5} = 20\%$
\$1800–\$1899	5	$\frac{5}{20} = \frac{1}{4} = 25\%$
\$1900–\$1999	4	$\frac{4}{20} = \frac{1}{5} = 20\%$

The final calculation we will look at, **cumulative frequency**, is the sum of the frequency for a given class and all previous classes. The cumulative frequency of the last class equals the sample size.

Memory Booster!

Relative Frequency relates a class to the whole.

Cumulative Frequency is the running total of the amount of data.

Example 5

Calculate the cumulative frequency for each class in Example 1.

Solution:

Table 8 - Plasma Screen TV Prices

Class	Frequency	Cumulative Frequency
\$1500–\$1599	2	2
\$1600–\$1699	5	7 (2 + 5)
\$1700–\$1799	4	11 (2 + 5 + 4)
\$1800–\$1899	5	16 (2 + 5 + 4 + 5)
\$1900–\$1999	4	20 (2 + 5 + 4 + 5 + 4)

Now let's put all of these concepts together in a single example.

Example 6

Data collected on the number of miles that professors drive to work daily is listed below. Use this data to create a frequency table that includes the class boundaries, midpoint, relative frequency, and cumulative frequency. Use six classes. Be sure that your class limits have the same number of decimals as the data set.

Table 9 - Miles Professors Drive to Work Each Day

3.8	2.7	9.3	6.5	5.8	7
10.2	1	3.7	9.1	6.2	11
11.9	5.5	4.8	7.3	9.1	1.4

Solution:

Since this example calls for 6 classes, a good starting point for the class width is

$$\frac{11.9-1}{6}=1.81\bar{6}6\approx 1.8.$$

Because our data are in miles, a more sensible class width to use is 2.

Next, to choose the lower class limit of the first class, begin by considering the smallest piece of data, which is 1 mile. In this case, 1.0 is a reasonable place to begin our classes. Adding the class width of 2 gives us the table on the following page.

cont'd on the next page

Table 10 - Miles Professors Drive to Work Each Day Frequency Table	
Class	**Frequency**
1.0–2.9	
3.0–4.9	
5.0–6.9	
7.0–8.9	
9.0–10.9	
11.0–12.9	

Once again, note that all of the data values will fall within the range of the class limits. So, no adjustments in the classes are necessary.

The upper class boundary and midpoint of the first class are as follows:

$$\text{Upper Boundary Class 1: } \frac{2.9+3.0}{2}=2.95,$$

$$\text{Midpoint Class 1: } \frac{1+2.9}{2}=1.95.$$

Use the class width to find the other class boundaries and midpoints. The frequency, relative frequency, and cumulative frequency are calculated as in previous examples.

Table 11 - Miles Professors Drive to Work Each Day Frequency Table

Class	Freq.	Class Boundaries	Midpoint	Relative Frequency	Cumulative Frequency
1.0–2.9	3	0.95–2.95	1.95	$\frac{3}{18}=\frac{1}{6}\approx 17\%$	3
3.0–4.9	3	2.95–4.95	3.95	$\frac{3}{18}=\frac{1}{6}\approx 17\%$	6
5.0–6.9	4	4.95–6.95	5.95	$\frac{4}{18}=\frac{2}{9}\approx 22\%$	10
7.0–8.9	2	6.95–8.95	7.95	$\frac{2}{18}=\frac{1}{9}\approx 11\%$	12
9.0–10.9	4	8.95–10.95	9.95	$\frac{4}{18}=\frac{2}{9}\approx 22\%$	16
11.0–12.9	2	10.95–12.95	11.95	$\frac{2}{18}=\frac{1}{9}\approx 11\%$	18

Exercises

2.1

1. For the frequency table below, compute the **class boundaries** between each class.

Table for Exercise 1 - Age of Taste Test Participants (in Years)	
Class	**Frequency**
15–19	7
20–24	8
25–29	10
30–34	2
35–39	3

2. For the frequency table below, compute the **class boundaries** between each class.

Table for Exercise 2 - Braking Time for Vehicles (in Minutes)	
Class	**Frequency**
0.05–0.07	12
0.08–0.10	15
0.11–0.13	14
0.14–0.16	15
0.17–0.19	14

3. For the frequency table below, identify the **class width** of each class.

Table for Exercise 3 - Age at First Marriage (in Years)	
Class	**Frequency**
15–18	2
19–22	5
23–26	4
27–30	5
31–34	4

4. For the frequency table below, identify the **class width** of each class.

Table for Exercise 4 - Hourly Wage at First Job (in Dollars)	
Class	**Frequency**
7.5–8.4	12
8.5–9.4	50
9.5–10.4	48
10.5–11.4	45
11.5–12.4	34

5. For the frequency table below, identify the **midpoint** of each class.

Table for Exercise 5 - Age of Survey Participants (in Years)	
Class	**Frequency**
15–24	9
25–34	8
35–44	12
45–54	1
55–64	3

6. For the frequency table below, identify the **midpoint** of each class.

Table for Exercise 6 - Cost of a 12 oz. Soda (in Dollars)	
Class	**Frequency**
0.25–0.49	2
0.50–0.74	15
0.75–0.99	12
1.00–1.24	5
1.25–1.49	9

7. For the frequency table below, calculate the **relative frequency** for each class.

Table for Exercise 7 - Age of Purchasing First Home (in Years)	
Class	**Frequency**
15–18	2
19–22	7
23–26	4
27–30	15
31–34	3

8. For the frequency table below, calculate the **relative frequency** for each class.

Table for Exercise 8 - Hourly Wages of Surveillance Operators (in Dollars)	
Class	**Frequency**
7.5–8.4	92
8.5–9.4	78
9.5–10.4	68
10.5–11.4	45
11.5–12.4	34

9. For the frequency table below, calculate the **cumulative frequency** for each class.

Table for Exercise 9 - Age of Purchasing First Car (in Years)	
Class	**Frequency**
15–19	12
20–24	8
25–29	15
30–34	12
35–39	9

10. For the frequency table below, calculate the **cumulative frequency** for each class.

Table for Exercise 10 - Grades on a Difficult Test	
Class	**Frequency**
A	2
B	5
C	7
D	13
F	10

Directions: For each of the following sets of data, a frequency table has been started. Complete the frequency table for that data.

11. The following data describes the heights of 30 men, in inches, entering the military in April 2006.

72.8	71.2	70.3	73.4	72.6	74.1
70.9	71.6	72.1	74.6	75.0	72.0
69.1	69.5	72.6	72.4	73.6	75.1
71.8	71.6	71.9	70.9	70.2	69.3
72.1	72.3	72.5	73.4	74.0	75.0

Table for Exercise 11 - Heights of Men (in Inches)	
Class	**Frequency**
69.0–69.9	3
70.0–70.9	
71.0–71.9	
72.0–72.9	
73.0–73.9	
74.0–74.9	
75.0–75.9	

12. The following data represents the number of exercises at the end of various sections in a traditional college algebra textbook.

145	137	138	112	137	100	78	127	97
70	143	133	150	124	115	110	45	141
119	92	84	94	105	71	95	117	104

Table for Exercise 12 - Number of Section Exercises	
Class	**Frequency**
40–59	
60–79	
80–99	
100–119	
120–139	
140–159	

13. At a state fair, one game involves guessing the number of marbles in a glass jar. The following data represents the guesses that people made during one hour at the state fair.

1234	1645	1469	1467	1549	1348	1671	1300	1200	1199
1621	1547	1501	1410	1487	1299	1500	1688	1301	1399

Table for Exercise 13 - Number of Marbles in the Jar	
Class	**Frequency**
1100–1199	
1200–	
1300–1399	4
–1499	4
1500–1599	
1600–	4

14. The following data represents the amount of pocket change, in dollars, in the pockets of a sample of men in an office building.

0.23	0.52	0.76	0.79	0.8	0.21	0.13
1.05	1.24	1.15	1.10	0.98	0.28	0.64
1.34	0.38	0.31	0.42	0.41	0.24	1.42

Table for Exercise 14 - Amount of Pocket Change (in Dollars)	
Class	**Frequency**
–0.24	4
0.25–0.49	
0.50–	2
0.75–0.99	
–1.24	4
1.25–1.49	

Directions: For each data set below, create a frequency table with the indicated number of classes. Include the frequency, class boundaries, the midpoint, the relative frequency, and the cumulative frequency of each class.

15. The following data represents the number of curl-ups in 60 seconds for a group of sixteen 8-year-old boys. Use 6 classes that have a class width of 5. Begin with a lower limit of 15.

31	34	41	36	27	29	18	33
31	28	34	22	26	28	36	42

16. The following data represents times, in minutes, for completing a one mile Run-Walk from a group of twenty-four 17 year old girls. Use 6 classes that have a class width of 2.00. Begin with a lower limit of 6.00.

15.23	13.52	11.35	11.15	12.20	9.90	10.37	14.05
10.02	17.35	8.33	8.05	9.87	9.28	10.62	6.65
9.55	10.23	13.93	10.97	9.75	12.85	12.82	10.93

17. The following data represents the caloric intake for a group of fifteen men between the ages of 20 and 39. Use 5 classes that have a class width of 400. Begin with a lower limit of 1800.

2700	2200	2500	2800	2600
3000	2600	2200	3100	2800
1800	3500	2500	3000	2900

18. The following data represents normal monthly precipitation in inches for the month of September in Alaska. Use 6 classes that have a class width of 3.50. Begin with a lower limit of 0.

2.7	1.72	1.39	6.88	2.59	2.04	2.43
9.28	1.06	3.29	1.57	4.23	0.95	8.37
0.6	4.41	6.73	1.92	2.28	2.74	18.65

Graphical Displays of Data

2.2

Sometimes it is easier to view the data we gather as a graph rather than a table. A graph is a snapshot that allows us to view patterns at a glance without undergoing lengthy analysis of the data. Also, graphs are much more visually appealing than a table or list. A graph should be able to stand alone, without the original data. To do this, the graph must be given a title, as well as labels for both axes. When appropriate a legend, source, and date should be included as well.

Pie Charts

Let's begin by looking at graphs which display qualitative data. When the categories of data are parts of a whole, such as parts of a budget, a pie chart can be created. A **pie chart** shows how large each category is in relation to the whole; it is created from a frequency distribution by using the relative frequencies. The size, or central angle, of each wedge in the pie chart is calculated by multiplying 360° by the relative frequency of each class and rounding to the nearest whole degree.

Rounding Rule

When constructing a pie chart, round each angle to the nearest whole degree.

Example 7

Create a pie chart from the following data describing the distribution of housing for students in a statistics class. Calculate the size of each wedge in the pie chart to the nearest whole degree.

Table 12 - Types of Housing

Types of Housing	Number of Students
Apartment	20
Dorm	15
House	9
Sorority/ Fraternity House	5

Solution:

	Frequency	Angles
Apartment	$= \frac{20}{49} \approx 0.41 = 41\%$	$\frac{20}{49} \cdot 360° \approx 147°$
Dorm	$= \frac{15}{49} \approx 0.31 = 31\%$	$\frac{15}{49} \cdot 360° \approx 110°$
House	$= \frac{9}{49} \approx 0.18 = 18\%$	$\frac{9}{49} \cdot 360° \approx 66°$
Sorority/Fraternity House	$= \frac{5}{49} \approx 0.10 = \underline{10\%}$	$\frac{5}{49} \cdot 360° \approx \underline{37°}$
	100%	360°

cont'd on the next page

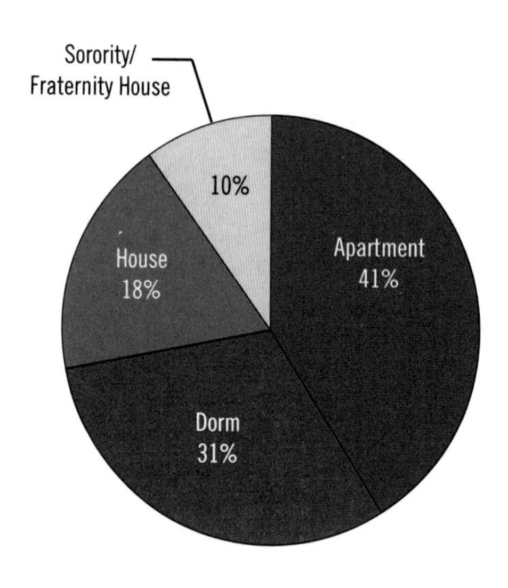

✦ Bar Graphs

Another way to display qualitative data is with a bar graph. **Bar graphs** are used to represent categorical data. The height of the bar represents the amount of data in that category. The horizontal axis contains the qualitative categories, and the vertical axis represents the frequency of each category. Because the bars represent categories, the width of each bar is meaningless, and the bars usually do not touch. However, to avoid misrepresenting the data, the bars should be of uniform width.

Example 8

Create a bar graph of the following data describing the distribution of housing for students in a statistics class.

Table 13 - Types of Housing

Types of Housing	Number of Students
Apartment	20
Dorm	15
House	9
Sorority/ Fraternity House	5

Solution:

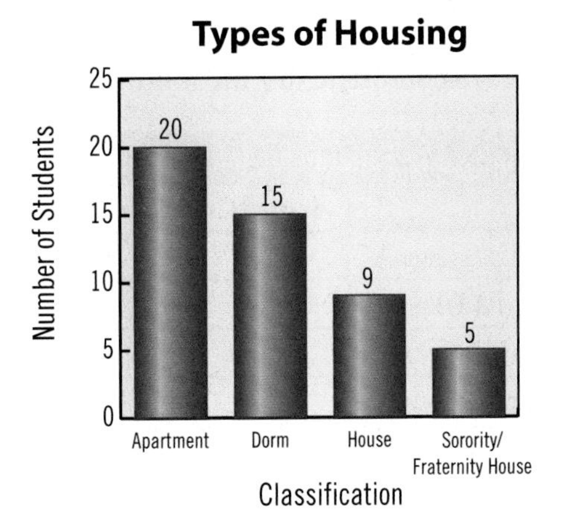

Notice that the bars in the previous example are in order from largest to smallest. This special type of bar graph with the bars in descending order is called a **pareto chart**. Pareto charts are typically used with nominal data. The reason for this is that if a pareto chart were created from ordinal or quantitative data, the values on the x-axis might seem out of order after the bars were rearranged from largest to smallest.

Consider the two bar graphs below showing the frequency of various shoe sizes in a women's sorority. In Figure 2, the bar graph has the bars arranged according to shoe size, with the first bar representing the smallest shoe size and the last bar representing the largest shoe size. Figure 3 is a pareto chart of the same data. Look at the labels on the x-axis. The shoe sizes seem out of order because they represent ordinal data which has a natural order to it and therefore does not lend itself to being rearranged in a pareto chart.

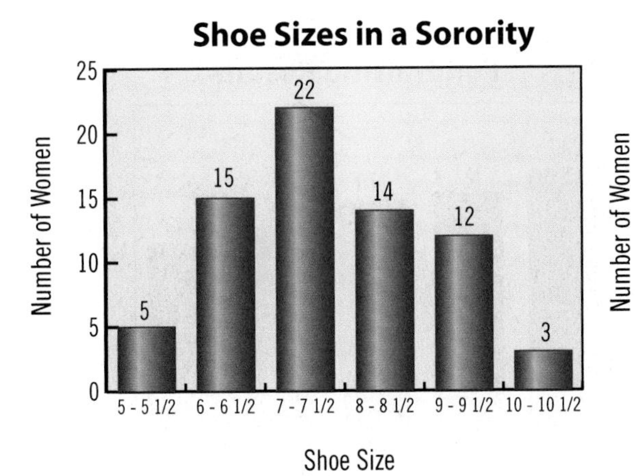

Figure 2 – Bar Graph According to Shoe Size

Appropriate

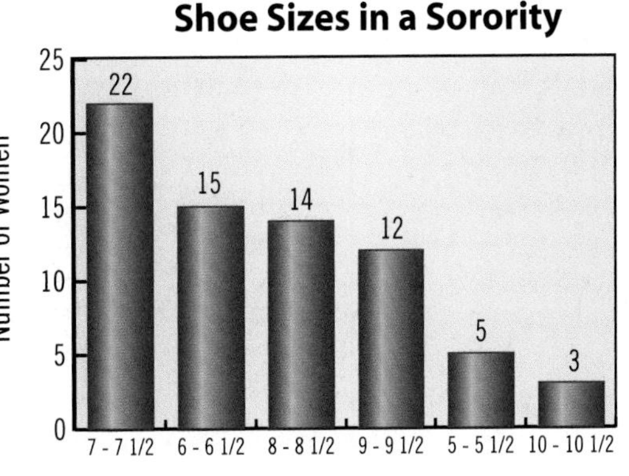

Figure 3 – Pareto Chart

Not Appropriate

Example 9

Create a pareto chart, if appropriate, for the following data:

Table 14 - Fundraising Information per Class

Class	Number of Fundraiser Items Sold
Ms. Boyd's Class	200
Ms. Willard's Class	157
Mr. Smith's Class	176
Mr. Carter's Class	181
Ms. Meredith's Class	143

Solution:

This data is nominal; therefore, it would be appropriate to create a pareto chart from this data. Begin by rearranging the data from the largest frequency to the smallest frequency.

Table 15 - Ordered Fundraising Information per Class

Class	Number of Fundraiser Items Sold
Ms. Boyd's Class	200
Mr. Carter's Class	181
Mr. Smith's Class	176
Ms. Willard's Class	157
Ms. Meredith's Class	143

Next, create a bar graph of the ordered data. The resulting pareto chart is shown below.

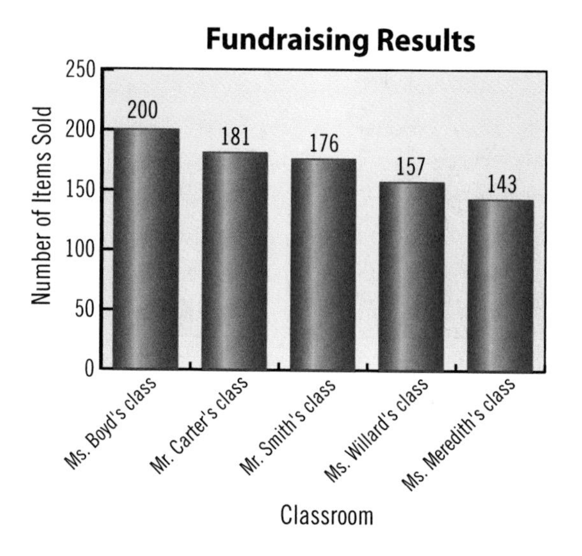

When we want to create a bar graph that compares different groups, we use a **side-by-side bar graph**. In Example 7, we looked at the housing data from one statistics class. Let's compare this to the data from another class. To do so, create a bar for each class and for each category. Identify the bars in some way, such as different colors, to denote which bars represent a given class. In this type of graph, it is important to include a legend that denotes which color represents which category.

Example 10

Create a side-by-side bar graph of the following data describing the distribution of student housing in two statistics classes.

Table 16 - Housing Information from Classes A & B

Types of Housing	Number of Students from Class A	Number of Students from Class B
Apartment	20	13
Dorm	15	24
House	9	6
Sorority/ Fraternity House	5	7

Solution:

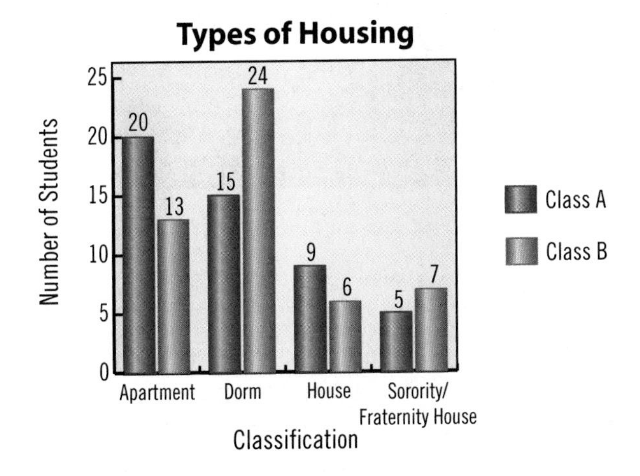

With a side-by-side bar graph, it is very easy to compare the housing characteristics of two classes. For instance, Class A has more students that live in apartments, and Class B has more students who live in dorms. Unfortunately, side-by-side bar graphs are not very convenient if you want to know how many students from both classes live in a certain type of housing. A more efficient graph for displaying the data for this purpose is the **stacked bar graph**. Imagine picking up each bar for Class B and placing it on top of the bar for Class A. Your result would be a stacked bar graph. Here too, a legend is essential.

Example 11

Create a stacked bar graph for the data in Example 10.

Solution:

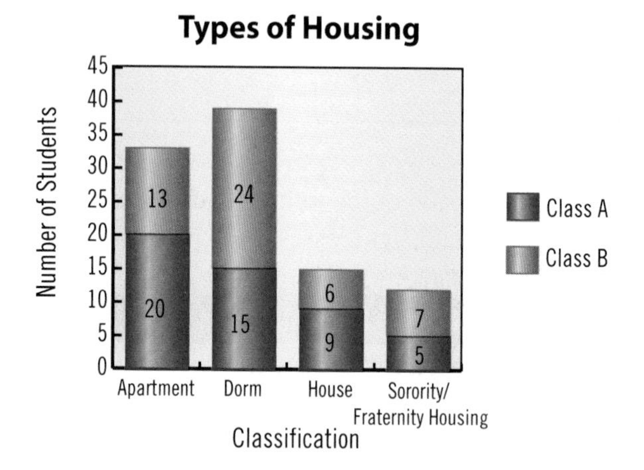

With the stacked bar graph, it is easier to see that more students live in the dorms than in apartments.

✦ Histograms

A bar graph of a frequency distribution is called a **frequency histogram**. To construct a histogram, we begin by finding the class boundaries of the frequency distribution. The horizontal axis of a histogram is a real number line on which you mark the class boundaries of every class. The width of the bars represents the width of each class. Since the upper class boundary of one class is the same as the lower class boundary of the next class, the bars should touch. The bars should be uniform in width; thus, histograms are only appropriate for frequency distributions that have classes of uniform width. Also, no classes can have an undetermined width, such as "15 or greater". The height of each bar represents the frequency of the class; thus, frequency is graphed on the vertical axis. Although we use class boundaries to draw the histogram, it is appropriate to use either the class boundaries or the midpoints when labeling the *x*-axis of a frequency histogram. We will show both in the next example.

Example 12

Construct a histogram of the plasma screen TV prices from the previous section. The frequency table of the data is reprinted here.

Table 17 - Plasma Screen TV Prices

Class	Frequency	Midpoint	Class Boundaries
$1500–$1599	2	1549.5	1499.5–1599.5
$1600–$1699	5	1649.5	1599.5–1699.5
$1700–$1799	4	1749.5	1699.5–1799.5
$1800–$1899	5	1849.5	1799.5–1899.5
$1900–$1999	4	1949.5	1899.5–1999.5

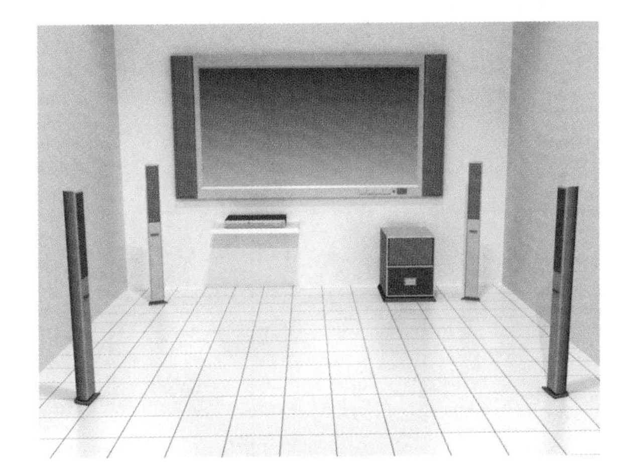

Solution:

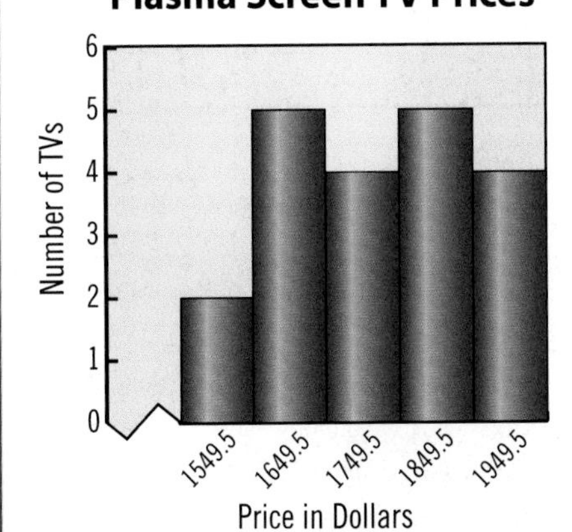

With Midpoints

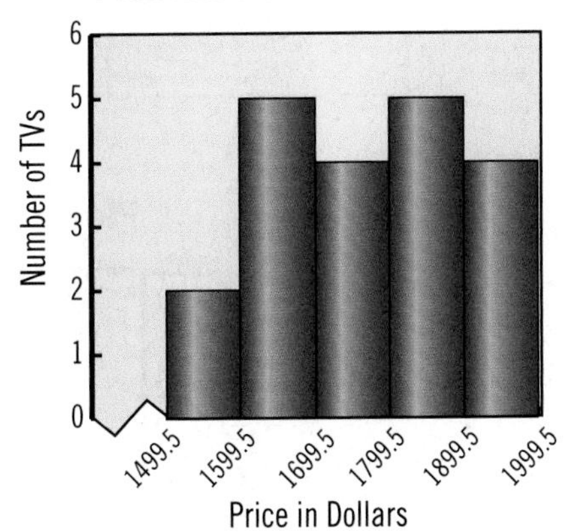

With Class Boundaries

✦ Relative Frequency Histogram

There are times at which it is beneficial to display the relative frequency of a distribution. A **relative frequency histogram** is identical to a regular histogram, except that the heights of the bars represent the relative frequencies of each class rather than simply the frequencies. It is appropriate to label a relative frequency histogram with either decimals or percentages. The following example shows a relative frequency histogram drawn for the previous example of plasma screen TVs.

Example 13

Construct a relative frequency histogram of the plasma screen TV prices from the previous section. The frequency table of the data is reprinted here.

Table 18 - Plasma Screen TV Prices

Class	Frequency	Relative Frequencies
\$1500–\$1599	2	$\frac{2}{20}=\frac{1}{10}=10\%$
\$1600–\$1699	5	$\frac{5}{20}=\frac{1}{4}=25\%$
\$1700–\$1799	4	$\frac{4}{20}=\frac{1}{5}=20\%$
\$1800–\$1899	5	$\frac{5}{20}=\frac{1}{4}=25\%$
\$1900–\$1999	4	$\frac{4}{20}=\frac{1}{5}=20\%$

Solution:

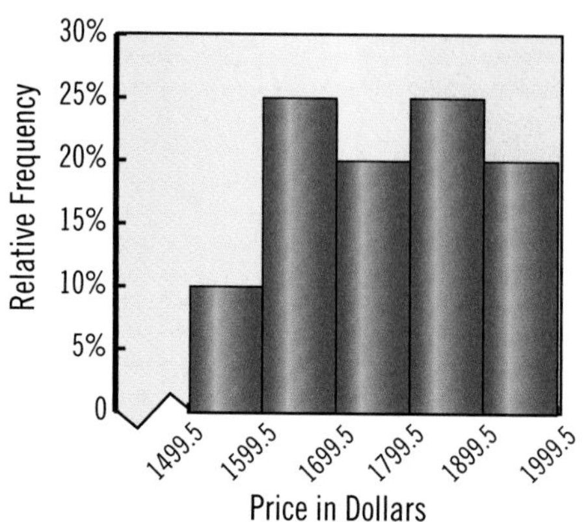

Using the class midpoints, we can also construct what is called a **frequency polygon**. A frequency polygon is a visual display of the frequencies of each class using the midpoints from the histogram.

Steps for Constructing a Frequency Polygon

1. *Mark the class boundaries on the x-axis and the frequencies on the y-axis*. Note, extra classes at the lower and upper ends will be added, each having a frequency of 0. In our plasma TV price example, These classes will be 1400 – 1499 at the lower end and 2000 – 2099 at the upper end.

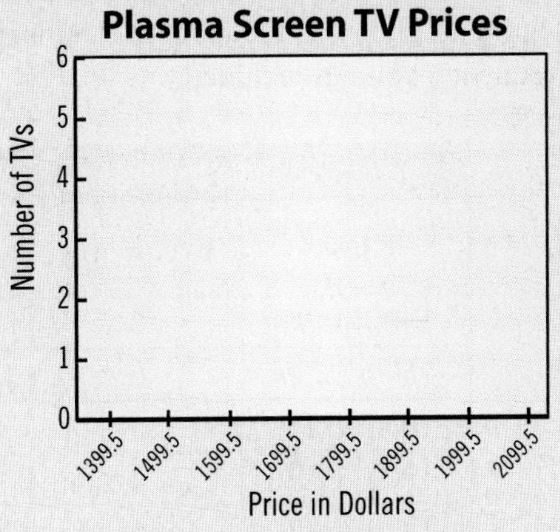

2. *Add the midpoints to the x-axis and plot a point at the frequency of each class directly above its midpoint*. Notice that the class boundaries on the x-axis have been lightened. This is due to the fact that frequency polygons represent the midpoints of the data.

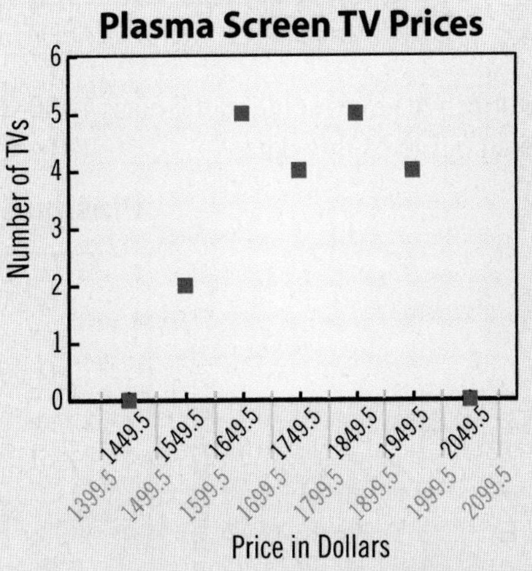

3. *Join each point to the next with a line segment*. Notice, this is not a smooth curve, but a polygon.

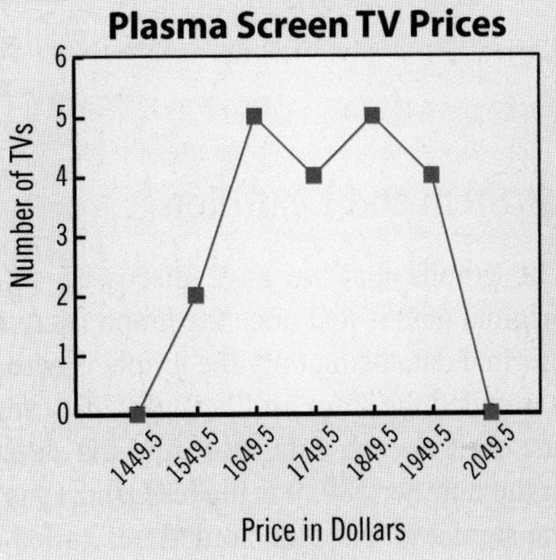

An **ogive** (pronounced "oh-jive") is another type of line graph which depicts *cumulative frequency* of each class from a frequency table. Begin by tabulating the cumulative frequencies for each class. Unlike creating a frequency polygon, we only include an extra class at the lower end for this graph, giving it a frequency of 0. Next, plot a point at the cumulative frequency for each class directly above its *upper class boundary*. The ogive is created by joining the points together with line segments. Below is the table of the Plasma Screen TV Prices with the cumulative frequency column included.

Table 19 - Plasma Screen TV Prices (in Dollars)

Class	Frequency	Cumulative Frequency	Class Boundaries
1400–1499	0	0	1399.5–1499.5
1500–1599	2	2	1499.5–1599.5
1600–1699	5	7	1599.5–1699.5
1700–1799	4	11	1699.5–1799.5
1800–1899	5	16	1799.5–1899.5
1900–1999	4	20	1899.5–1999.5

Notice we have included the class 1400–1499 with a frequency of 0. The following is an ogive of this data.

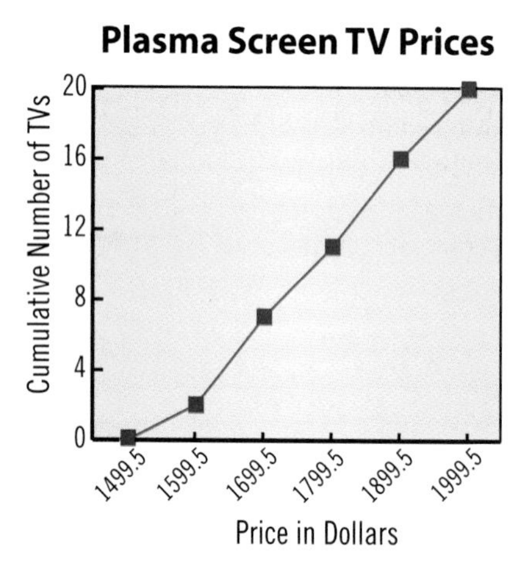

✦ Stem and Leaf Plots

The graphs that we have discussed so far all have one thing in common: the original data is lost once the graph is created, meaning that you cannot recreate the original data using only the graph. There is a graph that retains the original data—it is called the **stem and leaf plot**. In a stem and leaf plot, the leaves are usually the last digit in each data value and the stems are the remaining digits. For example, in the number 189, 9 is the leaf, and 18 is the stem. Be sure to include a legend for the stem and leaf plot, sometimes called a key, so that the reader can interpret the information. For example, $4|1 = 41$.

Steps for Constructing a Stem and Leaf Plot:

1. *Create two columns, one on the left for stems and one on the right for leaves.*

2. *List each of the stems that occurs in the data set in numerical order.* Each stem is normally listed only once; however, the stems are sometimes listed two or more times if splitting the leaves would make the data set's features clearer.

3. *List each leaf next to its stem.* Each leaf will be listed as many times as it occurs in the original data set. There should be as many leaves as there are data values. Be sure to line the leaves up in straight columns so that the table is visually accurate.

4. Create a key to guide interpretation of the stem and leaf plot.

5. The leaves may then be put in order, if desired, to create an **ordered stem and leaf plot**.

Memory Booster!

Although this is not a correct way to draw a stem and leaf plot, this figure will help you remember the concepts associated with a stem and leaf plot.

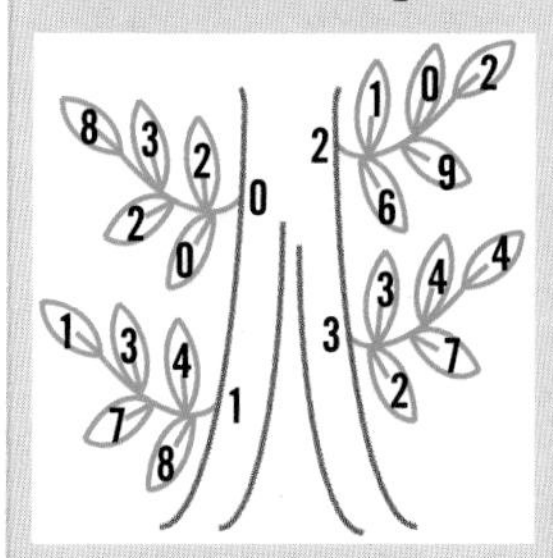

Example 14

Create a stem and leaf plot of the following ACT test scores from a group of entering college freshmen:

ACT Scores

18	23	24	31	19
27	26	22	32	18
35	27	29	24	20
18	17	21	25	26

Solution:

ACT Scores

Stem	Leaves
1	8 9 8 8 7
2	3 4 7 6 2 7 9 4 0 1 5 6
3	1 2 5

Key: 1|8 = 18

The ordered stem and leaf plot from the last example is:

ACT Scores	
Stem	Leaves
1	7 8 8 8 9
2	0 1 2 3 4 4 5 6 6 7 7 9
3	1 2 5

Key: 1|8 = 18

From this stem and leaf plot we can see that the majority of students scored in the 20's. But with only three categories, patterns are difficult to see. We can divide the leaves into two groups, 0 – 4 and 5 – 9. To do this write each stem *twice* instead of just once. Write any leaves from 0 to 4 with the first stem and any leaves 5 to 9 with the second stem. The above solution would then appear as follows:

ACT Scores	
Stem	Leaves
1	
1	7 8 8 8 9
2	0 1 2 3 4 4
2	5 6 6 7 7 9
3	1 2
3	5

Key: 1|7 = 17

Now, turn your book so that the stems are along the bottom of the table. Notice that if you look at the stem and leaf plot this way, it resembles a bar graph where the height of the bars is equal to the number of leaves. Thus a stem and leaf plot is similar to a histogram.

Exercises

2.2

1. Given the following data, construct a bar graph. Would it be appropriate to create a pareto chart from this data? If so, create a pareto chart as well.

Table for Exercise 1 - Math Grades on Test 1	
Grade	**Number of Students**
A	30
B	56
C	47
D or F	12

2. Given the following data, construct a bar graph. Would it be appropriate to create a pareto chart from this data? If so, create a pareto chart as well.

Table for Exercise 2 - Votes for President	
Candidate	**Number of Votes**
Bush	51
Kerry	46
Nader	2
Other	1

3. Given the following data, construct a bar graph. Would it be appropriate to create a pareto chart from this data? If so, create a pareto chart as well.

Table for Exercise 3 - Value-Added Tax	
Country	**Percent Tax**
Spain	16%
Canada	7%
Norway	25%
Japan	5%
United Kingdom	17.5%

Source: wikipedia.org

4. Given the following data, construct a side-by-side bar graph and a stacked bar graph; then answer the questions below.

Table for Exercise 4 - Math Attendance		
Math Class	**Freshmen**	**Sophomores**
Statistics	147	45
Algebra	160	73
Calculus	23	92
Quantitative Reasoning	12	120

a. Which course has the most students enrolled? Which graph did you use to answer this question?

b. Which course has the most sophomores enrolled? Which graph did you use to answer this question?

c. Which course has the most freshmen enrolled? Which graph did you use to answer this question?

d. Which course has the least students enrolled? Which graph did you use to answer this question?

5. Given the following data, construct a side-by-side bar graph and a stacked bar graph; then answer the questions below.

Table for Exercise 5 - Types of Pets Seen By the Vet		
Type of Animal	**Number seen by Dr. Warren**	**Number seen by Dr. Campbell**
Cats	47	59
Dogs	56	37
Reptiles	13	6
Birds	28	30

a. Which veterinarian saw more dogs this month? Which graph did you use to answer this question?
b. Which type of animal was seen the least this month? Which graph did you use to answer this question?
c. Which veterinarian saw more cats this month? Which graph did you use to answer this question?
d. Which animal did Dr. Warren see the least this month? Which graph did you use to answer this question?

6. Given the following data, calculate the angles needed to construct a pie chart. Round your answers to the nearest whole degree.

Table for Exercise 6 - Attitude Toward Math	
Attitude	**Number of Students**
Love	23
Like	46
Indifferent	9
Hate	12

7. Given the following data, calculate the angles needed to construct a pie chart. Round your answers to the nearest whole degree.

Table for Exercise 7 - College Majors	
Major	**Number of Students**
English	171
Business	569
Education	346
Science	285

8. For the frequency table below, construct a histogram.

Table for Exercise 8 - Ages of Taste Test Participants (in Years)	
Class	**Frequency**
15–19	7
20–24	8
25–29	10
30–34	2
35–39	3

9. For the frequency table below, construct a histogram.

Table for Exercise 9 - Braking Time for Vehicles (in Minutes)	
Class	**Frequency**
0.05–0.07	12
0.08–0.10	15
0.11–0.13	14
0.14–0.16	15
0.17–0.19	14

10. For the frequency table below, construct a histogram.

Table for Exercise 10 - Age at First Marriage (in Years)	
Class	**Frequency**
15–18	2
19–22	5
23–26	4
27–30	5
31–34	4

11. For the frequency table below, construct a histogram.

Table for Exercise 11 - Hourly Wage at First Job (in Dollars)	
Class	**Frequency**
7.5–8.4	12
8.5–9.4	50
9.5–10.4	48
10.5–11.4	45
11.5–12.4	34

12. Use the data from Exercise 8 to create a frequency polygon.

13. Use the data from Exercise 9 to create a frequency polygon.

14. Use the data from Exercise 10 to create an ogive.

15. Use the data from Exercise 11 to create an ogive.

16. The following data represents the number of curl ups completed in 60 seconds by a group of sixteen 8 year old boys. Create a stem and leaf plot for this data.

31	34	41	36
27	29	18	33
31	28	34	22
26	28	36	42

17. The following data represents the caloric intake for a group of fifteen men between the ages of 20 and 39. Create a stem and leaf plot for this data.

2700	2200	2500	2800	2600
3000	2600	2200	3100	2800
1800	3500	2500	3000	2900

18. The following data represents times in minutes for completing a one mile run from a group of twenty-four 17 year old girls. Create an **ordered** stem and leaf plot for this data.

12.4	12.3	11.1	11.9	12.1	9.5	11.6	10.8
10.2	9.3	10.1	11.2	8.2	9.3	9.5	12.5
9.4	9.7	10.7	10.9	9.3	10.4	12.9	10.6

19. The following data represents normal monthly precipitation in inches for the month of September in Alaska. Create an **ordered** stem and leaf plot for this data.

2.73	2.81	2.54	2.59	2.70	2.88	2.64
2.55	2.86	2.68	2.77	2.61	2.56	2.62
2.78	2.64	2.50	2.67	2.89	2.74	2.81

20. Graphs of Canadian gas prices (adjusted to U.S. dollars per gallon from Canadian dollars per liter) are shown below. Use these graphs to answer the following questions.

a. In what region(s) is gas the most expensive?

b. What retailer(s) offers the cheapest prices on gas?

c. What is the cheapest price for gas in a given region?

d. What is the most expensive price a retailer charges for gas?

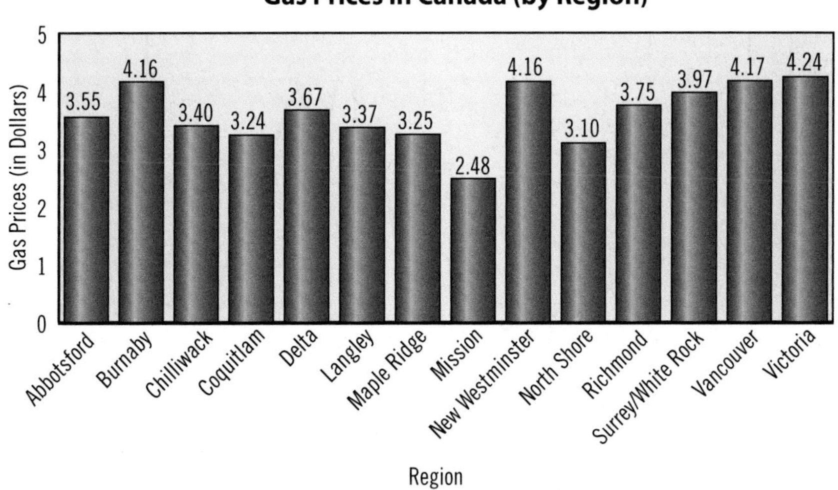

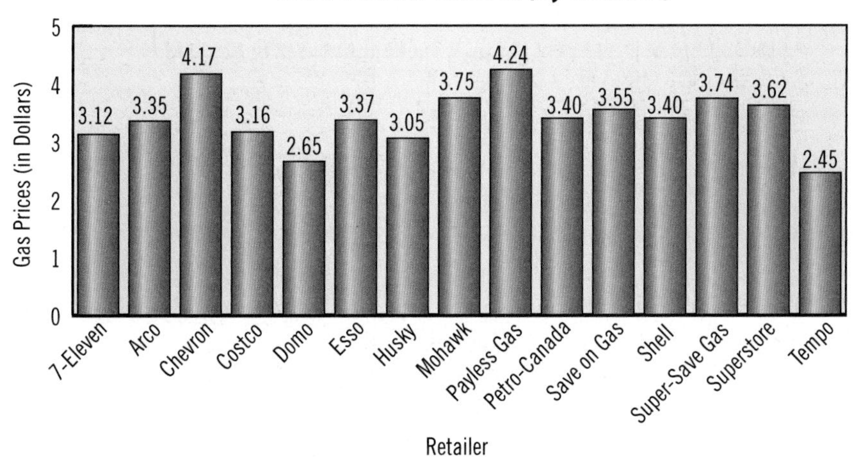

Source: www.gasolinewars.com

21. Below is a line graph depicting the spot and retail gasoline prices in the U.S. for the year 2005. Use this graph to answer the questions below.

a. Approximately, what was the highest spot price for gasoline?

b. Approximately, what was the lowest retail price for gasoline?

c. Approximately, when did the lowest spot price for gasoline occur?

d. Approximately, when did the highest retail price for gasoline occur?

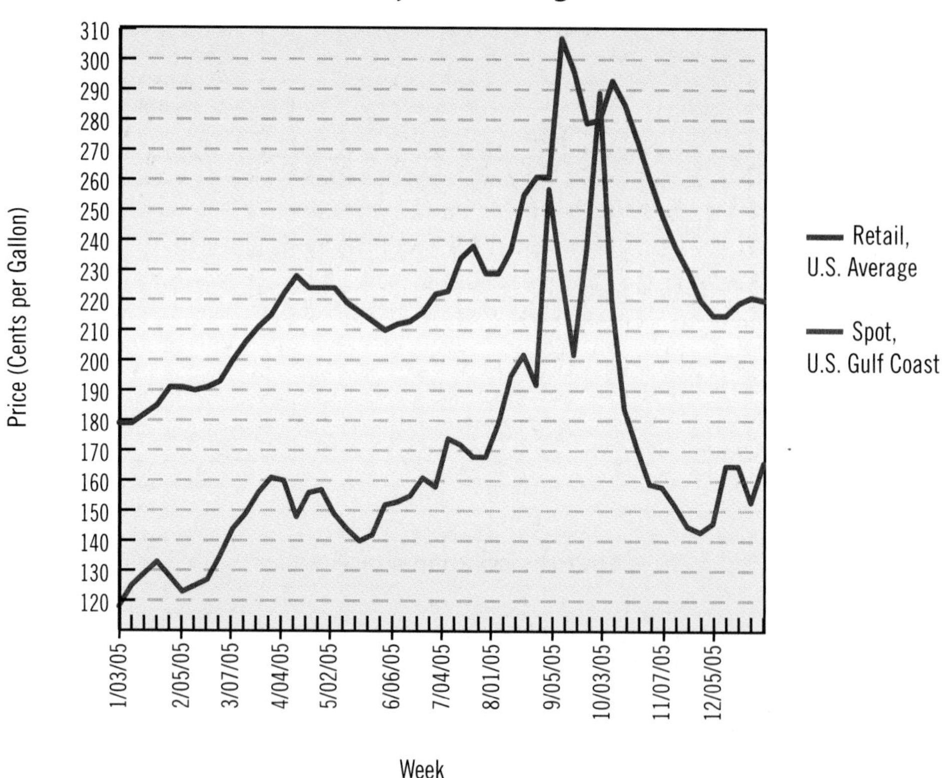

Source: Department of Energy

22. The pie chart on the following page depicts Shelly's monthly budget. Use this pie chart to answer the questions that follow.

a. What does Shelly spend the most money on each month?

b. What does Shelly spend the least on each month?

c. What percentage of Shelly's budget is spent on household bills? (Housing, Utilities, Cable, and Cell Phone)

d. What percentage of her monthly budget does Shelly spend on her car? (Car, Insurance, and Gas)

e. Disposable income is the money remaining after all *essentials* have been paid. What percentage of Shelly's budget is disposable income? (Entertainment, Cable with Internet, Miscellaneous, Cell Phone)

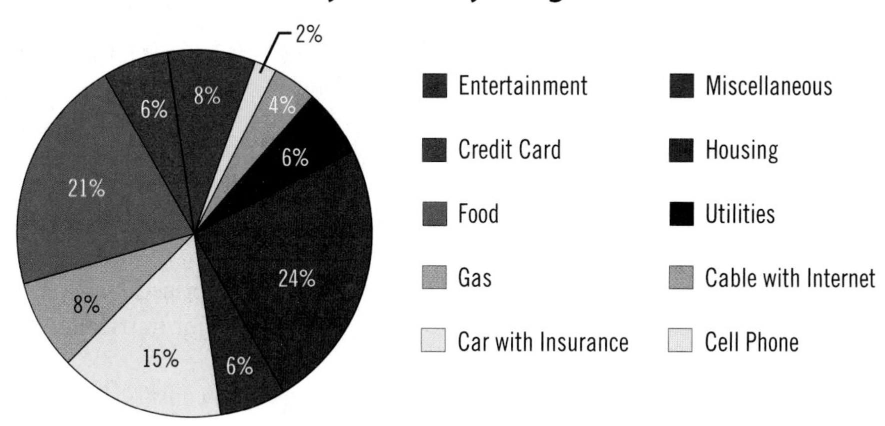

23. Below is a stem and leaf plot of ACT math scores for a group of college music students. Use this information to answer the questions below.

a. What was the lowest math score a student in this class received on the ACT?

b. What was the highest math score a student in this class received on the ACT?

c. Which math score occurred the most often?

d. How many students are represented by this information?

ACT Math Scores from Students in a College Music Class

Stem	Leaves
1	2 3 3 4
1	5 7 8 8 8 8 9 9
2	0 0 0 1 1 2 2 2 3
2	5 5 5 6 7
3	2 3
3	4

Key: 1|2 = 12

24. Meteorologists categorize hurricanes according to their maximum sustained wind speed using the Saffir-Simpson scale. This scale divides hurricanes into five categories, with Category 1 hurricanes having maximum sustained winds between 74 and 95 knots. The bar graph below depicts the number of hurricanes along the U.S. coast from 1851 to 2005. Each bar in the graph represents one of the categories used to classify hurricanes. Use this graph to answer the following questions:

a. How many Category 3 hurricanes, with wind speeds of 111 to 130 knots, have hit the U.S. coast between 1851 and 2005?

b. A major hurricane is considered a hurricane in categories 3, 4, or 5 (i.e. wind speeds greater than 111 knots.) How many major hurricanes have hit the U.S. coast between 1851 and 2005?

c. What is the total number of hurricanes to hit the U.S. coast between 1851 and 2005?

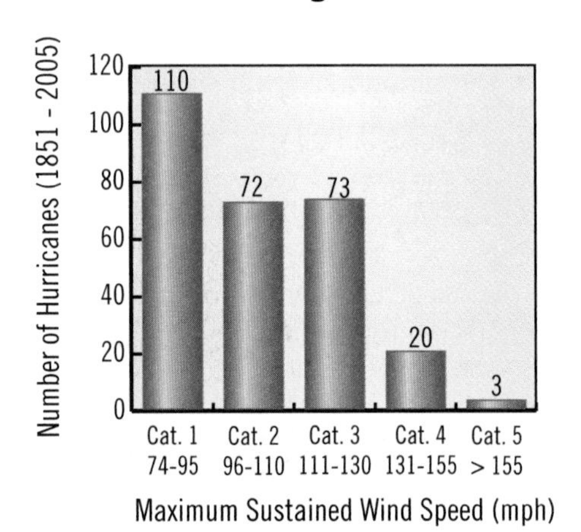

Source: National Oceanic & Atmospheric Administration

Analyzing Graphs

2.3

When you are analyzing a graph, you are trying to first determine the overall pattern of the data. Is it symmetrical? Does the majority of the data lie to one side or the other? Is the frequency the same for all categories? The answers to these questions tell us the basic shape of the distribution. There are four basic shapes that you should recognize.

Shapes of Distributions

1. **Uniform** – the frequency of each class is relatively the same. The distribution will have a rectangular shape.

2. **Symmetrical** – the data lies evenly on both sides of the distribution.

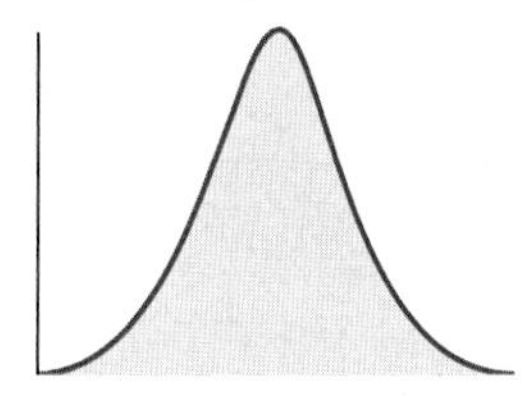

3. **Skewed to the Right** – the majority of the data falls on the left of the distribution.

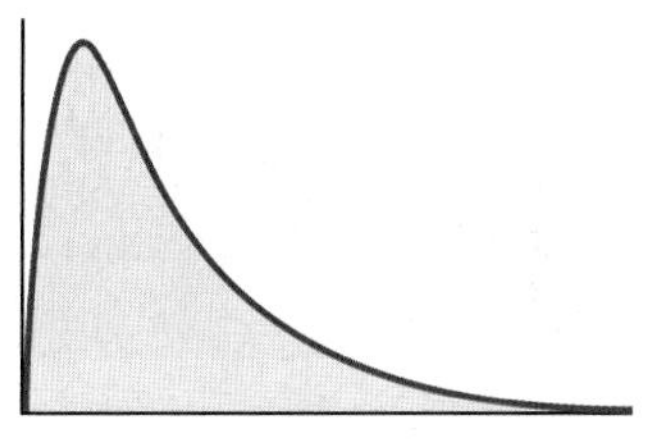

4. **Skewed to the Left** – the majority of the data falls on the right of the distribution.

Notice that the definitions of skewed to the right and skewed to the left seem backwards. This is because the names are based on what happens to the mean of the distribution, not on where the majority of the data lies.

Example 15

Describe the overall shape of the following distribution.

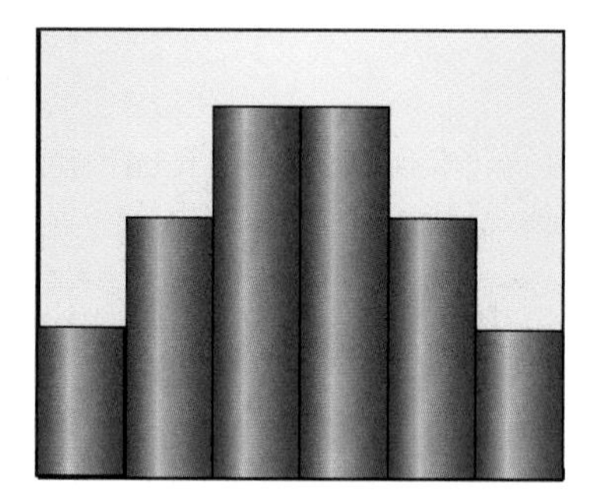

Solution:

Notice that if we draw a smooth curve skimming the top of the histogram, we begin to see a curve similar to the shape of the symmetric curve. To be symmetric, the left and right sides of the graph should be close to mirror images. Drawing a line down the center of the graph, we can see that both sides of the graph are indeed mirror images of each other.

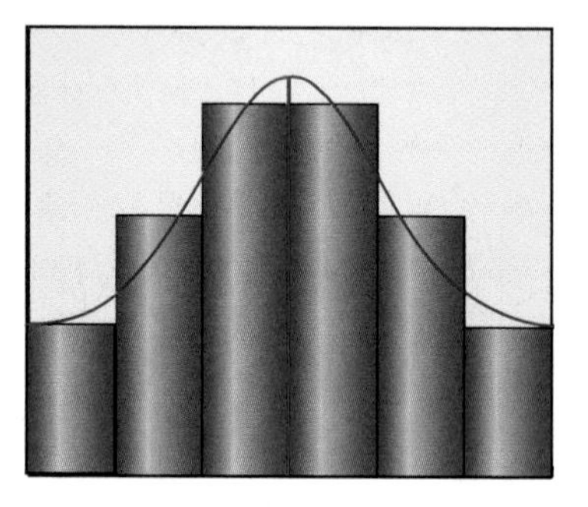

Thus this histogram has a symmetric shape.

Another item to look for when analyzing a graph is an outlier. An **outlier** is a data value that falls outside the normal shape of the distribution. Outliers indicate an unusual data value, which could be a mistake in the data collection. However, they should not be discarded! We will talk more about the effect of an outlier on the shape of a distribution in Chapter 3.

We also need to determine whether the graph is a time-series or a cross-sectional graph. A **time-series** graph is a picture of how data changes over time. Consider the following table of Consumer Price Indices between the years 1920 and 1990. We can use one common type of time-series graph, a line graph, to depict the change in the value of the CPI over that time span. A **line graph** is constructed by joining the data points in order with line segments.

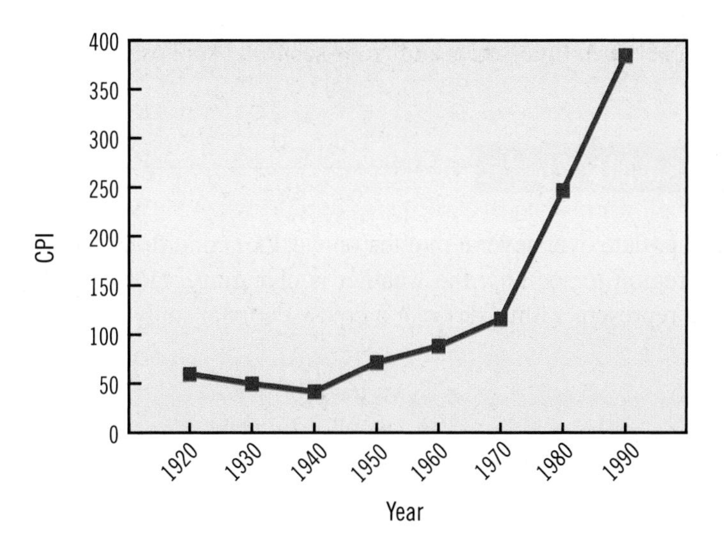

Table 20 - Consumer Price Index	
Year	**CPI**
1920	60.0
1930	50.0
1940	42.0
1950	72.1
1960	88.7
1970	116.3
1980	247.0
1990	384.4

From this time-series graph, we see that the CPI was at its lowest around 1940 and that it has risen dramatically since 1970.

In contrast to a time-series graph, a **cross-sectional** graph is a picture of the data at a given moment in time. Consider the following table which gives a breakdown of one shopper's bill at a local grocery store.

Table 21 - Grocery Store Bill	
Item	**Cost**
Produce	$12.00
Frozen Items	$23.00
Dairy	$31.50
Dry Goods	$9.00

We can see in this cross-sectional graph the that most money was spent on dairy products on this particular shopping trip. Let's look at an example of how to distinguish between time-series and cross-sectional studies.

Example 16

You gather data over several months on weather conditions in your region to see how the weather is changing. Does this data represent a time-series or a cross-sectional study?

Solution:

The data was gathered over several months, so it is a time-series.

Lastly, you must be careful when looking at graphs, because sometimes they can be misleading. If you stretch or shrink the scale on the y-axis, the shape of the graph may change dramatically. A line that rises gently on one scale might look very steep with a different scale. Choose a scale that best represents the data. If there are large differences between the data values, then the graph should accurately reflect the differences. On the other hand, if the difference in the data values is small, then the graph should reflect this as well.

Example 17

Consider the graph below on minimum wage as reported by the Associated Press. What errors can you find in the graph? How should they be fixed?

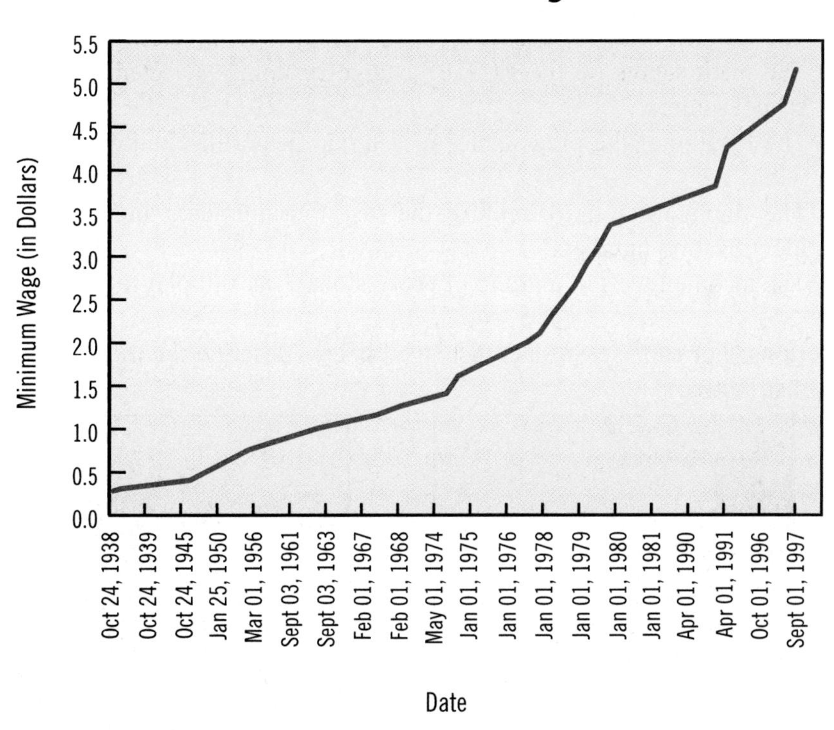

Source: Department of Labor

Solution:

Notice that the x-axis does not have a consistent scale. The years are sometimes one year apart and sometimes even nine years apart. To correct this graph, the x-axis needs to be adjusted to use a consistent scale. The corrected graph can be found in Exercise 10 in the Chapter Exercises.

Exercises

2.3

Discussion Questions

Directions: For each set of data described below, discuss the most likely shape of its distribution.

1. The weights of the defensive linemen on Big Ten University football teams.
2. The math scores on the ACT of a large randomly selected group of high school seniors.
3. The lengths of the pregnancies of a group of gorillas being studied in the wild.
4. The digits used in compiling the last four numbers in social security identification numbers.
5. The income levels of a group of professional baseball players.

Directions: For each set of data displayed below, describe the most likely shape of its distribution.

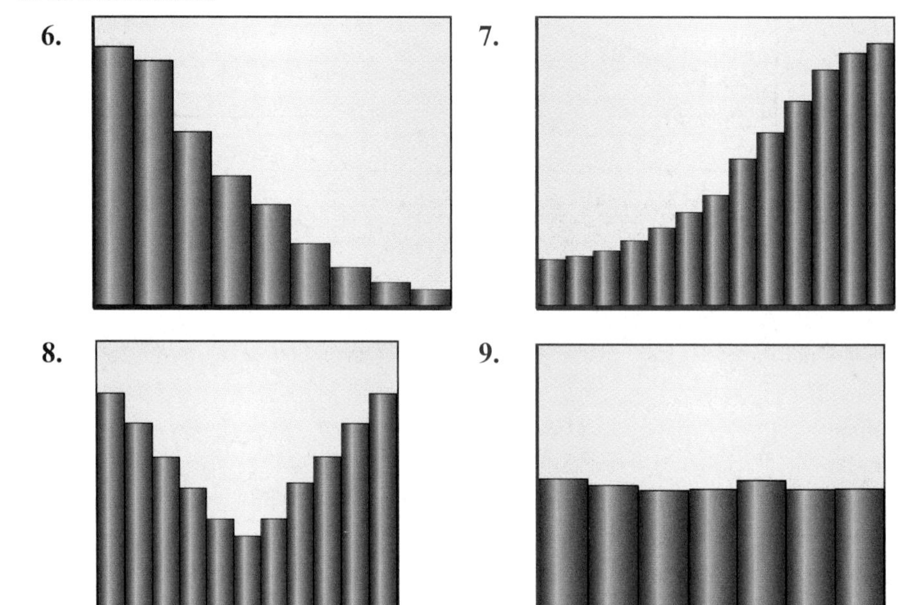

Directions: For each graph described below, decide if it is a time-series or a cross-sectional study.

10. A map of Louisiana with the individual parishes shaded to represent the average income level of each parish.
11. A line graph depicting the change in the minimum wage since 1965.
12. A bar graph showing the average tax rate for different countries around the world.
13. A stem and leaf plot displaying the number of home runs hit by each player on the Yankees Ball Club.
14. A bar chart of the average SAT score each year for the last five years.

15. For the graph shown below, decide if the scale on the graph shows an accurate depiction of the situation.

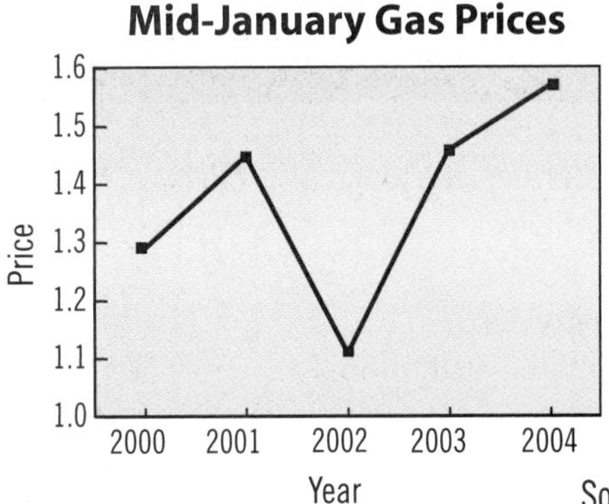

Source: Department of Energy

16. Look at the two graphs shown below depicting the *same* data on people's overall satisfaction level with the care they received at their local hospital. Which of these two graphs shows the more accurate picture of hospital satisfaction? Has hospital satisfaction increased? Are people satisfied with the care at their local hospital according to these graphs?

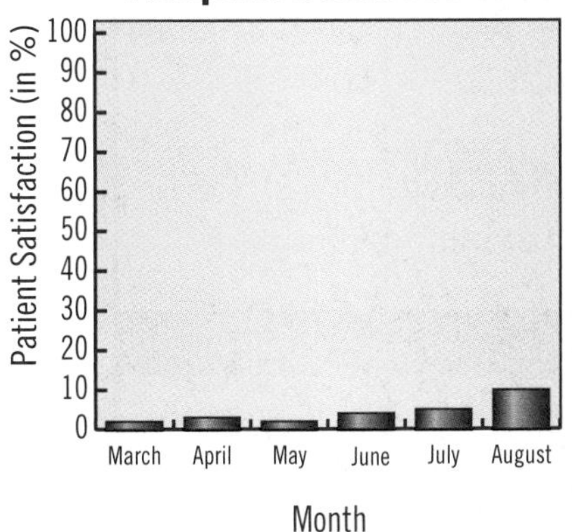

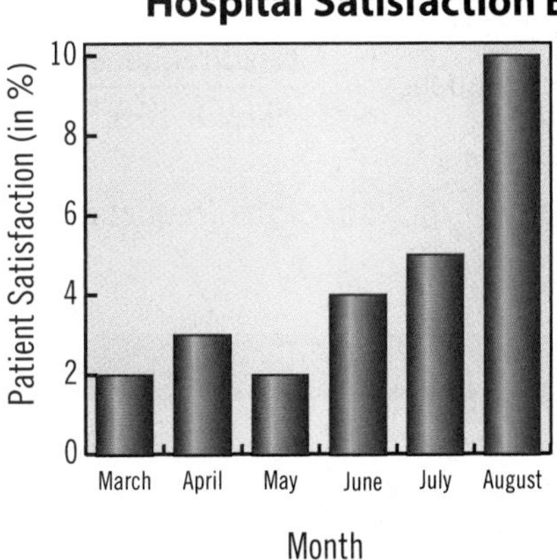

17. What errors occur in the following histogram?

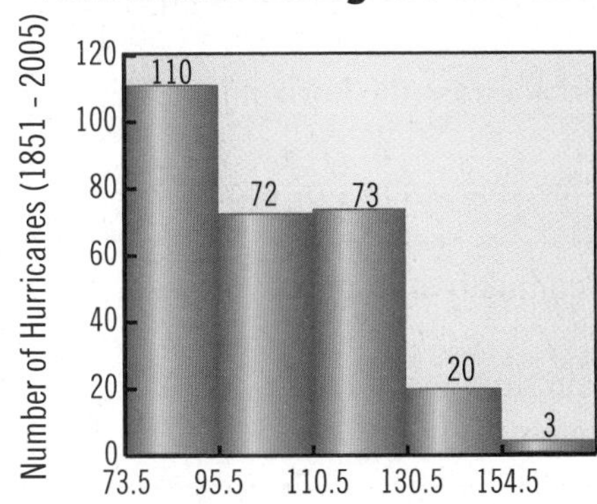

Source: National Oceanic & Atmospheric Administration

Chapter Review

Section 2.1 - Frequency Distributions	Pages
• **Frequency distribution** – a table that divides the data into classes and shows how many data values occur in each class	p. 44 – 52
• **Steps in constructing a frequency distribution:** **1.** Decide how many classes should be in the distribution. **2.** Choose an appropriate class width. **3.** Find the class limits. **4.** Determine the frequency of each class.	p. 46
• Other calculations commonly displayed in a frequency distribution: **1. Class boundaries** – dividing lines between classes **2. Midpoint** – $\text{Midpoint} = \dfrac{\text{Upper Limit} + \text{Lower Limit}}{2}$ **3. Relative frequency** – $\text{Relative Frequency} = \dfrac{\text{Class Frequency}}{\text{Sample Size}} = \dfrac{f}{n}$ **4. Cumulative frequency** – the sum of the frequency for that class and all previous classes	p. 48 – 52

Section 2.2 - Graphical Displays of Data	Pages
• **Definitions**	
Pie chart – displays each category in relation to the whole	p. 59 – 60
Bar graph– typically used to display qualitative data	p. 60 – 64
Pareto chart – a bar graph placed in descending order of frequency	
Side-by-side bar graph – a bar graph used to compare different groups	
Stacked bar graph – a bar graph used to compare different groups as well as category totals	
Histogram – a bar graph of a frequency distribution; the horizontal axis is a number line	p. 64 – 66
Frequency polygon – a visual display of the frequencies of each class using the midpoints from the histogram	p. 67
Ogive – a type of line graph which depicts cumulative frequency of each class	p. 68
Stem and leaf plot – retains the original data; leaves are the last digit in each data value and the stems are the remaining digits	p. 68 – 70

Section 2.3 - Analyzing Graphs	Pages
• **Shapes of Distributions:** **1. Uniform** – the frequency of each class is relatively the same. The distribution will have a rectangular shape. **2. Symmetrical** – the data lies evenly on both sides of the distribution. **3. Skewed to the Right** – the majority of the data falls on the left of the distribution. **4. Skewed to the Left** – the majority of the data falls on the right of the distribution.	p. 79
• **Definitions** **Outlier** – a data value that falls outside the normal shape of the distribution **Time-series graph** – a picture of how data changes over time **Cross-sectional graph** – a picture of data at a given moment in time	p. 81

Chapter Exercises

Directions: For the data sets below, construct a frequency table with the indicated number of classes. Include the frequency, the class boundaries, the midpoint, the relative frequency, and the cumulative frequency of each class.

1. The following data represents the number of days absent from school in one school year for each of the 24 students in Ms. Jinn's fourth grade class. Use 6 classes, each having a class width of 5 days. Begin with a lower limit of zero.

17	8	12	3	0	5	13	12
25	10	6	8	11	0	1	4
19	21	22	9	16	9	3	2

2. The following data represents the weight, in pounds, of twenty-four collegiate offensive linemen in a particular state. Use 6 classes each having a class width of 25 pounds. Begin with a lower limit of 175.

195	210	255	267	231	229	301	199
178	281	245	256	278	205	217	223
279	196	235	248	262	291	302	189

Directions: Create a stem and leaf plot for each set of data. Be sure to include a key to your stem and leaf plot.

3. The following data represents the saline concentration in a salt-water aquarium on various days. Order your leaves, but do not split the stems.

1.022	1.021	1.022	1.023	1.019	1.021	1.022	1.017	1.024
1.021	1.022	1.022	1.023	1.020	1.019	1.023	1.025	1.018

4. The following data represents the number of grams of fat per serving of a representative sample of various foods found in someone's kitchen pantry. Split the stems in this exercise to better reveal the patterns in the data. Do you believe that this person is on a low-fat diet? Explain your reasoning using the data as your evidence.

0	2	0	6	8	1	10	2
3	14	4	21	13	7	9	17

5. Determine which graph would most clearly depict the information gathered in each of the following studies:

a. The number of tickets sold at one theater over the course of a year

b. The number of tickets sold at one theater for each movie showing during one week

c. The number of tickets sold for each movie this week, specifically comparing the movie choices of 18-35 year olds to 36 and older

d. Ticket prices for all theaters across the country

6. Ages of students volunteering at a local animal shelter:

Table for Exercise 6

Class	Frequency
10–12	5
13–15	9
16–18	2
19–21	11
22–24	6

a. What is the relative frequency for the 4th class?

b. What is the cumulative frequency for ages 18 and under?

c. What is the upper class boundary for the 2nd class?

d. What is the class width?

e. What percentage of the student volunteers were between 19 and 24?

f. How many students were surveyed?

g. What was the youngest age surveyed?

7. Answer each question using the stacked bar graph below.

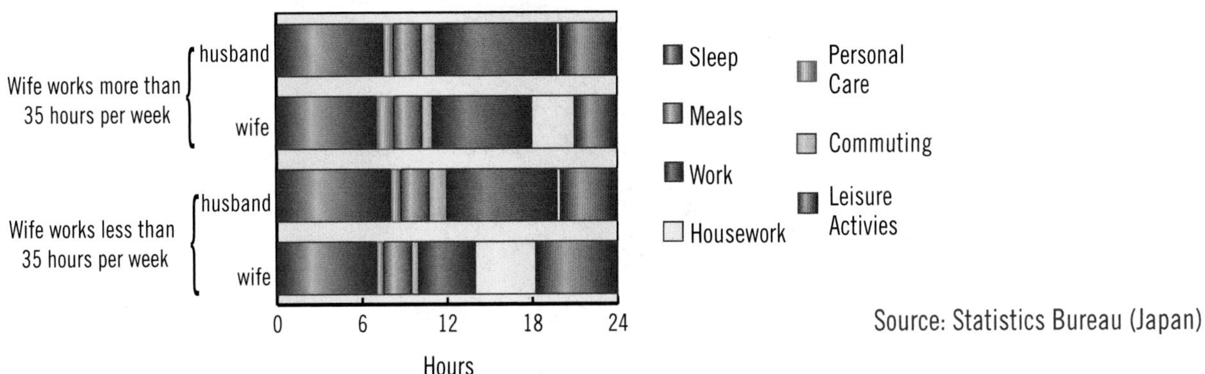

a. For wives that spend more than 35 hours working, how many hours on average are spent each week sleeping?
b. For husbands whose wives work fewer than 35 hours per week, how many hours on average are spent each week on personal care? How accurate do you feel your answer is?
c. Compare the number of hours spent on leisure activities for wives in each category.
d. How easy did you find it to interpret this graph? What type of graph could have been used to represent this data in a clearer fashion?

8. Answer each question using the side-by-side bar graph below.

Location of Internet Access by Age Group

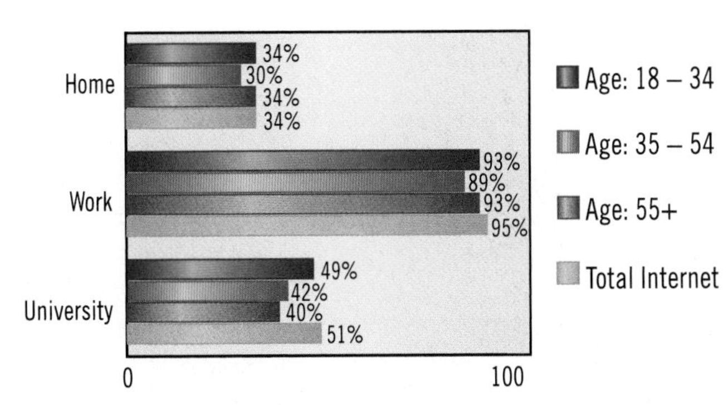

Source: Media Metrix October 2003

a. What percentage of people age 35-54 use the internet at work?
b. In the age category 18-34, where do most people use the internet?
c. What age group is most likely to use the internet at the university?
d. Add the percentages together for the 18-34 age group in all three categories. How is it possible that you get an answer over 100%? Can you think of the how the construction of the survey might have produced such results?

9. Answer each question using the pie chart below.

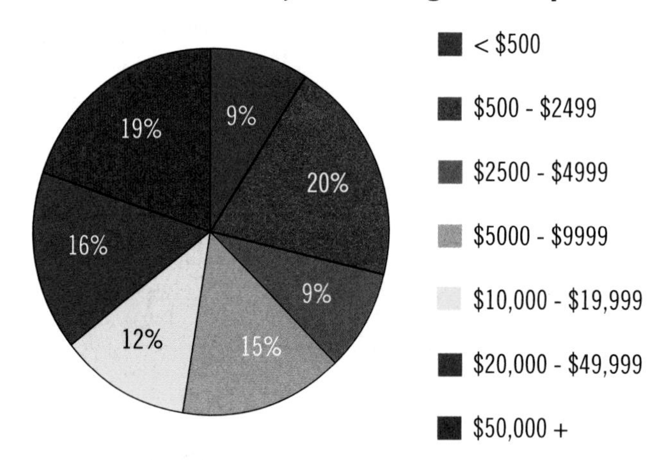

a. What percentage of respondents initially invested over $50,000?
b. What percentage of respondents initially invested less than $10,000?
c. What percentage of respondents initially invested between $5000 and $19,999?
d. What percentage of respondents initially invested $10,000 or more?

10. Use the line graph of the (unadjusted) federal hourly minimum wage to answer the questions below. (Note: The dollar values in the graph represent current dollar values as they have not been adjusted for inflation.)

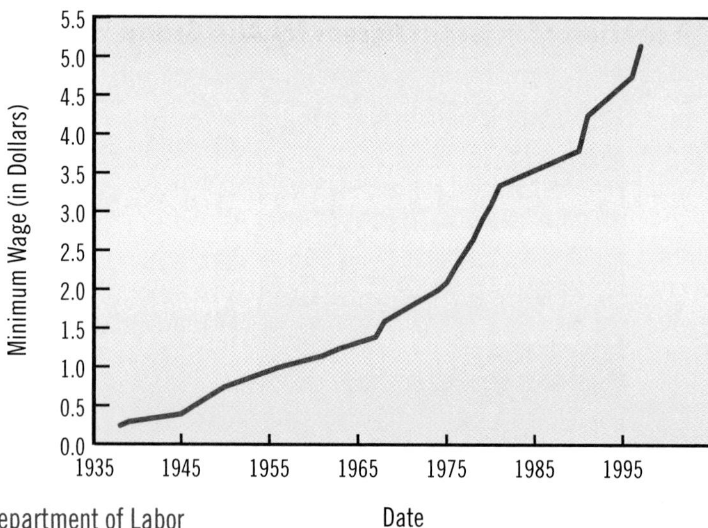

Source: U.S.Department of Labor

a. What is the overall trend in the value of minimum wage?
b. When was the lowest minimum wage reported?
c. Has there been any time period where the minimum wage decreased?

11. A pre-test and a post-test were administered to one class at the beginning and end of two weeks of classes. The scores for each of the tests are shown graphically below.

Pre-Test Scores

-0.5 9.5 19.5 29.5 39.5 49.5 59.5 69.5 79.5 89.5 99.5

Scores

Post-Test Scores

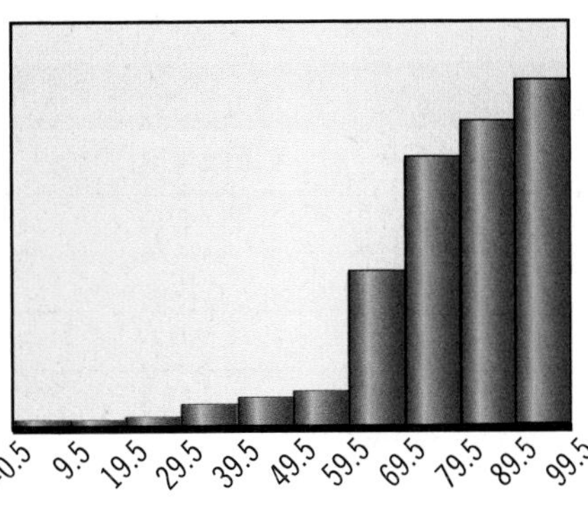

Scores

a. Identify the general shape of the pre-test scores.
b. Identify the general shape of the post-test scores.
c. Why do you think that the pre-test scores are distributed the way they are?
d. Why do you think that the post-test scores are distributed the way they are?

12. Use the line graph of the federal hourly minimum wage adjusted for inflation to answer the questions below.

Federal Minimum Wage Rates (adjusted for inflation)

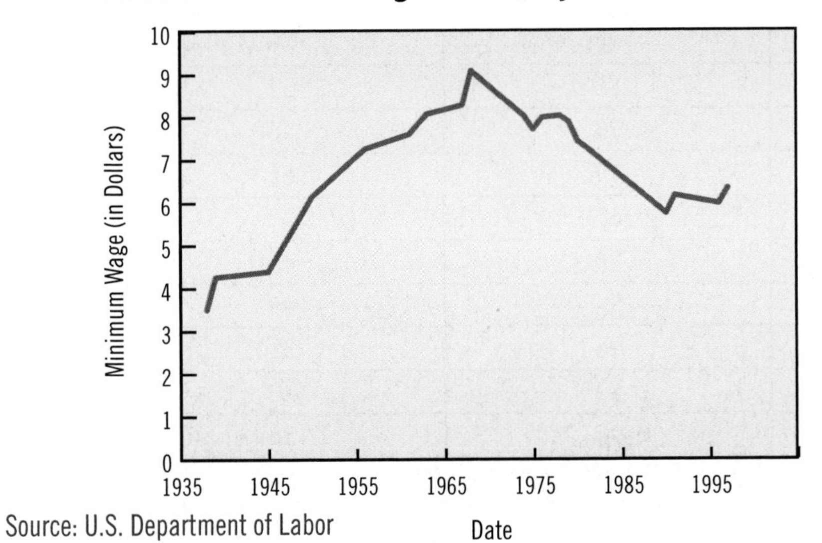

Source: U.S. Department of Labor

a. What is the overall trend in the value of minimum wage? How does this trend compare to the trend in the unadjusted minimum wage graph?
b. When was the lowest minimum wage reported?
c. Has there been any time period when the minimum wage decreased?

Chapter Project: – Fast Food Comparison

For three fast food restaurants, a random sample of service times (in seconds) was taken. The results are given below.

Project Table - Service Times

Kim's Kajun Kitchen	Chez Carolyn	Emily's Eatery
111	109	99
94	84	63
57	93	53
80	123	82
78	97	75
109	56	112
92	79	65
34	57	55
67	32	65
122	68	86
95	45	94
46	99	87
103	82	62
110	76	68
93	54	61
86	84	49
75	86	37
49	73	46
61	79	65
82	67	86
76	96	48
92	112	95
65	94	33
92	82	49
94	96	76
106	125	35
112	98	68
83	120	57
72	78	41
119	63	49

1. For each restaurant's service times, do the following:
   a. Create a frequency distribution with the given classes. Include the frequency, class boundaries, relative frequency, and cumulative frequency of each class.

Class
30–49
50–69
70–89
90–109
110–129

   b. Use the frequency table to construct a histogram of the data.
   (Hint: It would be a good idea to use Microsoft Excel to create both the frequency tables and the histograms at the same time.)

2. Based on the graphical descriptions of the data you created in Part 1, which fast food restaurant do you believe consistently has the shortest service times? Explain why.

Chapter Technology – Creating Graphs

TI-84 PLUS

The graphing function on the TI-84 Plus calculator will allow you to create many of the graphs we have discussed in this chapter. We will look at two of these graphs, the histogram and the line graph.

Histogram

Let's begin by creating a histogram. First, you must enter your data into the calculator. To do so, press STAT, then choose EDIT, and type in your data in the first list, L1. Your screen should now look similar to the following screenshot.

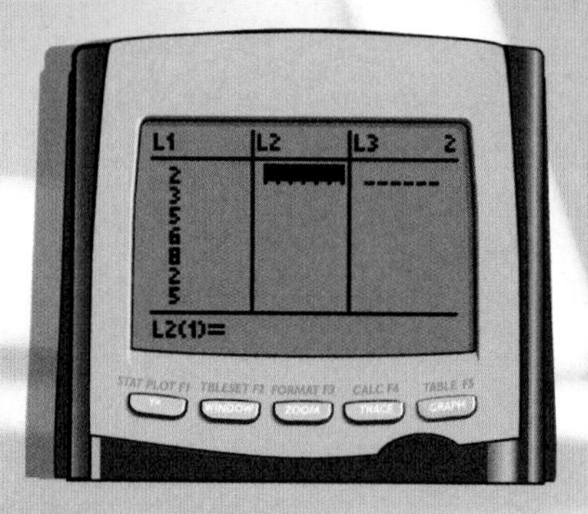

Once your data is entered, you need to set the specifications for the graph. Press 2ND and then press Y=. This displays the STAT PLOTS menu. Press ENTER. This brings up the Graph Settings menu below.

Choose ON, highlight the type of graph you want to create, and press ENTER. The histogram is the third option. Now, when you press GRAPH, you should see a histogram of your data. If your graph does not appear or it seems strange, you can modify the screen window by pressing ZOOM and then choosing option 9: *ZoomStat*. This will resize the screen so that it is appropriate for a statistical graph.

Example T.1

Use your TI-84 Plus to recreate the histogram of our plasma screen TV prices. The prices are reprinted below for your convenience.

Table T.1 - Plasma Screen TV Prices				
1595	1599	1685	1699	1699
1699	1699	1757	1787	1799
1799	1885	1888	1899	1899
1899	1984	1999	1999	1999

Solution:

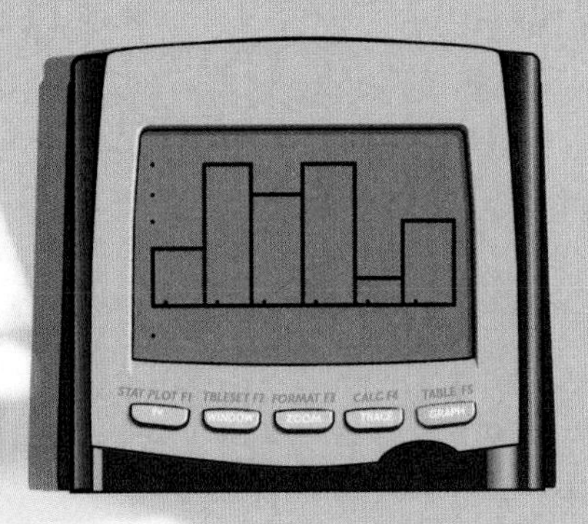

1. Enter your data into the List Editor by pressing STAT and then choosing EDIT.
2. Once the data has been entered, press 2ND and then press Y= to bring up the STAT PLOTS menu.
3. Press ENTER. Set the Menu so that the graph will display a histogram. Press GRAPH. Your screen should look like the one at right. If it does not appear that way at first, pressing ZOOM and then choosing option 9: *ZoomStat* should correct the problem.

Line Graph

The process for creating a line graph is similar to constructing a histogram; however, to create a line graph we need both x and y values. This means we will need to enter data into Lists 1 and 2. If your data is not in these lists, you can change the XList and/or YList in the STAT PLOTS menu.

Example T.2

Use your TI-84 Plus to recreate the line graph of CPI values from Section 2.3. The data is reproduced below.

Table T.2 - Comsumer Price Index

Year	CPI
1920	60
1930	50
1940	42
1950	72.1
1960	88.7
1970	116.3
1980	247.0
1990	384.4

Solution:

Press STAT, choose EDIT, and enter the years in List 1 and the CPI values in List 2. You should have the following screen.

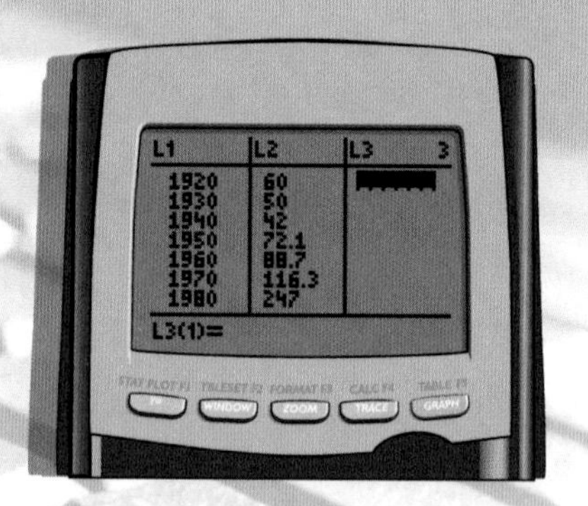

Now set up the STAT PLOTS menu to produce a line graph by selecting the second graph type in the menu. Press GRAPH. If you do not see a line graph at first, press ZOOM and choose option 9: *ZoomStat*. You should then see the line graph shown on the right.

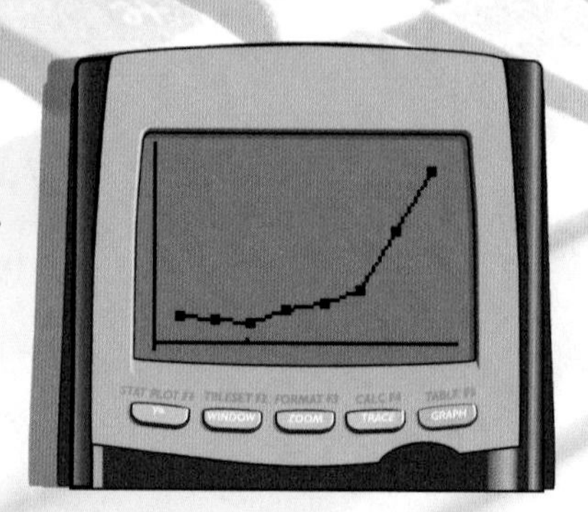

MICROSOFT EXCEL

Basic Charts

You can create many more graphs in Microsoft Excel using the Chart Wizard than you can on your TI-84 Plus. Chart Wizard can be used to create bar graphs, pie charts, and line graphs, to name a few. Begin by entering your data into the worksheet. Then, click on the **Chart Wizard** button in the toolbar.

Σ ▾ 100% ▾

This Wizard will walk you through the steps used to create a graph. Simply follow the steps in the wizard, clicking on the next button. The wizard takes you through the steps of choosing the type of chart, locating your sources of data, giving your chart a title, and creating the final output.

Example T.3

Use Microsoft Excel to recreate the pie chart of the housing data from Section 2.2. The data is reproduced below.

Table T.3 - Types of Housing

Types of Housing	Number of Students
Apartment	20
Dorm	15
House	9
Sorority/ Fraternity House	5

Solution:

Begin by entering the type of housing in cells A1 through A4. Then, enter the number of students in cells B1 through B4. Your Excel worksheet should look like the one below.

	A	B
1	Apartment	20
2	Dorm	15
3	House	9
4	Sorority/Fraternity	5

Now click on the *Chart Wizard* button in the toolbar. Choose *Pie* in the Chart Type box then click *Next*. In the *Data Range* box, type in $A$1:$B$4 which is where your data is located in the worksheet, then click *Next*. On the next screen, enter a *Chart Title* such as Housing Data then click *Next*. The last step is to choose the location for the output. Decide whether you want the pie chart in its own worksheet or on the same one with your data. Once you have made your choice, click *Finish*. Now you have a professional pie chart!

Histograms and Frequency Tables

The Chart Wizard is wonderful for creating a wide variety of very nice graphs; however, if you want to create a histogram, you would not use the Chart Wizard. You do not need to create a frequency table first either, as Excel does both at the same time. Simply enter your data in one or more columns in the worksheet. In another column, enter the upper class boundaries you would like Excel to use; otherwise, the class boundaries will be chosen for you and may not be the best choice.

Once the data is entered into the worksheet, go to TOOLS and select DATA ANALYSIS. (If DATA ANALYSIS does not appear in your Excel menu, you can easily add this function by looking under TOOLS. Click on Add-ins and select Analysis Toolpack. When finished, it is a good idea to remove the tool pack, as it slows Excel down.) Choose *Histogram* from the menu. The *Input Range* is where your data is located. The *Bin Range* is where your upper class boundaries have been entered. Click on *Labels* if the first cell in your Input Range is a data label. Finally, click on *Chart Output* to create a histogram instead of just a frequency table. Then click on OK to generate the output.

Example T.4

Let's recreate the histogram of plasma TV prices using Microsoft Excel. The data is again reprinted below.

Table T.4 - Plasma Screen TV Prices				
1595	1599	1685	1699	1699
1699	1699	1757	1787	1799
1799	1885	1888	1899	1899
1899	1984	1999	1999	1999

Solution:

Begin by entering the data in Column A in Excel. Next, enter the upper boundaries of the classes you want to use in Column B. Once your data is entered, go to TOOLS, then DATA ANALYSIS. Choose *Histogram* from the menu. The *Input Range* is where your data is located. The *Bin Range* is where your upper class boundaries have been entered. Click on *Labels* if the first cell in your Input Range is a data label. Next, click on *Chart Output* to create a histogram instead of just a frequency table. Finally, click on OK to generate the output.

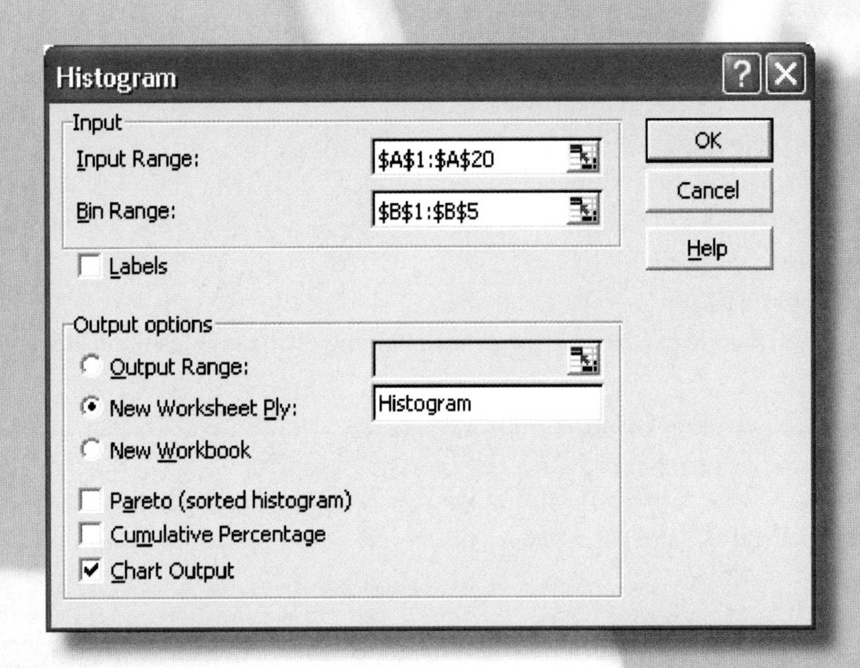

When your histogram has been created, you will notice that the bars do not touch. To correct this error, double click on one of the bars in the histogram. Click on *Options* and set the *Gap Width* to 0. You will also notice that the class boundaries are appearing on the *x*-axis where the midpoints should be, and there is a 'more' category on the graph. To correct these errors, return to the Excel sheet with the data. Change the bins to be the midpoints and delete the 'more' entry from the spreadsheet. Once these adjustments are made, the graph will be a correctly formatted histogram like the one shown below.

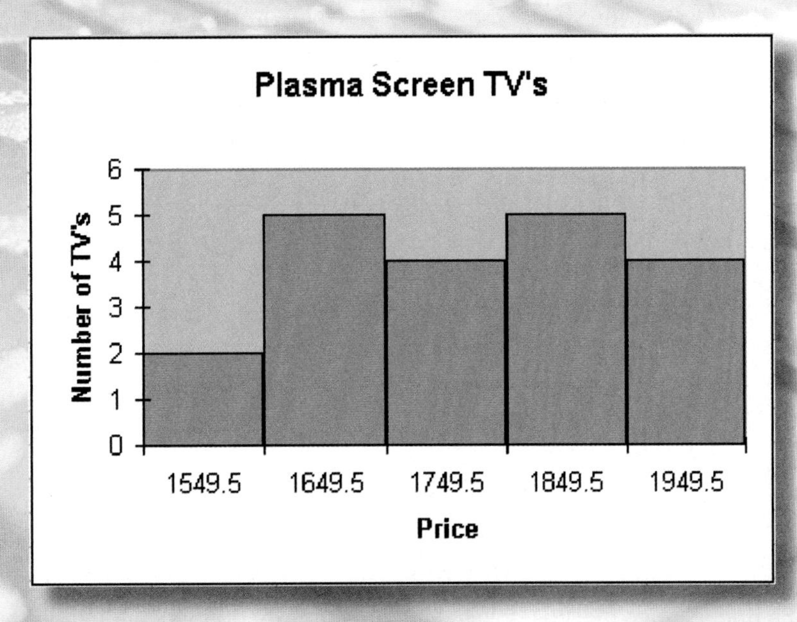

MINITAB

You can also create many statistical graphs in Minitab. All of the graphs shown in this chapter are possible in Minitab. A few examples are shown below.

Histogram

Example T.5

Create a histogram in Minitab using the data containing the heart rates of 50 students, given below.

77	84	79	80	67	84	82	84	69	81
94	68	65	86	78	79	83	83	84	82
93	80	81	80	62	98	77	93	82	80
82	73	77	79	81	70	72	85	84	80
83	77	80	70	75	74	85	87	79	88

Solution:

Begin by entering the data in column 1 in Minitab. Next, enter the class boundaries you want to use in column 2. Once your data is entered, go to GRAPH then choose HISTOGRAM. When the histogram menu appears, select C1 as the **Graph variables**. Then select OPTIONS and choose **CutPoints** under "Type of Intervals" and select C2 as the **Midpoint/cutpoint positions** under "Definition of Intervals". If you want the histogram to only mark the midpoints, choose **MidPoint** under "Type of Intervals" and enter 5 for the **Number of intervals** under "Definition of Intervals". Once you are finished, hit OK twice to generate the graph.

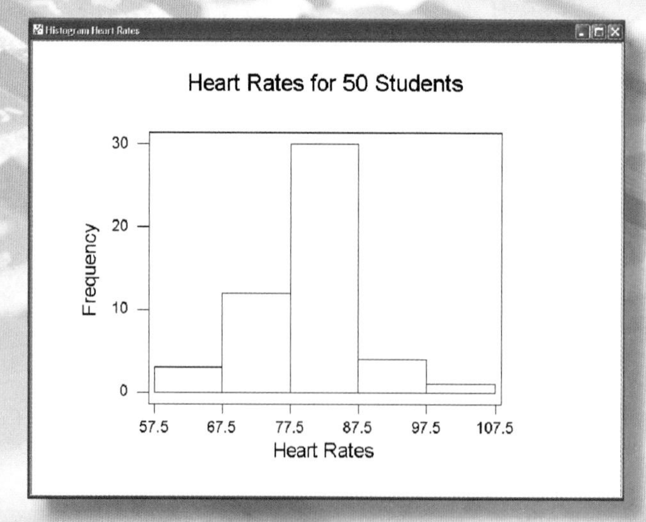

You can add a title, change the axis labels, and add color by double clicking on the graph.

Stem and Leaf Plots

Example T.6

Using the data from Example T.5, create a stem and leaf plot in Minitab.

Solution:

Begin by entering the data in column 1 in Minitab. Next, choose GRAPH and select STEM-AND-LEAF. Select C1for the **Variables** and enter 10 for the **Increment**. Select OK to display the following stem and leaf plot.

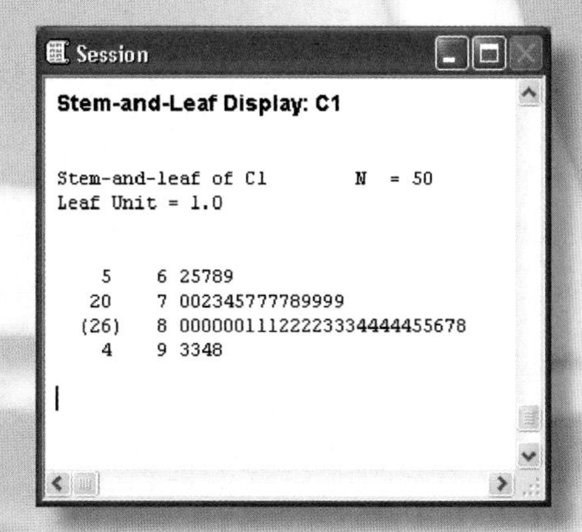

Numerical Descriptions of Data

Sections

Objectives

The objectives of this chapter are as follows:

1. Calculate the mean, median, and mode
2. Determine the most appropriate measure of center
3. Compute the range, variance, and standard deviation
4. Determine the percentiles, quartiles, and five number summary of a data set
5. Construct a box plot

Chapter 3

Introduction

Do statistics always report data accurately? A study of geography majors at the University of North Carolina demonstrates why it is important to rely on not one, but several, measures of central tendency.

According to the findings of the study, geography majors at the University of North Carolina have an average starting salary significantly higher than geography majors from other colleges and universities. With such a superior mean value, it is tempting to attribute the success of these students to factors relating to the school itself. Perhaps UNC has a top notch geography program or draws better students with its reputation. It's even possible that their instructors are more qualified.

However, what we often overlook when considering the mean of a set of data is its vulnerability to the effects of an outlier. What is a potential outlier in this study? As some of you might guess, the starting salary of one UNC geography major in particular, Michael Jordan, is a bit higher than the rest of the study's population. Consequently, it is essential that we evaluate other statistical measures such as median and mode to arrive at a true representation of the facts.

Measures of Center

3.1

Memory Booster!

Quantitative data are counts and measures.

We have seen how to describe data using tables and graphs; now we will learn how to summarize data using numbers. In order to summarize data numerically we will use the center, dispersion, and shape of the data. Let's begin with the center of the data.

The center of the data is the single value that best describes a typical value in the data set. For quantitative data, it also describes the location of the data on the real number line. Thus the center of the data is the "average" value of the data. In statistics; however, "average" can be calculated many ways.

✦ Mean

Math Symbols!

$\sum$ - Greek "sigma"; indicates to sum up what follows

The general usage of the word average is to indicate the **arithmetic mean**, $\bar{x}$. The mean is the sum of all of the data values divided by the number of data values. The formula is

$$\bar{x} = \frac{x_1 + x_2 + \cdots + x_n}{n} = \frac{\sum x_i}{n}$$

where x_i is the ith data value and n is the number of data values in the sample.

This formula is actually the **sample mean** because we are calculating the mean of a set of sample values. If we are calculating the mean of all the values in a population, we call it the **population mean**, μ. The procedure is the same: divide the sum of all the population values by the number of values in the population, N. The formula is written in a slightly different manner,

$$\mu = \frac{x_1 + x_2 + \cdots + x_N}{N} = \frac{\sum x_i}{N}$$

where x_i is the ith value in the population, and N is the number of values in the population.

Note that the mean might not be a value in the data set.

Math Symbols!

μ, population mean; Greek symbol mu

$\bar{x}$, sample mean; read as "x-bar"

Example 1

Calculate the mean of the following sample data.

5, 7, 8, 10, 4, 3, 12

Solution:

We have values from a sample, so we are calculating the sample mean.

$$\bar{x} = \frac{5+7+8+10+4+3+12}{7} = 7.0$$

Rounding Rule

When calculating the mean, round to one more decimal place than is given.

For certain data sets, the formula above for calculating a mean value is not sufficient. For example, suppose that Walter wants to calculate his overall average in his U.S. History course. If the course syllabus states that the final grade is determined by tests (40%), homework (20%), quizzes (10%) and final exam (30%), how should Walter calculate his grade?

When each data value in the set does not hold the same relative importance, a **weighted mean** must be used. To calculate a weighted mean for a sample, first multiply each value by its respective weight. Then divide the sum of these products by the sum of the weights to obtain the mean. The mathematical formula is given as follows:

$$\text{Weighted Mean, } \bar{x} = \frac{\sum (x_i \cdot w_i)}{\sum w_i}$$

where x_i is the ith data value, and w_i is the weight of the ith data value.

The procedure for calculating the weighted mean for a population is the same, but with the notation μ rather than $\bar{x}$.

Example 2

The syllabus in Walter's U.S. History class states that the final grade is determined by tests (40%), homework (20%), quizzes (10%) and a final exam (30%). Calculate Walter's final grade using the following scores:

Tests: 83
Homework: 98
Quizzes: 90
Final Exam: 87

Memory Booster!

When calculating a weighted mean, the weights do not change from person to person.

Solution:

First, let's determine which numbers are values of x and which are weights. A helpful hint for identifying the weights is that the weights do not change from person to person. In this case the weights must be the percentages for each category. The values for x are then each category's average. The weighted mean is calculated as follows:

$$\begin{aligned}\bar{x} &= \frac{83(0.4)+98(0.2)+90(0.1)+87(0.3)}{0.4+0.2+0.1+0.3} \\ &= \frac{87.9}{1} \\ &= 87.9\end{aligned}$$

✦ Median

Another measure of center is the median. The **median** is the middle value in an ordered array of data.

To find the median of a data set:

1. List the data in ascending (or descending) order.
2. If the data set has an ODD number of values, the median is the middle value.
3. If the data set has an EVEN number of values, the median is the average (mean) of the two middle values.

Note that this means the median may not be a value in the data set.

Example 3

Find the median of the following sets:

a. 3, 4, 6, 7, 2, 8, 9 **b.** 5, 7, 8, 1, 4, 9, 8, 9

Solution:

a. First, put the data in order: 2, 3, 4, **6**, 7, 8, 9. Since there are an odd number of values, the median is the number in the middle. Median = 6.

b. First, put the data in order: 1, 4, 5, **7**, **8**, 8, 9, 9. Since there are an even number of values, the median is the average of the two numbers in the middle.

$$\text{Median} = \frac{7+8}{2} = 7.5$$

✦ Mode

The last measure of center we will discuss is the mode. The **mode** is the value in the data set that occurs most frequently. If all of the data values occur only once, or they each occur an equal number of times, we say that there is **no mode**. If only one value occurs the most, then the data set is said to be **unimodal**. If exactly two values occur equally often, then the data set is said to be **bimodal**. If more than two values occur equally often, the data set is **multimodal**. Note that if there is a mode, it will always be a value in the data set.

Example 4

Find the mode of each of the following data sets. State if the data set is unimodal, bimodal, or neither.

a. 6, 4, 6, 1, 7, 8, 7, 2, 5, 7
b. 3, 4, 7, 8, 1, 6, 9
c. 2, 5, 7, 2, 8, 7, 9, 3

Solution:

a. The number 7 occurs more than any other value, so the mode is 7. This data set is unimodal.

b. The values all occur only once, so there is no mode. This data set is neither unimodal nor bimodal.

c. The values 2 and 7 both occur an equal number of times; thus they are both modes. This data set is bimodal.

✦ Choosing an Appropriate Measure of Center

We now know how to calculate three measures of center: mean, median, and mode. For a given data set, which measure of center is best? To answer this question, first consider what we might mean by the "average" college student. The "average" in this sense would refer to the most typical college student. Thus it describes the most frequently occurring characteristics of college students; such as gender or ethnicity. These are both nominal data sets. Thus for nominal data, the mode should be used. The mode is also the best choice for ordinal data. (Note, however, that in our college student example, the changing demographics of college students make the concept of an "average" college student meaningless.)

This answers the question for qualitative data, but what about quantitative data? Recall from Chapter 2 that an outlier is an extreme data value. Because it is much larger or much smaller than the rest of the data, it could affect the center. Let's see how an outlier affects the various measures of center in quantitative data sets.

Example 5

Calculate the mean, median, and mode of the following data:

84, 80, 82, 77, 78, 80, 79, 42

Solution:

Mean: Remember, the mean is the sum of all the data points divided by the number of points.

$$\bar{x} = \frac{84+80+82+77+78+80+79+42}{8} \approx 75.3$$

Median: We have an even number of values, so we will need the average of the middle two values.

42, 77, 78, **79**, **80**, 80, 82, 84

$$\text{Median} = \frac{79+80}{2} = 79.5$$

Mode: The number 80 occurs more than any other number, so it is our mode.

Note that the median and mode of the previous example are close to the same value, but the mean is smaller. This is because the outlier, 42, is much smaller than the rest of the data. The value for the mean will be pulled toward any outlier; thus the mean is pulled towards the tail of a skewed distribution. For this reason, if a quantitative data set has an outlier or is skewed, you should use the median. In summary, here are a few guidelines for choosing the most appropriate measure of center for a given data set.

Determining the Most Appropriate Measure of Center:

1. For qualitative data, the mode should be used.
2. For quantitative data the mean should be used, unless the data set contains outliers or is skewed.
3. For quantitative data sets that are skewed or contain outliers, the median should be used.

Example 6

Choose the best measure of center for the following data sets:

a. The average T-shirt size (S, M, L, XL) of American women.
b. The average salary for a professional team of baseball players.
c. The average price of homes in a subdivision of similar homes.

Solution:

a. This is qualitative data; thus the mode is the most appropriate choice.
b. This is quantitative data that has outliers since the superstars on the team make substantially more than the typical players. Therefore, the median is the best choice.
c. This is quantitative data that does not have outliers since the homes are similar. Therefore, the mean is the best choice.

✦ Graphs and Measures of Center

When looking at the graph of a distribution, it is often possible to estimate certain measures of center. For example, the mode is the number that occurs most frequently. Graphically, a mode is the highest peak of the distribution. Since the median is the number in the middle, on a distribution it will be the number that divides the area of the distribution in half. As we just stated, the mean of a distribution will be pulled toward any outliers. For example, if the distribution is skewed to the right, the mean of the distribution will be shifted to the right. If all three measures of center are plotted on a distribution that is skewed to the right, then the mean is the measure of center farthest to the right. Using this information, we can determine the general location (and therefore an approximate value) of each measure of center.

Example 7

Determine which letter represents the mean, the median, and the mode in the graph below.

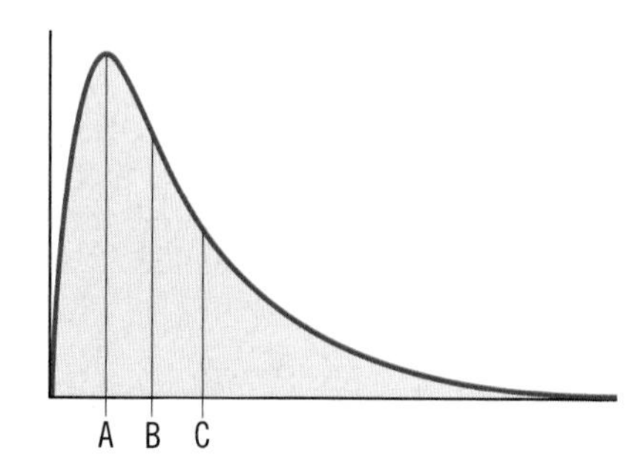

Solution:

Mean: This distribution is skewed to the right, so the mean will be the measure of center farthest to the right. Here it is represented by line C.

Median: The median is the value that divides the area of the distribution in half. Here it is represented by line B.

Mode: The mode is always located at the peak of a distribution, so it is line A.

Exercises

3.1

Directions: Calculate the mean, median, and mode for the following data sets, and when applicable state whether the data set is unimodal, bimodal, or multimodal. Remember to round the mean to one decimal place more than is given in the data set.

1. 8, 9, 10, 12, 4, 3, 5, 7, 10
2. 5, 5, 5, 5, 5, 5, 5
3. 5, 6, 3, 5, 2, 7, 8, 35
4. 17, 19, 15, 16, 20, 19, 17, 3
5. 1, 2, 3, 4, 5, 7, 8, 11, 12, 15

6. 2, 5, 4, 6, 7, 2, 4, 3

7. 3, 4, 10, 9, 5, 7, 1

8. 0.5, 0.6, 1.7, 0.5, 1.3, 1.0, 1.3, 0.5

9. 1, 1, 2, 5, 3, 5, 6, 7, 9, 11

10. 2, 5, 6, 8, 10, 12, 2, 13, 4

11. Weights of newborn babies (in pounds).

6.8	9.1	8.7	7.5	8.2
5.4	6.5	8.5	7.3	6.6
5.9	7.3	9.3	7.5	7.8

12. Batting averages for a sample of professional baseball players.

.275	.265	.333	.244	.279
.250	.288	.292	.370	.243
.236	.321	.305	.292	.250

13. Lengths of long distance phone calls (in minutes).

0.8	42.2	20.6	2.8
36.7	18.6	23.3	11.5
3.7	14.9	9.4	1.5
14.9	31.1	23.5	9.5

14. High temperatures for cities in the Southeast (in degrees Fahrenheit).

85	82	93	88
92	79	84	90
77	83	91	89
90	85	87	91
89	92	95	88

15. Ages of 20 American entrepreneurs.

28	39	43	53
35	32	34	29
33	31	32	31
25	22	30	29
41	36	23	47

16. Marquis is calculating his GPA. His grades are as follows: A (15 hours), B (18 hours), C (8 hours), and D (3 hours). Note that an A is equivalent to a 4.0, a B is equivalent to a 3.0, a C is equivalent to a 2.0, and a D is equivalent to a 1.0. What is Marquis' GPA?

17. Beth is calculating her GPA. Her grades are as follows: A (12 hours), B (22 hours), C (14 hours), and F (3 hours). Note that an A is equivalent to a 4.0, a B is equivalent to a 3.0, a C is equivalent to a 2.0, a D is equivalent to a 1.0, and an F is equivalent to 0. What is Beth's GPA?

18. The following table gives the average balance for one bank customer for the months of October through December.

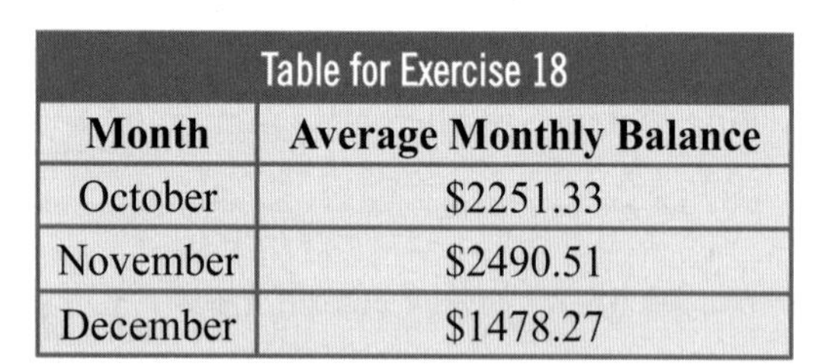

Table for Exercise 18	
Month	**Average Monthly Balance**
October	$2251.33
November	$2490.51
December	$1478.27

Calculate the average balance for the months of October through December. Remember: the average balance for each month must be weighted by the number of days in that month.

19. The following table gives the average balance for one bank customer for the months of July through September.

Table for Exercise 19	
Month	**Average Monthly Balance**
July	$402.45
August	$322.97
September	$298.64

Calculate the average balance for the months of July through September. Remember: the average balance for each month must be weighted by the number of days in that month.

20. Susan is calculating her final grade in biology. The grade is broken down as follows: tests (50%), lab (30%), and final exam (20%). Susan's category averages are:

Tests: 82
Lab: 78
Final Exam: 86

What is her final grade? Round your answer to one decimal place.

Directions: For each of the following data sets, determine the most appropriate measure of center.

21. The average hairstyle of female students on a given college campus (length, color, straight/curly, etc.).

22. The average salary of professional football players.

23. The average number of minutes students spend completing a homework assignment in an honors class.

24. The average price of similar sofas at different furniture stores.

Directions: For each of the two graphs below, determine which letter represents (a) the mean, (b) the median, and (c) the mode.

25.

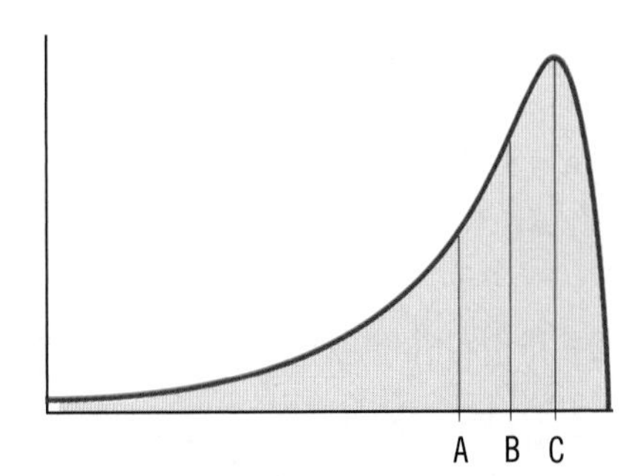

26.

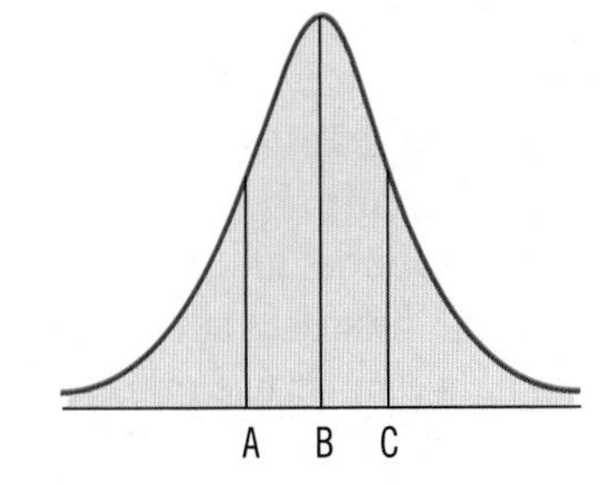

Measures of Dispersion

3.2

In the previous section, we calculated various measures of center. These numerical descriptions tell us where the data is located on a number line, but they tell us nothing about the shape of the data. Is this data spread out or are all of the values very close together? How much of the data is close to the mean? How close is the data to the mean? To answer these questions, we need to learn about several other numerical descriptions, called measures of dispersion.

Range

The easiest measure of dispersion to calculate is the **range**. The range of a data set is the difference between the largest and smallest values in the data set. The formula for finding the range is

$$\text{Range} = \text{Maximum} - \text{Minimum}.$$

Example 8

Find the range of the following data sets:

a. 4, 3, 2, 7, 6, 5, 8

b. 2, 2, 8, 2, 2, 2, 2

Solution:

a. The maximum value is 8 and the minimum value is 2, so the range is

$$\begin{aligned}\text{Range} &= 8-2\\ &= 6\end{aligned}$$

b. The maximum value for the data set is also 8 and the minimum value is also 2, so the range is also

$$\begin{aligned}\text{Range} &= 8-2\\ &= 6\end{aligned}$$

Calculating the range is very easy; however, the range is not as descriptive as other measures of dispersion. Consider the two data sets in the above example. Notice that both data sets have the same range. However, almost all of the values in the second data set are the same while the values in the first data set are more spread out. To distinguish between these two situations, we must use other measures of dispersion.

✦ Variance

Rather than simply looking at the difference between the largest and smallest data values, we might instead look in more detail at how much each piece of data differs from the mean. The difference is called the **deviation from the mean** and is calculated by

$$\text{Deviation from the mean} = x_i - \bar{x}.$$

It may seem natural to use the 'average' deviation from the mean as a measure of dispersion. However, the sum of the deviations from the mean for any data set is always equal to zero because the positive deviations cancel out the negative deviations. Instead, we remove this cancellation effect by first squaring the deviations. The average squared deviation from the population mean is called the **population variance,** σ^2.

$$\sigma^2 = \frac{\sum (x_i - \mu)^2}{N}$$

Unlike the formulas for population and sample mean, the formula for sample variance is different from the population version. The **sample variance,** s^2, is given by

$$s^2 = \frac{\sum (x_i - \bar{x})^2}{n-1}.$$

Should you desire to calculate variance by hand, especially with large data sets, it is helpful to use a table as illustrated in the following example. However, using a calculator or spreadsheet program can greatly simplify this task, as demonstrated in the technology section at the end of this chapter.

Math Symbols!

σ^2 stands for the population variance

s^2 represents sample variance

Rounding Rule

Round the standard deviation and variance to one more decimal place than what is given in the data set.

Example 9

Calculate the sample variance for the sample: 5, 8, 7, 6, 9.

Solution:

First, note that the mean of the data is 7, and the sample size is 5. ($\bar{x} = 7$, $n = 5$)

Table 1 - Sample Variance

x_i	$(x_i - \bar{x})$	$(x_i - \bar{x})^2$
5	−2	4
8	1	1
7	0	0
6	−1	1
9	2	4

cont'd on the next page

Thus the sample variance is

$$s^2 = \frac{\sum(x_i - \bar{x})^2}{n-1} = \frac{10}{5-1} = 2.5.$$

✦ Standard Deviation

Math Symbols!

σ represents the population standard deviation

s represents sample standard deviation

Although the variance has important theoretical properties, the units of the variance are the square of the units in which the data is measured. By taking the square root of the variance, we return to a quantity which is in the same units as the data. This closely related measure of dispersion is called the **standard deviation**, σ, given by the formula

$$\sigma = \sqrt{\sigma^2} = \sqrt{\frac{\sum(x_i - \mu)^2}{N}}.$$

Similarly, $s = \sqrt{s^2} = \sqrt{\frac{\sum(x_i - \bar{x})^2}{n-1}}$.

Example 10

Calculate the sample standard deviation of the data from the previous example.

Solution:

Since the sample variance equals 2.5, the sample standard deviation is equal to the square root of 2.5, namely

$$s = \sqrt{2.5} \approx 1.6.$$

The standard deviation provides a measure of how much we might expect a typical member of the data set to differ from the mean. The greater the standard deviation, the more the data is "spread out." Note that, by definition, the standard deviation cannot be negative. If the standard deviation is 0, then all of the data values must be the same.

The standard deviation allows us to interpret differences from the mean with some sense of scale. For instance, if a data set consisted of house prices, then differences of thousands of dollars as opposed to tens of thousands of dollars might be considered small. However, if a data set consisted of gas prices in various towns, a difference of even a single dollar would be considered very large. As we shall see, the standard deviation allows us to make such judgments of whether a difference is large or small, in a systematic way.

Which of the two graphs has the larger standard deviation?

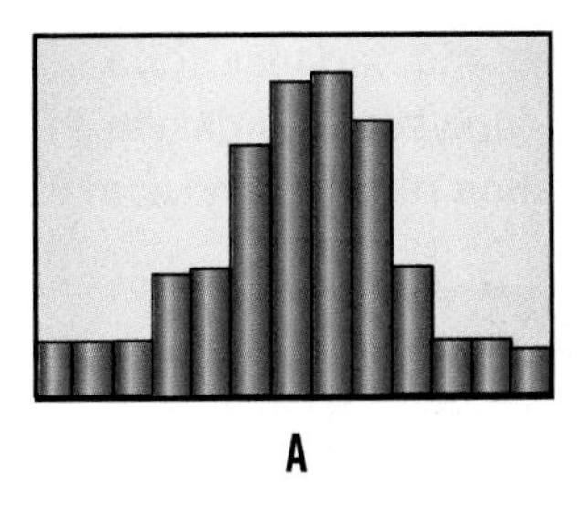

A

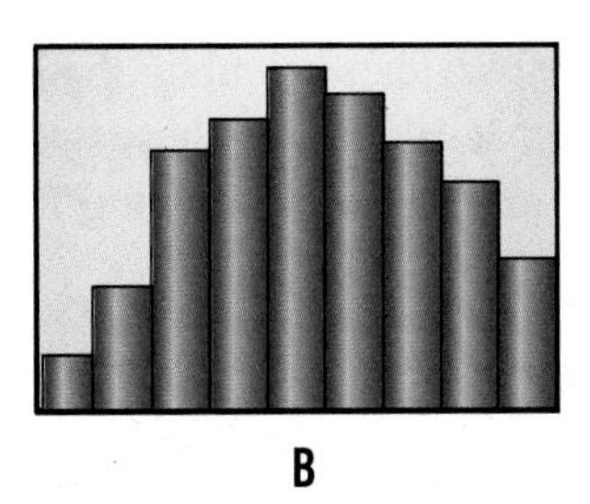

B

We know that the graph with the larger standard deviation should appear more spread out; but without the numerical labeling for these graphs, one cannot tell which graph has the largest spread. In fact, if these graphs are from two different types of data, they may not have the same units of measurement, and so comparison is meaningless. However, we can do this comparison mathematically by using the **coefficient of variation, CV**. The CV is the ratio of the standard deviation to the mean as a percentage.

$$\text{CV} = \frac{s}{\bar{x}} \cdot 100\%$$

It allows us to compare the spread of data from two different sources. For example, the mean and standard deviation of Data Set A from the graphs above are 23 inches and 4 inches respectively; whereas, the mean and standard deviation of Data Set B are $117,000 and $1596.03, respectively.

$$\begin{aligned}\text{The CV of Data Set A } &= \frac{4}{23} \cdot 100\% \\ &\approx 17.4\%\end{aligned}$$

$$\begin{aligned}\text{The CV of Data Set B } &= \frac{1596.03}{117{,}000} \cdot 100\% \\ &\approx 1.4\%\end{aligned}$$

Now, it is clear to see that Data Set A has the larger standard deviation.

✦ Standard Deviation and Variance of Grouped Data

Now that we can calculate the standard deviation and variance for a set of data, let's look at calculating these two measures for grouped data, or data in a frequency table. Consider the following frequency distribution of final grades in a statistics class.

Table 2 - Final Grades	
Grade	**Frequency (f)**
94 – 100	5
87 – 93	8
80 – 86	12
73 – 79	7
66 – 72	4

Since we are not given the original data, we do not know the exact values of the five data points in the first class. All we know for certain is that these five values are all between 94 and 100. Since we need an estimate of each data value in order to calculate standard deviation, the class midpoint is a good average to use for each data point in that class. For example, the class midpoint for the first class is 97. We then estimate that the first class contains five values that are each approximately equal to 97. Using this same reasoning, the second class contains eight values that are each approximately equal to 90. Therefore, for grouped data the formula for standard deviation is

$$s=\sqrt{\frac{n\left[\sum\left(f\cdot x^{2}\right)\right]-\left[\sum\left(f\cdot x\right)\right]^{2}}{n(n-1)}},$$

where n = sample size,
f = frequency, and
x = the midpoint.

To help us apply this formula, we can extend the frequency table to include three more columns: 1) the midpoint, x, 2) the frequency times the midpoint, $f\cdot x$, and 3) the frequency times the midpoint squared, $f\cdot x^2$. The sums of these columns are then used in the standard deviation formula.

Let's now create a table with these summations and use them to calculate the standard deviation of the grades given above.

Example 11

Calculate the standard deviation for the frequency distribution of test scores given on the previous page.

Solution:

Table 3 - Final Grades

Grade	Frequency (f)	Class Midpoint, x	$f \cdot x$	$f \cdot x^2$
94 – 100	5	97	485	47,045
87 – 93	8	90	720	64,800
80 – 86	12	83	996	82,668
73 – 79	7	76	532	40,432
66 – 72	4	69	276	19,044
	$n = \sum f = 36$		$\sum(f \cdot x) = 3009$	$\sum(f \cdot x^2) = 253{,}989$

We can now substitute the approximated values into the formula as follows:

$$s = \sqrt{\frac{36(253{,}989) - (3009)^2}{36(35)}}$$
$$\approx 8.4$$

Thus the standard deviation of the grades is approximately 8.4 points.

The variance for grouped data can easily be estimated using the relationship between variance and standard deviation. Recall that the variance is the standard deviation squared. Thus the variance for our example is approximately $(8.4)^2 = 70.56$. Again, this is just an approximation since we are not working with the original data set.

Memory Booster!

Remember, calculating the standard deviation from a frequency distribution will only produce an estimate of the standard deviation of the original data.

Example 12

Calculate the sample standard deviation and variance for the following distribution of gas prices at the stations in a certain city.

Table 4 - Gas Prices	
Gas Price	**Frequency (f)**
1.95 – 1.99	1
2.00 – 2.04	3
2.05 – 2.09	5
2.10 – 2.14	6
2.15 – 2.19	2
2.20 – 2.24	1

Solution:

Begin by extending the frequency table to include columns for the midpoint, the frequency multiplied by the midpoint, and the frequency multiplied by the midpoint squared. Next, calculate the values for each cell in the table and total the columns. The resulting frequency table is shown below.

Table 5 - Final Grades				
Gas Price	**Frequency (f)**	**Class Midpoint, x**	$f \cdot x$	$f \cdot x^2$
1.95 – 1.99	1	1.97	1.97	3.8809
2.00 – 2.04	3	2.02	6.06	12.2412
2.05 – 2.09	5	2.07	10.35	21.4245
2.10 – 2.14	6	2.12	12.72	26.9664
2.15 – 2.19	2	2.17	4.34	9.4178
2.20 – 2.24	1	2.22	2.22	4.9284
	$n=\sum f=18$		$\sum(f \cdot x)=37.66$	$\sum(f \cdot x^2)=78.8592$

Notice that no rounding occurred in the table. All resulting decimal places were recorded. Substituting these values into the formula for standard deviation of grouped data yields the following:

$$s=\sqrt{\frac{18(78.8592)-(37.66)^2}{18(17)}}\approx 0.062$$

Thus the standard deviation for gas prices in the city is approximately 0.062 dollars. To calculate the variance, square the value of the standard deviation. Thus

$$s^2=(0.062)^2\approx 0.004.$$

✦ Empirical Rule

When a set of data is approximately bell-shaped, the empirical rule can be used to estimate the percentage of values within a few standard deviations of the mean. The empirical rule says the following:

Empirical Rule ✦

- Approximately 68% of the data lies within 1 standard deviation of the mean.
- Approximately 95% of the data lies within 2 standard deviations of the mean.
- Approximately 99.7% of the data lies within 3 standard deviations of the mean.

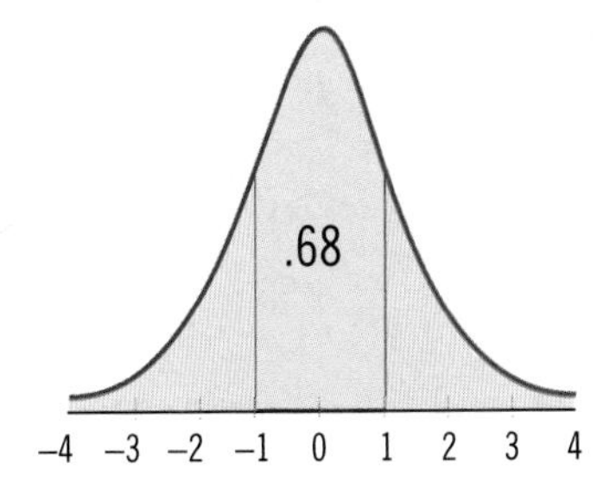

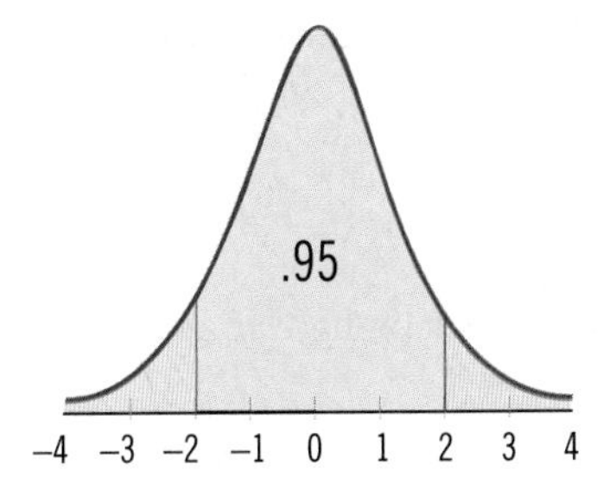

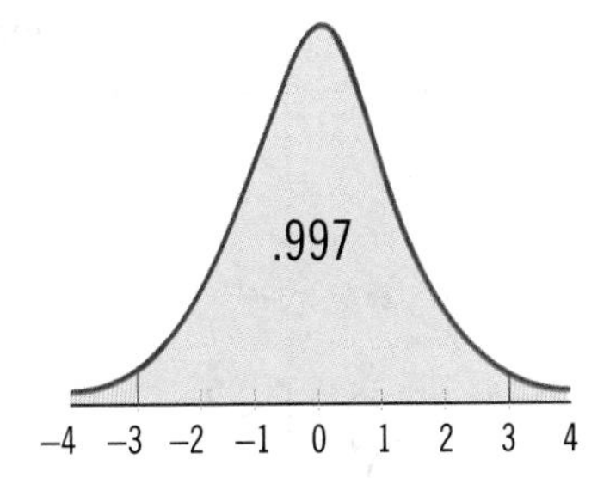

Figure 1 – Graphical Depiction of the Empirical Rule

Example 13

The average weight of newborn babies is bell-shaped with a mean of 3325 grams and standard deviation of 571 grams.

a. What percentage of newborn babies weigh between 2183 and 4467 grams?
b. What percentage of newborn babies weigh less than 3896 grams?

Solution:

a. Since we know that the data is bell-shaped, we can apply the empirical rule. We need to know how many standard deviations 2183 and 4467 are from the mean. By subtracting, we can find how far each of these figures is from the mean. Then, dividing by the standard deviation, we can convert these differences into numbers of standard deviations.

cont'd on the next page

Here are the calculations:

$$3325 - 2183 = 1142 \quad \text{and} \quad 4467 - 3325 = 1142$$

$$\frac{1142}{571} = 2 \qquad\qquad \frac{1142}{571} = 2$$

Thus these weights lie two standard deviations above and below the mean. According to the empirical rule, approximately 95% of values lie within two standard deviations of the mean. Therefore, we can say approximately 95% of newborn babies weigh between 2183 and 4467 grams.

b. To begin, let's find out how many standard deviations a weight of 3896 is away from the mean by performing the same calculation as before.

$$3896 - 3325 = 571$$

Thus it is one standard deviation above the mean. The empirical rule says that 68% of data lies within one standard deviation of the mean. Because of the symmetry of the distribution, half of this 68% is above the mean and half is below. Putting the upper 34% together with the 50% of data that is below the mean, we have that

$$50\% + 34\% = 84\%$$

of babies weigh less than 3896 grams.

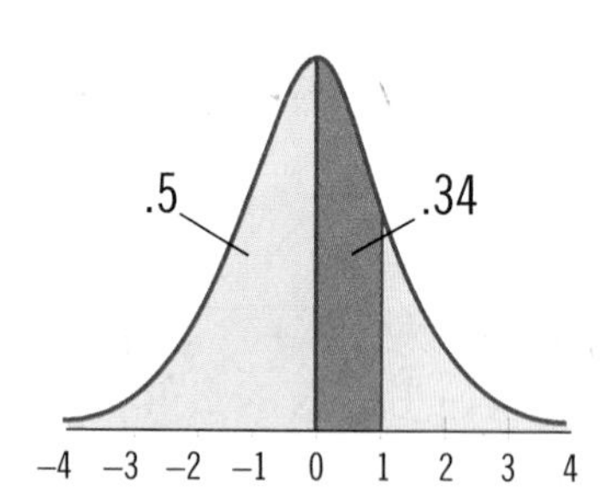

✦ Chebyshev's Theorem

Although the empirical rule is handy for bell-shaped data, it cannot be applied to other distributions. Chebyshev's Theorem is helpful when the empirical rule cannot be used. However, Chebyshev's Theorem simply gives a minimum estimate, but is NOT exact.

Chebyshev's Theorem

The proportion of data that lie within K standard deviations of the mean is at least $1-\frac{1}{K^2}$, for $K > 1$. When $K = 2$ and $K = 3$, Chebyshev's Theorem says:

- $K = 2$: At least $1-\frac{1}{2^2}=\frac{3}{4}=75\%$ of the data lies within two standard deviations of the mean.
- $K = 3$: At least $1-\frac{1}{3^2}=\frac{8}{9}=88.9\%$ of the data lies within three standard deviations of the mean.

Example 14

Suppose that in one small town the average income is \$34,200 with a standard deviation of \$2200. What percentage of households earn between \$29,800 and \$38,600?

Solution:

Since we do not know that the data is bell-shaped, we cannot apply the empirical rule here. However, we can apply Chebyshev's Theorem to find a minimum estimate. In order to do so, we need to know how many standard deviations \$29,800 and \$38,600 are from the mean. By subtracting, we can find how far each of these figures is from the mean. Then, dividing by the standard deviation, we can convert these differences into numbers of standard deviations. Here are the calculations:

$$\$34,200-\$29,800=\$4400 \quad \text{and} \quad \$38,600-\$34,200=\$4400$$

$$\frac{\$4400}{\$2200}=2 \qquad\qquad \frac{\$4400}{\$2200}=2$$

Thus these incomes lie two standard deviations above and below the mean. Chebyshev's Theorem can be applied for $K = 2$. Using the calculation previously shown in the box above, we can say that at least 75% of household incomes lie within this range.

Exercises

3.2

Directions: Calculate the range, population variance, and population standard deviation of the following samples.

1. 8, 12, 10, 12, 4, 13, 5, 17, 10
2. 5, 5, 5, 5, 5, 5, 5
3. 5, 6, 3, 5, 7, 7, 8, 25
4. 17, 19, 21, 16, 20, 19, 17, 13
5. 1, 2, 3, 4, 5, 7, 8, 9, 10

Directions: Calculate the range, sample variance, and sample standard deviation of the following samples.

6. 8.1, 9.5, 10.7, 12.3, 4.4, 3.9, 5.1, 7.3, 10.1
7. 4, 4, 4, 4, 4, 4, 4
8. 1, 2, 3, 2, 3, 4, 5, 35
9. 2.0, 1.7, 1.9, 1.5, 1.6, 1.9, 1.7, 3.0
10. 15, 12, 13, 14, 15, 17, 18, 11, 12, 15

Directions: Calculate the sample standard deviation for each of the following data sets:

11. Weights of newborn babies (in pounds).

6.8	9.1	8.7	7.5	8.2
5.4	6.5	8.5	7.3	6.6
5.9	7.3	9.3	7.5	7.8

12. Batting averages for a sample of professional baseball players.

.275	.265	.333	.244	.279
.250	.288	.292	.370	.243
.236	.321	.305	.292	.250

13. Lengths of long distance phone calls (in minutes).

0.8	42.2	20.6	2.8
36.7	18.6	23.3	11.5
3.7	14.9	9.4	1.5
14.9	31.1	23.5	9.5

14. High temperatures for cities in the Southeast (in degrees Fahrenheit).

85	82	93	88
92	79	84	90
77	83	91	89
90	85	87	91
89	92	95	88

15. Ages of 20 American entrepreneurs.

28	39	43	53
35	32	34	29
33	31	32	31
25	22	30	29
41	36	23	47

Directions: Decide if each statement is true or false.

16. If the standard deviation of a data set is zero, then all entries in the data set equal zero.

17. The population variance and sample variance are the same value for the same set of data.

18. It is possible to have a standard deviation of –3 for some data set.

19. It is possible to have a standard deviation of 435,000 for some data set.

Directions: Calculate the sample standard deviation and sample variance of each frequency distribution.

20.

Table for Exercise 20- Ages of Taste Test Participants	
Class	**Frequency**
15 – 19	7
20 – 24	8
25 – 29	10
30 – 34	2
35 – 39	3

21.

Table for Exercise 21 - Braking Time for Vehicles in Minutes	
Class	**Frequency**
0.05 – 0.07	12
0.08 – 0.10	15
0.11 – 0.13	14
0.14 – 0.16	15
0.17 – 0.19	14

22.

Table for Exercise 22 - Age at First Marriage	
Class	**Frequency**
15 – 18	2
19 – 22	5
23 – 26	4
27 – 30	5
31 – 34	4

23.

Table for Exercise 23 - Hourly Wage at First Job	
Class	**Frequency**
7.5 – 8.4	12
8.5 – 9.4	50
9.5 – 10.4	48
10.5 – 11.4	45
11.5 – 12.4	34

24. For the following sets of data, calculate the coefficient of variation, CV. Which of the 2 data sets has the *larger* spread?
    A: 31, 33, 35, 39, 31, 32, 30, 40, 13, 41, 38, 32, 37, 33
    B: 2.3, 3.5, 4.3, 2.1, 4.8, 3.9, 2.0, 3.3, 4.0, 1.6, 2.2

25. For the following sets of data, calculate the coefficient of variation, CV. Which of the 2 data sets has the *larger* spread?
    A: $1.23, $1.55, $1.01, $1.89, $2.35, $2.56, $2.71, $1.75, $2.01, $1.59
    B: 119, 145, 97, 121, 118, 98, 102, 114, 118, 99, 100, 113, 116, 111

26. For the following sets of data, calculate the coefficient of variation, CV. Which of the 2 data sets has the *smaller* spread?
    A: 21,568 20,888 20,037 21,932 22,000 21,123 21,567 22,298
    B: 4.5, 4.3, 6.5, 3.3, 4.7, 3.67, 4.01, 3.89, 4.4, 2.99, 4.88, 3.77

27. For the following sets of data, calculate the coefficient of variation, CV. Which of the 2 data sets has the *smaller* spread?
    A: 328, 444, 283, 289, 345, 327, 298, 277, 419, 402, 399, 418, 401
    B: 78, 73, 92, 89, 74, 88, 91, 71, 70, 89, 81, 83, 84, 93, 94, 100

Directions: Use the empirical rule to answer the following questions.

28. Suppose that starting salaries for graduates of one university have a bell-shaped distribution with a mean of $25,400 and standard deviation of $1300. What percentage of graduates have starting salaries between $24,100 and $26,700?

29. Suppose that starting salaries for graduates of one university have a bell-shaped distribution with a mean of $25,400 and standard deviation of $1300. What percentage of graduates have starting salaries between $22,800 and $28,000?

30. Suppose that electric bills for the month of May in one city have a bell-shaped distribution, with a mean of 119 units and standard deviation of 22 units. What percentage of electric bills are greater than 97 units?

31. Suppose that electric bills for the month of May in one city have a bell-shaped distribution, with a mean of 119 units and standard deviation of 22 units. What percentage of electric bills are less than 163 units?

32. Suppose it is known that verbal SAT scores have a bell-shaped distribution with a mean of 500 and a standard deviation of 100. What percentage of SAT scores are no more than 600?

33. Suppose it is known that verbal SAT scores have a bell-shaped distribution with a mean of 500 and a standard deviation of 100. What percentage of SAT scores are at least 300?

Directions: Use Chebyshev's Theorem to answer the following questions.

34. Suppose that salaries for associate mathematics professors at one university have a mean of \$64,900 with a standard deviation of \$9400. What is the minimum percentage of associate professors with starting salaries between \$46,100 and \$83,700?

35. Suppose that household electric bills for the months of May through August in a city in Florida have a mean of 230 units with a standard deviation of 58 units. What is the minimum percentage of electric bills between 56 units and 404 units?

36. Car insurance premiums in one region have a quarterly mean of \$246 with a standard deviation of \$31. What is the minimum percentage of car insurance premiums between \$184 and \$308?

Measures of Relative Position

3.3

In addition to measuring the center of a data set and how spread out the data values are, we can also measure the relative position of values in the data set. Often, it is useful to know exactly where a particular value is located within a set, or the number of values that are above or below it. For example, the median is the number whose position is in the middle of the data set.

Quartiles

If we divide a data set into four parts, the numbers that form the divisions are called the **quartiles**. If we wanted to divide a line segment into four parts, we would draw three dividing lines. Similarly, when dividing a data set into four parts, we use three quartiles.

Q_1 = First Quartile = the number greater than or equal to 25% of the data
Q_2 = Second Quartile = the number greater than or equal to 50% of the data
Q_3 = Third Quartile = the number greater than or equal to 75% of the data

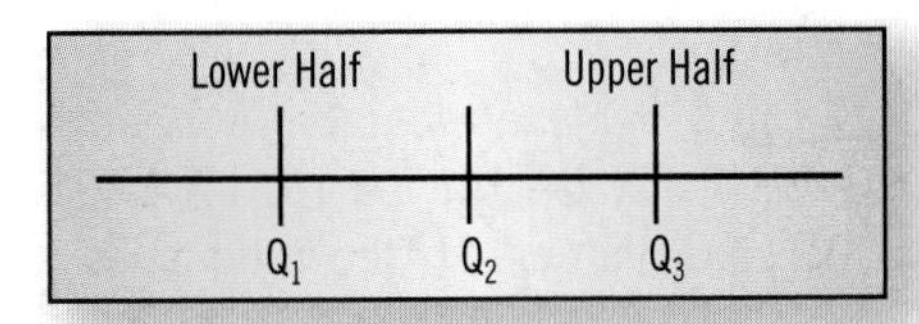

To find the quartiles, first note that the second quartile is the middle of the data set; thus it is the median of the data. Therefore, we begin finding quartiles by ordering the data set and finding the median, Q_2. Next, we use the median to divide the data into an upper half and a lower half. For an odd number of data values, include the median in each half. If there were an even number of data values, then the median would not be a value in the data set, so we do not include it in either the upper or lower half of the data. The first quartile is the median of the lower half of the data. The third quartile is the median of the upper half of the data.

Example 15

Find the three quartiles for the data set 8, 9, 10, 2, 5, 3, 7, 12, 15.

Solution:

First, put the data in order: 2, 3, 5, 7, 8, 9, 10, 12, 15.
Next, find the median. There are 9 data values, so the fifth value is the median, which is 8. Thus $Q_2 = 8$. Now we need to divide the data into two halves. Since we have an odd number of data values, include the median in each half. The lower half of the data is 2, 3, 5, 7, 8, which has a median of 5. Thus $Q_1 = 5$. The upper half of the data is 8, 9, 10, 12, 15, which has a median of 10. Thus $Q_3 = 10$. In summary, $Q_1 = 5$, $Q_2 = 8$, and $Q_3 = 10$.

If you try to work the previous example on your TI-84 Plus calculator, you will notice that the solutions given in the above example do not match the values for Q_1 and Q_3 given by the calculator. The reason is that the TI-84 Plus calculator does not include the median in the upper or lower half of the data set. However, by not including the median you are calculating a completely different value, created by statistician John Tukey, called a **hinge**. The lower hinge is a rough approximation of the first quartile, and the upper hinge is a rough approximation of the third quartile. To complicate matters further, for some data sets the values of the hinges and the quartiles are the same. So some textbooks (and the TI-84 Plus) do not differentiate between quartiles and hinges. The solutions given in this textbook will use the method of calculating quartiles described in Example 14. So be aware that if you use a TI-84 Plus calculator to find quartiles, the values given by the calculator are hinges, not quartiles, and though they may be the same numbers for some data sets, these two numbers will not always be the same.

Example 16

Find the three quartiles for the data set 10, 12, 14, 15, 14, 16, 17, 18, 10, 19, 17, 17.

Solution:

First, put the data in order: 10, 10, 12, 14, 14, 15, 16, 17, 17, 17, 18, 19. Next, find the median. There are 12 data values, so the median is the mean of the middle two values.

$$\begin{aligned}\text{Median} &= \frac{15+16}{2}\\ &= 15.5.\end{aligned}$$

Thus $Q_2 = 15.5$. Next we need to divide our data into two parts. Since we have an even number of data values the median is not a value in the data set, so we will not include it in the upper or lower half of the data. The lower half of the data is 10, 10, 12, 14, 14, 15. The median of the lower half of the data is the mean of the middle two values, which is

$$\frac{12+14}{2} = 13.$$

Thus $Q_1 = 13$. The upper half of the data is 16, 17, 17, 17, 18, 19. The median of the upper half of the data is the mean of the middle two values, which is

$$\frac{17+17}{2} = 17.$$

Thus $Q_3 = 17$. In summary, $Q_1 = 13$, $Q_2 = 15.5$, and $Q_3 = 17$.

✦ Five-Number Summary

Quartiles are used in a numerical description, aptly called the five number summary because it contains five numbers: the minimum value; the first quartile, Q_1; median or second quartile, Q_2; the third quartile, Q_3; and the maximum value. The five-number summary is made up of these five numbers listed in order from smallest to largest.

Example 17

Write the five-number summary for the data from Example 15. The data was 8, 9, 10, 2, 5, 3, 7, 12, 15.

Solution:

The minimum value is 2, the maximum value is 15, and we have previously determined that the quartiles are 5, 8, and 10. Thus the five-number summary is 2, 5, 8, 10, 15.

✦ Box Plots

A **box plot** is a graphical representation of a five-number summary. It is sometimes referred to as a "Box and Whiskers plot." The "box" refers to the rectangle that is created from joining the lines representing the first and third quartile. The "whiskers" are the lines that extend to reach the minimum and maximum values.

Creating a Box Plot:

1. Begin with a horizontal (or vertical) number line.
2. Draw a small line segment above (or next to) the number line to represent each of the numbers in the five-number summary.
3. Connect the line segment that represents the first quartile to the line segment representing the third quartile, forming a box with the median's line segment in the middle.
4. Connect the "box" to the line segments representing the minimum and maximum values to form the "whiskers".

Example 18

Draw a box plot to represent the five number summary: 2, 6, 8, 10, 14.

Solution:

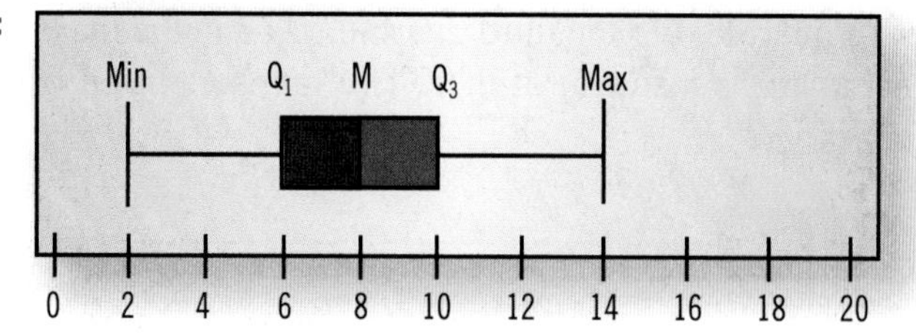

By drawing several box plots next to each other, different data sets may be easily compared. The box plots show the distribution of the values from each data set; consequently, comparisons between median values, maximum values, and minimum values are a simple task.

Example 19

The box plots below show the distribution of ACT scores for three groups of seniors.

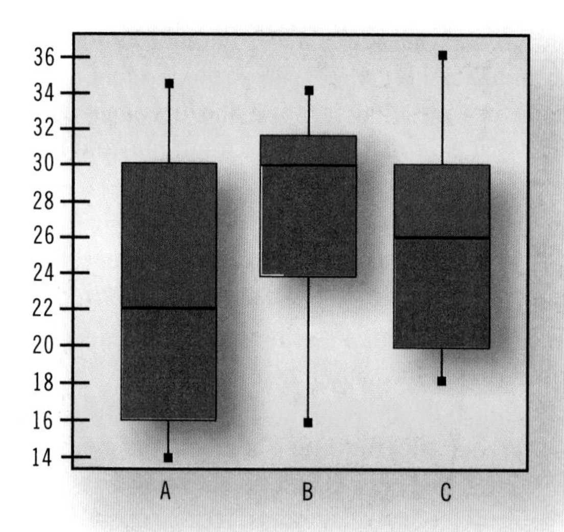

Answer the following questions:

a. Which group has the best score?
b. Which group has the lowest score?
c. Which group has the highest median?
d. Which group has the smallest deviation?
e. Which group did better overall?

Solution:

a. The maximum value, 36, is in group C.
b. Group A has the lowest score, 14.
c. The highest median is 30 in group B.
d. Group C has the smallest range; therefore, it has the smallest deviation.
e. Group B did better overall. Despite having a minimum value smaller than group C's, group B has a Q_1, median, and Q_3 larger than those from any other group.

✦ Percentiles

Instead of dividing the data into just 4 parts, we may choose to divide the data up into any number of parts. Data divided into 10 parts is called deciles and into 8 parts is called octiles. Let's consider dividing the data into 100 parts. These divisions are called **percentiles**. Percentiles are commonly seen in standardized testing and medical records. They tell you what percentage of the data lies *at or below* a given value. For instance, if your ACT score falls in the 63rd percentile, this indicates that 63% of ACT scores are less than or equal to yours.

One difficulty when it comes to calculating percentiles is that there is little agreement on exactly how to do it. Various statistical software packages, such as SAS, actually allow you to choose the method you want to use to calculate the value of a given percentile. Each of the methods could result in a different answer, depending on the size and variation of your data set.

Despite the controversy surrounding how best to calculate percentiles, we must choose a method. Therefore, we will use the following formula to calculate the value of a certain percentile in a given data set.

$$l = n \cdot \frac{p}{100}$$

where l = location of the pth percentile, n = the sample size, and p stands for the pth percentile.

When using this formula to find the location of the percentile's value in the data set, you must make sure that you follow these two rules:

1. If the formula results in a decimal value for l, the location is the next *largest* integer.
2. If the formula results in a whole number, the percentile's value is the average of the value in that location and the one in the next largest location.

Example 20

A class of 25 students had the following ACT scores:

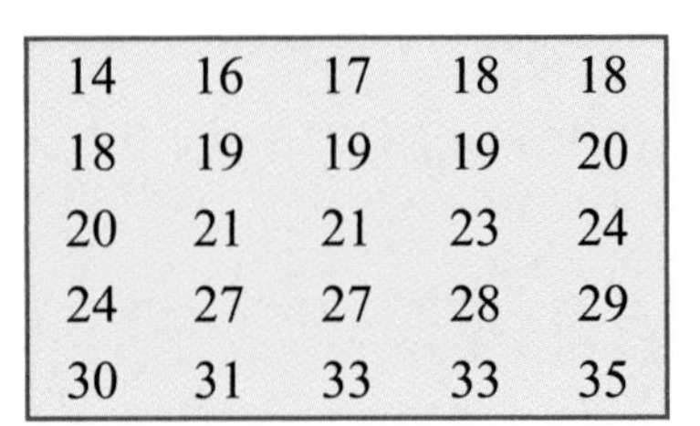

14	16	17	18	18
18	19	19	19	20
20	21	21	23	24
24	27	27	28	29
30	31	33	33	35

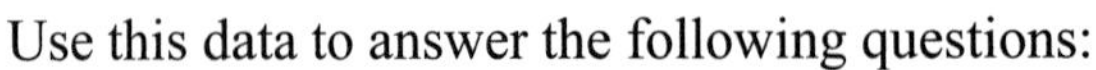

Use this data to answer the following questions:

a. What is the value of the 10th percentile?

b. What is the value of the 40th percentile?

Solution:

a. The sample size for our data set is $n = 25$. Since we want the 10th percentile, $p = 10$. In order to find the value of the 10th percentile, we need to find its location in the data set. Using the formula, we get

$$l = 25 \cdot \frac{10}{100} = 2.5.$$

Because the formula resulted in a decimal value, we round up to the next largest integer, in this case, 3. So the 10th percentile is the number in the 3rd position, which is 17. Thus the 10th percentile is 17.

b. Again, $n = 25$. We want the value of the 40th percentile, so $p = 40$. Using the formula, we get

$$l = 25 \cdot \frac{40}{100} = 10.$$

Since the formula resulted in a whole number, the percentile's value is between that location and the next. Thus it is the number exactly between the 10th and 11th value in the data set. The 10th value is 20. The 11th value is 20 also. So the number exactly between them is

$$\frac{20 + 20}{2} = 20.$$

Therefore, the value of the 40th percentile is 20.

We can also find the percentile of a value in a data set; however, we need to use a slightly different formula. Begin as you did when finding the percentile's value above by ordering the data set from smallest to largest. Next, apply the formula

$$p = \frac{x}{n} \cdot 100$$

where p = the percentile expressed as a whole number, x = the number of values in the data set *less than* the given value, and n = the sample size.

Rounding Rule

When calculating a percentile, always round up to the next integer.

Example 21

Using the data from the previous example, Deya's score was a 23. In what percentile is her score?

Solution:

Deya's score is the 14th in the ordered list of 25 scores. There are 13 scores less than Deya's; thus, $x = 13$. The sample size is 25, so $n = 25$. Plugging the known values into our formula and solving for p gives us

$$p = \frac{13}{25} \cdot 100$$
$$= 52.$$

Thus Deya's score is in the 52nd percentile. In other words, 52% of the ACT scores were less than or equal to Deya's.

✦ Standard Scores

We can compare two scores from different populations by comparing their respective percentiles. For example, a score in the 56th percentile is not as high as a score in the 87th percentile. We could also determine how the values relate to the respective means of their data set. This measure is called a **standard score**, or ***z*-score**. A standard score tells us how far a value is from the mean, specifically, how many standard deviations it is away from the mean. The formula is

$$z = \frac{x - \mu}{\sigma} \quad \text{(population)}$$

or

$$z = \frac{x - \bar{x}}{s} \quad \text{(sample).}$$

Standard scores are useful in comparing data values from populations with different means and standard deviations. For example, they could be used to determine if you scored better on the ACT exam or the SAT exam (assuming you took both).

Rounding Rule

When calculating a z-score in this text, round your answer to two decimal places.

Example 22

If the average score on the math section of the SAT test is a 500 with a standard deviation of 150 points, what is the standard score for a student who scored a 630?

Solution:

$\mu = 500$ and $\sigma = 150$. The value of interest is $x = 630$, so we have:

$$z = \frac{630 - 500}{150} \approx 0.87.$$

Thus a student with a score of 630 would be 0.87 standard deviations above the mean.

Example 23

Jodi scored an 87 on her calculus test and was bragging to her best friend how well she had done. She said that her class had an average of 80 with a standard deviation of 5; therefore, she had done better than the class average. Her best friend, Ashley, was disappointed. She had scored only an 82 on her calculus test. The class average for her class was a 73 with a standard deviation of 6.

Who *really* did better on their test, in their respective classes, Jodi or Ashley?

Solution:

Let's calculate each student's standard score.

For Jodi,

$$z = \frac{87 - 80}{5} = 1.4.$$

For Ashley,

$$z = \frac{82 - 73}{6} = 1.5.$$

Thus Ashley actually did better on her calculus test with respect to her class, despite the fact that Jodi had the higher score, because Ashley's score was more standard deviations above her class average.

Exercises

3.3

Directions: Calculate the five-number summary of the following populations.

1. 8, 12, 10, 12, 4, 13, 5, 17, 10

2. 5, 6, 3, 5, 7, 7, 8, 25

3. 17, 19, 21, 16, 20, 19, 17, 13

4. 1, 2, 3, 4, 5, 7, 8, 9, 10

Directions: Draw a box and whiskers plot for each set of data on the same graph (as in Example 19). Use your box and whiskers to answer the following questions:

a. Which data set has the largest value?
b. Which data set has the smallest value?
c. Which data set has the largest median?

5. Data Set A: 2, 3, 5, 5, 5, 6, 6, 6, 7, 7, 8
   Data Set B: 3, 3, 4, 4, 4, 5, 6, 6, 9, 12

6. Data Set A: 10, 12, 13, 14, 15, 16, 17, 18
   Data Set B: 11, 15, 17, 18, 19, 19, 20

7. Test Scores for Class A: 45, 60, 57, 83, 72, 93, 87, 73, 92
   Test Scores for Class B: 23, 88, 67, 89, 91, 76, 72, 100, 95, 35

8. Diameter of Cans (in cm) from Assembly Line A:
   5.6, 5.7, 5.1, 5.7, 5.5, 5.9, 5.7, 5.5, 5.6, 5.6
   Diameter of Cans (in cm) from Assembly Line B:
   5.4, 5.7, 5.6, 5.5, 5.6, 5.7, 5.7, 5.8, 5.6, 5.5

Directions: Answer the questions that follow each set of data:

9. Weights of newborn babies (in pounds):

6.8	9.1	8.7	7.5	8.2
5.4	6.5	8.5	7.3	6.6
5.9	7.3	9.3	7.4	7.8

a. Which weight represents the 50th percentile?
b. What is the percentile of a weight of 8.2 pounds?

10. Batting averages for a sample of professional baseball players:

.275	.265	.333	.244	.279
.250	.288	.292	.370	.243
.236	.321	.305	.292	.250

a. Which batting average represents the 35th percentile?
b. What is the percentile of an average of .244?

11. Lengths of long distance phone calls (in minutes):

0.8	42.2	20.6	2.8
36.7	18.6	23.3	11.5
3.7	14.9	9.4	1.5
14.9	31.1	23.5	9.5

a. Which time represents the 82nd percentile?
b. What is the percentile of a time of 11.5 minutes?

12. High temperatures for cities in the Southeast (in degrees Fahrenheit):

85	82	93	88
92	79	84	90
77	83	91	89
90	85	87	91
89	92	95	88

a. Which temperature represents the 75th percentile?
b. What is the percentile of a temperature of 93 degrees?

13. Ages of 20 American entrepreneurs:

28	39	43	53
35	32	34	29
33	31	32	31
25	22	30	29
41	36	23	47

a. Which age represents the 25th percentile?
b. In what percentile is an age of 28 years?

Directions: For the values listed below, calculate the standard score. Round your answer to two decimal places.

14. $\mu = 25$, $\sigma = 3$, and $x = 27$

15. $\bar{x} = 37$, $s = 8$, and $x = 34$

16. $\mu = 0.32$, $\sigma = 0.01$, and $x = 0.29$

17. $\mu = 2$, $\sigma = 0.5$, and $x = 1.8$

18. $\bar{x} = 180$, $s = 10$, and $x = 210$

19. Carlita scored a 32 on her ACT and a 730 on her SAT examinations. Given that the average ACT score is 18 with a standard deviation of 6 and the SAT has an average score of 500 with a standard deviation of 150, which exam did Carlita perform better on?

20. A manufacturer makes aluminum cans and long neck bottles. The average diameter on an aluminum can is supposed to be 4.2 inches with an allowable deviation of 0.01 inches. The average diameter on a long neck bottle is supposed to be 3.8 inches with an allowable standard deviation of 0.02 inches. A factory worker randomly selects a can from the assembly line and it has a diameter of 4.3 inches. The worker then selects a bottle from the assembly line and it has a diameter of 3.75 inches. Which assembly line is closest to specifications?

21. Don played in a local golf tournament for charity and scored a round of 63 while the average round for the day was a 74 with a standard deviation of 3 strokes. Later that week, Don played in a pro-am tournament and scored a 65 while the average score for the day was a 79 with a standard deviation of 4 strokes. Which was Don's better round of golf in comparison to the competition? (Remember, in golf, lower scores are better!)

Directions: For each of the following graphs, where the mean score is marked, what is a likely z-score for the indicated value? (a) –1.3, (b) 0, or (c) 1.7.

22.

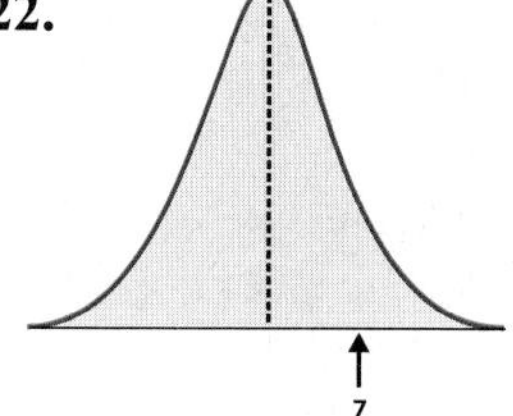

23.

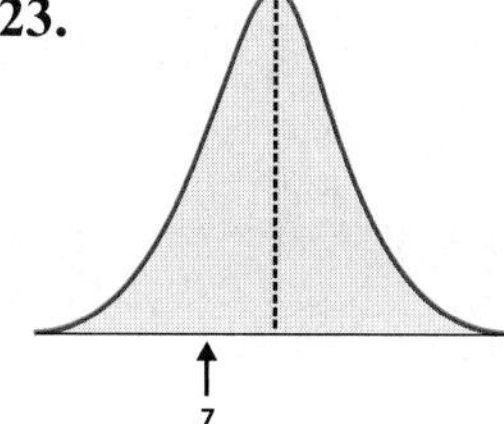

24.

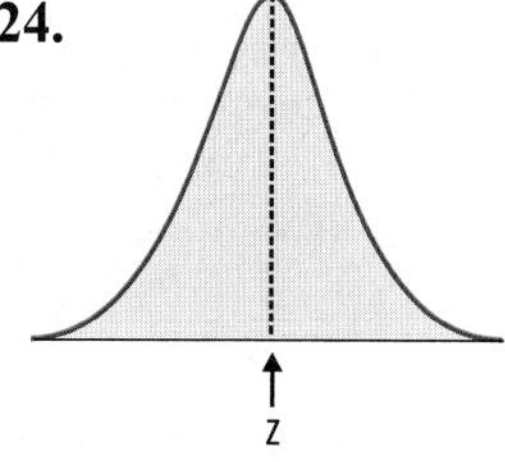

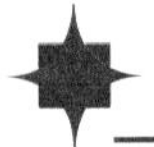

Chapter Review

Section 3.1 - Measures of Center	Pages
• The three measures of center are the mean, median, and mode.	p. 104 – 107
• **Properties of the Mean:** **1.** Most familiar and widely used **2.** Its value is affected by EVERY value in the data set **3.** Is not necessarily a value in the data set **4.** Appropriate choice for quantitative data with no outliers	p. 104 – 106
• **Weighted Mean:** $\bar{x} = \frac{\sum (x_i \cdot w_i)}{\sum w_i}$	p. 105 – 106
• **Properties of the Median:** **1.** Easy to compute by hand **2.** The middle number of the ordered data set **3.** Only determined by middle values of a data set, and not affected by extreme numbers **4.** Useful measure of center for skewed distributions **5.** Is not necessarily a value in the data set	p. 106 – 107
• **Properties of the Mode:** **1.** A data set does not have to have a mode. **2.** A data set can have more than one mode. **3.** If a mode exists for a data set, the mode is a value in the data set. **4.** Not affected by outliers in the data set **5.** Only measure of center appropriate for qualitative data	p. 107
• **Determining the Most Appropriate Measure of Center:** **1.** For qualitative data, the mode should be used. **2.** For quantitative data, the mean should be used, unless the data set contains outliers or is skewed. **3.** For quantitative data sets that contain outliers or are skewed, the median should be used.	p. 108 – 109

Section 3.2 - Measures of Dispersion	Pages
• The three measures of dispersion are the range, variance, and standard deviation.	p. 114 – 117
• **Properties of the Range:** **1.** Easiest measure of dispersion to compute **2.** Only affected by the largest and smallest value, so it can be misleading	p. 114

Section 3.2 - Measures of Dispersion (cont.)	Pages
• **Properties of Variance:** **1.** Easily computed using a calculator or computer **2.** Affected by every value in the data set **3.** Population variance and sample variance formulas yield different results. **4.** Difficult to interpret because of its unusual units **5.** Equal to the standard deviation squared	p. 115 – 116
• **Properties of Standard Deviation:** **1.** Easily computed using a calculator or computer **2.** Affected by every value in the data set **3.** Population standard deviation and sample standard deviation formulas yield different results. **4.** Interpreted as the average distance a data value is from the mean; thus it cannot take on negative values **5.** Equal to the square root of the variance	p. 116 – 117
• **Coefficient of Variation:** $\text{CV} = \frac{s}{\bar{x}} \cdot 100\%$	p. 117
• **Standard Deviation and Variance of Grouped Data:** $s = \sqrt{\frac{n\left[\sum(f \cdot x^2)\right] - \left[\sum(f \cdot x)\right]^2}{n(n-1)}}$	p. 118 – 121
• **Empirical Rule:** • Approximately 68% of the data lies within 1 standard deviation of the mean. • Approximately 95% of the data lies within 2 standard deviations of the mean. • Approximately 99.7% of the data lies within 3 standard deviations of the mean.	p. 121 – 122
• **Chebyshev's Theorem:** The proportion of data that lie within K standard deviations of the mean is at least $1 - \frac{1}{K^2}$, for $K > 1$. • $K = 2$: at least 75% of the data lies within two standard deviations of the mean. • $K = 3$: at least 88.9% of the data lies within three standard deviations of the mean.	p. 122 – 123

Section 3.3 - Measures of Relative Position	Pages
• **Quartiles:** Q_1 = First Quartile = the number greater than or equal to 25% of the data Q_2 = Second Quartile = the number greater than or equal to 50% of the data Q_3 = Third Quartile = the number greater than or equal to 75% of the data	p. 129 – 130

Section 3.3 - Measures of Relative Position (cont.)	Pages
• **Five-Number Summary:** the minimum value, Q_1, median (Q_2), Q_3, and the maximum value	p. 131
• **Box Plot:** a graphical depiction of a five-number summary	p. 131 – 133
• **Percentiles:** $l = n \cdot \frac{p}{100}$ and $p = \frac{x}{n} \cdot 100$	p. 133 – 135
• **Standard Scores:** $z = \frac{x - \mu}{\sigma}$ (population) or $z = \frac{x - \overline{x}}{s}$ (sample)	p. 135 – 136

Chapter Exercises

1. Four friends went out to eat one evening, and the mean price of their dinners was \$10.54. If the prices of the first three meals were \$9.62, \$11.59, and \$10.03, what was the cost of the fourth meal?

2. Suppose that a list of company CEOs is compiled, and 37% are over the age of 45. True or False: Q_3 must be greater than 45.

3. Suppose that 200 employees at a major theme park are surveyed, and 54% are under the age of 25.
   a. True or False: Of those surveyed, the mean age must be under 25.
   b. True or False: Of those surveyed, the median age must be under 25.

4. Give one example of a data set with a large variation and one example of a data set with a small variation.

5. The mean grade for Test 2 in Marquetta's biology class is 73, and the standard deviation is 8 points. If the standard score for Marquetta's test is 1.25, what is her grade?

6. Neikia wants to determine the average amount of money moviegoers typically spend on snacks at the theater. He surveys 150 people leaving the theater one evening and finds that 79 did not spend any money on refreshments. The maximum amount spent was \$22.50.
   a. What is the largest possible value for the mean?
   b. What is the median amount spent on refreshments?

7. Suppose that Julie's height has a standard score of 0.44. True or False: Julie is taller than average.

Chapter Project A: Olympic Gold

The following tables give the years of the Winter Olympics and the number of gold medals won in each year by the United States.

Table Project A - USA Gold Medal Count

Year	1928	1932	1936	1948	1952	1956	1960	1964	1968	1972
Number of Gold Medals	2	6	1	3	4	2	3	1	1	3
Number of Events	14	14	17	22	22	25	28	34	35	35

Year	1976	1980	1984	1988	1992	1994	1998	2002	2006
Number of Gold Medals	3	6	4	2	5	6	6	10	?
Number of Events	37	39	38	46	57	61	68	78	84

Note that the number of events is not the same for every year. When analyzing the data, it will become necessary to take into account the number of gold medals as a percentage of the number of events. Now, we are ready to estimate the expected number of gold medals for the U.S. in the 2006 winter games.

1. Begin by calculating the number of gold medals won each year as a percentage of the number of events. To do this, divide the number of gold medals by the number of events and then multiply by 100. Round your answers to the nearest whole percentage. Create a table similar to the one above with the percentage of gold medals won in each year.
2. Calculate the range of the number of gold medals and the range in the percentage of gold medals.
3. Calculate the median number of gold medals and the median percentage of gold medals.
4. Calculate the mode of the number of gold medals and the mode of the percentage of gold medals.
5. Calculate the mean number of gold medals won.
6. Calculate the mean percentage of gold medals won by adding the total number of gold medals won in all years and dividing by the total numbers of events in all years then multiplying by 100. Round your answer to the nearest whole percentage. (Remember, you cannot average percentages. This is why we have to go back to the number of gold medals and the number of events.)
7. Calculate the five-number summary for the number of gold medals won.
8. Draw a box plot for the number of gold medals won.
9. Using the fact that there were 84 events in the 2006 Olympics; estimate the expected number of gold medals by multiplying 84 by the median percentage of gold medals won by the USA.
10. The US actually won 9 gold medals in the 2006 Winter games in Torino, Italy. How does your calculation compare to the actual number?

Source: www.olympic.org

Chapter Project B: Where would you invest your money?

Let's look at several sets of stock prices. The following prices were obtained in the second half of 2001 and are listed in dollars per share.

Table Project B - Stock Prices

Netflix	Wal-Mart	Continental Airlines
18	46	28
14	53	32
13	54	35
14	55	28
10	51	25
5	56	22
3	54	15
9	55	10
11	54	10
13	50	5
11	51	4
12	50	6
16	53	8

1. Find the mean of each set of stocks.
2. Find the median of each set of stocks.
3. Calculate the variance and standard deviation for each set of stocks.
4. If you have $10,000 to invest, what stock would you buy under the following circumstances? Justify your reasoning.
   a. You are nearing retirement and need a stable investment for the future.
   b. You are a wealthy entrepreneur hoping to make a large profit in a short amount of time.

Chapter Technology – Calculating Descriptive Statistics

TI-84 PLUS

You can use your TI-84 Plus calculator to generate all of the numerical summaries at once using the STAT menu. Press the STAT button, then choose EDIT and enter your data in List 1. Then press STAT again and now choose CALC (calculate). Choose option 1 for 1-Var Stats and press ENTER. If your data is in List 1, press ENTER again since L1 is the default list. If you did not type your data in List 1, enter the List your data is in, such as L3 or L5. (These list names are in blue, above the numeric keys.)

The values shown on the screen will be, in order:

Table T.1 - Calculator Values

Value	Definition
$\bar{x}$	the sample mean
$\sum x$	the sum of the values
$\sum x^2$	the sum of the square of each value
Sx	the sample standard deviation
σx	the population standard deviation
n	the number of data values
$minX$	the minimum value
Q_1	the first hinge (often used for the first quartile)
Med	the median
Q_3	the third hinge (often used for the third quartile)
$maxX$	the maximum value

Example T.1

Use your TI-84 Plus calculator to calculate the list below of numerical summaries for the data values:

12, 14, 15, 19, 12, 10, 13, 19, 20, 12, 23.

Solution:

Begin by entering your data in the first list, L1. Press STAT, then choose CALC, and option 1: 1-Var-Stats. Press ENTER.

The beginning of the list is shown to the right.

Box Plots

The TI-84 Plus can also produce box plots. Begin by entering your data in the first list, L1. Next, go to the STAT PLOT screen and turn on Plot 1. Then choose the second of the two box plot options. This will produce a standard box plot as described earlier. Remember to press ZOOM and choose option 9: ZoomStat if your graph does not appear as you would like.

Example T.2

Create a box plot, using your TI-84 Plus calculator, from the following data values:

12, 14, 15, 19, 12, 10, 13, 19, 20, 12, 23.

Solution:

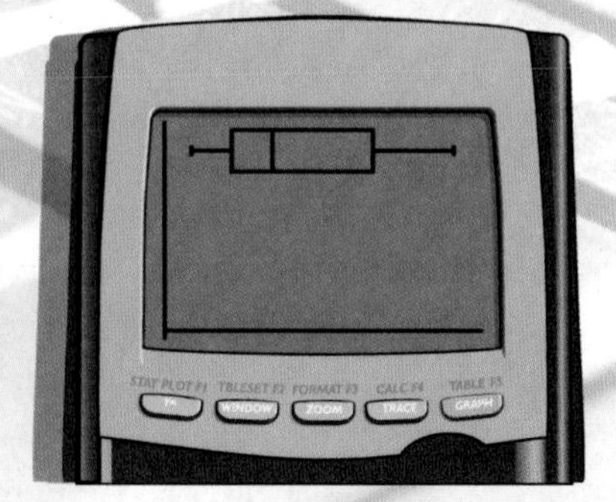

Enter the data in List 1. Go to the STAT PLOT screen by pressing 2ND Y=. Turn on Plot 1 and choose the second box plot option. Next enter L1 in the Xlist. Then press GRAPH. You should see the box plot shown to the right. If it does not appear that way at first, pressing ZOOM and choosing option 9: *ZoomStat* should correct the problem.

MICROSOFT EXCEL

Excel can calculate the values of the descriptive statistics individually or collectively. Individually, the commands are:

Table T.2 - Excel Commands

Calculation	Command
Mean	=AVERAGE(*cell range*)
Median	=MEDIAN(*cell range*)
Mode	=MODE(*cell range*)
Population Variance	=VARP(*cell range*)
Sample Variance	=VAR(*cell range*)
Population Standard Deviation	=STDEVP(*cell range*)
Sample Standard Deviation	=STDEV(*cell range*)
Minimum Value	=MIN(*cell range*)
First Quartile	=QUARTILE(*cell range*, 1)
Third Quartile	=QUARTILE(*cell range*, 3)
Maximum	=MAX(*cell range*)

You can also use the DATA ANALYSIS menu under TOOLS to create a list of numerical summaries. Go to TOOLS then DATA ANALYSIS then choose *Descriptive Statistics*. In the *Input Data* box, enter the cells where your data is located. Click on the box in front of *Summary Statistics*, then click OK. The output will produce many values we have not discussed as well as ones we have studied. The values you should recognize are the Mean, Median, Mode, Standard Deviation (for a sample), Sample Variance, Range, Minimum, Maximum, and Count, which is the number of data values.

Example T.3

Use the Data Analysis menu in Microsoft Excel to calculate the list of descriptive statistics for the data values:

12, 14, 15, 19, 12, 10, 13, 19, 20, 12, 23.

Solution:

Begin by typing in the data in Column A. Go to TOOLS, then DATA ANALYSIS, then choose *Descriptive Statistics*, and click OK. In the *Input Data* box, enter the cells where your data is located. Select 'New Worksheet Ply' and type in Descriptive Statistics. Click on the box in front of *Summary Statistics*, then click OK. The Descriptive Statistics menu should look similar to the following:

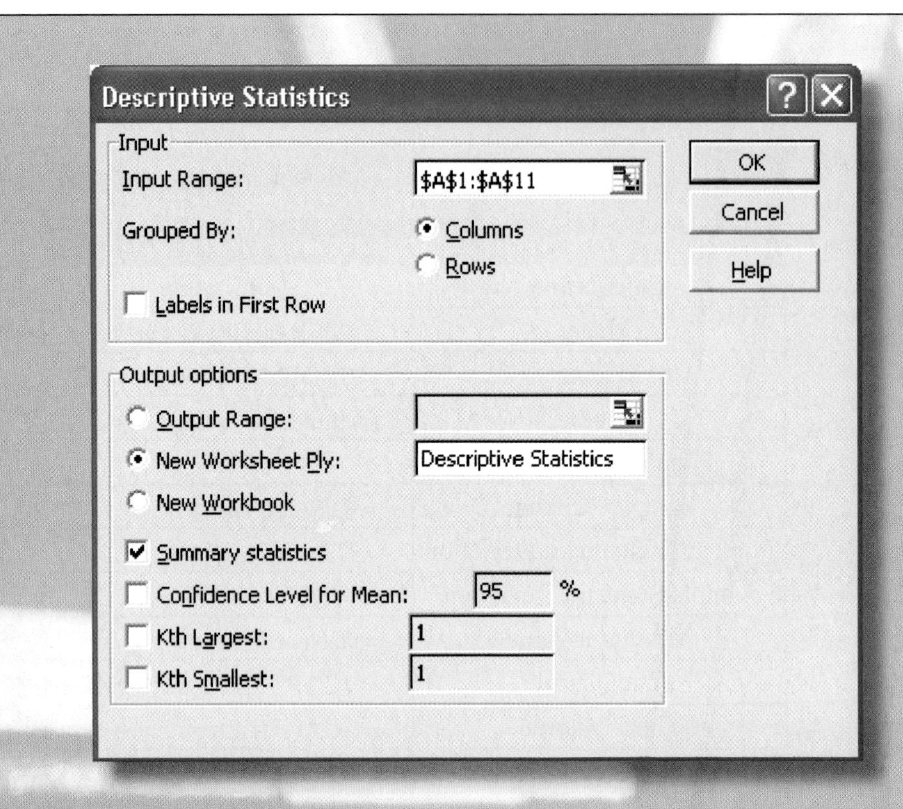

Clicking OK should produce the following list of descriptive statistics:

	A	B
1	*Column1*	
2		
3	Mean	15.36364
4	Standard Error	1.26687
5	Median	14
6	Mode	12
7	Standard Deviation	4.201731
8	Sample Variance	17.65455
9	Kurtosis	-0.95587
10	Skewness	0.588007
11	Range	13
12	Minimum	10
13	Maximum	23
14	Sum	169
15	Count	11

MINITAB

Sample Mean and Standard Deviation

Minitab can be used to find the mean, median, and standard deviation of a data set. It can also give the values for the five number summary.

Example T.4

Suppose six people participated in a 1000 meter run. Their times, measured in minutes, are given below.

4, 10, 9, 11, 9, 7

Use the Display Descriptive Statistics option in Minitab to calculate the list of descriptive statistics for the data values.

Solution:

Begin by typing in the data in column C1. Go to STAT then BASIC STATISTICS and then choose *Display Descriptive Statistics*. In the dialog box, input "C1" under **Variables**. Once this is done, select OK. Observe the output screen for the descriptive statistics.

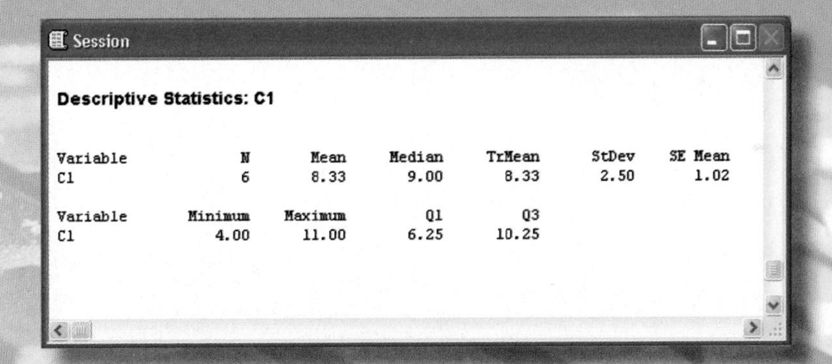
Session

Descriptive Statistics: C1

Variable	N	Mean	Median	TrMean	StDev	SE Mean
C1	6	8.33	9.00	8.33	2.50	1.02

Variable	Minimum	Maximum	Q1	Q3
C1	4.00	11.00	6.25	10.25

Probability, Randomness, & Uncertainty

Sections

4.1 Classical Probability

4.2 Probability Rules

4.3 Basic Counting Rules

4.4 Additional Counting Techniques

Objectives

The objectives of this chapter are as follows:

1. Identify the sample space of a probability event
2. Calculate basic probabilities
3. Determine if two events are mutually exclusive
4. Determine if two events are independent
5. Use the Addition Rule and Multiplication Rule to calculate probability
6. Calculate permutations and combinations
7. Use basic counting rules to calculate probability

Chapter 4

Introduction

Assessing how likely things are to happen is a useful tool in everyday life. In statistics, a measure of this likelihood is called probability. How would it affect your wager if you could *know* that you have only a 6% chance of winning in a game of cards? Or, what if instead you could know your chance of winning the hand is 58%? In both cases, in order to calculate your probabilities of winning, you have to know which cards are left in the deck. That is, you have to be able to count all of the possible outcomes that could be dealt.

Casinos know all too well what happens when players use statistical techniques to help them beat the odds. These players are called card-counters, and casinos do not like them. By keeping up with the cards that have already been dealt, card-counters are able to determine which cards are left in the deck, and therefore can predict when the high cards are more likely to turn up. Casinos make money because the odds of each game are stacked slightly in their favor, and the element of chance is essential.

By using probability to help them decide how to place their bets, card-counters are able to walk away with winnings that are unlikely to occur simply by chance. For example, in the early 1990's, a group of MIT students studied card counting techniques in order to win big in blackjack at the Las Vegas casinos. Win big they did, as their winnings peaked at $4 million before the casinos caught on. Unfortunately for the students, not only were they unable to hold onto their money, but they were also quickly blacklisted from every casino in the state!

Classical Probability

4.1

Be Careful!

The sample space is the *set* of outcomes for an experiment, not the *number* of outcomes.

In this chapter, our primary goal is to introduce the basic techniques and terminology that will allow us to calculate probabilities. First, a **probability experiment**, or *trial*, is any process in which the result is random in nature. Examples of probability experiments include flipping a coin, tossing a pair of dice, or drawing a raffle ticket. In each of these examples, there is more than one possible result and that result is determined at random. In a given probability experiment, each individual result that is possible is called an **outcome**. The set of all possible outcomes for a given probability experiment is called the **sample space**. An **event** is a subset of outcomes from the sample space.

For example, consider the experiment of throwing a die. There are six possible outcomes: namely the numbers 1 through 6. The sample space is the set of all outcomes, which in this case is simply $\{1, 2, 3, 4, 5, 6\}$. On the other hand, the event 'throwing an even number' is the subset of outcomes $\{2, 4, 6\}$.

Example 1

Consider an experiment in which a coin is tossed and then a 6-sided die is rolled.

a. List the sample space for the experiment.
b. List the outcomes in the event 'tossing a tail then rolling an odd number'.

Solution:

a. Each outcome consists of a coin toss and a die roll. For example, heads and a 3 would be denoted as H3. Thus

$$\text{Sample space} = \begin{Bmatrix} \text{H1} & \text{T1} \\ \text{H2} & \text{T2} \\ \text{H3} & \text{T3} \\ \text{H4} & \text{T4} \\ \text{H5} & \text{T5} \\ \text{H6} & \text{T6} \end{Bmatrix}.$$

b. Choosing the members of the sample space which fit the event 'tossing a tail then rolling an odd number' gives

$$\{\text{T1}, \text{T3}, \text{T5}\}.$$

Sometimes it is easy to list the outcomes for an experiment, as in the previous example. However, at times, the sample space is large, and special techniques are needed to ensure that no outcomes are omitted. An experiment may have a particular pattern that enables you to more easily organize the outcomes. For example, a pattern was used to create the sample space in Example 1. The first column shows each possibility beginning with heads, and the second column shows the possibilities beginning with tails. By using this pattern, we can easily see that every possible outcome is accounted for.

Example 2

Consider the experiment in which a red die and a blue die are rolled together. Use a pattern to help list the sample space.

Solution:

We begin by listing all outcomes in which a 1 is rolled on the red die and the blue die is allowed to vary from 1 to 6. There are 6 of these outcomes: 11, 12, 13, 14, 15, and 16. Note, 12 does not denote the number "twelve" but the outcome in which a one is rolled on the red die and a two is rolled on the blue die. Next, we list all outcomes in which a 2 is rolled on the red die. Again, there are 6 outcomes of this form. The pattern continues until all 36 outcomes are listed.

cont'd on the next page

$$\text{Sample space} = \begin{Bmatrix} 11 & 12 & 13 & 14 & 15 & 16 \\ 21 & 22 & 23 & 24 & 25 & 26 \\ 31 & 32 & 33 & 34 & 35 & 36 \\ 41 & 42 & 43 & 44 & 45 & 46 \\ 51 & 52 & 53 & 54 & 55 & 56 \\ 61 & 62 & 63 & 64 & 65 & 66 \end{Bmatrix}$$

Note that rolling a 1 on the red die and a 2 on the blue die, denoted 12, is a different outcome than rolling a 2 on the red die and a 1 on the blue die, here denoted 21.

When an experiment consists of several stages, a **tree diagram** allows the outcomes to be organized in a systematic manner. The tree begins with the possible outcomes for the first stage and then branches for each additional possibility. Each of the elements of the last row in the tree diagram represents a unique outcome in the sample space. To identify the specific outcomes, list the sequence of events that describe a path from the top of the tree to the bottom of the tree. Notice that the number of possibilities in the bottom row of the tree is equal to the *number* of outcomes in the sample space. The tree diagram method is demonstrated in the following example.

Example 3

Consider a family with 3 children. Use a tree diagram to find the sample space for the gender of each child in regard to birth order.

Solution:

The tree begins with the 2 possibilities for the first child—girl or boy. It then branches for each of the other 2 births in the family as shown below.

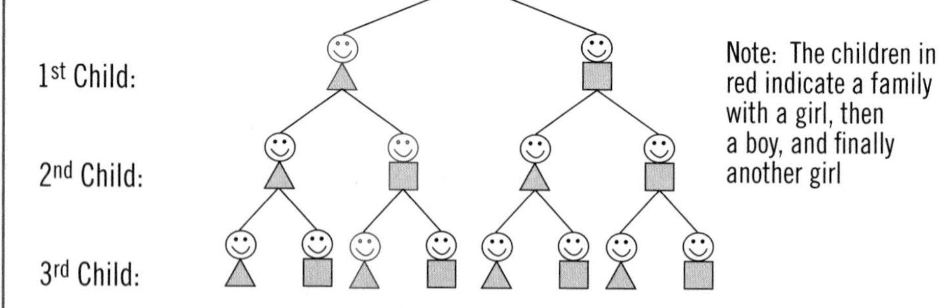

Figure 1 – Tree Diagram of a Three Child Family

Using the tree diagram as a guide, the sample space can be written as follows:

$$\text{Sample space} = \begin{Bmatrix} \text{GBB} & \text{GBG} & \text{GGB} & \text{GGG} \\ \text{BBB} & \text{BBG} & \text{BGB} & \text{BGG} \end{Bmatrix}$$

We have established how to identify all of the possible outcomes and can now look at three methods for calculating the probability of each outcome. The least precise type of probability is called subjective probability. **Subjective probability** is simply an educated guess regarding the chance that an event will occur. For example, the local meteorologist gives a subjective probability when he predicts the chance of rain. Clearly, subjective probability is only as good as the expertise of the person giving that probability. If the weatherman says there is a 90% chance of rain, you will probably take your umbrella with you to work. However, if your roommate looks out the window and guesses a 90% chance of rain, you won't be so quick to believe him. For this reason, you should always be cautious with subjective probabilities.

Rounding Rule

When calculating probability, either give the exact fraction or decimal rounded to three decimal places. If the probability is extremely small, it is permissible to round the decimal to the first nonzero digit.

Empirical probability, on the other hand, is found by experiment. Specifically, the empirical probability of an event is calculated by dividing the number of times an event occurs by the total number of trials performed.

Empirical Probability

Empirical Probability:
If E is an event, then $P(E)$, read "the probability that E occurs", is as follows:

$$P(E) = \frac{\text{frequency of event } E}{\text{total number of times the experiment is performed}} = \frac{f}{n}.$$

For example, suppose a fisherman wants to know the probability that a fish he catches in his favorite fishing hole is a catfish. For his experiment, he records the number and types of fish he catches one week. Suppose that during this week he catches 92 fish, 43 of which are catfish. He can then calculate the empirical probability of catching a catfish as follows:

$$P(\text{catfish}) = \frac{\text{number of catfish caught}}{\text{total number of fish caught}} = \frac{43}{92} \approx 0.467$$

The fisherman has obtained a reasonable estimate based on one week's worth of fishing. How do you think his estimate of the probability of catching a catfish would change if instead he recorded his catches for a month? Two months? A year? The **Law of Large Numbers** says that the greater the number of trials, the closer the empirical probability will be to the true probability.

What, then, is the true probability? **Classical probability**, also called *theoretical probability*, is the most precise type of probability. This is because classical probability is calculated by taking *all* possible outcomes for an experiment into account. In the previous example about the pond, the only way that the classical probability of catching a catfish could be calculated is if the exact number of catfish and total fish living in the pond were both known.

Let's give a proper definition for classical probability. Classical probability states that if all outcomes are equally likely, the probability of an event is equal to the number of outcomes included in the event (E) divided by the total number of outcomes in the sample space (S). Classical probability is written mathematically in the following box.

Classical Probability

Classical Probability:
If all outcomes are equally likely to occur, then $P(E)$, read "the probability that E occurs", is

$$P(E)=\frac{n(E)}{n(S)},$$

where $n(E)$ = the number of outcomes in the event, and
$n(S)$ = the number of outcomes in the sample space.

Note that probability may be written as either a fraction, decimal, or percentage. For instance, we may say the probability of the fisherman catching a catfish is $\frac{43}{92}$, 0.467, or 46.7 %. Let's look at a few examples of how to distinguish between the three different types of probability.

Example 4

Determine whether each of the following probabilities is subjective, empirical, or classical.

a. The probability of selecting the queen of spades out of a standard deck of cards.

b. An economist predicts a 20% chance that technology stocks will decrease in value over the next year.

c. A police officer wishes to know the probability that a driver chosen at random will be driving under the influence of alcohol on a Friday night. At a roadblock, he records the number of drivers and the number of drivers driving with more than the legal blood alcohol limit. He determines that the probability is 3%.

Solution:

a. All 52 outcomes are known and are equally likely to occur, so this is an example of classical probability.

b. The economist is giving an educated guess; therefore, this is an example of subjective probability.

c. The police officer's probability is based on the evidence gathered from the roadblock. Because all drivers were not included, the probability is empirical.

Let's consider the experiment of tossing a coin. We know the probability of getting a head is $\frac{1}{2}$. Let's verify that number with the classical probability formula. There are two outcomes in the sample space, heads and tails (and both outcomes are assumed to be equally likely), thus $n(S) = 2$. There is one way to get heads, so $n(E) = 1$.

So we have,

$$P(\text{heads}) = \frac{n(E)}{n(S)} = \frac{1}{2} = 0.5.$$

Consider the probability of rolling a 6-sided die and obtaining an even number. There are 3 outcomes in the event: rolling a 2, 4, or 6, so $n(E) = 3$. There are 6 outcomes in the sample space (and all outcomes are equally likely); thus $n(S) = 6$. So we calculate the probability to be

$$P(\text{even}) = \frac{n(E)}{n(S)} = \frac{3}{6} = \frac{1}{2} = 0.5.$$

It's important to note that this form of classical probability applies only when the outcomes are equally likely. There is an obvious fallacy in the statement "There is a 50% chance I will be hit by lightning tomorrow." Even though there are only two outcomes in the sample space {hit by lightning, not hit by lightning}, the probability of being struck by lightning is **not** $\frac{1}{2} = 50\%$ because the outcomes are not equally likely. (Thank goodness!)

Example 5

Consider drawing a card from a standard 52-card deck.

a. What is the probability of drawing a heart?

b. What is the probability of drawing a face card (king, queen, or jack)?

Solution:

a. There are 13 hearts in the deck, so the probability of choosing a heart is

$$P(\text{heart}) = \frac{n(E)}{n(S)} = \frac{13}{52} = \frac{1}{4} = 0.25.$$

b. There are 3 face cards for each of the 4 suits, so there are 12 face cards in the deck. The probability of choosing a face card is then

$$P(\text{face card}) = \frac{n(E)}{n(S)} = \frac{12}{52} = \frac{3}{13} \approx 0.231.$$

Example 6

Consider a family with 3 children. What is the probability that all 3 children are boys?

Solution:

The numerator of our probability fraction will be easy, because we know that there is only one way that the family can have all boys. To determine how many outcomes are in the sample space, we can use either a tree diagram or a pattern. From Example 3, we know there are 8 possible outcomes. The probability of having 3 boys is

$$P(\text{all boys}) = \frac{n(E)}{n(S)} = \frac{1}{8} = 0.125.$$

Example 7

Consider a family with 6 boys. What is the probability that the 7th child will also be a boy?

Solution:

Many people would say that this family is due to have a girl, but let's consider the mathematics. Suppose we took the time to write down every possible combination for birth orders of 7 children. Because the family already has 6 boys, we would have to mark out all combinations that contain a girl in the first 6 children. The only two outcomes left in the sample space would be BBBBBBG and BBBBBBB. The probability of baby number 7 being a boy is then $P(\text{all boys}) = \frac{n(E)}{n(S)} = \frac{1}{2} = 0.5$, just as with any other pregnancy!

Example 8

What is the probability of flipping 4 coins and obtaining exactly one head?

Solution:

First, let's determine the total number of outcomes in the sample space. Using a pattern, we see there are 16 possible outcomes. (Verify this yourself.) Next, we must determine the number of ways that exactly one head can be thrown. Let's list all the possibilities:

$$\begin{Bmatrix} \text{HTTT} & \text{TTHT} \\ \text{THTT} & \text{TTTH} \end{Bmatrix}$$

There are 4 outcomes in which exactly one head appears, so the probability for this event is $P(1 \text{ head}) = \frac{n(E)}{n(S)} = \frac{4}{16} = \frac{1}{4} = 0.25.$

Exercises

4.1

Sample Spaces
Directions: Find the sample space for the given experiments.

1. Two coins are tossed.

2. A child's board game contains a spinner with 3 colors: orange, yellow, and green. Give the sample space for two consecutive spins.

3. A family has four children. Give the sample space in regard to the sex of the children.

4. A bag contains four marbles: one each of red, blue, green, and violet. Two marbles are drawn from the bag. Assume that the first marble is not put back in the bag before drawing the second marble.

5. When buying a new car, you've narrowed your choices to three colors: red, black, or silver. You also need to decide whether to have a sunroof or not, and whether you want leather or cloth interior.

6. When ordering a pizza with a coupon, you can have a choice of crusts: thin, hand-tossed, or stuffed. You can also choose one topping from the following: pepperoni, ham, sausage, onion, bell pepper, or olives.

7. When choosing your seat at the opera, you can choose from 3 levels and then whether you want an aisle seat or not.

8. When building your new house, you have a choice of flooring for the kitchen: tile, concrete or wood. You must also choose the counter tops from the following: granite, concrete, wood, or stainless steel.

Directions: Determine whether each of the probabilities is subjective, empirical, or classical.

9. Jeff wants to know whether a certain coin is fair or not. He flips the coin 100 times and obtains tails 61 times. He calculates that the probability of obtaining a tail with his coin is 61%.

10. Caroline estimates that there is only a 10% chance that they will have a quiz in biology.

11. Each of Mr. Dorrough's 18 students has dropped their name in the hat for a prize drawing. Stephanie calculates that she has a 1/18 chance of winning.

12. On a game show, the contestant must choose one of three doors, behind one of which is a new car. He has a one-in-three chance of winning the car.

13. A computer manufacturer brags that there is less than a 5% chance that their computers will need to be serviced within the first year.

Directions: Calculate each empirical probability.

14. A very large bag contains more coins than you are willing to count. Instead, you draw some coins out at random, replacing each coin before the next draw. You record the picks in the following table:

Table for Exercise 14

Quarters	Nickels	Dimes	Pennies
23	17	29	38

a. What is the probability that on your next draw you will obtain a nickel?
b. What is the probability that on your next draw you will obtain a penny?
c. What is the probability that on your next draw you will obtain either a quarter or a dime?

15. A telemarketer's computer selects phone numbers at random. The telemarketer has recorded the number of respondents in each age bracket for one evening in the following table:

Table for Exercise 15

18–25	26–35	36–45	Over 45
29	40	55	51

a. What is the probability that the next respondent will be over 45?
b. What is the probability that the next respondent will be between 26 and 35?
c. What is the probability that the next respondent will be at least 36?

Directions: Use classical probability techniques to determine the following probabilities. Assume individual outcomes are equally likely.

16. Martha has a box full of 17 different CDs: 5 rock, 3 blues, 6 pop, and 3 R & B. If she randomly pulls out a CD, what is the probability that it is a blues CD?

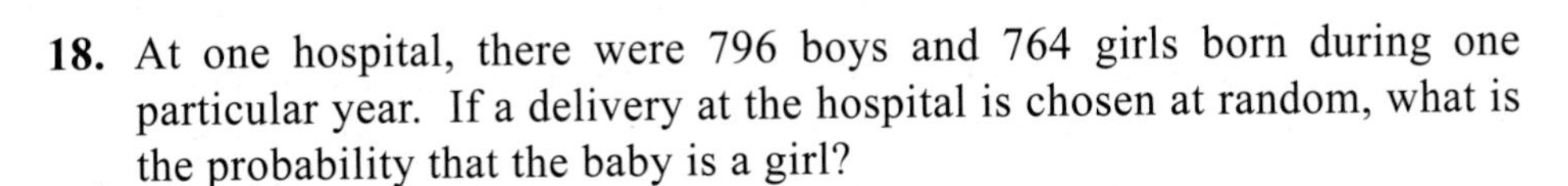

17. Chloë puts a coin into a gumball machine that contains 12 blue, 15 pink, 9 orange, 16 yellow, and 14 white gumballs. What is the probability that Chloë gets a yellow gumball?

18. At one hospital, there were 796 boys and 764 girls born during one particular year. If a delivery at the hospital is chosen at random, what is the probability that the baby is a girl?

19. What is the probability that a person selected at random will have a March birthday? Assume that every day of the year contains an equal number of birthdays, and they were not born in a leap year.

20. If Mark grabs a utensil out of a drawer without looking, what is the probability that he grabs a fork if there are 7 knives, 9 spoons, and 6 forks in the drawer?

21. If Timmy is fishing in his newly stocked pond and knows that there are 200 bream and 150 bass in it, what is the probability that the first fish he catches will be a bass?

22. John is at a cookout and wants to get a drink from the cooler. If there are 12 colas, 10 bottles of water, and 5 root beers in the chest, what is the probability that he randomly grabs a root beer?

23. A college algebra class has 14 freshmen, 21 sophomores, 9 juniors and a senior enrolled. What is the probability that the professor randomly selects the senior to answer a question?

24. Mary Ann is sewing and needs the spool of white thread. Her basket of sewing supplies is sitting next to her, and it contains 26 different colors of thread, including the white spool she needs. If she grabs one spool without looking, what is the probability that she has chosen the white spool of thread?

25. For a school fundraiser, 1000 raffle tickets are sold for $5 each. Each ticket is assigned a 3-digit number using the digits 0–9. What is the probability that the winning ticket will be one with three repeating digits?

Probability Rules

4.2

In order to calculate a probability, there are a number of properties and rules we need to know. First, we need to know the basic properties of probability and the concept of the complement. Next, we must be able to recognize mutually exclusive events and independent events. Finally, we will look at the addition rule for probability, the multiplication rule for probability, and conditional probability.

We begin by discussing some basic facts about probability itself. First, probability is always a number between 0 and 1. That is, $0 \leq P(E) \leq 1$. If the probability of an event is 0, then the event will not occur. If the probability of an event is 1, then the event is absolutely certain to occur. The closer a probability is to 0, the less likely the event is to happen, and the closer a probability is to 1, the more likely it is to happen.

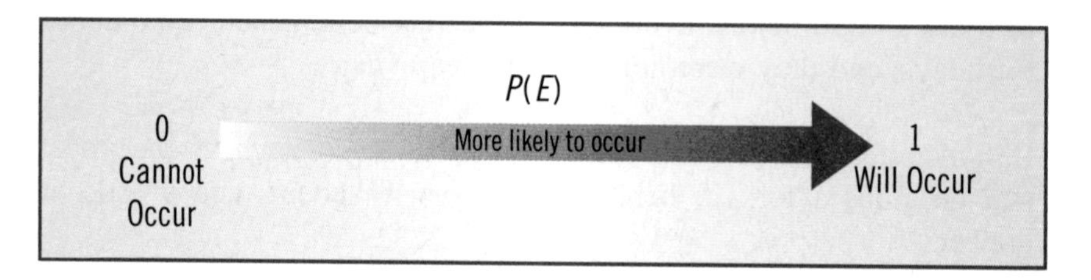

Figure 2 – Properties of Probability

Also, for a given experiment, one of the outcomes in the sample space has to occur. If the sample space is the list of all possible outcomes, then when the experiment is performed, one of them must occur. Mathematically, we write this fact as $P(S) = 1$. Along those same lines, it is impossible for none of the outcomes in the sample space to occur. We can write this second fact as $P(\varnothing) = 0$, where Ø is the "empty set", which is the set of no outcomes.

Facts about Probability:

1. $0 \leq P(E) \leq 1$

2. $P(S) = 1$

3. $P(\varnothing) = 0$

✦ The Complement

Consider the sample space for a probability experiment, and one event, E, in that sample space. The **complement** for E, denoted E^c, consists of all outcomes in the sample space that are not in E. For example, let's pretend that the experiment is rolling a 6-sided die. If E is the event of rolling an even number $\{2, 4, 6\}$, then its complement is rolling an odd number $\{1, 3, 5\}$. When you put an event and its complement together, you will always have the entire sample space. An illustration of the die example is given in Figure 3.

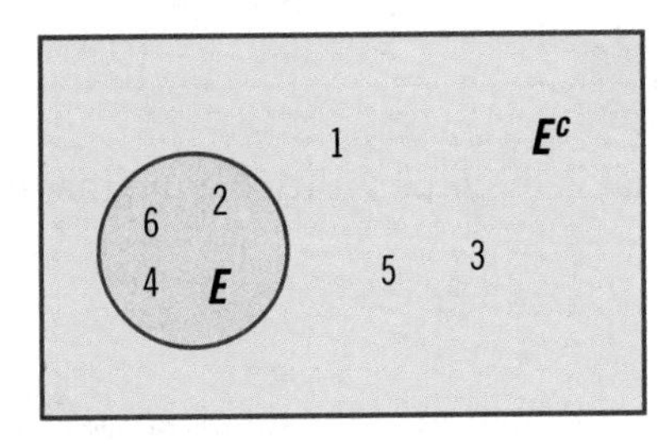

Figure 3 – Venn Diagram Illustrating the Complement of a Set

Example 9

Describe the complement for each of the following events.

a. Choose a red card out of a standard deck of cards.
b. Out of 31 students in your statistics class, 15 are out sick with the flu.
c. In your area, 91% of phone customers use PhoneSouth.

Solution:

a. A standard deck of cards contains 26 red and 26 black cards. Since all 26 red cards are contained in the event, the complement contains all 26 black cards.
b. Since 15 students are out with the flu, the complement contains the other 16 students that are not sick.
c. Since the event contains all PhoneSouth customers, everyone who is not a PhoneSouth customer is in the complement. Thus the complement contains the other 9% of customers in your area.

Because combining the two subsets E and E^c gives us the entire sample space, it is true that the sum of their probabilities is always one. Written mathematically, this says

$$P(E) + P(E^c) = 1.$$

This fact is convenient, because for some problems the probability of E^c is easier to calculate than the probability of E. For these problems, you can calculate $P(E^c)$ and subtract that value from 1 to obtain $P(E)$, written mathematically as

$$P(E) = 1 - P(E^c).$$

Probability Rule for the Complement

For an event E and its complement E^c:

$$P(E) = 1 - P(E^c)$$

Example 10

Roll a pair of dice. What is the probability that neither die is a 3?

Solution:

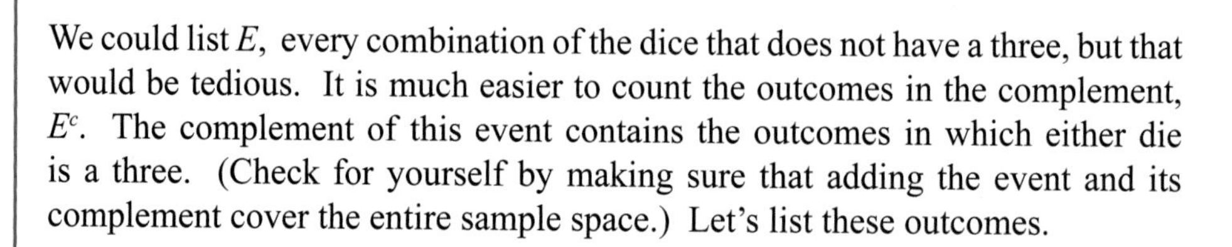

We could list E, every combination of the dice that does not have a three, but that would be tedious. It is much easier to count the outcomes in the complement, E^c. The complement of this event contains the outcomes in which either die is a three. (Check for yourself by making sure that adding the event and its complement cover the entire sample space.) Let's list these outcomes.

$$\begin{array}{ccc} \{3, 1\} & & \{1, 3\} \\ \{3, 2\} & & \{2, 3\} \\ & \{3, 3\} & \\ \{3, 4\} & & \{4, 3\} \\ \{3, 5\} & & \{5, 3\} \\ \{3, 6\} & & \{6, 3\} \end{array}$$

There are 11 outcomes where at least one of the dice is a 3. Thus $P(E^c) = \frac{11}{36}$. Subtracting this probability from 1 gives us $P(E) = 1 - \frac{11}{36} = \frac{25}{36} \approx 0.694$.

Therefore, the probability that neither die is a 3 is approximately 0.694.

✦ Addition Rule for Probability

So far in this chapter we have discussed probabilities that involve a single event, such as drawing a heart from a deck of cards or rolling seven with a pair of dice, but now we want to discuss how to find probabilities that involve combinations of events. The first type of combination we will consider is the probability that one event **or** another event occurs. For example, what is the probability of drawing a heart **or** a queen from a deck of cards? There are 52 cards in a deck and 13 of these cards are hearts, so the probability of drawing a heart is $\frac{13}{52}$. Similarly, there are 4 queens in a deck, so the probability of drawing a queen is $\frac{4}{52}$. But what is the probability of drawing one or the other? Wouldn't it be convenient if we could add the previous two probabilities together? Let's look at a picture illustrating this situation to help us see if adding these probabilities will give us the correct solution.

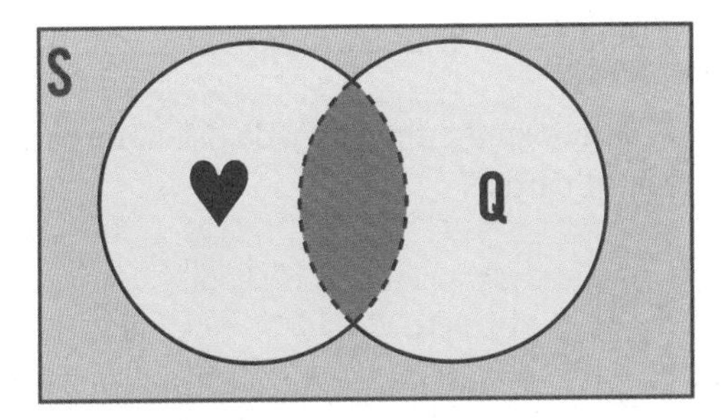

Figure 4 – Venn Diagram Illustrating Probability of Choosing Either a Heart or a Queen

The rectangle represents the entire sample space, all 52 cards in the deck. The circle on the left represents choosing one of the 13 hearts, and the circle on the right represents choosing one of the four queens. The shaded area in the middle represents any cards that are both hearts and queens, namely the queen of hearts. Figure 4 shows a problem with our proposed solution—the overlapping region has been counted twice! The queen of hearts was counted as a heart and again as a queen. Because its probability was counted both for the hearts and the queens, we must subtract it in order to obtain the correct probability. Therefore, the probability of choosing a heart or a queen is

$$\begin{aligned} P(\text{heart or queen}) &= P(\text{heart}) + P(\text{queen}) - P(\text{heart and queen}) \\ &= \frac{13}{52} + \frac{4}{52} - \frac{1}{52} \\ &= \frac{16}{52} \\ &= \frac{4}{13} \\ &\approx 0.308. \end{aligned}$$

We can generalize this concept to any probability of one event or another occurring as shown in the formula, often called the **Addition Rule for Probability**, in the box below.

Addition Rule for Probability
For two events, E and F,

$$P(E \textbf{ or } F) = P(E) + P(F) - P(E \text{ and } F).$$

Example 11

Find the probability of choosing either a spade or a face card (king, queen, jack) out of a standard deck of cards.

Solution:

The key word here is "or". Using the Addition Rule for Probability, we have:

$$\begin{aligned}P(\text{spade or face card}) &= P(\text{spade}) + P(\text{face card}) - P(\text{spade and face card})\\ &= \frac{13}{52} + \frac{12}{52} - \frac{3}{52}\\ &= \frac{22}{52}\\ &= \frac{11}{26}\\ &\approx 0.423\end{aligned}$$

Example 12

Roll a pair of dice. What is the probability of rolling an even total or a total greater than 9?

Solution:

The key word is again "or". Using the Addition Rule for Probability, we have

$$P(\text{even or} > 9) = P(\text{even}) + P(> 9) - P(\text{even and} > 9).$$

We know from previous examples that there are 36 outcomes in the sample space of rolling a pair of dice. The only tricky part is then to determine the number of outcomes that give an even total, and the number of outcomes that give a total greater than nine. Let's list these outcomes.

Even totals:
2 – (1, 1)
4 – (1, 3); (3, 1); (2, 2)
6 – (1, 5); (5, 1); (2, 4); (4, 2); (3, 3)
8 – (2, 6); (6, 2); (3, 5); (5, 3); (4, 4)
10 – (4, 6); (6, 4); (5, 5)
12 – (6, 6)

Totals greater than 9:
10 – (4, 6); (6, 4); (5, 5)
11 – (5, 6); (6, 5)
12 – (6, 6)

Thus there are 18 even totals and 6 totals greater than nine. We can also look through the list and see that 4 of the outcomes are even and greater than nine. Using these values, we can now fill in the probabilities in the formula.

$$\begin{aligned} P(\text{even or} > 9) &= P(\text{even}) + P(>9) - P(\text{even and} > 9) \\ &= \frac{18}{36} + \frac{6}{36} - \frac{4}{36} \\ &= \frac{20}{36} \\ &= \frac{5}{9} \\ &\approx 0.556 \end{aligned}$$

Some events do not have any outcomes in common, which makes computing their combined probabilities even easier. Events that share no outcomes are called **mutually exclusive** events. For example, consider the events of choosing a seven or a face card out of a standard deck of cards. Is it possible to choose a card that is both a face card and a seven? No. These two events have no outcomes in common, so they are mutually exclusive. Figure 5 illustrates this relationship.

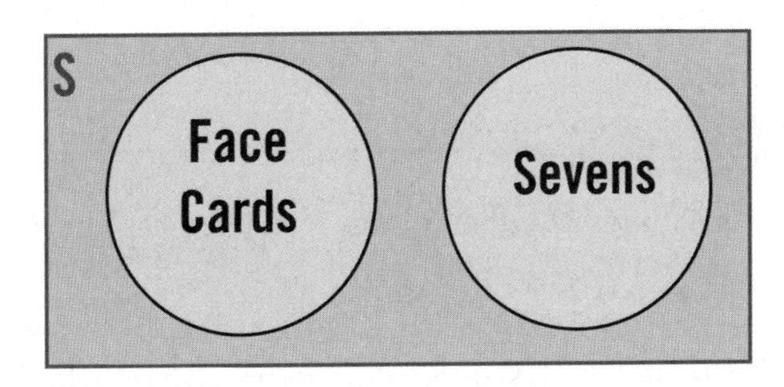

Figure 5 – Venn Diagram Illustrating Mutually Exclusive Events

Notice that since these two events have no outcomes in common, the probability of both events occurring at the same time is zero. In fact, this will be true for any two events that are mutually exclusive. We can therefore shorten the formula for the Addition Rule of Probability for mutually exclusive events to the following:

Addition Rule for Mutually Exclusive Events:

If two events, E and F, are mutually exclusive, then

$$P(E \textbf{ or } F) = P(E) + P(F).$$

Example 13

What is the probability of drawing a face card or a seven from a standard deck of cards?

Solution:

These two events are mutually exclusive, so we can use the shorter version of the Addition Rule. There are 12 face cards in a deck and 4 sevens in a deck, so the probabilities are as follows:

$$\begin{aligned} P(\text{face card or } 7) &= P(\text{face card}) + P(7) \\ &= \frac{12}{52} + \frac{4}{52} \\ &= \frac{16}{52} \\ &= \frac{4}{13} \\ &\approx 0.308 \end{aligned}$$

Example 14

What is the probability that a family with 3 children has either all girls or all boys?

Solution:

These two events are mutually exclusive because a family with all girls cannot also have all boys. Thus we can use the Addition Rule for mutually exclusive events. There are 8 outcomes in the sample space for a family with three children with respect to birth order. Of these eight outcomes, there is one that is all girls and one that is all boys. Using this information, we obtain the following solution:

$$\begin{aligned} P(GGG \text{ or } BBB) &= \frac{1}{8} + \frac{1}{8} \\ &= \frac{2}{8} \\ &= \frac{1}{4} \\ &= 0.25 \end{aligned}$$

✦ Multiplication Rule for Probability

Now let's consider the probability of two events both happening, instead of the probability of one or the other happening. For example, what is the probability of choosing a queen and then a king from a deck of cards? The key word in these problems is the word "and."

Before we can calculate these types of probabilities, we need to introduce a few new terms. The phrase **with repetition** means that outcomes may be repeated. For example, if you are choosing numbers *with repetition* for a bank PIN then you are allowed to repeat numbers, such as in the PIN 1231. Similarly, the phrase **with replacement** refers to placing objects back into consideration, such as choosing a card from a deck and then returning it to the deck and shuffling the deck for the next choice. On the other hand, the phrase **without repetition** means that outcomes may not be repeated. For example, if you are choosing a bank PIN *without repetition* then you are not allowed to repeat numbers, such as in the PIN 1234. In the same manner, the phrase **without replacement** means your first choice is not put back in for consideration, such as drawing a card and then drawing a second card from those left over. These four phrases affect the number of possible outcomes in the sample space.

Lastly, two events are **independent** if one event happening does not influence the probability of the other event happening. For example, if after drawing a card you replace the card drawn—and shuffle the deck—then the probability of the next card drawn is not affected by what was picked first. Therefore, these two selections of a card are independent events.

When two events are independent, the probability that both events occur can easily be calculated by multiplying their respective probabilities together. Written mathematically, the formula referred to as the **Multiplication Rule for Probability** is shown in the box below.

Multiplication Rule for Probability:

For two independent events, E and F,

$$P(E\ \textbf{and}\ F) = P(E)P(F).$$

Example 15

Choose two cards from a standard deck, with replacement. What is the probability of choosing a king and then a queen?

Solution:

Because the cards are replaced after each draw, the two events are independent. Using the Multiplication Rule for Probability, we have

$$\begin{aligned} P(\text{king and queen}) &= P(\text{king})P(\text{queen}) \\ &= \left(\frac{4}{52}\right)\left(\frac{4}{52}\right) \\ &\approx 0.006. \end{aligned}$$

When two events are not independent, it means that the outcome of one influences the probability of the other. For example, consider drawing two cards from a standard deck *without replacement*. If a king is drawn first, what is the probability that you will draw a king again on your second pick? This is called **conditional probability**, and is denoted as $P(2^{\text{nd}}\text{ king}|1^{\text{st}}\text{ king})$. This notation is read "the probability of drawing a king second, given that a king was drawn first." To find this probability, let's think about how many kings are available in the deck for the second draw. One king has been taken out, so there are three kings left. How many cards do we have left in the deck from which to choose? One card was removed, so there are now just 51 cards in the deck. Thus

$$P(2^{\text{nd}}\text{ king}|1^{\text{st}}\text{ king}) = \frac{3}{51} \approx 0.059.$$

> **Memory Booster!**
>
> **Mutually Exclusive:** consider outcomes from a SINGLE trial
>
> VS.
>
> **Independent:** consider outcomes from CONSECUTIVE trials

Example 16

What is the probability of choosing a red card from the deck, given that the first card was a diamond? Assume the cards are chosen without replacement.

Solution:

First, we must determine how many red cards are left in the deck after the first pick. Because the first card was a diamond (which is a red card), there are only 25 red cards left in the deck instead of 26. There are also only 51 cards total left in the deck. Thus the conditional probability is

$$P(\text{red}|\text{diamond}) = \frac{25}{51} \approx 0.490.$$

Conditional probability is necessary in order to find the probability of two events happening consecutively that are not independent. In this case, we say that the two events are **dependent**. For example, suppose we wanted to calculate the probability of choosing two face cards in a row, without replacement. For these types of problems, we multiply the probability of one event by the conditional probability of the other event as shown in the formula below.

Multiplication Rule for Dependent Events:
For two dependent events, E and F,

$$P(E \text{ and } F) = P(E)P(F|E).$$

Example 17

What is the probability of choosing two face cards in a row? Assume that the cards are chosen without replacement.

Solution:

We are dealing with dependent events, so we must use the Multiplication Rule for Dependent Events. When the first card is picked, all 12 face cards are available out of 52 cards. When the second card is drawn, there are only 11 face cards left out of 51 cards left in the deck. Thus we have

$$\begin{aligned} P(\text{face card and face card}) &= P(1^{\text{st}} \text{ face card})P(2^{\text{nd}} \text{ face card}|1^{\text{st}} \text{ face card}) \\ &= \left(\frac{12}{52}\right)\left(\frac{11}{51}\right) \\ &= \frac{132}{2652} \\ &\approx 0.050. \end{aligned}$$

Exercises 4.2

Complement
Directions: Describe the complement for each of the following events.

1. Out of 253 apple trees in the orchard, just 42 are ready for harvesting.
2. In the graduating class at one large high school, 39% of the graduates plan to attend college out-of-state.
3. Out of 21 players on the baseball team, 4 are left-handed.
4. The adult literacy rate for the United States is 97%.
5. A local news station claims that 70% of their viewers are over the age of 30.

Complement Rule

Directions: Use the complement rule to find the following probabilities.

6. A food distributor estimates that 2% of the eggs it delivers to grocery stores are cracked. What is the probability that an egg selected at random will not be cracked?

7. Every eleventh box of Sugar Pops cereal that is produced contains a special toy. If you buy a box of Sugar Pops, what is the probability that your box does not contain a toy?

8. Find the probability of rolling two dice and not getting doubles.

9. Suppose that a family has five children. What is the probability that the family has at least one girl?

10. Given that every fifth person in line will get a coupon for a free box of popcorn at the movies, what is the probability that you don't get a coupon when you're in line?

Mutually Exclusive Events

Directions: Determine whether the following events are mutually exclusive.

11. Out of a standard deck, choose a jack or choose a diamond.

12. Out of a standard deck, choose a red card or a club.

13. For a survey on campus, a sophomore or a history major is chosen.

14. When determining the schedule for his spring semester, Troy must decide whether to take a history class that meets at 9:00 a.m. on Mondays, Wednesdays, and Fridays or a science class which meets Mondays and Wednesdays at 9:30 a.m. (Assume that each class meets for 50 minutes.)

15. When assigning parts in a high school play, a senior or someone who's been in at least 2 other plays can play the lead role.

16. Choose a diet cola or a bottle of water out of the refrigerator.

17. Rent a movie that is appropriate for a 3-year-old child, or rent a movie that is rated R.

Independent Events

Directions: Determine whether the following events are independent.

18. Finding the batteries in your calculator dead. Finding the battery in your car dead.

19. Getting a letter from your aunt. Getting a bank statement in the mail.

20. Claire's name being drawn in the school raffle, without replacement. Ben's name being drawn in the school raffle.

21. Buying a new shirt on sale. Having enough money for lunch that day.

22. Lois ordering a steak. Ann ordering a salad.

23. Winning the Mississippi state lottery. Winning the Florida state lottery.

Directions: Calculate the following probabilities.

24. Suppose you draw two cards out of a standard deck with replacement. What is the probability that one of the cards is a jack and the other is a club?

25. What is the probability of choosing a jack and then a club out of a standard deck of cards with replacement?

26. Suppose that 4 out of the 15 doctors in a hospital are female, 11 out of the 15 are under the age of 45, and 2 are both female and under the age of 45. What is the probability that you are randomly assigned a female doctor or a doctor under the age of 45?

27. There are 15 different crayons in a new box. What is the probability that the orange and then the green will be chosen at random, without replacement?

28. In the teacher's pencil box, 13 out of the 20 pencils have no eraser, and 12 out of the 20 are not mechanical pencils. There are 4 pencils that are both mechanical and have erasers on them. What is the probability that Ava chooses a pencil at random from the box and gets either a mechanical pencil, or a pencil with a good eraser?

29. Ashley's internet service is terribly unreliable. In fact, on any given day, there is a 15% chance that her internet connection will be lost at some point that day. What is the probability that her internet service is not broken for five days in a row?

30. The local arbor society is giving out free trees to the public. They have 200 pine trees, 100 dogwoods, 125 oaks, and 200 birch trees. If you are the first in line to receive your free tree, what is the probability that you randomly get either an oak or a dogwood, since you cannot choose what type of tree you are given?

31. What's the probability of rolling at most one even number with two dice?

32. On a five-day vacation, the forecast is a 50% chance of rain every day. What's the probability that it rains every day?

33. On awards day at the end of the year, Jasmine has an 85% chance of winning the top award in English and a 4 out of 5 chance of winning an award for athletics. What's the probability that Jasmine wins both awards?

Directions: Calculate each of the following conditional probabilities.

34. Consider a standard deck of 52 cards. What is the probability that you draw without replacement the queen of hearts and then another heart?

35. Consider a standard deck of 52 cards. What is the probability that you draw without replacement a jack and then a face card?

36. Mrs. Harvey's algebra class contains 42 students, classified by academic year and gender as follows:

Table for Exercise 36

Class	Males	Females
Freshman	9	13
Sophomore	4	5
Junior	4	2
Senior	2	3

Mrs. Harvey must choose two students at random for a special project.

a. What is the probability that a sophomore boy and then a freshman girl are chosen?

b. What is the probability that two boys are chosen?

37. David likes to keep a jar of change on his desk. Right now his jar contains the following:

26 pennies
16 quarters
11 dimes
19 nickels

What is the probability that David reaches in and randomly grabs a quarter and then a nickel?

Basic Counting Rules

4.3

All of the probability calculations throughout the previous sections have depended on being able to count numbers of outcomes in events and in sample spaces. To do this more effectively, we need to introduce more sophisticated counting techniques.

The first counting technique is called the **Fundamental Counting Principle**. Simply put, it states that you can multiply together the number of possible outcomes for each stage in an experiment in order to obtain the total number of outcomes for that experiment. As an example of how this rule works, let us revisit the example of rolling a red die and then a blue die. We could call the first stage of the experiment "rolling the red die" and the second stage "rolling the blue die." There are 6 possible outcomes for stage 1, and for each of the 6 outcomes in stage 1, there are 6 possible outcomes for stage 2. We can multiply the number of outcomes from stage 1 by the number of outcomes from stage 2 in order to obtain the total number of outcomes for our experiment. In this case, (6)(6) = 36 total outcomes for the die rolling experiment. We know that this is the correct number because we have methodically listed all 36 outcomes in Example 2 earlier in this chapter.

Fundamental Counting Principle

For a sequence of n experiments where the first experiment has k_1 outcomes, the second experiment has k_2 outcomes, the third experiment has k_3 outcomes, and so forth, the total number of possible outcomes for the sequence of experiments is $k_1 \cdot k_2 \cdot k_3 \cdot \ldots \cdot k_n$.

A helpful method for setting up a fundamental counting problem is to think of each stage as a "slot." First, determine the number of slots that must be filled. Then decide how many outcomes are possible for each slot. For example, suppose one university issues student identification numbers that consist of two letters followed by three digits (0–9). How can we determine how many unique ID numbers the school can assign? The first step is to notice that there are 5 ID number "slots" to be filled.

	a b c ... *x y z*	*a b c* ... *x y z*	0 1 2 ... 8 9	0 1 2 ... 8 9	0 1 2 ... 8 9
	Slot 1	Slot 2	Slot 3	Slot 4	Slot 5
Number of Choices:	26 ×	26 ×	10 ×	10 ×	10

The first two slots contain 26 outcomes each, one outcome for each letter of the alphabet. The other 3 slots contain 10 possibilities each, one for each digit 0–9. After determining this information, we simply multiply the number of possibilities for each slot together and obtain (26)(26)(10)(10)(10) = 676,000 unique ID numbers.

Example 18

Suppose that Blake is ordering a banana split with 3 scoops of ice cream. If there are 31 flavors, and he wants all three flavors to be different, how many different ways can his banana split be made? For now, let's assume that the *order* in which the scoops are placed in the dish is important to Blake.

Solution:

31 30 29

There are 3 slots to fill—one for each different scoop of ice cream. Blake can pick any of the 31 flavors for his first scoop, so there are 31 choices for the first slot. For his second pick, he can choose any flavor except the one he has already chosen. This leaves only 30 choices for the second slot. For the last scoop, Blake can pick any flavor except the flavors he chose first and second, which leaves 29 choices for the last slot. As a result, we multiply (31)(30)(29) to obtain 26,970 different ways to make Blake's banana split.

Example 19

Robin is preparing a snack for her twins, Matthew and Lainey. She wants to give each child one item. She has the following snacks on hand: carrots, raisins, crackers, grapes, apples, yogurt, and granola bars. How many different ways can she prepare a snack for her twins?

Solution:

There are two slots to be filled—one for each twin. Because there is no requirement that the twins have different snacks, there are 7 possibilities for each. Therefore, there are (7)(7) = 49 possibilities for the way she can prepare the snacks.

✦ Factorial

An important mathematical expression used in many probability calculations is a factorial. A **factorial** is the product of all positive integers less than or equal to n. Symbolically, factorials are written as

$$n! = n(n-1)(n-2)\cdots(2)(1).$$

Thus 5! is the product of all positive integers less than or equal to 5, which equals (5)(4)(3)(2)(1) = 120. One important factorial expression to memorize is 0!, which appears in the denominator of many probability calculations. By special definition, the value of 0! is defined as 1.

$$0! = 1$$

Since factorials equal the product of a string of positive numbers, their values get very big, very quickly. While it was easy to calculate 5! without a calculator, expressions such as 100! are much too large even for ordinary calculators to compute.

Math Symbols!

$n! = n(n-1)(n-2) \cdots (2)(1)$

$0! = 1$

Example 20

Calculate the following factorial expressions.

a. $7!$ **b.** $\frac{4!}{0!}$ **c.** $\frac{95!}{93!}$ **d.** $\frac{5!}{(5-3)!}$ **e.** $\frac{6!}{2!(6-2)!}$

Solution:

a. Multiply each positive integer less than or equal to 7.

$$7! = (7)(6)(5)(4)(3)(2)(1) = 5040$$

b. Calculate each factorial then divide. Remember that 0! = 1.

$$\frac{4!}{0!} = \frac{24}{1} = 24$$

c. It would be very cumbersome to multiply out 95! and 93! and then divide. Instead, we will cancel first,

$$\frac{95!}{93!} = \frac{(95)(94)\cancel{(93)}\cancel{(92)}\cdots\cancel{(2)}\cancel{(1)}}{\cancel{(93)}\cancel{(92)}\cdots\cancel{(2)}\cancel{(1)}} = 95 \cdot 94 = 8930$$

d. You must begin by subtracting the expression in parentheses. Next, calculate each factorial expression and then divide.

$$\frac{5!}{(5-3)!} = \frac{5!}{2!} = \frac{120}{2!} = \frac{120}{2} = 60$$

e. Make sure that you begin by subtracting 6 – 2.

$$\frac{6!}{2!(6-2)!} = \frac{6!}{2!4!} = \frac{720}{2(24)} = 15$$

✦ Permutations and Combinations

When you need to count the number of ways objects can be chosen out of a group, then the problem you are dealing with is either a permutation or a combination. The way that you distinguish between the two calculations is that for a **permutation**, the order in which the objects are chosen is *important*. That is, the

Memory Booster!

Permutation:
order matters

Combination:
order does not matter

members of the group are picked out in a particular order. For example, suppose that a group of ten runners is racing. The first three runners to finish are in order, Chelsie, Chris, then Katie. This is different than them finishing Chris, Katie, then Chelsie, especially to Chelsie! Although they are the same three people, these are two different outcomes for the race. However, if you were trying to choose a team of three runners for a race, the team would be the same no matter what order they are listed in. In this case we would say that order is not important. When the order in which the objects are chosen is *not important*, we use a **combination**.

Once you have determined whether the order in which objects are chosen is important in your problem, you apply one of the following rules to calculate the total number of possibilities.

The Number of Ways of Choosing *r* Objects from *n* Distinct Objects:

When order is **important**:

$${}_nP_r = \frac{n!}{(n-r)!} \qquad \text{(Permutation)}$$

When order is **not important**:

$${}_nC_r = \frac{n!}{r!(n-r)!} \qquad \text{(Combination)}$$

The number of combinations might also be written as C_r^n or $\binom{n}{r}$. A permutation is read "n things permuted r at a time", whereas, a combination can be read in the shorter form "n choose r".

Example 21

A class of 18 fifth graders is holding elections for class president, vice-president, and secretary. How many different ways can the officers be elected?

Solution:

First, note that the order of the students chosen is important, i.e., it is different if someone is elected for president rather than vice-president. Therefore, this is a permutation. The sample size is 18, so $n = 18$. Because 3 students are to be elected, $r = 3$. Thus

$${}_{18}P_3 = \frac{18!}{(18-3)!} = \frac{18!}{15!} = \frac{18\cdot 17\cdot 16\cdot \cancel{15\cdots 1}}{\cancel{15\cdots 1}} = 18\cdot 17\cdot 16 = 4896.$$

So there are 4896 ways the class officers can be elected.

Example 22

Suppose that a baseball coach is putting the 9 starting baseball players in a batting order for the big game. How many different ways can he order those 9 players?

Solution:

Again, the order is important, so we know that this is a permutation problem. Because there are 9 players, $n = 9$. However, for this problem, r is also 9, because all 9 players are to be chosen for the line-up. Filling in the permutation formula then gives us

$${}_9P_9 = \frac{9!}{(9-9)!} = \frac{9!}{0!} = \frac{9!}{1} = 362{,}880.$$

There are 362,880 possible batting orders from which the coach can choose.

Example 23

Consider that a cafeteria is serving the following vegetables for lunch one weekday: carrots, green beans, lima beans, celery, corn, broccoli, and spinach. Suppose now that Bill wishes to order a vegetable plate with 3 different vegetables. How many ways can his plate be prepared?

Solution:

For this problem, we are choosing 3 vegetables out of a list of 7. It makes no difference what order the vegetables are placed on Bill's plate, so this must be a combination. Fill in the combination formula using $n = 7$ and $r = 3$.

$${}_7C_3 = \frac{7!}{3!(7-3)!} = \frac{7!}{3!4!} = 35.$$

Therefore, we have 35 different ways to prepare Bill's vegetable plate.

✦ Special Permutations

Finally, let's consider the special permutations in which some of the objects being counted are identical. For example, consider the permutation on the letters in the word Mississippi. Because the letters 'i', 's', and 'p' are *repeated*, we must use a different permutation formula which takes into consideration the repeated objects, or in this case, the repeated letters.

The Number of Permutations of n Objects, in which k_1 are all alike, k_2 are all alike, etc., is given by:

$$\frac{n!}{k_1!k_2!\cdots k_p!}, \text{ where } k_1 + k_2 + \cdots + k_p = n$$

Example 24

How many different ways can you arrange the letters in the word Tennessee?

Solution:

Because we can make no distinction between each 'e', 'n', or 's' in the word, we need to group the letters together. The letters of Tennessee are grouped as follows:

$$\begin{aligned} T &= 1 \\ E &= 4 \\ N &= 2 \\ S &= 2 \end{aligned}$$

Since there is a total of 9 letters in Tennessee, substituting these values into the the special permutation formula gives

$$\frac{9!}{1!4!2!2!} = 3780.$$

Thus there are 3780 ways to arrange the letters in the word Tennessee.

Let's apply these counting techniques to the probability methods we learned in the first two sections.

Example 25

Maya has a bag of 15 blocks, each of which is a different color including red, blue, and yellow. Maya reaches into the bag and pulls out 3 blocks. What is the probability that the blocks she has chosen are red, blue, and yellow?

Solution:

Let's first determine the number of outcomes in the sample space. We want to count the number of ways that 3 blocks can be drawn. This is a combination because the order of the colors is not important. The number of ways to choose 3 blocks out of a group of 15 is

$$_{15}C_3 = \frac{15!}{3!(15-3)!} = 455.$$

How many combinations contain the red, blue, and yellow blocks? Order does not matter, so there is only one way to choose these colors. Thus the probability that Abby chooses red, blue, and yellow is

$$P(\text{red, blue, yellow}) = \frac{1}{455} \approx 0.002.$$

Exercises 4.3

Fundamental Counting Principle

Directions: Use the Fundamental Counting Principle to determine the total number of outcomes.

1. Determine the number of 5-digit zip codes that can be made from the digits 0–9. Assume that digits may repeat.

2. Henry is setting a 6-character password on his computer. He is told that the first two characters must be letters and the last four must be numbers. No character may be used twice. How many choices does Henry have for his password?

3. There are six children, three girls and three boys, singing in the school program. As they line up to perform, the choral director insists that the first person be a girl and the last a boy. How many ways can the children line up?

4. How many even 4-digit pin numbers can be created from the digits 0–9?

5. How many odd 6-digit pin numbers can be made from the digits 0–9?

6. In the game of Clue, the guilty person can be chosen from 6 people, using 6 different possible weapons, in 9 possible rooms. How many possible crime scenarios are there?

7. When ordering his new computer, Joe must choose one of 5 monitors, one of 4 printers, and one of 6 scanners. How many different computer systems can he choose?

8. As Candy is trying to decide what to wear, she can choose from 8 blouses, 3 skirts and 4 pairs of shoes. Assuming everything coordinates, how many different possible outfits does she have?

9. A couple having twins is deciding on names. If they narrowed their choices to 5 boy names and 7 girl names, how many different pairs of names could their children have if they have one child of each sex?

Factorial
Directions: Evaluate the following factorial expressions.

10. $6!$ **11.** $8!$ **12.** $\frac{6!}{4!}$ **13.** $\frac{8!}{5!}$ **14.** $\frac{6!}{4!2!}$

15. $\frac{8!}{5!3!}$ **16.** $\frac{6!}{4!(6-4)!}$ **17.** $\frac{8!}{5!(8-5)!}$ **18.** $0!$ **19.** $\frac{7!}{0!}$

Permutations and Combinations
Directions: Find the following permutations or combinations.

20. ${}_7P_4$ **21.** ${}_5P_3$ **22.** ${}_5P_4$ **23.** ${}_7P_6$ **24.** ${}_6P_1$

25. ${}_8P_1$ **26.** ${}_9P_9$ **27.** ${}_3P_3$ **28.** ${}_5C_2$ **29.** ${}_8C_5$

30. ${}_7C_7$ **31.** ${}_4C_4$ **32.** ${}_5C_1$ **33.** ${}_{12}C_1$ **34.** $\frac{{}_3C_2}{{}_3P_1}$

35. $\frac{{}_{10}P_2}{{}_{10}C_2}$ **36.** ${}_6C_4 + {}_6C_3 + {}_6C_2 + {}_6C_1$ **37.** ${}_5P_4 + {}_5P_3 + {}_5P_2 + {}_5P_1$

38. Farmer John has nine prize-winning cows. How many ways can he choose three of his cows to show at the state fair?

39. There are twelve members in the local garden club. In how many ways can a president and secretary be chosen? (Assume that no member can hold both positions at the same time.)

40. A teacher must choose parts for the upcoming Thanksgiving play from her class of 18 students. How many ways can she choose the parts of pilgrim, native american, and turkey?

41. A teacher must choose parts for the upcoming Thanksgiving play from her class of 18 students. She needs a group of four students to serve as program attendants before the start of the play. How many ways can this group be chosen?

42. There are eight people hosting a party. Three people are needed to stay and clean up after the party is over. How many ways can the clean-up crew be chosen?

43. There are eight people hosting a party. One person must set up the catering, another must bring flowers, and someone else needs to bring drinks. In how many ways can these tasks be assigned?

44. In assigning seats for a classroom, how many ways can a teacher place 8 students on the front row from her roll of 35?

45. In how many ways can a masters student fulfill the degree requirements in mathematics if 10 classes are needed from a choice of 15 classes?

46. In how many ways can 1st, 2nd, and 3rd place prizes be awarded in a local science fair, if there are 25 participants?

47. In how many ways can a task force of 4 people be chosen from a group of 12 employees?

48. If 3 people need to serve as chaperones on a school trip, how many ways can they be chosen from the parents of the 20 students? (Assume each child has 2 parents available.)

49. The Seago family is planning their vacation. Each of the five family members is allowed to nominate three places they would like to visit. If they want to visit 4 different places during the trip, in how many ways can they plan their road trip assuming no family members choose the same place?

50. When painting a plate at a paint your own pottery studio, Sara wants to use 3 different colors from the palette of 30 colors. How many different color combinations can she choose from?

51. In how many ways can the letters in the word 'statistics' be arranged?

52. In how many ways can the letters in the word 'probability' be arranged?

Directions: Use counting techniques to compute the following classical probabilities.

53. At a carnival entrance tickets are assigned 5-digit numbers using the digits 0–9. If one ticket number is chosen randomly for a prize, what is the probability that every number on the ticket is even?

54. Suppose that your boss must choose 3 employees in your office to attend a conference in Jamaica. Because all 20 of you want to go, he decides that the only fair way is to draw names out of a hat. What is the probability that you, Suzanne, and Alex are chosen?

55. Rhonda and Laura are planning to watch two movies over the weekend from Laura's collection of 24 DVDs. Rhonda has two favorites among the collection. What is the probability that the girls would randomly choose those two movies to watch?

56. Bill is planting tulip bulbs in the front yard. There are 2 white bulbs and 2 red bulbs mixed together in a bucket. What is the probability that Bill plants the 4 bulbs in a row so that they are alternating in color?

57. Kim is playing Scrabble. What is the probability that she chooses the tiles with the letters of her name, in order, when she draws three tiles from the bag? Assume when she begins there is one tile of each letter in the alphabet left in the bag.

58. Stephanie has 11 different outfits in her closet that she wears on Sundays. Ann has 14 different outfits, of which 3 are the same as Stephanie. What is the probability that they wear the same outfit on a particular Sunday?

Challenge Exercises

Directions: Find the simplified expression for each of the following.

59. ${}_nC_n$ **60.** ${}_nC_1$ **61.** ${}_nP_1$ **62.** ${}_nP_n$ **63.** ${}_nP_{n-1}$

Additional Counting Techniques

4.4

There are many other key terms that you may see in counting problems. These terms are important because they define the method that must be used in order to obtain the correct answer. Words to look for include *at least*, *at most*, *greater than*, *less than*, *between*, etc. A few examples of the many different types of problems you may encounter are given below.

Example 26

A group of 12 tourists is visiting London. At one particular museum, a discounted admission is given to groups of at least ten. How many combinations of tourists can be made for the museum visit so that the group receives the discounted rate?

Solution:

The key words for this problem are *at least*. If at least 10 are required, then the group will get the discount rate if 10, 11, or 12 tourists go to the museum. We must then calculate the number of combinations for each of the three possibilities. Note that these are combinations, not permutations, because the order of tourists chosen to go to the museum is irrelevant.

Number of groups of 10 tourists:

$$_{12}C_{10} = \frac{12!}{10!(12-10)!} = 66$$

Number of groups of 11 tourists:

$$_{12}C_{11} = \frac{12!}{11!(12-11)!} = 12$$

Number of groups of 12 tourists:

$$_{12}C_{12} = \frac{12!}{12!(12-12)!} = 1$$

To find the total number of groups, we add the combinations together:

$$66 + 12 + 1 = 79$$

There are 79 different groups that can be formed to tour the museum at the discounted rate.

Example 27

Jack is setting a password on his computer. He is told that his password must contain at least 3 but no more than 5 characters. He may use either letters or numbers (0–9). How many different possibilities are there for his password if each character can only be used once?

Solution:

First, notice the key words *at least* and *no more than*. These words tell us that Jack's password can be 3, 4, or 5 characters in length. Also, note that for a password the order of characters is important, so this must be a permutation problem. There are 26 letters and 10 digits to choose from, so we must calculate the number of permutations possible from taking groups of 3, 4, or 5 characters out of 36 possibilities.

Number of permutations of 3 characters:

$$ {}_{36}P_3 = \frac{36!}{(36-3)!} = 42,840 $$

Number of permutations of 4 characters:

$$ {}_{36}P_4 = \frac{36!}{(36-4)!} = 1,413,720 $$

Number of permutations of 5 characters:

$$ {}_{36}P_5 = \frac{36!}{(36-5)!} = 45,239,040 $$

To find the total number of passwords, we add all three of the above permutations together and obtain 46,695,600 passwords:

$$ 42,840 + 1,413,720 + 45,239,040 = 46,695,600 $$

(Imagine how many possibilities there would be if you could use up to eight or nine characters!)

✦ Combining Counting Techniques

At times it is necessary to **combine** counting techniques in order to find all possibilities for a given problem. A common scenario that occurs is when a combination or permutation rule must be used in order to determine the number of outcomes for each slot of a fundamental counting principle problem. Solving this type of problem is best learned by example.

Example 28

Tina is packing her suitcase to go on a weekend trip. She wants to pack 3 shirts, 2 pairs of pants, and 2 pairs of shoes. She has 9 shirts, 5 pairs of pants, and 4 pairs of shoes to choose from. How many ways can Tina pack her suitcase? (We will pretend everything matches.)

Solution:

Begin by thinking of this problem as a fundamental counting principle problem. There are 3 suitcase slots for Tina to fill—a slot for shirts, a slot for pants, and a slot for shoes. To fill these slots, the combination formula must be used because we are choosing some items out of her closet, and the order of the choosing does not matter. Let's first calculate these combinations:

Number of combinations of shirts:

$$_9C_3 = \frac{9!}{3!(9-3)!} = 84$$

Number of combinations of pants:

$$_5C_2 = \frac{5!}{2!(5-2)!} = 10$$

Number of combinations of shoes:

$$_4C_2 = \frac{4!}{2!(4-2)!} = 6$$

The final step in this problem is different than the final step in the last two problems, because the fundamental counting principle tells us we have to *multiply*, rather than add, the combinations together to get the final result. There are then

$$(84)(10)(6) = 5040$$

different ways Tina can pack for the weekend.

Example 29

An elementary school principal is putting together a committee of 6 teachers to head up the spring festival. There are 8 first grade, 9 second grade, and 7 third grade teachers at the school.

a. How many ways can the committee be formed?

b. How many ways can the committee be formed if there must be 2 teachers chosen from each grade?

Solution:

a. There are no stipulations as to what grades the teachers are from, so we need to simply calculate the number of combinations by choosing 6 committee members from the 24 teachers.

$$_{24}C_6 = \frac{24!}{6!(24-6)!} = 134{,}596$$

There are then 134,596 ways to choose the spring festival committee.

b. For this problem we need to combine the fundamental counting principle and combination rule. First, we see that there are 3 slots to fill: one first grade slot, one second grade slot, and one third grade slot. Each slot is made up of 2 teachers, and we must use the combination rule to determine how many ways these slots can be filled.

Number of first-grade combinations:

$$_{8}C_2 = \frac{8!}{2!(8-2)!} = 28$$

Number of second-grade combinations:

$$_{9}C_2 = \frac{9!}{2!(9-2)!} = 36$$

Number of third-grade combinations:

$$_{7}C_2 = \frac{7!}{2!(7-2)!} = 21$$

Finally, the fundamental counting principle says that we have to multiply together the outcomes for all three slots in order to obtain the total number of combinations. Thus there are

$$(28)(36)(21) = 21{,}168$$

ways to choose the committee.

Exercises

4.4

Directions: Use the fundamental counting principle, combination formula, permutation formula, or a combination of the methods to solve each problem.

1. License tags in a particular area of the state must begin with one of the following letters: L, D, Y, or K. The rest of the tag must contain two letters followed by three digits (0–9). If characters cannot repeat, how many unique tags can be made?

2. License tags in another state are made up of three letters followed by three digits (0 – 9), none of which can repeat. Additionally, each tag number may be assigned to a regular tag, a wildlife conservation tag, or a veterans tag. How many unique tags can be made?

3. How many 4-digit numbers can be created from the digits 0–9 if the first and last digits must be odd and no digit can repeat?

4. How many 4-digit numbers can be made from the digits 0–9 if each number created must be greater than 5000?

5. How many 5-card hands containing exactly 2 hearts and 3 clubs can be drawn without replacement from a standard deck of cards?

6. How many 5-card hands containing at least 4 face cards can be drawn without replacement from a standard deck of cards?

7. Virginia's Veggie Café offers 5 types of homemade bread, 10 toppings, and 6 different condiments. How many different super sandwiches can be made if a super sandwich consists of 6 different toppings and 2 different condiments?

8. Mysti is picking out material for her new quilt. There are 12 possible plaids, 8 different solids, and 4 floral prints that she can choose from. If she needs 3 plaids, 2 solids, and 2 floral prints for her quilt, how many different ways can she choose the materials?

9. Because Tristan has diabetes, he must make sure that his daily diet includes 2 vegetables, 3 fruits and 2 breads. When going to the grocery store, he has a choice of 20 vegetables, 8 fruits, and 5 breads. In how many ways can he make up his daily requirements if he doesn't like to eat 2 helpings of the same thing?

10. How many different teams of 4 can be chosen from a group of 20 girls and 15 boys if each team must have at least one boy on it?

11. A football coach needs to choose 11 players to start on offense. There are 6 freshmen, 6 sophomores, 8 juniors, and 7 seniors on the team. How many ways can the starting 11 be chosen if the coach wants all 7 seniors to play?

12. Lindsay is checking out books at the library, and she is primarily interested in mysteries and nonfiction. She has narrowed her selections down to 7 mysteries and 8 nonfiction books.

a. How many combinations of books can she check out if she is only allowed three books at a time?

b. How many combinations of books can she check out if she is only allowed three books at a time, and she wants at least one mystery?

Challenge Problem

13. In choosing what music to play at a charity fund raising event, Marlow needs to have an equal number of symphonies from Mozart, Haydn, and Schubert. If he is setting up a schedule of the 12 songs to be played during the show, how many different schedules are possible if he has 41 Mozart, 104 Haydn, and 8 Schubert symphonies from which to choose?

14. A baseball coach needs to choose 9 players to be in the batting lineup for the first game of the season. There are 5 freshmen, 4 sophomores, 7 juniors, and 4 seniors on the team.

a. How many ways can the batting order be chosen if the coach wants no more than 2 freshmen to play?

b. How many ways can the batting order be chosen if the coach wants all 4 seniors to play?

Chapter Review

Section 4.1 - Classical Probability	Pages
• **Probability experiment or trial** – any process in which the result is random in nature	p. 152
• **Outcome** – each individual result that is possible for a given experiment	p. 152
• **Sample Space** – the set of all possible outcomes for a given experiment	p. 152
• **Event** – a subset of the sample space	p. 152
• **Subjective Probability** – an educated guess regarding the chance that an event will occur	p. 155–156
• **Empirical Probability** – if all outcomes are based on experiment, $P(E)=\frac{\text{number of times the event } E \text{ occurs}}{\text{total number of times the experiment is performed}}=\frac{f}{n}$	p. 155–156
• **Law of Large Numbers** – the greater the number of trials, the closer the empirical probability will be to the true probability	p. 155
• **Classical Probability** – if all outcomes are equally likely, $P(E)=\frac{n(E)}{n(S)}$	p. 155 – 156

Section 4.2 - Probability Rules	Pages
• **Complement** – all outcomes in the sample space that are not in the event, $P(E) = 1 - P(E^c)$	p. 162 – 164
• **Addition Rule for Probability** – probability that one event happens or the other event happens, $P(E \textbf{ or } F) = P(E) + P(F) - P(E \text{ and } F)$	p. 164 – 168
• **Mutually exclusive** – events that share no outcomes	p. 167 – 168
• **With repetition** – outcomes may be repeated	p. 169 – 171
• **Without repetition** – outcomes may not be repeated	p. 169 – 171
• **With replacement** – objects are placed back into consideration for the following choice	p. 169 – 171
• **Without replacement** – objects are not placed back into consideration for the following choice	p. 169 – 171

cont'd on the next page

Section 4.2 - Probability Rules (cont.)	Pages
• **Multiplication Rule for Probability** – for two independent events, E and F, $P(E \textbf{ and } F) = P(E)P(F)$	p. 169 – 171
• **Independent Events** – if one event happening does not influence the probability of the other event happening	p. 169
• **Conditional Probability** – when two events are not independent, the outcome of one influences the probability of the other	p. 170
• **Multiplication Rule for Dependent Events** – for two dependent events, E and F, $P(E \text{ and } F) = P(E)P(F \mid E)$	p. 170 – 171

Section 4.3 - Basic Counting Rules	Pages
• **Fundamental Counting Principle** – for a sequence of n experiments where the first experiment has k_1 outcomes, the second experiment has k_2 outcomes, the third experiment has k_3 outcomes, and so forth. The total number of possible outcomes for the sequence of experiments is $k_1 \cdot k_2 \cdot k_3 \cdot \ldots \cdot k_n$.	p. 175 – 176
• **Factorial** – the product of all positive integers less than or equal to n $n! = n(n-1)(n-2) \cdots (2)(1)$ Remember $0! = 1$	p. 176 – 177
• **Permutation** – when order is important, ${}_nP_r = \dfrac{n!}{(n-r)!}$	p. 177 – 179
• **Combination** – when order is not important, ${}_nC_r = \dfrac{n!}{r!(n-r)!}$	p. 177 – 179
• **Special Permutations** – the number of permutations of n objects, in which k_1 are all alike, k_2 are all alike, etc., is given by $\dfrac{n!}{k_1!k_2!\cdots k_p!}$	p. 179 – 181

Section 4.4 - Additional Counting Techniques	Pages
• **Combining Counting Techniques**	p. 185 – 188

Chapter Exercises

Directions: Find the probability for each of the following scenarios. Use the most appropriate method.

1. If 56% of adults attend church services in a typical month, 46% listen to a Christian radio broadcast in a typical month and 39% do both, what's the probability that a random adult will neither attend church nor listen to a Christian radio broadcast in a typical month?

2. Find the probability of being dealt 5 hearts from a standard deck of playing cards.

3. Find the probability that out of the 1319 college freshmen on campus, you are randomly chosen by the yearbook staff to give a quote about your first year at college which will be published in the yearbook.

4. In the dice game Yahtzee you achieve the most points by throwing a Yahtzee, which is throwing the same number on 5 different dice in one roll. Find the probability that you don't roll a Yahtzee.

5. In a large doll store, 47% of the dolls blink, 56% of the dolls have moveable legs and arms and 23% of the dolls neither blink nor have moveable limbs. Find the probability that a doll chosen at random will both blink and have moveable limbs.

6. In a class of 87 people, 27 wear glasses, 32 are blonde, and 38 are neither blonde nor wear glasses. Find the probability that a student chosen at random will have blonde hair and wear glasses.

7. Joe is planting tulip bulbs in his garden, and he has 5 red bulbs and 5 yellow bulbs to plant in one row. What is the probability that he randomly plants the bulbs so that the colors alternate?

8. Emily has a bad habit of losing her keys. This time, she estimates that there is a 0.2 chance that she left them in her office, a 0.35 chance that she left them in her bag, and a 0.15 chance that she left them on the kitchen counter. What is the probability that her keys will be in none of these places?

9. Robert is ordering dessert, and he wants the fudge pie. The waitress tells him that he can have up to four toppings: ice cream, chocolate sauce, whipped cream, and a cherry. Since he cannot decide how many of the toppings he wants, he tells the waitress to surprise him. If the waitress randomly chooses which toppings to add, what is the probability that Robert gets just chocolate sauce and a cherry?

10. Out of 62 people surveyed, 22 own a laptop computer, 39 own a desktop computer, and 9 own no computer at all. What is the probability that a person selected at random from those surveyed owns both a laptop and a desktop computer?

11. Find the probability of drawing 4 cards from a deck of 52 cards and not getting all aces.

12. What's the probability of rolling a total of at least 11 with two dice?

13. Raffle tickets for a trip to Miami are assigned 3-digit numbers using the digits 0–9. What is the probability that the number on the winning ticket is either even or less than 300?

Directions: Answer the following questions.

14. Suppose you have money to buy 5 new downloads for your iPod. If there are 12 new downloads you'd like to have, how many different ways could you purchase your five? Assume that the order of songs does not matter.

15. In planning the teaching assignments for next semester, Mr. Hinton must have a teacher in each of the 7 grades during each of the 6 periods of the day. If he has 10 teachers to choose from, which allows 3 teachers to be on break at a time, how many different teaching schedules could he come up with?

16. The student council at one high school must choose two representatives from each of the sophomore, junior, and senior classes to attend the annual student council convention. If there are 6 sophomores, 5 juniors, and 7 seniors on the student council, in how many ways can the group be chosen for the convention?

17. The Pancake House offers hash browns with up to five different toppings: cheese, ham, bacon, onions, and mushrooms. How many different ways can the hash browns be served?

18. Richard is assigned the task of setting the passwords for every computer at his office. Different computers have different guidelines as to how the password may be set. How many different 4-digit passwords using the digits 0–9 can he create under each of the following guidelines?

a. The passwords must be odd and greater than 3000.
b. The passwords must be even and greater than 4000.
c. The passwords must be even and less than 5000.

19. Suppose that there are 8 employees at a café in downtown Jackson, MS. The boss needs one employee to serve as host or hostess, one to clean tables, and one to supervise the kitchen. How many ways can these tasks be assigned?

20. Calculate the probability of choosing, without replacement, a spade and then a heart out of a standard deck of cards.

21. Calculate the probability of drawing two cards from a standard deck, with replacement, and obtaining a five and an ace.

22. Chandra has 10 pieces of jewelry in her jewelry case, and she wants to take 4 pieces with her on vacation. In how many ways can she select the jewelry for her trip?

23. A homeowner wants to plant some new trees in her yard. If she has 6 pink crepe myrtles, 4 red oak trees and 5 yellow poplar trees to plant down the drive, in how many ways can she plant them?

24. In planning the room assignments for the overnight field trip, Mrs. Viant needs to put three girls and one chaperone in each room. If there are 12 girls and 4 chaperones on the trip, how many room assignments can she possibly have for the rooms 100, 101, 102, and 103?

Chapter Project: Law of Large Numbers

Directions: This project is designed to be completed in groups, but may be completed by an individual. Each group is to follow the steps below using a 6-sided die.

Step 1: Calculate the probability of rolling a single die and getting a 4.

Probability = ☐

Step 2: Each group member is to roll a die 10 times and record the outcomes. Compile the outcomes for all group members, and compute the percentage of 4's rolled by the group.

Percentage of 4's = ☐

Step 3: This time, each group member is to roll a die an additional 40 times, for a total of 50 rolls per person. Again, combine the outcomes for all group members, and compute the percentage of 4's rolled by the group.

Percentage of 4's = ☐

Step 4: Let's combine the information from all groups in order to look at the class as a whole. On the board, make a chart with one column for "number of fours" and one for "number of rolls." Fill in the chart with the information from each group. Using this information, calculate the percentage of fours rolled by the class.

Number of 4's Rolled	Total Number of Rolls	Percentage of 4's Rolled

Step 5: Let's evaluate the results of this experiment. Compare the percentages calculated in steps 2, 3, and 4. You should see that the more times the dice were tossed, the closer the percentage of fours rolled should come to the probability calculated in step 1. This is precisely what the law of large numbers says: the greater the number of trials, the closer the empirical probability comes to the theoretical probability. In fact, if it were possible to complete an infinite number of die tosses, the percentage of fours rolled would indeed be equal to the theoretical probability.

Chapter Technology: Calculating Probabilities

TI-84 PLUS

Factorial

To calculate a factorial, enter the n value in the calculator, press MATH, then scroll over to the probability menu PRB. The factorial is option 4: !. Select option 4 and ! will appear on the screen after the value you entered. To obtain the answer, press ENTER.

Example T.1

Calculate 9!

Solution:

In your graphing calculator, press 9, then MATH, PRB, and option 4. Press ENTER.

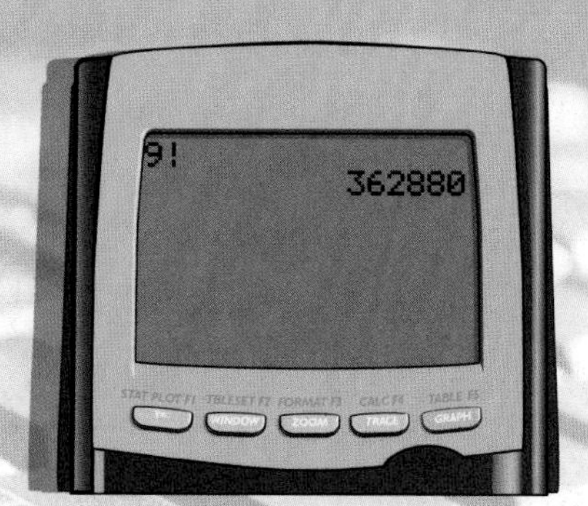

As you can see in the screen on the right, 9! = 362,880.

Permutation

To calculate a permutation, enter the n value in the calculator, press MATH, then scroll over to the probability menu PRB. Permutation is option 2: nPr. Select option 2 and nPr will appear on the screen after the value you entered. Enter the value for r and press ENTER.

Example T.2

Calculate ${}_{11}P_4$.

Solution:

First, type 11 and then press the MATH button. Next, scroll over to PRB and choose option 2. Then type 4 and press ENTER.

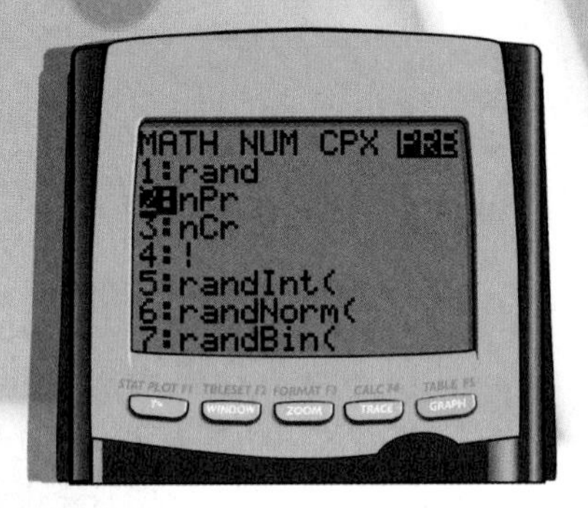

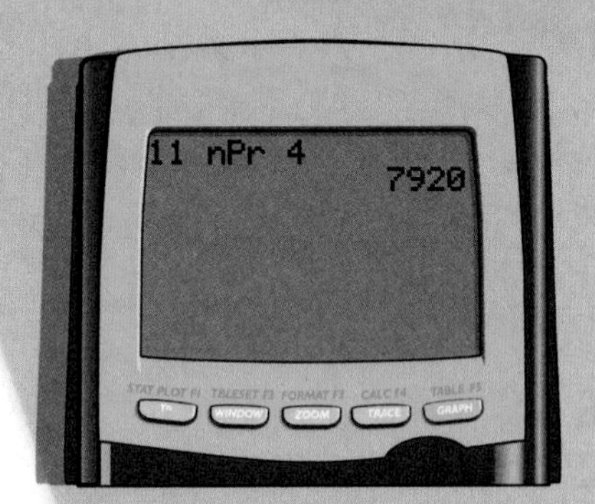

Thus ${}_{11}P_4 = 7920$.

Combination

To calculate a combination, enter the n value in the calculator, press MATH, then scroll over to the probability menu PRB. Combination is option 3: nCr. Select option 3 and nCr will appear on the screen after the value you entered. Enter the value for r and press ENTER.

Example T.3

Calculate ${}_{15}C_9$.

Solution:

Type in 15, then press the MATH. Next, scroll over to PRB and choose option 3. Then type 9 and press ENTER.

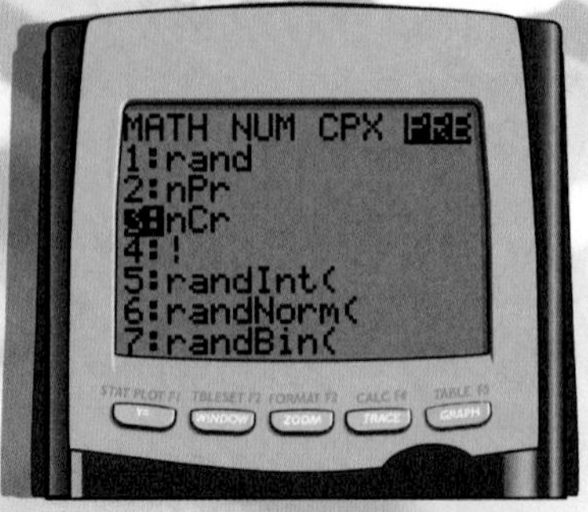

Thus ${}_{15}C_9 = 5005$.

MICROSOFT EXCEL

Probability

Microsoft Excel can calculate many mathematical functions used in probability. Individually, the commands are:

Table T.1 - Excel Commands

Calculation	Command
Factorial	=FACT(*value*)
Permutation	=PERMUT(n, r)
Combination	=COMBIN(n, r)

Excel is also a great tool to use when trying to calculate the probability of several independent events as shown in the following example.

Example T.4

A coin is flipped, a die is rolled, and a card is drawn from a deck. Find the probability of getting a tail on the coin, a 5 on the die, and drawing a heart from the deck of cards.

Solution:

First fill out the data in a worksheet as shown below.

	A	B	C	D	E
1		Coin	Die	Card	
2	Chance of Success	1	1	13	
3	Possible Outcomes	2	6	52	
4					
5					

Then, in cell B4, divide B2 by B3. To do this, type the formula "=B2/B3". Copy the formula from cell B4 into C4 and D4. Now the probabilities for each event are listed in Row 4. Remember, when events are independent we multiply the probabilities of each event occurring with each other. Therefore, in cell E4, multiply the cells B4 through D4 by typing "=B4*C4*D4". The result, shown below, will be the probability of getting a tail on the coin, a 5 on the die, and drawing a heart from the deck of cards.

	A	B	C	D	E
1		Coin	Die	Card	
2	Chance of Success	1	1	13	
3	Possible Outcomes	2	6	52	
4		0.5	0.166667	0.25	0.020833
5					

Probability Distributions

Sections

Objectives

The objectives of this chapter are as follows:

1. Calculate the expected value of a probability distribution
2. Calculate the variance and the standard deviation of a probability distribution
3. Identify a distribution as binomial, Poisson, or hypergeometric
4. Calculate probabilities using the binomial, Poisson, or hypergeometric distribution

Chapter 5

Introduction

In the previous chapter, we looked at methods for computing simple probabilities. For example, we learned how to find the probability that you win the local school raffle or the probability that a student randomly—and unfortunately—sits down to one of the two broken computers in the computer lab on campus.

Suppose instead that you wanted to know a more complicated probability, one that cannot easily be solved by using classical probability techniques alone. For instance, consider the scenario in which a university randomly selects sixty students to complete a survey regarding the cost of tuition. What is the probability that all sixty students chosen are freshmen and male? Or, suppose that an investor wants to compare the risks of two different stock options for his portfolio. Is there a way to estimate the long-term benefit for each investment option? Let's look at some alternate probability techniques to solve these types of problems.

Expected Value

5.1

Let's suppose that you and some friends go to a fall carnival. One game at the carnival is a coin toss challenge. For a dollar you get a chance to flip three coins. If all 3 coins are the same (heads or tails), then you win three dollars. If any other combination of coins turns up, you don't win anything. Let's consider for a moment the probabilities associated with this carnival game.

The number of coins that turn up heads each time you play the game will be some number between 0 and 3, which is completely determined by chance. A variable like this is called a **random variable** and is usually denoted by a capital letter, such as X, Y, or Z. For example, if X = the number of coins showing heads, then an outcome of three coins showing two heads and one tail gives X the value of 2, while an outcome of three coins showing heads would make $X = 3$.

2 Heads, 1 Tail
$X = 2$

3 Heads
$X = 3$

In our example above, notice that, although the values of X are determined by chance, the values of X are not all equally likely. Of the eight possible outcomes, only *one* gives a value of $X = 3$. Therefore, the probability of getting three heads is

$$P(X = 3) = \frac{1}{8}.$$

On the other hand, there are 3 throws that give $X = 2$, so

$$P(X=2)=\frac{3}{8}.$$

A table or formula that lists the probabilities for each outcome of the random variable, X, is called a **probability distribution**. Consider the probability distribution for our example.

Table 1 - Number of Heads

x	0	1	2	3
$P(X=x)$	$\frac{1}{8}$	$\frac{3}{8}$	$\frac{3}{8}$	$\frac{1}{8}$

Just as data may be discrete or continuous, random variables may also be discrete or continuous. A **discrete random variable** may take on either finitely many values or have infinitely many values that are determined by a counting process. Our example of counting heads in a coin toss is an example of a discrete random variable, and its table is a **discrete probability distribution**.

Definitions

A **random variable** is a variable whose numeric value is determined by the outcome of a random experiment.

A **probability distribution** is a table or formula that lists the probabilities for each outcome of the random variable, X.

Example 1

Create a probability distribution for X, the sum of 2 rolled dice.

Solution:

To begin, let's list all of the possible values for X.
When rolling two dice, there are 36 possible rolls, each giving a sum between 2 and 12, inclusive. To find the probability distribution, we need to calculate the probability for each outcome.

$P(X=2)=\frac{1}{36}$ because there is only one way to get a sum of 2: a one on each die.

$P(X=3)=\frac{2}{36}$ because you may get the sum of 3 in two ways: 1, 2 or 2, 1.

Continuing this process will give us the probability distribution shown on the following page.

cont'd on the next page

Table 2 - Rolling Two Dice											
x	2	3	4	5	6	7	8	9	10	11	12
$P(X=x)$	$\frac{1}{36}$	$\frac{2}{36}$	$\frac{3}{36}$	$\frac{4}{36}$	$\frac{5}{36}$	$\frac{6}{36}$	$\frac{5}{36}$	$\frac{4}{36}$	$\frac{3}{36}$	$\frac{2}{36}$	$\frac{1}{36}$

Check for yourself that the probabilities listed above are the true values for the probability distribution of X, the sum of 2 rolled dice.

Probability distributions are important when calculating the average outcome for an experiment. Let's go back to our original example of the coin toss challenge. Now, if you play the game a number of times, are you more likely to come out winning or losing money? The solution to this problem cannot be found simply by using one of the patterns we have seen in the previous sections. In order to calculate this probability, you need to be able to find something called the *expected value* of a random variable. Also, the random variable that we need to use is not X, the number of heads, but a new random variable W = the money won playing a single game.

There are two possible outcomes for the coin toss game—you win \$2 or lose \$1. (The net amount you win is just \$2, considering it costs a dollar to play the game in the first place.) Then, the possible values for the discrete random variable W are 2 and -1. (A loss is considered a negative value.) To calculate the expected value for our example, we need to calculate the discrete probability distribution for W. We then need to know the probability for each value of W. There are 2 ways for all three coins to be the same—all heads or all tails—and 8 ways to flip three coins. The probability of all three coins being the same is then $\frac{2}{8}=\frac{1}{4}$. By the complement rule, we know that the probability of not getting all three alike must be $1-\frac{1}{4}=\frac{3}{4}$. Let's draw the discrete probability distribution of W.

Table 3 - Coin Toss Game	
w	$P(W=w)$
2	$\frac{1}{4}$
-1	$\frac{3}{4}$

The **expected value**, $E(X)$, that we referred to earlier is the value that you "expect" the random variable to have on average. Formally, it is the mean of the probability distribution, i.e., $E(X) = \mu$. To calculate this expected value, multiply each value in the distribution by its probability. The expected value is then the sum of the multiplied values. The formula for expected value is as follows:

Expected Value, *E(X)*, for a Discrete Probability Distribution

$$E(X)=\mu=\sum\left[x_i \cdot P(X=x_i)\right]$$

We calculate the expected value as follows:

$$E(W)=2\cdot P(W=2)+(-1)\cdot P(W=-1)$$

$$\begin{aligned}E(W)&=(2)\left(\frac{1}{4}\right)+(-1)\left(\frac{3}{4}\right)\\&=\frac{2}{4}-\frac{3}{4}\\&=-\frac{1}{4}\\&=-\$0.25\end{aligned}$$

Now that we know the expected value of the random variable W, we can say whether we are more likely to come out winning or losing money after having played the game a number of times. The expected value of the coin toss challenge is –\$0.25. This tells us that if you were to play the game a large number of times, you would expect an average loss of 25 cents for each game you play. Notice that the expected value gives us a *long-term average*. For any given try, a loss of a quarter is not an option. (Remember, you either win \$2 or lose \$1.) However, if you repeat the game many times, an average loss of a quarter per game is to be expected. So, what is the answer to our original question? Are you more likely to come out winning or losing money after playing the game over a period of time? Although it is possible for you to win every time, the expected value tells us that over time you are more likely to lose money on the coin toss game (which is why the game is profitable for the carnival company).

Example 2

Peyton is trying to decide between two different investment options. The two plans are summarized in the table below. The left-hand column for each plan gives the potential profits, and the right-hand columns give their respective probabilities. Which plan should he choose?

Table 4 - Investment Plans

Investment Plan A		Investment Plan B	
\$1200	$P=0.1$	\$1500	$P=0.3$
\$950	$P=0.2$	\$800	$P=0.1$
\$130	$P=0.4$	–\$100	$P=0.2$
–\$575	$P=0.1$	–\$250	$P=0.2$
–\$1400	$P=0.2$	–\$690	$P=0.2$

cont'd on the next page

Solution:

It is difficult to determine which plan will yield the higher return simply by looking at the table on the previous page. Let's use expected value to compare the plans.

For Investment Plan A:

$$\begin{aligned} E(X) &= (1200)(0.1)+(950)(0.2)+(130)(0.4)+(-575)(0.1)+(-1400)(0.2) \\ &= 120+190+52-57.50-280 \\ &= 24.50 \end{aligned}$$

For Investment Plan B:

$$\begin{aligned} E(X) &= (1500)(0.3)+(800)(0.1)+(-100)(0.2)+(-250)(0.2)+(-690)(0.2) \\ &= 450+80-20-50-138 \\ &= 322 \end{aligned}$$

From these calculations, we see that the expected value of Plan A is \$24.50, and the expected value of Plan B is \$322. Over time, Plan B appears to be the wiser investment option for Peyton.

Rounding Rule

Round the standard deviation and variance to one more decimal place than what is given in the data set.

At times it is also important to have a measure of the variation of a discrete probability distribution. This is especially important for problems such as the previous investment example, where a high variation in profit would make for a riskier investment. The variance and standard deviation of a discrete probability distribution are found using the following formulas.

Variance for a Discrete Probability Distribution

$$\sigma^2 = \sum\left[x^2 \cdot P(X=x)\right] - \mu^2$$

Standard Deviation for a Discrete Probability Distribution

$$\sqrt{\sigma^2} = \sigma = \sqrt{\sum\left[x^2 \cdot P(X=x)\right] - \mu^2}$$

Example 3

Which of the investment plans in the previous example carries more risk, Plan A or Plan B?

Solution:

To decide which plan carries more risk, we need to look at their variances. Let's calculate the variance separately for each investment plan. To do this, multiply each value of x by its respective probability. Add these values together to obtain the expected value. Remember that $E(x) = \mu$. Next, multiply the square of each value of x by its respective probability; then find the sum of the results. The variance is then the expected value squared subtracted from this sum. We will use a table to organize our data as we calculate the variance. We will use the expected value we calculated in the previous example as the mean.

For Investment A:

Table 5 - Investment A			
x	$P(X=x)$	$x \cdot P(X=x)$	$x^2 \cdot P(X=x)$
\$1200	$P=0.1$	120	144,000
\$950	$P=0.2$	190	180,500
\$130	$P=0.4$	52	6760
–\$575	$P=0.1$	–57.5	33,062.5
–\$1400	$P=0.2$	–280	392,000
		$\mu=\sum[x_i \cdot P(X=x_i)]=24.5$	$\sum[x_i^2 \cdot P(X=x_i)]=756{,}322.5$

$$\sigma^2 = \sum[x^2 \cdot P(X=x)] - \mu^2 = 755{,}722.25$$

$$\sigma = \sqrt{755{,}722.25} = \$869.32$$

For Investment B:

Table 6 - Investment B			
x	$P(X=x)$	$x \cdot P(X=x)$	$x^2 \cdot P(X=x)$
\$1500	$P=0.3$	450	675,000
\$800	$P=0.1$	80	64,000
–\$100	$P=0.2$	–20	2000
–\$250	$P=0.2$	–50	12,500
–\$690	$P=0.2$	–138	95,220
		$\mu=\sum[x_i \cdot P(X=x_i)]=322$	$\sum[x_i^2 \cdot P(X=x_i)]=848{,}720$

$$\sigma^2 = \sum[x^2 \cdot P(X=x)] - \mu^2 = 745{,}036$$

$$\sigma = \sqrt{745{,}036} = \$863.15$$

What do these results tell us? Comparing the standard deviations, we see that not only does Plan B have a higher expected value, but its profits vary slightly less than those of Plan A. We may conclude that Plan B carries a slightly lower amount of risk than Plan A.

Example 4

Suppose that Randall and Blake decide to make a friendly wager on the football game they are watching one afternoon. For every kick the kicker makes, Randall wins $30. For every kick the kicker misses, Randall has to pay Blake $40. Prior to this game, the kicker has made 18 of his past 23 field goals this season.

a. What is the expected value for Randall's bet?

b. Suppose that the kicker attempts 4 field goals during the game. How much total should Randall expect to win?

Solution:

a. There are two possible outcomes for this bet: Randall wins $30 ($x = 30$) or Randall loses $40 ($x = -40$). If the kicker has made 18 of his past 23 field goals, then we assume that the probability that he will make a field goal—and that Randall will win the bet—is $\frac{18}{23}$. By the complement rule, the probability that the kicker will miss—and Randall will lose the bet—is $1-\frac{18}{23}=\frac{5}{23}$. Then we calculate the expected value as follows:

$$\begin{aligned} E(X) &= (30)\left(\frac{18}{23}\right)-(40)\left(\frac{5}{23}\right) \\ &= \frac{540}{23}-\frac{200}{23} \\ &= \frac{340}{23} \\ &\approx 14.78 \end{aligned}$$

We see that the expected value of the wager is $14.78. Randall should expect that if the same bet were made many times, he would win an average of $14.78 per bet.

b. We know that over the long-term Randall would win an average of $14.78 per bet. So $14.78 \cdot 4 = 59.12$. If he and Blake place 4 bets, then Randall can expect to win approximately $59.12.

Exercises

5.1

Directions: Create the probability distribution for each scenario.

1. The number of tails when flipping 4 coins.

2. The number of even numbers showing when a pair of dice are rolled.

3. The difference of the two numbers when a pair of dice are rolled. (Largest value – smallest value)

4. The number of heads in 5 tosses of a coin.

Directions: Find the expected value of each of the following discrete probability distributions.

5.

x	$P(X=x)$
15	0.6
22	0.4

6.

x	$P(X=x)$
–55	0.45
30	0.55

7.

x	$P(X=x)$
14	0.3
21.5	0.4
–2	0.3

8.

x	$P(X=x)$
–$1.50	0.3
0	0.5
$2.75	0.1
$5.00	0.1

Directions: Determine the expected values for each scenario.

9. Suppose that you and a friend are playing cards and you decide to make a friendly wager. The bet is that you will draw two cards without replacement from a standard deck. If both cards are hearts, you win $25. Otherwise, you have to pay your friend $5.
   a. What is the expected value of the bet?
   b. If this same bet is made 100 times, how much would you expect to win or lose?

10. Taranique loves to play the floating duck game at the carnival. For \$2.00 per try, you get to choose one duck out of the 50 swimming in the water. If Taranique is lucky, she will pick one of the 8 winning ducks and go home with a pink teddy bear.

   a. What is the expected value of the game, if the value of the prize is \$5.00?

   b. If Taranique plays the game 10 times, how much can she expect to win or lose?

11. Scott likes to day-trade on the Internet. On a good day, he averages a \$2200 gain. On a bad day, he averages a \$1600 loss. Suppose that he has good days 25% of the time, bad days 35% of the time, and the rest of the time he breaks even.

   a. What is the expected value of Scott's day-trading hobby?

   b. If Scott day-trades every day for three weeks, how much money should he expect to win or lose?

   c. What is the variance of Scott's hobby?

12. Mike's older brother, Jeff, bets him that he can't roll two dice and get doubles three times in a row. If Mike does it, Jeff will give him \$100. Otherwise, Mike has to give Jeff \$5.

   a. What is the expected value of Mike's bet?

   b. What is the expected value of Jeff's bet?

   c. If Mike and Jeff make the same bet 30 times, how much can Mike expect to win or lose?

13. An insurance company offers Mississippi adults between the ages of 25 and 34 a \$100,000 life insurance policy for \$18 a month. They use the fact that Mississippi has a yearly death rate of 172.8 per 100,000 residents 25-34 years of age.

   a. Find the expected value for the insurance company at the end of one year per customer.

   b. If the insurance company has 10,000 customers in Mississippi, what is their revenue at the end of the year?

14. Suppose the same insurance company as in the previous question insures adults ages 25 to 34 in California for the same amount of money per month, but offers a \$175,000 policy for that amount of money. The reason for the differences in the payout is that the death rate in California for that age group is 81.6 per 100,000 residents.

   a. Find the expected value for the insurance company at the end of one year for California per customer.

   b. If the insurance company has 10,000 customers in California, what is their revenue at the end of the year?

   c. Which state is more profitable for the insurance company (as compared to Mississippi in the previous problem)?

15. A church in town is raffling off $50. You can buy one ticket for $1.00, three tickets for $2.50 or five tickets for $4.00. Assume that the church sells 100 tickets.

  a. Find the expected value for each ticket option.

  b. Should you buy one, three, or five tickets in order to maximize your expected winnings?

16. A department store is running a promotion one Saturday by giving out coupons for $10 of free merchandise. Based on data collected in the past, only one-fourth of customers who shop on that Saturday use the coupon, but do not purchase any other merchandise. However, one-third of customers spend $40, and one-third spend $75. The remainder of customers who come in the store do not take advantage of the promotion at all.

  a. Find the expected value for the department store on their promotion.

  b. If the store has 712 customers on the promotional Saturday, what is their net expected profit at the end of the day?

17. A car dealership is offering an interesting incentive in order to get people to come and test drive their new sports cars. Everyone who agrees to a test drive gets to choose a key. There are 75 car keys in the bag, and 4 of them unlock a sports car. For the customers who choose a winning key, the dealership agrees to knock $1000 off of the price if they buy a new car. (Assume that each key is returned after being drawn.)

  a. From the perspective of the car dealership, what is the expected value of the incentive?

  b. If 90 customers come in and choose a key, how much can the dealership expect to give up in sales?

Binomial Distribution

5.2

The **binomial distribution** is a special discrete probability function for problems with a fixed number of trials, where each trial has only two possible outcomes and one of these outcomes is counted. For example, a common binomial experiment is that of flipping a coin. The coin is flipped a fixed number of times, and the only possible outcomes are heads and tails. We can then choose to count either the number of heads or the number of tails obtained for an experiment. The outcome that is counted is called a **success**.

Let's say that you want to determine the probability of getting less than two heads when three coins are tossed. We want to count the number of heads obtained in three coin tosses, so getting a head will be considered the success. We can calculate this probability by using classical techniques from section 4.1. First, let's find the sample space.

HHH HHT HTH THH

TTT TTH THT HTT

If we want the probability of getting fewer than two heads, then we need to count the outcomes that have either exactly one head or no heads. There are four outcomes that meet these criteria, so the probability of getting less than two heads is $\frac{4}{8}=\frac{1}{2}$.

That's easy enough. But what if we want to find the probability of getting exactly six heads in ten trials? Or fifteen heads in twenty trials? The sample spaces for these problems are much too large to list by hand. We must find a better way to calculate these types of probabilities.

The binomial distribution can be used to calculate more complicated binomial probabilities. First, let's see when it would be appropriate to use the binomial distribution. All of the following guidelines must be met for an experiment to be classified as a binomial distribution.

Binomial Distribution Guidelines

1. The experiment consists of a fixed number of identical trials, n.
2. Each trial is independent of the others.
3. For each trial, there are only two possible outcomes. For counting purposes, one outcome is labeled a success, the other a failure.
4. For every trial, the probability of getting a success is called p. The probability of getting a failure is then $1-p$.
5. The binomial random variable, X, is the number of successes in n trials.

When all of these rules are satisfied, the following formula can be used to determine the probability of obtaining x successes out of n trials.

Probability for the Binomial Distribution

$$P(X=x) = {}_nC_x \cdot p^x (1-p)^{(n-x)}$$

where x = the number of successes,
n = the number of trials, and
p = the probability of getting a success on any trial

Example 5

What is the probability of getting exactly 6 heads in 10 coin tosses?

Solution:

We showed earlier that coin tosses meet the criteria of the binomial distribution. For this problem, there are 10 coin tosses, so $n = 10$. We will say that a success is getting a head. We want 6 successes, so $x = 6$. The probability of flipping a head is 0.5, which means that $p = 0.5$. Substituting these values into the binomial formula gives us

$$\begin{aligned} P(X=6) &= {}_{10}C_6 (0.5)^6 (1-0.5)^{(10-6)} \\ &= \frac{10!}{4!6!}(0.5)^6 (0.5)^4 \\ &= (210)(0.015625)(0.0625) \\ &\approx 0.205. \end{aligned}$$

Though the binomial formula is not an overly complex formula to use, many graphing calculators can calculate a binomial probability directly. In this text, we will use a TI-84 Plus calculator to calculate the necessary probabilities. The following examples continue to demonstrate the use of the binomial formula as well as describe the two different types of binomial calculations that can be performed on a TI-84 Plus. For specific directions on how to use your TI-84 Plus, or other available technology such as Microsoft Excel and Minitab, see the technology section at the end of this chapter.

Example 6

A quality control expert at a large factory estimates that 10% of all batteries produced are defective. If he takes a sample of 15 batteries, what is the probability that exactly 2 are defective?

Solution:

First, we know that we are testing 15 batteries, so $n = 15$. Second, let's consider a defective battery to be a success, so the probability of getting an individual

cont'd on the next page

success is $p = 0.1$. We are looking for the probability that **two** are defective, so we want $P(X = 2)$.

Using the binomial formula, our solution would require us to calculate the following expression:

$$\begin{aligned} P(X=2) &= {}_{15}C_2 (0.1)^2 (0.9)^{13} \\ &\approx 0.267. \end{aligned}$$

However, our TI-84 Plus calculator can calculate $P(X = x)$ directly. On your calculator, press 2ND, then VARS to access the Distributions menu. Choose option 0:binompdf(. The format for entering the statistics is binompdf(n, p, x). Thus using a TI-84 Plus, we would enter

$$\begin{aligned} P(X=2) &= \text{binompdf}(15, 0.1, 2) \\ &\approx 0.267. \end{aligned}$$

Therefore, the probability that exactly two of the batteries are defective is 0.267.

Example 7

A quality control expert at a large factory estimates that 10% of all the batteries produced at the factory are defective. If he takes a sample of 15 batteries, what is the probability that no more than 2 are defective?

Solution:

This scenario is the same as the previous example; therefore, we know that $n = 15$ and $p = 0.1$. This time we want the probability that **no more than two** are defective, which is $P(X \le 2)$. Thus we are looking for the probability that $x = 0$, or $x = 1$, or $x = 2$. We can find $P(X \le 2)$ by adding these three individual probabilities.

$$P(X \le 2) = P(X = 0) + P(X = 1) + P(X = 2)$$

Using the binomial formula, our solution would require us to calculate the following expression

$$\begin{aligned} P(X \le 2) &= P(X=0) + P(X=1) + P(X=2) \\ &= {}_{15}C_0 (0.1)^0 (0.9)^{15} + {}_{15}C_1 (0.1)^1 (0.9)^{14} + {}_{15}C_2 (0.1)^2 (0.9)^{13} \\ &\approx 0.816. \end{aligned}$$

A TI-84 Plus calculator can calculate $P(X = x)$ directly as seen in the previous example. Thus using a TI-84 Plus, we would enter

$$\begin{aligned}P(X \leq 2) &= P(X=0)+P(X=1)+P(X=2)\\ &= \text{binompdf}(15,0.1,0)+\text{binompdf}(15,0.1,1)+\text{binompdf}(15,0.1,2)\\ &\approx 0.816.\end{aligned}$$

However, our TI-84 Plus calculator allows us to use an even more efficient method for this particular problem, as it will also directly calculate a cumulative probability, $P(X \leq x)$. Press 2ND, then VARS to access the Distributions menu again. This time choose option A:binomcdf(. The format for entering the statistics is the same as described above. So, using this more efficient method, we would enter

$$\begin{aligned}P(X \leq 2) &= \text{binomcdf}(15,0.1,2)\\ &\approx 0.816.\end{aligned}$$

Therefore, the probability that no more than two of the batteries are defective is 0.816.

Example 8

What is the probability that a family with 6 children has more than 2 girls? Assume that the gender of one child is independent of the gender of any of the other children.

Solution:

We are choosing from six children, so $n = 6$. Let's define a success as having a girl. Thus the probability of obtaining a success is $p = 0.5$. We are considering the event of having **more than two** girls, $X > 2$. This is the complement to the event of having no more than two girls, $X \leq 2$. Thus

$$P(X > 2) = 1 - P(X \leq 2).$$

Using the binomial formula for this problem would be cumbersome, so let's use a TI-84 Plus. We would enter the statistics on the calculator as follows:

$$\begin{aligned}P(X > 2) &= 1 - P(X \leq 2)\\ &= 1 - \text{binomcdf}(6,0.5,2)\\ &\approx 0.656.\end{aligned}$$

Therefore, the probability that the family with 6 children has more than 2 girls is 0.656.

Example 9

Suppose that 20% of the programs at a local sporting event contain a special discount coupon. If all 8 friends in your group bought programs, what is the probability that at least half of your friends received the discount coupon?

Solution:

We are looking in 8 programs; thus $n = 8$. If we define a success to be obtaining a discount coupon, then the probability of obtaining a success would be $p = 0.2$. We are interested in the probability that **at least half of the 8** friends gets a discount coupon, so at least 4 out of the 8, or $P(X \geq 4)$. As in the previous example, we will have to use the complement to solve the problem. So,

$$P(X \geq 4) = 1 - P(X < 4).$$

This is still not exactly what we need because the TI-84 Plus calculator can only calculate cumulative probabilities of the form $P(X \leq x)$. Fortunately, this situation is not too difficult to deal with due to one of the characteristics of the binomial distribution. The value for x must be a whole number; therefore, $P(X < 4) = P(X \leq 3)$. Using all of this information and a TI-84 Plus calculator, we calculate the probability as follows

$$\begin{aligned} P(X \geq 4) &= 1 - P(X < 4) \\ &= 1 - P(X \leq 3) \\ &= 1 - \text{binomcdf}(8, 0.2, 3) \\ &\approx 0.056. \end{aligned}$$

Therefore, the probability that at least half of the eight friends find a discount coupon in their program is 0.056, which is not very likely.

While the binomial formula and different technologies are nice tools for finding binomial probabilities, binomial tables can be used as well. One drawback to these tables is that there are a limited number of probability values available. Thus you must still use the formula or appropriate technology if the necessary probability is not in the table. We will look briefly at two binomial tables, the standard binomial table and the cumulative binomial table. The standard binomial table gives the probability for a specific number of successes. For example, if we want to know the probability of getting exactly 6 heads in 10 coin tosses (as in Example 5), we could use the standard binomial table, found in Appendix A.

In contrast, the cumulative binomial table gives the cumulative probability up to and including a certain number of successes. You can think of the cumulative binomial table as $\leq$. For example, suppose we want to know the probability of getting less than or equal to 6 heads in 10 tosses. For this problem, we need to use the cumulative binomial table, because it will give the cumulative probability up to and including 6 successes. In other words, it gives $P(X \leq 6)$.

Both types of binomial tables are read the same way, so after you have decided which table is best for the problem, you need to know how to read the table. There is a table for each value of n, so the first step is to choose the table corresponding to the value of n for your problem. Next, find the probability p across the top of the table and the number of successes x down the left-hand side. The cell where these two columns intersect is the probability of obtaining x successes in n trials.

Let's walk through one complete example of using a binomial table. Take, for instance, the probability of rolling a die ten times and obtaining odd digits on 8 of the rolls. We see for this problem that there are 10 trials ($n = 10$) and 8 successes ($x = 8$), since we consider rolling an odd digit to be a success. Three of the six numbers on the die are odd, so the probability of rolling an odd digit is $\frac{3}{6} = \frac{1}{2} = 0.5$. Thus $p = 0.5$. We have the three values we need, so we are now ready to find the probability in the table. Which table should we use? Because we are given a specific value for x, rather than a range of values, we need to use the standard binomial table. On the table, we choose $n = 10$, and then look for the cell where the $p = 0.5$ column and $x = 8$ row meet. We then see that the probability of getting 8 odd numbers in 10 rolls is 0.044.

n	X	0.3	0.4	0.5	0.6
10	0	0.028	0.006	0.001	0.000
10	1	0.121	0.040	0.010	0.002
10	2	0.233	0.121	0.044	0.011
10	3	0.267	0.215	0.117	0.042
10	4	0.200	0.251	0.205	0.111
10	5	0.103	0.201	0.246	0.201
10	6	0.037	0.111	0.205	0.251
10	7	0.009	0.042	0.117	0.215
10	8	0.001	0.011	0.044	0.121
10	9	0.000	0.002	0.010	0.040
10	10	0.000	0.000	0.001	0.006

Exercises

5.2

Directions: Determine whether or not the given procedure results in a binomial distribution. If not, identify at least one requirement that is not satisfied.

1. A survey of college students rating the food in the campus dining hall on a scale from 1-10.

2. Recording the number of times an interception is made during the playoff game if there are 35 passes thrown.

3. Spinning a roulette wheel 9 times and recording the score.

4. Spinning a roulette wheel 9 times and recording the number of times you land on black.

5. Surveying 124 people and recording their "no" responses to the question: "Do you think that Internet sites should be federally regulated?'

Directions: Assume the random variable X has a binomial distribution with the given probability of obtaining a success. Find the following probabilities, given the sample size.

6. $P(X = 3)$, $n = 5$, $p = 0.40$

7. $P(X = 4)$, $n = 10$, $p = 0.30$

8. $P(X \leq 8)$, $n = 12$, $p = 0.10$

9. $P(X \leq 2)$, $n = 3$, $p = 0.90$

10. $P(X < 7)$, $n = 18$, $p = 0.40$

11. $P(X < 9)$, $n = 17$, $p = 0.50$

12. $P(X > 3)$, $n = 4$, $p = 0.80$

13. $P(X > 5)$, $n = 10$, $p = 0.70$

14. $P(X \geq 6)$, $n = 7$, $p = 0.20$

15. $P(X \geq 8)$, $n = 15$, $p = 0.60$

Directions: Find the probabilities in each of the following scenarios.

16. Suppose that the probability of David making a free throw in the championship basketball game is 60%, and each throw is independent of his last throw. Assume that David attempts seven free throws during the game.

a. What is the probability that he will make more than four of his free throws?

b. What is the probability that he will make all of his free throws?

17. At one university, freshmen account for 30% of the student body.

a. If a group of 12 students is randomly chosen by the school newspaper to comment on textbook prices, what is the probability that less than 3 of the students are freshmen?

b. If a group of 10 students is randomly chosen by the school newspaper to comment on textbook prices, what is the probability that more than 3 of the students are freshmen?

18. The SugarBear Candy Factory makes two types of chocolate candy bars – milk chocolate and milk chocolate with almonds. In a typical day, 60% of the candy bars being made are milk chocolate, and the rest are milk chocolate with almonds. At the end of the day, if a quality control expert randomly chooses 14 chocolate bars, what is the probability that:

a. half of them contain almonds?

b. at least half of them contain almonds?

19. Suppose that the probability of your favorite baseball player getting a hit at each at bat is 0.30. Assume that each at bat is independent of any other at bat.

a. What is the probability that he bats six times and gets less than two hits?

b. What is the probability that he bats nine times and gets at most four hits?

20. Suppose Carlos is taking a multiple-choice test where there are 5 possible answers on each question. If he randomly guesses on 4 questions, what's the probability that he gets

a. exactly 3 of those questions correct?

b. at least 1 out of the 4 questions correct?

c. none of the 4 questions correct?

21. In a pediatrician's office, the probability of a "no show" for a check-up appointment on any given day is 1 out of 10. Find the probability that out of 18 appointments,

a. less than 4 don't show.

b. at least 2 don't show.

c. the doctor sees every patient scheduled.

22. The probability of a plant surviving in Kerry's garden is 0.8. If she plants 19 new plants this year, what's the probability that
    a. at least 5 of them survive?
    b. more than 3/4 of them survive?
    c. she has a bad year and none of them survive?

23. In a national park in Alaska, there are 100 polar bears. As part of the monitoring of the park, the first year the rangers caught 20 bears, tagged them and released them back into the park. A year later they caught 13 bears, what's the probability that
    a. 6 are tagged?
    b. at least 10 are tagged?
    c. none are tagged?

24. In a standard deck of 52 cards, 12 are face cards (king, queen, jack). Assume that five cards are selected with replacement out of a well-shuffled deck.
    a. What is the probability of getting exactly three face cards?
    b. What is the probability of getting at least one face card?

25. Underneath the cap of cola bottles is a chance to win a free cola. If the probability of winning is 1 out of 11, what's the probability that if you buy 16 colas, you
    a. win at least one time?
    b. don't win at all?
    c. win with half of the bottles?

26. Ronnie owns a fireworks stand and knows that in the fireworks business, 1 out of every 13 fireworks are duds. If Juanita buys 10 firecrakers at Ronnie's stand, what's the probability that
    a. no more than 3 are duds?
    b. she has a perfect display with no duds?
    c. more than half are duds?

Poisson Distribution

5.3

The **Poisson distribution** is a peculiar, though interesting, distribution. The Poisson distribution is similar to the binomial in that for the Poisson distribution we are again looking to count the number of "successes" obtained. The major difference between the Poisson distribution and the binomial distribution is that for the Poisson distribution, there is not a set number of trials in which the successes must occur. Instead, Poisson distribution problems are generally looking for the successes to occur within some window of time or measurement. For example, a Poisson distribution could be used for problems involving such scenarios as the number of defects on a length of copper wiring or the number of calls into one company's computer tech support line one evening. Thus a random variable distributed according to the Poisson distribution may take on infinitely many values, but these values are determined by counting. Therefore, the Poisson distribution is a discrete probability distribution.

Poisson Distribution Guidelines

1. The successes must occur one at a time.
2. Each success must be independent of any other successes.

Math Symbols

x = number of successes

λ = average number of successes each period

If these conditions are met, then the following Poisson distribution formula can be used to find the probability of x number of successes occurring during the given period.

Probability for the Poisson Distribution

$$P(X=x)=\frac{e^{-\lambda}\lambda^{x}}{x!}$$

where $e \approx 2.71828$ and
λ = the average number of successes for each period

Memory Booster!

Make sure that your calculation for λ is in the same unit of measurement as the question.

Notice that there is only one parameter that must be determined for a Poisson distribution, and it is λ (pronounced "lambda"). The parameter λ gives the average number of successes for the given period. One caution when you are working Poisson problems is that you must make sure that your calculation for λ is in the same unit of measurement as the question. For example, suppose you are told that the local barber finishes 1 haircut every 15 minutes. If you are asked to find the probability that he will finish 6 haircuts in one hour, you must calculate the average per hour, not per minute. In this case, you must convert haircuts per minute to haircuts per hour as follows:

$$\frac{1\text{ haircut}}{15\text{ minutes}}\cdot\frac{60\text{ minutes}}{1\text{ hour}}=\frac{4\text{ haircuts}}{1\text{ hour}}.$$

Thus $\lambda = 4$.

Rounding Rule

Using the Poisson Distribution, we will round answers to four decimal places.

Example 10

Calculate the probability that our barber is feeling extra-speedy one day and finishes 6 haircuts in one hour. Recall that we mentioned he usually finishes 1 haircut every 15 minutes.

Solution:

Each haircut is considered a success, and we are looking for $x = 6$. We already determined that the barber averages 4 haircuts every hour and that $\lambda = 4$. We have the two values needed for the Poisson formula, so we can substitute as follows:

$$\begin{aligned} P(X=6) &= \frac{e^{-4} \cdot 4^6}{6!} \\ &\approx \frac{(0.0183156)(4096)}{720} \\ &\approx 0.1042. \end{aligned}$$

We see that the probability that he finishes 6 haircuts in one hour is 10.42%.

Just as we learned in the previous section, using a TI-84 Plus calculator can simplify the process of calculating probabilities. The technology section at the end of this chapter describes how to use various technologies to calculate Poisson probabilities, but in these next few examples, we will use a TI-84 Plus calculator to perform our calculations.

Example 11

A popular accounting office takes in on average 3 new tax returns a day during tax season. What is the probability that on a given day they will take in just one new tax return? Assume that the number of tax returns follows a Poisson distribution.

Solution:

We will consider obtaining a new tax return to be a success. Because we are looking for the probability that exactly 1 new return will come in, we are looking for one success, $x = 1$. The business averages 3 new tax returns each day, so $\lambda = 3$. Thus we need to find $P(X = 1)$. Using the Poisson formula, we have

$$\begin{aligned} P(X=1) &= \frac{e^{-3} \cdot 3^1}{1!} \\ &\approx 0.1494. \end{aligned}$$

Using a TI-84 Plus, begin by pressing 2ND, then VARS to access the Distributions menu. Choose option B:poissonpdf(. The format for entering the statistics is poissonpdf(λ, x). Thus

$$\begin{aligned} P(X=1) &= \text{poissonpdf}(3,1) \\ &\approx 0.1494. \end{aligned}$$

Thus the probability of the business getting just one new tax return is 0.1494.

Example 12

Suppose that the dial-up Internet connection at your house goes out an average of 0.9 times every hour. If you plan to be connected to the Internet for 3 hours one afternoon, what is the probability that you will stay connected the entire time? Assume that the dial-up disconnections follow a Poisson distribution.

Solution:

We will need to consider a disconnection to be a success for the purpose of this problem. We are looking for the probability that no successes occur over the course of 3 hours; thus $x = 0$. What is λ? The dial-up connection averages 0.9 disconnects each hour, so for 3 hours we multiply to get $\lambda = (0.9)(3) = 2.7$. We can substitute the values for x and λ into the Poisson formula as follows:

$$\begin{aligned} P(X=0) &= \frac{e^{-2.7}(2.7)^0}{0!} \\ &\approx 0.0672. \end{aligned}$$

Using a TI-84 Plus calculator, we would enter the values as shown below.

$$\begin{aligned} P(X=0) &= \text{poissonpdf}(2.7,0) \\ &\approx 0.0672 \end{aligned}$$

Thus the probability of staying connected for all 3 hours is 0.0672 or 6.72%.

Example 13

A fast food restaurant averages 1 incorrect order every 3 hours. What is the probability that they will get no more than 3 orders wrong on any given day between 5 and 11 PM? Assume that fast food errors follow a Poisson distribution.

Solution:

We will consider a wrong order a success for this example. We are looking for the probability of getting no more than 3 successes, which we can write as $P(X \le 3)$. To find λ, we need to calculate the average number of errors that occur in a 6-hour period. This is twice that of a 3-hour period, so $\lambda = (2)(1) = 2$. Using the Poisson formula, we must find the probability that x equals 0, 1, 2, and 3 and add these probabilities together. Thus

$$\begin{aligned} P(X \le 3) &= P(X=0) + P(X=1) + P(X=2) + P(X=3) \\ &= \frac{e^{-2}(2)^0}{0!} + \frac{e^{-2}(2)^1}{1!} + \frac{e^{-2}(2)^2}{2!} + \frac{e^{-2}(2)^3}{3!} \\ &\approx 0.8571. \end{aligned}$$

Using our TI-84 Plus, we would enter

$$\begin{aligned} P(X \le 3) &= P(X=0) + P(X=1) + P(X=2) + P(X=3) \\ &= \text{poissonpdf}(2,0) + \text{poissonpdf}(2,1) \\ &\quad + \text{poissonpdf}(2,2) + \text{poissonpdf}(2,3) \\ &\approx 0.8571. \end{aligned}$$

However, just as the TI-84 Plus can calculate a cumulative binomial probability, it can calculate a cumulative Poisson probability as well. Press 2ND, then VARS to access the Distribution menu. Choose option C:poissoncdf(. The format for entering the statistics for the cumulative Poisson distribution is the same as before. Using the cumulative Poisson distribution on the TI-84 Plus simplifies our calculation as follows

$$\begin{aligned} P(X \le 3) &= \text{poissoncdf}(2,3) \\ &\approx 0.8571. \end{aligned}$$

Thus the fast food restaurant has an 85.71% chance of getting no more than 3 orders wrong.

Example 14

A math professor averages grading 20 exams an hour. What is the probability that she grades more than 35 of her 60 statistics exams during her uninterrupted hour and a half between classes?

Solution:

Let's define a success to be grading an exam. We want to find the probability that more than 35 successes occur. This probability can be written as $P(X > 35)$. We will have to use the complement rule to find the probability that we need, $P(X > 35) = 1 - P(X \le 35)$. Next, we need to find λ. We know that the average for one hour is 20, so the average for 1.5 hours is $\lambda = (1.5)(20) = 30$. Using the Poisson formula to calculate each of the necessary probabilities would be very time-consuming, so let's just use a TI-84 Plus calculator as shown below.

$$\begin{aligned} P(X > 35) &= 1 - P(X \le 35) \\ &= 1 - \text{poissoncdf}(30,35) \\ &\approx 0.1574 \end{aligned}$$

Thus the math professor has a 15.74% chance of getting more than 35 statistics exams graded between classes.

Example 15

A typist averages 4 typographical errors per paragraph. If he is typing a 5 paragraph document, what is the probability that he will make less than 10 mistakes?

Solution:

Let's define a success as making a mistake. (Yes, it does sound strange, but it is the best way to solve the problem!) If the typist averages 4 mistakes per paragraph, his average for 5 paragraphs is (5)(4) = 20. Thus $\lambda = 20$. We are looking for the probability of less than 10 mistakes, $P(X < 10)$. We need to rewrite the probability as $P(X < 10) = P(X \le 9)$ to use the cumulative Poisson distribution on our calculator. So we enter

$$\begin{aligned} P(X < 10) &= P(X \le 9) \\ &= \text{poissoncdf}(20,9) \\ &\approx 0.0050. \end{aligned}$$

Thus the typist has a 0.0050 probability of making fewer than 10 mistakes.

As with the binomial distribution, a Poisson distribution table can be used for certain values of λ. Otherwise, the Poisson distribution formula or technology must be used. To use the table, you must simply know the values of λ and x. The probability you are looking for will be located in the cell where the λ column and x row meet. The Poisson distribution table can be found in Appendix A.

For example, suppose that a length of copper wiring averages one defect every 200 feet. What is the probability that one 300-foot stretch will have no defects? To solve this problem, we must first determine λ and x. If there is a defect on average every 200 feet, then we can expect 1.5 defects for a 300-foot stretch; thus $\lambda = 1.5$. Because we are looking for the probability of seeing no defects, $x = 0$. Using the Poisson table, we find that the probability that corresponds with these values of $\lambda = 1.5$ and $x = 0$ is 0.2231. We can say that there is a 22.31% chance of finding no defect in that section of wire.

	λ				
x	1.50	1.60	1.70	1.80	1.90
0	0.2231	0.2019	0.1827	0.1653	0.1496
1	0.3347	0.3230	0.3106	0.2975	0.2842
2	0.2510	0.2584	0.2640	0.2678	0.2700
3	0.1255	0.1378	0.1496	0.1607	0.1710

Exercises 5.3

Directions: For the following scenarios, determine the value of λ.

1. A 5th grader averages 3 grammatical errors per paragraph. What is his average for 4 paragraphs?

2. A surveillance officer reports 2 incidents of shoplifting per month, on average. How many incidents of shoplifting on average are reported per year?

3. A baggage handler at an airport moves an average of 500 bags on an eight-hour shift. How many bags does she average per hour?

4. An assembly line, on average, produces 1 defective part for every 100 parts that roll off the line. What is the average number of defects for a group of 20 parts?

5. You average 70 heartbeats per minute. What is the average number of heartbeats you have in 10 seconds?

Directions: Use the Poisson table to determine the following probabilities. Assume that the random variable X follows a Poisson distribution with the given value of λ.

6. $P(X = 2),\ \lambda = 1.8$

7. $P(X = 0),\ \lambda = 2.6$

8. $P(X = 13),\ \lambda = 6.5$

9. $P(X = 8),\ \lambda = 9.3$

10. $P(X \leq 2),\ \lambda = 3.8$

11. $P(X \leq 5),\ \lambda = 7.1$

12. $P(X < 4),\ \lambda = 8.3$

13. $P(X < 8),\ \lambda = 2.9$

14. $P(X \geq 3),\ \lambda = 4.9$

15. $P(X \geq 12),\ \lambda = 9.6$

Directions: Use the Poisson distribution to find the following probabilities.

16. The Oxford Gift Shop averages 4 sales each hour. Betty is scheduled to work the cash register from 1:00-3:00 on Saturday afternoon.
- **a.** What is the probability that Betty rings up exactly 10 customers?
- **b.** What is the probability that Betty rings up more than 10 customers?

17. The Pancake House is so popular that they boast to sell a stack of pancakes every 2 minutes.
- **a.** What is the probability that there is a 10-minute interval in which no pancakes are sold?
- **b.** What is the probability that there is a 5-minute interval in which less than 3 stacks of pancakes are sold?

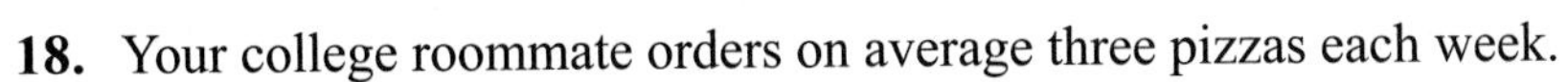

18. Your college roommate orders on average three pizzas each week.
- **a.** In any given week, what is the probability that your roommate orders at least four pizzas?
- **b.** In any given week, what is the probability that your roommate orders more than four pizzas?

19. Rob is a busy physician in the emergency room. He sees an average of 4 major trauma patients each night.

a. What is the probability that less than three major trauma patients will be admitted on any given night?

b. What is the probability that no more than five major trauma patients will be admitted on any given night?

20. Suppose a teller in a bank drive-through services customers at a rate of 12 cars every hour.

a. What is the probability that the teller will service less than 5 customers in 30 minutes?

b. What is the probability that the teller will service 3 customers in 15 minutes?

21. On average, Patrick sees a spider in his home once a month.

a. What is the probability that Patrick sees 2 spiders in a month and a half?

b. What is the probability that Patrick sees no more than one spider in a month and a half?

22. An artist cannot sell a finished piece of pottery if she discovers the clay has a defect in it. Suppose that she has to discard 2 pieces for every 56 pieces she makes.

a. What is the probability that in 14 pieces of pottery, just one piece is defective?

b. What is the probability that in 28 pieces of pottery, at least one piece is defective?

23. Suppose Kenny reads an average of one book a week.

a. What is the probability that he reads 2 books in two days?

b. What is the probability that he reads at most 3 books in 4 weeks?

24. A landscape architect knows that in the cable that he lays for landscape lighting he can expect one defect in 300 yards of cable.

a. What is the probability that in 100 yards of cable, he would find 2 defects?

b. What is the probability that in 200 yards of cable, he would find less than 3 defects?

Hypergeometric Distribution

5.4

As with the Poisson distribution, the hypergeometric distribution is also similar to the binomial distribution. The **hypergeometric distribution** is characterized by a given number of trials, n, and a specified number of countable successes, x, that occur within those n trials, in the same manner as the binomial distribution. However, the hypergeometric distribution is distinguished because it deals with *dependent* trials rather than independent trials.

Recall that two events, or trials, are dependent if one affects the other. A common example of dependent events is that of choosing cards without replacement. For example, suppose that a king is choosen out of a standard deck of cards and not replaced. The probability that a king will be choosen again on the second draw is affected by the first card choice. In fact, the chances of getting a king a second time are lowered because there is one less king in the stack. Choosing cards without replacement is just one example of dependent events. Any other experiment in which there is no replacement or repetition of objects will be also be *dependent*.

What does it take for an experiment to qualify as a hypergeometric distribution? The following properties are true for all hypergeometric problems.

Properties of a Hypergeometric Distribution

1. Each trial consists of selecting one of the N items in the population and results in either a *success* or a *failure*.
2. The experiment consists of n trials.
3. The total number of possible successes in the entire population is k.
4. The trials are dependent. (i.e., selections are made without replacement)

If a given scenario displays all of the properties above, then the following formula can be used to determine the probability of x successes occurring in n trials.

Probability for the Hypergeometric Distribution

$$P(X=x)=\frac{\left({}_kC_x\right)\left({}_{N-k}C_{n-x}\right)}{\left({}_NC_n\right)},$$

where N = the number of items in the entire population,
n = the number of trials performed,
k = the number of successes in the entire population, and
x = the number of successes obtained in n trials.

For example, let's suppose that you bring a cooler full of soft drinks to a football tailgate party. You have randomly packed 12 cans of regular sodas and 6 cans of diet soda in the ice chest. If you grab 3 drinks for your friends without looking, what is the probability that exactly two of the drinks will be regular sodas?

Notice that this is a hypergeometric problem because you are selecting from a finite population without replacement. That is, the drink you pull out first affects the probability of which drink you will pull out second. We now need to determine all of the necessary variables. There are 18 drinks in the ice chest, so the entire population is $N = 18$. Just 3 drinks are actually taken out of the ice chest, making the number of trials $n = 3$. For this example, we will consider a regular soda to be a success. Because there are 12 regular sodas in the ice chest, there are 12 successes in the entire population, and $k = 12$. We are looking for the probability of getting exactly 2 regular sodas, so $x = 2$. Now we simply substitute these values in the hypergeometric formula as follows:

$$\begin{aligned} P(X=2) &= \frac{\left({}_{12}C_2\right)\left({}_{18-12}C_{3-2}\right)}{\left({}_{18}C_3\right)} \\ &= \frac{\left({}_{12}C_2\right)\left({}_{6}C_1\right)}{\left({}_{18}C_3\right)} \\ &= \frac{(66)(6)}{816} \\ &= \frac{396}{816} \\ &\approx 0.4853. \end{aligned}$$

We see that there is a 48.53% chance that you will pull exactly two regular sodas out of the ice chest.

Example 16

At the local grocery store there are 20 boxes of cereal on one shelf, half of which contain a prize. Suppose that you buy 3 boxes of cereal. What is the probability that all 3 boxes contain a prize?

Solution:

There are 20 boxes total, so the entire population contains $N = 20$ boxes. Three boxes are purchased and each is considered a trial, so $n = 3$. A box with a prize is considered a success, and we are looking for the probability that all 3 trials are successes, so $x = 3$. Because half of the boxes have prizes, there are 10 possible successes in the entire population, and $k = 10$. Substituting into the formula, we have

$$P(X=3) = \frac{\left({}_{10}C_3\right)\left({}_{20-10}C_{3-3}\right)}{\left({}_{20}C_3\right)}$$

$$= \frac{({}_{10}C_3)({}_{10}C_0)}{({}_{20}C_3)}$$
$$= \frac{(120)(1)}{1140}$$
$$\approx 0.1053.$$

You have a 10.53% chance of getting a prize in all 3 boxes of cereal.

Example 17

There are 7 yellow and 9 green marbles in a bag. If 5 marbles are chosen at random, what is the probability that exactly 3 of the marbles chosen will be yellow?

Solution:

There are 16 total marbles in the bag, so the entire population contains $N = 16$ marbles. Because 5 marbles are to be taken out and their color noted, the number of trials is $n = 5$. We will consider a success to be obtaining a yellow marble. Out of the entire population, 7 marbles are yellow, so the total number of successes in the entire population is $k = 7$. We want to calculate the probability of getting exactly 3 yellow marbles, so $x = 3$. We then have

$$P(X = 3) = \frac{({}_{7}C_3)({}_{16-7}C_{5-3})}{({}_{16}C_5)}$$
$$= \frac{({}_{7}C_3)({}_{9}C_2)}{({}_{16}C_5)}$$
$$= \frac{(35)(36)}{4368}$$
$$= \frac{1260}{4368}$$
$$\approx 0.2885.$$

The probability of getting exactly 3 yellow marbles is 28.85%.

Example 18

A shipment of 25 light bulbs contains 3 defective bulbs. If 5 bulbs are selected randomly, what is the probability that less than 2 of the bulbs chosen are defective?

Solution:

There are 25 light bulbs in the entire population, so $N = 25$. Five bulbs are to be selected and checked for defects, making $n = 5$. A success for this problem is a defective light bulb, and since there are 3 defects in the whole shipment, $k = 3$. Because we are looking for the probability that less than 2 successes occur, $x < 2$. This example is different from the others because we need to calculate the probability that less than 2 lightbulbs chosen are defective. To do this, we must remember that successes have to be whole numbers. This means that $P(X < 2) = P(X = 0) + P(X = 1)$. We use the hypergeometric formula to obtain the following:

$$\begin{aligned}P(X<2) &= P(X=0)+P(X=1)\\ &= \frac{({}_{3}C_{0})({}_{25-3}C_{5-0})}{({}_{25}C_{5})}+\frac{({}_{3}C_{1})({}_{25-3}C_{5-1})}{({}_{25}C_{5})}\\ &= \frac{({}_{3}C_{0})({}_{22}C_{5})}{({}_{25}C_{5})}+\frac{({}_{3}C_{1})({}_{22}C_{4})}{({}_{25}C_{5})}\\ &= \frac{(1)(26334)}{53130}+\frac{(3)(7315)}{53130}\\ &= \frac{48279}{53130}\\ &\approx 0.9087.\end{aligned}$$

The probability of getting less than 2 defective light bulbs is 90.87%.

Example 19

A produce distributor is carrying 9 boxes of Granny Smith apples and 8 boxes of Golden Delicious apples. If 4 boxes are randomly delivered to one local market, what is the probability that at least 3 of the boxes delivered contain Golden Delicious apples?

Solution:

The truck is carrying a total of 17 boxes, which means that the entire population contains $N = 17$ boxes. Four of these boxes are actually delivered, so $n = 4$. We will consider a Golden Delicious box of apples to be a success, and there are $k = 8$ successes in the entire population. We want to calculate the probability that at least 3 successes occur, so $x \geq 3$. However, if at least 3 of the 4 boxes delivered are successes, then it is possible that either 3 or 4 successes occur.

We then have,

$$\begin{aligned} P(X \geq 3) &= P(X=3) + P(X=4) \\ &= \frac{({}_8C_3)({}_{17-8}C_{4-3})}{({}_{17}C_4)} + \frac{({}_8C_4)({}_{17-8}C_{4-4})}{({}_{17}C_4)} \\ &= \frac{({}_8C_3)({}_9C_1)}{({}_{17}C_4)} + \frac{({}_8C_4)({}_9C_0)}{({}_{17}C_4)} \\ &= \frac{(56)(9)+(70)(1)}{2380} \\ &= \frac{574}{2380} \\ &\approx 0.2412. \end{aligned}$$

The probability that at least 3 boxes of Golden Delicious apples are delivered to the market is 24.12%.

A summary of the three distributions discussed in this chapter is shown below.

Table 7 - Comparing the Binomial, Poisson, and Hypergeometric Distributions

	Binomial	Poisson	Hypergeometric
Discrete Distributions	✓	✓	✓
Independent Trials	✓	✓	✗
Fixed Number of Trials	✓	✗	✓
Formula	$P(X=x) = {}_nC_x \cdot p^x(1-p)^{(n-x)}$ x = the number of successes, n = the number of trials, and p = the probability of getting a success on any trial	$P(X=x) = \frac{e^{-\lambda}\lambda^x}{x!}$ $e \approx 2.71828$ and λ = the average number of successes for each period	$P(X=x) = \frac{({}_kC_x)({}_{N-k}C_{n-x})}{({}_NC_n)}$ N = the number of items in the entire population, n = the number of trials performed, k = the number of successes in the entire population, and x = the number of successes obtained in n trials.
Description	Use for problems with a fixed number of trials, where each trial is independent and only has two possible outcomes.	Use for problems with independent trials where successes occur during some window of time or measurement.	Use for problems with a fixed number of trials, but unlike Binomial distributions each trial is dependent.

Exercises

5.4

Directions: Find the following probabilities.

1. In a standard deck of 52 cards, 13 are hearts. Assume that five cards are selected without replacement out of a well-shuffled deck.
   a. What is the probability of getting exactly two hearts?
   b. What is the probability that all five cards will be hearts?

2. Suppose that one Christmas Abby and Andrew's mother forgot to label their gifts. Out of 10 wrapped presents, 5 are for Abby and 5 are for Andrew.
   a. What is the probability that exactly one of the first four presents is for Abby?
   b. What is the probability that at most three of the first five gifts opened are for Andrew?

3. Suppose that 12 of the 20 azaleas for sale at a large nursery have pink flowers and the rest have red flowers. Because it is early in the season, they have not begun to bloom and you cannot yet tell what color each plant will be.
   a. If 8 azaleas are chosen at random, what is the probability that exactly 6 will be pink?
   b. If 5 azaleas are chosen at random, what is the probability that none of the azaleas are pink?

4. Eloise loves jellybeans, and the yellow ones are her favorite. One afternoon she is snacking on a bag of 18 jellybeans, five of which are yellow. She grabs a handful of six jellybeans.
   a. What is the probability that more than half of the jellybeans are yellow?
   b. What is the probability that less than two of the jellybeans are yellow?

5. The manager of a furniture store has just received a shipment of sofas and recliners. He knows that the order contains 5 sofas and 9 recliners.
   a. What is the probability that the first three items brought into the store are recliners?
   b. What is the probability that out of the first seven items brought into the store, no more than two are sofas?

6. Suppose Audrey received a box of chocolates for Valentine's Day. Just after opening the box, she lost the paper which had the description of each chocolate on it. However, she knows that there were 6 truffles and 5 caramel candies left.
   a. What is the probability that the first 2 chocolates she eats are both truffles?
   b. What is the probability that at least 1 of the first 3 chocolates she eats is caramel?

7. Karen has 20 squares of material to use for her quilt, 8 of which are polka-dotted and the rest are floral.

a. What's the probability that she randomly uses all floral squares for the first 5 pieces of the quilt?

b. What's the probability that 2 out of the first 6 pieces she randomly chooses are polka-dotted?

8. Jay has 10 pieces of mail to open, 4 of which are junk mail.

a. What is the probability that he randomly opens 2 pieces of mail and they are both junk mail?

b. What is the probability that he randomly opens 3 pieces of mail and at least two of them are junk mail?

9. Grab bags at the town festival are filled with either a coupon for a free hamburger or a coupon for a free order of french fries. Suppose there are 22 hamburger coupons left and 18 french fry coupons left.

a. What is the probability that you and your friend each get bags with hamburger coupons in them?

b. Given that you both got hamburger coupons, what is the probability that if you each choose again, you both get french fry coupons?

10. Suppose there are 8 green tags, 12 white tags and 4 red tags left to use as name tags at a conference. If tags are given out randomly at the registration desk,

a. What is the probability that the first two tags given out are red?

b. What is the probability that if part **a.** did happen, then more than 3 out of the next 5 tags would be white?

Chapter Review

Section 5.1 - Expected Value	Pages
• **Definitions** **Random Variable** – a variable whose numeric value is determined by the outcome of a random experiment **Probability Distribution** – a table or formula that lists the probabilities for each outcome of the random variable, X **Discrete Random Variable** – a variable that may take on either finitely many values, or have infinitely many values that are determined by a counting process **Discrete Probability Distribution** – a table or formula that lists the probabilities for each outcome of the discrete random variable, x	p. 202 – 203
• **Expected Value for a Discrete Probability Distribution:** The mean of a probability distribution, $E(X)=\mu=\sum\left[x_i \cdot P(X=x_i)\right]$	p. 204 – 209
• **Variance for a Discrete Probability Distribution:** $\sigma^2=\sum\left[x^2 \cdot P(X=x)\right]-\mu^2$	p. 206
• **Standard Deviation for a Discrete Probability Distribution:** $\sqrt{\sigma^2}=\sigma=\sqrt{\sum\left[x^2 \cdot P(X=x)\right]-\mu^2}$	p. 206

Section 5.2 - Binomial Distribution	Pages
• **Binomial Distribution** – A special discrete probability function for problems with a fixed number of trials, where each trial has only two possible outcomes, and one of these outcomes is counted, $P(X=x)={}_nC_x \cdot p^x(1-p)^{(n-x)}$.	p. 212 – 217
• **Binomial Distribution Guidelines:** **1.** The experiment consists of n, a fixed number, identical trials. **2.** Each trial is independent of the others. **3.** For each trial, there are only two possible outcomes. For counting purposes, one outcome is labeled a success, the other a failure. **4.** For every trial, the probability of getting a success is called p. The probability of getting a failure is then $1-p$. **5.** The binomial random variable, X, is the number of successes in n trials.	p. 212

Section 5.3 - Poisson Distribution	Pages
• **Poisson Distribution** – A discrete probability distribution that uses a fixed interval of time or space in which the number of successes are recorded, $P(X=x)=\frac{e^{-\lambda}\lambda^{x}}{x!}$.	p. 221 – 226
• **Poisson Distribution Guidelines:** **1.** The successes must occur one at a time. **2.** Each success must be independent of any other successes.	p. 221

Section 5.4 - Hypergeometric Distribution	Pages
• **Hypergeometric Distribution** – A special discrete probability function for problems with a fixed number of dependent trials and a specified number of countable successes, $P(X=x)=\frac{\left({}_{k}C_{x}\right)\left({}_{N-k}C_{n-x}\right)}{\left({}_{N}C_{n}\right)}$.	p. 229 – 233
• **Hypergeometric Distribution Guidelines:** **1.** Each trial consists of selecting one of the N items in the population, and results in either a *success* or a *failure.* **2.** The experiment consists of n trials. **3.** The total number of possible successes in the entire population is k. **4.** The trials are dependent.	p. 229

Chapter Exercises

Directions: Find the probability for each of the following scenarios. Use the most appropriate method.

1. A survey indicates that among college females, 35% of them say shopping is their favorite pastime. Suppose you randomly select 5 females on your campus.

  a. What is the probability that 2 of them say shopping is their favorite pastime?

  b. What is the probability that at least 4 of them say shopping is not their favorite pastime?

2. On a small university campus, research indicates that the proportion of people who use an Apple computer is about 3 out of 100. Suppose you randomly choose 7 people on this campus. What's the probability that at most 2 of them are Mac users?

3. If, in a typical work day, the president of the company is interrupted twice, what is the probability that in a 5 day work week she will be interrupted more than 12 times?

4. One day while cleaning out your junk drawer at home, you put all of the loose batteries in a pile. Realizing later that 2 out of the 8 were actually new batteries, it's necessary to test each battery before throwing it away. Find the probability that you pull out the two new batteries first among the first four.

5. A gumball machine at the local pizza place is filled with plastic toys. The machine is filled with 28 rings, 35 bouncy balls, 18 rubber spiders and 41 tattoos. Suppose you and three friends want to get four toys from the machine, but would like 2 rings and 2 tattoos. Find the probability that you get what you want on the first 4 tries. (This requires some thought. Don't undercount!)

6. The male to female ratio at one large university is 5:4. If 8 students are selected at random for a survey regarding student housing, what is the probability that at least 6 of them are girls?

7. Philip is running late for work, and he needs his black socks. Unfortunately, Philip never matches his socks, so there are 10 loose white and 2 loose black socks in his drawer. If he reaches in and randomly grabs two socks, what is the probability that he gets the black socks he needs?

8. Every time Brietta goes to the mall, she has a 0.6 chance of forgetting where she parked her car. If she goes to the mall 5 times this month, what is the probability that she will forget where she has parked less than 2 times?

9. Your boss has ordered lunch for everyone at the office. There are 12 chicken salad plates and 6 tuna salad plates to choose from. If you randomly select three plates for yourself and two co-workers, what is the probability that they all contain chicken salad?

10. You forgot to study for your statistics quiz, and you don't know the answers to any of the 10 questions on the page. If each question has 4 multiple choice options (one of which is correct), what is the probability that you will guess the right answer on at least one question?

11. One southern city experiences an ice storm on average once every eight years. Calculate the probability that there is an ice storm in the city twice in the next three years.

12. Suppose a friend is helping put on a fundraiser for the local animal shelter. One activity is a game using a bowl that contains 6 green marbles and 8 blue marbles. To play the game, each person draws 2 marbles without replacement (and without looking). If both marbles are green, the player wins $25. If not, the player must donate $10 to the animal shelter. The marbles are then replaced for the next player.

a. Calculate the expected value of the game.

b. Suppose that the marble game is played 350 times. How much money would you expect to be donated to the animal shelter?

Chapter Project : Playing Roulette

We all dream of winning big, becoming an instant millionaire; but how likely is that? Let's say we decide to pursue our goal of winning big money by going to a casino and continually playing what we think will be an easy game: roulette. Can we expect to win big in the long run? Is one bet better than another? How do the casinos make so much money anyway? If we are betting against the casino, how do they make sure that they always win? This project will help you answer these questions.

Let's begin with a lesson in roulette. Roulette is a casino game that involves spinning a marble on a wheel that is marked with numbered squares that are either red, black, or green. Half of the numbers 1–36 are colored red and half are black and the numbers 0 and 00 are green. Each number occurs only once on the wheel.

We can make many different types of bets, but two of the most common are to bet on a single number or to bet on a color (either red or black). These will be the two bets we will consider in this project. After all players place their bets on the table, the wheel is spun and the marble tossed onto the wheel. Where the marble lands on the wheel determines the winner. The marble can land on only one color and number at a time.

We begin by placing a bet on a number between 1 and 36. This bet pays 36 to 1 in most casinos, which means we will be paid $36 for each $1 we bet on the winning number. If we lose, we simply lose however much money we bet.

1. Calculate the probability that we will win on a single spin of the wheel.

2. Calculate the probability that we will lose.

3. If we bet $8 on the winning number, how much money can we expect to win?

4. What is the expected value for betting on a single number if we bet $1?

5. For a $5 bet, what is the expected value for betting on a single number?

6. What is the expected value for betting on a single number, if we bet $10?

7. Do you see a pattern in the answers to the last three questions?

We decide that we can certainly increase our chances of winning if we bet on a color instead of a number. Roulette allows us to bet on either red or black and if the number is that color, we win. This bet pays even money in most casinos. This means that for each dollar we bet, we will win $1 for choosing the winning color. So, if we bet $5 and win, we would keep our $5 and win $5 more. If we lose, we lose however much money we bet, just as before.

8. What is the probability that we will win on a single spin if we bet on red?

9. What is the probability that we will lose on a single spin if we bet on red?

10. If we bet $60 on the winning color, how much money can we expect to win? Is this more or less than we can expect to win by betting $8 on our favorite number? Explain why.

11. What is the expected value for betting on red if we bet $1?

12. For a $5 bet, what is the expected value for betting on red?

13. What is the expected value for betting on red, if we bet $10?

14. Do you see a pattern in the answers to the last three questions?

15. How does the expected value of betting on a number compare to the expected value of betting on a color? Is one bet more profitable than another?

16. If our goal was to play roulette so that we can "win it big," what does the expected value of the game tell us about our chances of winning a large amount of money?

17. Are the casinos really gambling when we place a bet against them? Explain.

Chapter Technology – Calculating Probabilities

TI-84 PLUS

Binomial Distribution

To calculate the probability $P(X = x)$ for x successes out of n trials using a TI-84 Plus, use option 0: binompdf(under the DISTR menu. Press 2ND, then VARS to access the menu and then scroll down to option 0. The format for entering the statistics is binompdf(numtrials, p, x), where "numtrials" is the number of trials, n, p is the probability of a success on each trial, and x is the number of successes.

Calculating the cumulative probability, $P(X \leq x)$, using a TI-84 Plus is similar. Use option A: binomcdf(under the DISTR menu. Press 2ND, then VARS to access the menu and then scroll down to option A. The format for entering the statistics is binomcdf(numtrials, p, x), where numtrials, p, and x are defined as before.

Poisson Distribution

To use a TI-84 Plus to calculate the value of a Poisson distribution, use option B:poissonpdf (under the DISTR menu). Press 2ND, then VARS to access the menu and then scroll down to option B. The format for entering the statistics is poissonpdf(μ, x), where μ is the value λ, and x represents the number of successes.

Your TI-84 Plus can also calculate the value of a cumulative Poisson distribution, $P(X \leq x)$, which we do not calculate directly in this text. Use option C:poissoncdf(under the DISTR menu. Press 2ND, then VARS to access the menu and then scroll down to option C. The format for entering the statistics is poissoncdf(μ, x). Again μ is the value λ, and x represents the number of successes.

Example T.1

A popular accounting office takes in on average 2.78 new tax returns a day during tax season. What is the probability that on a given day they will take in 4 new tax returns? Assume that the number of tax returns follows a Poisson distribution. Use your TI-84 Plus to calculate this value.

Solution:

We need the value of the Poisson distribution for $\lambda = 2.78$ and $x = 4$. Since we want exactly 4 successes, we will use option B: poissonpdf(. Press 2ND, then VARS to access the menu and then scroll down to option B. Enter 2.78 for μ and 4 for x. Thus $P(X = 4) \approx 0.1544$.

MICROSOFT EXCEL

Binomial Distribution

The formula for calculating the probability of getting x successes out of n trials in Microsoft Excel is =BINOMDIST(number_s, trials, probability_s, cumulative) where "number_s" is the number of successes, "trials" is the number of trials, "probability_s" is the probability of choosing a success, and "cumulative" is TRUE if we want the probability of getting at most x successes and FALSE if we want the probability of getting exactly x successes.

Example T.2

What is the probability of getting exactly 6 heads in 10 tosses of a fair coin? Use Microsoft Excel to calculate your answer.

Solution:

If we define "getting a head" as a success, then we want 6 successes. There are 10 trials and the probability of getting a success on a given trial is 0.5. Since we want exactly 6 successes, we do not want the cumulative probability, so we will let cumulative be FALSE. So we would type in the following formula:

= BINOMDIST(6, 10, 0.5, FALSE)

This formula returns the value 0.205; thus the probability of getting exactly 6 heads in 10 tosses of a fair coin is 0.205.

Poisson Distribution

The formula for calculating the probability using the Poisson distribution in Microsoft Excel is =POISSON(x, mean, cumulative) where x is the number of events, "mean" is the expected numerical value, and "cumulative" is TRUE if we want the probability of getting less than or equal to x and FALSE if we want the probability of getting exactly x. You may enter a 1 in place of TRUE or a 0 in place of FALSE.

Hypergeometric Distribution

The formula for calculating the probability using the hypergeometric distribution in Microsoft Excel is =HYPGEOMDIST(sample_s, number_sample, population_s, number_pop) where "sample_s" is the number of successes in the sample, "number_sample" is the sample size, "population_s" is the number of successes in the population, and "number_pop" is in the population size.

Example T.3

At the local grocery there are 20 boxes of cereal on one shelf, half of which contain a prize. Suppose that you buy 3 boxes of cereal. What is the probability that all 3 boxes contain a prize?

Solution:

The sample size is 3. Since we want each box in the sample to be a success, the number of successes in the sample is also 3. The population size is 20. Half of the population are successes, so there are 10 successes in the population. Thus we enter the following formula: =HYPGEOMDIST(3, 3, 10, 20). Thus the probability is 0.1053.

MINITAB

Minitab will calculate probabilities using a wide variety of distributions including binomial, Poisson, hypergeometric, and discrete distributions expressed by the user. In fact, any distribution you will encounter in this text is available by selecting CALC then choosing PROBABILITY DISTRIBUTIONS. Options include Probability, Cumulative probability, and Inverse cumulative probability (introduced in Chapter 6).

Example T.4

A local pizza place offers free large pizzas on Thursday nights. The offer is good provided that the customer correctly chooses a coin toss when the pizza is delivered. Also, the offer can be used a maximum of 4 times per customer. John, Cristobal, and Rex arrive at the restaurant and realize they only have enough money to purchase one pizza. What is the probability they will be able to get at least three pizzas for free?

Solution:

First, state the problem: we want to know the probability of success – at least three successful coin flips, where each trial (coin flip) has a .50 probability of success and the maximum number of trials is 4. This is the cumulative probability of 2 successes in 4 trials. Choose CALC, then PROBABILITY DISTRIBUTIONS and *Binomial*. Enter 4 trials with a 0.5 chance of success and enter 2 as the input constant. Make sure Cumulative probability is selected and choose OK. This probability is displayed in the Session window. Since this is the probability of failure, $1 - 0.6875 = 0.3125$ is the probability of getting at least three pizzas for free. The Binomial Distribution dialog box is shown below.

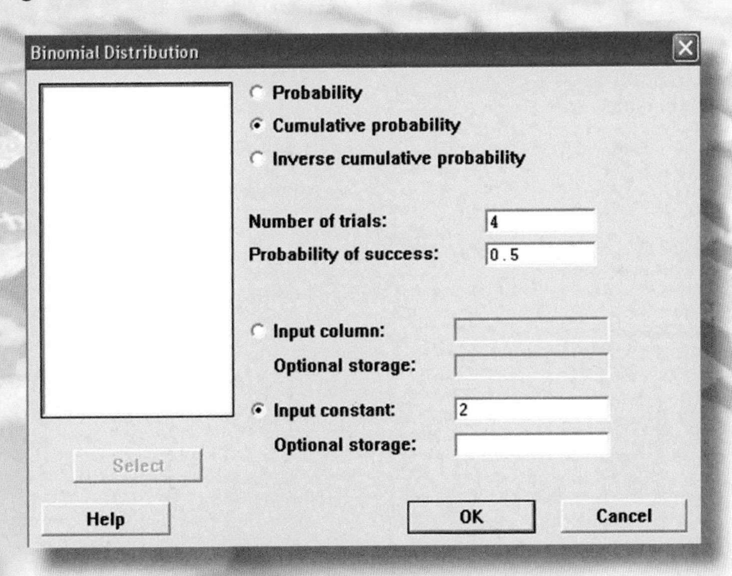

Continuous Random Variables

Chapter 6

Sections

Objectives

The objectives of this chapter are as follows:

1. Identify the properties of a normal distribution
2. Find areas under the standard normal distribution
3. Calculate probability using the normal distribution
4. Identify z-values for given areas under the standard normal curve
5. Learn how to use a Student t-distribution table

Introduction

Suppose that a prize of $500 is being offered for the biggest pumpkin entered at the state fair. You have worked hard in the garden all summer long, and you are certain that your pumpkin will win the grand prize. Upon arriving at the fair, you are shocked to see a pumpkin that appears to be as big as yours. How can that be? You wait anxiously as the judges measure both pumpkins in order to determine the winner.

Four feet and ten inches in diameter for both pumpkins, the measurements are identical! Determined to find a winner, the judges take a second look. This time they measure to the nearest half-inch, and find that both pumpkins are about 4 feet and 10 and a half inches in diameter. Because the measurements are still equal, a more precise measuring tool must be used. Now the judges use a scale of 1/8, and find that the pumpkins each measure 4 feet and 10 and 3/8 inches. The smallest scale on the judges' yardstick has been used, but still there is no contest winner.

Imagine that the judges are so determined to find the winner that they continue to use more and more precise tools to measure each of the pumpkins. In theory, how many measurements could be taken? The measuring process could theoretically go on forever, because the measurement of length is continuous in nature. That is, for any two numbers in an interval, there is always a number between them. Unfortunately, this means that the state fair judges may be in for a very long day. And if it were *my* pumpkin, I believe that I would just split the prize money and call it a day!

Introduction to the Normal Curve

6.1

In Chapter 2, we discussed different types and shapes of distributions of data. In Chapters 4 and 5, we discussed probability distributions for both discrete and continuous random variables. In this chapter, we will focus on two types of continuous probability distributions that occur frequently, the normal distribution and the Student *t*-distribution.

The most prevalent distribution is the normal distribution. A **normal distribution** is a continuous probability distribution for a given random variable, X, that is completely defined by its mean and standard deviation. A graphical representation of a normal distribution is a symmetric, bell-shaped curve centered above the mean of the distribution. Figure 1 shows an example of a normal curve.

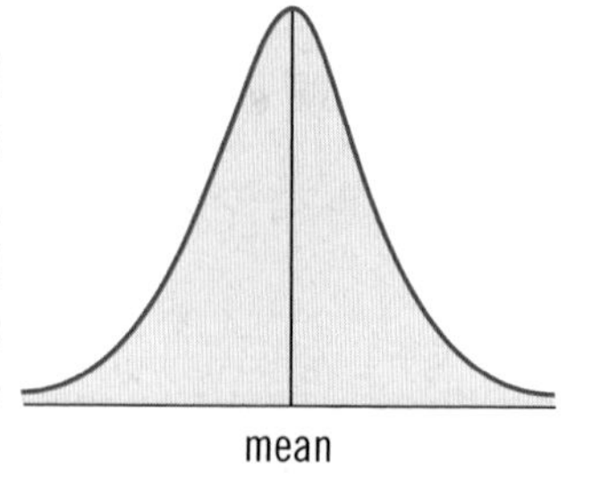

Figure 1 – Normal Curve

An example of a data set that would produce a normal distribution is the height of 500 randomly selected men, such as the distribution in Figure 2. The random variable in this example is men's heights. The heights are approximately normally distributed with a mean close to 69.2 inches. Heights of men produce a normal distribution because most men are fairly close to the same height, give or take a few inches. Very tall and very short men are rare. Some other examples of data that are normally distributed over a large randomly selected sample are shoe size, weight, and pregnancy duration.

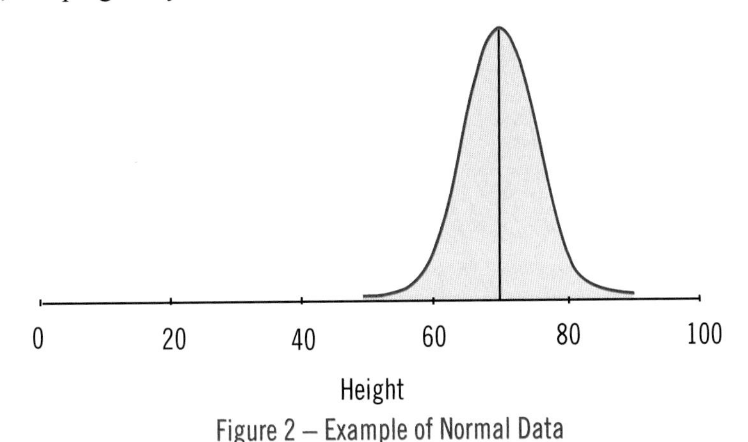

Figure 2 – Example of Normal Data

Let's consider some characteristics of a normal distribution. First, a normal distribution is symmetric and bell-shaped. The symmetry of the curve means that if you cut the curve in half, the left and right sides are mirror images. We say that the line of symmetry is $x = \mu$. The bell shape of the curve means that the majority of the data is in the middle of the distribution, and the amount of data tapers off evenly in both directions from the center.

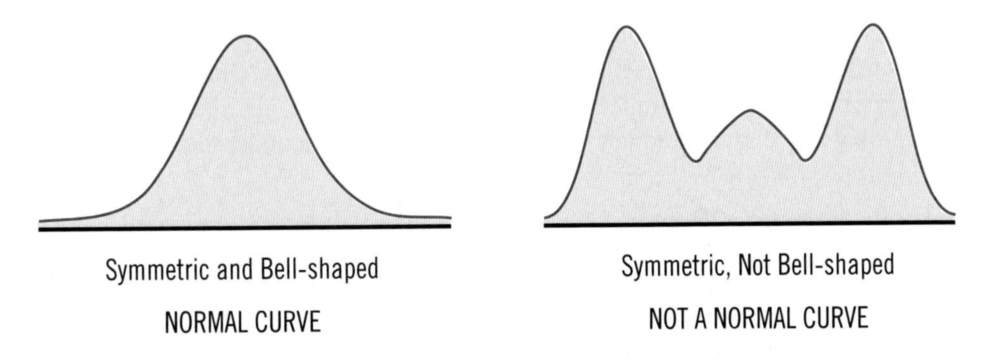

Figure 3 – General Shape of a Normal Curve

The fact that a normal distribution is symmetric and bell-shaped tells us many other facts about the distribution. First, there is only one mode, and it is located at the center of the distribution. Second, recall from Chapter 3 that a symmetric bell-shaped curve has the property that its mean equals its median and its mode. For a normal distribution, the mean, median, and mode are all the same value.

The shape of a normal distribution is determined by its standard deviation. The inflection points on either side of a normal distribution are at $\mu - \sigma$ and $\mu + \sigma$. An **inflection point** is a point on the curve where the curvature of the line changes. In a normal curve, the distance from the mean to one of the inflection points is equal to one standard deviation. The *larger* the standard deviation, the *more* area there will be in the tails of the distribution. Therefore, the curve will appear flatter. The *smaller* the standard deviation, the *less* area there will be in the tails of the distribution. Therefore, the curve will appear more narrow. Since the mean is the center, and the standard deviation determines the shape, a normal curve is completely defined by its mean and standard deviation.

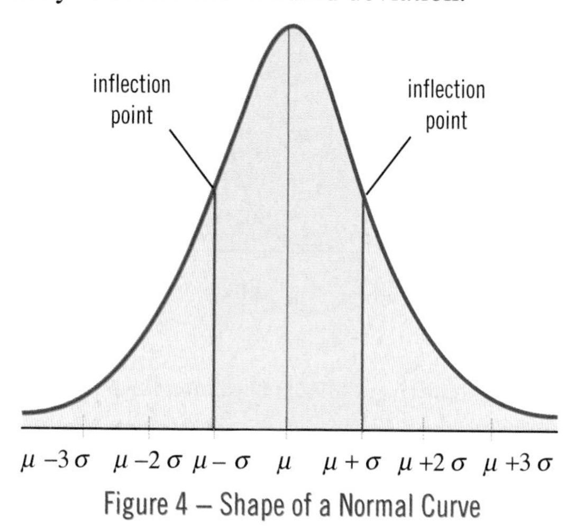

Figure 4 – Shape of a Normal Curve

Another important property of a normal distribution is that the total area under the curve of a normal distribution is equal to 1. This value is derived from the interpretation of the area under the curve. The normal distribution is a continuous probability distribution; thus the area under the curve to the left of a specific value of the random variable, x, equals the probability that a randomly chosen value will be less than x, i.e., $P(X < x)$. Therefore, the total area under the curve is equivalent to the probability of randomly choosing any value that the distribution can take. This probability certainly equals 1; therefore, the total area under the curve equals one.

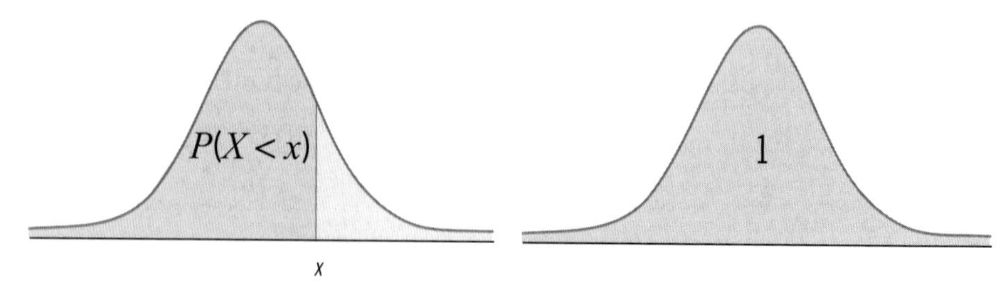

Figure 5 – Area Under the Normal Curve

In addition, the x-axis is a horizontal asymptote for the normal distribution. This fact is derived from the mathematical definition of the normal curve's probability density function, which can never equal zero. For us, this says that the normal curve will approach the x-axis on both ends, but will never touch or cross it.

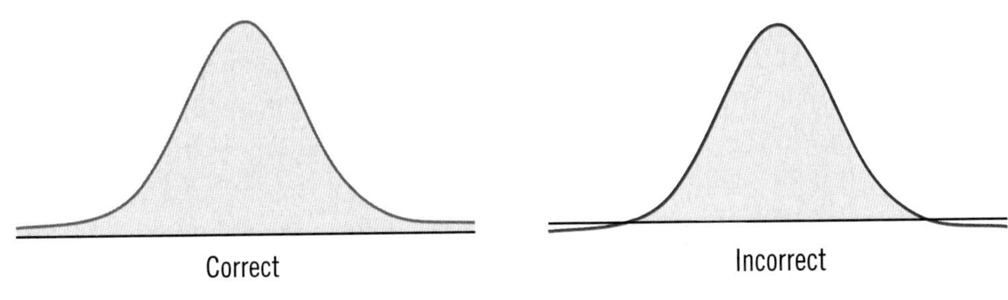

Figure 6 – Horizontal Asymptote of a Normal Curve

Properties of a Normal Distribution

1. A normal curve is symmetric and bell-shaped.
2. A normal curve is completely defined by its mean, μ, and standard deviation, σ.
3. The total area under a normal curve equals 1.
4. The x-axis is a horizontal asymptote for a normal curve.

Now that we've looked at properties of normal distributions, let's consider how to determine if a distribution could be classified as normal. Several obvious conditions must be met first. The data must be continuous, not discrete. In addition, most data values must lie close to the mean of the distribution, and data values much smaller or larger than the mean are less likely to occur (i.e., bell-shaped).

Consider the heights of the players in the men's collegiate basketball league. First of all, the data is *approximately* continuous in nature. In addition, there are both shorter people and extremely tall people in the set, but the majority of their heights cluster around the mean height. Because the data meets these conditions, it is likely to be a normal distribution.

However, consider the letter grades on a Biology final. Although there are likely to be grades at both ends of the scale, the data is not continuous. It is discrete, and in fact, not even quantitative data. Therefore, one would be less likely to call the distribution normal.

Let's consider another set of data, since the previous distribution fails to be normal for a fairly obvious reason. What if we look at the percentages of the test grades on the Putnam Exam? This exam, taken by several thousand mathematics students a year, produces results which are continuous but fail the test for normality in another way. Because of the extreme difficulty of this test, both the mean and the median of the results are 0. As we know from Chapter 3, this distribution would be skewed to the right because of all the 0's, making it unlikely to be normal.

The William Lowell Putnam Competition

The competition began in 1938 and is designed to stimulate a healthy rivalry in mathematical studies in the colleges and universities of the United States and Canada. It exists because Mr. William Lowell Putnam had a profound conviction in the value of organized team competition in regular college studies. Mr. Putnam, a member of the Harvard class of 1882, wrote an article for the December 1921 issue of the Harvard Graduates' Magazine in which he described the merits of an intellectual intercollegiate competition. To establish such a competition, his widow, Elizabeth Lowell Putnam, in 1927 created a trust fund known as the William Lowell Putnam Intercollegiate Memorial Fund. The first competition supported by this fund was in the field of English, and a few years later, a second experimental competition was held, this time in mathematics between two institutions. It was not until after Mrs. Putnam's death in 1935 that the examination assumed its present form and was placed under the administration of the Mathematical Association of America.

Source: http://math.scu.edu/putnam/historyJan.html

How many normal curves are there? Look again for a moment at the second property of normal curves, which says that a normal curve is completely defined by its mean and standard deviation. Because there are an infinite number of possibilities for μ and σ, there are an infinite number of normal curves.

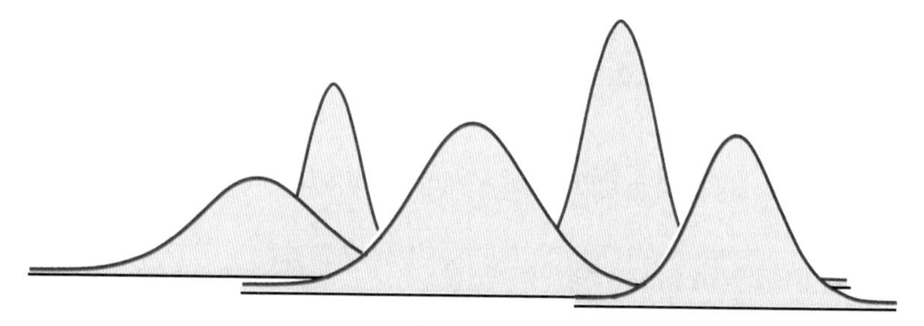

Figure 7 – Various Normal Curves

✦ The Standard Normal Distribution

As we just noted, there are infinitely many normal curves, each with a different mean and standard deviation. However, it is simpler to study a single standard curve instead of an unlimited number of normal curves. To do this, we need to standardize normal curves using standard scores. We can then reference this **standard normal curve** rather than having to compute the area under each different normal curve. The standard normal curve has all of the properties of a normal curve, and in addition, the mean is always 0 and the standard deviation is always 1.

Properties of the Standard Normal Distribution

1. The standard normal curve is symmetric and bell-shaped.
2. The standard normal curve is completely defined by its mean, $\mu = 0$, and standard deviation, $\sigma = 1$.
3. The total area under the standard normal curve equals 1.
4. The x-axis is a horizontal asymptote for the standard normal curve.

To standardize a normal curve to the standard normal curve, we convert each x-value to a standard score, z, using the formula

$$z = \frac{x - \mu}{\sigma}.$$

For the remainder of this book, we will reference a standard score in several different ways, as is the common practice among statisticians. A standard score may also be referred to as a z-score or a z-value.

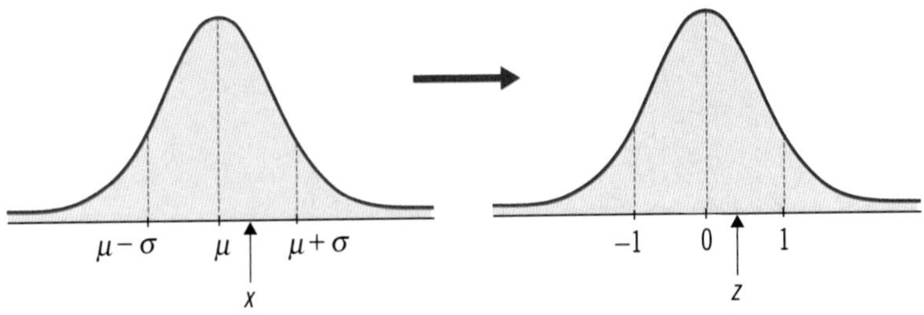

Figure 8 – Converting to the Standard Normal Curve

Let's look at how we convert data values to z-scores. Below is a normal curve with a mean of 25 and a standard deviation of 3.2, where an x-value of 27 is indicated with the arrow. If we standardize this curve, we have the standard normal curve which always has a mean of 0 and a standard deviation of 1.

To obtain the corresponding z-value for an x-value of 27, we substitute the mean and standard deviation of the normal curve into the standard scores formula as follows:

$$
\begin{aligned}
z &= \frac{x-\mu}{\sigma} \\
&= \frac{27-25}{3.2} \\
&\approx 0.63
\end{aligned}
$$

Rounding Rule

Round z-values to 2 decimal places.

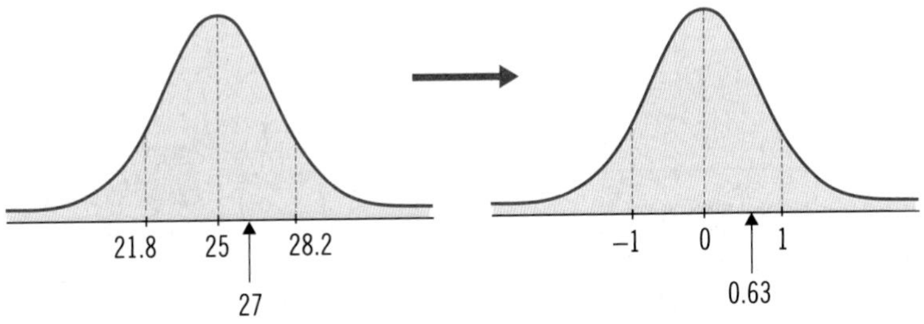

21.8 25 28.2 27 −1 0 1 0.63

Figure 9 – Converting to the Standard Normal Curve

Example 1

Given a normal curve with $\mu = 48$ and $\sigma = 5$, convert to a standard normal curve and indicate where a score of $x = 45$ would be on each curve.

Solution:

Begin by finding the z-score for $x = 45$ as shown on the following page.

cont'd on the next page

$$
\begin{aligned}
z &= \frac{x-\mu}{\sigma} \\
&= \frac{45-48}{5} \\
&= -0.6
\end{aligned}
$$

Now draw each of the distributions, marking a standard score of $z = -0.6$ on the appropriate distribution.

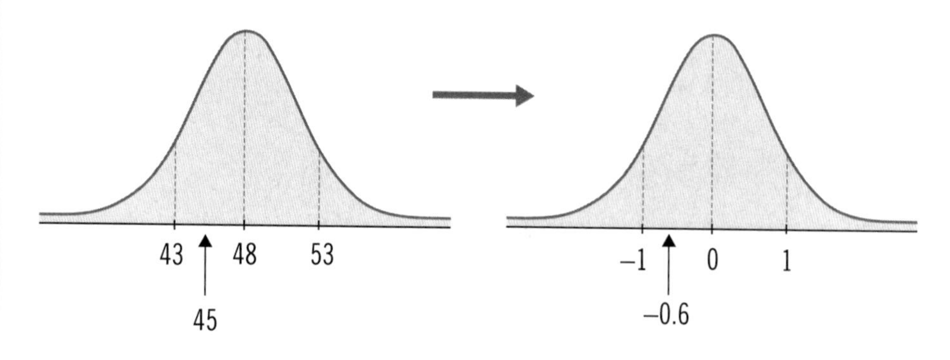

The distribution on the left is a normal curve with a mean of 48 and a standard deviation of 5. The distribution on the right is a standard normal curve with a standard score of $z = -0.6$ indicated.

Exercises

6.1

Properties of Normal Distributions
Directions: Determine if the following statements are true or false. If a statement is false, state the reason why.

1. There are a limited number of normal distributions.

2. There is only one standard normal distribution.

3. The mean of a normal distribution is always 0.

4. The mean of the standard normal distribution is always 0.

5. The standard deviation of the standard normal distribution is always 0.

6. For any normal distribution, the mean, median, and mode are equal.

7. The line of symmetry for a normal distribution is $x = \mu$.

8. The y-axis is a vertical asymptote for all normal distributions.

9. The inflection points for any normal distribution are one standard deviation on either side of the mean.

10. Normal distributions are symmetric, but do not have to be bell-shaped.

Recognizing Normal Distributions

Directions: For the following distributions, determine if the distribution is likely to be normal. If not, explain why.

11. The weight of 500 American men.

12. The age at death of 200 gorillas studied in the wild.

13. The grades for 100 students on a relatively easy exam.

14. The length of 300 babies born in a Las Vegas hospital.

15. The frequency of the outcomes on a roulette wheel.

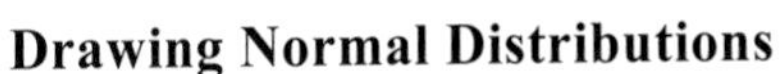

Drawing Normal Distributions

Directions: For each problem below, draw a normal curve with the given characteristics. Indicate the mean and the inflection points on the graph, as well as where the given x-value would be.

16. $\mu = 65$ and $\sigma = 20$; $x = 40$

17. $\mu = 5$ and $\sigma = 0.25$; $x = 4.8$

18. $\mu = 15$ and $\sigma = 2$; $x = 19$

19. $\mu = 0.023$ and $\sigma = 0.001$; $x = 0.02$

20. $\mu = 12{,}000$ and $\sigma = 2000$; $x = 10{,}750$

Converting to the Standard Normal Distribution

Directions: Draw the standard normal curve. For each problem below, calculate the standard score of the given x-value. Indicate where the z-value would be.

21. $\mu = 65$ and $\sigma = 20$; $x = 40$

22. $\mu = 5$ and $\sigma = 0.25$; $x = 4.8$

23. $\mu = 15$ and $\sigma = 2$; $x = 19$

24. $\mu = 0.023$ and $\sigma = 0.001$; $x = 0.02$

25. $\mu = 12{,}000$ and $\sigma = 2000$; $x = 10{,}750$

Reading a Normal Curve Table

6.2

Many of the distributions in which statisticians are interested are normal distributions. One important application of a normal distribution is that the *area* under any part of the normal curve is equal to the *probability* of the random variable falling within that region. Recall from the previous section that the area to the left of a specific value, x, of the random variable is equal to $P(X < x)$. Similarly, the area to the right of x equals $P(X > x)$. (Notice that the inequality symbol points in the direction of the area!) Furthermore, we do not talk about finding the probability that x is a specific value because of the fact that the normal curve is a continuous probability distribution. Instead, we refer to the probability that x is within a range of values. Thus choosing to include the endpoint of our range does not change the value of the probability. So, we can say that $P(X < x) = P(X \leq x)$.

To demonstrate how area and probability are related, let's look at an example. Suppose the mean score on Test 1 in your statistics class was 75 with a standard deviation of 5, and the distribution of test scores was normal. What is the probability that a randomly selected test score, such as yours, was better than 80? Look at the normal distribution of test scores shown in Figure 10 below. Notice that the mean is 75, and because the standard deviation is 5, a score of 80 is one standard deviation above the mean. The area under the normal curve above one standard deviation above the mean is equal to 0.1587 (we will show how to find this number in a minute). If the area containing all scores better than 80 equals 0.1587, then 15.87% of all of the test scores are better than 80, and thus the probability of randomly choosing a score greater than 80 is 15.87%.

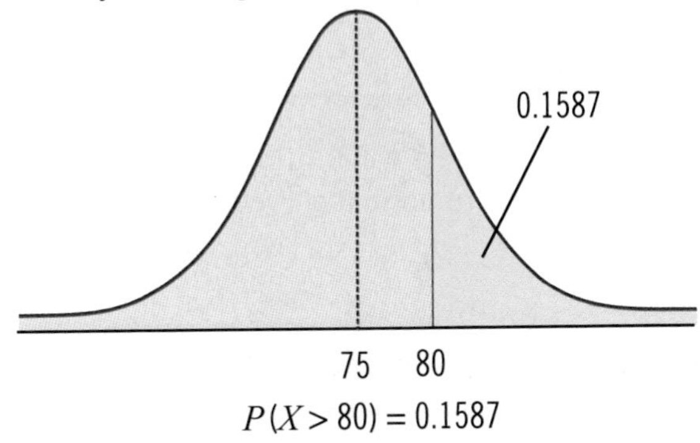

$P(X > 80) = 0.1587$

Figure 10 – Normal Distribution of Test Scores

Now that we have demonstrated the usefulness of the area under a normal curve, we need to be able to calculate the area under the normal curve in any given region. The area under a curve can be found using integration, which is a calculus technique. However, because there are an infinite number of normal curves, it would be tedious to have to calculate the area for each unique curve. Because the area under any given part of the *standard* normal curve is fixed, tables can be created so that one can to look up the area under the standard normal curve without relying on calculus.

There are many different standard normal curve tables that statisticians use, each one giving us a different method for finding the same area under the normal curve. Although you can use different tables to produce the same results, we will use the cumulative normal curve table, found in Appendix A, which gives the area under the standard normal curve to the left of the given z-value. A small excerpt of that table is shown below.

z	0.00	0.01	0.02	0.03	0.04
0.0	0.5000	0.5040	0.5080	0.5120	0.5160
0.1	0.5398	0.5438	0.5478	0.5517	0.5557
0.2	0.5793	0.5832	0.5871	0.5910	0.5948
0.3	0.6179	0.6217	0.6255	0.6293	0.6331
0.4	0.6554	0.6591	0.6628	0.6664	0.6700
0.5	0.6915	0.6950	0.6985	0.7019	0.7054
0.6	0.7257	0.7291	0.7324	0.7357	0.7389
0.7	0.7580	0.7611	0.7642	0.7673	0.7704
0.8	0.7881	0.7910	0.7939	0.7967	0.7995

Figure 11 – Excerpt from the Cumulative Normal Curve Table

Since we will be using z-values rounded to two decimal places, our cumulative normal table reflects just that. The first decimal place of the z-value is listed down the left-hand column, with the second decimal place across the top row. Where the appropriate row and column intersect, we find the amount of area under the standard normal curve to the *left* of that particular z-value. Let's now look at an example of finding the area to the left of a particular z.

Example 2

Area to the Left

Find the area under the standard normal curve to the left of $z = 1.37$.

Solution:

The first part of the z-value is 1.3. The second part is 0.07. So, look across the row labeled 1.3 and down the column labeled 0.07. This row and column intersect at 0.9147. Thus the area under the normal curve to the left of $z = 1.37$ is 0.9147.

z	0.05	0.06	0.07	0.08	0.09
1.0	0.8531	0.8554	0.8577	0.8599	0.8621
1.1	0.8749	0.8770	0.8790	0.8810	0.8830
1.2	0.8944	0.8962	0.8980	0.8997	0.9015
1.3	0.9115	0.9131	0.9147	0.9162	0.9177
1.4	0.9265	0.9279	0.9292	0.9306	0.9319
1.5	0.9394	0.9406	0.9418	0.9429	0.9441

The cumulative normal curve table we are using only gives the area to the *left* of a given z-value, but we can use the table, along with the properties of the standard normal distribution, to find other areas as well.

First, let's consider the area to the right of a given z-score. To obtain this area, remember that the total area under the standard normal curve is 1. So, if the table gives us the area to the left of z, then subtracting that area from 1 gives us the area to the *right* of z.

A shortcut to finding the area to the right of a given z is to use the symmetry of the curve. In terms of area under the curve, this means that the area to the right of z is equal to the area to the left of $-z$. To find the area to the right of z, instead of looking up the area to the left of z and subtracting that area from 1, you can simply look up the area to the left of $-z$.

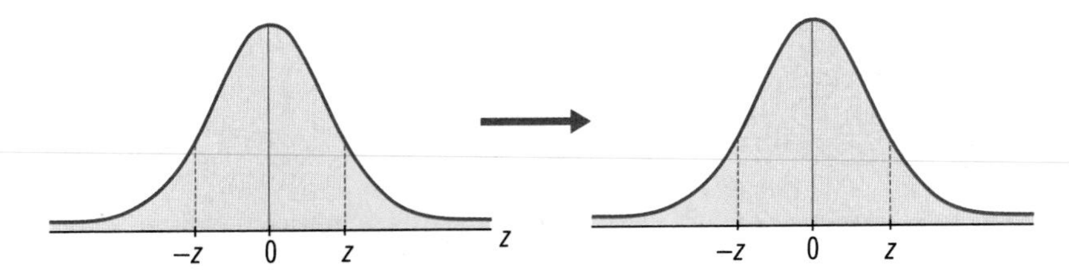

Figure 12 – Calculating Area to the Right of z

Example 3

Area to the Right

Find the area under the standard normal curve to the right of $z = 1.37$.

Solution:

- Method 1 – From the previous example, we know that the area to the left of $z = 1.37$ is 0.9147. So, the area to the right of $z = 1.37$ is $1 - 0.9147 = 0.0853$.

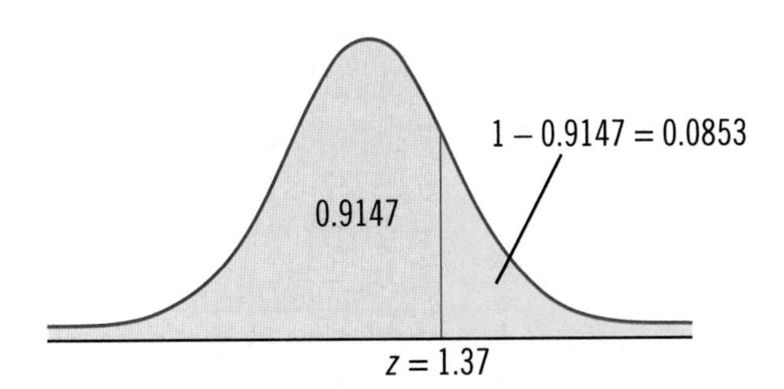

cont'd on the next page

- Method 2 – We can look up $z = -1.37$, which also gives us 0.0853.

	0.09	0.08	0.07	0.06	0.05
-1.7	0.0367	0.0375	0.0384	0.0392	0.0401
-1.6	0.0455	0.0465	0.0475	0.0485	0.0495
-1.5	0.0559	0.0571	0.0582	0.0594	0.0606
-1.4	0.0681	0.0694	0.0708	0.0721	0.0735
-1.3	0.0823	0.0838	0.0853	0.0869	0.0885
-1.2	0.0985	0.1003	0.1020	0.1038	0.1056

Let's consider the area between two z-scores. To find the area between two values of z, use the table to look up the area to the left of each z-value and then subtract the smaller area from the larger area.

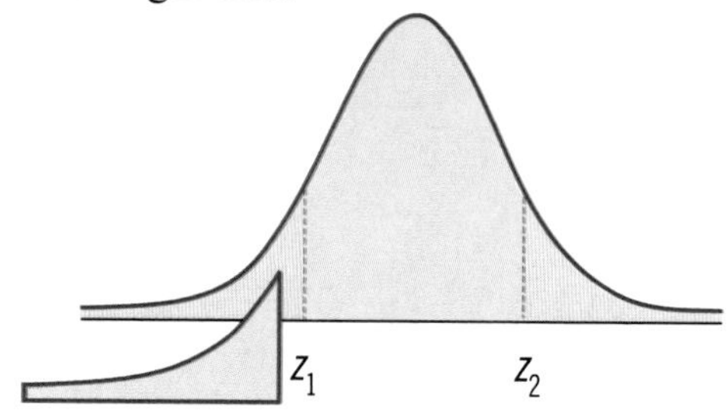

Figure 13 – Calculating Area in Between

Example 4

Area Between

Find the area under the standard normal curve between $z = -1.68$ and $z = 2$.

Solution:

First, look up the area to the left of $z = -1.68$, which gives you 0.0465. Second, look up the area to the left of $z = 2$, which gives you 0.9772.

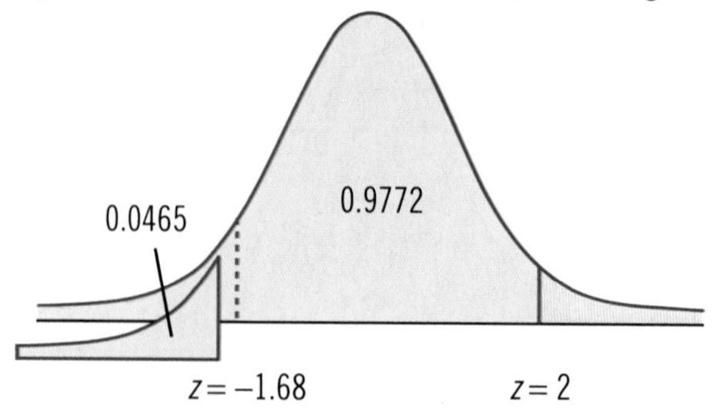

Last, subtract $0.9772 - 0.0465 = 0.9307$. Thus the area between the two z-values is 0.9307.

Example 5

Area in the Tails

Find the area under the standard normal curve to the left of $z = -2.5$ and to the right of $z = 3$.

Solution:

There are two areas that we must find. Thus the total area we are interested in is the sum of these two areas. Let's begin by finding the area to the left of $z = -2.5$. Look up $z = -2.5$ in the normal curve table. The area equals 0.0062. Next, we need to find the area to the right of $z = 3$. Using the symmetric property of the normal curve, look up $z = -3$ in the cumulative normal curve table. The area is 0.0013. Thus the sum of the two areas is $0.0062 + 0.0013 = 0.0075$.

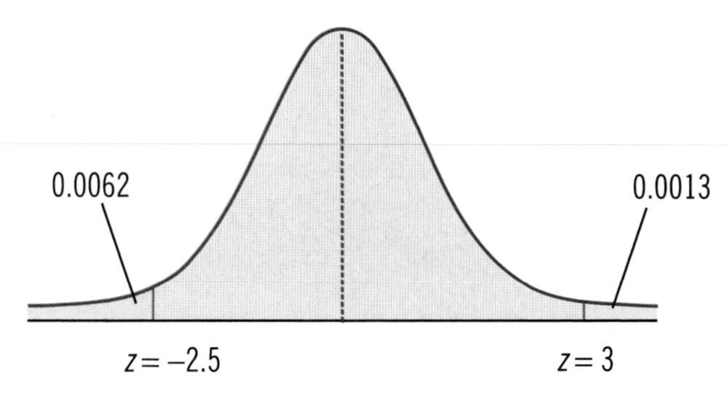

Example 6

Area in the Tails

Find the area under the standard normal curve to the left of $z = -1.23$ and to the right of $z = 1.23$.

Solution:

Notice that the absolute value of each z is the same. Thus the area in the tails of the distribution will be the same because of the symmeteric property of the normal curve. So to find the area in the two tails, we only need to look up the negative tail and multiply the answer by 2.

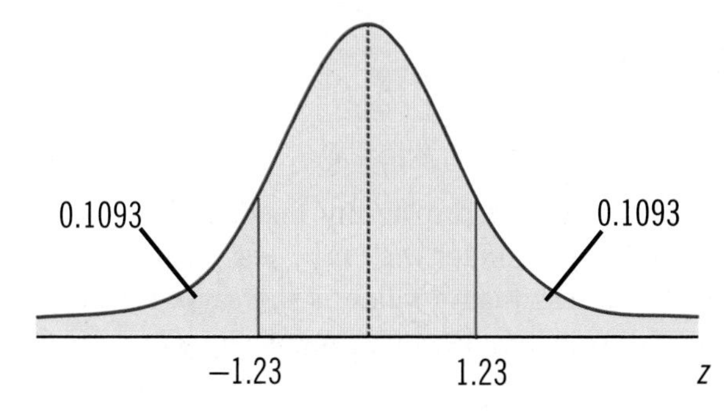

cont'd on the next page

Use the normal curve table to find the area to the left of $z = -1.23$. This gives us a value of 0.1093. Multiply this area by two in order to obtain the combined area in the tails. Thus $(0.1093)(2) = 0.2186$. So the area is 0.2186.

Memory Booster!

$P(X)$ = Area under the curve

Now that we know how to find the area under the normal curve, we know how to find the probability that a randomly chosen value is within a specific range of values. Thus finding the area to the left of $z = 1.37$ is the same as calculating $P(z < 1.37)$. Notice that the inequality sign points the same direction in which we shaded the area under the curve. Also recall that $P(z < 1.37) = P(z \le 1.37)$ because whether or not we include the endpoint does not change the value of the probability. The following table summarizes the basic rules we use for each of the four general types of area problems.

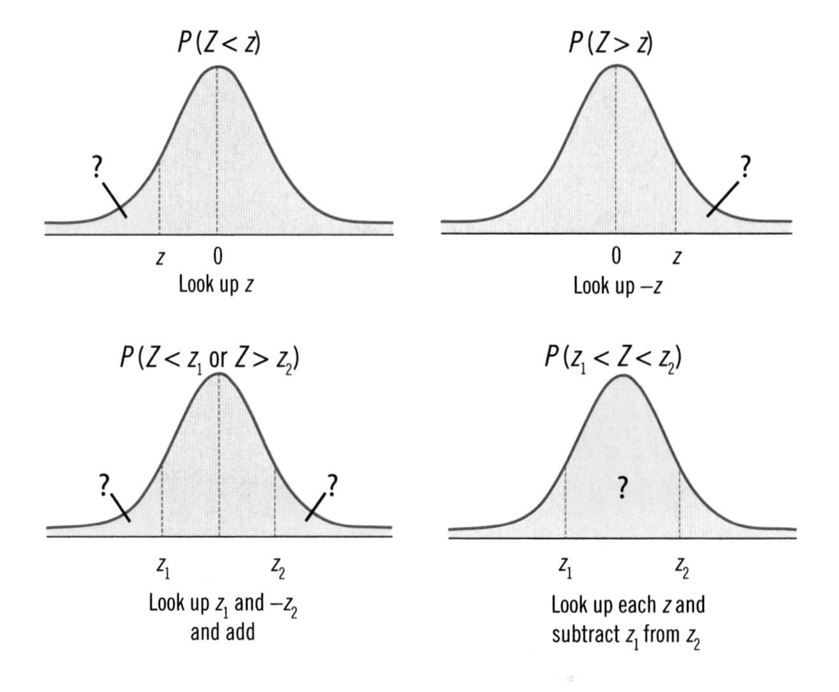

Figure 14 – Guidelines for Using the Normal Curve Table Found in Appendix A

Example 7

Intrepret $P(z \le -2.67)$.

Solution:

$P(z \le -2.67)$ stands for the probability that z is less than or equal to -2.67. This is equal to the area under the curve less than $z = -2.67$. Note that the probability that z is less than a value is the same as the probability that z is less than or equal to that value, or symbolically $P(z < -2.67) = P(z \le -2.67)$.

Example 8

Find the following probabilities.

a. $P(z < 1.45)$ b. $P(z \geq -1.37)$

c. $P(1.25 < z < 2.31)$ d. $P(z < -2.5 \text{ or } z > 2.5)$

e. $P(z < -4.01)$ f. $P(z \leq 3.98)$

Solution:

a. $P(z < 1.45)$ is the area under the curve to the left of $z = 1.45$. Look up $z = 1.45$. The area is 0.9265.

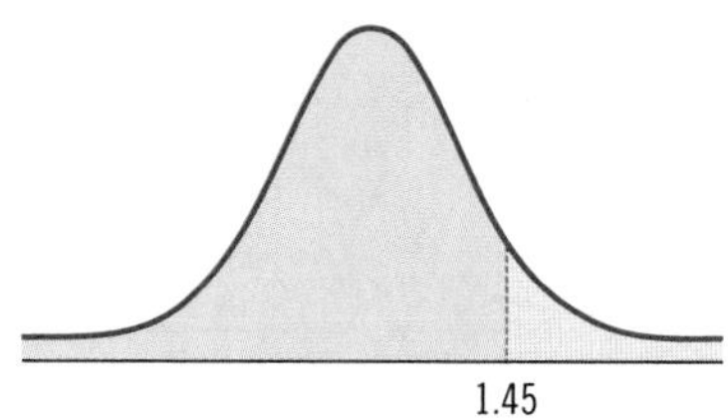

b. $P(z \geq -1.37)$ is the area under the curve to the right of $z = -1.37$. Use the symmetry property of the normal curve and look up $z = 1.37$. The area to the right of $z \geq -1.37$ is 0.9147.

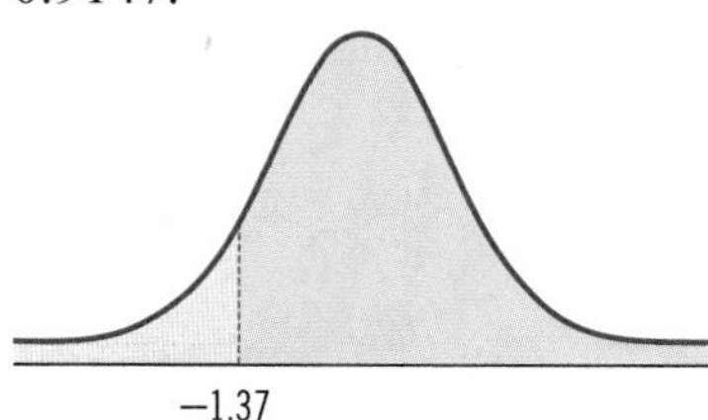

c. $P(1.25 < z < 2.31)$ is the area between $z = 1.25$ and $z = 2.31$. Look up each value in the normal curve table. The area to the left of $z = 1.25$ is 0.8944. The area to the left of $z = 2.31$ is 0.9896. The area between $z = 1.25$ and $z = 2.31$ is the difference of these two areas. Thus, the area is $0.9896 - 0.8944 = 0.0952$.

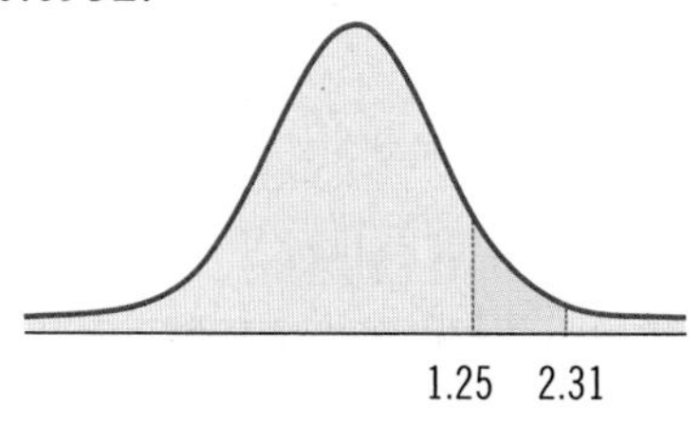

d. $P(z < -2.5 \text{ or } z > 2.5)$ is the area to the left of $z = -2.5$ or to the right of $z = 2.5$. Since the normal curve is symmetric, these two areas are the same; therefore, look up the area to the left of $z = -2.5$ in the table and multiply the area by 2. The area to the left of $z = -2.5$ is 0.0062. Thus the area we are interested in equals $(0.0062)(2) = 0.0124$.

cont'd on the next page

−2.5 2.5

e. $P(z < -4.01)$ is the area under the normal curve to the left of $z = -4.01$. Notice that $z = -4.01$ is not in normal curve table. It is smaller than the values in the table, which means that it is further to the left. Thus the area under the curve is smaller than all of the areas listed in the table. Thus $P(z < -4.01)$ is approximately 0.

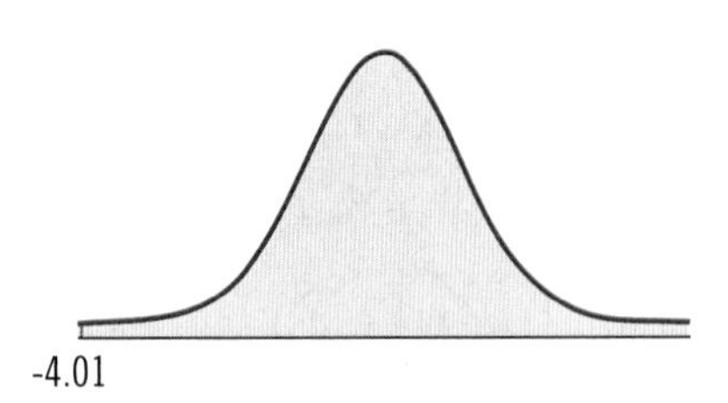

f. $P(z \leq 3.98)$ is the area under the normal curve to the left of $z = 3.98$. Notice that $z = 3.98$ is not in normal curve table. It is larger than the values in the table, which means that it is further to the right. Thus the area under the curve is larger than all of the areas listed in the table. Thus $P(z \leq 3.98)$ is approximately 1.

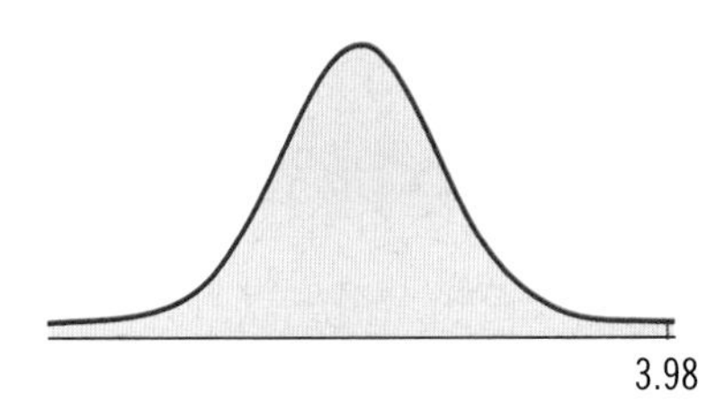

Exercises

6.2

Area Under the Standard Normal Curve

Directions: Find the area to the left of the given z-value. Note: Answers found using technology will vary slightly.

1. $z = 2.35$

2. $z = 1.78$

3. $z = -1.25$

4. $z = -0.19$

5. $z = 2$

6. $z = 1.3$

7. $z = 3.57$

8. $z = -4.12$

Directions: Find the area to the right of the given z-value. Note: Answers found using technology will vary slightly.

9. $z = 1.35$ **10.** $z = 1.7$ **11.** $z = -2.51$

12. $z = -0.39$ **13.** $z = -1$ **14.** $z = 3.68$

Directions: Find the area between the given z-values.

15. $z = 0.35$ and $z = 1.85$ **16.** $z = -1.25$ and $z = 2.16$

17. $z = -1.78$ and $z = -0.95$ **18.** $z = -0.19$ and $z = 1$

19. $z = -3.57$ and $z = 1.85$

Directions: Find the area less than $-z$ and greater than z for each value of z given.

20. $z = 1.46$ **21.** $z = 2.11$ **22.** $z = 3.05$

Directions: Find the area to the left of z_1 and the right of z_2.

23. $z_1 = -2.31, z_2 = 1.67$ **24.** $z_1 = -1.75, z_2 = 1.89$ **25.** $z_1 = -2, z_2 = 1$

26. $z = -3.81, z_2 = 2.37$ **27.** $z_1 = 1.31, z_2 = 1.93$ **28.** $z_1 = 0.35, z_2 = 1.75$

29. $z_1 = -2.31, z_2 = -1.67$ **30.** $z_1 = -3, z_2 = -2$

Finding Probabilities Using the Normal Curve Table
Directions: Find the given probabilities.

31. $P(z < -3.14)$ **32.** $P(z < 1.43)$ **33.** $P(z > 2.72)$

34. $P(z > -0.81)$ **35.** $P(-1.86 < z < 3.14)$ **36.** $P(0.78 < z < 2.64)$

37. $P(0 < z < 2.78)$ **38.** $P(-2.81 < z < -1.14)$

39. $P(z < -1.26 \text{ or } z > 1.26)$ **40.** $P(z < -2.39 \text{ or } z > 2.39)$

Finding Probability Using the Normal Curve

6.3

Recall that one of the key properties of the normal curve is that the area under any region is equivalent to the probability of the random variable falling within that region. In the previous section, we learned how to use the standard normal curve to find area under the curve. But what about the area under a non-standard normal curve? We can find the area and, therefore, the probability under any normal curve by first converting the x-values to standard scores and then using the standard normal curve.

Recall from the previous section that there are four basic types of probability problems:

1. Probability less than some value
2. Probability greater than some value
3. Probability between two values
4. The probability less than one value and greater than another value

Let's look at an example of each type of probability problem.

Example 9

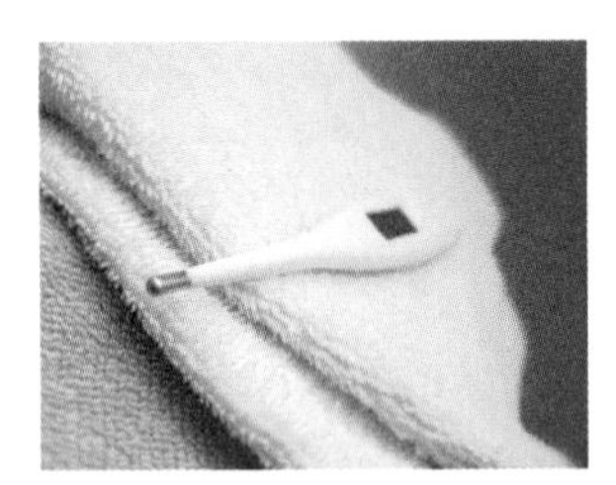

Probability Less Than

Body temperatures of adults are normally distributed with a mean of 98.6° F and a standard deviation of 0.73° F. What is the probability of having a normal body temperature **less than** 96.9° F?

Solution:

We are interested in finding the probability that a person's body temperature, X, is less than 96.9° F, written mathematically as $P(X < 96.9)$. The picture on the left illustrates this. First, we need to convert 96.9 to a standard score.

$$\begin{aligned} z &= \frac{x-\mu}{\sigma} \\ &= \frac{96.9-98.6}{0.73} \\ &\approx -2.33 \end{aligned}$$

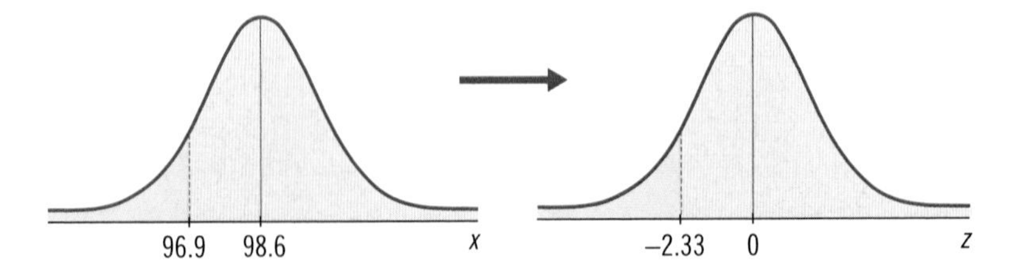

Second, we need to find the area to the left of $z = -2.33$, written mathematically as $P(z < -2.33)$. The picture on the right shows this portion of the standard normal curve. Looking up -2.33 in the table, we see that it has an area of 0.0099. Thus the probability of having a body temperature less than 96.9 degrees F is 0.0099.

$$P(X < 96.9) = P(z < -2.33) = 0.0099$$

Example 10

Probability Greater Than

Body temperatures of adults are normally distributed with a mean of 98.6° F and a standard deviation of 0.73° F. What is the probability of having a normal body temperature **greater than** 100° F?

Solution:

This time we are interested in finding the probability that a person's body temperature, X, is greater than 100° F, written mathematically as $P(X > 100)$. The picture on the left illustrates this. First, convert 100 to a standard score.

$$\begin{aligned} z &= \frac{x-\mu}{\sigma} \\ &= \frac{100-98.6}{0.73} \\ &\approx 1.92 \end{aligned}$$

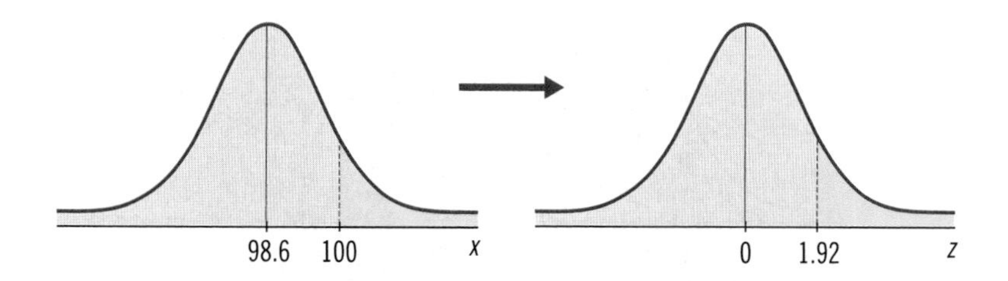

Second, we need to calculate the area in the standard normal curve to the right of $z = 1.92$, which we denote mathematically as $P(z > 1.92)$. The picture on the right shows this region of the curve. Because we need the area to the right of $z = 1.92$, look up the opposite sign of z, which is -1.92. This area is 0.0274. Hence, the probability that a person has a normal body temperature greater than 100 ° F is 0.0274.

$$P(X > 100) = P(z > 1.92) = 0.0274$$

Example 11

Probability Between

Body temperatures of adults are normally distributed with a mean of 98.6° F and a standard deviation of 0.73° F. What is the probability of having a normal body temperature **between** 98° F and 99° F?

Solution:

We are now interested in finding the probability that a person's body temperature, X, is between 98° F and 99° F, written mathematically as $P(98 < X < 99)$. The picture on the left below illustrates this region. First, convert 98 and 99 each to standard scores.

For 98° F:

$$\begin{aligned} z &= \frac{x-\mu}{\sigma} \\ &= \frac{98-98.6}{0.73} \\ &\approx -0.82 \end{aligned}$$

For 99° F:

$$\begin{aligned} z &= \frac{x-\mu}{\sigma} \\ &= \frac{99-98.6}{0.73} \\ &\approx 0.55 \end{aligned}$$

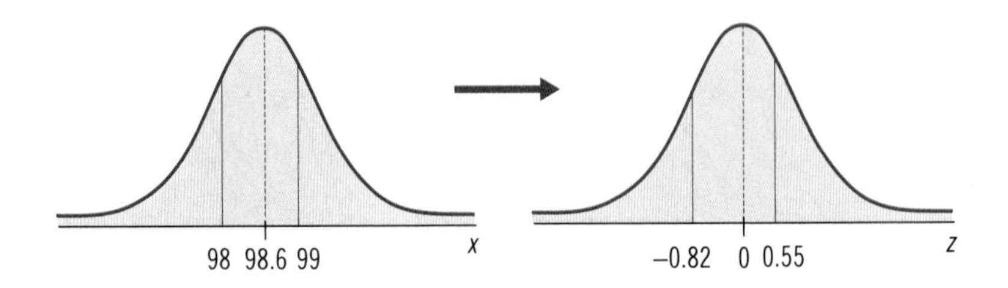

We now need to calculate the area between $z = -0.82$ and $z = 0.55$, written mathematically as $P(-0.82 < z < 0.55)$. The picture on the right illustrates this area under the standard normal curve. Finally, remember that to find area between two z-scores you must find both areas and subtract. In the table, we see that $z = -0.82$ has an area of 0.2061 to its left and $z = 0.55$ has an area of 0.7088 to its left. Subtracting the two areas gives us $0.7088 - 0.2061 = 0.5027$. Thus the probability of having a body temperature between 98° F and 99° F is 0.5027.

$$P(98 < X < 99) = P(-0.82 < z < 0.55) = 0.5027$$

Example 12

Probability in the Tails

Body temperatures of adults are normally distributed with a mean of 98.6° F and a standard deviation of 0.73° F. What is the probability of having a normal body temperature **less than** 97.6° F or **greater than** 99.6° F?

Solution:

We are interested in finding the probability that a person's body temperature, X, is less than 97.6° F or greater than 99.6° F. We can denote this mathematically as $P(X < 97.6 \text{ or } X > 99.6)$. The picture on the left below shows the area under the curve in which we are interested. First, convert each temperature to a standard score.

For 97.6° F:

$$\begin{aligned} z &= \frac{x-\mu}{\sigma} \\ &= \frac{97.6-98.6}{0.73} \\ &\approx -1.37 \end{aligned}$$

For 99.6° F:

$$\begin{aligned} z &= \frac{x-\mu}{\sigma} \\ &= \frac{99.6-98.6}{0.73} \\ &\approx 1.37 \end{aligned}$$

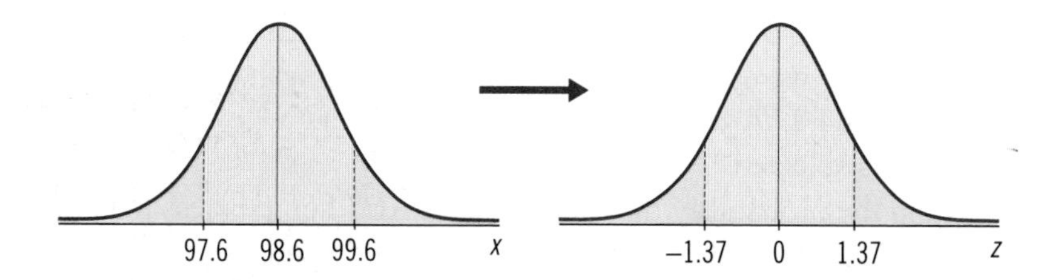

We now need to calculate the area to the left of $z = -1.37$ and to the right of $z = 1.37$. This may be denoted mathematically as $P(z < -1.37 \text{ or } z > 1.37)$. Notice that these z-scores are equal distances from the mean. Because of the symmetry of the curve, the shaded areas must then be equal. The easiest way to calculate the total area is to find the area to the left of $z = -1.37$ and simply double it. We then have $(0.0853)(2) = 0.1706$. Thus the probability of having a normal body temperature either less than 97.6° F or greater than 99.6° F is 0.1706.

$$P(X < 97.6 \text{ or } X > 99.6) = P(z < -1.37 \text{ or } z > 1.37) = 0.1706$$

Exercises

6.3

Directions: Answer each of the following questions. Remember to round your answer to two decimal places for percentages and four decimal places for probabilities. Note: Answers found using technology will vary slightly.

1. Deviation IQ Scores, sometimes called Wechsler IQ scores, have a mean of 100 and a standard deviation of 15.
   a. What percentage of the general population have deviation IQs lower than 85?
   b. What percentage of the general population have deviation IQs larger than 130?
   c. What percentage of the general population have deviation IQs between 90 and 110?
   d. What percentage of the general population have deviation IQs less than 90 or greater than 110?

2. Replacement times for CD players are normally distributed with a mean of 7.1 years and a standard deviation of 1.4 years. (Based on data from "Getting Things Fixed", Consumer Reports.)
   a. Find the probability that a randomly selected CD player will have a replacement time of less than 8.0 years.
   b. Find the probability that a randomly selected CD player will have a replacement time of more than 9.0 years.
   c. Find the probability that a randomly selected CD player will have a replacement time between 5 and 8 years.
   d. Find the probability that a randomly selected CD player will have a replacement time of less than 5.1 or greater than 9.1 years.

3. In a recent year, the ACT scores for high school students with an A or B grade point average were normally distributed, with a mean of 24.2 and a standard deviation of 4.2. A student with an A or B average who took the ACT during this time is selected.
   a. Find the probability that the student's ACT score is less than 20.
   b. Find the probability that the student's ACT score is greater than 31.
   c. Find the probability that the student's ACT score is between 25 and 32.
   d. Find the probability that the student's ACT score is less than 23.2 or greater than 25.2.

4. The heights of American women ages 18 to 24 are normally distributed with a mean of 65 inches and a standard deviation of 2.4 inches. An American woman in this age bracket is chosen at random.
   a. What is the probability that she is more than 68 inches tall?
   b. What is the probability that she is less than 70 inches tall?
   c. What is the probability that she is between 64 and 69 inches tall?
   d. What is the probability that she is less than 59 or greater than 71 inches tall?

5. According to the data released by the Chamber of Commerce of a certain city, the weekly wages of office workers are normally distributed with a mean of $700 and a standard deviation of $50. Consider a worker chosen at random from this city.

  a. What is the probability that the worker makes a weekly wage of less than $600?

  b. What is the probability that the worker makes a weekly wage of more than $810?

  c. What is the probability that the worker makes a weekly wage of between $620 and $770?

  d. What is the probability that the worker makes a weekly wage of less than $620 or greater than $780?

6. The serum cholesterol levels in milligrams/deciliter (mg/dL) in a certain Mediterranean population are found to be normally distributed with a mean of 160 and a standard deviation of 50. Scientists of the National Heart, Lung and Blood Institute consider this pattern ideal for a minimal risk of heart attacks.

  a. Find the percentage of the population who have blood cholesterol levels less than 150 mg/dL.

  b. Find the percentage of the population who have blood cholesterol levels which exceed the ideal level by at least 10 mg/dL.

  c. Find the percentage of the population who have blood cholesterol levels between 150 and 200 mg/dL.

  d. Find the percentage of the population who have blood cholesterol levels less than 100 mg/dL or greater than 220 mg/dL.

7. Monthly telephone bills in one region are normally distributed with a mean of $72 and a standard deviation of $14.

  a. What is the probability that a phone bill chosen at random is less than $75?

  b. What is the probability that a phone bill chosen at random is at least $90?

  c. What is the probability that a phone bill chosen at random is between $80 and $100?

  d. What is the probability that a phone bill chosen at random is less than $58 or greater than $86?

8. Scores on a midterm exam are normally distributed with a mean of 73 and a standard deviation of 9 points.

  a. What is the probability that a student chosen at random has a score between 80 and 90?

  b. What is the probability that a student chosen at random has a score of at most 85?

  c. What is the probability that a student chosen at random has a score greater than 95?

  d. What is the probability that a student chosen at random has a score less than 55 or greater than 91?

9. The average lifetime for a car battery is 148 weeks with a standard deviation of 8 weeks.

a. If a company guarantees its battery for 3 years, what percentage of the batteries sold would you expect to be returned before the end of the warranty period? Assume a normal distribution.

b. Imagine you were the CEO of the battery company. Evaluate the warranty offer and list any changes you would make as the CEO.

10. Ella refuses to tell you her ACT score. However, she does tell you that her score is above the mean. Which of the following z-scores is possible for her ACT score?

a. -0.08

b. 1.43

c. 0

d. Not enough information

11. The average wage of first year graduates from nursing school is $34,000. At Olivia's job interview, she found out that she would make at most the mean starting salary for first year graduates. Which of the following z-scores are possible for her wage?

a. 0

b. -1.42

c. 0.78

d. Not enough information

Finding z-Values Using the Normal Curve

6.4

In the previous section, we used the properties of the standard normal distribution to solve probability problems. In this section we will use these same properties but in reverse. Specifically, if we know a probability, we can use the table in reverse to find the corresponding z-value.

The first problem type we will look at is when you are asked to find the z-value for a given area to the *left* of z. Recall that the cumulative standard normal table gives area to the left of z. Therefore, to find z when you are given cumulative area to the left, simply scan through the body of the table until you find the area you need. The answer, then, is simply the z-value that corresponds with that area.

Helpful Hint!

Not sure about your answer? Draw a sketch to make sure your answer makes sense.

Example 13

Area to the Left

What z-value has an area of 0.7357 to its left?

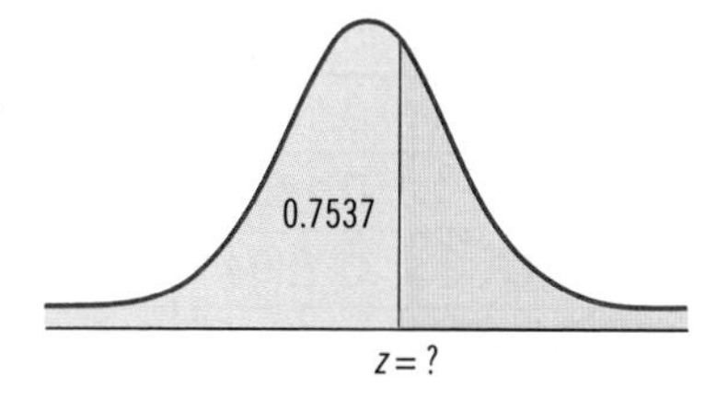

Solution:

Scan through the *interior* of the table for an area of 0.7357. Look to the left and up to find the corresponding z-value. Looking at the row and column titles, you will see that $z = 0.63$.

z	0.00	0.01	0.02	0.03	0.04	0.05
0.0	0.5000	0.5040	0.5080	0.5120	0.5160	0.5199
0.1	0.5398	0.5438	0.5478	0.5517	0.5557	0.5596
0.2	0.5793	0.5832	0.5871	0.5910	0.5948	0.5987
0.3	0.6179	0.6217	0.6255	0.6293	0.6331	0.6368
0.4	0.6554	0.6591	0.6628	0.6664	0.6700	0.6736
0.5	0.6915	0.6950	0.6985	0.7019	0.7054	0.7088
0.6	0.7257	0.7291	0.7324	0.7357	0.7389	0.7422
0.7	0.7580	0.7611	0.7642	0.7673	0.7704	0.7734

Example 14

Find the value of z such that the area to the left of z is 0.2000.

Solution:

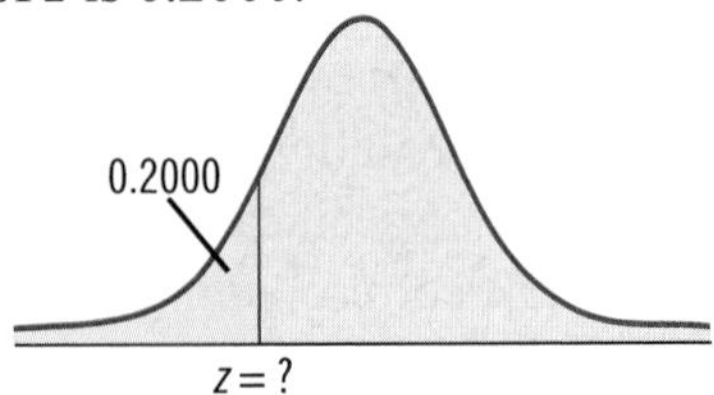

First, find 0.2000 in the body of the normal curve table. Notice that none of the areas is exactly 0.2000. Our area falls between the areas of 0.2005 and 0.1977. One method for finding the z-value for an area not in the table is interpolation; however, for our purposes, we will use the area closest to the one we want. The value of 0.2005 is the closest one to 0.2000. Its z-value is –0.84. So the z-score for an area of 0.2000 is approximately $z = -0.84$.

z	0.08	0.07	0.06	0.05	0.04	0.03
–1.1	0.1190	0.1210	0.1230	0.1251	0.1271	0.1292
–1.0	0.1401	0.1423	0.1446	0.1469	0.1492	0.1515
–0.9	0.1635	0.1660	0.1685	0.1711	0.1736	0.1762
–0.8	0.1894	0.1922	0.1949	0.1977	0.2005	0.2033
–0.7	0.2177	0.2206	0.2236	0.2266	0.2296	0.2327
–0.6	0.2483	0.2514	0.2546	0.2578	0.2611	0.2643

Next, let's consider a problem in which the area given lies to the *right* of z. The method for finding these z-scores is to first look up the given area in the table and find its corresponding z-value. However, you must then switch the sign of z to obtain the answer you need. (The reasoning is that by the symmetry of the curve, the area to the right of z is equal to the area to the left of $-z$.)

Example 15

Area to the Right

What z-value has an area of 0.0096 to its right?

Solution:

Scan through the *interior* of the table for an area of 0.0096. Look to the left and up to find the corresponding z-value. What we find is that $z = -2.34$. However, remember that the table assumes that the area is to the *left* of z. Because the area is to the right of z, and by the symmetry of the normal curve, we can simply change the sign of the z-value to obtain the correct answer. The z-value that has 0.0096 of the area to its right is $z = 2.34$.

z	0.08	0.07	0.06	0.05	0.04	0.03
–2.7	0.0027	0.0028	0.0029	0.0030	0.0031	0.0032
–2.6	0.0037	0.0038	0.0039	0.0040	0.0041	0.0043
–2.5	0.0049	0.0051	0.0052	0.0054	0.0055	0.0057
–2.4	0.0066	0.0068	0.0069	0.0071	0.0073	0.0075
–2.3	0.0087	0.0089	0.0091	0.0094	0.0096	0.0099
–2.2	0.0113	0.0116	0.0119	0.0122	0.0125	0.0129

The third problem type, in which the area given lies between $-z$ and z, requires more work than the previous two examples. First, find the area remaining in the two tails by subtracting the given area from the total area of one. Next, divide by two to find the area in just the left tail. Use this area to obtain $-z$ from the table. Finally, because the question asks for the positive value of z, simply switch the sign of z to obtain the correct answer.

Example 16

Area Between

Find the value of z such that the area between $-z$ and z is 0.90.

Solution:

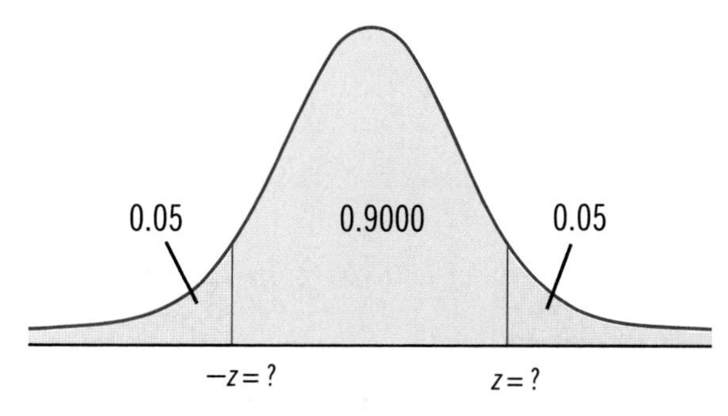

If the area between $-z$ and z is 0.90, then the area in the tails must be $1-0.90=0.10$. Because the curve is symmetric, this area is divided equally between each tail. So, half of that area is in the left-hand tail, which is $\frac{0.10}{2}=0.0500$. Because our z-table gives area to the left, we can look up this value.

z	0.08	0.07	0.06	0.05	0.04	0.03
–1.8	0.0301	0.0307	0.0314	0.0322	0.0329	0.0336
–1.7	0.0375	0.0384	0.0392	0.0401	0.0409	0.0418
–1.6	0.0465	0.0475	0.0485	0.0495	0.0505	0.0516
–1.5	0.0571	0.0582	0.0594	0.0606	0.0618	0.0630
–1.4	0.0694	0.0708	0.0721	0.0735	0.0749	0.0764

cont'd on the next page

Look in the interior of the table for the area closest to 0.0500. We find two areas that are equally close to 0.0500, so use the z-value exactly between those two z-values. The z-value halfway between $z = -1.64$ and $z = -1.65$ is -1.645. Thus the z-value such that the area between $-z$ and z is 0.90 is $z = 1.645$.

The final problem type is a problem in which the area given is contained in the tails of the curve. Because the curve is symmetric, divide the area in half to find the amount of area in the left tail. Look in the table to find the z-value that corresponds with this area. Finally, as in the previous example, the question asks for the positive value of z. Simply switch the sign of z to obtain the correct answer.

Example 17

Area in the Tails

Find the value of z such that the area to the left of $-z$ plus the area to the right of z is 0.1616.

Solution:

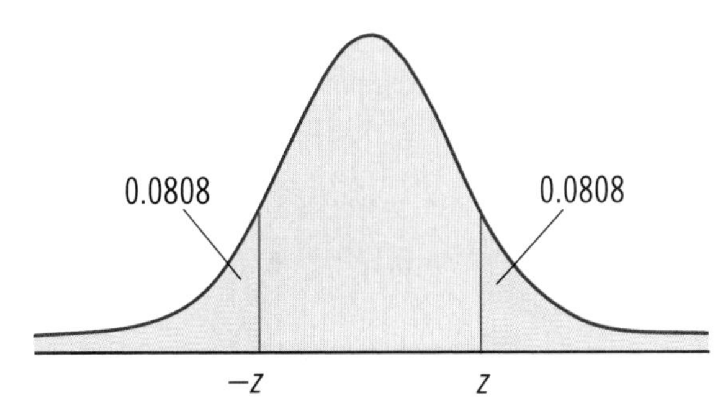

If the area in both tails together is 0.1616, then the area in one of the tails must be half of that or $\frac{0.1616}{2} = 0.0808$.

z	0.05	0.04	0.03	0.02	0.01	0.00
−2.0	0.0202	0.0207	0.0212	0.0217	0.0222	0.0228
−1.9	0.0256	0.0262	0.0268	0.0274	0.0281	0.0287
−1.8	0.0322	0.0329	0.0336	0.0344	0.0351	0.0359
−1.7	0.0401	0.0409	0.0418	0.0427	0.0436	0.0446
−1.6	0.0495	0.0505	0.0516	0.0526	0.0537	0.0548
−1.5	0.0606	0.0618	0.0630	0.0643	0.0655	0.0668
−1.4	0.0735	0.0749	0.0764	0.0778	0.0793	0.0808
−1.3	0.0885	0.0901	0.0918	0.0934	0.0951	0.0968

Scan through the interior of the table for an area of 0.0808. The corresponding z-value is $z = -1.40$. Then the z-value such that the area to the left of $-z$ plus the area to the right of z is 0.1616 is $z = 1.40$.

Example 18

What z-value represents the 90th percentile?

Solution:

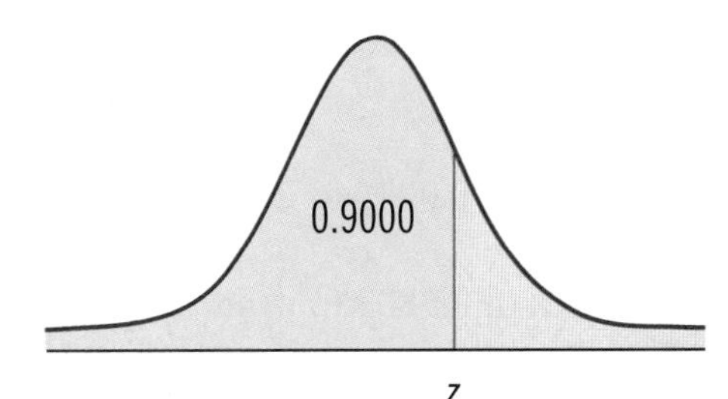

The 90th percentile is the z-value for which 90% of the area under the standard normal curve is to the left of z. So, we need to find the value of z that has 0.9000 to its left. Looking for 0.9000 (or extremely close to it) we find 0.8997, which corresponds to a z-value of 1.28. Thus $z = 1.28$ represents the 90th percentile.

Memory Booster!

By rearranging the standard score formula to solve for x, we have

$$x = z\sigma + \mu.$$

Now that we can find z for any given area, we can use it to help us find particular values of x. This is extremely helpful when we want to know data values that represent cutoff points or mark certain intervals in our data set. Consider the following example:

Example 19

The body temperatures of adults are normally distributed with a mean of 98.6° F and a standard deviation of 0.73° F. What temperature represents the 90th percentile?

Solution:

In order to determine the temperature that represents the 90th percentile, we first need to find the z-value that represents the 90th percentile. We found this value in the previous example to be $z = 1.28$. Once we have the value of z, we can substitute z, μ, and σ into the standard score formula and solve for x.

$$\begin{aligned} z &= \frac{x-\mu}{\sigma} \\ 1.28 &= \frac{x-98.6}{0.73} \\ (1.28)(0.73) &= x - 98.6 \\ 0.9344 + 98.6 &= x \\ 99.5344 &= x \end{aligned}$$

Thus a temperature of 99.53° F represents the 90th percentile.

Exercises

6.4

Finding z Using the Normal Curve Table

Directions: Find the indicated z-values. Note: Answers found using technology will vary slightly.

1. What z-value has an area of 0.0038 to its left?
2. What z-value has an area of 0.9803 to its left?
3. What z-value has an area of 0.0212 to its left?
4. What z-value has an area of 0.0838 to its right?
5. What z-value has an area of 0.9706 to its right?
6. What z-value has an area of 0.5987 to its right?
7. Find the value of z such that the area between $-z$ and z is 0.99.
8. Find the value of z such that the area between $-z$ and z is 0.80.
9. Find the value of z such that the area between $-z$ and z is 0.95.
10. Find the value of z such that the area to the left of $-z$ plus the area to the right of z is 0.5686.
11. Find the value of z such that the area to the left of $-z$ plus the area to the right of z is 0.0286.
12. Find the value of z such that the area to the left of $-z$ plus the area to the right of z is 0.0040.
13. What z-value represents the 95th percentile?
14. What z-value represents the 30th percentile?
15. What z-value represents the 75th percentile?
16. What z-value represents the third quartile?
17. What z-value represents the first quartile?

Application Problems

Directions: Solve the following problems. Note: Answers found using technology will vary slightly.

18. The body temperatures of adults are normally distributed with a mean of 98.6° F and a standard deviation of 0.73° F. What temperature represents the 85th percentile?

19. Assume that the salaries of elementary school teachers in the United States are normally distributed with a mean of $27,000 and a standard deviation of $6000. What is the cutoff salary for teachers in the top 10%?

20. The Verbal Reasoning, Biological Sciences, and Physical Sciences sections of the MCAT are scored on a 1 to 15 scale with a mean and standard deviation of 8 and 2.5, respectively. If the medical school you are applying to only takes students who score in the top 5%, what is the lowest score you could make and still be considered for acceptance?

21. The weights of college female students are normally distributed with a mean of 150 pounds and a standard deviation of 20 pounds. What weight represents the cutoff for the first quartile of female students on campus?

22. Scores on a midterm exam are normally distributed with a mean of 73 and a standard deviation of 9 points. What score is at the 90th percentile?

23. The numbers of monthly cell phone minutes used by students at one university are normally distributed with a mean of 110 minutes and standard deviation of 33 minutes. What is the third quartile for cell phone usage at this university?

24. The numbers of monthly cell phone minutes used by students at one university are normally distributed with a mean of 110 minutes and standard deviation of 33 minutes. What number represents the 30th percentile for cell phone usage at this university?

25. A local firehouse receives on average 45 calls per week with a standard deviation of 6 calls. Suppose that the firefighters are anticipating an unusually busy week. How many calls should they prepare for if they anticipate that this week will be busier than 85% of their weeks?

26. A local firehouse receives on average 45 calls per week with a standard deviation of 6 calls. Suppose that the firefighters are anticipating an unusually slow week. How many calls should they prepare for if they anticipate that this week will be busier than just 15% of their weeks?

Finding t-Values Using the Student t-Distribution

6.5

There are many distributions which are useful in statistical analysis other than the normal distribution. Another important continuous distribution, which we shall explore in more detail later, is the Student t-distribution. For now, let's look at its properties. The t-distribution is similar to the normal distribution in shape. It, too, is bell-shaped, but with more area under the tails, or "fatter" tails, on each end. Also, where the normal distribution has two parameters, the mean and standard deviation, the t-distribution has only one parameter, the number of degrees of freedom. Figure 15 is a diagram comparing the two distributions.

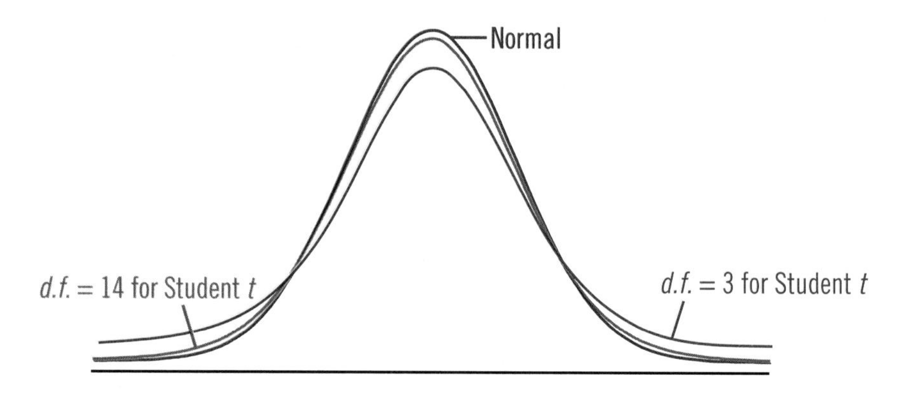

Figure 15 – Comparison of the Normal and Student t-Distributions

Properties of the t-Distribution

1. A t-curve is symmetrical and bell-shaped, centered about 0.

2. A t-curve is completely defined by its number of degrees of freedom, $d.f.$

3. The total area under a t-curve equals 1.

4. The x-axis is a horizontal asymptote for a t-curve.

Just as there are infinitely many normal curves, there are also infinitely many possible t-curves. In contrast to the normal curve, there is no "standard" t-distribution that we can use to calculate probabilities. Instead, each t-curve must be looked at individually. Tables, similar to the z-value table, are used to calculate probabilities for the t-distribution. In this text, we combine these tables into one, which contains the most commonly used probability values. Figure 16 shows a small portion of the Student t-distribution table. Note that the interior of the Student t-distribution table gives the t-scores associated with particular areas of the curve. This is the reverse of the Standard Normal distribution, which gives areas.

d.f.	0.100	0.050	0.025	0.010	0.005
1	3.078	6.314	12.706	31.821	63.656
2	1.886	2.920	4.303	6.965	9.925
3	1.638	2.353	3.182	4.541	5.841
4	1.533	2.132	2.776	3.747	4.604
5	1.476	2.015	2.571	3.365	4.032

Figure 16 – Excerpt from the t Table

As the number of degrees of freedom increases, the "fatter" tails of the t-distribution become thinner. They begin to appear more like the tails of a normal curve until, at infinitely many degrees of freedom, the t-distribution matches the standard normal distribution. Consequently, as you look down each column in the table the t-values decrease. A full t-distribution table can be found in Appendix A. Notice that the last row of that table has some familiar numbers in it. Can you figure out why?

In general, the value of t such that an area of α is to the right of t is denoted by t_α, as shown in Figure 17.

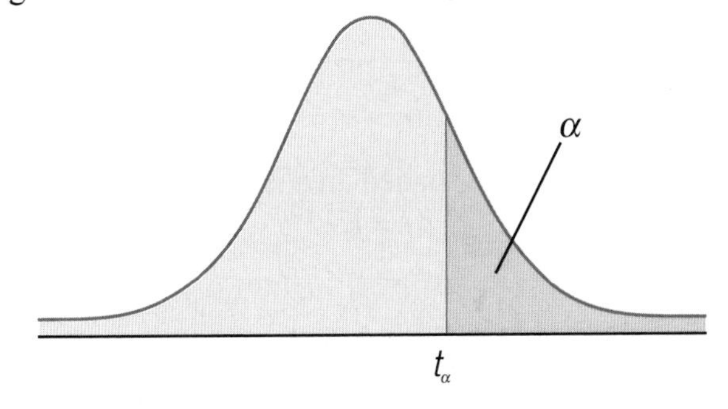

Figure 17 – Area to the Right of t_α

Rounding Rule

Round t-values to 3 decimal places.

Example 20

Find the value of $t_{0.025}$ with 25 degrees of freedom.

Solution:

The number of degrees of freedom is listed in the first column of the t-distribution table. Since our example has 25 degrees of freedom, the value we need lies on that row. Looking at the area of the shaded region, 0.025, tells us which column to use. This row and column intersect at 2.060. Thus $t_{0.025} = 2.060$.

d.f.	0.100	0.050	0.025	0.010	0.005
23	1.319	1.714	2.069	2.500	2.807
24	1.318	1.711	2.064	2.492	2.797
25	1.316	1.708	2.060	2.485	2.787
26	1.315	1.706	2.056	2.479	2.779

Example 21

How many degrees of freedom make $t_{0.05} = 1.86$?

Solution:

Finding 1.86 in the 0.05 column will give us the corresponding degrees of freedom of 8.

d.f.	0.100	0.050	0.025	0.010	0.005
6	1.440	1.943	2.447	3.143	3.707
7	1.415	1.895	2.365	2.998	3.499
8	1.397	1.860	2.306	2.896	3.355
9	1.383	1.833	2.262	2.821	3.250
10	1.372	1.812	2.228	2.764	3.169

Example 22

Area to the Right

Find the value of t such that the shaded area is 0.1 for 17 degrees of freedom.

d.f. = 17

0.10

t_α

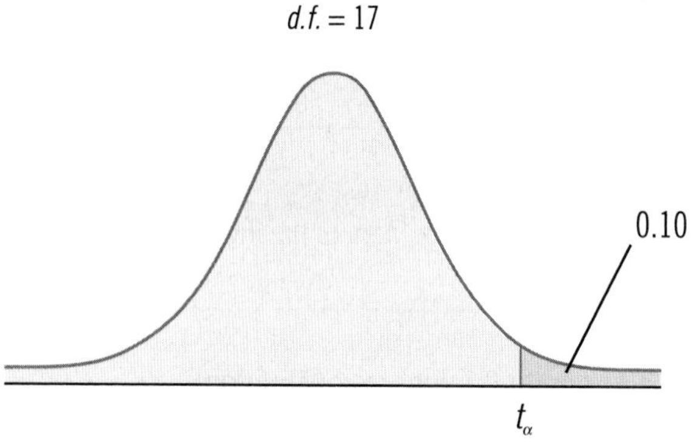

Solution:

Looking across the row for 17 degrees of freedom to the column of 0.100, we can see that the t-value for the shaded area is 1.333.

d.f.	0.100	0.050	0.025	0.010	0.005
15	1.341	1.753	2.131	2.602	2.947
16	1.337	1.746	2.120	2.583	2.921
17	1.333	1.740	2.110	2.567	2.898
18	1.330	1.734	2.101	2.552	2.878
19	1.328	1.729	2.093	2.539	2.861

Example 23

Area to the Left

Find the value of t such that the shaded area is 0.05 for a t-distribution. Assume 11 degrees of freedom.

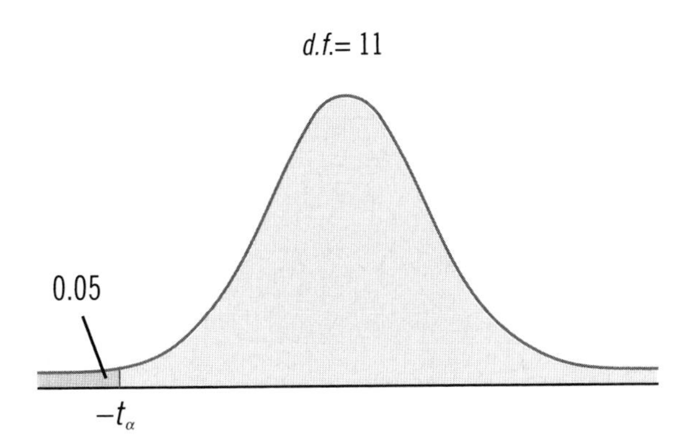

Solution:

Because the t-distribution is symmetrical, we can look up the t-value for 0.05 to the right. Using the table, we get $t = 1.796$. However, since the area is to the left, the t-value needs to be negative. So, in our example, $t_{0.05} = -1.796$.

d.f.	0.100	0.050	0.025	0.010	0.005
9	1.383	1.833	2.262	2.821	3.250
10	1.372	1.812	2.228	2.764	3.169
11	1.363	1.796	2.201	2.718	3.106
12	1.356	1.782	2.179	2.681	3.055
13	1.350	1.771	2.160	2.650	3.012

Sometimes the area that we are concerned with lies in both the right and left tails. When this happens, we call this type of problem *two-tailed.* In general, the value of t such that an area of α is divided equally between both tails is denoted by $t_{\alpha/2}$. With this type of problem, the area α will need to be divided by 2, so we can find the corresponding t-value. You will see an example of a two-tailed problem on the next page.

Example 24

Area in the Tails

Find the value of t such that the shaded area is 0.02. Assume 7 degrees of freedom.

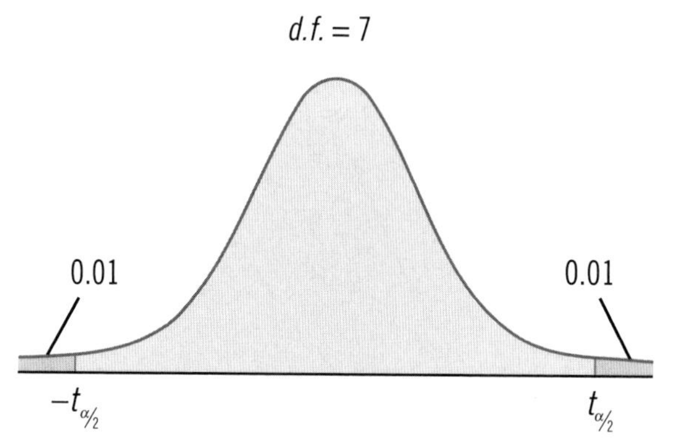

Solution:

Since the graph indicates that the area is divided between both ends, we must divide the area in half before we look up α. Therefore, $\frac{\alpha}{2} = \frac{0.02}{2} = 0.01$. Now we find t just as in the other examples. So the critical value for t is 2.998.

Another place we use the concept of two-tailed is when the area we want to know lies between two t-values. In this case, we need to do a little more manipulation with the original data in order to know which α column to look in. Consider the picture in Figure 18.

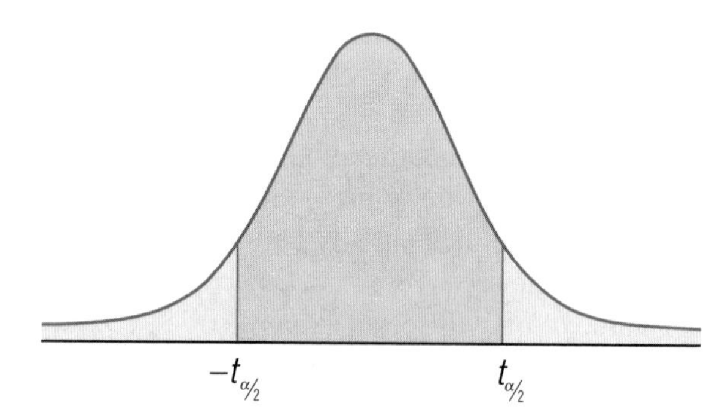

Figure 18 – Area Between Two t-Values

Suppose we are looking for the t-value that has 95% of the curve's area in between the positive and negative values of t. Since our Critical t-Value Table deals only with the outside tails, what we need to consider is the amount of area which remains in the tails. If 95% of the area is on the inside, then 5% remains in the tails. When divided equally, that gives us 2.5%, or 0.025, on each end. So the α column we would use is the 0.025 column.

Example 25

Area Between
Find the critical t-value such that the area in between $-t$ and t is 99%. Assume 29 degrees of freedom.

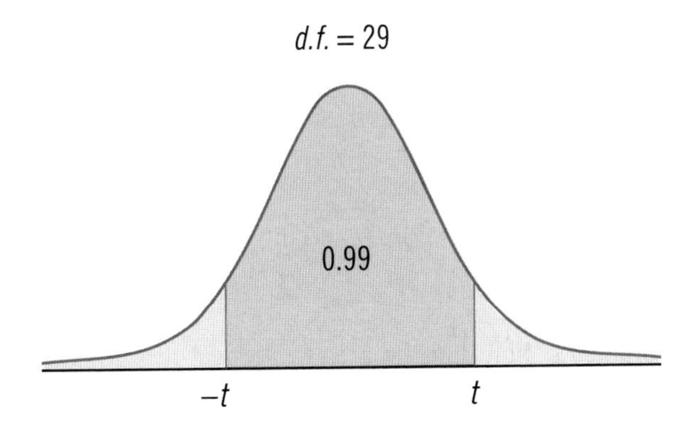

Solution:

Since 99% of the area of the curve is in the middle, that leaves 1%, or 0.01 of the area on the outside. Dividing that in two, we have that $\frac{\alpha}{2} = \frac{0.01}{2} = 0.005$. Using 29 degrees of freedom, the critical value for our example is $t = 2.756$.

Exercises 6.5

Directions: Find the appropriate t-value.

1. Find the value of $t_{0.05}$ with 15 degrees of freedom.
2. Find the value of $t_{0.01}$ with 8 degrees of freedom.
3. Find the value of $t_{0.005}$ with 26 degrees of freedom.
4. Find the value of $t_{0.1}$ with 20 degrees of freedom.
5. Find the value of $t_{0.01}$ with 29 degrees of freedom.
6. Find the value of $t_{0.025}$ with 6 degrees of freedom.
7. How many degrees of freedom make $t_{0.025} = 2.074$?

8. How many degrees of freedom make $t_{0.005} = 3.106$?

9. Find the value of t such that the area to the right of t equals 0.05. Assume 9 degrees of freedom.

10. Find the value of t such that the area to the right of t equals 0.025. Assume 11 degrees of freedom.

11. Find the value of t such that the area to the right of t equals 0.01. Assume 5 degrees of freedom.

12. Find the value of t such that the area to the right of t equals 0.005. Assume 14 degrees of freedom.

13. Find the value of t such that the area to the left of t equals 0.05. Assume 10 degrees of freedom.

14. Find the value of t such that the area to the left of t equals 0.10. Assume 3 degrees of freedom.

15. Find the value of t such that the area to the left of t equals 0.005. Assume 12 degrees of freedom.

16. Find the value of t such that the area to the left of t equals 0.025. Assume 27 degrees of freedom.

17. Find the value of t such that the area to the left of $-t$ and the right of t equals 0.01. Assume 13 degrees of freedom.

18. Find the value of t such that the area to the left of $-t$ and the right of t equals 0.20. Assume 23 degrees of freedom.

19. Find the value of t such that the area to the left of $-t$ and the right of t equals 0.02. Assume 12 degrees of freedom.

20. Find the value of t such that the area to the left of $-t$ and the right of t equals 0.05. Assume 18 degrees of freedom.

21. Find the value of t such that the area in between $-t$ and t equals 95%. Assume 24 degrees of freedom.

22. Find the value of t such that the area in between $-t$ and t equals 98%. Assume 4 degrees of freedom.

23. Find the value of t such that the area in between $-t$ and t equals 90%. Assume 25 degrees of freedom.

Chapter Review

Section 6.1 - Introduction to the Normal Curve	Pages
• **Normal distribution** – a continuous probability distribution for a given random variable X	p. 248 – 251
• **Properties of a Normal Distribution:** **1.** A normal curve is symmetric and bell-shaped. **2.** A normal curve is completely defined by its mean and standard deviation. **3.** The total area under the curve always equals 1. **4.** The x-axis is a horizontal asymptote for the normal curve.	p. 251
• **Standard Normal Curve** – has the same properties as the normal distribution; in addition, it has a mean of 0 and standard deviation of 1	p. 252 – 254
• **Standard Score Formula:** $z = \frac{x - \mu}{\sigma}$	p. 252 – 254

Section 6.2 - Reading a Normal Curve Table	Pages
• **Cumulative normal table**– table that gives area used to find the area to the *left* of a given z-value	p. 258 – 262
• **Rules for Finding Area Under the Normal Curve Using a Cumulative Distribution Table:** **1.** For area to the left of z: look up z. **2.** For area to the right of z: look up $-z$. **3.** For area between two z-values: look up each z and subtract. **4.** For area to the left of z_1 plus the area to the right of z_2: look up z_1 and $-z_2$ and add.	p. 262

Section 6.3 - Finding Probability Using the Normal Curve	Pages
• Area under the normal curve = probability of a random variable falling within that region	p. 266
• **Method for Finding Probability for Any Normal Curve:** **1.** Convert the x-value to a standard score (z). **2.** Use the standard normal table as outlined in 6.2.	p. 266 – 269

Section 6.4 - Finding z-Values Using the Normal Curve	Pages
• When given a probability, the cumulative standard normal table may be used in reverse in order to find the corresponding standard score.	p. 273 – 277

Section 6.5 - Finding t-Values Using the Student t-Distribution	Pages
• **Properties of the t-Distribution:** **1.** A t-curve is symmetrical and bell-shaped, centered about 0. **2.** A t-curve is completely defined by its number of degrees of freedom. **3.** The total area under a t-curve equals 1. **4.** The x-axis is a horizontal asymptote for a t-curve.	p. 280 – 285

Chapter Exercises

Directions: Fill in the blanks to make the statements true.

1. The total area under the *t*-curve equals __________. (Enter a number value.)

2. A normal curve is single-peaked which tells us it has __________ mode(s). (Enter the number of modes.)

3. The normal curve is symmetric about the mean, which tells us that the value of the mean equals the value of the __________ equals the value of the __________.

4. The normal curve is __________-shaped.

5. You can completely describe a normal distribution by two numbers, the __________ and the __________.

6. The total area under the normal curve equals __________. (Enter a number value.)

7. You can completely describe a *t*-curve by the __________.

Directions: Answer the following questions. Remember to round your answer to two decimal places for percentages and four decimal places for probabilities. (Hint: Recall from Chapter 3 that a *z*-score is the number of standard deviations above or below the mean.)

8. What percentage of the values in a normal distribution are more than one standard deviation away from the mean?

9. What percentage of the values in a normal distribution are greater than the mean, but less than 2 standard deviations above the mean?

10. What is the probability that a single-value from a normal distribution is greater than 3 standard deviations above the mean?

11. What is the area under the curve of a normal distribution between 1 and 3 standard deviations below the mean?

12. What is the relative frequency of the data greater than 2 standard deviations above the mean?

13. What percentage of the values in a normal distribution are less than the first quartile?

14. Kanuk, the polar bear, is pregnant. If polar bears have an average gestation period of 32 weeks with a standard deviation of 1.5 weeks, then what is the likelihood that Kanuk's baby will be born before 30 weeks?

15. Candy was surprised when the doctor told her that her baby was in the 85th percentile on the weight table. Which of the following would be a reasonable standard score for her baby's weight?

a. 85
b. −2.75
c. 1.04
d. 0
e. Not enough information

16. Find the value of t such that the area to the right of t equals 0.10. Assume 22 degrees of freedom.

17. Find the value of t such that the area to the left of t equals 0.01. Assume 16 degrees of freedom.

18. Find the value of t such that the area to the left of $-t$ and the right of t equals 0.10. Assume 28 degrees of freedom.

19. Find the value of t such that the area in between $-t$ and t equals 80%. Assume 17 degrees of freedom.

20. Look up the z-value for a right tailed probability of 0.1. Now look up the t-value for a right tailed probability of 0.1 with 17 degrees of freedom. Explain using the shape of the t-curve why the t-value is greater than the z-value.

Chapter Project: Curving Grades Using the Normal Curve

Dr. Smith, a biology professor at the University of XYZ has decided to give his classes a final from the University of ABC's biology department. He would like to compare the results of his students to those from the other university. After years of using this final, the University of ABC knows that the test results look like the following normal curve:

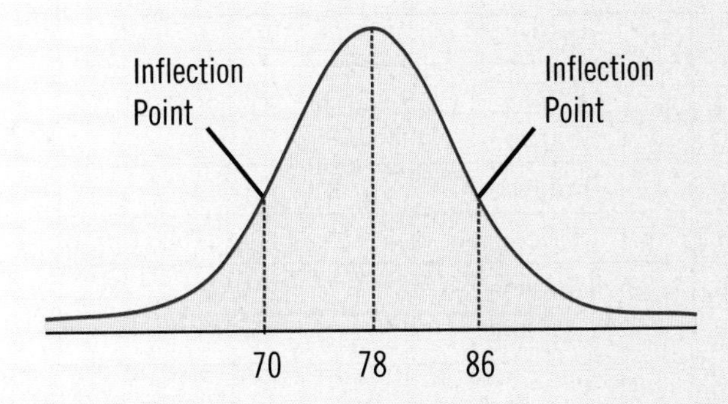

1. State the mean of the distribution of the Biology Final.

2. State the standard deviation of the Biology Final.

Grading Curve Option I

Originally, Dr. Smith decides to curve his students' grades as follows:

- Students at or above the 90th percentile will receive an A,
- Students in the 80th – 89th percentiles will receive a B,
- Students in the 70th – 79th percentiles will receive a C,
- Students in the 60th – 69th percentiles will receive a D,
- Students below the 60th percentile will receive an F.

3. Find the *z*-scores that correspond to the following percentiles:

- 90th percentile
- 80th percentile
- 70th percentile
- 60th percentile

4. Using that information, find the scores that correspond to the curved grading scale: (Round to the nearest whole number.)

- A's: __________ – 100
- B's: __________ – __________
- C's: __________ – __________
- D's: __________ – __________
- F's: 0 – __________

5. The following is a partial list of grades for students in Dr. Smith's class. Using the grading scale you just created, find the new curved **letter grade** that the students will receive on their test given their raw score.

Table P1 - List of Student's Grades

Name	Raw Score / Grade
Adam	82 / B
Bill	77 / C
Susie	91 / A
Troy	86 / B
Sharon	75 / C
Laura	66 / D
Eric	88 / B
Marcus	69 / D
Stephanie	79 / C

Grading Curve Option II

After reviewing the results, Dr. Smith decides to consider an alternate curving method. He decides to assign grades as follows:

- A: Any student scoring at least 2 standard deviations above the mean of the standardized test.
- B: Any student scoring from 1 up to 2 standard deviations above the mean of the standardized test.
- C: Any student scoring from 1 standard deviation below the mean up to 1 standard deviation above the mean of the standardized test.
- D: Any student scoring from 2 standard deviations below the mean up to 1 standard deviation below the mean of the standardized test.
- F: Any student scoring less than 2 standard deviations below the mean of the standardized test.

6. Using the above information, create Dr. Smith's new grading scale. (Round to the nearest whole number.)

- A's: __________ – 100
- B's: __________ – __________
- C's: __________ – __________
- D's: __________ – __________
- F's: 0 – __________

7. Using the grading scale you just created, return to the partial list of grades and find the new curved **letter grade** that the students receive on their test given their raw score.

8. Review the grades each student received using the two grading scales. Which grading scale do you feel is more fair? Explain why.

Chapter Technology – Finding Areas

NOTE: When using technology to find the area under the normal curve, the values may be slightly different than the values found using the normal curve tables. The technology value is the more accurate of the two values.

TI-84 PLUS

Finding the Area Under the Curve

To use your TI-84 Plus to find the area under any normal curve, use option 2: normalcdf(under the DISTR menu. Press 2ND, then VARS to access the menu and then scroll down to option 2. The format for entering the statistics is normalcdf(lower bound, upper bound, μ, σ). If you want to find the area under the standard normal curve, you do not need to enter μ or σ. It is sufficient to enter normalcdf(lower bound, upper bound).

Example T.1

Find the indicated area under the following normal curves:

a. Area between $z = -1.23$ and $z = 2.04$.

b. Area to the left of $z = 2.37$.

c. Area to the right of $z = -1.09$.

d. Area under a normal curve with a mean of 64, a standard deviation of 5.3, to the left of $x = 58$.

Solution:

a. Press 2ND, then VARS to access the menu and then scroll down to option 2. Since this is the area under the standard normal curve, we do not need to enter μ or σ. Enter normalcdf(−1.23, 2.04). The area is 0.86998.

b. Press 2ND, then VARS to access the menu and then scroll down to option 2. Since this is the area under the standard normal curve, we do not need to enter μ or σ. We want the area to the left, so the lower bound would be $-\infty$. We cannot enter $-\infty$, so we will enter a very small value for the lower endpoint, such as -10^{99}. This number appears as −1E99 when entered correctly into the calculator. To enter −1E99, press (−), 1, 2ND, ,, 9, 9. Enter normalcdf(−1E99, 2.37). The area is 0.9911.

cont'd on the next page

c. Press 2ND , then VARS to access the menu and then scroll down to option 2. Again, this is the area under the standard normal curve, so we do not need to enter μ or σ. We want the area to the right, so the upper bound would be ∞. We cannot enter ∞, so we will enter a very large value for the upper endpoint, such as 10^{99}. This number appears as 1E99 when entered correctly into the calculator. To enter 1E99, press 1 , 2ND , , , 9 , 9 . Enter normalcdf(−1.09, 1E99). The area is 0.8621.

d. Press 2ND , then VARS to access the menu and then scroll down to option 2. This is the area under a normal curve, so we will enter $\mu = 64$ and $\sigma = 5.3$. We want the area to the left, so we will enter a very small value for the lower endpoint, such as -10^{99}. Enter normalcdf(−1E99, 58, 64, 5.3). The area is 0.1288.

Finding *z*-Values

The TI-84 Plus calculator can find *z* or *x*-values given the probability also. To find an *x* or *z*-value, use option 3: invNorm(under the DISTR menu. Press 2ND , then VARS to access the menu and then scroll down to option 3. The format for entering the statistics is invNorm(area, μ, σ). If you want to find *z* instead of *x*, you do not need to enter μ or σ. It is sufficient to enter invNorm(area). The "area" is the area to the left of *x* or *z*, only. If you know the area to the right, or the area between, you must first determine the corresponding area to the left.

Example T.2

Find the indicated value:

a. The value of *z* with an area of 0.6781 to the left.
b. The value of *z* with an area of 0.2500 to the right.
c. The value of *z* such that 90% of the area is between *z* and $-z$.
d. A given distribution has a mean of 25 and a standard deviation of 2.45. What value of *x* has an area of 0.3761 to the left?

Solution:

a. Press 2ND , then VARS to access the menu and then scroll down to option 3. Enter invNorm(0.6781). The value of $z = 0.46$.

b. Since the area given is the area to the right, we must first determine the area to the left. The area to the left is $1 - 0.2500 = 0.7500$. Press 2ND , then VARS to access the menu and then scroll down to option 3. Enter invNorm(0.7500). The value of $z = 0.67$.

c. Since the area given is between two values of z, $1 - 0.90 = 0.10$ is the area in the tails. Thus, half of that area, or 0.05, is the area to the left of $-z$. Press 2ND, then VARS to access the menu and then scroll down to option 3. Enter invNorm(0.05). The value returned by the calculator is −1.6448. Note that the value we obtained earlier in this chapter using the normal curve table was 1.645.

d. This is not the standard normal distribution, so will need to enter $\mu = 25$ and $\sigma = 2.45$. The area given is to the left, so we can enter it directly. Press 2ND, then VARS to access the menu and then scroll down to option 3. Enter invNorm(0.3761, 25, 2.45). So the value of x is 24.23.

Finding *t*-Values

To use your TI-84 Plus to find the area under the Student t-distribution, use option 5: tcdf(under the DISTR menu. Press 2ND, then VARS to access the menu and then scroll down to option 5. The format for entering the statistics is tcdf(lower bound, upper bound, *d.f.*). Just as we did with the normal distribution, if we want the area to the left, we will enter −1E99 for the lower bound. If we want the area to the right, we will enter 1E99 for the upper bound. The "*d.f.*" stands for degrees of freedom.

Example T.3

Find the area to the left of $t = -1.638$ with 3 degrees of freedom.

Solution:

Press 2ND, then VARS to access the distribution menu then choose option 5:tcdf. Since we want the area to the left, the lower bound is $-\infty$, which we will enter as −1E99. Enter tcdf(−1E99, −1.638, 3). Thus the area to the left of $t = -1.638$ with 3 degrees of freedom is 0.09997.

MICROSOFT EXCEL

Finding the Area Under the Curve

The formula to find the area under the normal curve in Microsoft Excel is =NORMDIST(x, mean, standard_dev, cumulative). The value you are interested in is x, "mean" is the mean of the distribution, "standard_dev" is the standard deviation of the distribution, and "cumulative" is TRUE, which can be entered as the number 1, so that we will get the area to the left, just as in the cumulative normal distribution table. Unlike with the TI-84 Plus, you must enter a mean of 0 and a standard deviation of 1 to find the area under the standard normal curve. Further, Microsoft Excel only calculates the area to the left of the given value of x. To find the area between two values or to the right of a value, you must use the rules discussed in this chapter.

Finding z-Values

The formula to find z or x-values given the probability in Microsoft Excel is =NORMINV(probability, mean, standard_dev). The probability is the area to the left of our value; thus if the area we are given is to the right or the area in between, we must first determine the area to the left, just as we did with the TI-84 Plus. The "mean" is the mean of the distribution, "standard_dev" is the standard deviation of the distribution.

Finding t-Values

In this chapter, we actually calculated the inverse t-distribution, not the area under the t-distribution. To calculate the inverse t, we can use Microsoft Excel. The formula is =TINV(probability, deg_freedom). This formula is actually for the area in two-tails, so if we know the area in one-tail, the probability we would enter would be the area times 2.

Example T.4

Find the value of $t_{0.05}$ with 15 degrees of freedom.

Solution:

We know the degree of freedom is 15. The area 0.05 is the area to the right of t, which is a one-tailed test. $0.05 \cdot 2 = 0.10$, so we will enter a probability of 0.10. Using Microsoft Excel, enter "=TINV(0.10, 15)". The value displayed will be 1.753.

MINITAB

Finding the Area Under the Curve

To find the area under the curve in Minitab, enter the x-value which you have been given in the first column and row. Once your data is entered, go to CALC, choose PROBABILITY DISTRIBUTION, and then select *Normal*. When the normal distribution menu appears, make sure **Cumulative probability** is selected and enter your **Mean** and **Standard deviation**. Select C1 as the **Input Column**. Once you are finished, hit OK, and the probability will appear. Like Excel, Minitab only calculates the area to the left of the given value of x. To find the area between two values or to the right of a value, you must use the rules discussed in this chapter.

Finding z-Values

To find z or x-values given the probability in Minitab, enter the probability given in the first column and row. Once your data is entered, go to CALC, choose PROBABILITY DISTRIBUTION, and then select *Normal*. When the normal distribution menu appears, make sure **Inverse cumulative probability** is selected and enter your **Mean** and **Standard deviation**. Select C1 as the **Input Column**. Once you are finished, hit OK, and the z or x-value will appear. The probability is the area to the left of our value, Thus if the area we are given is to the right or the area in between, we must first determine the area to the left, just as we did with the TI-84 Plus.

Example T.5

Find the value of z given the probability to the left of z is 0.9265

Solution:

Enter 0.9265 in the first column, row and then go to CALC, choose PROBABILITY DISTRIBUTION, and then select NORMAL. Make sure **Inverse cumulative probability** is selected, the mean = 0 and the standard deviation = 1. The answer which appears will be $z = 1.4502$.

Finding t-Values

To find t-values given the area in Minitab, enter the area given in the first column and row. Once your data is entered, go to CALC, choose PROBABILITY DISTRIBUTION, and then select t. When the t-distribution menu appears, make sure **Inverse cumulative probability** is selected, and enter your **Degrees of freedom**. Select C1 as the **Input Column**. Once you are finished, hit OK, and the t-value will appear. If the area we are given is to the right or the area in between, we must first determine the area to the left, just as we did with the TI-84 Plus.

Sampling Distributions

Sections

Objectives

The objectives of this chapter are as follows:

1. Use the Central Limit Theorem to identify characteristics of a sampling distribution
2. Use the properties of the Central Limit Theorem to estimate the mean and standard deviation of a sampling distribution of sample means
3. Use the Central Limit Theorem to find probabilities for population means
4. Use the Central Limit Theorem to find probabilities for population proportions
5. Calculate probabilities by using the normal distribution as an approximation to the binomial distribution

Chapter 7

Introduction

Imagine that it is a cold, rainy day—the perfect kind of day for a bowl of hot vegetable soup. Fortunately, you have what you need on hand, and soon your soup is simmering away on the stove. The only ingredients left to add are the seasonings, and you want to make sure that the flavor is just right. How will you test to see if the seasonings are correct? Is it necessary to eat the entire pot of soup? Certainly not! Most likely, you would try just one or two spoonfuls in order to decide whether or not the soup is ready to eat.

Little did you know that something as simple as testing soup is actually an example of statistical sampling. Just as you estimated the seasoning of the whole batch from a bite or two, often information of an entire population can be obtained from one well-chosen sample. On a small scale, you are using statistical sampling when you simply poll your friends about what TV shows they watch. On a large scale, the Nielsen Company is using statistical sampling when it surveys families across the country to find out the same information. Indeed, the statistical sampling process is everywhere.

Central Limit Theorem

7.1

We have looked at using the normal curve to find the probability that an individual has some characteristic of a population, such as the probability of a single student scoring below 65 on an exam. Until now, we have had to be assured that the population is normally distributed. What if the population has a skewed distribution? Can we find a way to use the concepts from the normal curve? The answer is yes—if we apply a valuable statistical tool called the Central Limit Theorem.

Before we can understand the significance of the Central Limit Theorem, we need to introduce the term **sampling distribution**. Suppose we have a population distribution of any shape – such as skewed, uniform, or symmetric. We can take a sample from that distribution and calculate the mean of that sample. If we continue taking samples until all possible samples of a certain size have been selected, then the distribution created from the means of each of these samples would be called a **sampling distribution for sample means**. Instead of the sample mean, we could also calculate the sample proportion, sample variance, or sample standard deviation for each sample. This would give us other sampling distributions, such as the sampling distribution for sample proportions.

A few logical questions at this point would be, "How does the mean of the sampling distribution relate to the mean of the population? What about the standard deviation? How is it related to the population's standard deviation?". We can answer these questions by using the Central Limit Theorem.

The Central Limit Theorem

For any given population with mean μ and standard deviation σ, a sampling distribution of the sample mean, with sample sizes of at least 30, will have the following three characteristics:

1. The sampling distribution will approximate a normal distribution, regardless of the shape of the original distribution. Larger sample sizes will produce better approximations.

2. The mean of a sampling distribution, $\mu_{\bar{x}}$, equals the mean of the population.

$$\mu_{\bar{x}} = \mu$$

3. The standard deviation of a sampling distribution, $\sigma_{\bar{x}}$, equals the standard deviation of the population divided by the square root of the sample size.

$$\sigma_{\bar{x}} = \frac{\sigma}{\sqrt{n}}$$

The first statement concerns the shape of the sampling distribution of the sample mean. In plain English, it says that no matter what the original distribution looks like, a sampling distribution of the sample means will appear normal if large enough samples are chosen. The larger the sample size used to create the sampling distribution, the closer the distribution will be to the normal curve. Note, however, that if the original population is normally distributed, the distribution of sample means will be normal for any sample size.

Example 1

The following histogram represents the population distribution of the weights of horse jockeys. If a sampling distribution with samples of size 45 is constructed, what would be the shape of the sampling distribution?

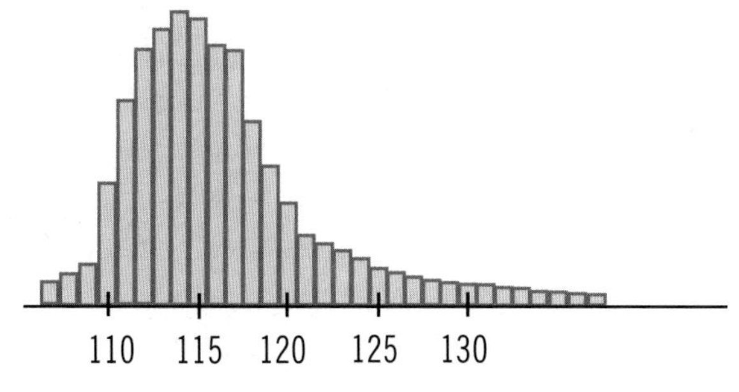

Solution:

Since the sample size is more than 30, its shape would approximate a normal curve.

The second statement describes the relationship between the mean of the population and the mean of a sampling distribution. Simply stated, the means are equal, $\mu_{\bar{x}} = \mu$. In fact, this condition is true for any sample size.

Example 2

If the mean of a given sampling distribution is $\mu_{\bar{x}} = 85$, what is the mean of the population?

Solution:

Since we know the mean of a sampling distribution is the same as that of its population, the population mean is $\mu = 85$.

The last statement compares the standard deviation of a sampling distribution to that of its population. The standard deviation of a sampling distribution can be found by dividing the standard deviation of the population by the square root of the sample size.

This property can be denoted $\sigma_{\bar{x}} = \dfrac{\sigma}{\sqrt{n}}$. As with the mean, the formula for standard deviation of a sampling distribution is true for any sample size.

Formulas for the Central Limit Theorem

1. $\mu_{\bar{x}} = \mu$
2. $\sigma_{\bar{x}} = \dfrac{\sigma}{\sqrt{n}}$

Rounding Rule

Remember to round standard deviation to one more decimal place than what is given in the data set.

Example 3

If the standard deviation of a given population is $\sigma = 9$, and a sampling distribution is created from the population with sample sizes of $n = 100$, what is the standard deviation of the sampling distribution?

Solution:

$$\sigma_{\bar{x}} = \frac{\sigma}{\sqrt{n}}$$

$$\sigma_{\bar{x}} = \frac{9}{\sqrt{100}}$$

$$\sigma_{\bar{x}} = 0.9$$

Therefore, the sampling distribution's standard deviation would be $\sigma_{\bar{x}} = 0.9$.

Example 4

Based on information from the National Association of Theater Owners, the standard deviation of ticket prices is $0.72. If samples of size 52 are taken, what would be the standard deviation of the sampling distribution?

Solution:

$$\begin{aligned}\sigma_{\bar{x}} &= \frac{\sigma}{\sqrt{n}} \\ &= \frac{0.72}{\sqrt{52}} \\ &\approx 0.09984 \\ &\approx 0.100\end{aligned}$$

So, the sampling distribution's standard deviation would be $\sigma_{\bar{x}} = 0.100$.

Exercises

7.1

Directions: Determine if the following statements are true or false. If a statement is false, state the reason why.

1. A sampling distribution refers to individuals rather than groups.

2. The shape of a sampling distribution that follows the requirements of the Central Limit Theorem will be approximately bell-shaped.

3. A sampling distribution has a standard deviation equal to $\frac{\sigma}{\sqrt{n}}$.

4. A sampling distribution has a mean equal to $\frac{\mu}{\sqrt{n}}$.

Directions: For each set of population information below, find the mean and standard deviation of the sampling distribution.

5. $\mu = 35$; $n = 64$; $\sigma = 9$

6. $\mu = 28$; $n = 81$; $\sigma = 6$

7. $\mu = 12$; $n = 36$; $\sigma = 2.3$

8. $\mu = 52$; $n = 100$; $\sigma = 7$

9. $\mu = 9.5$; $n = 39$; $\sigma = 10$

10. $\mu = 582$; $n = 1201$; $\sigma = 23.6$

Directions: Answer each of the following questions.

11. According to a Portland Press Herald Report, the average price of heating oil in Maine in December 2004 was $1.87 per gallon. If 100 samples of size 37 were collected from around Maine during that time, what would you expect the average of the sampling distribution to be?

12. According to a local school district, middle school students are assigned an average of 2.5 hours of homework per night. If 144 samples of size 50 are collected from this school district, what would you expect the average of the sampling distribution to be?

13. Some health reports claim that the average cold lasts 7 days. If 120 samples of size 100 are taken from across the U.S., what would you expect the average of the sampling distribution to be?

14. An internet source shows that the average one-way fare for business travel is $217, the lowest in five years. If 215 samples of size 45 are collected from across the United States, what would you expect the average of the sampling distribution to be?

15. The Federal Reserve Bank of New York conducted a study and claims that the inflation rates of American households had a population standard deviation of 0.2 percentage points in 1996. If a sampling distribution is created using samples of size 78, what would be the standard deviation of the sampling distribution?

Source: "Inflation Inequality in the United States." Federal Reserve Bank of New York, Staff Report no. 173, October 2003.

16. A study of elementary school students reports that children begin reading at age 5.7 years on average, with a standard deviation of 1.1 years. If a sampling distribution is created using samples of size 55, what would be the standard deviation of the sampling distribution?

17. A study on the latest fad diet claimed that the amount of weight lost by all people on this diet varied by a standard deviation of 5.8 pounds. If a sampling distribution is created using samples of size 100, what would be the standard deviation of the sampling distribution?

18. The Council of Christian Colleges and Universities tuition survey claims that in 2003-04 the population standard deviation of CCCU tuition and fees is $3048. If a sampling distribution is created using samples of size 35, what would be the standard deviation of the sampling distribution?

19. Researchers French and Jones report that the standard deviation of health care costs is \$4271 per year for people younger than 65 years old. If a sampling distribution is created using samples of size 500, what would be the standard deviation of the sampling distribution?
Source: French, E. & Jones, J. B., (2003) On the distributions and dynamics of health care costs. Available online at http://www.albany.edu/ jbjones/healc32.pdf.

20. Researchers French and Jones report that the standard deviation of health care costs is \$6072 per year for people older than 65 years old. If a sampling distribution is created using samples of size 350, what would be the standard deviation of the sampling distribution?
Source: French, E. & Jones, J. B., (2003) On the distributions and dynamics of health care costs. Available online at http://www.albany.edu/ jbjones/healc32.pdf.

Central Limit Theorem with Population Means

7.2

Recall that the procedure for solving normal curve word problems is as follows:

1. Draw a picture that describes the information in the question.
2. Convert the values in the problem to standard scores, z-values.
3. Use the normal curve table and the z-values to find the area under the curve.

When solving word problems involving sampling distributions, the same procedure is applied. The only difference is the formula for the z-value. We now must use the sampling distribution's mean and standard deviation instead of the population's. Thus the z-score formula for population means becomes

$$z = \frac{\bar{x} - \mu_{\bar{x}}}{\sigma_{\bar{x}}} \qquad \text{or} \qquad z = \frac{\bar{x} - \mu}{\left(\frac{\sigma}{\sqrt{n}}\right)}.$$

Example 5

The body temperatures of adults are normally distributed with a mean of 98.6° F and a standard deviation of 0.73° F. What is the probability of a sample of 36 adults having an average normal body temperature less than 98.3° F?

Solution:

The picture on the left shows the area under the normal curve in which we are interested. Notice that we can standardize the sampling distribution of interest and get the standard normal curve on the right.

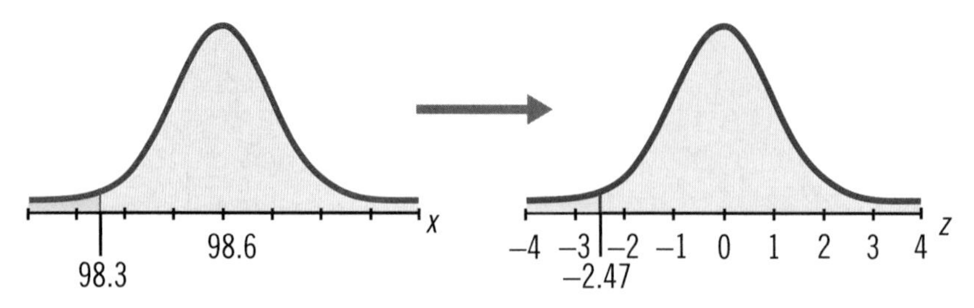

To do this, convert 98.3° F to a standard score using

$$z = \frac{\bar{x} - \mu}{\left(\frac{\sigma}{\sqrt{n}}\right)}$$

$$z = \frac{98.3 - 98.6}{\left(\frac{0.73}{\sqrt{36}}\right)}$$

$$z \approx -2.47.$$

Now we want to find the area to the left of $z = -2.47$, so look up $z = -2.47$ in the cumulative standard normal curve table. This gives us an area to the left of $z = -2.47$ equal to 0.0068. Thus the probability of a sample of 36 adults having an average normal body temperature less than 98.3° F is 0.0068.

Example 6

The body temperatures of adults are normally distributed with a mean of 98.6° F and a standard deviation of 0.73° F. What is the probability of a sample of 40 adults having an average normal body temperature greater than 99° F?

Solution:

The picture on the left shows the area under the normal curve in which we are interested. Notice that we can standardize the normal distribution of interest and get the standard normal curve on the right.

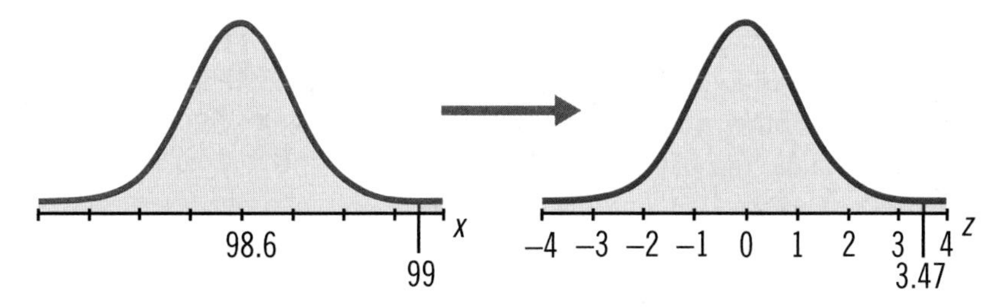

To do this, convert 99° F to a standard score.

$$z = \frac{\bar{x} - \mu}{\left(\frac{\sigma}{\sqrt{n}}\right)}$$

$$z = \frac{99 - 98.6}{\left(\frac{0.73}{\sqrt{40}}\right)}$$

$$z \approx 3.47$$

Now we want to find the area to the right of $z = 3.47$. We can use the normal curve's symmetric property and look up $z = -3.47$ in the table. This gives an area to the right of $z = 3.47$ equal to 0.0003. Thus the probability of a sample of 40 adults having an average normal body temperature greater than 99° F is 0.0003.

Example 7

In 2006, prices of women's athletic shoes were normally distributed with a mean of \$75.15 and a standard deviation of \$17.89. What is the probability that the average price of a sample of 50 pairs of women's athletic shoes will differ from the population mean by less than \$5.00?

Solution:

By subtracting and adding \$5.00 from the mean, we find that the area under the normal curve in which we are interested is **between** \$70.15 and \$80.15.

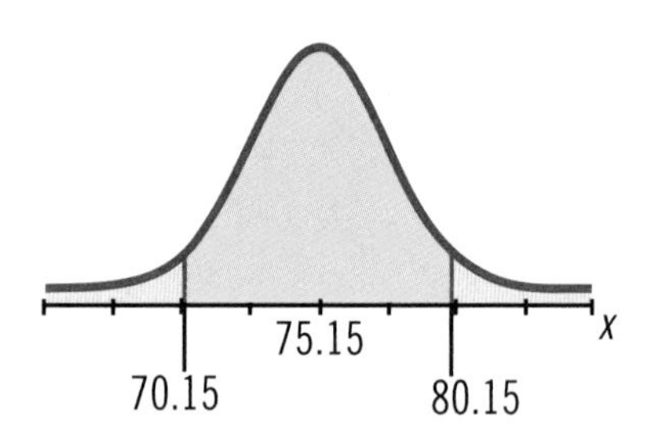

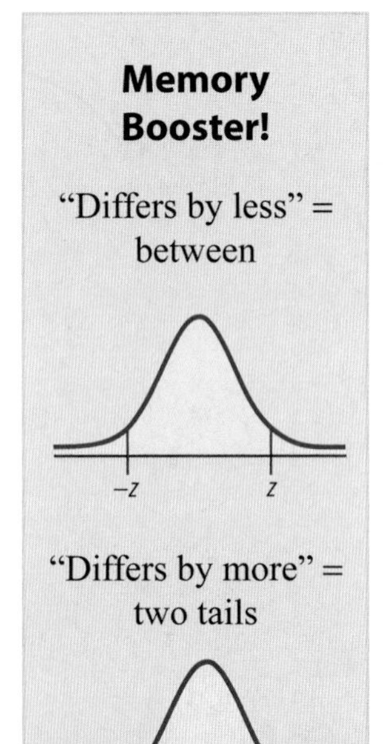

By converting to z-scores, we can standardize this normal distribution. Notice that because the distance between both of the endpoints of interest and the mean is 5, we do not need to find two distinct z-scores. They will be the same, but one positive and the other negative. To calculate the z-score, the distance given is substituted into the top of the formula, giving

$$z = \frac{\bar{x} - \mu}{\left(\frac{\sigma}{\sqrt{n}}\right)}$$

$$z = \frac{5.00}{\left(\frac{17.89}{\sqrt{50}}\right)}$$

$$z \approx 1.98.$$

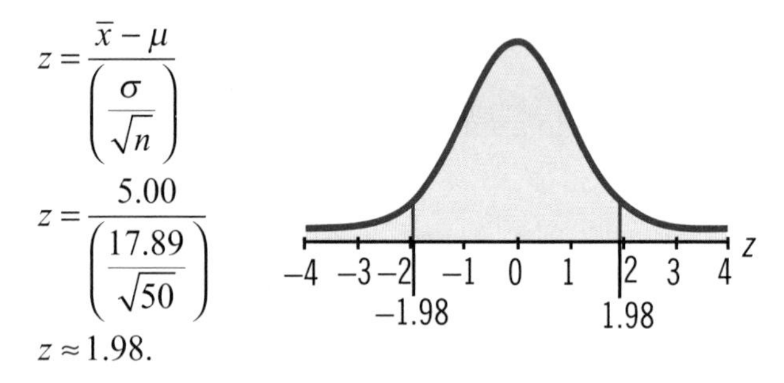

Because we want to find the area between two z-scores, recall from the previous chapter that we look up both z-values and subtract. Looking up $z = -1.98$ gives an area equal to 0.0239. Looking up $z = 1.98$ gives an area equal to 0.9761. Subtracting the two areas, we get

$$0.9761 - 0.0239 = 0.9522.$$

Thus the probability of a sample of 50 pairs of women's athletic shoes having an average price that differs from the population mean by less than \$5.00 is 0.9522.

Example 8

In 2006, prices of women's athletic shoes were normally distributed with a mean of \$75.15 and a standard deviation of \$17.89. What is the probability that the average price of a sample of 30 pairs of women's athletic shoes will differ from the population mean by more than \$3.00?

Solution:

By subtracting and adding \$3.00 from the mean, we find that the area under the normal curve in which we are interested is **less than** \$72.15 and **greater than** \$78.15.

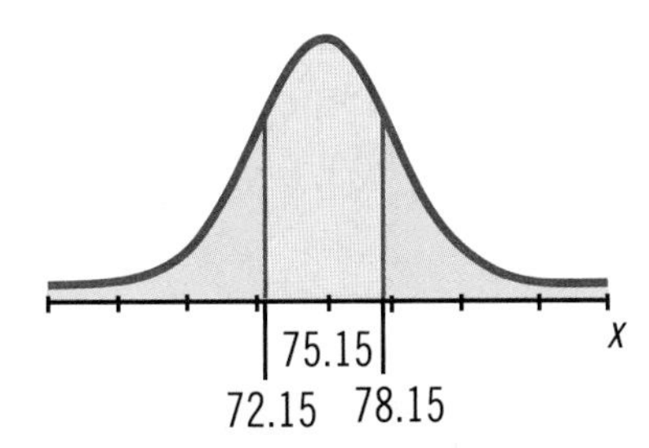

By converting to z-scores, we can standardize the normal distribution. Notice that because the distance between both of the endpoints of interest and the mean is 3, we do not need to find two distinct z-scores. They will be the same, but one positive and the other negative. To calculate the z-score, the distance given is substituted into the top of the formula, giving

$$z = \frac{\bar{x} - \mu}{\left(\frac{\sigma}{\sqrt{n}}\right)}$$

$$z = \frac{3.00}{\left(\frac{17.89}{\sqrt{30}}\right)}$$

$$z \approx 0.92.$$

−4 −3 −2 −1 0 1 2 3 4 z

−0.92 0.92

Because we want to find the area in the tails, recall from the previous chapter that we look up the area for both regions and add. However, because of the symmetry of the curve, looking up $z = -0.92$ and multiplying by 2 will be sufficient. Looking up $z = -0.92$ gives an area equal to 0.1788. Multiplying by two, we get

$$(0.1788)(2) = 0.3576.$$

Thus the probability of a sample of 30 pairs of women's athletic shoes having an average price that differs from the population mean by more than \$3.00 is 0.3576.

Exercises

7.2

Directions: For the data given below calculate the value of z using the Central Limit Theorem formula. Round your answers to 2 decimal places.

1. $\bar{x} = 38, n = 73, \mu = 36, \sigma = 39$

2. $\bar{x} = 10, n = 100, \mu = 11, \sigma = 10.2$

3. $\bar{x} = 0.55, n = 17, \mu = 1.1, \sigma = 1.1$

4. $\bar{x} = 0.52, n = 1500, \mu = 0.49, \sigma = 0.6$

5. $\bar{x} = 110{,}045, n = 100, \mu = 114{,}023, \sigma = 30{,}000$

6. $\bar{x} = 513, n = 1000, \mu = 514.2, \sigma = 30.2$

Directions: Answer each of the following questions. Use the Central Limit Theorem only where appropriate.

7. The walking gait of adult males is normally distributed with a mean of 2.4 feet and a standard deviation of 0.3 feet. A sample of 34 men's strides is taken.
   a. Find the probability that an individual man's stride is less than 2.1 feet.
   b. Find the probability that the mean of the sample taken is less than 2.1 feet.
   c. Find the probability that the mean of the sample taken is more than 2.5 feet.
   d. Find the probability that the sample mean differs from the population mean by more than 0.06 feet.

8. The distribution of the number of people in line at a grocery store check out has a mean of 3 and a variance of 9. A sample of 50 grocery lines is taken.
   a. Calculate the probability that the sample mean of the line length is more than 4.
   b. Calculate the probability that the sample mean of the line length is less than 2.5.
   c. Calculate the probability that the sample mean differs from the population mean by less than 0.5.

9. Intelligence is often cited as being normally distributed with $\mu = 100.0$ and $\sigma = 15.0$.

- **a.** What is the probability of a random person on the street having an intelligence level of less than 95?
- **b.** If a random sample of 50 people is taken, what is the probability that their mean intelligence level will be less than 95?
- **c.** If a random sample of 50 people is taken, what is the probability that their mean intelligence level will be more than 95?
- **d.** If a random sample of 50 people is taken, what is the probability that their mean intelligence level will be more than 105?
- **e.** If a random sample of 50 people is taken, what is the probability that their mean will differ from the population mean by more than 5?

10. If the diameter of oak trees is normally distributed with a mean of 4 feet and a standard deviation of 0.375 feet,

- **a.** what is the probability of walking down the street and finding an oak tree with a diameter of more than 5 feet?
- **b.** what is the probability of sampling a set of 87 oak trees and finding their mean to be more than 4.1 feet in diameter?
- **c.** what is the probability of sampling a set of 87 oak trees and finding their mean to be less than 3.92 feet in diameter?
- **d.** what is the probability of sampling a set of 87 oak trees and finding their mean to differ from the population mean by less than 0.1 foot in diameter?

11. A glass tube maker claims that his tubes have a length which is normally distributed with a mean of 9 cm and a variance of 0.25 cm.

- **a.** What is the probability that a quality control regulator will pull a tube off the assembly line which has a length between 8.6 and 9 cm?
- **b.** What is the probability that a random sample of 40 tubes will have a mean of less than 8.8 cm?
- **c.** What is the probability that a random sample of 35 tubes will have a mean of more than 9.2 cm?
- **d.** What is the probability that a random sample of 75 tubes will have a mean that differs from the population mean by more than 0.1 cm?

12. A pencil manufacturer claims that their pencils have a length which is normally distributed with a mean of 6 inches and a variance of 0.2 inches. What is the probability that a randomly chosen pencil will have a length of more than 6.4 inches?

13. A vineyard claims that the average berry count per grape cluster is 43 with a standard deviation of 3.6 berries. What is the probability that in next year's crop, his average berry count from a sample of 50 clusters is between 42 and 44?

14. A book publisher claims that its average book size is 250 pages long with a standard deviation of 70. As a reviewer, you get paid per book, and not per page, to read over the manuscript. What is the probability that you are randomly given a book of less than 200 pages to proof?

15. Airlines predict that there are on average 10 "no shows" per 100 seats on each flight with a standard deviation of 3.4. What is the probability that a random sample of 45 flights has an average of more than 12 "no shows" per 100 seats?

16. In a Scrabble® tournament the average score was 420.2 points with a standard deviation of 105. What is the probability that the score of a random competitor differs from the average score by less than 50 points?

17. A tea bag manufacturer needs to place 2 g of tea in each bag. If the machinery places an average of 2.6 g of tea in each bag with a standard deviation of 0.3 g, what is the probability that a randomly chosen bag will have between 2 and 2.8 g of tea?

18. A pollster claims that the average American family spends $927 on Christmas gifts with a standard deviation of $200. What is the probability that the average Christmas spending for a sample of 500 families differs from the average by less than $15?

19. A medical journal lists the average fetal heart rate at 140 beats per minute (bpm) with a standard deviation of 12 bpm. What is the probability that a fetal heart rate differs from the mean by more than 25 bpm?

20. A local sports magazine lists the average length of a baseball game in minutes to be 175.9 minutes with a standard deviation of 27 minutes. In a random sample of 30 games, what is the probability that the mean game length is at most 170 minutes?

21. The average teenager spends 7.5 hours per week playing computer and video games. In a random sample of 110 teenagers, what is the probability that their mean time for playing games is more than 8 hours per week? Let $\sigma = 3$.

22. The average wait time in a drive-thru chain is 193.2 seconds with a standard deviation of 29.5 seconds. What is the probability that in a random sample of 45 wait times, the mean is between 185.7 and 206.5 seconds?

23. The average amount spent per order at one fast food restaurant is $8.43 with a standard deviation of $1.52. What is the probability that the total for the first randomly selected order will be between $7 and $9?

24. The average hourly rate for babysitters in one town is \$7.05 with a standard deviation of \$0.55. What is the probability that a babysitter chosen at random will charge an hourly rate that differs from the mean by less than \$1.00?

25. The average sale price for a piece of art from members of a large artists' guild is \$545 with a standard deviation of \$76. If, at one show, 45 pieces are for sale, what is the probability that the average sale price for the show will be higher than \$575?

26. The average cost for plumbing repairs in one area is \$208 with a standard deviation of \$64. If 31 homeowners in that area are surveyed, what is the probability that the average cost for their plumbing repairs will be over \$200?

27. One pediatric clinic sees, on average, 42 patients per day with a standard deviation of 3.4 patients. What is the probability that the average patients per day for March (31 days) will be between 41 and 44 patients?

28. At a large university, the average student spends \$38.90 each month for cellular phone service with a standard deviation of \$3.64. Consider a group of 44 randomly chosen university students. What is the probability that their average cell phone bill differs from the mean by more than \$1?

29. A large bakery sells on average 11 dozen cookies per day with a standard deviation of 1.1 dozen cookies. Consider the average number of cookies sold in March (31 days) of one year. What is the probability that the mean differs from the population mean by less than 0.5 dozen?

Central Limit Theorem with Population Proportions 7.3

How often have you read in the newspapers that 45% of the population would do "this" or favor "that"? Have you ever wondered how researchers come up with these figures? We're now in a position to explain perhaps the most commonly encountered use of statistics – proportions. Of course, we all understand that only by conducting a census could a researcher establish the true percentage of the population who favor a particular issue. However, by applying the Central Limit Theorem to sample proportions in a manner similar to sample means, we can *estimate* the population proportion. A **population proportion** is the percentage of the population that has a certain characteristic. That percentage might be written as a percentage, fraction, or decimal. For example, if a census showed that 9 out of every 10 people carry a cell phone, then the population proportion is $\frac{9}{10}$, more often written 0.90 or 90%. We denote population proportion with the symbol p.

Memory Booster!

p = population proportion

$\hat{p}$ = sample proportion

In the same manner, the **sample proportion** is the percentage of the sample that has a certain characteristic. We denote the sample proportion as $\hat{p}$, which is read "p hat". For example, if we ask a sample of 10 people if they carry a cell phone and 8 of them say yes, then $\hat{p} = \frac{8}{10} = 0.80 = 80\%$.

In order to use the Central Limit Theorem for sample proportions, we must make sure that the following conditions have been met:

$$np \geq 5 \text{ and } n(1-p) \geq 5.$$

In this section you can safely assume that these conditions have been met for each example and exercise. (You may certainly verify them for yourself, if you wish!) When these conditions have been met, the sampling distribution of sample proportions can be assumed to be normal, thus allowing us to use the normal distribution and z-scores to calculate probability for population proportions.

When using the normal curve and Central Limit Theorem to discuss population proportions, the z-score formula changes slightly. The formula for z-scores for population proportions is

$$z = \frac{\hat{p} - p}{\sqrt{\frac{p(1-p)}{n}}}.^{*}$$

We can use this z-score formula along with our traditional procedures for using the normal curve table to calculate probabilities involving proportions.

**Footnote: Formally,* $z = \frac{\hat{p} - p}{\sigma_{\hat{p}}}$, *where* $\sigma_{\hat{p}} = \sqrt{\frac{p(1-p)}{n}}$.

Example 9

In a certain conservative precinct, 79% of the voters are registered Republicans. What is the probability that in a sample of 100 voters from this precinct more than 68 of the voters would vote for the Republican candidate?

Solution:

To begin, we must calculate the z-score. To do this, we need to use $p = 0.79$ and $n = 100$. We also need to find $\hat{p}$. $\hat{p} = \frac{68}{100} = 0.68$. We can now calculate z.

$$\begin{aligned} z &= \frac{\hat{p} - p}{\sqrt{\frac{p(1-p)}{n}}} \\ &= \frac{0.68 - 0.79}{\sqrt{\frac{0.79(1-0.79)}{100}}} \\ &\approx \frac{-0.11}{0.0407} \\ &\approx -2.70 \end{aligned}$$

Since we want the probability of getting a value greater than z, we can look up negative z which would be $-z = 2.70$ in the normal curve table. The area is 0.9965. Thus the probability of more than 68 voters choosing the Republican candidate is 0.9965.

Rounding Rule

When dealing with intermediate calculation, do not round the intermediate steps. For example, although $\hat{p}$ is normally rounded to three decimal places, for intermediate calculation it is best to keep $\hat{p}$ out to six decimal places in order to not affect the remaining calculations with a rounding error.

Example 10

In another precinct across town, the population is very different. In this precinct, 81% of the voters are registered Democrats. What is the probability that, in a sample of 100 voters from this precinct, less than 80 of the voters would vote for the Democratic candidate?

Solution:

To begin, we must calculate the z-score. To do this, we need to use $p = 0.81$ and $n = 100$. We also need to find $\hat{p}$. $\hat{p} = \frac{80}{100} = 0.80$. We can now calculate z.

cont'd on the next page

$$\begin{aligned} z &= \frac{\hat{p}-p}{\sqrt{\frac{p(1-p)}{n}}} \\ &= \frac{0.80-0.81}{\sqrt{\frac{0.81(1-0.81)}{100}}} \\ &\approx \frac{-0.01}{0.03923} \\ &\approx -0.25 \end{aligned}$$

Since we want the probability of getting a value less than z, we can look up $z = -0.25$ in the normal curve table. The area is 0.4013. Thus the probability of less than 80 voters choosing the Democratic candidate is 0.4013.

Example 11

It is estimated that 7% of all Americans have diabetes. Suppose that a sample of 74 Americans is taken. What is the probability that the proportion of the sample that is diabetic differs from the population proportion by less than 1%?

Solution:

By subtracting and adding 0.01, we find that the area under the normal curve in which we are interested is **between** 0.06 and 0.08.

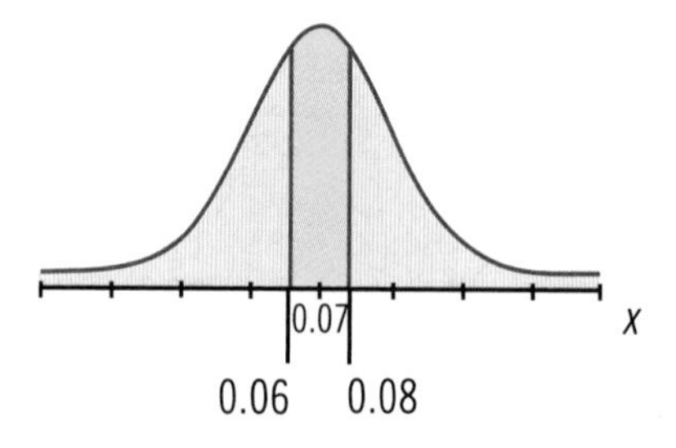

By converting to z-scores, we can standardize the normal distribution of interest. Recall that we do not need to find two distinct z-scores. They will be the same, but one positive and the other negative. To calculate one z-score, the distance given is substituted into the top of the formula, giving

$$z = \frac{\hat{p} - p}{\sqrt{\frac{p(1-p)}{n}}}$$

$$= \frac{0.01}{\sqrt{\frac{0.07(1-0.07)}{74}}}$$

$$\approx 0.34.$$

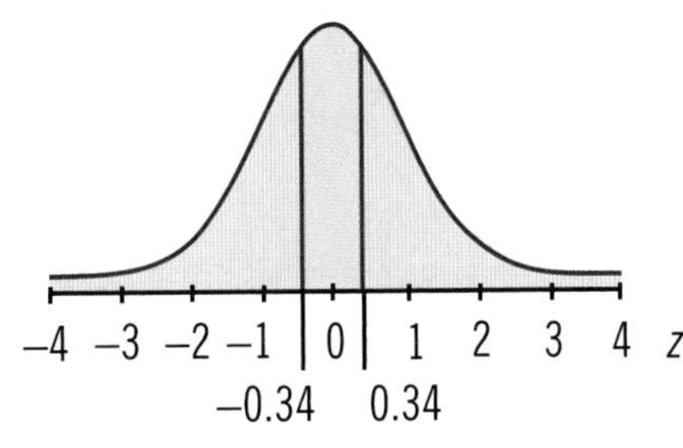

Because we want to find the area between two z-scores, we look up both z-values and subtract. Looking up $z = -0.34$ gives an area equal to 0.3669. Looking up $z = 0.34$ gives an area equal to 0.6331. Subtracting the two areas, we get

$$0.6331 - 0.3669 = 0.2662.$$

Thus the probability that the proportion of Americans in the sample with diabetes differs from the population proportion by less than 1% is 0.2662.

Example 12

It is estimated that 7% of all Americans have diabetes. Suppose that a sample of 91 Americans is taken. What is the probability that the proportion of the sample that is diabetic differs from the population proportion by more than 2%?

Solution:

By subtracting and adding 0.02, we find that the area under the normal curve in which we are interested is **less than** 0.05 and **greater than** 0.09.

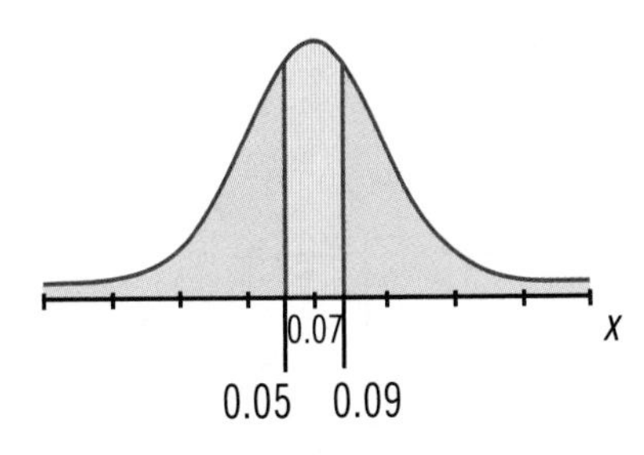

By converting to z-scores, we can standardize the normal distribution of interest. Recall that we do not need to find two distinct z-scores. They will be the same, but one positive and the other negative. To calculate one z-score, the distance given is substituted into the top of the formula, giving

cont'd on the next page

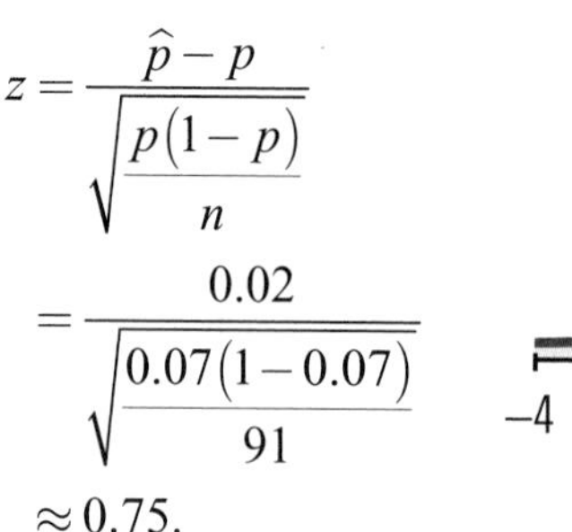

$$z=\frac{\hat{p}-p}{\sqrt{\frac{p(1-p)}{n}}}=\frac{0.02}{\sqrt{\frac{0.07(1-0.07)}{91}}}\approx 0.75.$$

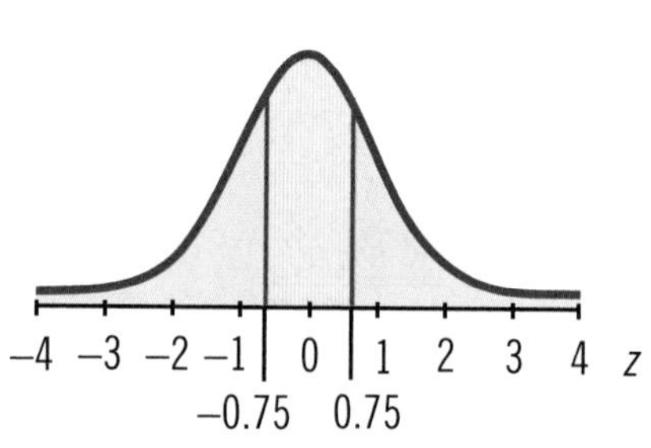

Because we want to find the area in the tails of two z-scores, we look up the area for both regions and add. However, because of the symmetry of the curve, looking up $z = -0.75$ and multiplying by 2 will be sufficient. Looking up $z = -0.75$ gives an area equal to 0.2266. Multiplying by two, we get

$$(0.2266)(2) = 0.4532.$$

Thus the probability that the sample proportion differs from the population proportion by more than 2% is 0.4532.

Exercises

7.3

Directions: For the data given below, a) calculate the sample proportion and b) calculate the value of z using the Central Limit Theorem formula. Round both answers to two decimal places using the value you found for $\hat{p}$ in the Central Limit Theorem formula.

1. $p = 0.34, x = 35, n = 100$

2. $p = 0.54, x = 563, n = 1000$

3. $p = 0.8, x = 24, n = 29$

4. $p = 0.61, x = 66, n = 98$

5. $p = 0.15, x = 1152, n = 12{,}500$

6. $p = 0.93, x = 987, n = 1012$

Directions: Answer each of the following questions.

7. A large car dealership claims that 47% of their customers are looking to buy a sport utility vehicle (SUV). A sample of 61 customers is surveyed. What is the probability that less than 40% are looking to buy an SUV?

8. The local nursery is waiting for its spring annuals to be delivered, and 20% of the plants ordered are petunias. If the first truck contains 120 plants packed at random, what is the probability that no more than 30 of the plants are petunias?

9. A hometown car lot sells pre-owned cars. Their sales manager reports that 78% of the cars they sell cost less than $10,000. What is the probability that in a sample of 45 cars, more than 75% cost less than $10,000?

10. A consumer report stated that 65% of the vehicles sold nationally were SUVs. A sample of 100 vehicles sold in the last month is taken. What is the probability that at least 70 of the vehicles sold are SUVs?

11. A local florist says that 85% of all flowers sold on Valentine's Day are roses. Consider a sample of 150 orders for flowers. What is the probability that less than 80% of the orders in the sample are for roses?

12. At one private college, 34% of students are business majors. Suppose that 260 students are randomly selected from a list in the registrar's office. What is the probability that the proportion of business students in the sample differs from the population proportion by less than 2%?

13. At a large grocery store, 72% of shoppers are women. In order to obtain information about spending habits, 40 shoppers are randomly chosen for a survey. What is the probability that the proportion of women in the sample differs from the mean by more than 3%?

14. A major appliance retailer claims that 18% of all appliances sold are dishwashers. If 112 appliances are sold one day, what is the probability that the proportion of dishwashers sold differs from the population mean by over 2%?

15. A popular restaurant downtown says that 20% of diners order the daily special. Consider an evening in which 128 people dine at the restaurant. What is the probability that the proportion of diners that order the special differs from the population mean by less than 1%?

Approximating the Binomial Distribution Using the Normal Distribution

7.4

In Section 5.2 we learned how to calculate probabilities using the binomial distribution. However, we restricted our discussion to small values for the random variable. It is often necessary, though, to calculate probabilities that involve large values for the binomial random variable. In this section, we will discuss a technique for approximating these probabilities using the normal distribution.

Let's begin by reviewing the properties of the binomial distribution. Remember that a binomial distribution requires that:

- The experiment consists of n identical trials.
- Each trial is independent of the others.
- For each trial, there are only two possible outcomes. For counting purposes, one outcome is labeled a *success*, the other a *failure.*
- For every trial, the probability of getting a success is called p. The probability of getting a failure is then $1-p$.
- The binomial random variable, X, is the number of successes in n trials.

Memory Booster!

Remember:
x = the number of successes
n = the number of trials
p = the probability of getting a success

In order to calculate the probability $P(X)$ for a binomial random variable, we learned to use technology, the binomial tables in the appendix (Table D), or the binomial formula, $P(X=x)={}_nC_x\cdot p^x(1-p)^{(n-x)}$. However, we cannot use the table for values of n larger than 20. Nor can calculators compute combinations for large values of n, thus making the formula impractical for very large values of n.

Now consider the first property of the Central Limit Theorem. This property states that under certain conditions, a sampling distribution created from a distribution of any shape will approximate a normal distribution. If a binomial distribution meets the conditions that $np \geq 5$ and $n(1-p) \geq 5$, then a sampling distribution can be created with a mean equal to $\mu = np$ and a standard deviation equal to $\sigma=\sqrt{np(1-p)}$ that will be approximately normal. To summarize:

Normal Distribution Approximation of a Binomial Distribution

If the conditions that $np \geq 5$ and $n(1-p) \geq 5$ are met for a given binomial distribution, then a normal distribution can be used to approximate its probability distribution with the given mean and standard deviation

$$\mu = np$$

$$\sigma=\sqrt{np(1-p)}.$$

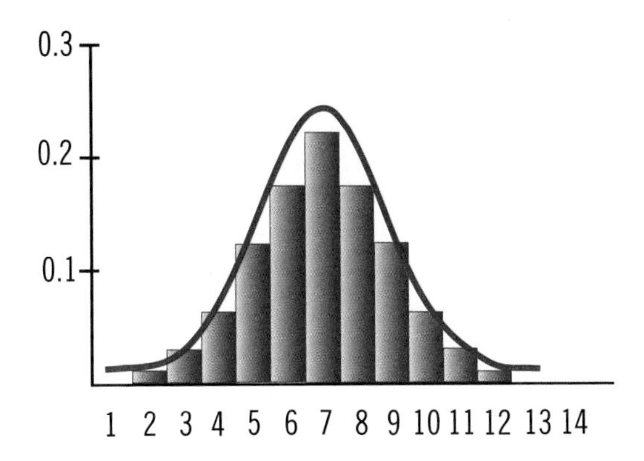

Figure 1 - Binomial Distribution with Normal Curve

Thus when we need to calculate a probability for a large value of a binomial random variable, we do not need to use Table D or the binomial formula. Instead, we can use the normal distribution.

Remember that the basic steps in finding a probability using the normal distribution are:

1. Draw a normal curve labeled with the information in the problem.
2. Convert the value of the random variable(s) to a z-value(s).
3. Use the normal curve table to find the appropriate area under the curve.

These steps do not change when we are approximating a binomial distribution; the first step is simply more involved. Let's look at calculating the mean and standard deviation in more depth.

When using the normal distribution to approximate the binomial distribution, begin by determining the values of n and p. These values are necessary to verify that the conditions $np \geq 5$ and $n(1-p) \geq 5$ are met. Without these conditions being satisfied, the normal curve cannot be used as an appropriate approximation. Once the conditions have been verified, calculate the values of the mean and standard deviation using the formulas $\mu = np$ and $\sigma = \sqrt{np(1-p)}$. Draw a normal curve and mark the value of the mean in the center. Finding the value of x used in the probability calculation involves using a continuity correction which we will discuss in detail next.

We are trying to use the normal distribution, which is *continuous*, to approximate the binomial distribution, which is *discrete*. Therefore, a **continuity correction** must be used to convert the whole number value of the discrete binomial random variable to an interval range of the continuous normal random variable. To make this correction, determine the value of the binomial random variable, x, and convert it to the interval from $x - 0.5$ to $x + 0.5$.

Continuity Correction

A **continuity correction** is a correction factor employed when using a continuous distribution to approximate a discrete distribution.

Let's practice using the continuity correction before solving any probability problems.

Example 13

Use the continuity correction factor to describe the area under the normal curve that approximates the probability that at least 2 people in a statistics class of 50 cheated on the last test. Assume that the number of people who cheated is a binomial distribution with a mean of 5 and a standard deviation of 2.12.

Solution:

Begin by converting the discrete number 2 into an interval by adding and subtracting 0.5 from the number 2. The discrete number 2 is changed to the continuous interval from 1.5 to 2.5. Now, draw a normal curve and indicate the interval from 1.5 to 2.5 to represent the number 2. Next, shade the area corresponding to the phrase *at least 2*. This would be the area greater than or equal to 2. Thus the area corresponding to *at least 2* would include the interval from 1.5 to 2.5 and all values to the right of 2.5.

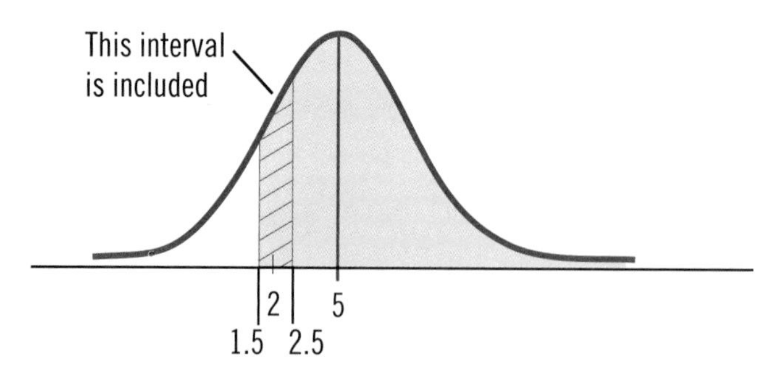

Thus the area under the normal curve approximating the probability that at least 2 people in your statistics class cheated on your last test would be the area to the right of 1.5.

Example 14

Use the continuity correction factor to describe the area under the normal curve that approximates the probability that less than 5 of the sitcoms playing on TV tonight are reruns. Assume that the number of reruns on TV tonight is a binomial distribution with a mean of 7 and a standard deviation of 2.16.

Solution:

To begin we need to convert the discrete number 5 to the continuous interval from 4.5 to 5.5 by adding and subtracting 0.5 to 5. Draw a normal curve with a mean of 7 and a standard deviation of 2.16. Indicate the interval from 4.5 to 5.5 on the curve. Next, shade the area *less than 5*. We need the area to the left of our interval. Since the phrase *less than 5* does not include the number 5, we do *not* shade the interval from 4.5 to 5.5.

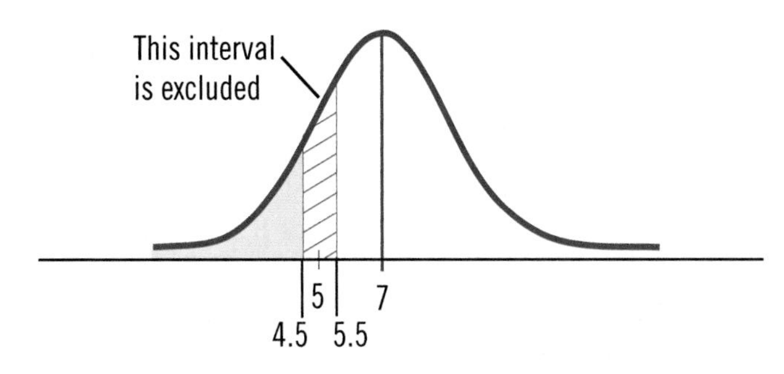

Thus the area under the normal curve corresponding to the probability that less than 5 sitcoms are reruns is the area to the left of 4.5.

To help you use the continuity correction properly, refer to the following table for some examples.

Table 1 - Examples of the Continuity Correction		
Statement	**Symbolically**	**Area**
At least 45, or no less than 45	≥ 45	Area to the right of 44.5
More than 45, or greater than 45	> 45	Area to the right of 45.5
At most 45, or no more than 45	≤ 45	Area to the left of 45.5
Less than 45, or fewer than 45	< 45	Area to the left of 44.5
Exactly 45, or equal to 45	$= 45$	Area between 44.5 and 45.5

Normal Curves

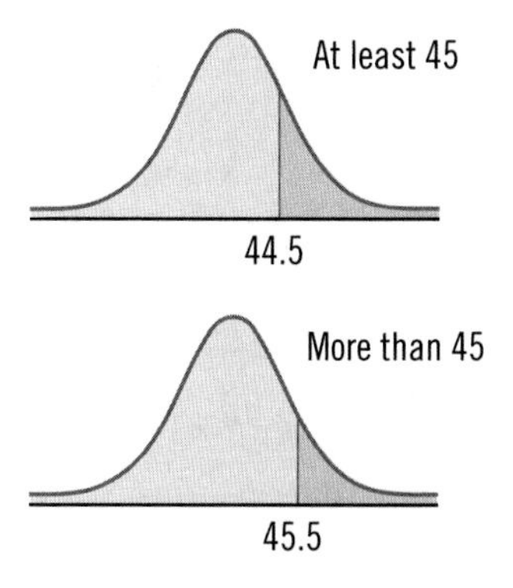

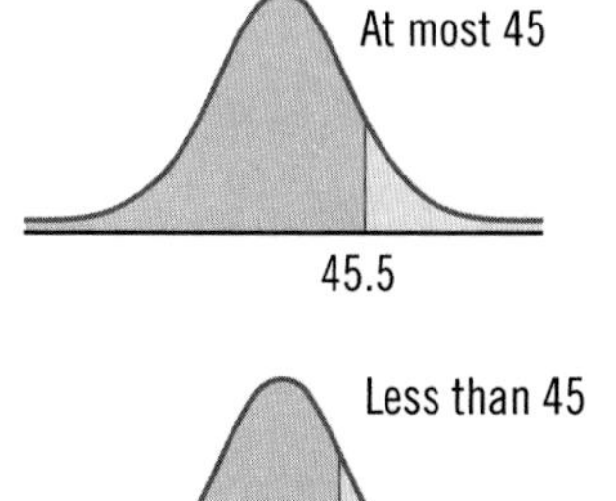

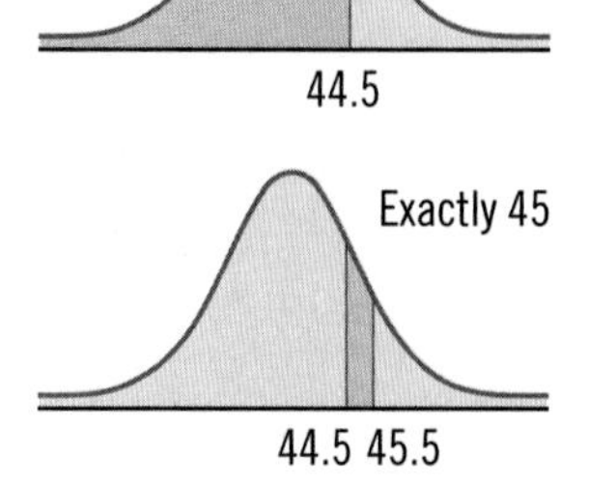

Now that we have the tools to translate a discrete value into a continuous interval, let's look at some examples of probability questions. Our process for using the normal curve to approximate the binomial distribution is as follows:

1. Determine the values of n and p.
2. Verify that the conditions $np \geq 5$ and $n(1-p) \geq 5$ are met.
3. Calculate the values of the mean and standard deviation using the formulas $\mu = np$ and $\sigma = \sqrt{np(1-p)}$.
4. Use a continuity correction to determine the interval corresponding to the value of x.
5. Draw a normal curve labeled with the information in the problem.
6. Convert the value of the random variable(s) to a z-value(s).
7. Use the normal curve table to find the appropriate area under the curve.

Example 15

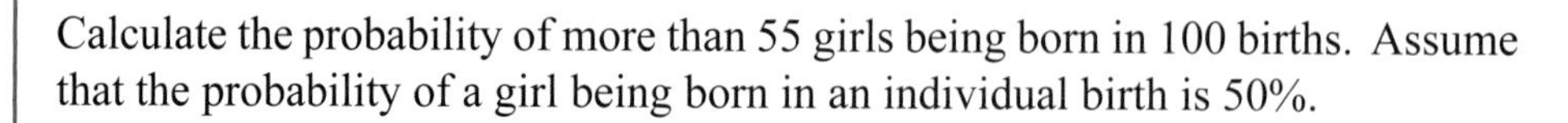

Calculate the probability of more than 55 girls being born in 100 births. Assume that the probability of a girl being born in an individual birth is 50%.

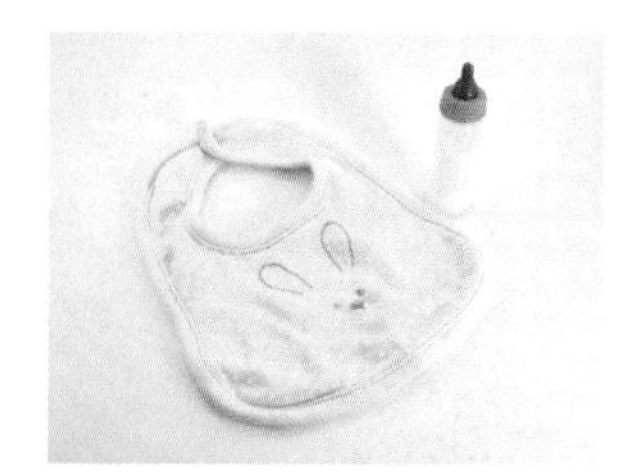

Solution:

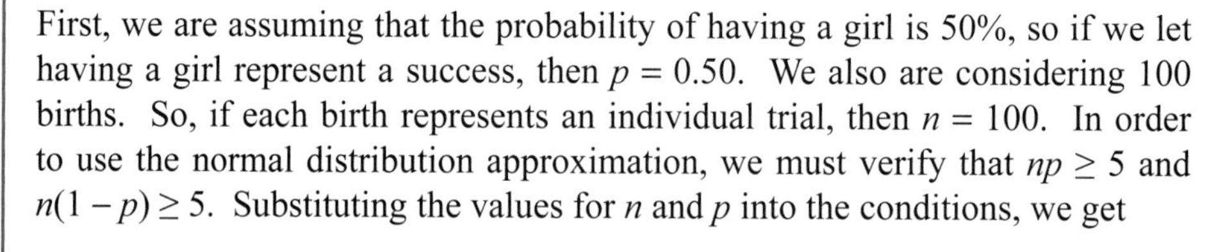

First, we are assuming that the probability of having a girl is 50%, so if we let having a girl represent a success, then $p = 0.50$. We also are considering 100 births. So, if each birth represents an individual trial, then $n = 100$. In order to use the normal distribution approximation, we must verify that $np \geq 5$ and $n(1-p) \geq 5$. Substituting the values for n and p into the conditions, we get

$$np = 100(0.5) = 50 \geq 5, \text{ as necessary}$$

and

$$n(1-p) = 100(1-0.5) = 50 \geq 5, \text{ as necessary.}$$

Thus the conditions are met and we can use the normal distribution approximation.

Next, we must calculate the mean and standard deviation. Substituting the values for n and p into the formulas, we get

$$\mu = np$$
$$\mu = 100(0.5)$$
$$\mu = 50$$

and

$$\sigma = \sqrt{np(1-p)}$$
$$\sigma = \sqrt{100(0.5)(1-0.5)}$$
$$\sigma = \sqrt{25}$$
$$\sigma = 5.$$

Therefore, the mean is 50 and the standard deviation is 5.

We now need to use the continuity correction to determine the interval corresponding to our discrete x-value of 55. By adding and subtracting 0.5, we get the interval from 54.5 to 55.5. Now we can draw our normal curve, with a mean of 50 and a standard deviation of 5. Mark the interval from 54.5 to 55.5 on the curve. We are asked for the probability of obtaining *more than* 55 girls, so we want the area to the *right* of the interval, but *not* including it. So, we want the area to the right of 55.5.

Memory Booster!

Remember that using the normal distribution to *approximate* a binomial distribution results in an APPROXIMATE answer.

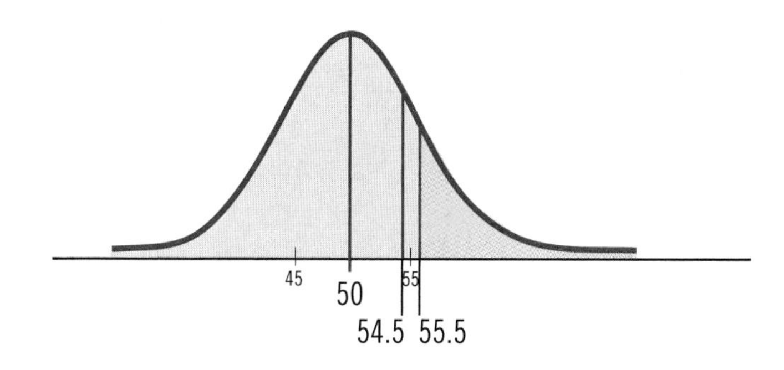

Next, convert the value 55.5 to a z-value using the z formula.

$$z = \frac{55.5-50}{5}$$
$$z = 1.10$$

Lastly, use the normal curve table to find the area to the right of $z = 1.10$. Using the symmetry property of the normal curve, look up the area to the left of $z = -1.10$ instead. This gives us an area of 0.1357.

Therefore, the probability of having more than 55 girls out of 100 births is approximately 0.1357.

Example 16

After many hours of studying for your statistics test, you believe that you have a 90% probability of answering any given question correctly. Your test includes 50 true/false questions. What is the probability that you will miss no more than 4 questions?

Solution:

If a trial is a single question on the test, then we have 50 trials, so $n = 50$. Since we are asked about missing questions on the test, we should define a success as *missing* a question. This definition might seem "backwards", but it will result in a more straightforward solution to the problem. If a success is missing a question, and we have a 90% chance of getting it right, then we have a 10% chance of missing the question. Thus $p = 0.10$.

Next, we need to verify the conditions that allow us to use the normal curve approximation. Substituting the values $n = 50$ and $p = 0.10$ into the conditions, we get

$$np \geq 5$$
$$50(0.1) = 5 \geq 5$$

and

$$n(1-p) \geq 5$$
$$50(1-0.1) = 45 \geq 5.$$

Thus both conditions are met.

Now we need to calculate the mean and standard deviation. Substituting the values for n and p into the formulas, we get

$$\mu = np$$
$$\mu = 50(0.1)$$
$$\mu = 5$$

and

$$\sigma = \sqrt{np(1-p)}$$
$$\sigma = \sqrt{50(0.1)(1-0.1)}$$
$$\sigma = \sqrt{4.5}$$
$$\sigma \approx 2.121$$

Therefore, the mean is 5 and the standard deviation is 2.121.

Using a continuity correction to determine the interval corresponding to our discrete x-value of 4, add and subtract 0.5 to 4 to get the interval from 3.5 to 4.5. Draw a normal curve with a mean of 5 and a standard deviation of 2.121 and mark the interval from 3.5 to 4.5 on the curve. We are asked for the probability of missing *no more than* 4 questions, so we want the area to the *left* of the interval and *including* the interval. So we want the area to the left of 4.5.

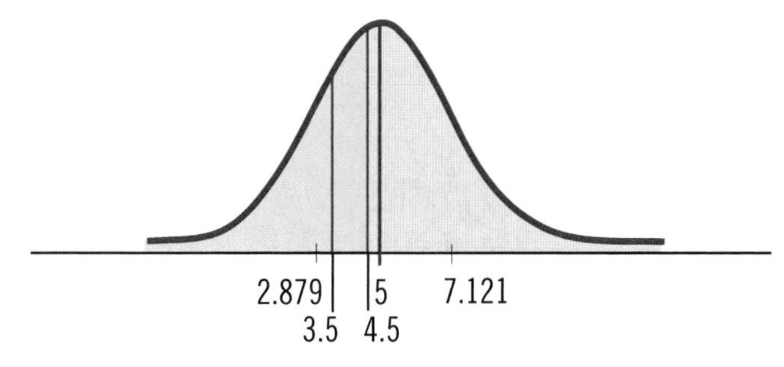

Next, convert the value 4.5 to a z-value using the z formula.

$$z = \frac{4.5 - 5}{2.121}$$
$$z \approx -0.24$$

Lastly, use the normal curve table to find the area to the left of $z = -0.24$. This gives us an area of 0.4052. Therefore, the probability of missing no more than 4 questions out of 50 is approximately 0.4052.

Example 17

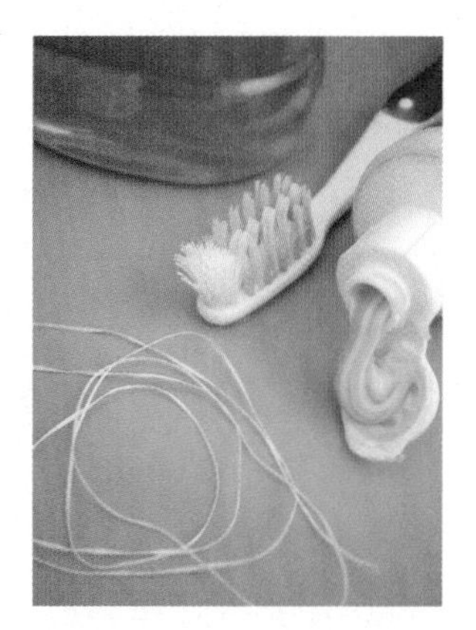

Many toothpaste commercials advertise that 3 out of 4 dentists recommend their brand of toothpaste. What is the probability that out of a random survey of 400 dentists, 300 will have recommended Brand X toothpaste? Assume that the commercials are correct, and therefore, there is a 75% chance that any given dentist will recommend Brand X toothpaste.

Solution:

Let's define a success to be a dentist who recommends Brand X toothpaste. Then the probability of obtaining a success is $p = 0.75$. Since we are surveying 400 dentists, $n = 400$. Substitute these values into the conditions that $np \geq 5$ and $n(1 - p) \geq 5$. Substituting the values for n and p into the formulas, we get

cont'd on the next page

$$np = 400(0.75) = 300 \geq 5, \text{ as necessary}$$

and

$$n(1 - p) = 400(1 - 0.75) = 100 \geq 5, \text{ as necessary.}$$

Thus the conditions are met and we can use the normal distribution approximation.

Using these values again, we calculate that the mean is 300 and the standard deviation is 8.66 as follows:

$$\mu = np$$
$$\mu = 400(0.75)$$
$$\mu = 300$$

and

$$\sigma = \sqrt{np(1-p)}$$
$$\sigma = \sqrt{400(0.75)(1-0.75)}$$
$$\sigma = \sqrt{75}$$
$$\sigma \approx 8.660.$$

Using the continuity correction, we add and subtract 0.5 to our discrete x-value of 300. Thus our interval is 299.5 to 300.5. Draw a normal curve and mark the center, as well as the interval, on the curve.

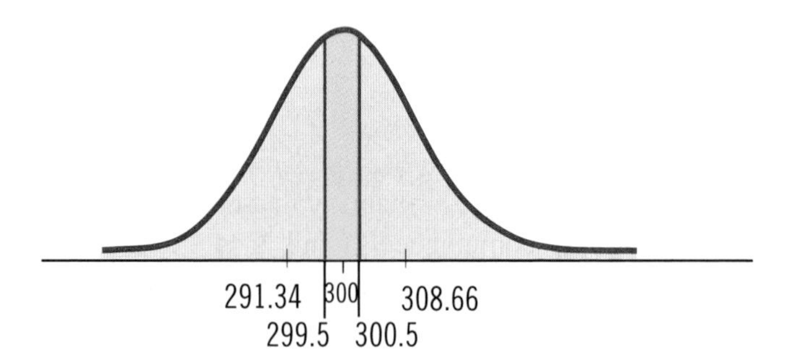

We are interested in the probability that $x = 300$, so we only want the area in our interval. So we will need to convert both 299.5 and 300.5 to standard scores and then find the area between them. First, converting 299.5 to a standard score we obtain

$$z = \frac{299.5 - 300}{8.660}$$
$$z \approx -0.06.$$

Converting 300.5 to a standard score we get

$$z = \frac{300.5 - 300}{8.660}$$
$$z \approx 0.06.$$

Using the cumulative normal curve table to find the area to the left of each z-score, we find that the areas are 0.4761 and 0.5239, respectively. Subtracting the smaller from the larger, we find that the area within our interval is

$$0.5239 - 0.4761 = 0.0478.$$

Thus the probability of exactly 300 out of a sample of 400 dentists recommending Brand X toothpaste is approximately 0.0478.

Exercises

7.4

Directions: Describe the area under the normal curve that would be used to approximate the following binomial probabilities.

1. Consider the probability that at least 40 out of 232 planes will malfunction on the runway.

2. Consider the probability that at least 100 out of 250 new mothers received prenatal care.

3. Consider the probability that more than 500 out of 12,000 tax returns were filed incorrectly.

4. Consider the probability that more than 180 out of a sample of 1200 elderly people in the United States will have the flu this winter.

5. Consider the probability that at most 30 out of 534 DVD players on the assembly line are defective.

6. Consider the probability that at most 15 out of 400 high school seniors will not pass their senior year.

7. Consider the probability that fewer than 12 out of 367 fourth-graders will not pass their state placement test.

8. Consider the probability that 70 out of 100 trees planted in an orchard will live to maturity.

9. Consider the probability that 35 out of 200 registered voters will not vote in the election.

10. Consider the probability that at least 225 out of a sample of 1200 elderly in the United States will have the flu this winter.

Directions: Determine whether the normal curve can be used as an approximation to the following binomial probabilities by verifying the conditions.

11. Consider the probability that fewer than 15 out of the 123 people watching a movie have already read the book. Assume that the probability of a given person having read the book is 40%.

12. Consider the probability that more than 100 out of 238 fifth-graders have seen all of the Harry Potter movies. Assume that the probability of a given fifth-grader having seen all of the Harry Potter movies is 54%.

13. Consider the probability that at most 2 out of 30 television sets on an assembly line are defective. Assume that the probability of a given television set being defective is 5%.

14. Consider the probability that no more than 5 out of 120 teenage girls become pregnant before finishing high school. Assume that the teen pregnancy rate is 4%.

Directions: Approximate the following binomial probabilities using the normal distribution. You can safely assume that the conditions for using the normal distribution have been met in each of the following scenarios.

15. What is the probability that more than 150 out of 230 eighth graders at a local middle school have been exposed to drugs? Assume that a previous study at this school reported that the probability of an individual eighth grade student having been exposed to drugs is 63%.

16. What is the probability that more than 100 out of 300 elections will contain voter fraud? One report suggests that there is a 32% chance of an individual election containing voter fraud.

17. What is the probability that more than 20 out of a class of 347 high school seniors will drive under the influence of alcohol on prom night? The local chapter of MADD fears that the probability of a high school senior drinking and driving on prom night is 38%.

18. What is the probability that more than 200 out of the 248 hunters staying at a hunting club this season will obtain their game of choice? Club records indicate that hunters have an 83% chance of obtaining their desired game animal.

19. What is the probability that at least 67 out of 100 cars stopped at a roadblock will not be given a ticket? Local authorities report that tickets usually are given to 23% of cars stopped.

20. What is the probability that at least 25 out of a survey of 200 Cajuns do not actually like Cajun food? A regional food critic believes that the probability of a Cajun not liking Cajun food is 14%.

21. What is the probability that at least 140 out of 200 drivers surveyed will report that they typically speed? The state's highway patrol estimates that 78% of drivers typically exceed the speed limit.

22. What is the probability that at least 130 out of 145 preschoolers watch more than 4 hours of television per night? One previous study indicates that the probability that a child watches more than 4 hours of television per night is 84%.

23. What is the probability that no more than 32 out of 150 vehicles inspected at a service station will fail their yearly inspection and not receive an inspection sticker? One service station's records indicate that the probability of a car failing its inspection is 18%.

24. What is the probability that no more than 130 out of the 2300 tax returns filed at a local CPA's office will be inaccurate? Previous records indicate only a 7% probability that a given tax return from this office is incorrect.

25. What is the probability that no more than 5 out of 250 puppies born to a well-respected dog breeder will have birth defects? This dog breeder usually averages only 2 birth defects in 50 births.

26. What is the probability that no more than 50 out of 150 former smokers will resume smoking within 6 months of quitting? Assume that the probability of a former smoker resuming smoking is 35%.

27. What is the probability that fewer than 100 homes out of a sample of 1200 homes will not have a TV? A study indicates that 9 out of 10 homes have televisions.

28. What is the probability that fewer than 100 out of 320 high school seniors will not attend college? The registrar's office at a local college estimates that the probability of a high school student going on to college is 76%.

29. What is the probability that 125 out of 200 people are overweight? Estimates show that the probability of someone being overweight is 65%.

30. What is the probability that out of 350 entering college freshmen, 50 will need to take developmental mathematics? Research shows that there is a 13% chance of a college freshman needing to take developmental mathematics.

Chapter Review

Section 7.1 - Central Limit Theorem	Pages
• **Central Limit Theorem** For any given population with mean μ and standard deviation σ, a sampling distribution of the sample mean, with sample sizes of at least 30, will have the following three characteristics: **1.** The sampling distribution will approximate a normal distribution, regardless of the shape of the original distribution. Larger sample sizes will produce better approximations. **2.** The mean of a sampling distribution, $\mu_{\bar{x}}$, equals the mean of the population, i.e. $\mu_{\bar{x}} = \mu$. **3.** The standard deviation of a sampling distribution, $\sigma_{\bar{x}}$, equals the standard deviation of the population divided by the square root of the sample size, i.e. $\sigma_{\bar{x}} = \frac{\sigma}{\sqrt{n}}$.	p. 300 – 303

Section 7.2 - Central Limit Theorem with Population Means	Pages
• **Formula for z using the Central Limit Theorem for population means:** $$z = \frac{\bar{x} - \mu}{\left(\frac{\sigma}{\sqrt{n}}\right)}.$$	p. 306 – 309

Section 7.3 - Central Limit Theorem with Population Proportions	Pages
• **Definitions** **Population Proportion** – the percentage of the population that has a certain characteristic (p) **Sample Proportion** – the percentage of the sample that has a certain characteristic ($\hat{p}$)	p. 314
• **Formula for z using the Central Limit Theorem for population proportions:** $$z = \frac{\hat{p} - p}{\sqrt{\frac{p(1-p)}{n}}}$$	p. 314 – 318

Section 7.4 - Approximating the Binomial Distribution Using the Normal Distribution	Pages
• **Continuity Correction** – a correction factor employed when using a continuous distribution to approximate a discrete distribution	p. 321 – 323
• **Process for Using the Normal Curve to Approximate the Binomial Distribution:** **1.** Determine the values of n and p. **2.** Verify that the conditions $np \geq 5$ and $n(1-p) \geq 5$ are met. **3.** Calculate the values of the mean and standard deviation using the formulas $\mu = np$ and $\sigma = \sqrt{np(1-p)}$. **4.** Use a continuity correction to determine the interval corresponding to the value of x. **5.** Draw a normal curve labeled with the information in the problem. **6.** Convert the value of the random variable(s) to a z-value(s). **7.** Use the normal curve table to find the appropriate area under the curve.	p. 324 – 329

Chapter Exercises

Directions: Answer the following questions.

1. The average lifetime of laptop computers produced by one company is 4.6 years with a standard deviation of 1.1 years. Consider samples of laptop computers of size 55. What is the standard deviation of the sampling distribution?

2. Suppose that the average 18 – 25 year old male watches 19 football games each football season with a standard deviation of 3.7 games. If 31 men in this age range are surveyed, what is the probability that they watch on average no more than 17 games a season?

3. In one city, 83% of eligible voters are registered to vote. If 425 eligible voters are surveyed, what is the probability that at least 82% are registered to vote?

4. A phone company reports that 51% of its customers subscribe to their high-speed internet service. If 180 customers are randomly selected for a survey, what is the probability that more than 94 of them subscribe to the high-speed internet service?

5. A local fast food restaurant reports that, on average, they sell 235 hamburgers a day with a standard deviation of 11.3. Consider a sample of 31 days. What is the probability that the average number of hamburgers sold will differ from the mean by more than 3 hamburgers?

6. The average size of candy bars from one candy maker is 39 grams with a standard deviation of 2.3 grams. If you produce a sampling distribution with samples of size 40 from the candy production line, what would you expect the mean of the sampling distribution to be?

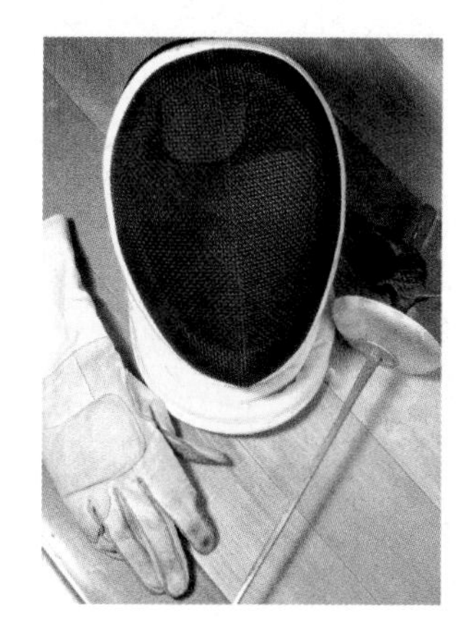

7. The average length of a rapier, which is a fencing sword, is 90 cm with a standard deviation of 4.2 cm. If 38 rapiers are sampled, what is the probability that their average length is less than 88 cm?

8. The average length of an adult blue whale is 30 meters with a standard deviation of 5.7 meters. If a sample of 31 whales is taken, what is the probability that their average length differs from the mean by less than 2 meters?

9. What is the probability that more than 35 out of 119 college algebra students will attend a late afternoon study session provided by their professor? Previous experience tells the professor that there is a 40% probability that a given student will attend the study session.

10. What is the probability that less than 45 calculus students will fail their calculus course this semester? There are 210 students currently enrolled in the class and previous semesters indicate that there is a 28% chance of an individual student failing this course.

11. Suppose that for a particular make and model of car the average sale price for car dealerships nationwide is \$24,560 with a standard deviation of \$390. What is the probability that the sale price at the car dealership in your area will be no more than \$24,000? Assume that the car prices are normally distributed.

12. One brand of light bulbs has an average lifetime of 400 hours with a standard deviation of 28 hours. What is the probability that the lifetime of one randomly chosen bulb will differ from the mean by over 50 hours? Assume that the population is normally distributed.

13. At one regional grocery store chain, the average customer spends \$7.29 on produce each visit with a standard deviation of \$2.33. Find the probability that the average amount spent on produce for 55 shoppers will be less than \$7.

14. At a local elementary school, the average student misses 4.1 days of school each year with a standard deviation of 1.1 days. What is the probability that the average days missed for a class of 30 students will be no more than 3.5?

15. At one theater, the average movie-goer spends \$6.32 on refreshments with a standard deviation of \$2.55. What is the probability that a random sample of 50 movie goers will spend an average of between \$6 and \$7 on refreshments?

16. At a local accounting firm, the average rate for completing an individual's tax return is \$225 with a standard deviation of \$28. Suppose that Alex, one of the employees, completes 33 individual tax returns one month. What is the probability that the average amount billed differs from the mean by more than \$5?

17. At one supermarket, patrons purchase an average of 17 items each visit with a standard deviation of 8.2 items. Suppose that 53 customers are checked out one morning. What is the probability that the average number of items purchased by the sample differs from the population mean by less than 2 items?

18. Salaries of employees at a large factory are normally distributed with a mean of \$27,500 and a standard deviation of \$2900. Consider the mean salary of a group of 8 employees and the mean salary of a group of 15 employees. Which group's mean would you expect to be closer to \$27,500?

Chapter Project A: Central Limit Theorem Experiment

Directions: You will need a die and at least 6 sets of data to complete this project.

Consider the distribution of all possible outcomes from rolling a single die. Let's use this distribution as our population distribution. We want to use this population distribution to explore the properties of the Central Limit Theorem. We must begin by determining the shape, center, and dispersion of our original distribution.

1. What would you expect the distribution of the outcomes from a roll of a single die to look like, in other words, what is its shape? (Hint: What is the probability of getting each value?)
Shape = ______________________________

2. Calculate the mean of our population. (Hint: What is the mean outcome for rolling a single die?)
$\mu =$ ______________________________

3. Calculate the standard deviation of our population. (Hint: What is the standard deviation of all possible outcomes from rolling a single die?)
$\sigma =$ ______________________________

In order to explore the properties of the Central Limit Theorem, let's begin by creating a distribution of the original population. To do so, follow these steps:

1. Roll your die 50 times and record each outcome.

2. Combine your results with at least 2 other students and tally the frequency of each roll of the dice from the combined results. Record your results in a table similar to the following:

Outcome	Frequency
1	
2	
3	
4	
5	
6	

3. Draw a bar graph of these frequencies.

4. Does the distribution appear to be a normal distribution? Is this what you expected from question 1?

Return to your original data. After each set of 10 rolls, calculate the mean of that sample. Round your answers to one decimal place. (You should have 5 sample means.)

5. Combine your sample means with that of as many of your classmates as you can. Record the sample means of each of your classmates' 5 samples. (For this experiment to work the best, you need to find at least 6 classmates.)

6. Tally the frequency of the sample means from your combined results in a table like the one that follows.

Sample Mean	Frequency
1.0 – 1.3	
1.4 – 1.7	
1.8 – 2.1	
2.2 – 2.5	
2.6 – 2.9	
3.0 – 3.3	
3.4 – 3.7	
3.8 – 4.1	
4.2 – 4.5	
4.6 – 4.9	
5.0 – 5.3	
5.4 – 5.7	
5.8 – 6.0	

7. Draw a histogram of the sample means.

8. Does the distribution appear to be a normal distribution?

9. What is the mean of your sample means? (Hint: Use the raw data you collected.) $\mu_{\bar{x}} =$ ______________
How does $\mu_{\bar{x}}$ compare to μ from question 2?

10. What is the standard deviation of the sample means? (Again, go back to your raw data and use a calculator or software package.) $\sigma_{\bar{x}} =$______________. How does $\frac{\sigma}{\sqrt{n}}$ compare to $\sigma_{\bar{x}}$ from question 3?

11. Do your results seem to verify the three conclusions of the Central Limit Theorem?

Chapter Project B: Sampling Distribution Simulation

Using the Hawkes Learning Systems Software: Statistics, open the lesson titled Sampling Distribution Simulation. The Sampling Distribution Simulation is designed to help you better understand sampling distributions as well as the Central Limit Theorem. Begin the simulation by choosing a parent distribution from the pull-down menu at the top of the screen. If you do not have a preference, the computer will automatically begin with a uniform parent distribution. Press SIMULATE to start the first iteration of data from that distribution. For each iteration, the computer randomly chooses 30 numbers from the parent distribution and displays them in the parent histogram. At the same time, the computer chooses samples of size 5, 15, and 30 from those same 30 numbers. The mean of each sample is calculated and then displayed in its respective histogram. Answer the following questions:

1. How many numbers are displayed in the parent histogram?

2. How many numbers are displayed in each of the sampling distributions?

Press NEXT to obtain a second iteration.

3. After the second iteration, how many numbers are displayed in the parent histogram?

4. After the second iteration, how many numbers are displayed in each of the sampling distributions?

Now let's see what happens after many iterations. Press AUTO and let the simulation run until about 100 iterations have passed. Press STOP again to stop the process.

5. How many numbers are displayed in the parent histogram? (This number will vary, depending on how many iterations have passed.)

6. How many numbers are displayed in each of the sampling distributions?

7. Which, if any, of the sampling distributions appear to have a normal shape?

Finally, press AUTO again and allow the program to process at least 1500 iterations.

8. Which of the sampling distributions appear to have a normal shape?

9. Compare the means of sample means listed in the table. Do these numbers behave as you would expect according to the Central Limit Theorem?

10. Compare the standard deviations of sample means listed in the table. Do these numbers behave as you would expect according to the Central Limit Theorem?

Take some time to explore the other parent distributions available. Although the shape of the parent histogram will differ with each parent used, the Central Limit Theorem principles will always remain constant.

Chapter Technology – Calculating z using the Central Limit Theorem

TI-84 PLUS

This chapter extends on the concepts discussed in the previous chapter. In this chapter, the methods for calculating the probability are the same as those discussed in Chapter 6. The difference is that we now have introduced new formulas for z that are needed when calculating the probability of a sample mean, or sample proportion. To learn how to use Microsoft Excel or a TI-84 Plus calculator to find a given probability, refer to the Chapter Technology section in Chapter 6. In this Chapter Technology section, we will focus on how to use your TI-84 Plus calculator to find the value of z.

Example T.1

Find the value of z using the formula for the Central Limit Theorem given that $\bar{x} = 34$, $\mu = 35$, $\sigma = 5$, and $n = 100$.

Solution:

The equation we need is the z-value for population means using the Central Limit Theorem. Let's begin by substituting our values into the equation.

$$z = \frac{\bar{x} - \mu}{\frac{\sigma}{\sqrt{n}}} = \frac{34 - 35}{\frac{5}{\sqrt{100}}}$$

Now we need to enter this into the calculator. We must make sure that we put parentheses around the numerator and the denominator. Enter the following into your calculator $(34-35)/\left(5/\sqrt{(100)}\right)$. Press ENTER. Thus $z = -2$.

Example T.2

Find the value of z using the formula for the Central Limit Theorem given that $\hat{p} = 0.56$, $p = 0.54$, and $n = 81$.

Solution:

The equation we need is the z-value for population proportions using the Central Limit Theorem. Let's begin by substituting our values into the equation.

$$z = \frac{\hat{p} - p}{\sqrt{\frac{p(1-p)}{n}}} = \frac{0.56 - 0.54}{\sqrt{\frac{0.54(1-0.54)}{81}}}$$

Now we need to enter this into the calculator. We must make sure that we put parentheses around the numerator and the denominator. Enter the following into your calculator $(.56 - .54)/\sqrt{(.54(1 - .54)/81)}$ and press ENTER. Thus $z = 0.36$.

MICROSOFT EXCEL

We can also find the values for z from the previous examples using Microsoft Excel.

Example T.3

Find the value of z using the formula for the Central Limit Theorem given that $x = 34$, $\mu = 35$, $\sigma = 5$, and $n = 100$.

Solution:

Recall that the Central Limit Theorem states that the standard deviation of a sampling distribution, $\sigma_{\bar{x}}$, equals the standard deviation of the population divided by the square root of the sample size. The formula for calculating the value of z is =STANDARDIZE(*x, mean, standard_dev*). Using the Central Limit Theorem, we input =STANDARDIZE(34, 35, 5/SQRT(100)); again, we see $z = -2$.

Example T.4

Find the value of z using the formula for the Central Limit Theorem given that $\hat{p} = 0.56$, $p = 0.54$, and $n = 81$.

Solution:

Here we will use $\sigma_{\hat{p}} = \sqrt{\dfrac{p(1-p)}{n}}$. Enter "=STANDARDIZE(0.56, 0.54, SQRT(0.56*(1-0.56)/81)) into Excel. The answer, 0.36262, is displayed.

MINITAB

Minitab can calculate a column of z-scores with a single operation.

Example T.5

An emergency dispatch processing center believes its call processing time has a population mean of 45 seconds and a standard deviation of 50 seconds. The operators are evaluated one day and their total incidents and average processing times are displayed in the table below. Find their z-scores.

n	98	85	105	110	91
Sample Mean	52	48	63	45	55

Solution:

First, enter the data into the worksheet. The first column is n, the number of incidents, and the second column contains each operator's sample mean. Select CALC then CALCULATOR and enter the following expression: (C2-45)/(50/SQRT(C1)). Choose to store the result in column C3 and select enter. The dialog box is shown to the right.

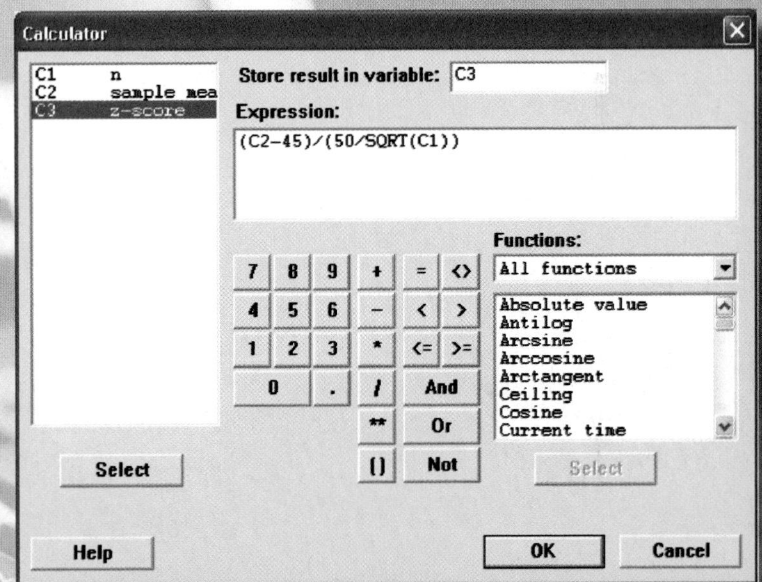

The column produced contains the z-scores for each operator.

Worksheet 1 ***

↓	C1	C2	C3
	n	sample mean	z-score
1	98	52	1.38593
2	82	48	0.54332
3	105	63	3.68890
4	110	45	0.00000
5	91	55	1.90788

Confidence Intervals

Sections

Objectives

The objectives of this chapter are as follows:

1. Determine the best point estimate for population parameters
2. Calculate the margin of error for confidence intervals for population parameters
3. Create confidence intervals for population parameters
4. Determine the minimum sample size required to obtain the desired level of precision for estimation

Chapter 8

Introduction

The battle for the Presidency of the United States in 2004 was fierce. Polls leading up to Election Day showed incumbent George W. Bush and challenger John Kerry in a statistical dead heat. Media outlets worked themselves into a frenzy making predictions and spinning the issues, while voters in key battleground states such as Florida, Ohio, and Pennsylvania basked in the attention lavished upon them by the dueling candidates. The presidency hinged on these key states, and Americans anticipated a tight race.

Election Day dawned, and data released from exit polls throughout the afternoon painted quite a different picture. Democrat John Kerry was projected to comfortably win in most of the battleground states, and some political commentators went so far as to predict that he would win the White House as well. However, when the votes were finally tallied late into the night, it was Bush who actually won many of these key states and subsequently the presidency. The exit polls had been wrong.

Many Americans were puzzled at the discrepancies in the information given to them by various media sources. How is it possible that the most reputable polling companies in America could be wrong in so many of their predictions? Were the exit polls biased? Or, is it possible that the results obtained by statisticians were within the bounds of statistical error?

Introduction to Estimating Population Means 8.1

The major purpose of statistics is to provide information so that informed decisions can be made. Statisticians are asked to provide information about a variety of population parameters. For example, a television network may want to know how many Americans watch their TV shows each week. It is impossible to survey each and every household, yet the network needs the best information available in order to make programming decisions. This is where inferential statistics play an important role.

Recall that inferential statistics is the branch that uses sample statistics to provide estimates for population parameters. For a particular parameter, there may be several sample statistics that could be used to provide estimates. However, the best estimate for a population parameter is one that is *unbiased*—that is, a sample statistic that does not consistently underestimate or overestimate the population parameter. When the estimate for a population is a single value, it is called a **point estimate**.

Point Estimate

A **point estimate** is a single number estimate of a population parameter.

In this section, we will concentrate on estimating the population mean. Three obvious statistics that could be used to estimate the population mean are the sample mean, median, and mode. It is not necessarily true of sample modes or medians that their values will be centered on the mean of the population. However, recall from our discussion of the Central Limit Theorem in section 7.1 that the mean of a sampling distribution of sample means *is* equal to the mean of the population. Thus, the sample mean does not consistently underestimate or overestimate the population mean and is actually the best point estimate of the population mean.

Memory Booster!

The best point estimate of the population mean, μ, is the sample mean $\bar{x}$.

Example 1

Find the best point estimate for the population mean of test scores on a standardized biology final exam. The following is a simple random sample taken from these test scores.

45	68	72	91	100	71
69	83	86	55	89	97
76	68	92	75	84	70
81	90	85	74	88	99
76	91	93	85	96	100

Solution:

The best point estimate for a population mean is the sample mean. The sample mean for the given sample of test scores is $\bar{x} \approx 81.6$. Thus the best point estimate for the population mean on this standardized exam is 81.6.

As you might imagine, using only one value for estimating the population mean, or any other population parameter, leaves much room for error. In fact, we have no way of knowing the likelihood that we have chosen the exact value of the parameter, or that we are even close. It would be much better to provide a range of values rather than just one number to estimate the parameter. This range of possible values is called an **interval estimate**.

Notice that an interval estimate is still an estimate. Just as we couldn't be certain that the point estimate was exact, we also cannot be certain that the interval estimate contains the true population parameter. Although not certain, how confident can we be? The probability that the interval actually contains the true population parameter is called the **level of confidence**, and is represented by the symbol c. Thus if we are 95% confident that the interval estimate contains the population parameter, then $c = 0.95$ and the interval estimate is a 95% confidence interval. An interval estimate associated with a certain level of confidence is called a **confidence interval**.

Definitions

An **interval estimate** is a range of possible values for the population parameter.

The **level of confidence** is the degree of certainty that the interval estimate contains the population parameter.

A **confidence interval** is an interval estimate associated with a certain level of confidence.

In general, how do we begin to find an confidence interval for a population parameter? Obviously, the confidence interval should be constructed around the best point estimate we can find. But, how big do we make the interval? The largest possible distance away from the point estimate that the confidence interval will cover is called the **margin of error,** or maximum error of estimate, E. To construct the confidence interval for population parameters, you simply add and subtract E from the point estimate.

Margin of Error

The **margin of error**, or *maximum error of estimate*, E, is the largest possible distance from the point estimate that a confidence interval will cover.

When estimating the population mean, begin with the best point estimate, $\bar{x}$, then add and subtract E, giving a lower endpoint of $\bar{x} - E$ and an upper endpoint of $\bar{x} + E$.

Confidence Interval for Population Means:

$$\bar{x} - E < \mu < \bar{x} + E$$
$$\text{or } (\bar{x} - E, \bar{x} + E)$$

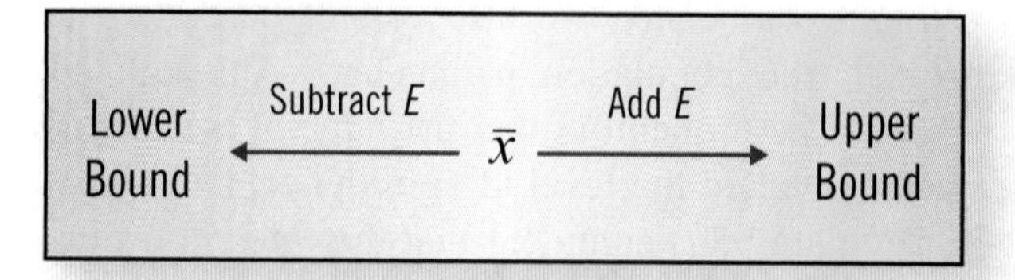

Figure 1 – Constructing a Confidence Interval for Means

Let us stop at this point and emphasize how to interpret confidence intervals. We have no way of being certain, short of knowing the actual parameter, if a particular interval estimate actually contains the population mean—just as we had no way of knowing how close the point estimate was to the true population mean. Furthermore, the population parameter is a fixed value. It is already determined—we just don't know its value. So when we interpret a confidence interval, we do *not* want to say that there is a 95% chance of the population mean falling within our confidence interval. We *can* say that we are 95% confident that the confidence interval contains the population mean, because this statement correctly implies that the population mean is fixed and it is either in the interval, or not.

Example 2

A college student researching study habits collects data from a random sample of 250 college students on her campus and calculates that the sample mean is $\bar{x} = 15.7$ hours per week. If the margin of error for her data using a 95% level of confidence is $E = 2.2$ hours, construct a 95% confidence interval for her data. Interpret your results.

Solution:

The best point estimate for the population mean is the sample mean, so use $\bar{x} = 15.7$ as the point estimate for this population parameter. We are given the value of the margin of error for our confidence interval, $E = 2.2$. To find the lower endpoint, subtract the margin of error from the sample mean. Thus the lower endpoint is

$$15.7 - 2.2 = 13.5 \text{ hours.}$$

To find the upper endpoint, add the margin of error to the sample mean. Thus the upper endpoint is

$$15.7 + 2.2 = 17.9 \text{ hours.}$$

Therefore, the confidence interval ranges from 13.5 to 17.9. The confidence interval can be written mathematically as

$$13.5 < \mu < 17.9,$$

or as the interval (13.5, 17.9).

The interpretation of our confidence interval is that we are 95% confident that the true population mean for the number of hours students on this campus study per week is between 13.5 and 17.9 hours.

So far, constructing a confidence interval for a population mean does not seem much more complicated than finding the point estimate for the population mean. However, when calculating the margin of error it is important to choose the right technique. The flow chart on the following page provides a simple way for determining the appropriate method.

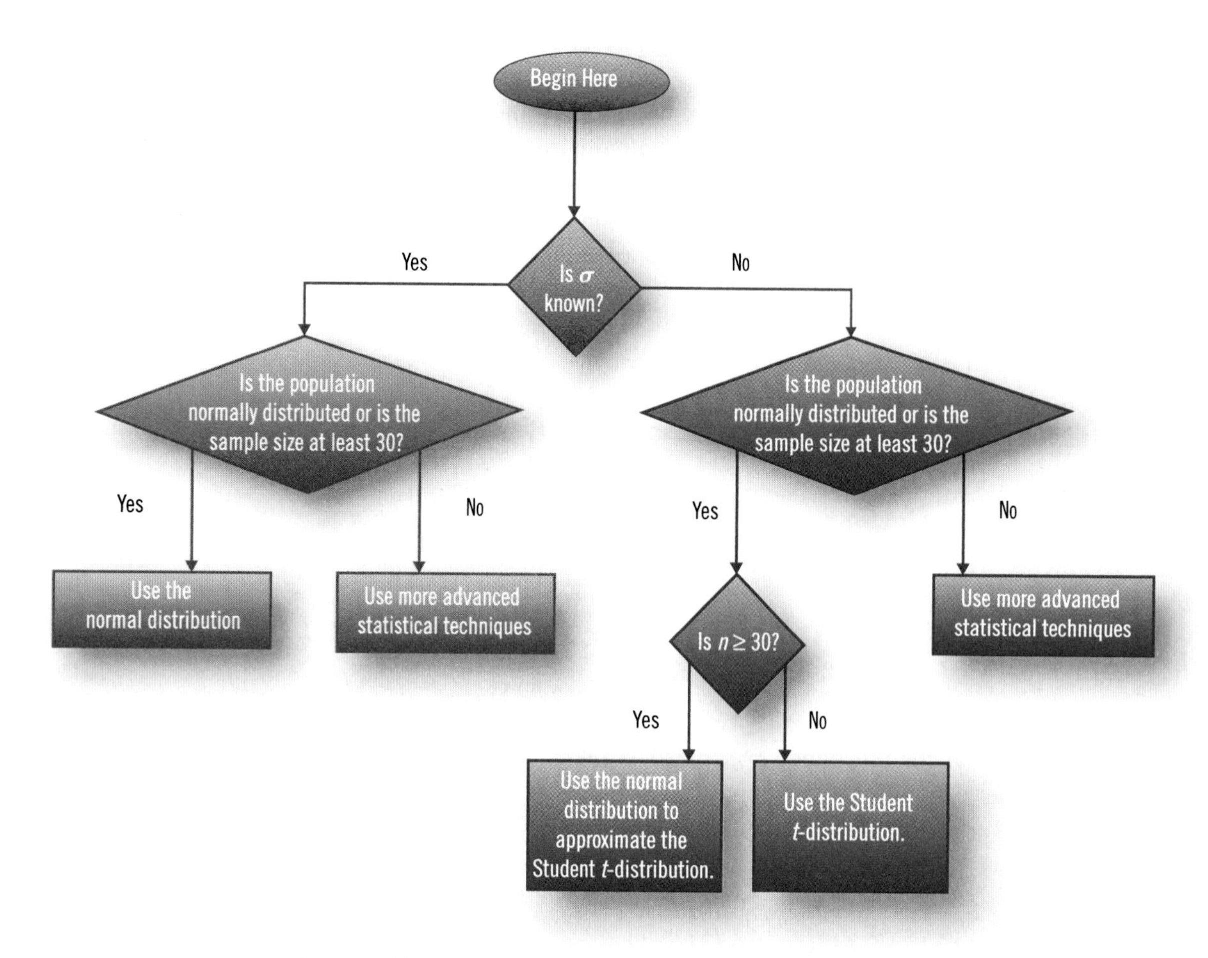

Figure 2 – How to choose z or t when using charts within this text

Notice that we begin by determining whether the population's standard deviation, σ, is known. If σ is known, then either the normal distribution or more advanced statistical techniques must be used. However, essentially the only way you could know the population's standard deviation is to know every value in the population. With this knowledge you could calculate *any* statistic you needed—there would be no need to estimate anything! Though moving down the left side of the flow chart is the best choice if σ is known, in practice it would be a very rare event to truly know the population's standard deviation. The left hand side of the chart is shaded in green as a reminder that, though this side accurately reflects statistical theory, it is seldom used in practice.

Rarely do we actually know the population standard deviation. As a result, two facts are crucial: the distribution of the population and the size of the sample. If the population is not normally distributed and the sample size is less than 30, then advanced statistical techniques that are beyond the scope of this book are required to estimate the population mean. However, if either the population is *normally distributed* or the sample size is *at least 30*, then the Student t-distribution may be used to calculate the margin of error for population means.

When the sample size is at least 30, the normal distribution may be used as an approximation to the Student t-distribution. To explain why this is true, recall for a moment that the Student t-distribution has only one parameter, the *degrees of freedom*. When using the t-distribution to calculate a confidence interval for population means, the degree of freedom is one less than the sample size, $n - 1$. Table C lists the values that we need for all sample sizes less than 30. However, individual sample sizes greater than 30 are not listed. The reason for this is that as the sample size tends toward infinity, the shape of the Student t-distribution approaches the normal distribution. Therefore, if the sample size is at least 30, we can use the normal distribution tables. In summary, for sample sizes less than 30, we use the Student t-distribution with $n - 1$ degrees of freedom; and for sample sizes of at least 30, we use the normal distribution as a close approximation to the t-distribution.

Example 3

a. An amusement company wants to estimate the average amount that each state fair visitor spends on tickets, games, and food. They choose, at random, 55 state fair visitors one afternoon and ask each to record the amount of money spent that afternoon before leaving the park. If the distribution of the population is skewed to the left and the population standard deviation is unknown, which technique should be used to calculate a confidence interval for this population mean?

b. The sportscasters at a small news station wish to estimate the number of hours that men spend watching sporting events each week. They plan to survey 25 men locally. From previous studies, they believe that the population has a skewed distribution with a standard deviation of 3.1 hours. Which technique should be used to calculate a confidence interval for this population mean?

Solution:

a. First, note that σ is unknown for this example. Next, look to the shape of the distribution and the sample size. The population is not normally distributed, but the sample size is greater than 30, so the normal distribution may be used to approximate the t-distribution when calculating this confidence interval.

b. In this example, we are told that $\sigma = 3.1$, so it is certainly known. Next look at the shape of the distribution and the sample size. Because the population is not normally distributed nor is the sample size large enough, the normal distribution cannot be used here. Other more advanced techniques would be needed to calculate this confidence interval.

Since, in practice, the population standard deviation is rarely known, this textbook will restrict its focus to estimating population means when σ is unknown. As we have seen, the technique to use in this case depends on the population's distribution and the size of the sample. We will discuss these techniques in detail in the next two sections.

Exercises

8.1

Directions: Answer each of the following questions.

1. A survey of 42 teachers finds that they spend on average \$18 a week on lunch. What is the best point estimate for the mean spent on lunch for all teachers?

2. The average number of pets per student for a sample of 47 students at Brown Elementary is 2.5 pets. What is the best point estimate for the mean number of pets per student for all students at Brown Elementary?

3. The sample mean of passing yards for a sample of college quarterbacks is 177 yards per game. What is the best point estimate for the mean number of yards passed per game for all college quarterbacks?

4. The average electric bill for 50 residents of Country Meadows is \$212 per month. What is the best point estimate for the mean monthly electric bill for all Country Meadows' residents?

5. A survey of teachers reports a point estimate for the mean amount of money spent each week on lunch is \$18. If the margin of error for a 95% confidence interval for the mean amount of money spent each week on lunch by all teachers is \$1.70, construct a 95% confidence interval for the mean amount of money spent each week on lunch by all teachers.

6. The mean batting average for a sample of 35 professional baseball players is 0.283. If the margin of error for the population mean with a 99% level of confidence is 0.051, construct a 99% confidence interval for the mean batting average for professional baseball players.

7. A sample of 581 teenagers revealed that the mean number of hours they sleep per night is 10.5 hours. If the margin of error for the population mean with a 98% confidence interval is 1.3 hours, construct a 98% confidence interval for the mean number of hours teenagers sleep per night.

8. The mean number of rushing yards for a sample of running backs is 78 yards per game. If the margin of error for the population mean with a 95% confidence interval is 9.1 yards, construct a 95% confidence interval for the mean number of rushing yards per game for all running backs.

Directions: For studies with the following characteristics, determine which of the following methods would be most appropriate when calculating the margin of error for the population mean: 1) using z, 2) using t, or 3) using more advanced techniques.

9. σ is unknown; $n = 14$; the population is normally distributed

10. σ is known; $n = 21$; the population is normally distributed

11. σ is unknown; $n = 45$; the population is uniformly distributed

12. σ is unknown; $n = 26$; the population is skewed to the right

13. σ is known; $n = 57$; the population is skewed to the left

14. σ is unknown; $n = 6$; the population is normally distributed

15. σ is known; $n = 11$; the population is uniformly distributed

16. σ is unknown; $n = 85$; the population is normally distributed

Directions: For the following studies, determine which of the following methods would be most appropriate when calculating the margin of error for the population mean: 1) using z, 2) using t, or 3) using more advanced techniques.

17. Amanda is studying trends in the mortgage lending industry. She wishes to estimate the average amount borrowed by a first-time homeowner. She obtains information from 61 first-time homeowners and finds that they borrowed an average of $134,600. Amanda believes that the distribution of home loans is skewed to the right. What distribution should Amanda use?

18. One of the assignments for Norie's summer internship is to estimate the number of people who respond to advertisements in the newspaper for the real estate firm where she is working. From 9 ads that the firm placed during the month of June, she found an average response rate of 12.3 people per ad. If the distribution of responses in the population is normal, what distribution should Norie use when making her estimate?

19. A boutique in a college town wants to estimate the average amount of sales they receive on weekends in which there is a home football game. They calculate the average amount of sales for 7 home football game weekends, and find the sample mean to be $2788. They assume that the population is skewed right, because sales are much higher the week of Homecoming. What distribution should they use to make their estimate?

20. An auto manufacturer wants to estimate the number of miles usually put on their vehicles before a visit to the mechanic is necessary. They obtain information from 59 of their vehicles and find that the average number of miles at their first visit to the shop is 41,300. The manufacturer believes that the standard deviation for the population is 5600 miles. What distribution should the auto manufacturer use?

21. A researcher for a special show on the cooking station wants to estimate the number of times that Americans dine out each week. The researcher interviews 25 Americans at random and finds that they dine out on average 1.6 times a week. If the population is distributed normally, what distribution should the researcher use to make his estimate?

22. An environmentalist wants to estimate the average amount of trash, in pounds, that households in America produce each week. She surveys 28 households and finds that they throw away on average 17.8 pounds of trash each week. If previous studies show that the population is normally distributed with a standard deviation of 2.9 pounds, which distribution should the environmentalist use?

Estimating Population Means (Large Samples)

8.2

In the previous section we discussed two ways to estimate a population parameter, a point estimate and a confidence interval. Constructing a confidence interval involves using a point estimate, and a margin of error for a specific level of confidence. In all of our previous examples, the margin of error was calculated for us since we did not yet have the tools to calculate this value. In this section we will see how to calculate the margin of error for population means when sufficiently large samples have been taken.

As we have seen, the methods used to calculate the margin of error vary depending on the characteristics of the study. Here, we will focus on estimating the population mean when the following criteria are true:

- All possible samples of a given size have an equal probability of being chosen,
- the size of the sample is at least 30 ($n \geq 30$) and
- the population's standard deviation is unknown.

When all of the above conditions are met, then the distribution used to calculate the margin of error for the population mean is the Student t-distribution. However, when $n \geq 30$, the critical values for the t-distribution are almost identical to the critical values for the normal distribution at corresponding levels of confidence. For this reason, we can use the normal distribution, or z-distribution, to approximate the t-distribution. Because all of the problems we look at in this section will have sample sizes of at least 30, we will use the normal distribution to calculate the margin of error for the mean of each population.

The first step in calculating the margin of error for confidence intervals for population means is to determine the critical z-value at your particular level of confidence. We call z_c the **critical value** for a level of confidence, c. This z_c has the property that the area under the standard normal curve between $-z_c$ and z_c is equal to c. For example, to find the critical value for a 95% confidence interval, we need to find the values of $-z_{0.95}$ and $z_{0.95}$ which have an area of 0.95 between them as shown in Figure 3 on the next page. Using our knowledge of the normal curve, an area of 0.05 is split equally between the two tails, so each tail has an area of 0.025. The standard normal z table tells us that the z-value with 0.025 of the distribution's area to its left is $-z_{0.95} = -1.96$. By the symmetry of the normal curve, the other critical value must be $z_{0.95} = 1.96$.

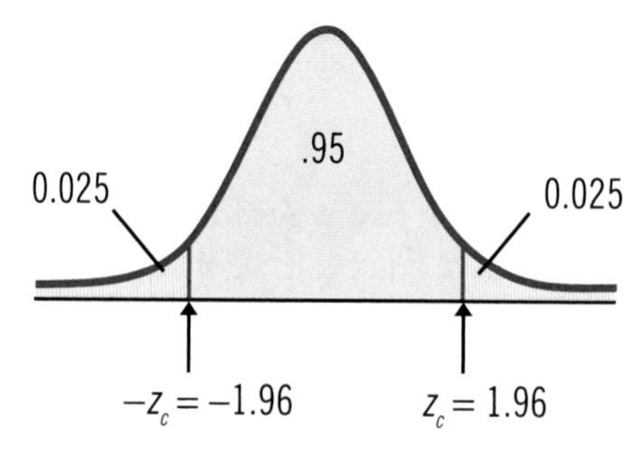

Figure 3 – Finding a Critical z-Value

The critical values for any given level of confidence may be calculated using the same method. Notice that the critical value is based only on the level of confidence and not on the size of the sample or any of the population's parameters. Hence, the critical values for a 95% level of confidence for population means using the normal curve remain the same for every problem. Commonly used levels of confidence include 90%, 95%, and 99%. The Critical z-Values Table, shown below, may be used as a reference.

Table 1 - Critical z-Values for Confidence Intervals

Level of Confidence, c	z_c
0.80	1.28
0.85	1.44
0.90	1.645
0.95	1.96
0.98	2.33
0.99	2.575

Rounding Rule

When calculating the margin of error, round to one more decimal place than the original data, or the same number of places as the standard deviation.

Note that critical values can be written using different symbols. The symbol z_c can also be written as $z_{\alpha/2}$, where $\alpha = 1 - c$, since $z_{\alpha/2}$ marks the place on the standard normal curve at which $\frac{\alpha}{2}$ of the distribution's area is to its right. Thus $z_{0.95}$ is also called $z_{0.05/2}$ or $z_{0.025}$.

Once we know the critical value for z, we can calculate the margin of error, E, using the following formula:

$$E = z_c \frac{s}{\sqrt{n}}.^*$$

Footnote: From the Central Limit Theorem, we can derive the theoretical formula for E which is* $E = z_c\sigma_{\bar{x}}$, *or equivalently* $E = z_c \frac{\sigma}{\sqrt{n}}$. *However, since* σ *is unknown, we use the sample standard deviation* **s, *a point estimate for* σ.

Example 4

Find the margin of error for a 99% confidence interval, given a sample of size 100 with a sample standard deviation of 15.50. Round your answer to two decimal places.

Solution:

First, refer to the table to find the critical value for $c = 0.99$. In the table we see that $z_{0.99} = 2.575$. Substituting these values into the formula for the margin of error, we get:

$$\begin{aligned} E &= z_c \frac{s}{\sqrt{n}} \\ &= 2.575 \frac{15.50}{\sqrt{100}} \\ &\approx 3.99. \end{aligned}$$

So the margin of error for this 99% confidence interval is 3.99.

Recall from the previous section that once the value for the margin of error for a particular confidence interval has been calculated, we simply add and subtract the margin of error to the point estimate to construct the confidence interval.

Confidence Interval for Population Means:

$$\bar{x} - E < \mu < \bar{x} + E, \text{ or}$$
$$(\bar{x} - E, \bar{x} + E)$$

Example 5

Construct a 99% confidence interval from the data in the above example, given that the sample mean equals 25.99.

Solution:

From the previous example, we have a margin of error of 3.99. The best point estimate for the population mean is the sample mean, 25.99. To calculate the lower endpoint of the confidence interval, subtract the margin of error from the point estimate:

$$25.99 - 3.99 = 22.00.$$

To calculate the upper endpoint of the confidence interval, add the margin of error to the point estimate:

$$25.99 + 3.99 = 29.98.$$

cont'd on the next page

Thus the confidence interval ranges from

$$22.00 \text{ to } 29.98,$$

$$\text{or } 22.00 < \mu < 29.98,$$

$$\text{or } (22.00, 29.98).$$

Example 6

A toy company wants to know the average number of new toys bought for children each year. Marketing strategists at the toy company collect data from the parents of 1842 children. The sample mean is found to be 4.7 toys with a sample standard deviation of 1.9 toys. Construct a 99% confidence interval for the number of new toys purchased per child each year.

Solution:

To begin, we need to calculate the margin of error. Since we want a 99% confidence interval, the critical z-value is 2.575. Substituting the values we have into the formula for margin of error, we get the following:

$$\begin{aligned} E &= z_c \frac{s}{\sqrt{n}} \\ &= 2.575 \frac{1.9}{\sqrt{1842}} \\ &\approx 0.1 \end{aligned}$$

To calculate the lower endpoint of the confidence interval, subtract the margin of error from the best point estimate, i.e. the sample mean,

$$4.7 - 0.1 = 4.6.$$

To calculate the upper endpoint of the confidence interval, add the margin of error to the point estimate:

$$4.7 + 0.1 = 4.8.$$

Thus the confidence interval ranges from

4.6 to 4.8 new toys per child per year,

or $4.6 < \mu < 4.8$,

or $(4.6, 4.8)$.

✦ Finding the Minimum Sample Size for Means

Suppose that a researcher wants to conduct a statistical study with a high level of confidence. It makes sense that if you want to increase the level of confidence, then you need to base your estimates on more data, i.e. larger sample sizes. A logical question at this point might be, "How large a sample size do I need?" The answer is that it depends on the level of confidence and margin of error that you plan to use. It would be nice to have very large samples with thousands of data values, but time and money will usually prevent us from such luxuries. So, what is the fewest number of data values that we need to obtain the desired level of confidence with a particular margin of error? That is, what is the minimum sample size we must have to obtain our requirements? Rather than simply using a very large sample, we can calculate the minimum sample size necessary using the following formula:

Rounding Rule

Always round sample sizes up to the next whole number.

$$n = \left(\frac{z_c \sigma}{E}\right)^2.$$

As we have repeatedly emphasized, knowing the population's standard deviation, σ, is very unlikely. So we must once again determine a way to estimate σ. One method is to use the sample standard deviation, s, a point estimate for σ. We can also use estimates for the population standard deviation from similar studies on the same population or from pilot studies on the population. Another method, if the range of the population is known, is to divide the range of the population by 6 and use this value as an estimate for the population's standard deviation.

When calculating the minimum sample size, remember that you are trying to determine the number of data values you need, which is a whole number. So if the formula results in a decimal value, you should always *round up* to the next whole number. For example, suppose we needed to choose a sample of students for our study, and using the formula for the minimum sample size resulted in the answer 59.07. How many students do we need in our sample? Well, 59 students would be too few. Since 59 is less than 59.07 and we cannot sample a 0.07 of a student, we must have at least 60 students in our sample. However, remember that this is just the *minimum* sample size needed in order to achieve the requirements. In reality, you should always try to sample as many as possible.

Example 7

Determine the minimum sample size needed if you wish to be 90% confident that the sample mean is within two units of the population mean. An estimate for the population's standard deviation of 8.4 is available from a previous study.

Solution:

From the information that we are given, we know the following values: $c = 0.90$, $E = 2$, and $\sigma \approx 8.4$. Since we desire a 90% level of confidence, we can use our Critical z-Value Table to determine that $z_{0.90} = 1.645$. Substituting these values into our formula for minimum sample size, we obtain the following:

$$\begin{aligned} n &= \left(\frac{z_c \sigma}{E}\right)^2 \\ &= \left(\frac{1.645 \cdot 8.4}{2}\right)^2 \\ &\approx 47.7342 \end{aligned}$$

So the size of the sample that we need to construct a 90% confidence interval for the population mean with the desired margin of error is at least 48.

Exercises 8.2

Note: For all exercises in this section, you may assume that the requirements mentioned in this section are met, namely, all samples of a certain size have an equal chance of being chosen, all sample sizes are at least 30, and the population's standard deviation is unknown.

Directions: Calculate the margin of error given the following information.

1. $n = 56$, $s = 3.14$, $c = 0.90$

2. $n = 81$, $s = 2.45$, $c = 0.95$

3. $n = 93$, $s = 1.25$, $c = 0.95$

4. $n = 134$, $s = 0.27$, $c = 0.99$

Directions: Construct a confidence interval given the following information.

5. $n = 89, s = 2.01, c = 0.95, \bar{x} = 45$

6. $n = 64, s = 8.01, c = 0.90, \bar{x} = 90.4$

7. $n = 607, s = 1.92, c = 0.99, \bar{x} = 18.45$

8. $n = 1123, s = 7.31, c = 0.95, \bar{x} = 87.12$

Directions: Calculate the minimum sample size needed to construct a confidence interval with the desired characteristics. The value of the population's standard deviation is an estimate based on a previous, reliable study.

9. $E = 0.5, \sigma = 5.25, c = 0.95$

10. $E = 2, \sigma = 12.10, c = 0.90$

11. $E = 1.5, \sigma = 4.75, c = 0.99$

12. $E = 3, \sigma = 15.03, c = 0.95$

Directions: Answer each of the following questions.

13. A sample of 78 students averaged 15 hours of studying per week with a standard deviation of 2.3 hours. Calculate the margin of error for a 90% confidence interval.

14. A sample of 56 teachers averaged 8 hours per week working at home after school with a standard deviation of 1.5 hours. Calculate the margin of error for a 95% confidence interval.

15. A sample of 120 American households has a mean computer usage time of 19.2 hours per week with a standard deviation of 3.3 hours. Calculate the margin of error for the mean computer usage of all American households. Use a 95% level of confidence.

16. A survey of 85 homeowners finds that they spend on average $67 a month on home maintenance with a standard deviation of $14. Calculate the margin of error for the mean amount spent on home maintenance by all homeowners. Use a 99% level of confidence.

17. A survey of 85 homeowners finds that they spend on average $67 a month on home maintenance with a standard deviation of $14. Find the 99% confidence interval for the mean amount spent on home maintenance by all homeowners.

18. A survey of 97 homeowners found that the average amount spent on lawn service per year was $720 per year with a standard deviation of $123. Calculate a 98% confidence interval for the average amount spent on lawn service per household each year.

19. A sample of 140 dieters reported that they "cheated" on their diet an average of 7 times a week with a standard deviation of 1.5. Calculate a 99% confidence interval for the average number of times dieters "cheat" on their diets each week.

20. A sample of 120 American households has a mean computer usage time of 19.2 hours per week with a standard deviation of 3.3 hours. Find the 90% confidence interval for the mean computer usage of all American households.

21. You would like to estimate the average number of credit cards college students have in their wallets. You would like to create a 98% confidence interval with a maximum error of 1 card. Assuming a standard deviation of 3.25, what is the minimum number of college students you must include in your sample?

22. A social worker is concerned about the number of prescriptions her elderly patients have. She would like to create a 99% confidence interval with a maximum error of 2 prescriptions. Assuming a standard deviation of 5.2, what is the minimum number of patients she must sample?

23. Suppose you are interested in determining the average number of hours students spend studying each week. You want a 95% level of confidence and a maximum error of 0.5 hours. Assuming the standard deviation is 2.5 hours, what is the minimum number of students you must include in your sample?

24. For a psychology experiment, Emma is assigned the task of finding the average number of hours students sleep per night. Her results must be at the 99% level of confidence with a maximum error of 0.25 hours. Assuming the standard deviation is 1.4 hours, how many students must Emma survey?

Estimating Population Means (Small Samples)

8.3

In the previous section, we were able to find confidence intervals for the population mean so long as the sample size was at least 30 by using the normal distribution as an approximation for the *t*-distribution. In reality, though, cost or time often prohibits researchers from sampling even thirty members of the population. Take, for example, the world of dentistry. When experimenting with new methods of treatment, the cost of a single trial can easily reach \$5000. You can see that testing 30 participants could get unreasonably costly—very quickly.

In this section, we will focus on the methods used to estimate a population mean when the sample obtained is less than 30. In this case, it is essential that the distribution of the population is approximately normal. Therefore, in this section we will focus on estimating the population mean when the following criteria are true:

- all possible samples of a given size have an equal probability of being chosen
- the size of the sample is less than 30 ($n < 30$)
- the distribution of the population is approximately normal
- the population's standard deviation is unknown

When all of the above conditions are met, the distribution used to calculate the margin of error for the confidence interval for the population mean is the Student *t*-distribution. The formula for the margin of error using the Student *t*-distribution is

$$E = t_{\alpha/2} \frac{s}{\sqrt{n}}.$$

Recall from Chapter 6 that to find the value for $t_{\alpha/2}$, you use the Critical *t*-Value Table with $n - 1$ degrees of freedom. Remember also that $\alpha = 1 - c$, where c is the level of confidence.

Memory Booster!

d.f. = degrees of freedom

Rounding Rule

When calculating the margin of error, round to one more decimal place than the original data, or the same number of places as the standard deviation.

Example 8

Assuming the population is approximately normal and all samples have an equal probability of being chosen, find the margin of error for a sample size of 10 given that $s = 15.5$ and the level of confidence is 95%.

Solution:

Since the question tells us the values for s and n, the only missing value in the calculation of E is $t_{\alpha/2}$. First we need to find α. Since the level of confidence is 95%, $\alpha = 1 - 0.95 = 0.05$. Therefore, $t_{\alpha/2} = t_{0.05/2} = t_{0.025}$.

cont'd on the next page

Since the sample size is 10, the number of degrees of freedom is 9. Therefore, $t_{0.025} = 2.262$ which gives us:

$$E = 2.262\frac{15.5}{\sqrt{10}}$$
$$\approx 11.1.$$

So the margin of error is 11.1.

Once we have calculated E, the confidence interval is constructed in the same way as before. We simply add and subtract the margin of error from the sample mean to find the endpoints of the interval.

Confidence Interval for Population Means

$$\bar{x} - E < \mu < \bar{x} + E, \text{ or}$$
$$(\bar{x} - E, \bar{x} + E)$$

Example 9

Construct a 95% confidence interval from the data in the above example, given that $\bar{x} = 187.7$.

Solution:

From the previous example, we know that $E = 11.1$. Therefore, to find the lower endpoint of the confidence interval, subtract 11.1 from the sample mean of 187.7, obtaining a lower endpoint of

$$187.7 - 11.1 = 176.6.$$

To find the upper endpoint of the confidence interval, add 11.1 to the sample mean of 187.7, obtaining an upper endpoint of

$$187.7 + 11.1 = 198.8.$$

Therefore the confidence interval is

$$176.6 < \mu < 198.8, \text{ or}$$

$$(176.6, 198.8).$$

Example 10

A student records the repair cost for 20 randomly selected computers. A sample mean of \$216.53 and standard deviation of \$15.86 are subsequently computed. Determine the 98% confidence interval for the mean repair cost for all computers. Assume the criteria for this section are met.

Solution:

To begin, we need to calculate $t_{\alpha/2}$ to be used in calculating the margin of error. The level of confidence is $c = 0.98$, so $\frac{\alpha}{2} = \frac{1-0.98}{2} = 0.01$. There are 20 computers in the sample, so the number of degrees of freedom is $d.f. = 19$. Using the Critical t-Value Table, we find that $t_{0.01} = 2.539$.

Next substitute these values into the margin of error formula.

$$\begin{aligned} E &= 2.539\frac{15.86}{\sqrt{20}} \\ &\approx 9.00 \end{aligned}$$

Therefore, subtracting the margin of error from the sample mean gives us a lower endpoint for the confidence interval of

$$216.53 - 9.00 = \$207.53.$$

By adding our margin of error to the sample mean, we obtain the upper endpoint of the confidence interval as follows

$$216.53 + 9.00 = \$225.53.$$

Thus the confidence interval for the population mean is

$$\$207.53 < \mu < \$225.53, \text{ or}$$

$$(\$207.53, \$225.53).$$

Exercises 8.3

Note: For all problems in this set of exercises, assume all of the following conditions are met:

- all possible samples of a given size have an equal probability of being chosen
- the size of the sample is less than 30
- the distribution of the population is approximately normal
- the population's standard deviation is unknown

Directions: Given the following information, find the value of $t_{\alpha/2}$ using the Critical t-Value Table.

1. $n = 13$, level of confidence is 95%

2. $n = 25$, level of confidence is 90%

3. $n = 18$, level of confidence is 99%

4. $n = 21$, level of confidence is 98%

Directions: Given the following information, calculate the margin of error.

5. $n = 14$, $s = 0.03$, level of confidence is 99%

6. $n = 20$, $s = 0.57$, level of confidence is 90%

7. $n = 27$, $s = 1.25$, level of confidence is 95%

8. $n = 12$, $s = 2.93$, level of confidence is 98%

Directions: Given the following information, construct confidence intervals.

9. $n = 14$, $\bar{x} = 95$, $s = 4.8$, level of confidence is 95%

10. $n = 25$, $\bar{x} = 56$, $s = 8$, level of confidence is 90%

11. $n = 8$, $\bar{x} = 7$, $s = 1.2$, level of confidence is 99%

12. $n = 13$, $\bar{x} = 1.97$, $s = 0.03$, level of confidence is 98%

Directions: Answer each of the following questions.

13. The average number of miles commuters drove to work each day was estimated to be 40.8 miles from a sample of 15 commuters. The standard deviation is 5.8 miles. Calculate the margin of error for a 95% confidence interval.

14. The average amount of money a sample of 25 drivers spent a week on gas was found to be $57 with a standard deviation of $2.36. Calculate the margin of error for a 90% confidence interval.

15. Wildlife conservationists studying grizzly bears in the United States found that the average weight of 25 adult males was 600 lbs. with a standard deviation of 90 lbs. Find the margin of error to construct a 98% confidence interval for the average weight of all adult, male grizzly bears.

16. The average length of 12 newborn iguanas is seven inches long with a standard deviation of 0.75 inches. Find the margin of error needed to construct a 90% confidence interval for the average length of all newborn iguanas.

17. A sample of 4 plane crashes finds that the average number of deaths was 49 with a standard deviation of 15. Find a 99% confidence interval for the average number of deaths per plane crash.

18. Suppose you sample 19 high school baseball pitchers and find that they have an average pitching speed of 87 miles per hour with a standard deviation of 0.98 mph. Find a 95% confidence interval for the average pitching speed of all high school baseball pitchers.

19. After looking at the attendance record of 28 men's college basketball games, the average number of fans was 4125 with a standard deviation of 741. In order to predict ticket sales for next year, the athletic office needs a confidence interval for attendance at these games. Using a confidence level of 95%, construct a confidence interval for the average number of fans at men's college basketball games.

20. Suppose you are thinking of becoming a pet owner and want to know the amount of time people spend caring for their pets. You survey 20 pet owners and find that the average amount of time they spend caring for the animals was 108 minutes per day. If the standard deviation you find was 17 minutes, construct a 98% confidence interval for the amount of time pet owners spend on their animals per day.

Estimating Population Proportions

8.4

Another parameter often estimated is the population proportion. Recall that a **population proportion**, denoted p, is the fraction of a population that has a certain characteristic. For example, the proportion of women around the world is approximately $p \approx \frac{1}{2} = 0.5$. Of course, when a population is very large it is impractical or impossible to find p. A sample must be taken and the sample information used to analyze the population. The **sample proportion**, denoted $\hat{p}$, is the fractional part of a sample that has a certain characteristic. For example, if in a sample of 135 teachers, 100 of the teachers are women, then the sample proportion of women is $\hat{p} = \frac{100}{135} \approx 0.741 = 74.1\%$.

The best **point estimate** used for estimating the population proportion is the sample proportion, $\hat{p}$. Let's look at an example of finding the best point estimate for a population proportion.

Rounding Rule

Round proportions to three decimal places.

Example 11

A graduate student wishes to know the proportion of American adults who speak two or more languages. He surveys 565 American adults and finds that 226 speak two or more languages. Estimate the proportion of all American adults that speak two or more languages.

Solution:

Use $\hat{p}$ as a point estimate for p.

$$\begin{aligned} \hat{p} &= \frac{226}{565} \\ &= 0.400 \end{aligned}$$

We estimate the population proportion to be 40.0%.

Rounding Rule

Round the margin of error for proportions to three decimal places.

As we've noted before, using a range of values rather than a point estimate increases the likelihood of estimating the true population proportion. Constructing a confidence interval for the population proportion is similar to constructing a confidence interval for the population mean. The first step in constructing a confidence interval is to calculate the margin of error. The formula for the margin of error of population proportions is

$$E = z_c \sqrt{\frac{\hat{p}\left(1-\hat{p}\right)}{n}}.$$

The margin of error formula uses the critical value of z_c, which can be found using the same table of critical z-values that we used when calculating margin of error for population means. The table of critical values is given below.

Table 2 - Critical z-Values for Confidence Intervals	
Level of Confidence, c	z_c
0.80	1.28
0.85	1.44
0.90	1.645
0.95	1.96
0.98	2.33
0.99	2.575

Once the margin of error is calculated, find the confidence interval for p by adding and subtracting E from $\hat{p}$. Thus a confidence interval for population proportions can be written as shown in the box below.

Confidence Interval for Population Proportions

$$\hat{p} - E < p < \hat{p} + E \text{, or}$$
$$(\hat{p} - E,\ \hat{p} + E)$$

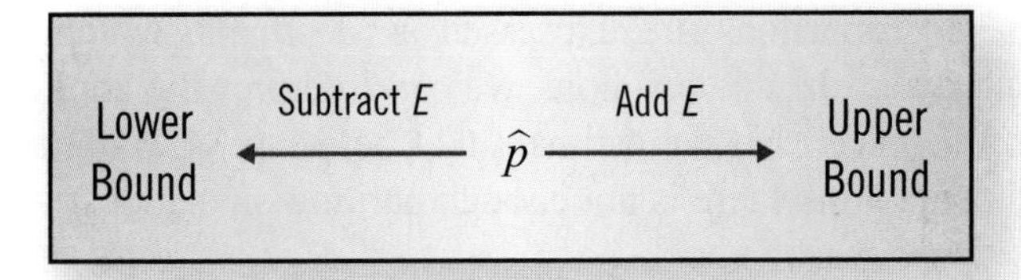

Figure 4 – Constructing a Confidence Interval for Proportions

Example 12

A survey of 345 university students found that 301 students think that there is not enough parking. Find the 90% confidence interval for the proportion of all university students who think that there is not enough parking on campus.

Solution:

First calculate $\hat{p}$.

$$\hat{p} = \frac{301}{345}$$
$$\approx 0.872463768116$$
$$\approx 0.872$$

cont'd on the next page

Rounding Rule

When dealing with intermediate calculation, do not round the intermediate steps. For example, although $\hat{p}$ is normally rounded to three decimal places, for intermediate calculation it is best to keep $\hat{p}$ out to six decimal places in order to not affect the remaining calculations with a rounding error.

Now, refer to the table to find $z_{0.90} = 1.645$.

Calculate E.

$$\begin{aligned} E &= z_{0.90}\sqrt{\frac{\hat{p}(1-\hat{p})}{n}} \\ &= 1.645\sqrt{\frac{0.872464(1-0.872464)}{345}} \\ &\approx 0.029542 \\ &\approx 0.030 \end{aligned}$$

Therefore subtracting the margin of error from the sample proportion gives us a lower endpoint for the confidence interval of

$$0.872 - 0.030 = 0.842.$$

By adding the margin of error to the sample proportion, we obtain the upper endpoint for the confidence interval as follows

$$0.872 + 0.030 = 0.902.$$

Thus the confidence interval for the population proportion is

$$0.842 < p < 0.902, \text{ or}$$
$$(0.842, 0.902).$$

Note that these proportions are expressed as decimals, which is typical for confidence intervals. In this example, we can interpret the confidence interval (0.842, 0.902) to mean that we estimate that between 84.2% and 90.2% of all university students think there is not enough parking on campus.

✦ Finding the Minimum Sample Size for Confidence Intervals Involving Proportions

Rounding Rule

Always round the minimum sample size up to the next whole number.

As we saw when estimating population means, it is often desirable to know the fewest number of data values that we need to obtain a certain level of confidence. The minimum sample size required for estimating population proportions may be calculated using the following formula:

$$n = \hat{p}(1-\hat{p})\left(\frac{z_c}{E}\right)^2$$

Typically, $\hat{p}$ is an estimate for p or a value from a previously conducted study. Note that because we are calculating the number of data values needed, you must always **round up** to the next whole number. Again, this is a *minimum* number of data values you should sample. It is always a good idea to sample as many as possible.

Example 13

What is the minimum sample size needed for a 99% confidence interval for the population proportion if $\hat{p} = 0.54$ and $E = 2\%$?

Solution:

To begin, we must look up $z_{0.99}$ in the Critical z-Value Table. According to the table, $z_{0.99} = 2.575$. Note that we must convert E to a decimal before substiting it into the formula.

$$\begin{aligned} n &= \hat{p}(1-\hat{p})\left(\frac{z_c}{E}\right)^2 \\ &= 0.54(1-0.54)\left(\frac{2.575}{0.02}\right)^2 \\ &\approx 4117.62 \end{aligned}$$

Thus we need a sample of size 4118 or larger.

Example 14

The state education commission wants to estimate the fraction of tenth-grade students who read at or below the eighth-grade level. The commissioner of education believes that the fraction is around 0.21.

a. How large a sample would be required to estimate the fraction of all tenth-graders who read at or below the eighth-grade level with an 85% level of confidence and an error of at most 0.03?

b. Suppose a sample of minimum size is chosen. Of these students, 84 read at or below the eighth-grade level. Using this data, estimate the sample proportion of all tenth-graders reading at or below the eighth-grade level.

c. Construct the 85% confidence interval for the population proportion of tenth-graders reading at or below the eighth-grade level.

Solution:

a. From the problem, we know that $E = 0.03$ and $\hat{p} = 0.21$. Using the Critical z-Value Table, we have that $z_{0.85} = 1.44$. We can now use this information to calculate the minimum sample size.

$$\begin{aligned} n &= \hat{p}(1-\hat{p})\left(\frac{z_c}{E}\right)^2 \\ &= 0.21(1-0.21)\left(\frac{1.44}{0.03}\right)^2 \\ &\approx 382.23 \end{aligned}$$

Thus the minimum sample size should be 383.

cont'd on the next page

b. To estimate the sample proportion, we divide the number reading at or below the eighth grade level by the sample size.

$$\begin{aligned}\hat{p} &= \frac{84}{383} \\ &\approx 0.219321 \\ &\approx 0.219\end{aligned}$$

c. To construct the confidence interval, we need to calculate the margin of error. Substituting the appropriate values into the formula, we have

$$\begin{aligned}E &= z_c\sqrt{\frac{\hat{p}(1-\hat{p})}{n}} \\ &= 1.44\sqrt{\frac{0.219321(1-0.219321)}{383}} \\ &\approx 0.030.\end{aligned}$$

Therefore, subtracting the margin of error from the sample proportion gives us a lower endpoint for the confidence interval of

$$0.219 - 0.030 = 0.189.$$

By adding the margin of error to the sample proportion, we obtain the upper endpoint for the confidence interval as follows

$$0.219 + 0.03 = 0.249.$$

Thus the confidence interval for the population proportion is

$$0.189 < p < 0.249, \text{ or}$$

$$(0.189, 0.249).$$

Exercises

8.4

Directions: Answer each of the following questions.

1. Out of 50 students surveyed, 41 felt that parking was a serious problem on campus. Give the point estimate for the proportion of all students who feel that parking is a serious problem.

2. Out of 130 students, only 7 had been placed in the wrong math class. Give the point estimate for the population proportion of all students who have been placed in the wrong math class.

3. 48 out of 112 homeowners pay for lawn service every month. Give a point estimate for the population proportion of all homeowners who pay for lawn service every month.

4. Out of 210 registered voters polled, 107 declared that they intended to vote for the incumbent in the upcoming election. Give the point estimate for the proportion of all registered voters who intend to vote for the incumbent.

5. Out of 50 students surveyed, 41 felt that parking was a serious problem on campus. Using a margin of error of 2%, give the interval estimate for the percentage of all students who feel that parking is a serious problem.

6. Out of 210 registered voters polled, 107 declared that they intended to vote for the incumbent in the upcoming election. Using a margin of error of 3%, give the interval estimate for the percentage of all registered voters who intend to vote for the incumbent.

7. 13 out of 147 faculty members surveyed at a community college know sign language. Calculate a margin of error for a 90% confidence interval for the proportion of faculty members that know sign language.

8. A sample of 200 computer chips is obtained and 4% are found to be defective. Using a 95% level of confidence, calculate the margin of error for the percentage of all computer chips that are defective.

9. Out of 140 kindergarteners surveyed, 32 say that pancakes are their favorite breakfast food. Using a 90% level of confidence, calculate the margin of error for the percentage of all kindergarteners who say pancakes are their favorite breakfast food.

10. Out of 54 patients surveyed at a local hospital, 49 reported that they were satisfied with the care they received. Calculate a margin of error for a 95% confidence interval of the proportion of patients satisfied with their care.

11. A survey of 145 students showed that only 87 checked their campus email account on a regular basis. Construct a 90% confidence interval for the percentage of students who do *not* check their email account on a regular basis.

12. A sample of 200 computer chips is obtained and 4% are found to be defective. Construct a 99% confidence interval for the percentage of all computer chips that are *not* defective.

13. Out of 190 adults surveyed, 71 exercise on a regular basis. Construct a 95% confidence interval for the percentage of all adults who exercise on a regular basis.

14. A survey of 47 members of a health club reported that 68% prefer walking or running on a treadmill to any other type of aerobic exercise. Construct a 95% confidence interval for the percentage of health club members who prefer walking/running on a treadmill.

15. Suppose you wish to determine the proportion of college students nationally who receive some form of financial aid. You want to be 98% confident of your results and have a maximum error of 5%. Calculate the minimum sample size that you must have to meet these requirements, given that the financial aid office at a local institution estimates the percentage to be 78%.

16. Suppose that you wish to determine the proportion of college students that own a cell phone. You want to be 95% confident in your results and have a maximum error of 3%. Calculate the minimum number of college students that you must survey to meet these requirements, given that previous estimates show that 68% of all college students own a cell phone.

17. A market researcher wishes to determine the proportion of women who shop online. The results must be accurate at the 90% level of confidence with a maximum error of 2%. Calculate the minimum sample size needed to meet these requirements, given that previous studies estimate that 59% of women shop online.

18. Suppose a researcher needs to verify the proportion of patients who will have mild side effects to a recently approved drug. The researcher must be 99% sure of her results with a maximum error of 1%. Calculate the minimum sample size the researcher needs to meet her requirements, given that the studies used to get the drug approved reported that only 12% of patients experienced mild side effects.

Estimating Population Variance

8.5

The final parameter that we will estimate is the population variance. Up to this point, the applications for creating confidence intervals for parameters such as the mean and proportion were obvious. Less obvious perhaps is the need to estimate the *variance* of a population. However, there are times when it is important for products or items to be identical. For example, an auto parts manufacturer must test to ensure that each auto part that comes off the assembly line matches the necessary specifications. The consistency of quality in the auto parts is essential. For quality control, the manufacturer would likely estimate the variation of all parts by using a sample of parts on the assembly line. If the estimate of the variance is too high, then adjustments in the manufacturing process must be made.

Let's begin as before with simply estimating the parameter from a single source. Once again the best single estimate for a population parameter is a sample statistic. In this case the best point estimate for the population variance, σ^2, is the sample variance, s^2. Likewise, the point estimate for the population standard deviation is the sample standard deviation. The following example illustrates how a point estimate may be used to estimate the variance and standard deviation of a population.

Rounding Rule

Recall that variance and standard deviation are rounded to one more decimal place than the original data.

Example 15

General Auto is testing the variance in the length of its windshield wiper blades. A sample of 12 windshield wiper blades is randomly selected, and the following lengths are measured in inches.

22.1	22.0	22.1	22.4	22.3	22.5
22.3	22.1	22.2	22.6	22.5	22.7

a. Give a point estimate for the population variance.

b. Give a point estimate for the population standard deviation.

Solution:

a. The sample variance, s^2, is the best point estimate for σ^2. Using a calculator, we calculate $s^2 = 0.05$. For review, see the technology for Chapter 3 (page 146). The point estimate for the population variance is then 0.05.

b. The sample standard deviation, s, is the best point estimate for σ. According to the calculator, $s = 0.22$. The point estimate for the population standard deviation is then 0.22.

As in previous sections, developing an interval estimate, or confidence interval, to help us in estimating the population variance is more accurate than a point estimate. However, confidence intervals for variance and standard deviation are constructed differently than the other methods we have used in this chapter. First, in order to calculate a confidence interval for a variance or standard deviation, the population must be normally distributed. We will assume that all examples and exercises meet this requirement so that we may concentrate on the applications here, not on verifying normality. Second, the formula for calculating the lower and upper bounds of the confidence interval requires that we use the chi-square test statistic, denoted χ^2. The χ^2 distribution is one that is completely defined by its degrees of freedom, $n - 1$. We will discuss the χ^2 distribution in more detail in Chapter 10. For now, a basic knowledge of how to read the χ^2 distribution table will be sufficient.

Rounding Rule

Round confidence intervals for variance and standard deviation to one more decimal place than the original data, or to the same number of decimal places as the point estimate.

Let's look at how to read the table. The table of critical values for the chi-square distribution can be found in Appendix A. Notice that the first column contains the degrees of freedom, *d.f.* For constructing variance estimates, the degrees of freedom will simply be $n - 1$. Across the top row of the table you will see ten commonly used values for χ^2_α. The column you will use depends on the level of confidence given in the problem, which dictates the corresponding level of α to choose. (Remember that $\alpha = 1 - c$.) The needed critical value is found in the cell where the row for degrees of freedom and the χ^2 column intersect.

The final way in which confidence intervals for variance and standard deviation are unique is the way in which the interval estimate is created. Rather than adding and subtracting a margin of error, we have a formula for calculating the bounds. The formula for a confidence interval for a population variance is as follows:

Confidence Intervals for Population Variance:

$$\frac{(n-1)s^2}{\chi^2_{\alpha/2}} < \sigma^2 < \frac{(n-1)s^2}{\chi^2_{(1-\alpha/2)}}$$

The formula for creating a confidence interval for a population standard deviation follows easily from this by simply taking the square root of each side to give the following:

Confidence Intervals for Population Standard Deviation:

$$\sqrt{\frac{(n-1)s^2}{\chi^2_{\alpha/2}}} < \sigma < \sqrt{\frac{(n-1)s^2}{\chi^2_{(1-\alpha/2)}}}$$

The steps for creating a confidence interval for a population variance (or standard deviation) are as follows:

1. From the sample data, calculate s^2.
2. Based on the level of confidence given, calculate $\frac{\alpha}{2}$ and $1 - \frac{\alpha}{2}$.
3. Use the χ^2 distribution table to find the critical values for $\frac{\alpha}{2}$ and $1 - \frac{\alpha}{2}$ with $n - 1$ degrees of freedom.
4. Substitute the necessary values into the confidence interval formula above.

Note that when the value of the degrees of freedom needed is not in the tables, use the closest value given. Let's look at an example of how this type of confidence interval is created.

Example 16

A commercial bakery is testing the variance in the weights of the cookies it produces. A sample of 15 cookies is randomly chosen and found to have a variance of 3.4 grams. Build a 95% confidence interval for the variance of the weights of all cookies produced by the company.

Solution:

Step 1: We are given in the problem $s^2 = 3.4$.

Step 2: Because $c = 0.95$, we know that $\alpha = 1 - 0.95 = 0.05$. Then $\frac{\alpha}{2} = \frac{0.05}{2} = 0.025$. Also, $1 - \frac{\alpha}{2} = 1 - 0.025 = 0.975$.

Step 3: Using the table with $15 - 1 = 14$ degrees of freedom, we see that $\chi^2_{0.025} = 26.119$ and $\chi^2_{0.975} = 5.629$.

Step 4: Substituting into the formula gives us the following:

$$\frac{(15-1)3.4}{26.119} < \sigma^2 < \frac{(15-1)3.4}{5.629}$$

$$1.8 < \sigma^2 < 8.5$$

The bakery estimates with 95% confidence that the variance in the weights of their cookies is between 1.8 and 8.5 grams.

Example 17

Consider again the bakery in the previous example. Suppose that the bakery needs to estimate the standard deviation of the weights of their cookies as well. Construct a 95% confidence interval for the standard deviation of the weights of all cookies produced.

Solution:

Remember the relationship between standard deviation and variance. To find the confidence interval for standard deviation, simply take the *square root* of the confidence interval for the variance. This gives us

$$\sqrt{1.8} < \sigma < \sqrt{8.5}$$

$$1.3 < \sigma < 2.9.$$

The bakery can be 95% confident that the standard deviation of the weights of all its cookies is between 1.3 and 2.9 grams.

✦ Minimum Sample Size

As in previous sections, there are times at which it is desirable to estimate how large a sample that must be chosen in order for the results of a study to have a certain level of precision. Unfortunately, calculating minimum sample size for estimating the variance of a population is not as simple as we have seen before. In fact, the best way to determine minimum sample size for these types of problems is to use a statistical package. For our purposes, an example of a table created with a statistical package is shown below.

Table 3 - Minimum Sample Sizes at 95% Confidence

s^2 is within the percentage of the value of σ^2	Minimum value of n needed	s is within the percentage of the value of σ	Minimum value of n needed
1%	77,209	1%	19,206
5%	3150	5%	769
10%	807	10%	193
20%	212	20%	49
30%	99	30%	22
40%	58	40%	13
50%	39	50%	9

Table 4 - Minimum Sample Sizes at 99% Confidence

s^2 is within the percentage of the value of σ^2	Minimum value of n needed	s is within the percentage of the value of σ	Minimum value of n needed
1%	133,450	1%	33,220
5%	5459	5%	1337
10%	1403	10%	337
20%	370	20%	86
30%	173	30%	39
40%	102	40%	23
50%	69	50%	15

Let's look at an example of how to use the table to find the minimum sample size necessary for a particular study.

Example 18

A market researcher wants to estimate the standard deviation of home prices in a metropolitan area in the Northeast. She needs to be 99% confident that her results are within 5% of the true standard deviation. Assuming that the housing prices in that area are normally distributed, what is the minimum number of housing prices she must acquire?

Solution:

According to the table, we see that for 99% confidence of being within 5% of the true standard deviation, the minimum sample size is 1337. Therefore, the market researcher must include at least 1337 housing prices in order for her study to have the level of precision that she needs.

Exercises

8.5

Directions: Find the following point estimates.

1. In a random sample of 112 adults, the mean height was found to be 72 inches with a standard deviation of 2.2 inches. Give a point estimate for the population variance of height for adults.

2. Consider a sample of caps for 3 inch pipes. Their diameters are measured and found to have a variance of 0.12 square inches. Give a point estimate for the population variance in diameter lengths of the caps.

3. A sample of 25 shoelaces was taken and their lengths measured. The laces were found to have a mean of 18.01 inches with a variance of 0.76 inches. Give a point estimate for the population variance in the lengths of shoelaces.

4. In a sample of 100 skateboards, the average length was found to be 33.4 inches with a standard deviation of 2.1 inches. Give a point estimate for the population variance of the length of skateboards.

5. What is the best point estimate for the population's standard deviation if the sample variance is 25?

6. What is the best point estimate for the population's variance if the sample standard deviation is 3.5?

7. What is the best point estimate for the population's variance if the sample variance is 25?

8. What is the best point estimate for the population's standard deviation if the sample standard deviation is 5.67?

Directions: Calculate the critical value for the left and right endpoints for a confidence interval for population variance given the following information.

9. $n = 25$, $\alpha = 0.05$

10. $n = 17$, $\alpha = 0.10$

11. $n = 22$, $c = 0.90$

12. $n = 10$, $c = 0.95$

Directions: Calculate the confidence interval for the population variance given the following information.

13. $n = 10$, $s^2 = 31.8$, $c = 0.95$

14. $n = 15$, $s^2 = 3.8$, $c = 0.90$

15. $n = 23$, $s^2 = 11.9$, $c = 0.99$

16. $n = 26$, $s^2 = 21.5$, $c = 0.98$

Directions: Use the information given to calculate the specified confidence interval for the population standard deviation.

17. $n = 63$, $s = 2.4$, $c = 0.99$

18. $n = 31$, $s = 0.9$, $c = 0.95$

19. $n = 40$, $s = 7.7$, $c = 0.98$

20. $n = 56$, $s = 3.5$, $c = 0.99$

Directions: Find the minimum sample size.

21. Find the minimum sample size needed to be 95% confident that the sample's variance is within 20% of the population's variance.

22. Find the minimum sample size needed to be 99% confident that the sample's variance is within 5% of the population's variance.

23. Find the minimum sample size needed to be 99% confident that the sample's standard deviation is within 10% of the population's standard deviation.

24. Find the minimum sample size needed to be 95% confident that the sample's standard deviation is within 1% of the population's standard deviation.

Directions: Construct confidence intervals given the following information:

25. Consider the following sample data.

3.2	3.6	2.9	3.0	3.0
3.1	3.2	3.3	2.9	3.3
2.9	3.1	3.4	3.3	3.0

Build a 99% confidence interval for the population variance.

26. The following sample of weights was taken from 14 boxes of crackers off the assembly line.

16.87	16.92	17.01	16.98	16.99	16.92	16.91
17.00	17.01	16.96	16.95	16.94	17.00	16.92

Build a 98% confidence interval for the population variance for all boxes of crackers that come off the assembly line.

27. The weights of 89 randomly selected new truck engines were found to have a standard deviation of 1.59 lbs. Calculate the 95% confidence interval for the population standard deviation of the weights of all new truck engines in this particular factory.

28. A butcher uses a machine that packages ground beef in one-pound portions. A sample of 52 packages of ground beef has a standard deviation of 0.2 pounds. Calculate a 99% confidence interval to estimate the standard deviation of the weights of all packages prepared by the machine.

Chapter Review

Section 8.1 - Introduction to Estimating Population Means

Section 8.1 - Introduction to Estimating Population Means	Pages
• **Definitions** **Point Estimate** – a single number estimate of a population parameter **Interval Estimate** – range of possible values for the parameter **Level of Confidence** – the degree of certainty that the interval estimate contains the population parameter **Confidence Interval** – an interval estimate associated with a certain level of confidence **Margin of Error** – largest possible distance from the point estimate that a confidence interval will cover	p. 346 – 348
• **Confidence Interval for Population Means** $\bar{x} - E < \mu < \bar{x} + E$ or $(\bar{x} - E, \bar{x} + E)$	p. 348

Section 8.2 - Estimating Population Means (Large Samples)

Critical z-Values — p. 356

Level of Confidence, c	z_c
0.80	1.28
0.85	1.44
0.90	1.645
0.95	1.96
0.98	2.33
0.99	2.575

• **Margin of Error for Means** — p. 356

$$E = z_c \frac{s}{\sqrt{n}}$$

• **Confidence Interval for Population Means** — p. 357

$$\bar{x} - E < \mu < \bar{x} + E$$

• **Minimum Sample Size for Means** — p. 359

$$n = \left(\frac{z_c \sigma}{E}\right)^2$$

Section 8.3 - Estimating Population Means (Small Samples)	Pages
• **Margin of Error for *t*-distribution** $E = t_{\alpha/2} \frac{s}{\sqrt{n}}$	p. 363
• **Confidence Interval for Population Means** $\bar{x} - E < \mu < \bar{x} + E$ or $(\bar{x} - E, \bar{x} + E)$	p. 364

Section 8.4 - Estimating Population Proportions	Pages
• **Margin of Error for Proportions** $E = z_c \sqrt{\frac{\hat{p}(1-\hat{p})}{n}}$	p. 368
• **Confidence Interval for Population Proportions** $\hat{p} - E < p < \hat{p} + E$ or $(\hat{p} - E, \hat{p} + E)$	p. 369
• **Minimum Sample Size for Proportions** $n = \hat{p}(1-\hat{p})\left(\frac{z_c}{E}\right)^2$	p. 370

Section 8.5 - Estimating Population Variance	Pages
• **Confidence Interval for Population Variance** $\frac{(n-1)s^2}{\chi^2_{\alpha/2}} < \sigma^2 < \frac{(n-1)s^2}{\chi^2_{(1-\alpha/2)}}$	p. 376
• **Confidence Interval for Population Standard Deviation** $\sqrt{\frac{(n-1)s^2}{\chi^2_{\alpha/2}}} < \sigma < \sqrt{\frac{(n-1)s^2}{\chi^2_{(1-\alpha/2)}}}$	p. 376

Chapter Exercises

Directions: Answer each of the questions below. You may assume that all samples of a given size had an equal probability of being chosen. Unless otherwise indicated, assume that the population's standard deviation is unknown.

1. A survey of 21 accountants finds that they spend on average \$45 a week on lunch with a standard deviation of \$3.70. Using a margin of error of \$2.00, give an interval estimate for the mean amount of money accountants spend on lunch each week. Assume that the population is normally distributed.

2. 19 of the 35 students in Mrs. Willard's Organic Chemistry class carried a cell phone to class today. Using Mrs. Willard's class as your sample with a margin of error of 3%, construct an interval estimate for the percentage of all students on campus who carry a cell phone to class.

3. Out of 300 randomly selected DVDs, a manufacturer found that only 2% were defective. Using a 95% level of confidence, calculate the margin of error for the proportion of all DVDs that are defective.

4. A sample of 253 working mothers was surveyed regarding the cost of healthcare. The mean amount of money spent on healthcare by these working mothers each year is \$1300 with a standard deviation of \$289. Calculate the margin of error for the population mean using a 98% level of confidence.

5. A sample of 15 dentists was timed brushing their own teeth. The mean amount of time spent brushing their teeth was 2.13 minutes with a standard deviation of 0.45 minutes. Calculate the margin of error for the mean amount of time all dentists spend brushing their teeth. Use a 99% level of confidence. Assume that the population is normally distributed.

6. 230 of the 600 bags of rice being transported in one boxcar were damaged in a train wreck. The insurance company needs to estimate the proportion of bags of rice on the entire train that were damaged in the wreck. Construct a 99% confidence interval for the percentage of all bags of rice that were damaged.

7. From a sample of 49 stray dogs, a veterinarian finds that their mean weight is 41 pounds with a standard deviation of 8.5 pounds. Construct a 98% confidence interval for the mean weight of all stray dogs in this area.

8. Although male polar bears weigh only about one pound at birth, the mean weight of a sample of 10 adult male polar bears is 1200 pounds with a standard deviation of 100 pounds. Construct a 98% confidence interval for the mean adult weight of all male polar bears. Assume that the population is normally distributed.

9. The following sample of the number of ounces in a soda can was taken from 14 cans coming off the assembly line.

12.01	12.02	11.95	11.99	11.94	12.01	12.03
11.98	12.00	12.03	11.98	12.05	11.93	11.98

Build a 95% confidence interval for the population variance for the volume of all the soda cans that come off the assembly line.

10. A researcher wants to interview military personnel returning from war regarding their experiences overseas. The researcher wants to know the proportion of soldiers that would not choose to return overseas if given the choice. If the researcher desires to construct a 90% confidence interval with a 5% margin of error, how many soldiers must the researcher interview? An estimate for the proportion of soldiers that would not return overseas if given the choice is 76%.

11. Juan is interning for the summer at a marketing firm. One of his assignments is to conduct research to estimate the percentage of Americans that attend after-Christmas sales. He needs his results to have a 99% level of confidence and a maximum error of 2%. If previous estimates have shown the percentage to be 20%, how many people must Juan survey?

Discussion Questions

Directions: Discuss each of the following questions with your classmates. Focus on the relationship between the parameters in the questions.

12. Lisa sets out to survey 500 people; however, she only receives responses from 387 people.
    a. How will this decrease in her sample size affect the margin of error for population means?
    b. How will this decrease in her sample size affect the width of her confidence interval for population means?

13. How will increasing the level of confidence without changing the sample size affect the width of the confidence interval for population means?

14. How will increasing the level of confidence without changing the sample size affect the margin of error for population means?

15. Which level of confidence will produce a wider confidence interval for population means: a 95% level of confidence or a 99% level of confidence?

16. If you decrease the sample size while keeping the margin of error for population means constant, what effect will this have on the level of confidence?

17. How will the width of the confidence interval for population means change if you increase the sample size?

Chapter Project: Creating Confidence Intervals

Create a Confidence Interval for Population Means

Choose one of the study questions below:

- How many hours per week do college students on your campus study?
- What is the average number of parking tickets college students on your campus receive per semester?
- What is the average price of rent paid per month by college students on your campus?

For the study question you chose, create a confidence interval for that parameter by doing each of the following:

1. Collect data on your study question from at least 30 students on your campus.
2. Calculate the sample mean and sample standard deviation of your sample data.
3. Calculate the margin of error for a 95% confidence interval using your sample statistics.
4. Calculate a 95% confidence interval using your sample statistics.
5. Write up a summary of your results.

Create a Confidence Interval for Population Proportions

Choose one of the study questions below:

- What percentage of college students on your campus own an MP3 player?
- What percentage of college students on your campus are seeking a degree in your field of study?
- What percentage of college students on your campus live in on-campus housing?

For the study question you chose, create a confidence interval for that parameter by doing each of the following:

1. Collect data on your study question from at least 30 students on your campus.
2. Calculate the sample proportion from your sample data.
3. Calculate the margin of error for a 90% confidence interval using your sample statistics.
4. Calculate a 90% confidence interval using your sample statistics.
5. Write up a summary of your results.

Chapter Technology – Confidence Intervals

TI-84 PLUS

Calculating a confidence interval on the TI-84 Plus calculator is trivial. There are three types of confidence intervals you can create using the TI-84 Plus:

1. A confidence interval for population means using z-values.
2. A confidence interval for population means using t-values.
3. A confidence interval for population proportions.

Confidence Intervals for Population Means with $n \geq 30$

When your sample size is at least 30, confidence intervals for the population mean are calculated using z-values. You need to know σ, $\bar{x}$, n, and your level of confidence, c. Since $n \geq 30$, you can use the sample standard deviation, s, as a point estimate for σ. Now, let's calculate a confidence interval on our TI-84 Plus.

To begin, press STAT, then scroll over and choose TESTS. From the TESTS menu, choose option 7: ZInterval. You should now see the screen shown here:

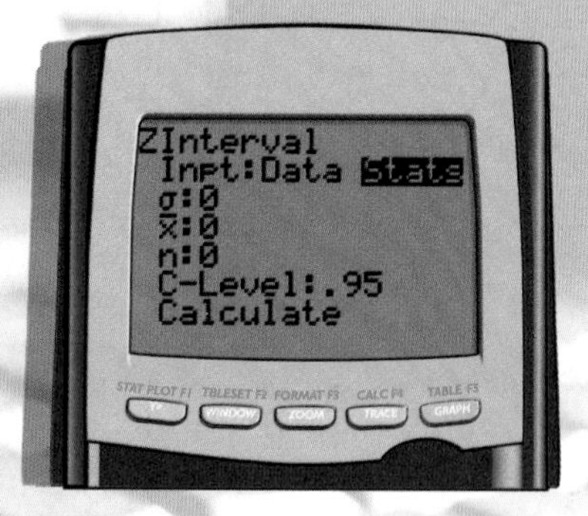

Notice that you can choose to calculate the confidence interval from a list of data or by entering in the values of the statistics. Let's assume we know the statistics. Choose Stats then press ENTER. Enter in the standard deviation for σ, the sample mean for $\bar{x}$, the sample size for n, and the level of confidence for *C-Level*. Highlight Calculate, then press ENTER. The confidence interval, along with the sample mean and sample size will be displayed, as shown in the screen shot here:

Example T.1

Use your TI-84 Plus to calculate an 85% confidence interval for the population mean amount teachers spend on lunch each week. A survey of 42 teachers finds that they spend on average $18 a week on lunch with a standard deviation of $2 a week.

Solution:

To begin, press STAT, then scroll over and choose TESTS. From the TESTS menu, choose option 7: ZInterval. Choose Stats, then press ENTER. Enter in 2 for σ, enter 18 for $\bar{x}$, a sample size of 42 for n, and 0.85 for the the level of confidence, *C-Level*. Highlight Calculate, then press ENTER. The screen will display the following results:

Thus the average that teachers spend on lunch each week is between $17.56 and $18.44.

Confidence Intervals for Population Means with $n < 30$

If you do not have a large sample size, $n < 30$, then you need to use t-values when calculating a confidence interval for population means. The procedure is the same as for z-values, except you choose option 8: TInterval instead of option 7. The statistics needed are a little different as well. For example, if you were given a sample size of 23 for the previous example, the data will be entered as follows:

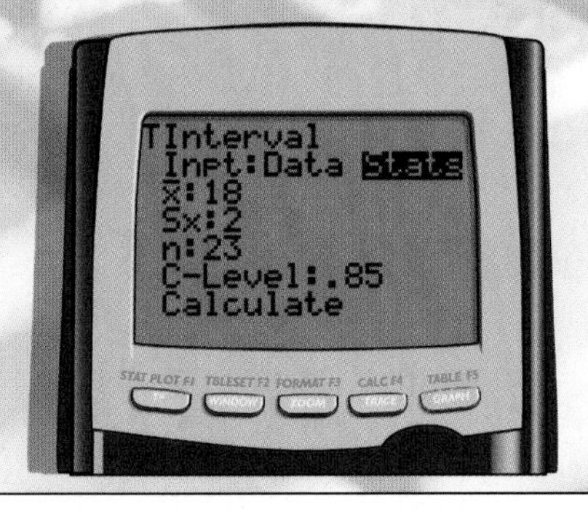

As you can see, you will need to enter the sample mean, $\bar{x}$, the sample standard deviation, *Sx*, the sample size, *n*, and the level of confidence, *C-Level*. Highlight Calculate, then press ENTER. The screen will display the following results:

Example T.2

Use your TI-84 Plus to calculate a 95% confidence interval for the population mean for the average amount teachers spend on lunch each week. A survey of 12 teachers finds that they spend on average \$20 a week on lunch with a standard deviation of \$3 a week.

Solution:

To begin, press STAT, then scroll over and choose TESTS. From the TESTS menu, choose option 8: TInterval. Choose Stats, then press ENTER. Enter in 20 for $\bar{x}$, enter 3 for *Sx*, a sample size of 12 for *n*, and 0.95 for the level of confidence, *C-Level*. Highlight Calculate, then press ENTER. The screen will display the following results:

Thus the average that teachers spend on lunch each week is between \$18.09 and \$21.91.

Confidence Intervals for Population Proportions

A TI-84 Plus can calculate confidence intervals for population proportions as well. To begin, press STAT and then scroll over and choose TESTS. Choose option A: 1-PropZInterval. All you need to calculate this confidence interval is the sample size, n; the number in the sample that have the characteristic in question, x; and the desired level of confidence, *C-Level*.

Example T.3

A survey of 345 university students found that 301 students think that there is not enough parking. Find the 90% confidence interval for the proportion of all university students who think that there is not enough parking on campus.

Solution:

Press STAT, then scroll over and choose TESTS. From the TESTS menu, choose option A: 1-PropZInterval. 301 students think that there is not enough parking, so enter 301 for x. Our sample size is 345, so enter 345 for n. Our level of confidence is 90%. This must be entered as a decimal, so enter 0.90 for *C-level*. Choose Calculate. The screen will display the following results:

Notice that the confidence interval is displayed as well as the value of the sample proportion, $\hat{p}$. Thus the 90% confidence interval for the population proportion is (0.843, 0.902).

MICROSOFT EXCEL

We can use Microsoft Excel to calculate the margin of error for large sample sizes directly, unlike the TI-84 Plus. To calculate the margin of error, use the formula =CONFIDENCE(alpha, standard_dev, size). "alpha" is the level of significance. It equals $1 - c$. "standard_dev" is the standard deviation of the population, and is assumed to be known. "size" is the sample size.

Example T.4

Calculate the margin of error for a 95% confidence interval given that $\sigma = 8.5$ and $n = 47$.

Solution:

Using the formula =CONFIDENCE(alpha, standard_dev, size), we can plug in the values. Remember that $\alpha = 1 - c$, so $\alpha = 1 - 0.95 = 0.05$. Enter =CONFIDENCE(0.05, 8.5, 47). Thus the margin of error is 2.43.

Important warning on using Microsoft Excel to calculate confidence intervals and margins of error:

1. The input template for this function says that it produces the confidence interval, but as we saw in the example, it only produces the value for the margin of error.
2. This calculation for margin of error is always based on the normal distribution, even when the sample size is less than 30.

Thus unless you are very familiar with the limitations of this function in Microsoft Excel, it is best to avoid using it.

Example T.5

Calculate the margin of error for a 95% confidence interval given that $s = 8.5$ and $n = 27$.

Solution:

Since our sample size is less than 30, we use a confidence interval built with the Student's t-distribution. A built in function for finding critical values of this distribution is TINV(probability, deg_freedom) where probability $= c$ and deg_freedom $= n - 1$. Using the calculation =TINV(0.05, 26)*8.5/SQRT(27)), we find the margin of error is 3.362.

Example T.6

During the 2004-2005 NBA season, LeBron James attempted 636 free throws and made 477 of these. Construct a 99% confidence interval for his true free throw percentage.

Solution:

The first step is calculating a point estimator for the interval. Enter the number of successes into the cell B1 and the total attempts into B2. We divide the number of successes by the total attempts. Enter =B1/B2 to calculate $\hat{p}$. The margin of error is calculated by entering =NORMS INV(.995)*SQRT(B3*(1-B3)/B2). Finally, subtracting the margin of error from $\hat{p}$ yields the left endpoint of the interval and adding the margin yields the right endpoint.

	A	B
1	x	477
2	n	636
3	sample p	0.75
4	E	0.04423
5	left endpoint	0.70577
6	right endpoint	0.79423

MINITAB

Minitab can produce a confidence interval for proportions in a single step.

Example T.7

A certain dosage of insecticide is applied to a sample of 200 roaches in a laboratory. The scientists have predicted that this dose should kill 50% of the roaches. When the results are tallied, 83 roaches are dead. Can the scientists make the claim with a confidence level of 95%?

Solution:

To produce the confidence interval for proportions dialog, select STAT, BASIC STATISTICS, and *1-Proportion*. Next, select **Summarized data** and enter the number of trials and successes. Then select Options to set the **Confidence level** and choose "Use test and interval based on normal distribution". Select OK on the options screen and OK on the main dialog screen and your results are displayed in the Session window. The fourth column displays the point estimator, $\hat{p}$, and the fifth contains the confidence interval.

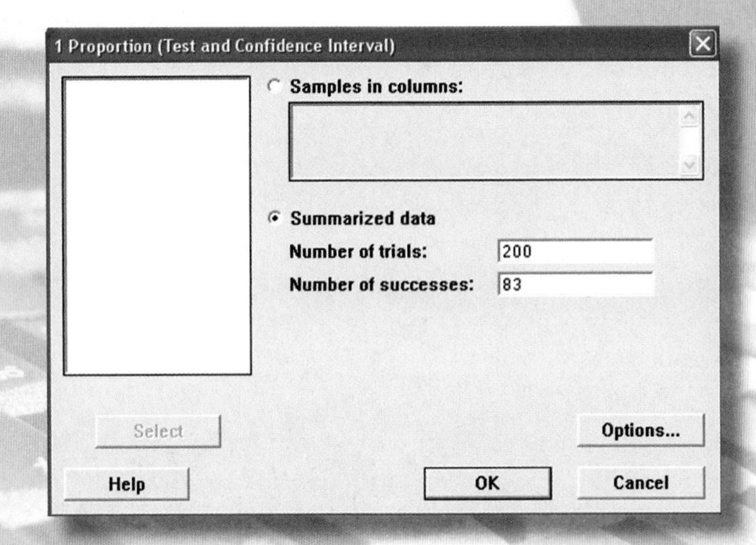

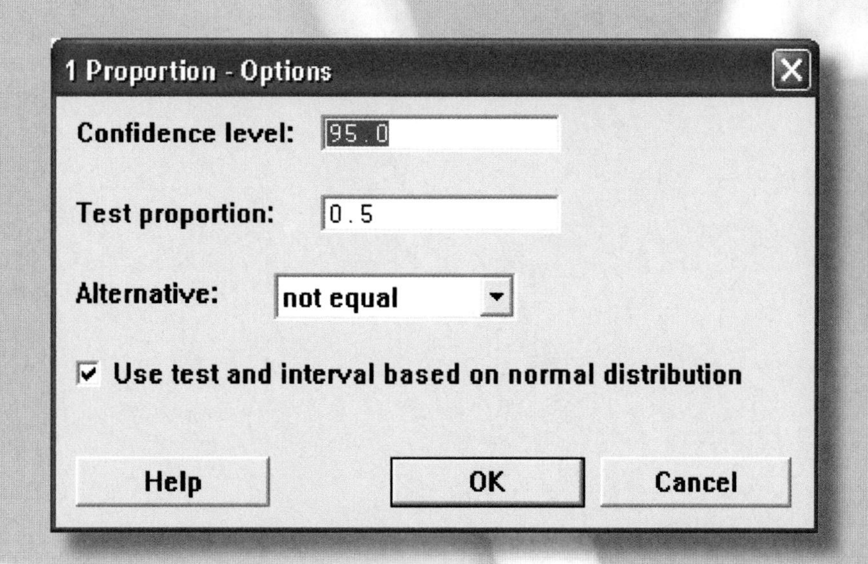

Clicking OK should produce the following screen:

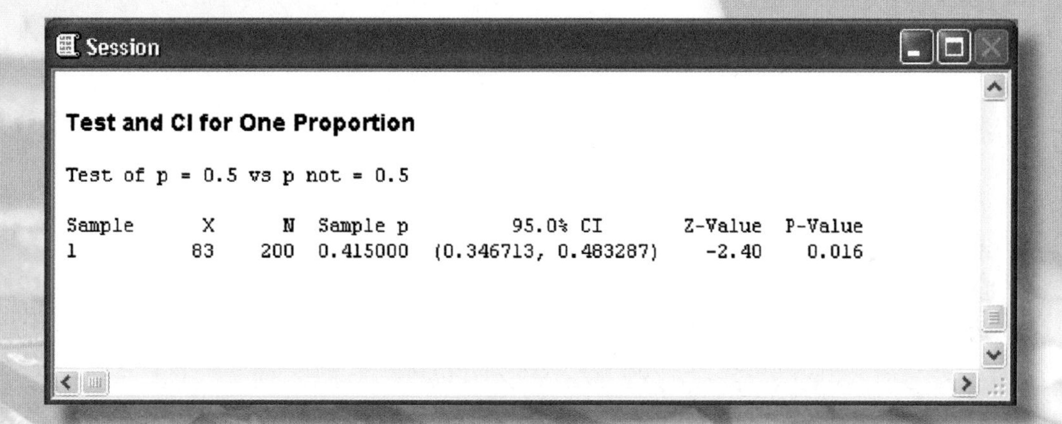

The confidence interval (0.346713, 0.483287) does not contain the claim of 0.50. Therefore, the scientists' claim is not supported at a confidence level of 95%.

Confidence Intervals for Two Samples

Sections

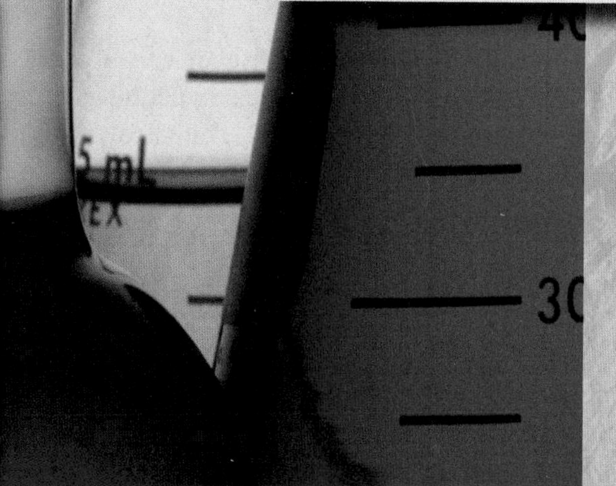

Objectives

The objectives of this chapter are as follows:

1. Calculate the confidence intervals for large independent samples, small independent samples, and dependent samples while comparing two means
2. Determine when a sample standard deviation is pooled
3. Calculate confidence intervals for the true difference in two population proportions

Chapter 9

Introduction

It seems these days that everyone is trying to do things better, faster—you name it. Each company wants to outdo the next in order to win not only the market, but bragging rights to being the best. Let's face it: our world is competitive, to say the least.

Take, for instance, the competition between cellular phone companies. Each company claims that their phone service is superior to the others. That is, each company wants to have the most reliable service covering the broadest territory. Is it possible to know exactly what percentage of the time *all* callers are able to connect for *each* phone company? Certainly not—especially when you consider the enormous volume of calls that are attempted even in a single hour.

Even though parameters such as these cannot be pinpointed, it is still possible to *compare* the percentage of calls completed for each company. Hey—the companies don't really care what percentage of calls are going through anyway, only if their percentage is higher than their competitor's, right? This is just one example of many cases in which knowing how to calculate a confidence interval for a difference in two population parameters comes in handy.

Comparing Two Means (Large, Independent Samples)

9.1

In the previous chapter we studied how to construct a confidence interval to estimate a single population parameter. However, it is often necessary to compare two populations, neither of whose population parameters is known. If we are considering two population parameters, sample data must be drawn from each population. There are then two sets of sample statistics instead of the single set used in the previous confidence interval formulas. Hence, the formulas from the previous chapter must be modified to include a second set of sample statistics, namely that from the second population.

Before we discuss confidence intervals for two populations in depth, one important distinction must be made regarding how the two populations are related. We must consider how the data from the two populations is connected, if at all. Two samples are said to be **independent** if the data from the first sample is not connected to the data from the second sample. For example, if we compare the average ACT score of high school seniors entering college to the average ACT of adults returning to college, then the data sets would be independent. If the two data sets are connected, then the samples are said to be **dependent**. Dependent data sets are also referred to as *matched data* or *paired data*. An example of dependent data sets may be a comparison of average grades on pre-tests and post-tests administered to the same group of students. Whether or not the two data sets under consideration are independent or dependent is one factor in determining the appropriate formulas to use for calculating confidence intervals.

Not only must we know whether the data sets are independent or not, but we must also consider whether the standard deviations of the populations are known values before choosing the appropriate method to use. Recall from section 8.1 that the standard deviation of the population is rarely known. If σ is known, then surely μ would be known as well, and there would be no need to estimate it. For this reason, in this text we will consider only the cases in which σ is unknown. When σ is unknown, we use the t-distribution for small samples ($n < 30$) and the normal distribution as an approximation for t whenever the samples chosen are at least 30.

Rounding Rule

Round E to one more decimal place than the original data, when given. Otherwise, round to the same number of decimal places as is given for the standard deviation or variance.

In this section we will consider the case in which we have two populations, from which large independent samples have been chosen. That is, the samples must meet the following criteria:

- Each sample must contain at least 30 values ($n \geq 30$).
- The data sets must be independent of each other.
- σ is unknown for both populations.

Recall from the last chapter the general process for creating a confidence interval, which contains two basic steps. First, calculate the margin of error for the desired level of precision. Second, add and subtract the margin of error from the point estimate to obtain the boundaries of the confidence interval.

The process for constructing a confidence interval with two samples is the same. However, we will use a new formula for the margin of error. The new formula that we use when dealing with two large independent samples is

$$E = z_c\sqrt{\frac{s_1^2}{n_1}+\frac{s_2^2}{n_2}}$$

where s_1 is the standard deviation of the first sample, n_1 is the size of the first sample, etc. The critical z-value, z_c, that is used depends on the level of confidence, c, that is desired. The table of critical values is listed below. It is the same table used in the previous chapter.

Table 1 - Critical z-Values for Confidence Intervals

Level of Confidence, c	z_c
0.80	1.28
0.85	1.44
0.90	1.645
0.95	1.96
0.98	2.33
0.99	2.575

The other change in this section is the value used for the point estimate. Since we are actually considering two populations, we need to consider a different point estimate than $\bar{x}$. Remember, we are not constructing a confidence interval for each sample separately, we are trying to *compare* the samples. We will then use the difference in the sample means as the point estimate, namely $\bar{x}_1 - \bar{x}_2$. Looking at the value of the difference will tell us if the means are the same, or if one is estimated to be greater than the other. Therefore, construct the confidence interval by adding and subtracting E from this difference.

Confidence Interval for Two Population Means:

$$(\bar{x}_1 - \bar{x}_2) - E < \mu_1 - \mu_2 < (\bar{x}_1 - \bar{x}_2) + E,$$
$$\text{or } \left((\bar{x}_1 - \bar{x}_2) - E,\ (\bar{x}_1 - \bar{x}_2) + E\right)$$

Example 1

A researcher is looking at the study habits of college students. A group of 42 freshmen reported an average study time per week of 15 hours with a standard deviation of 4.7 hours. A group of 39 seniors reported an average study time per week of 23 hours with a standard deviation of 5.2 hours. Construct a 95% confidence interval to estimate the true difference in the amount of time per week freshmen and seniors spend studying.

Solution:

Let Group 1 be freshmen and Group 2 be seniors. Begin by calculating our point estimate $\bar{x}_1 - \bar{x}_2$, which is 15 – 23 = –8. Next we calculate our margin of error using the formula for large, independent samples.

$$E = z_c\sqrt{\frac{s_1^2}{n_1} + \frac{s_2^2}{n_2}}$$

The sample sizes and standard deviations for each sample are given, so we only need to determine z_c before substituting the values into the formula. We need a 95% level of confidence, so $c = 0.95$. Looking up 0.95 in the table gives a critical value of 1.96. Thus,

$$E = 1.96\sqrt{\frac{4.7^2}{42} + \frac{5.2^2}{39}}$$
$$\approx 2.2.$$

Subtracting the margin of error from the point estimate gives a lower endpoint for the confidence interval of

$$-8 - 2.2 = -10.2.$$

By adding the margin of error to the point estimate, we obtain the upper endpoint of the confidence interval as follows

$$-8 + 2.2 = -5.8.$$

Thus the confidence interval for the difference in the two means is

$$-10.2 < \mu_1 - \mu_2 < -5.8 \text{, or}$$
$$(-10.2, -5.8).$$

The negative sign indicates that Group 1's average is less than Group 2's average. Therefore, freshmen study between 5.8 and 10.2 hours less than seniors.

Note that in the previous example we arbitrarily chose who would be in each group. If we had reversed the groups, the confidence interval would have been positive, indicating that group 1 (seniors) study more than group 2 (freshmen). The result is the same, just worded differently. Therefore, it does not matter how the groups are assigned. What matters is that you are consistent with the naming throughout the problem.

Example 2

A nursing instructor has created a prep course for the National Nursing Board Examination that she believes will result in nursing students performing better on the exam than if they had not taken the course. She collects data from 40 students who took her course and calculates that they had an average percentile score of 88 with a standard deviation of 4. She also collects data from 35 students who did not take her course and calculates that they had an average percentile score of 80 with a standard deviation of 6. Construct a 90% confidence interval to estimate the true difference in the average percentile score of students who took her course and those who did not.

Solution:

Let Group 1 be students who have taken the prep course and Group 2 be students who have not. Let's start by calculating our point estimate $\bar{x}_1 - \bar{x}_2$, which is $88 - 80 = 8$. We have the sample sizes and standard deviations for each sample, so we only need to determine z_c before substituting values into our margin of error formula. In this example we want a 90% level of confidence, so $c = 0.90$. Looking up 0.90 in our Critical z-Value Table gives us $z_c = 1.645$. Thus, for large, independent samples

$$E = 1.645\sqrt{\frac{4^2}{40} + \frac{6^2}{35}}$$
$$\approx 2.$$

Subtracting this margin of error from the point estimate gives a lower endpoint for the confidence interval of

$$8 - 2 = 6.$$

cont'd on the next page

By adding the margin of error to the point estimate, we obtain the upper endpoint of the confidence interval as follows

$$8+2=10.$$

The confidence interval for the difference of the two means is then

$$6<\mu_1-\mu_2<10\text{, or}$$
$$(6,\ 10).$$

Example 3

An automotive mechanic shop advertises that using a fuel additive will decrease the wear and tear on your engine. An independent researcher collects data from 50 cars that used the fuel additive and found that the average cost of engine repairs was \$3250 with a standard deviation of \$748.00. Data is then collected from 55 cars that did not use the fuel additive and finds that the average cost of engine repairs was \$3445 with a standard deviation of \$812.00. Construct an 85% confidence interval to estimate the true difference in the cost of engine repairs between cars using the fuel additive and those that do not.

Solution:

Let Group 1 be cars using the fuel additive and Group 2 be cars who do not. Start by calculating the point estimate $\bar{x}_1-\bar{x}_2$, which is $3250-3445=-195$. We have the sample sizes and standard deviations for each sample, so we only need to determine z_c. In this example we want an 85% level of confidence, so $c=0.85$ and $z_c=1.44$. Thus

$$E=1.44\sqrt{\frac{748.00^2}{50}+\frac{812.00^2}{55}}$$
$$\approx 219.$$

Subtract the margin of error from the point estimate to obtain a lower endpoint for the confidence interval of

$$-195-219=-414.$$

Add the margin of error to the point estimate to obtain the upper endpoint of the confidence interval as follows

$$-195+219=24.$$

The confidence interval for the difference of the two means is then

$$-414<\mu_1-\mu_2<24\text{, or}$$
$$(-414,\ 24).$$

Notice that in example 3 the confidence interval we calculated contained a range of values from a negative number to a positive number. Because the interval contains 0, you may interpret it to mean that it is possible there is no difference between the means of the 2 populations. However, if the interval is either all positive or all negative there likely is a difference in the population means.

Exercises

9.1

Directions: Find the point estimate, $\bar{x}_1 - \bar{x}_2$, for the true difference in population means for each of the following.

1. $\bar{x}_1 = 14$ and $\bar{x}_2 = 11$.

2. Sample 1's average is 45 and Sample 2's average is 56.

3. One mile running times for a group of 65 participants had an average of 8.45 minutes. The average time for second group of 77 participants is 8.25 minutes.

4. The average temperature in Group A is 72.4°F and the average temperature in Group B is 77.3°F.

Directions: Given the following information, calculate E.

5. $s_1 = 2.88,\ n_1 = 73,\ s_2 = 3.01,\ n_2 = 99,\ c = 0.99$

6. $s_1 = 5.3,\ n_1 = 31,\ s_2 = 4.9,\ n_2 = 40,\ c = 0.98$

7. $s_1 = 0.36,\ n_1 = 88,\ s_2 = 1.09,\ n_2 = 83,\ c = 0.95$

8. $s_1 = 11.54,\ n_1 = 115,\ s_2 = 10.65,\ n_2 = 122,\ c = 0.99$

Directions: Find the following confidence intervals. Assume that both σ's are unknown, and that there is independence between the populations being studied.

9. A local pizza delivery service claims to have a shorter delivery time than their national chain competitor. The manager of the local pizza store collects data on deliveries from two different drivers, one from each restaurant. Based on 48 deliveries, the local store's driver had an average delivery time of 29 minutes with a standard deviation of 8. Based on 52 deliveries, the national chain's driver had an average delivery time of 31 minutes with a standard deviation of 9 minutes. Construct a 90% confidence interval for the true difference in the mean delivery times between the two restaurants.

10. Dog-sled drivers use several different breeds of dogs to pull their sleds. One proponent of Siberian Huskies believes that sleds pulled by Huskies are faster than sleds pulled by other breeds. He times 35 Husky teams and they have an average time of 5.3 minutes with a standard deviation of 1.1 on a particular course. The average time for 32 non-Husky teams on the same course was 6.1 minutes with a standard deviation of 1.9. Construct a 99% confidence interval for the true difference in speed of a Husky team versus a non-Husky team.

11. There are two elementary schools in Caleb's community. Each school just took standardized exams. Forty-five of the third-graders from the first school scored an average of 75 on the exam with a standard deviation of 10. Thirty-nine of the third-graders from the second school scored an average of 78 with a standard deviation of 5. Construct a 95% confidence interval for the true difference in the average exam score between the two schools.

12. A contributor for the local newspaper is writing an article for the weekly fitness section. To prepare for the story, she conducts a study to compare the exercise habits of people who exercise in the morning to people who exercise in the afternoon or evening. She selects three different health centers from which to draw her sample. The 49 people she sampled who work out in the morning averaged 4.1 hours of exercise each week, with a standard deviation of 0.7 hours. The 54 people surveyed that exercise in the afternoon or evening averaged 3.7 hours of exercise each week, with a standard deviation of 0.5 hours. Construct a 95% confidence interval for the true difference in hours of exercise for morning versus afternoon or evening exercisers.

13. Russell is doing some research before buying his first house. He is looking at two different areas of the city, and he wants to know if there is a significant difference between the average prices of homes in the two areas. For the 34 homes he samples in the first area, the average home price is $168,500 with a standard deviation of $22,950. For the 39 homes he samples in the second area, the average home price is $171,800 to the nearest dollar with a standard deviation of $30,995. Construct the 90% confidence interval for the true difference between housing prices in the two areas.

14. A sports news station wants to know who are bigger sports fans, people who live in the north or the south. For their study, 121 southerners were surveyed and found to watch sports on average 4.1 hours a week, with a standard deviation of 1.8 hours. In the north, 137 people were surveyed and found to watch sports on average 3.9 hours a week with a standard deviation of 1.9 hours. Find the 99% confidence interval for the true difference between the hours of sports watched for the two regions.

Comparing Two Means (Small, Independent Samples)

9.2

Let's continue using confidence intervals to compare means from independent samples. However, we will restrict our focus in this section to cases where one or both sample sizes are small, $n < 30$. Our restrictions for this section include the following:

- The samples are independent
- Each population from which the samples were drawn has an approximately normal distribution
- $n < 30$ for one or both samples
- σ is unknown for both populations

When the above conditions are met, we begin the process of estimating the difference in the population means by finding a point estimate. The point estimate used for the difference in the sample means is the same as before, $\bar{x}_1 - \bar{x}_2$. However, the margin of error formula takes on a new form. First of all, because $n < 30$, we use the Student t-distribution in the formula just as we did when considering one sample. The critical values associated with a level of confidence, c, are found in Table C. Note that when the number of degrees of freedom is greater than 30, use the nearest number in the table.

When considering how to calculate the margin of error for small independent samples, we must consider whether or not the population variances are likely to be equal. Although neither population variance is actually known, in some cases it is reasonable to assume the variances are the same. For instance, a golf ball manufacturer claims that their ball travels further off the tee. If a player tests the new balls against other brands with the same club, it is reasonable to assume that the variance of the driving distances with the different balls will be equal, since the variance depends primarily on how consistently the player hits the ball, not on the balls themselves.

Let's first consider the case when the population variances are assumed to be the same. When this is true, the data from the samples can be **pooled**, or combined to give as good an estimate as possible for this common variance. The formula calculating the margin of error for pooled data is as follows:

$$E = t_{\alpha/2}\sqrt{\frac{(n_1-1)s_1^2+(n_2-1)s_2^2}{n_1+n_2-2}}\sqrt{\frac{1}{n_1}+\frac{1}{n_2}},$$

where the number of degrees of freedom is $d.f. = n_1 + n_2 - 2$.

Once the margin of error is calculated, adding and subtracting it from the point estimate gives the upper and lower bounds of the confidence interval.

Confidence Interval for Two Population Means:

$$(\bar{x}_1-\bar{x}_2)-E<\mu_1-\mu_2<(\bar{x}_1-\bar{x}_2)+E,$$
$$\text{or } \left((\bar{x}_1-\bar{x}_2)-E,\ (\bar{x}_1-\bar{x}_2)+E\right)$$

The following example illustrates how these types of confidence intervals are created.

Example 4

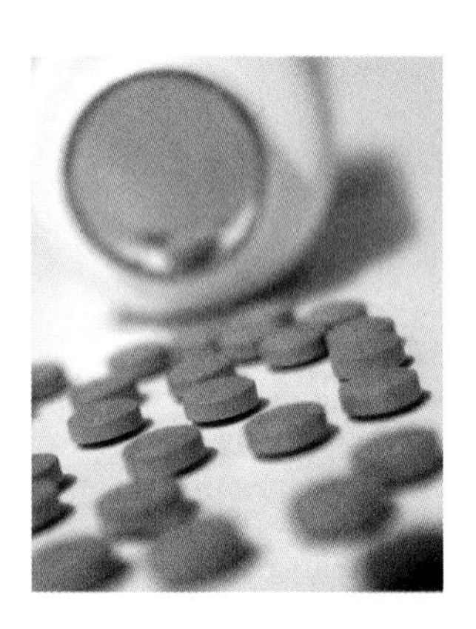

Obstetricians are concerned that a certain pain reliever may be causing lower birth weights. A medical researcher collects data from 12 mothers who took the pain reliever while pregnant and calculates that the average birth weight of their babies was 5.6 pounds with a standard deviation of 1.8. She also collects data from 20 mothers who did not take the pain reliever and calculates that the average birth weight of their babies was 6.3 pounds with a standard deviation of 2.1. Assuming that the population variances are the same, construct a 99% confidence interval for the true difference in the mean birth weights. Does the confidence interval suggest that there is a difference in the birth weights?

Solution:

Let Group 1 be the birth weights associated with taking the pain reliever and Group 2 be the birth weights associated with not taking the pain reliever. Begin by calculating the point estimate, $\bar{x}_1-\bar{x}_2$, which is $5.6-6.3=-0.7$.

Next calculate the margin of error using the formula

$$E=t_{\alpha/2}\sqrt{\frac{(n_1-1)s_1^2+(n_2-1)s_2^2}{n_1+n_2-2}}\sqrt{\frac{1}{n_1}+\frac{1}{n_2}}.$$

The sample sizes and standard deviations for each sample are given, so we only need to determine the critical t-value before substituting the values into the formula.

For populations where the variances are assumed to be the same, the number of degrees of freedom is

$$\begin{aligned} df &= n_1+n_2-2 \\ &= 12+20-2 \\ &= 30 \end{aligned}$$

Because $c=0.99$, we know that $\alpha=1-0.99=0.01$. Then $\frac{\alpha}{2}=\frac{0.01}{2}=0.005$.

With 30 degrees of freedom, $t_{0.005} = 2.750$.

Substituting into the margin of error formula gives us the following:

$$E = t_{\alpha/2}\sqrt{\frac{(n_1-1)s_1^2+(n_2-1)s_2^2}{n_1+n_2-2}}\sqrt{\frac{1}{n_1}+\frac{1}{n_2}}$$
$$= 2.750\sqrt{\frac{(12-1)(1.8)^2+(20-1)(2.1)^2}{12+20-2}}\sqrt{\frac{1}{12}+\frac{1}{20}}.$$

Since this is such a large formula, we will calculate it in stages, rounding to 6 decimal places.

$$\sqrt{\frac{(12-1)(1.8)^2+(20-1)(2.1)^2}{12+20-2}} \approx 1.995244 \text{ and } \sqrt{\frac{1}{12}+\frac{1}{20}} \approx 0.365148.$$

Which gives us

$$E = 2.750(1.995244)(0.365148)$$
$$E \approx 2.0$$

Subtracting the margin of error from the point estimate gives a lower endpoint for the confidence interval of

$$-0.7 - 2.0 = -2.7.$$

By adding the margin of error to the point estimate, we obtain the upper endpoint of the confidence interval as follows

$$-0.7 + 2.0 = 1.3.$$

Thus the confidence interval for the difference in the two means is

$$-2.7 < \mu_1 - \mu_2 < 1.3, \text{ or}$$
$$(-2.7, 1.3).$$

Since the confidence interval contains zero, it may be that there is no difference in the birth weights; certainly there is no statistical evidence to the contrary.

On the other hand, it may not be appropriate to assume that the two samples are drawn from populations with the same variance. For an illustration, let's return to our previous golf example. If, instead, different players were to drive the different balls, it would no longer be reasonable to assume a common variance.

If the variances are *not assumed to be equal* then the data **cannot be pooled** and the formula for the margin of error becomes

$$E = t_{\alpha/2}\sqrt{\frac{s_1^2}{n_1} + \frac{s_2^2}{n_2}},$$

where the number of degrees of freedom is the smaller of the values $n_1 - 1$ and $n_2 - 1$.

The following example illustrates how these types of confidence intervals are created.

Example 5

Unhappy with the class she is taking, Misty believes her difficulty in the class is the result of an inexperienced teacher. She believes that students in her class are performing more poorly on their exams than students in another class with a more experienced teacher. She collects exam scores from 11 of her classmates and calculates an exam average of 75 with a standard deviation of 8. Misty then collects data from 9 students in the other class and calculates that the exam average is 82 with a standard deviation of 5. Construct a 90% confidence interval for the true difference in the exam average between the two classes, assuming that the population variances are different. Does this confidence interval suggest that the inexperienced teacher's class is doing more poorly than the experienced teacher's class?

Solution:

Let Group 1 be the inexperienced teacher's class and Group 2 be the experienced teacher's class. Begin by calculating the point estimate, $\bar{x}_1 - \bar{x}_2$, which is

$$75 - 82 = -7.$$

Next calculate the margin of error using the formula

$$E = t_{\alpha/2}\sqrt{\frac{s_1^2}{n_1} + \frac{s_2^2}{n_2}}.$$

The sample sizes and standard deviations for each sample are given, so we only need to determine the critical t-value before substituting the values into the formula.

For populations where the variances are assumed to be different, the number of degrees of freedom is the smaller of $n_1 - 1 = 11 - 1 = 10$ and $n_2 - 1 = 9 - 1 = 8$. Thus

$$d.f. = 8.$$

Because $c = 0.90$, we know that $\alpha = 1 - 0.90 = 0.10$. Then $\frac{\alpha}{2} = \frac{0.10}{2} = 0.05$.

With 8 degrees of freedom, $t_{0.05} = 1.860$.

Substituting into the margin of error formula gives us the following:

$$\begin{aligned}
E &= t_{\alpha/2}\sqrt{\frac{s_1^2}{n_1} + \frac{s_2^2}{n_2}} \\
&= 1.860\sqrt{\frac{8^2}{11} + \frac{5^2}{9}} \\
&\approx 1.860(2.932) \\
&= 5.45352 \\
&\approx 5
\end{aligned}$$

Subtracting the margin of error from the point estimate gives a lower endpoint for the confidence interval of

$$-7 - 5 = -12.$$

By adding the margin of error to the point estimate, we obtain the upper endpoint of the confidence interval as follows

$$-7 + 5 = -2.$$

Thus the confidence interval for the difference in the two means is

$$-12 < \mu_1 - \mu_2 < -2, \text{ or}$$
$$(-12, -2).$$

By this, it appears that the true difference in the class means is a negative number, which in this scenario indicates that the second class did perform better than the first class. Therefore, we can be 90% confident that the inexperienced teacher's class is scoring between 2 and 12 points lower than the experienced teacher's class.

Margin of Error for Means (Small, Independent Samples)

Population variances assumed to be equal and data is pooled:

$$E = t_{\alpha/2}\sqrt{\frac{(n_1 - 1)s_1^2 + (n_2 - 1)s_2^2}{n_1 + n_2 - 2}}\sqrt{\frac{1}{n_1} + \frac{1}{n_2}},$$

where the number of degrees of freedom is $d.f. = n_1 + n_2 - 2$

Population variances assumed to be not equal and data is not pooled:

$$E = t_{\alpha/2}\sqrt{\frac{s_1^2}{n_1} + \frac{s_2^2}{n_2}},$$

where the number of degrees of freedom is the smaller of the values $n_1 - 1$ or $n_2 - 1$.

Exercises 9.2

Directions: Find the critical t-value for the following:

1. 95% level of confidence, population variances assumed to be equal, $n_1 = 11$, $n_2 = 20$.
2. 99% level of confidence, population variances assumed to be equal, $n_1 = 9$, $n_2 = 6$.
3. 90% level of confidence, population variances assumed to be not equal, $n_1 = 14$, $n_2 = 12$.
4. 95% level of confidence, population variances assumed to be not equal, $n_1 = 25$, $n_2 = 26$.
5. 95% level of confidence, population variances assumed to be not equal, $n_1 = 4$, $n_2 = 7$.
6. 99% level of confidence, population variances assumed to be equal, $n_1 = 22$, $n_2 = 18$.
7. 98% level of confidence, population variances assumed to be equal, $n_1 = 13$, $n_2 = 8$.

Directions: Calculate the margin of error, E, for the following scenarios.

8. Variances assumed to not be equal, $t_{0.01} = 2.764$, $n_1 = 16$, $n_2 = 11$, $s_1 = 6$, $s_2 = 5$.

9. Variances assumed to be equal, $t_{0.025} = 2.021$, $n_1 = 22$, $n_2 = 20$, $s_1 = 2$, $s_2 = 3$.

10. Variances assumed to be equal, $t_{0.005} = 2.807$, $n_1 = 15$, $n_2 = 10$, $s_1 = 10$, $s_2 = 13$.

11. Variances assumed to not be equal, $t_{0.05} = 2.132$, $n_1 = 6$, $n_2 = 5$, $s_1 = 2$, $s_2 = 5$.

Directions: Construct confidence intervals for each of the following scenarios. You can assume the following for each:

- The samples are independent.
- Each population from which the samples were drawn has a normal distribution.
- σ is unknown for both populations.

12. Different breeds of dogs produce different sized litters. Wesley believes that his coon hounds will have litters that are twice as big as his schnauzers. He has 10 female coon hounds and the average size of each of their last litters was 11 puppies with a standard deviation of 3. Wesley also has 9 female schnauzers, whose average litter size was 5 puppies with a standard deviation of 2. Assume that the population variances are not the same. Create a 90% confidence interval to estimate the true difference in the size of the two litters.

13. Midwives claim that a water birth reduces the amount of pain the mother perceives. To test this theory, one midwife asks 6 of her clients who had a water birth and 8 of her clients who did not have a water birth to rate their pain on a scale of 1 to 10. The average pain score of mothers who had a water birth was a 7 with a standard deviation of 1.2. The average pain score of mothers who did not have a water birth was 9 with a standard deviation of 0.9. Construct an 80% confidence interval for the true difference in the amount of pain perceived by mothers who have a water birth and mothers who do not. Assume that the variance of the two populations are not equal.

14. Steve believes his wife's cell phone battery does not last as long as his cell phone battery. On 8 different occasions, he measured the length of time his cell phone battery lasted, and calculated that the mean is 47 hours with a standard deviation of 6 hours. He measured the length of time his wife's cell phone battery lasted on 9 different occasions and calculated a mean of 45 hours with a standard deviation of 8 hours. Construct a 95% confidence interval of the true difference in battery life between Steve's cell phone and his wife's. Assume that the population variances are the same.

15. Dentists believe that a diet low in sugary foods can reduce the number of cavities in children. Ten children whose diet is believed to be high in sugar are examined and the average number of cavities is 3 with a standard deviation of 0.7. Twelve children whose diet is believed to be low in sugar are examined and the average number of cavities is 1.8 with a standard deviation of 0.6. Construct a 99% confidence interval for the true difference in the average number of cavities in children with a diet high in sugar and those whose diet is low in sugar. Assume that the variances in the two populations are the same.

16. A company that manufactures baseball bats believes that their new bat will allow you to hit the ball 30 feet farther than their current model. They hire a professional baseball player known for hitting home runs to hit 10 balls with each bat and they measure the distance each ball is hit to test their claim. The results of the batting experiment are shown below. Construct a 90% confidence interval for the true difference in the mean distance hit with the new model and the mean distance hit with the older model. Assume the variances of the two populations are the same. Remember to round the means and the standard deviations to one decimal place for this data.

Table for Exercise 16 - Distance in Feet	
New Model	**Old Model**
235	200
240	210
253	231
267	218
243	210
237	209
250	210
241	229
251	234
248	231

17. A marketing firm is doing research for an internet based company. They want to appeal to the age group who spends the most money. The company wants to know if there is a difference in the amount of money people spend on internet purchases per month depending on their age bracket. They looked at two age groups: 18-24 year olds and 25-30 year olds. The data they found is in the following table.

Table for Exercise 17		
	18-24 Year Olds	**25-30 Year Olds**
Mean Amount Spent	\$52.00	\$45.49
Standard Deviation	\$21.30	\$13.19
Sample Size	18	25

Assume the population variances are not the same. Calculate the 80% confidence interval to estimate the true difference in the amount of money these two age groups spend on internet purchases.

18. Is it worth pursuing a doctoral degree in education if you already have an undergraduate degree? One way to help make this decision is to look at the mean incomes of these two groups. 19 education majors were surveyed. Their mean salary was \$28,500 with a standard deviation of \$6800. Eleven people with doctorate degrees in education were found to have a mean salary of \$55,500 with a standard deviation of \$10,200. Assume the population variances are not the same. Calculate the 90% confidence interval to estimate the true difference in the salaries for graduate degrees and undergraduate degrees.

19. In evaluating your teaching, you decide to have your class of 15 rate you on a scale of 1-10 with 10 being excellent. After getting the results, a mean of 6.2 with a standard deviation of 1.5, you decide to make an effort to improve your teaching. Next semester with 13 students, the results were a mean of 7.5 with a standard deviation of 1.6. Use a 95% confidence interval to estimate the difference in your ratings from semester to semester. You can assume the population variances to be the same.

20. To conform to market trends, a prominent syrup manufacturer is changing the design of their syrup bottle from a tall slender bottle to a shorter rounder bottle that fits more easily in a microwave oven. The shareholders are concerned that the average amount of syrup in each bottle remains the same. A sample of 16 bottles of the older design holds on average 36.2 fluid ounces with a standard deviation of 0.90 fl. oz. and a sample of 20 bottles from the new design holds on average 35.9 fl. oz. with a standard deviation of 0.73 fl. oz. Because the same machine is being used to fill the bottles, you may assume the population variances to be the same. Construct a 90% confidence interval to estimate the difference in volume for the bottles.

21. A researcher is interested in exploring the relationship between calcium intake and weight loss. Two different groups, each with 25 dieters, are chosen for the study. Group A is required to follow a specific diet and exercise regimen and also take a 500 mg supplement of calcium each day. Group B is required to follow the same diet and exercise regimen, but with no supplemental calcium. After 6 months on the program, the members of Group A had lost an average of 12.7 pounds with a standard deviation of 2.2 pounds. The members of Group B had lost an average of 10.8 pounds with a standard deviation of 2.0 pounds during the same time period. Assume that the population variances are not the same. Create a 95% confidence interval to estimate the true difference between dieters who supplement with calcium and those who do not.

22. Gavin likes to grow tomatoes in the summer, and each year he experiments with ways to improve his crop. This summer he wants to determine whether the new fertilizer he has seen advertised will increase the number of tomatoes produced per plant. Being careful to control the conditions in his garden, he uses the old fertilizer on 12 tomato plants and the new fertilizer on the other 8 plants. Throughout the course of the growing season, he counts an average of 18.2 tomatoes per plant for the old fertilizer, with a standard deviation of 1.9 tomatoes. He counts an average of 21.4 tomatoes per plant for the new fertilizer, with a standard deviation of 2.6 tomatoes. Create a 90% confidence interval for the true difference in tomato production. Assume that the population variances are the same.

23. A retailer is interested in comparing the shopping habits of customers as different types of music play throughout the store. One day when the store played slow instrumental music, the 17 customers who made purchases spent on average $84 with a standard deviation of $20. The next day the store played upbeat instrumental music, and the 13 customers who made purchases spent on average $93 with a standard deviation of $22. Construct the 99% confidence interval for the true difference in the average amount spent while listening to different types of music. Assume that the population standard deviations are the same.

24. A researcher wants to know who reads more, women or men. She surveys 28 women and finds that they read on average 2.5 books a month, with a standard deviation of 0.9 books. Out of the 25 men she surveys, they read on average 2.1 books per month with a standard deviation of 0.8 books. Assuming that the population standard deviations are different, what is the 90% confidence interval for the true difference between the average number of books read each month for women and men?

Comparing Two Means (Dependent Samples)

9.3

So far in this chapter, we have looked at creating confidence intervals from independent samples, meaning that the data from the two samples have no influence on each other. However, sometimes situations arise when the data are **dependent**. Data may be dependent in a variety of ways. In this section we will consider how to deal with dependent data that are connected in a specific manner, or **paired**. To do this we will need the additional assumption that all the data are drawn from normal distributions.

Paired data might come from sources such as pre-test/post-test studies, a single population, or data might be matched based on certain characteristics. For instance, suppose researchers wanted to study whether a person's sleeping habits changed with a new drug. Data would be taken from a number of participants both before the drug was administered and after. The data from each participant would then be paired together. This allows the researcher to compare the sleeping habits while eliminating as many confounding variables as possible.

When working with paired data, we actually construct a confidence interval as if there were only one set of data values. The set of data that we use contains the differences between each pair of values. Thus the first step in computing a confidence interval for the difference in the means of two sets of paired data is to calculate the **paired differences**, *d*. The paired difference is found by simply subtracting each pair of values. Thus,

$$d = x_2 - x_1$$

for each pair of values in the data set.

Example 6

Given the following data, calculate the paired difference for the cost of utility bills in the months of March and April.

Table 2 - Utility Bills for Homes

	March	April
Home 1	\$119.75	\$121.06
Home 2	\$68.43	\$79.04
Home 3	\$202.39	\$189.55
Home 4	\$47.88	\$49.64
Home 5	\$66.01	\$68.52

cont'd on the next page

Solution:

Calculate the paired differences by subtracting across the rows.

Table 3 - Utility Bills for Homes			
	March	**April**	**Paired Difference, *d***
Home 1	\$119.75	\$121.06	\$1.31
Home 2	\$68.43	\$79.04	\$10.61
Home 3	\$202.39	\$189.55	–\$12.84
Home 4	\$47.88	\$49.64	\$1.76
Home 5	\$66.01	\$68.52	\$2.51

Rounding Rule

Round $\bar{d}$ to one more decimal place than the original data, when given.

The new statistic that we want to estimate is the difference in the means of the two populations of dependent data, $\mu_{\bar{d}}$. Recall that the best point estimate for a population mean is its sample mean. Therefore, the mean of the paired differences, $\bar{d}$, is the point estimate used here.

Example 7

Calculate the best point estimate for the difference in the means of the two populations given in Example 6.

Solution:

The mean of the paired differences is calculated by

$$\begin{aligned}\bar{d} &= \frac{\Sigma d}{n}\\ &= \frac{3.35}{5}\\ &= 0.67.\end{aligned}$$

Rounding Rule Reminder

Round E to one more decimal place than the original data, when given. Otherwise, round to the same number of decimal places as is given for the standard deviation or variance.

Now that we have the point estimate for the difference in the means of the two populations, we must calculate the margin of error. If there are n sets of paired data drawn from normal distributions, then the sampling distribution for the sample statistic $\bar{d}$ follows a t-distribution with $n - 1$ degrees of freedom. Hence, the formula for the margin of error becomes

$$E = t_{\alpha/2} \frac{s_d}{\sqrt{n}}$$

where the number of degrees of freedom is $n - 1$.

Once the margin of error is calculated, adding and subtracting E from the point estimate gives the upper and lower bounds of the confidence interval.

Confidence Interval for Two Dependent Means:

$$\bar{d} - E < \mu_d < \bar{d} + E,$$
$$\text{or } (\bar{d} - E,\ \bar{d} + E).$$

Example 8

In a CPR class, students are given a pre-test at the beginning of the class to determine their initial knowledge of CPR and then a post-test after the class to determine their new knowledge of CPR. Educational researchers are interested if the class increases a student's knowledge of CPR. Create a 95% confidence interval for the mean increase in the students' knowledge from taking the CPR class. Data from 10 students who took the CPR class are listed in the table below.

Table 4 - Students' CPR Scores

Student	1	2	3	4	5	6	7	8	9	10
Pre-Test Score	60	63	68	70	71	68	72	80	83	79
Post-Test Score	80	79	83	90	83	89	94	95	96	93

Solution:

The first step is to calculate the paired difference for each pair of data. Subtract the pre-test score from the post-test score.

Table 5 - Students' CPR Scores

Student	1	2	3	4	5	6	7	8	9	10
Pre-Test Score	60	63	68	70	71	68	72	80	83	79
Post-Test Score	80	79	83	90	83	89	94	95	96	93
d	20	16	15	20	12	21	22	15	13	14

Next, calculate the mean and sample standard deviation of the paired data. (Your calculator is the easiest way to do this.)

$$\bar{d} = 16.8$$
$$s_d = 3.6$$

The number of degrees of freedom is $d.f. = n - 1 = 10 - 1 = 9$.

Because $c = 0.95$, we know that $\alpha = 1 - 0.95 = 0.05$. Then $\frac{\alpha}{2} = \frac{0.05}{2} = 0.025$.

cont'd on the next page

With 9 degrees of freedom, $t_{0.025} = 2.262$.

Substituting into the margin of error formula gives us the following:

$$E = t_{\alpha/2}\frac{s_d}{\sqrt{n}}$$
$$= 2.262\frac{3.6}{\sqrt{10}}$$
$$= 2.575$$
$$\approx 2.6$$

Subtracting the margin of error from the point estimate gives a lower endpoint for the confidence interval of

$$16.8 - 2.6 = 14.2.$$

By adding the margin of error to the point estimate, we obtain the upper endpoint of the confidence interval as follows

$$16.8 + 2.6 = 19.4.$$

Thus the confidence interval for the difference in the two means is

$$14.2 < \mu_{\bar{d}} < 19.4, \text{ or}$$
$$(14.2, 19.4).$$

Exercises 9.3

Directions: Calculate $\bar{d}$ and s_d for each of the tables below.

1.

Table for Exercise 1

Participant	1	2	3	4	5	6	7	8	9
Sample A	22	21	19	20	17	18	20	19	22
Sample B	24	23	23	19	20	21	23	19	23

2.

Table for Exercise 2

Participant	1	2	3	4	5	6	7	8	9
Sample 1	2	1	1	2	1	1	2	1	2
Sample 2	4	3	2	1	2	2	2	1	3

3.

Table for Exercise 3								
Data	1	2	3	4	5	6	7	8
Red Group	78	79	91	83	79	92	45	68
Blue Group	67	85	90	84	86	96	51	66

4.

Table for Exercise 4									
Participant	1	2	3	4	5	6	7	8	9
Group A	18	23	24	19	17	22	18	16	20
Group B	24	23	23	19	20	21	23	19	23

Directions: In each of the following, calculate the confidence interval for the difference in the means of two sets of paired data.

5. $n = 41,\ \bar{d} = 2.23,\ \alpha = 0.01,\ s_d = 0.567$

6. $n = 30,\ \bar{d} = 12,\ \alpha = 0.05,\ s_d = 2.7$

7. $n = 5,\ \bar{d} = 1.14,\ \alpha = 0.10,\ s_d = 1.3$

8. $n = 25,\ \bar{d} = 3.4,\ \alpha = 0.05,\ s_d = 1.08$

Directions: Calculate confidence intervals for each of the following scenarios. Assume that the data are drawn from normal populations.

9. To determine if a new cold medicine works better than the most popular cold medicine, 16 people who had a cold volunteered for a study. The volunteers are matched based on age to create 8 pairs. A double-blind study is constructed where one of the volunteers in the pair is given the new cold medicine and the other member of the pair is given the traditional medicine. The duration of the cold (in days) is measured for each person and the results are shown in the table below. Construct a 99% confidence interval for the true difference in the mean duration of a cold between those taking the new medication and those taking the traditional medication.

Table for Exercise 9 - Duration of Cold Medicine (in Days)								
Pair	1	2	3	4	5	6	7	8
Duration (new drug)	4	5	3	6	3	4	5	7
Duration (other drug)	6	5	4	8	4	5	7	7

10. To determine if a new teaching method increases students' learning, the professor administers a pre-test to his class at the beginning of the semester and then a post-test at the end of the semester. The results from 15 randomly chosen students are given below. Construct a 95% confidence interval for the true difference in the mean scores to determine if the teaching method increased students' learning over the course of the semester.

Table for Exercise 10 - Test Scores		
Student	**Pre-Test Score**	**Post-Test Score**
1	60	80
2	61	87
3	65	91
4	71	97
5	68	89
6	67	86
7	65	85
8	62	83
9	63	89
10	68	93
11	69	94
12	70	99
13	65	92
14	62	91
15	57	83

11. A dietician wants to see how much weight a certain diet can help patients lose. Nine people agree to participate in her study. She weighs each patient at the beginning of the study and after 30 days of being on the diet. Her results are shown in the table below. Construct a 99% confidence interval for the true difference in the mean weight to determine the average amount of weight lost by going on this diet.

Table for Exercise 11 - Patients' Weights									
Participant	1	2	3	4	5	6	7	8	9
Starting Weight	180	176	152	230	183	214	250	197	201
Ending Weight	172	167	144	219	172	201	239	186	195

12. A workout coach believes that walking every day can produce the same health benefits as jogging. 10 volunteers are paired based on significant characteristics and half of the group is asked to walk everyday and half of the group is asked to jog everyday. The amount of weight lost in 30 days is recorded below. Construct a 95% confidence interval for the true difference in the amount of weight lost by walking and the amount of weight lost by jogging.

Table for Exercise 12 - Weight Loss					
Pair	1	2	3	4	5
Walking	8	9	10	7	9
Jogging	10	12	14	9	12

13. Researchers have developed a method to improve memory. To test their method, 12 participants are asked to memorize a list of words and the number of words remembered correctly is recorded. The participants are then taught the method to improve their memory and are asked to memorize another list of words and the number of words remembered correctly is again recorded. The results are shown in the table below. Construct a 98% confidence interval for the true difference between the number of words participants could memorize before and after learning the memory method.

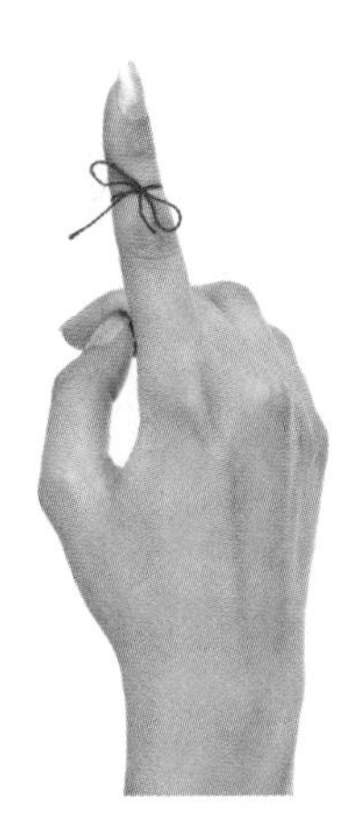

Table for Exercise 13 - Number of Words Memorized		
Participant	**Before**	**After**
1	8	13
2	5	16
3	6	12
4	7	16
5	6	15
6	3	13
7	8	17
8	10	20
9	12	19
10	6	17
11	8	18
12	5	13

14. A pharmaceutical company is running tests to see how well its new drug lowers cholesterol. A group of 10 adults volunteer to participate in the study. The cholesterol level of each participant is recorded once at the start of the study and then again after 3 months of taking the drug. The results are given in the table below. Construct a 99% confidence interval for the true difference in cholesterol levels after taking the new drug.

Table for Exercise 14 - Cholesterol Levels

Participant	Initial Level	Level After 3 Months
A	210	201
B	200	195
C	215	208
D	194	197
E	206	200
F	221	203
G	203	190
H	189	188
I	208	210
J	211	210

15. An infomercial claims that their new cooking device will dramatically cut the time you spend preparing meals. Wondering if their claim is true, you set out to determine how much time the new cooking device will really save you. Eight people who have purchased the item agree to participate in your study, and they each estimate the time they spent cooking dinner before they bought the new device and then after they began using it. Use the table of results to create a 90% confidence interval for the true amount of time that is saved by using the infomercial's item.

Table for Exercise 15 - Time Spent Preparing Meals

Participant	Time Spent without Service (in Minutes)	Time Spent with Service (in Minutes)
A	50	45
B	60	50
C	45	30
D	30	30
E	45	60
F	50	40
G	20	15
H	25	30

16. Philip wants to take a speed-reading course, but his wife thinks that it is a waste of time. To convince her that the course will really change the way that he reads, Philip decides to conduct an informal study. He polls 7 people, asking them to tell the number of pages they were able to read in an hour before and then after they took the course. The results he obtained are found below. Construct a 95% confidence interval for the true difference the speed-reading course makes.

Table for Exercise 16 - Number of Pages Read

Participant	A	B	C	D	E	F	G
Pages Read Pre-Course	35	50	45	50	60	70	65
Pages Read Post-Course	50	60	80	70	85	100	90

17. A home improvement show gives tips on ways to improve your house before it is listed for sale. The show claims that their tips will help your house sell faster. To test the show's claim, a researcher found 5 pairs of houses that were similar in condition, area, and asking price. The homeowners from one house in each pair were asked to follow the tips given in the show before putting their houses on the market. (No major renovations allowed.) The researchers then kept tabs on the subsequent time it took for each house to sell. Use the following results to find a 90% confidence interval for the true difference in time saved based on the show's tips.

Table for Exercise 17 - Number of Weeks to Sell a House

Pair	1	2	3	4	5
Weeks Until Sale Without Tips	9	10.5	3	5.5	14.5
Weeks Until Sale With Tips	7	6	5	1	4

18. During the fall semester of biology, the instructor teaching the class became ill and had to have another instructor stand in for him. After the students had taken a test under both instructors, the chairman of the biology department wanted to know if there was a significant difference in the performance of students under the different instructors. The results from each test are given in the table below. Construct a 90% confidence interval for the true difference in test scores for each student.

Table for Exercise 18 - Students' Grades

Student	1	2	3	4	5	6	7	8	9	10	11	12
Instructor A	58	82	78	91	83	77	45	87	94	92	68	77
Instructor B	52	60	83	90	87	70	48	81	90	90	71	70

19. At the beginning of an elementary physical education class, students are asked to do as many sit-ups as possible in a one-minute period. After working on technique for sit-ups, the students are once again timed to see if they have improved their ability to do sit-ups. Use the table of results to create a 95% confidence interval for the true difference the training made to the students' ability to do sit-ups.

Table for Exercise 19 - Number of Sit-Ups								
Student	A	B	C	D	E	F	G	H
Number of Sit-Ups Before	10	20	15	30	31	40	20	25
Number of Sit-Ups After	28	30	27	42	50	45	19	36

20. A typing program claims it can improve your typing skills. To test their claim and possibly help yourself out, you and three of your friends decide to try the program and see what happens. Use the table below to construct an 80% confidence interval for the true difference the typing course makes.

Table for Exercise 20 - Number of Words per Minute		
Participant	**Words per Minute Pre-Course**	**Words per Minute Post-Course**
A	42	54
B	50	52
C	37	41
D	22	30

21. After students were not doing so well in her math class, Ms. Comeaux decided to try a different approach and use verbal positive reinforcement at least once every hour. Use the following results to find a 99% confidence interval for the true difference in the students' performance with continuous positive verbal reinforcement.

Table for Exercise 21 - Students' Grades		
Student	**Without Reinforcement**	**With Reinforcement**
1	66	72
2	57	55
3	61	80
4	72	79
5	38	49
6	47	56

Comparing Two Proportions (Large, Independent Samples)

9.4

In this chapter we have seen how to compare population means for large and small independent samples as well as for dependent samples. Now we will turn our attention to comparing population proportions.

When we were comparing population means we constructed a confidence interval for the difference between the two population means. When comparing two population proportions, we will again use a difference for the point estimate, namely $\hat{p}_1 - \hat{p}_2$. Also, we will restrict our discussion to comparing population proportions using large, independent samples. If we use large, independent samples, then the sampling distribution for the difference in the sample proportions, $\hat{p}_1 - \hat{p}_2$, is a normal distribution. Thus we can use the normal distribution to calculate the margin of error and construct a confidence interval for the true difference in the population proportions.

Memory Booster!

Population Proportion

$$p = \frac{x}{N} = \frac{\text{\# of "successes"}}{\text{population size}}$$

Sample Proportion

$$\hat{p} = \frac{x}{n} = \frac{\text{\# of "successes"}}{\text{sample size}}$$

We mentioned that we need large, independent samples; but how large do the samples need to be in order to use the normal distribution? For comparing population means, "large" meant that the sample size was at least 30. When comparing population proportions, verifying that the samples are large enough is slightly more involved. To determine if the samples are large enough to use the normal distribution, make sure that the following products are all greater than or equal to 5:

$$n_1\hat{p}_1 \geq 5 \qquad n_2\hat{p}_2 \geq 5$$
$$n_1\left(1-\hat{p}_1\right) \geq 5 \qquad n_2\left(1-\hat{p}_2\right) \geq 5$$

Example 9

You wish to compare the proportion of women registered to vote in two different cities. You take a random sample of registered voters from each town and record the number of men and women in each sample. The following table summarizes your results.

Table 6 - Registered Voters

	Women	Men
City A	45	37
City B	32	28

Calculate the sample proportion for each sample and verify that the samples are large enough to use the normal distribution to compare the population proportions.

cont'd on the next page

Solution:

Begin by finding the sample proportion for each sample. The size of the sample is the sum of the number of men and women. For the first sample, from City A, the sample size is

$$n_1 = 45 + 37 = 82.$$

For the second sample, from City B, the sample size is

$$n_2 = 32 + 28 = 60.$$

The sample proportions are found by dividing the number of women in the sample by the sample size. Thus the sample proportions are as follows:

$$\hat{p}_1 = \frac{45}{82} \approx 0.549$$

and

$$\hat{p}_2 = \frac{32}{60} \approx 0.533.$$

Next, verify that these samples are large enough to use the normal distribution to compare the population proportions by making sure that the product of the sample sizes and the proportions are all greater than or equal to five. The products are as follows:

$$n_1 \hat{p}_1 = 82(0.549) = 45.018 \geq 5$$

$$n_1 \left(1 - \hat{p}_1\right) = 82(1 - 0.549) = 36.982 \geq 5$$

$$n_2 \hat{p}_2 = 60(0.533) = 31.98 \geq 5$$

$$n_2 \left(1 - \hat{p}_2\right) = 60(1 - 0.533) = 28.02 \geq 5$$

Thus all four products are greater than or equal to five and the normal distribution can be used.

For all exercises in this section, you can safely assume that the sample sizes given are large enough to justify using the normal distribution to compare the population proportions.

When comparing population proportions, the first step is to calculate the point estimate. The best point estimate for the difference in the population proportions is the difference in the sample proportions, $\hat{p}_1 - \hat{p}_2$.

The second step is to calculate the margin of error. The margin of error is found by multiplying the appropriate critical value for the sampling distribution by the standard deviation of the sampling distribution. Since the sampling distribution for the difference in the sample proportions is a normal distribution, we will use the critical value for z to calculate the margin of error. These are the same critical z-values we have used throughout this text.

Rounding Rule Reminder

Round proportions to three decimal places.

Table 7 - Critical z-Values for Confidence Intervals

Level of Confidence, c	z_c
0.80	1.28
0.85	1.44
0.90	1.645
0.95	1.96
0.98	2.33
0.99	2.575

Rounding Rule Reminder

Round the margin of error for proportions to three decimal places.

The margin of error for the difference in the population proportions is

$$E = z_c\sqrt{\frac{\hat{p}_1\left(1-\hat{p}_1\right)}{n_1}+\frac{\hat{p}_2\left(1-\hat{p}_2\right)}{n_2}}.$$

Thus, to construct a confidence interval for the true difference in the population proportions, add and subtract the margin of error from the point estimate.

Confidence Interval for Comparing Two Population Proportions

$$\left(\hat{p}_1-\hat{p}_2\right)-E<p_1-p_2<\left(\hat{p}_1-\hat{p}_2\right)+E,$$

$$\text{or } \left(\left(\hat{p}_1-\hat{p}_2\right)-E,\ \left(\hat{p}_1-\hat{p}_2\right)+E\right).$$

Example 10

In order to determine if a new instructional technology improves students' scores, a professor wants to know if a larger percentage of the class using the instructional technology passed than the class that did not use the new technology. Records show that 45 out of 50 students using the instructional technology passed the class and 38 out of 51 students who did not use the instructional technology passed the class. Construct a 95% confidence interval for the true difference between the proportion of students using the technology who passed and the proportion of students not using the technology who passed.

cont'd on the next page

Rounding Rule Reminder

When dealing with intermediate calculation, do not round the intermediate steps. For example, although $\hat{p}$ is normally rounded to three decimal places, for intermediate calculation it is best to keep $\hat{p}$ out to six decimal places in order to not affect the remaining calculations with a rounding error.

Solution:

The first step is to calculate the sample proportions. The sample proportion for sample 1 (using instructional technology) is

$$\hat{p}_1 = \frac{45}{50} = 0.900.$$

The sample proportion for sample 2 (without the instructional technology), is

$$\hat{p}_2 = \frac{38}{51} \approx 0.745098 \approx 0.745 .$$

The next step is to calculate the point estimate

$$\begin{aligned}\hat{p}_1 - \hat{p}_2 &= 0.900 - 0.745 \\ &= 0.155.\end{aligned}$$

Now we are ready to calculate the margin of error. The level of confidence is $c = 0.95$, so the $z_{0.95} = 1.96$. Substituting the values into the formula gives us

$$\begin{aligned}E &= z_c\sqrt{\frac{\hat{p}_1(1-\hat{p}_1)}{n_1} + \frac{\hat{p}_2(1-\hat{p}_2)}{n_2}} \\ &= 1.96\sqrt{\frac{0.900(1-0.900)}{50} + \frac{0.745098(1-0.745098)}{51}} \\ &\approx 0.146.\end{aligned}$$

Subtracting the margin of error from the point estimate gives a lower endpoint for the confidence interval of

$$0.155 - 0.146 = 0.009.$$

Adding the margin of error to the point estimate gives an upper endpoint of

$$0.155 + 0.146 = 0.301.$$

Thus the 95% confidence interval for the difference in the population proportions is

$$(0.009,\ 0.301), \text{ or}$$
$$0.009 < p_1 - p_2 < 0.301.$$

Example 11

School administrators want to know if there is a smaller percentage of students at a school in a lower economic district (School A) who carry cell phones than the percentage of students who carry a cell phone at a school in an upper economic district (School B). A survey of students is conducted at each school and the following results are tabulated:

Table 8 - Response to Survey		
	School A	**School B**
Carry a cell phone	45	53
Do not carry a cell phone	31	25

Construct a 90% confidence interval for the true difference in the proportion of students who carry a cell phone at the two schools.

Solution:

Here, we must calculate the sample sizes by adding the number of yes and no responses for each school. Thus the sample size for the first sample (School A) is

$$n_1 = 45 + 31 = 76$$

and the sample size for the second sample (School B) is

$$n_2 = 53 + 25 = 78.$$

Using these sample sizes to calculate the sample proportions gives us the following

$$\hat{p}_1 = \frac{45}{76} \approx 0.592105 \approx 0.592$$

and

$$\hat{p}_2 = \frac{53}{78} \approx 0.679487 \approx 0.679.$$

Subtracting these two sample proportions produces the point estimate.

$$\begin{aligned}\hat{p}_1 - \hat{p}_2 &= 0.592 - 0.679 \\ &= -0.087\end{aligned}$$

cont'd on the next page

Since we want to create a 90% confidence interval, we need to use $z_{0.90} = 1.645$ as the critical value. Thus the margin of error is

$$\begin{aligned} E &= z_c\sqrt{\frac{\hat{p}_1(1-\hat{p}_1)}{n_1}+\frac{\hat{p}_2(1-\hat{p}_2)}{n_2}} \\ &= 1.645\sqrt{\frac{0.592105(1-0.592105)}{76}+\frac{0.679487(1-0.679487)}{78}} \\ &\approx 0.127. \end{aligned}$$

Subtracting the margin of error from the point estimate gives a lower endpoint of

$$-0.087 - 0.127 = -0.214.$$

Adding the margin of error to the point estimate gives an upper endpoint for the confidence interval of

$$-0.087 + 0.127 = 0.040.$$

Therefore, the 90% confidence interval for the difference in the population proportions is

$$(-0.214,\ 0.040), \text{ or}$$
$$-0.214 < p_1 - p_2 < 0.040.$$

Because the confidence interval contains zero, there is no statistical evidence to suggest that there is a difference between the population proportions.

Exercises

9.4

Directions: Verify that the normal distribution can be used to compare the population proportions given the following information, or show how the conditions have not been met.

1. $n_1 = 130, n_2 = 200, \hat{p}_1 = 0.87, \hat{p}_2 = 0.72$

2. $n_1 = 1100, n_2 = 1200, \hat{p}_1 = 0.05, \hat{p}_2 = 0.01$

3. $n_1 = 23, n_2 = 14, \hat{p}_1 = 0.43, \hat{p}_2 = 0.72$

4. $n_1 = 19, n_2 = 12, \hat{p}_1 = 0.13, \hat{p}_2 = 0.26$

Directions: Calculate the following point estimates.

5. Sample 1 has 17 out of 97 'yes' responses, and Sample 2 has 46 out of 131 'yes' responses. Calculate the point estimate, $\hat{p}_1 - \hat{p}_2$, for the difference in the sample proportion of 'yes' responses given.

6. Records from public school A show that 61 out of 156 students start 1st grade having lost their first tooth, while at public school B, 46 out of 121 students have lost their first tooth when entering 1st grade. Find the point estimate, $\hat{p}_1 - \hat{p}_2$, for the difference in the sample proportion of students who have lost their first tooth before entering 1st grade at the two schools.

7. Given that the first sample of egg cartons had 15 broken eggs out of 360 eggs in it and that the second carton had 12 broken eggs out of 540 eggs, find the point estimate, $\hat{p}_1 - \hat{p}_2$, for the difference in the sample proportion of broken eggs.

8. Find the point estimate, $\hat{p}_1 - \hat{p}_2$, for the difference in the sample proportion of no votes given the following information. Sample A: 16 no, 32 yes. Sample B: 43 no, 55 yes.

Directions: Calculate the margin of error given the following information.

9. $n_1 = 130$, $n_2 = 200$, $\hat{p}_1 = 0.87$, $\hat{p}_2 = 0.72$, 95% level of confidence

10. $n_1 = 1100$, $n_2 = 1200$, $\hat{p}_1 = 0.05$, $\hat{p}_2 = 0.01$, 95% level of confidence

11. $n_1 = 88$, $n_2 = 74$, $\hat{p}_1 = 0.43$, $\hat{p}_2 = 0.61$, 99% level of confidence

12. $n_1 = 559$, $n_2 = 612$, $\hat{p}_1 = 0.13$, $\hat{p}_2 = 0.26$, 99% level of confidence

13. $n_1 = 108$, $n_2 = 104$, $\hat{p}_1 = 0.33$, $\hat{p}_2 = 0.65$, 90% level of confidence

14. $n_1 = 500$, $n_2 = 600$, $\hat{p}_1 = 0.73$, $\hat{p}_2 = 0.67$, 90% level of confidence

Directions: Answer each of the following questions.

15. Doctors at a fertility clinic wish to determine if taking fertility drugs increases the chances of a multiple birth (having 2 or more babies at once.) Doctors record the number of multiple births and the number of single births for a sample of patients taking fertility drugs and a sample of patients not taking fertility drugs. The data is shown below. Construct a 99% confidence interval for the true difference in the proportion of multiple births between women taking fertility drugs and those who are not.

Table for Exercise 15

	With Drugs	Without Drugs
Single Birth	32	43
Multiple Birth	12	13

16. Psychiatrists wish to determine if there is a higher incidence of divorce among couples where one of the spouses has suffered a serious head injury than among the general population. Of 51 couples who were married at the time one of the spouses had a serious head injury, 20 are still married. Of 50 randomly selected couples where no head injury occurred, 24 have remained married during this same time period. Construct a 95% confidence interval for the true difference in the divorce rate of couples where a head injury occurred and the general population.

17. Do a larger percentage of Southerners attend church on a weekly basis than do Northerners? A sample of Northerners and Southerners is interviewed about their church attendance and the results of the survey are shown below. Construct a 90% confidence interval for the true difference in the percentage of Northerners and Southerners who attend church on a weekly basis.

Table for Exercise 17

	Northern	Southern
Attend Church	12	35
Do Not Attend Church	34	16

18. Do more women than men exercise on a regular basis? To help answer this question, a survey was conducted asking men and women if they exercised on a regular basis, meaning 3 or more times a week. 49 out of the 100 women said yes. 52 out of 100 men said yes. Construct a 95% confidence interval for the true difference in the proportion of men and women who exercise on a regular basis.

19. For his senior psychology project, Robert wants to explore which gender is more honest, men or women. To do this, he leaves a wallet filled with money and identification cards in a local store and hides a video camera to record the results. He repeats the experiment 80 times. Of the 37 women who picked up the wallet, 30 returned it to Robert. Of the 43 men who found the wallet, 29 returned it to him. Construct a 90% confidence interval for the true difference between the proportion of men and women who will return a lost wallet.

20. Ann, a teacher, is concerned with the obsession many of her students have with video games. She is afraid that the video games have a negative impact on her students' performance in the classroom. To test her theory, Ann sends home a letter explaining the study, and 82 parents agree to let their children participate. Ann then randomly divides the students into two equal groups. Group A is required to play video games for two hours one evening, while Group B is not allowed any video game time. The following day, the students are given a review test over previously learned material. From Group A, 29 students pass the test. From Group B, 34 students pass the test. Construct a 95% confidence interval for the true difference between test results as affected by video games.

21. Frank owns a nursery, and he has had trouble in the past getting his gerbera daisies to bloom. This spring he has decided to experiment with a new, more expensive fertilizer called FertiGro. Of the 40 plants he treats with his regular fertilizer, 27 bloom. Of the 35 he treats with FertiGro, 29 bloom. Construct a 99% confidence interval for the true difference in the proportion of plants that bloom using FertiGro and those using regular fertilizer.

22. A state politician is interested in knowing how voters in rural areas and cities differ in their opinions about gun control. For his study, 75 rural voters were surveyed, and 41 were found to support gun control. Also included in the study were 75 voters from cities, and 53 of these voters were found to support gun control. Construct a 90% confidence interval for the true difference in the proportion of rural and city voters that favor gun control.

23. Aaliyah wants to know if there is a difference in the proportion of people who order water as opposed to soda at a restaurant. She went to 2 popular restaurants in town and found the following:

Table for Exercise 23

	Number of Customers	
	Soda	**Water**
Restaurant A	52	61
Restaurant B	72	68

Construct a 95% confidence interval for the true difference in the proportion of people who order water compared to soda.

24. A local city government is trying to promote road safety by encouraging drivers to buckle up. Their campaign director is trying to decide to which age group she should direct most of the promotions. She believes that fewer older adults buckle up, as they came from a generation without seatbelts. She surveyed 49 senior adults and found that 30 of them buckle up on a regular basis. She then surveyed 52 middle-aged adults and found that 42 of them buckle up on a regular basis. Construct an 90% confidence interval to estimate the true difference in proportion of senior adults who wear seatbelts compared to middle-aged adults.

25. Early childhood development studies indicate that the more a child is read to from birth, the earlier they begin to read. A local parent's group wants to test this theory and samples families with young children. They find the following results:

Table for Exercise 25		
	Read to more than 3 times per week	**Read to less than 3 times per week**
Started reading by age 4	46	31
Started reading after age 4	40	57

Construct a 98% confidence interval to estimate the true difference in the proportion of children who read at an early age when they are read to from birth compared to those who were not read to.

Comparing Two Population Variances

9.5

So far we have discussed comparing two population means and two population proportions. Now let's turn our attention to the problem of comparing two population variances.

In order to compare two population variances, we must ensure that the following two conditions are met:

- The two populations must be independent, not matched or paired in any way, and
- The two populations must be normally distributed.

These conditions are very strict; much more so than the conditions for other confidence intervals, as the methods used to compare population variances produce very poor results when these conditions have been violated, even to a small degree. You can safely assume that these conditions have been met for all exercises in this section.

To compare population means and proportions, we used the differences in the sample statistics as our point estimate. To compare population variances, we will use the ratio of the sample statistics as our point estimate. Thus the best point estimate for comparing population variances is $\frac{s_1^2}{s_2^2}$.

To simplify the remainder of the calculations, let us impose the rule that the larger of the two sample variances is s_1^2 and the smaller variance is s_2^2. This rule is not mathematically necessary; however, the calculations are more complicated if the ratio is less than 1, so we will consistently put the larger variance in the numerator.

Let's specify a few other notations we will use in this section. First, remember that the level of confidence is c and $\alpha = 1 - c$. Second, if s_1^2 is the larger sample variance, then n_1 is the size of the sample with the larger variance. The value s_2^2 is then the smaller sample variance, so n_2 is the size of the sample with the smaller variance.

Another major difference we encounter when comparing two population variances as opposed to population means is the distribution used for the critical value. When independent samples are drawn from normal distributions, the sampling distribution for the ratio of the sample variances is an F-distribution. For now, a few comments on the properties of the F-distribution and a brief discussion on how to read the F-distribution table should suffice.

- The F-distribution is skewed to the right,
- The values of F are always greater than 0, and
- The shape of the F-distribution is completely determined by its two parameters, the degrees of freedom of the numerator and the degrees of freedom of the denominator of the ratio.

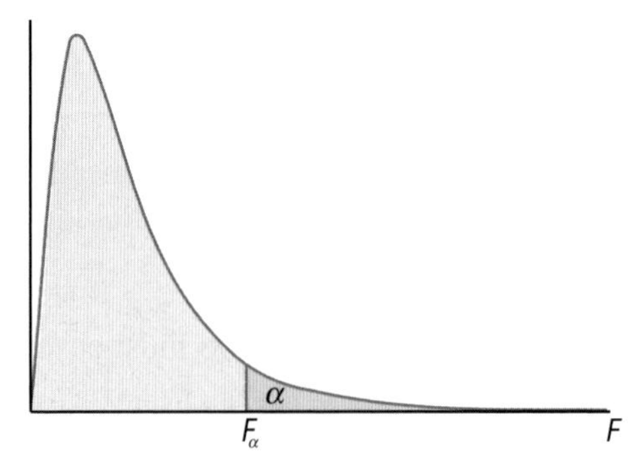

Figure 1 – F-Distribution

Rounding Rule

In order to be consistent with the F-distribution tables, we will round calculations to four decimal places.

Table H lists the critical values of the F-distribution for various degrees of freedom and levels of confidence. The number of degrees of freedom for the F-distribution is one less than the sample size, so the number of degrees of freedom for the numerator is $n_1 - 1$ and the number of degrees of freedom for the denominator is $n_2 - 1$. To find the critical values using the F-distribution table, first find the section of the table for the appropriate level of significance, α. Next, find the value in the table where the degrees of freedom for the numerator (across the top) and the degrees of freedom for the denominator (down the side) intersect.

Example 12

Consider the following information from two independent samples:

$$n_1 = 18 \qquad n_2 = 25$$
$$s_1^2 = 1.23 \qquad s_2^2 = 1.19$$

Using a 95% level of confidence, calculate the point estimate for comparing the population variances, $\frac{s_1^2}{s_2^2}$. Also, calculate the critical values $F_{\frac{\alpha}{2}}$ and $F_{1-\frac{\alpha}{2}}$.

Solution:

Let's begin by calculating the point estimate.

$$\frac{s_1^2}{s_2^2} = \frac{1.23}{1.19}$$
$$\approx 1.0336$$

Next, remember that the level of significance, α, equals $1 - c$. So $\alpha = 0.05$. To find the F-critical values, we also need the degrees of freedom:

$$\begin{aligned} d.f._1 &= n_1 - 1 \\ &= 18 - 1 \\ &= 17 \end{aligned}$$

and

$$\begin{aligned} d.f._2 &= n_2 - 1 \\ &= 25 - 1 \\ &= 24. \end{aligned}$$

Using the two degrees of freedom and $\alpha = 0.05$, find $F_{\frac{\alpha}{2}}$. Use the section of the table for 0.025 and find the value where the row for *d.f.* = 24 intersects with the column for *d.f.* = 17. Thus

$$F_{\frac{0.05}{2}} = 2.3865.$$

To find the critical value $F_{1-\frac{\alpha}{2}}$ we use similar methods. Thus

$$\begin{aligned} F_{1-\frac{0.05}{2}} &= F_{0.975} \\ &= 0.3907. \end{aligned}$$

Now that we can calculate the point estimate and the critical values, let's construct the confidence interval. Just as when constructing confidence intervals for the population variance with one sample, we cannot calculate a margin of error for the confidence interval for comparing population variances. We must calculate the upper and lower endpoints for the confidence interval separately. Thus the confidence interval is given in the box below.

Confidence Interval for Comparing Two Population Variances

$$\left(\frac{s_1^2}{s_2^2} \cdot \frac{1}{F_{\frac{\alpha}{2}}}\right) < \frac{\sigma_1^2}{\sigma_2^2} < \left(\frac{s_1^2}{s_2^2} \cdot \frac{1}{F_{1-\frac{\alpha}{2}}}\right)$$

Example 13

Suppose that a quality control inspector wants to make sure that two different machines are filling soda cans equally. Specifically, she wants to ensure that the population variance in the amount of soda per can is the same between the two machines. A sample of 16 cans from machine A shows a sample variance of 0.78, while a sample of 18 cans from machine B shows a sample variance of 0.65. Assuming that both populations are normally distributed, construct a 90% confidence interval for the ratio of the population variances.

Solution:

To begin, list the information that was given in the problem. We will call the sample from machine A sample 1 since its variance is larger. We can also include the degrees of freedom for each sample by subtracting 1 from the sample size.

$$\begin{aligned} n_1 &= 16 & n_2 &= 18 \\ s_1^2 &= 0.78 & s_2^2 &= 0.65 \\ d.f._1 &= 15 & d.f._2 &= 17 \end{aligned}$$

We know that $c = 0.90$, therefore $\alpha = 0.10$. The point estimate is the ratio of the sample variances

$$\begin{aligned} \frac{s_1^2}{s_2^2} &= \frac{0.78}{0.65} \\ &= 1.2. \end{aligned}$$

To calculate the left endpoint of the confidence interval, we need to find the critical value $F_{\frac{\alpha}{2}}$. Using the row for *d.f.* = 17 and the column for *d.f.* = 15 in the table for $\frac{\alpha}{2} = 0.05$ gives us

$$F_{0.05} = 2.3077.$$

Substituting these values into the formula for the left endpoint gives us

$$\begin{aligned} \frac{s_1^2}{s_2^2} \cdot \frac{1}{F_{0.05}} &= (1.2)\frac{1}{2.3077} \\ &\approx 0.5200. \end{aligned}$$

To calculate the right endpoint of the confidence interval, we need to find the critical value $F_{1-\frac{\alpha}{2}}$. Using the row for *d.f.* = 17 and the column for *d.f.* = 15 in the table for $1-\frac{\alpha}{2} = 0.95$ gives us

$$F_{0.95} = 0.4223.$$

Substituting these values into the formula for the right endpoint gives us

$$\frac{s_1^2}{s_2^2} \cdot \frac{1}{F_{0.05}} = (1.2)\frac{1}{0.4223}$$
$$\approx 2.8416.$$

Therefore, the 90% confidence interval for the ratio of the population variances is

$$0.5200 < \frac{\sigma_1^2}{\sigma_2^2} < 2.8416, \text{ or}$$
$$(0.5200, 2.8416).$$

The methods described in this section can be applied to population standard deviations in the same way they are applied to population variances. Use n_1 to represent the size of the sample with the larger standard deviation where s_1 is the standard deviation of this sample. Thus n_2 represents the size of the sample with the smaller standard deviation and s_2 is the standard deviation of this sample. The number of degrees of freedom is found in the same way, by subtracting one from the sample size. The F critical values are also found the same way. The formula for the confidence interval for the ratio of population standard deviations is given in the box below.

Confidence Interval for Comparing Two Population Standard Deviations

$$\left(\frac{s_1}{s_2} \cdot \frac{1}{\sqrt{F_{\frac{\alpha}{2}}}}\right) < \frac{\sigma_1}{\sigma_2} < \left(\frac{s_1}{s_2} \cdot \frac{1}{\sqrt{F_{1-\frac{\alpha}{2}}}}\right)$$

Exercises

9.5

Note: You may assume that all populations are normally distributed and that the samples chosen from each population are independent of each other.

Directions: For the following information, calculate the point estimate for the ratio of the population variances.

1. $n_1 = 13, n_2 = 17, s_1^2 = 3.467, s_2^2 = 2.903$

2. $n_1 = 23, n_2 = 20, s_1^2 = 0.689, s_2^2 = 0.542$

3. $n_1 = 8, n_2 = 10, s_1^2 = 12.103, s_2^2 = 10.874$

4. $n_1 = 26, n_2 = 28, s_1^2 = 1.472, s_2^2 = 1.327$

Directions: For the following information, find the left and right F critical values, $F_{\frac{\alpha}{2}}$ and $F_{1-\frac{\alpha}{2}}$, which would be needed to construct a confidence interval for the ratio of the population variances.

5. $n_1 = 18, n_2 = 23, s_1^2 = 12.470, s_2^2 = 12.205$, 95% level of confidence

6. $n_1 = 9, n_2 = 9, s_1^2 = 0.974, s_2^2 = 0.903$, 95% level of confidence

7. $n_1 = 21, n_2 = 20, s_1^2 = 134.943, s_2^2 = 125.908$, 90% level of confidence

8. $n_1 = 18, n_2 = 19, s_1^2 = 20.461, s_2^2 = 18.321$, 90% level of confidence

9. $n_1 = 12, n_2 = 13, s_1^2 = 5.100, s_2^2 = 5.098$, 99% level of confidence

10. $n_1 = 7, n_2 = 6, s_1^2 = 4.110, s_2^2 = 3.873$, 99% level of confidence

Directions: For the following information, construct the indicated confidence interval for the ratio of the population variances.

11. $\frac{s_1^2}{s_2^2} = 1.23,\ F_{\frac{\alpha}{2}} = 2.4499,\ F_{1-\frac{\alpha}{2}} = 0.3556$

12. $\frac{s_1^2}{s_2^2} = 1.67,\ F_{\frac{\alpha}{2}} = 5.4160,\ F_{1-\frac{\alpha}{2}} = 0.0688$

13. $n_1 = 13, n_2 = 12, \frac{s_1^2}{s_2^2} = 1.14, \alpha = 0.05$

14. $n_1 = 14, n_2 = 18, \frac{s_1^2}{s_2^2} = 2.01, \alpha = 0.05$

15. $n_1 = 9, n_2 = 12, \frac{s_1^2}{s_2^2} = 1.23, \alpha = 0.01$

16. $n_1 = 20, n_2 = 20, \frac{s_1^2}{s_2^2} = 1.87, \alpha = 0.10$

17. $n_1 = 15, n_2 = 16, s_1^2 = 3.007, s_2^2 = 2.897$, 95% level of confidence

18. $n_1 = 18, n_2 = 17, s_1^2 = 4.067, s_2^2 = 3.903$, 95% level of confidence

19. $n_1 = 20, n_2 = 23, s_1^2 = 124.940, s_2^2 = 123.980$, 90% level of confidence

20. $n_1 = 16, n_2 = 15, s_1^2 = 2.560, s_2^2 = 2.519$, 90% level of confidence

21. $n_1 = 11, n_2 = 10, s_1^2 = 0.584, s_2^2 = 0.499$, 99% level of confidence

22. $n_1 = 6, n_2 = 7, s_1^2 = 26.041, s_2^2 = 25.084$, 99% level of confidence

Directions: Answer the following questions.

23. A new brand of golf balls claims that using these golf balls can increase your precision. A golf pro tests this claim by hitting 20 golf balls with the old brand and 20 golf balls with the new brand. The sample variance for the old brand was 16.790 and the sample variance for the new brand was 15.910. Construct a 90% confidence interval for the ratio of the population variances.

24. A professor wants to make sure that two different versions of a test are equivalent. He decides to compare the variances of the test scores from each version. A sample of 19 scores on Version A has a sample variance of 2.450, while a sample of 20 scores from Version B has a sample variance of 2.391. Construct a 95% confidence interval for the ratio of the population variances of the two versions.

25. A graduate student walking around a large college campus notices that students, not faculty, often own the more expensive cars. He collects data from 15 professors and 14 students on the price of their cars and calculates the variance from each sample. The sample variance for the sample of professors' cars is 2456.45 and the sample variance for the sample of students' cars is 3567.90. Construct a 90% confidence interval for the ratio of the population variances of professors' and students' cars.

26. A researcher wants to compare the consistency with which two marksmen hit the bull's eye of a target. The first marksman misses an average of 4.5 bull's eyes per session with a variance of 2.78. The second marksman misses an average of 5.3 bull's eyes per session with a variance of 2.34. If 20 sessions were recorded for each marksman, construct a 95% confidence interval for the ratio of the population variances between these two marksmen.

27. A paint technician for an oil company has to make sure that the painted coating on the inside of the oil tankers has a consistent depth throughout. He measures the depth of the coating at 15 spots inside the first tanker and calculates a sample variance of 0.3918. He then measures the depth of the coating at 15 spots inside a second tanker sprayed using the same equipment and calculates a sample variance of 0.4231. Construct a 99% confidence interval for the ratio of the population variances to help determine if there was a difference in how consistently the equipment applied the coating in one tanker versus the other.

28. When shopping for a new car, Emily became interested in how consistently cars are priced in her area and was looking at buying either a Honda Accord or a Toyota Camry. She priced 10 different Accords with similar features at several dealerships in her area. The sample variance for the Accords was 273,529. Emily then priced 9 different Camrys with similar features at several different dealerships in her area. The sample variance for the Camrys was 231,361. Construct a 99% confidence interval for the ratio of the population variances for the two different brands.

Chapter Review

Section 9.1 - Comparing Two Means (Large, Independent Samples)	Pages
• **Definitions** **Independent** – samples are independent if the data from the first sample is not connected to the data from the second sample **Dependent** – samples are dependent if two data sets are connected; data are referred to as *paired data* when drawn from dependent sets	p. 398
• **Criteria for Comparing Two Population Means (Large, Independent Samples)** • Each sample must contain at least 30 values ($n \geq 30$). • The data sets must be independent of each other. • Each population from which the samples have been chosen is normally distributed. • σ is unknown for both populations.	p. 399
• **Margin of Error for Comparing Two Population Means (Large, Independent Samples)** $E = z_c\sqrt{\frac{s_1^2}{n_1}+\frac{s_2^2}{n_2}}$	p. 399
• **Point Estimate for Comparing Two Population Means** $\bar{x}_1 - \bar{x}_2$	p. 399 – 400
• **Confidence Interval for Comparing Two Population Means** $(\bar{x}_1-\bar{x}_2)-E<\mu_1-\mu_2<(\bar{x}_1-\bar{x}_2)+E$, or $((\bar{x}_1-\bar{x}_2)-E,\ (\bar{x}_1-\bar{x}_2)+E)$.	p. 400

Section 9.2 - Comparing Two Means (Small, Independent Samples)	Pages
• **Definitions** **Pooled** – combined; the data can be pooled if the population variances are assumed to be the same.	p. 405
• **Criteria for Comparing Two Population Means (Small, Independent Samples)** • The samples are independent. • Each population from which the samples were drawn has a normal distribution. • $n < 30$ for one or both samples. • σ is unknown for both populations.	p. 405
• **Point Estimate for Comparing Two Population Means** $\bar{x}_1 - \bar{x}_2$	p. 405

Section 9.2 - Comparing Two Means (Small, Independent Samples) cont.	Pages
• **Margin of Error for Comparing Two Population Means (Small, Independent Samples)**	
If the Population Variances are Assumed to be Equal $$E = t_{\alpha/2}\sqrt{\frac{(n_1 - 1)s_1^2 + (n_2 - 1)s_2^2}{n_1 + n_2 - 2}}\sqrt{\frac{1}{n_1} + \frac{1}{n_2}}$$ Degrees of Freedom: $n_1 + n_2 - 2$	p. 405
If the Population Variances are Assumed to be Not Equal $$E = t_{\alpha/2}\sqrt{\frac{s_1^2}{n_1} + \frac{s_2^2}{n_2}}$$ Degrees of Freedom: smaller of $n_1 - 1$ and $n_2 - 1$	p. 408
• **Confidence Interval for Two Poulation Means** $$(\bar{x}_1 - \bar{x}_2) - E < \mu_1 - \mu_2 < (\bar{x}_1 - \bar{x}_2) + E,$$ or $\left((\bar{x}_1 - \bar{x}_2) - E,\ (\bar{x}_1 - \bar{x}_2) + E\right)$.	p. 406

Section 9.3 - Comparing Two Means (Dependent Samples)	Pages
• **Definitions** **Paired Difference** – the difference between each pair of values in the data set, $d = x_2 - x_1$	p. 415
• **Point Estimate for Comparing the Difference in Two Population Means (Dependent Samples)** $$\bar{d}$$	p. 415
• **Mean of Paired Differences** $\bar{d} = \frac{\sum d}{n}$	p. 416
• **Margin of Error for Comparing Two Population Means (Dependent Samples)** $$E = t_{\alpha/2}\frac{s_d}{\sqrt{n}}$$	p. 416
• **Confidence Interval for Two Dependent Means** $$\bar{d} - E < \mu_d < \bar{d} + E,$$ or $(\bar{d} - E,\ \bar{d} + E)$	p. 417

Section 9.4 - Comparing Two Proportions (Large, Independent Samples)	Pages
• **Point Estimate for Comparing Two Population Proportions** $$\hat{p}_1 - \hat{p}_2$$	p. 425
• **Margin of Error for Comparing Two Population Proportions (Large, Independent Samples)** $$E = z_c\sqrt{\frac{\hat{p}_1(1-\hat{p}_1)}{n_1} + \frac{\hat{p}_2(1-\hat{p}_2)}{n_2}}$$	p. 427
• **Confidence Interval for Comparing Two Population Proportions** $$(\hat{p}_1 - \hat{p}_2) - E < p_1 - p_2 < (\hat{p}_1 - \hat{p}_2) + E,$$ or $$\left((\hat{p}_1 - \hat{p}_2) - E,\ (\hat{p}_1 - \hat{p}_2) + E\right)$$	p. 427

Section 9.5 - Comparing Two Population Variances	Pages
• **Point Estimate for Comparing Two Population Variances** $$\frac{s_1^2}{s_2^2}$$	p. 435
• **Critical Values for Comparing Two Population Variances** $F_{\frac{\alpha}{2}}$ and $F_{1-\frac{\alpha}{2}}$ with the number of degrees of freedom for the numerator is $n_1 - 1$ and the number of degrees of freedom for the denominator is $n_2 - 1$.	p. 436
• **Confidence Interval for the Ratio of Population Variances** $$\left(\frac{s_1^2}{s_2^2} \cdot \frac{1}{F_{\frac{\alpha}{2}}}\right) < \frac{\sigma_1^2}{\sigma_2^2} < \left(\frac{s_1^2}{s_2^2} \cdot \frac{1}{F_{1-\frac{\alpha}{2}}}\right)$$	p. 437
• **Confidence Interval for Comparing Two Population Standard Deviations** $$\left(\frac{s_1}{s_2} \cdot \frac{1}{\sqrt{F_{\frac{\alpha}{2}}}}\right) < \frac{\sigma_1}{\sigma_2} < \left(\frac{s_1}{s_2} \cdot \frac{1}{\sqrt{F_{1-\frac{\alpha}{2}}}}\right)$$	p. 439

Chapter Exercises

Directions: Answer each of the following questions.

1. Two math teachers each believe that their very large class is doing better than the other class. The teachers decided to determine whose class is doing better by looking at scores from a recent exam. A random sample of 40 scores from class A yields an average exam score of 88 with a standard deviation of 5.75. A random sample of 37 scores from class B yields an average exam score of 85 with a standard deviation of 6.23. Construct a 95% confidence interval for the true difference in the average exam score between the two classes.

2. Insurance companies claim that more accidents are caused by people who talk on the phone while driving than by people who do not. Cellular phone companies dispute this claim. Insurance companies from 2 areas compare the proportion of claims involving drivers who were talking on the phone at the time of the accident with those who were not. Here are the results.

Table for Exercise 2 - Number of Accidents		
	While on Phone	**While Not on Phone**
Area A	44	50
Area B	71	63

Construct a 95% confidence interval for the true difference in the proportion of accidents occuring while the driver was talking on the phone and those that did not involve a driver on a cell phone.

3. To determine how well a new drug reduces a fever, 10 adults running a fever in a hospital agree to take the new drug instead of the traditional treatment. Their temperature is taken before the drug is administered and 30 minutes after taking the drug. The results are shown below. Construct a 99% confidence interval for the decrease in the temperature of patients taking this new drug.

Table for Exercise 3 - Patient Temperatures										
Patient	1	2	3	4	5	6	7	8	9	10
First Temperature	100.1	101.3	102.1	102.7	101.9	100.8	103.1	102.5	103.5	101.7
Second Temperature	98.9	99.1	99.2	99.0	98.7	98.6	99.4	99.2	100.1	99.2

4. Voters in a dry county (meaning no alcohol is sold in the county) will soon decide whether the county should remain dry. Local business people fear that more people in the food industry favor the sale of alcohol than do people in the general population. A survey of local restaurant and convenience store owners is conducted as well as a survey of the general population. The results of the surveys are shown below. Construct a 90% confidence interval for the true difference in the population proportions of those favoring the sale of alcohol between the food industry owners and the general population.

Table for Exercise 4 - Results of Survey		
	Food Industry	**General Population**
Favor Alcohol Sale	54	32
Oppose Alcohol Sale	19	20

5. An apple a day keeps the doctor away? To test this theory, a pediatrician asks 12 children to be a part of a study. With parental consent, the number of doctor's visits from the previous 6 months is recorded for each child. The children are then asked to eat an apple each day for the next 6 months, after which the number of doctor's visits for that time period is again recorded for each child. Use the table below to construct a 95% confidence interval for the true difference in the number of doctor's visits that an apple a day makes.

Table for Exercise 5 - Number of Doctor's Visits												
Child	A	B	C	D	E	F	G	H	I	J	K	L
Without Apples	2	1	2	2	3	0	4	1	2	1	2	0
With Apples	1	1	0	2	2	1	1	2	0	0	2	1

6. Who makes better grades, men or women? To answer this question, you survey male and female students on your campus, asking each to confidentially give you their GPA. For the 28 men surveyed, you calculate their mean GPA to be 3.1 with a standard deviation of 0.7 points. For the 26 women surveyed, you calculate their mean GPA to be 3.3 with a standard deviation of 0.6 points. Assuming that the population variances are the same, construct a 90% confidence interval for the true difference between the mean GPA of men and women.

7. A popular magazine is conducting a study in order to rank each state according to the overall health of its citizens. One measure that is used is the average number of times each citizen exercises per week. Out of the 85 people surveyed in State A, they exercise on average 1.9 times a week with a standard deviation of 0.5 times. Out of the 91 people surveyed in State B, they exercise on average 2.3 times a week with a standard deviation of 1.1 times. Construct a 95% confidence interval for the true difference in exercises rates between the two states.

8. A research firm is hired by a snack food company to determine where they should focus their resources in their new advertising campaign. Specifically, they are interested in comparing the snack food preferences of adults and children. To do this, 100 adults and 100 children are surveyed. Among the adults, 62% say that they prefer salty over sweet snack food. Among the children, 54% say that they prefer salty over sweet snack food. Construct a 95% confidence interval for the true difference between the proportion of adults and children that prefer salty over sweet snack food.

9. Comparing results on standardized tests is a very common way to evaluate programs within certain university settings. The Institute for Higher Learning Board (IHL) often evaluates programs based on how students perform on these tests. Suppose that University A reported that 101 students who took the GED had an average percentile ranking of 81 with a standard deviation of 5. University B reported that 216 students took the GED and had an average percentile ranking of 86 with a standard deviation of 7. Construct a 90% confidence interval for the true difference in the population means on the GED between University A and University B.

10. A television station is interested in the number of televisions that the average family has in their home. They performed a survey in two different areas of town to see if there was a difference by area. After surveying 20 homes on one side of town, they found that the average number of televisions found in the home was 2.4 with a standard deviation of 0.5. On the other side of town, they found that after sampling 15 homes, the average number of televisions was 1.9 with a standard deviation of 0.7. Assume the population variances are not the same. Calculate the 90% confidence interval to estimate the true difference in the number of T.V.'s found in homes in the different areas.

11. One petroleum company claims that using their engine-cleaning product will improve a car's gas mileage. It was added to 6 cars, and here are the before and after gas mileages. Construct a 95% confidence interval for the true improvement in gas mileage for the cars.

Table for Exercise 11

Car	1	2	3	4	5	6
Without Engine Cleaner	29	48	61	35	21	19
With Engine Cleaner	31	50	66	43	29	20

12. Have you ever noticed that the Gulf Stream air current can speed up air travel in certain directions? With this in mind, a business-person wants to determine how much faster the return trip from Dallas to Memphis is than the initial trip from Memphis to Dallas. He records the time to and from Dallas on his next 7 flights and obtains the data below. Assuming the variance of the two populations is not the same, construct a 90% confidence interval for the true difference in the mean time to fly between Dallas and Memphis. (Hint: convert all times to minutes)

Table for Exercise 12

Memphis to Dallas	Dallas to Memphis
1 hr 49 min	1 hr 32 min
1 hr 51 min	1 hr 23 min
2 hr 03 min	1 hr 38 min
1 hr 47 min	1 hr 26 min
1 hr 53 min	1 hr 30 min
2 hr 01 min	1 hr 19 min
1 hr 45 min	1 hr 28 min

13. A coach wants to compare the consistency of the two best batters on his baseball team. The coach considers the number of times per game that each player gets on base. Out of a sample of 15 games, John's sample variance is 4.389 and Mark's sample variance is 3.981. Construct a 95% confidence interval for the ratio of the population variances for the two different batters.

14. To ensure that the braking time of a client's car meets manufacturer's specifications, a mechanic decides to compare the consistency of its braking distance to a car known to meet manufacturer's specifications. The mechanic measures the braking distance of the client's car 12 times and calculates a sample variance of 3136. He then measures the braking distance of the control car 12 times and calculates a sample variance of 3048. Construct a 99% confidence interval for the ratio of the population variances for the two different cars.

15. A professor wishes to verify that professors of various ranks are paid similarly throughout his region. He decides to compare the variance in the salaries of associate professors with the variance in the salaries of full professors. A sample of 15 associate professors reveals a sample variance of 609,961. A sample of 14 full professors reveals a sample variance of 674,041. Construct a 90% confidence interval for the ratio of the population variances for the different ranks of professors.

Chapter Project A: Battle of the Sexes

The age old Battle of the Sexes... who is smarter, better, etc. Well, let's see if we can answer that question once and for all!

Step 1 – The next page contains a short 10 question test. Feel free to photocopy that page for the purposes of this experiment. You will need to find 30 willing volunteers, approximately 15 male and 15 female (though it does not have to be evenly split) to take this short test. The answers to the test are provided for your benefit. Grade each person's test and record the number of answers correct.

Answers to Intelligence Test:

1. Eight
2. Benjamin Franklin
3. Red
4. 1440
5. A musical instrument
6. 6 cylinder engine in the shape of a 'V'
7. Aurora
8. British Petroleum
9. 206
10. Ruby

Step 2 – Divide the results into two groups, male and female, then calculate the mean number correct and the sample standard deviation for each group. Record your statistics below.

n_{male} = ____________________ n_{female} = ____________________

$\bar{x}_{\text{male}}$ = ____________________ $\bar{x}_{\text{female}}$ = ____________________

s_{male} = ____________________ s_{female} = ____________________

Step 3 – Construct a 95% confidence interval for the true difference in the number of correct items for males and females. Assume that the variances are not the same.

Step 4 – Write up your conclusion. Is one gender "smarter" than the other? Consider how you choose your sample, the validity of the intelligence test, and the limits in interpreting the confidence interval when drawing your final conclusion.

Intelligence Test for the Battle of the Sexes

Directions: Thank you for agreeing to help with this educational project. Please read and answer each of the 10 questions below. Do your best! Remember, you are representing your entire gender in this heated battle.

1. How many sides does a stop sign have?

2. Whose face is on the $100 bill?

3. Which color is on top on a stop light?

4. How many minutes are in a day?

5. What is a viola?

6. What does V6 stand for when referring to a car?

7. What is the name of the princess in the fairy tale Sleeping Beauty?

8. What does the BP stand for in the name of the national gas station chain?

9. How many bones are in the human body?

10. What is the birthstone for July?

Chapter Project B: Knowledge of Current Events

Many adults assume that college students do not pay attention to current events. Let's do an experiment to see if a simple crash course can improve college students' knowledge of current events. This experiment will involve some research and preparation on your part.

Step 1 – Begin by preparing two posters with pictures of the following people. Only label the posters with a letter next to each person's picture. Make sure that the pictures are not in the same order on the two posters and that the same letter is not used for the same person on the two posters. Using different pictures on the two posters would be a good idea.

- **A** The President of the United States
- **B** The Vice-President of the United States
- **C** The Secretary of State
- **D** The Chief Justice of the Supreme Court
- **E** The Prime Minister of Great Britian
- **F** The Pope

Step 2 – Next, you will need to find 10 willing college student volunteers. You will give them a pre-test and a post-test using the posters. For each volunteer, show them the first poster and ask them to name each person and state their title. Record how many answers they get correct out of 12 (6 names and 6 titles). After the pre-test, give them the answers. Coach them for 5 minutes on who each person is and their titles. Next, visit with them for a few minutes about the class, what you are doing in this project, the weather, your favorite sports team, etc. Then, show them the other poster and ask them to name each person and their title. Record how many answers they get correct out of 12. Thank them for their help with the project.

Step 3 – Once data has been collected from 10 college students, calculate the paired difference for each of the students by subtracting the pre-test score from the post-test score.

Step 4 – Calculate each of the following sample statistics:

$\bar{d} =$ ________________ $s_d =$ ________________

Step 5 – Use the t-distribution table to find $t_{\alpha/2}$ for a 95% confidence interval. $t_{\alpha/2} =$ ________________

Step 6 – Construct a 95% confidence interval for the improvement in college students' knowledge of people in current events.

Step 7 – Did your "crash course" on people in the news improve the students' knowledge? Consider how you chose your sample and if their initial knowledge of the people on your poster impacted your results.

Student	1	2	3	4	5	6	7	8	9	10
Pre-Test Score										
Post-Test Score										
d										

Chapter Technology – Confidence Intervals

TI-84 PLUS

Confidence Intervals for Mean (Large, Independent Samples)

Using a TI-84 Plus to calculate the confidence interval can be done from data or from statistics that have already been calculated. The process is similar to the way we calculated confidence intervals for one sample. For our examples, we will be calculating confidence intervals from given statistics.

Example T.1

A nursing instructor has created a prep course for the National Nursing Board Examination that she believes will result in nursing students performing better on the exam than if they had not taken the course. She collects data from 40 students who took her course and calculates that they had an average percentile score of 88 with a standard deviation of 4. She also collects data from 35 students who did not take her course and calculates that they had an average percentile score of 80 with a standard deviation of 6. Construct a 90% confidence interval to estimate the true difference in the average percentile score of students who took her course and those who did not.

Solution:

Since this confidence interval involves large sample sizes, we will use z to calculate the interval. Press STAT, then scroll to TESTS and choose option 9: 2-SampZInt, since we have two samples and are using z for large sample sizes. Next enter the statistics. Using the prep course results as group 1 and the non-prep course results as group 2, the statistics would be entered in as shown in the screen shots below.

Once you have entered the statistics, choose Calculate. The confidence interval is displayed, along with several of the entered statistics, as shown in the screen shot below on the right.

Thus the true difference in group 1 and group 2's mean score is between 6.034 and 9.966.

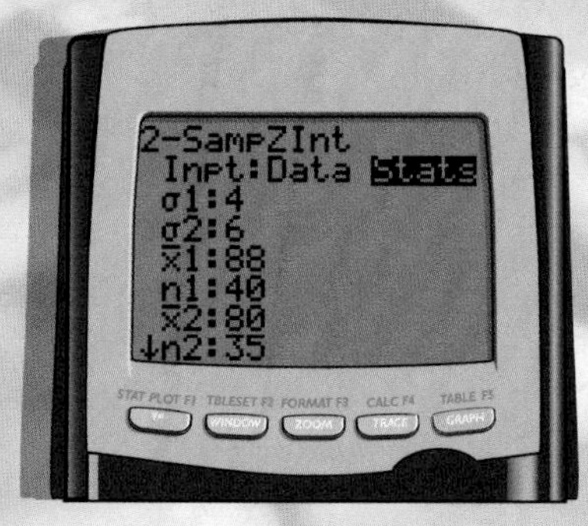

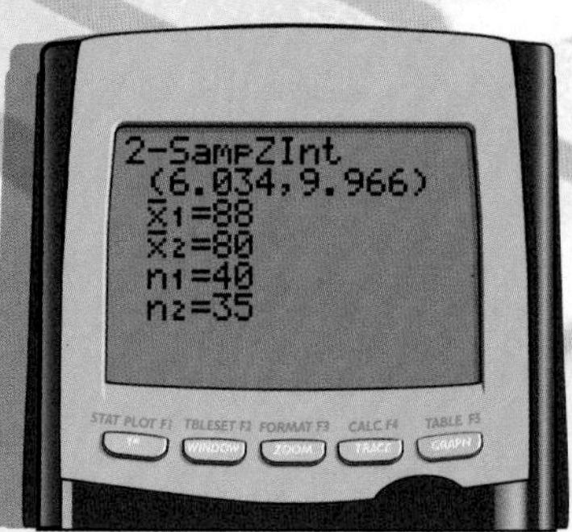

Confidence Intervals for Mean (Small, Independent Samples)

For confidence intervals using small sample sizes, we need to know if the variances are assumed to be the same. The TI-84 Plus uses the same test for both by allowing you to set whether the data is pooled (the variances are the same) or not pooled (the variances are different).

Example T.2

Obstetricians are concerned that a certain pain reliever may be causing lower birth weights. A medical researcher collects data from 12 mothers who took the pain reliever while pregnant and calculates that the average birth weight of their babies was 5.6 pounds with a standard deviation of 1.8. She also collects data from 20 mothers who did not take the pain reliever and calculates that the average birth weight of their babies was 6.3 pounds with a standard deviation of 2.1. Assuming that the population variances are the same, construct a 99% confidence interval for the true difference in the mean birth weights of the two groups

Solution:

Since this confidence interval involves small sample sizes, we will use t to calculate the interval. So press STAT, then scroll to TESTS and choose option 0: 2-SampTInt, since we have two samples and are using t for small sample sizes. Next enter the statistics. Using the results from the mothers taking the pain reliever as group 1 and the results from the mothers not using the pain reliever as group 2, the statistics would be entered in as shown in the screen shots below.

The confidence level is $c = 0.99$. Since the population variances are assumed to be the same, we want to pool the data, so we choose YES. Then press calculate.

The results shown in the screen on the bottom right indicate that the degrees of freedom for this example is 30. Also, the true difference in the means of the two groups is between –2.704 and 1.3035. Notice that these results are slightly different from the results we got earlier in the chapter, because these results are more accurate. The results in the chapter have errors due to rounding at several different stages of the calculation.

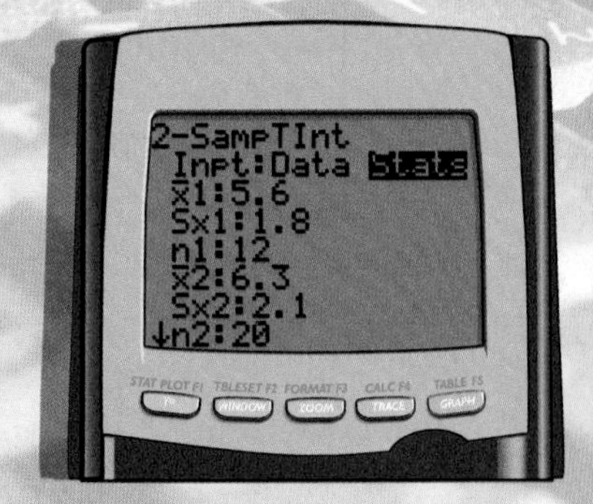

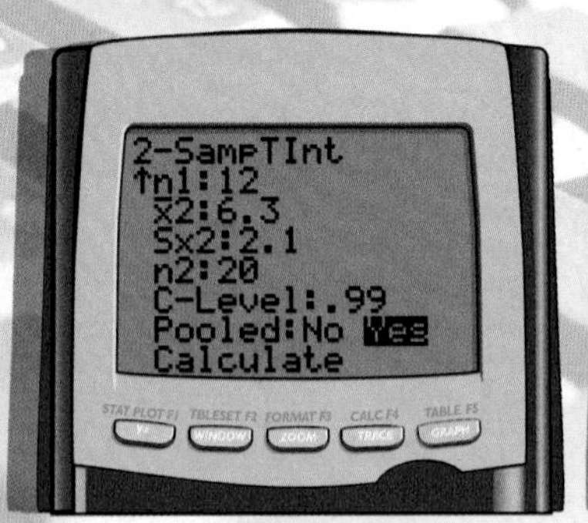

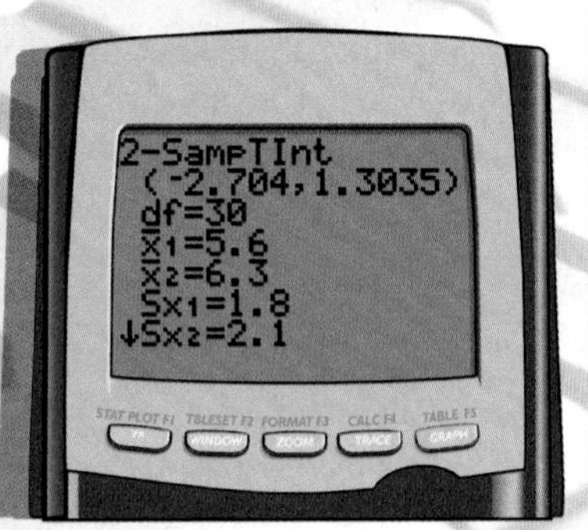

Confidence Intervals for Mean (Dependent Samples)

A TI-84 Plus does not directly do a calculation for the confidence interval for dependent samples, but we can do a confidence interval for a one sample *t*-test on the paired difference.

Example T.3

In a CPR class, students are given a pre-test at the beginning of the class to determine their initial knowledge of CPR and then a post-test after the class to determine their new knowledge of CPR. Educational researchers are interested if the class increases a student's knowledge of CPR. Create a 95% confidence interval for the mean increase in the students' knowledge from taking the CPR class. Data from 10 students who took the CPR class are listed in the table below.

Student	1	2	3	4	5	6	7	8	9	10
Pre-Test Score	60	63	68	70	71	68	72	80	83	79
Post-Test Score	80	79	83	90	93	89	94	95	96	93

Solution:

First, let's enter our pre-test scores in List 1 and and Post-Test scores in List 2. Next, we want to calculate the paired difference in List 3. To do so, Highlight L3 and enter the formula to subtract the pre-test scores from the post-test scores by pressing 2ND 2 – 2ND 1 then ENTER. This will produce the formula L2 – L1 and calculate the paired difference for each pair of data.

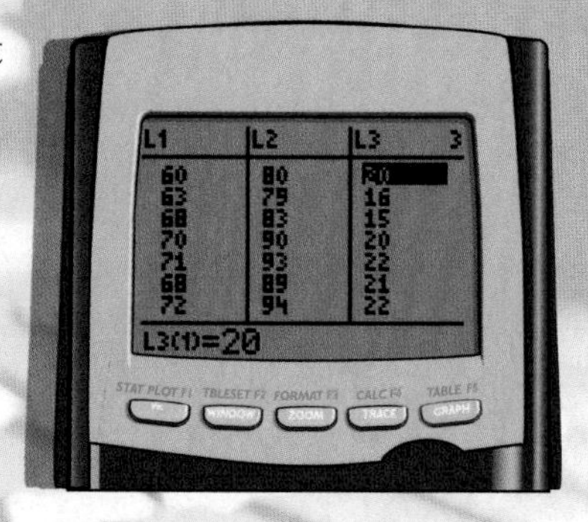

We now have the paired difference in List 3 and can create a one sample *t* confidence interval from the raw data. Press STAT and scroll to TESTS and choose option 8: TInterval. We want to calculate the confidence interval from the data, so choose the Data option. Our data is in List 3, so enter L3 by pressing 2ND then 3. The frequency of the data is the default, 1. Also, we want a 95% confidence interval, so enter 0.95 for the *C-Level*. The menu should then appear as it does in the screen shot to the right.

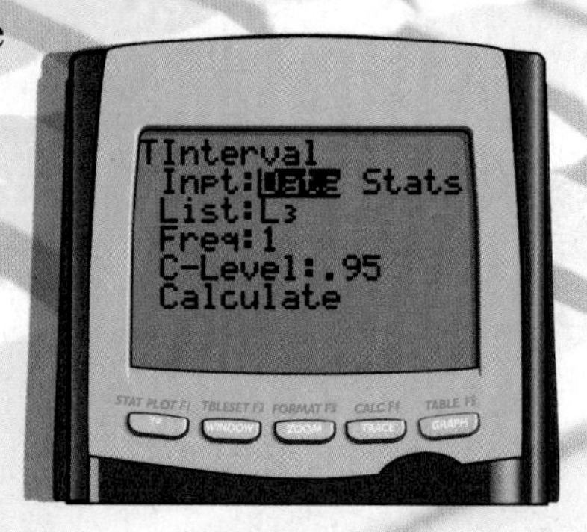

cont'd on the next page

Choose Calculate. The following results are generated:

These results show a confidence interval of (15.281, 20.319). This is the same results we got from calculating this problem by hand earlier in the chapter. Included with the confidence interval is also the mean of the paired differences, the sample standard deviation of the paired differences, and the sample size.

Comparing Two Proportions (Large Independent Samples)

Using a TI-84 Plus to construct a confidence interval to compare two proportions is very simple. We simply enter the number of successes and the sample size for each sample, as well as the level of confidence.

Example T.4

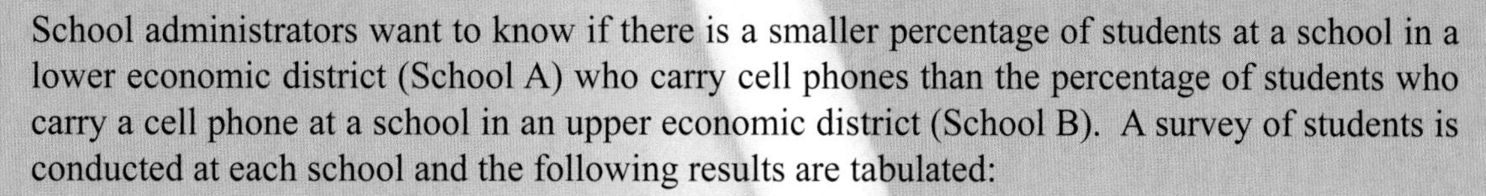

School administrators want to know if there is a smaller percentage of students at a school in a lower economic district (School A) who carry cell phones than the percentage of students who carry a cell phone at a school in an upper economic district (School B). A survey of students is conducted at each school and the following results are tabulated:

	School A	School B
Carry Cell Phones	45	53
Do Not Carry Cell Phones	31	25

Construct a 90% confidence interval for the true difference in the proportion of students who carry a cell phone at the two schools. Does this confidence interval suggest that fewer students carry a cell phone at School A than School B?

Solution:

Press STAT and scroll over to TESTS. Choose option B: 2-PropZInt. x_1 is the number of successes from the first sample, n_1 is the first sample's size. x_2 is the number of successes from the second sample, n_2 is the second sample's size. *C-Level* is the confidence level as a decimal. The data should be entered as shown in the screen shot to the right.

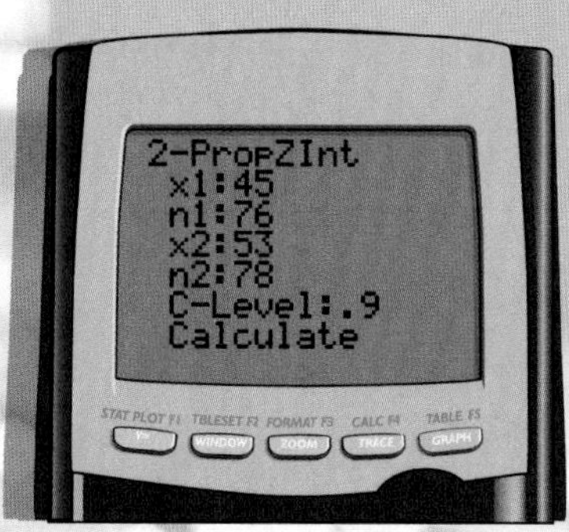

Press Calculate. The results are shown below.

Thus, the confidence interval is from –0.2145 to 0.03971. The sample proportions are also shown. The difference in the solution presented here and the solution presented earlier in this chapter is due to rounding.

MICROSOFT EXCEL

Confidence Intervals for Mean (Large, Independent Samples)

Example T.5

Francisco owns a small apple orchard and believes the mild winter he experienced last year decreased his trees' production. He harvested, weighed, and recorded his trees' production several days during the harvest–35 days the first year and 40 days the second year. Last year he averaged 240.6 pounds with a standard deviation of 5.2 pounds. This year he averaged 238.2 pounds with a standard deviation of 6.8 pounds. Construct a 99% confidence interval to estimate the true difference in average daily yield.

Solution:

Begin by calculating a point estimate, $\bar{x}_1 - \bar{x}_2$, which is 240.6 – 238.2 (cell B5). Next, we produce the margin of error using the critical value obtained using the NORMSINV function with probability $1 - \frac{\alpha}{2}$. Enter the calculation =NORMSINV(0.995)*SQRT(B3^2/B4+C3^2/C4) to produce $E = 3.6$. The interval, (–1.2, 6.0) contains the true difference in daily yield between the two years. The Excel sheet is shown below.

	A	B	C
1	Year	1	2
2	sample mean	240.6	238.2
3	standard deviation	5.2	6.8
4	n	35.0	40
5	point estimate	2.4	
6	E	3.6	
7	left endpoint	-1.2	
8	right endpoint	6.0	

Confidence Intervals for Mean (Small, Independent Samples)

Example T.6

Beth is a big hockey fan. One day she hears that the teams in the Eastern Conference score more than their counterparts in the Western Conference. The 15 teams in the east are averaging 3.159 goals per game with a standard deviation of 0.337. At the same time, the 15 western teams are averaging 3.063 goals per game with a standard deviation of 0.365. Assume that population variances are the same in order to construct a 90% confidence interval to estimate the true difference in the average goals per game of eastern teams compared to western teams.

Solution:

First, calculate a point estimate, $\bar{x}_1 - \bar{x}_2$, which is 3.159 – 3.063 = 0.096. Next, calculate the margin of error by using a *t*-interval with 28 degrees of freedom. Use =TINV(0.1, 28)*SQRT(((B4–1)*B3^2+(C4–1)*C3^2)/(B4+C4–2))*SQRT(1/B4+1/C4) to calculate *E*, the margin of error. Finally, subtract the margin of error from the point estimate to determine the left endpoint and add the margin of error to the point estimate to determine the right endpoint. Since the interval, (–0.122, 0.314), contains both positive and negative values, the claim of eastern scoring superiority is unsubstantiated at the 90% confidence level.

	A	B	C
1	Conference	Eastern	Western
2	sample mean	3.159	3.063
3	standard deviation	0.337	0.365
4	n	15	15
5	point estimate	0.096	
6	E	0.218	
7	left endpoint	-0.122	
8	right endpoint	0.314	

Confidence Intervals for Mean (Dependent Samples)

Example T.7

Abby thinks traffic is worse in the afternoon. She asks her coworkers to record their travel times to and from work one day. The following table lists the times they recorded.

	Darren	Jane	Justin	Kyia	Libby	Ron	Rusty	Siani
Morning	25	15	48	5	9	55	35	17
Afternoon	31	14	55	7	12	56	32	15

Construct a 95% confidence interval around the difference in morning and afternoon travel times.

Solution:

First, we must choose the correct procedure. Since the data is paired, we will construct an interval centered at the mean difference of pairs using the Student's t-distribution. We calculate the mean difference, count of pairs, and standard deviation of differences– $\bar{d}$, n, and s_d, respectively. (Use =AVERAGE(B4:I4), =COUNT(B4:I4), and =STDEV(B4:I4) to produce the mean, count and standard deviation of the paired differences.) The margin of error $E = 0.083$ is calculated as =TINV(.95, 7)*B7/SQRT(B5). Finally, subtracting and adding the margin of error to the mean difference yields the left and right endpoints. The Excel spreadsheet is shown below.

	A	B	C	D	E	F	G	H	I
1		Darren	Jane	Justin	Kyia	Libby	Ron	Rusty	Siani
2	Morning	25	15	48	5	9	55	35	17
3	Afternoon	31	14	55	7	12	56	32	15
4	Paired Difference, d	-6	1	-7	-2	-3	-1	3	2
5	n	8							
6	Mean of paired differences	-1.625							
7	Standard deviation of paired differences	3.623							
8	E	0.083							
9	Left endpoint	-1.708							
10	Right endpoint	-1.542							

Comparing Two Proportions (Large Independent Samples)

Example T.8

A city planner believes two separate districts have significantly different proportions of people below poverty level. She takes a sample from each district to justify her claim. Out of a sample of 252 persons, district 1 has 63 people below the poverty line. On the other hand, district 2 reports 45 persons below the poverty line with a sample size of 256. Construct a 95% confidence interval around the difference in proportion of persons below the poverty line.

Solution:

First, we calculate a point estimate, $\hat{p}_1 - \hat{p}_2 = 0.7$. Again, we find the critical value of the normal distribution using the NORMSINV function with a probability of $1 - \frac{\alpha}{2}$. Margin of error, E, is calculated with =NORMSINV(0.975)*SQRT((D2*(1-D2))/C2+D3*(1-D3)/C3). Subtracting and adding the margin of error to the point estimate yields the left and right endpoints. Since the entire interval, (0.003, 0.145), is positive, the claim is supported at the 95% confidence level. The Excel sheet is shown below.

	A	B	C	D
1	District	x	n	p-hat
2	1	63	252	0.25
3	2	45	256	0.175781
4				
5	difference of estimates		0.074	
6	E		0.071	
7	left endpoint		0.003	
8	right endpoint		0.145	

MINITAB

Confidence Intervals for Mean (Dependent Samples)
Minitab can construct a confidence interval around the mean of the difference of two paired means with a single instruction.

Example T.9

Geraldo's Tire Store claims to have better prices than a rival store. To prove this claim, the statistician at Geraldo's prices a random sample of 10 tire models at each store. The data is in the table below. Construct a 95% confidence interval for the true difference in tire prices at the two stores.

Geraldo's	45	75	80	65	73	109	125	78	152	73
Rick's	47	80	85	62	80	105	130	80	155	78

Solution:

First, enter the tire prices in two columns in the worksheet. Next, select STAT, select BASIC STATISTICS, and choose *Paired t*. The first sample is the column C1 and the second sample is column C2. Select Options in order to specify a confidence level of 95. Choose OK for both the options and main dialog windows and the confidence interval is produced. The dialog boxes are shown below and on the next page.

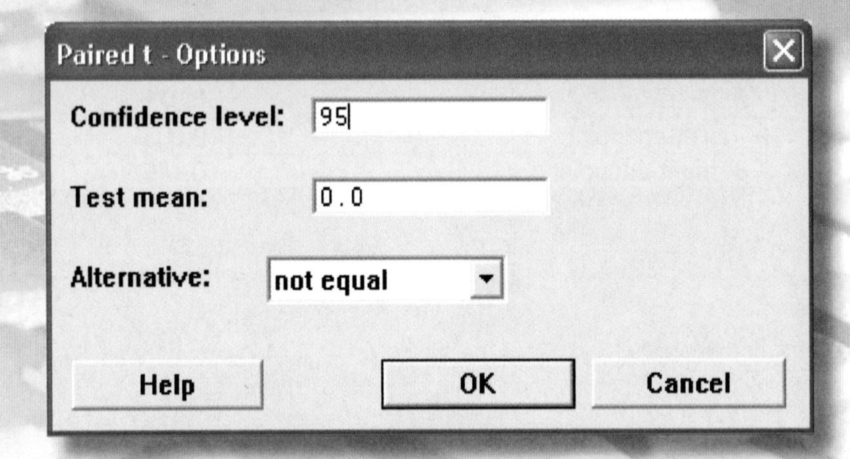

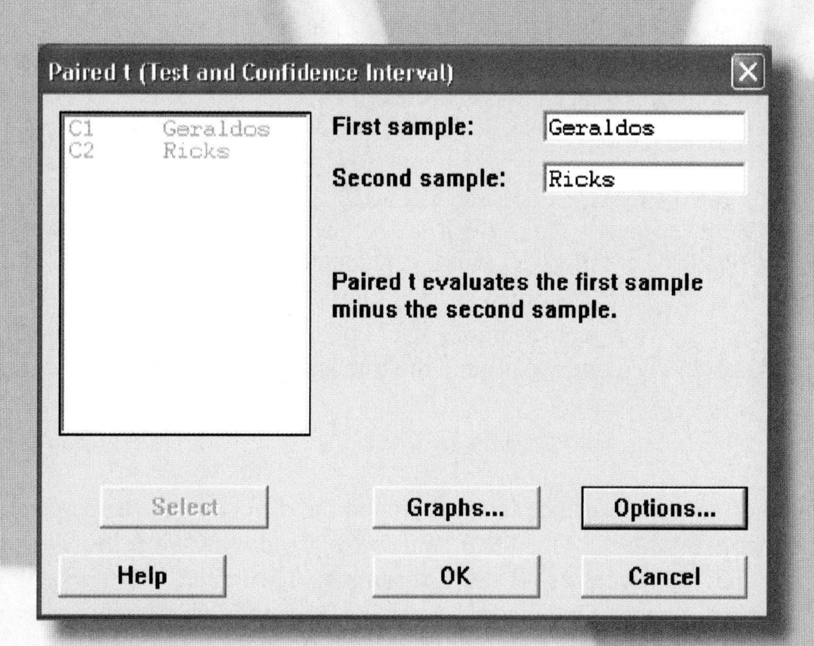

Clicking OK should produce the following screen:

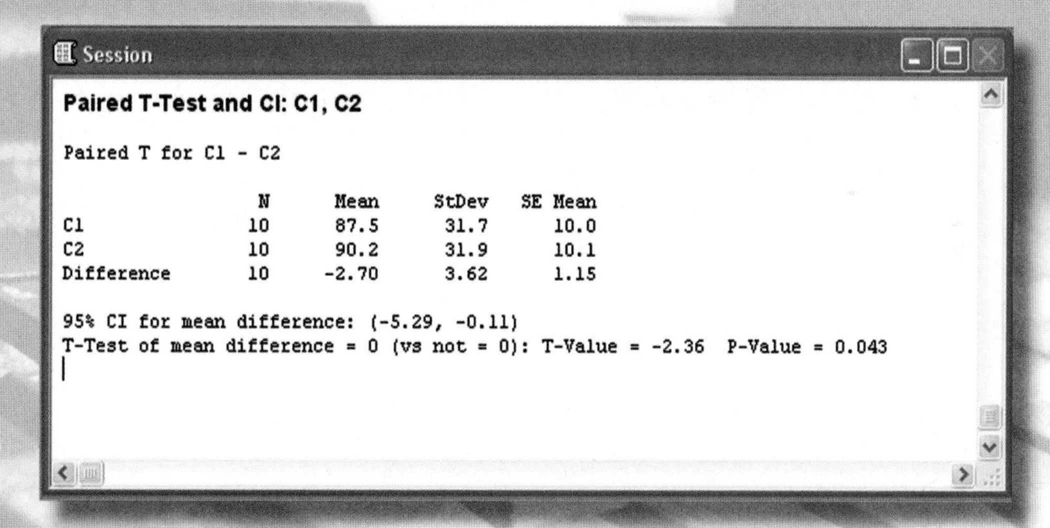

The interval produced, (–5.29, –0.11), contains only negative values. Therefore, Geraldo's can claim better prices than Rick's.

Comparing Two Proportions (Large Independent Samples)

Minitab can also construct a confidence interval around the mean of the difference of two proportions with a single instruction.

Example T.10

A local tourism board wants to know if the percentage of tourists visiting a site on labor day increased between two years. The first year, a sample of 150 people was taken of which 115 were tourists. The following year, a sample of 174 people was taken of which 120 were tourists. Construct a 90% confidence interval around the true change in percentage of tourists.

Solution:

We will construct a 90% confidence interval around the difference in the percentage of tourists. Select STAT, select BASIC STATISTICS, and *2 Proportions*. Choose **Summarized data** and enter 174 trials and 120 successes for the first sample. Then enter 150 trials and 115 successes for sample 2. Next, choose Options and enter 90 for Confidence level. Choose OK for the Options and main dialog and the results are displayed in the Session window. The dialog is shown below.

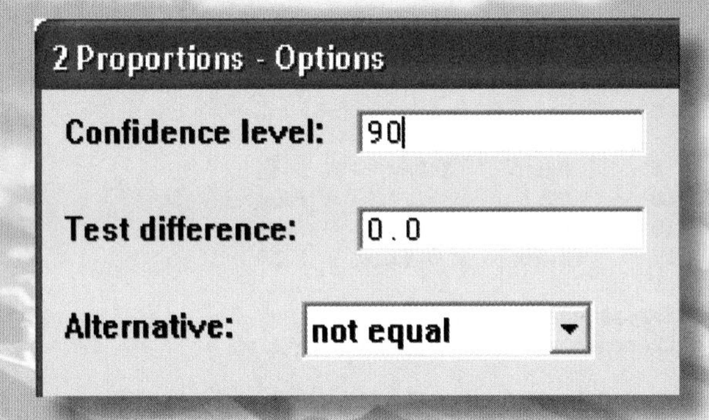

2 Proportions (Test and Confidence Interval)

Samples in one column:
Samples:
Subscripts:
Samples in different columns:
First:
Second:
Summarized data:

	Trials:	Successes:
First sample:	174	120
Second sample:	150	115

Select
Options...
Help
OK
Cancel

Clicking OK should produce the following screen:

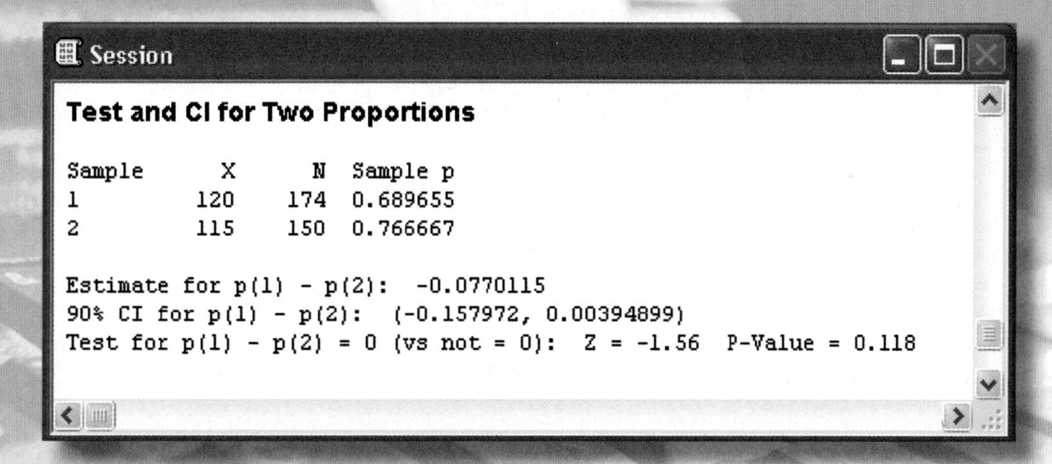

Session

Test and CI for Two Proportions

```
Sample       X      N  Sample p
1          120    174  0.689655
2          115    150  0.766667

Estimate for p(1) - p(2):  -0.0770115
90% CI for p(1) - p(2):  (-0.157972, 0.00394899)
Test for p(1) - p(2) = 0 (vs not = 0):  Z = -1.56  P-Value = 0.118
```

The confidence interval produced is (–0.157972, 0.00394899). With 90% certainty, the true percentage of tourists decreased–possibly as much as 16%.

Hypothesis Testing

Objectives

The objectives of this chapter are as follows:

1. State the null and alternative hypotheses
2. Choose the appropriate test statistic
3. Use rejection regions to draw a conclusion
4. Use *p*-values to draw a conclusion
5. Interpret conclusions to a hypothesis test
6. Describe Type I and Type II errors and recognize when they occur
7. Perform hypothesis tests for means, proportions, and variance
8. Perform a chi-square test for goodness of fit
9. Perform a chi-square test for association

Sections

Chapter 10

Introduction

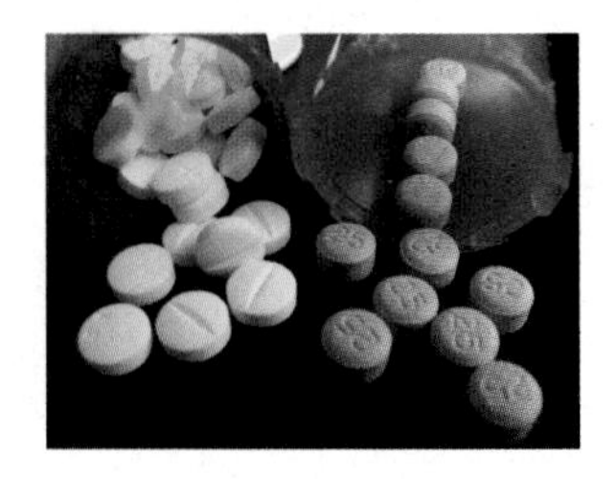

Anyone can make a claim about a population parameter. You have probably heard claims similar to the following: "Four out of five dentists prefer StarBrite toothpaste" and "Fewer than 1% of dogs attack people unprovoked." People make decisions everyday based on claims such as these. Sometimes the decisions are relatively insignificant, like what type of toothpaste to buy. Other times the decisions we make based on these claims are very important, like whether or not to use a new prescription drug that has come on the market. Or perhaps you may choose your next car based on reported safety ratings. In these cases, it is absolutely necessary that extensive testing be done to ensure that the information given to the public is as accurate as possible.

For example, the Food and Drug Administration has the daunting task of determining which drugs are safe and effective enough to be sold to the public. The process through which a drug obtains approval is grueling, and usually it takes years for the few drugs that pass inspection to be released. Even then, some drugs are later found to be dangerous enough that they are taken back off the market. For example, recently a popular arthritis drug was found to increase the risk of heart, stomach, and skin problems. For these reasons, the FDA decided that this drug's risks outweighed its benefits, and it was subsequently pulled from the market.

How do you begin to test claims made about population parameters? Where do we draw the line? Hypothesis testing is one statistical process that sets a uniform standard for evaluating these types of claims.

Fundamentals of Hypothesis Testing

10.1

"In any given one-year period, 9.5% of the population, or about 18.8 million American adults, suffers from a depressive illness…" – Reader's Digest
Source: http://www.rd.com/content/openContent.do?contentId=10671

"For cars, this Corporate Average Fuel Economy (CAFE) has been 27.5 miles per gallon since 1990. Truck fuel economy has gradually increased to 22.2 mpg for the 2007 model year…" – USA Today
Source: http:usatoday.com/money/autos/2006-05-02-mileage-usat_x.htm

You can find statements like these in almost every newspaper article today. However, these claims are not limited only to journalism. They appear in any number of scientific studies—studies that decide whether new drugs are safe, whether new public safety campaigns are working, or whether one or another candidate is likely to win an election. How, then, are statistics used to shed light on these important questions? In short, they all depend on a technique called **hypothesis testing**.

The first stage in hypothesis testing is to determine the hypotheses and express them mathematically. A **hypothesis** is a theory or premise—often it is the claim that someone is making that must be investigated. To begin, state this hypothesis in mathematical terms involving the population parameter. Then, write its mathematical opposite. These two hypotheses form a pair of logically opposite possibilities. For example, consider a claim made by the manager of a local candy manufacturing plant. Ever since his machinery was overhauled, he has claimed that the chocolate bars produced no longer have an average weight of 4 ounces, as they should. Since his claim refers to the value of the population mean of the chocolate bars, μ, it is expressed mathematically as $\mu \neq 4$. In this example, the mathematical opposite is $\mu = 4$.

A hypothesis test then compares the merit of these two competing hypotheses by examining the data that is collected. However, it does not treat the two hypotheses equally. One of the two hypotheses describes the currently accepted value for the population parameter. This hypothesis is called the **null hypothesis**, written as H_0. A hypothesis test examines whether or not the data collected is consistent with this null hypothesis. If the data disagrees "sufficiently" with the null hypothesis, then it is *rejected* in favor of the other hypothesis—which is appropriately named the **alternative hypothesis,** H_a. The alternative hypothesis is sometimes referred to as the *research hypothesis* because it is the hypothesis that leads to the research being conducted. In our example about the chocolate bars, the bars should be 4 ounces. Hence, the null hypothesis is that the population mean is four and the alternative hypothesis is that it is not four. Mathematically, this is written as

$$H_0: \mu = 4$$
$$H_a: \mu \neq 4.$$

Example 1

Determine the null and alternative hypotheses for the following:

It is generally accepted among leading educators that the average student studies no more than 15 hours per week. In the spring newsletter, a national student organization claims that its members study more per week than the average student.

Solution:

Notice that the claim refers to the population mean, μ. The organization's claim that its members study more than the average student is written mathematically as $\mu > 15$. The logical opposite of this claim is $\mu \leq 15$. Since this second inequality is the generally accepted statement among educators, it is the null hypothesis. Thus the two hypotheses are

$$H_0: \mu \leq 15$$
$$H_a: \mu > 15.$$

Example 2

Determine the null and alternative hypotheses for the following:

A leading news authority claims that the President's job approval rating has dropped over the past 3 months. Previous polls put the President's approval rating at a minimum rate of 56%. The president's chief of staff is concerned about this claim since it is an election year, and he wants to run a test on the claim.

Solution:

The news authority's claim refers to a population proportion, which we symbolically write as p. Their claim is that the President's approval rating has dropped from the generally accepted value, which is believed to be at least 56%. Thus their claim is that the President's approval rating is *less than* 56%. Written mathematically, their claim is that $p < 0.56$. The opposite of their claim is $p \geq 0.56$. Notice that this second inequality is the mathematical way of writing the currently accepted belief, so we call it the null hypothesis. Therefore, we have

$$H_0: p \geq 0.56$$
$$H_a: p < 0.56.$$

Example 3

Determine the null and alternative hypotheses for the following:

Leading authorities have stated that approximately 1% of dogs attack people even if the dogs are unprovoked. An animal rights organization believes that 1% is not accurate.

Solution:

The currently accepted statement in this scenario is that 1% of dogs attack people even if unprovoked. The animal rights organization believes that this percentage is wrong. Thus their claim is that $p \neq 0.01$. The logical opposite of this claim is that $p = 0.01$. This second equation is the null hypothesis because it states what is generally accepted by leading authorities. Thus the two hypotheses are

$$H_0: p = 0.01$$
$$H_a: p \neq 0.01.$$

Once the hypotheses have been clearly stated, we can begin to decide which of the two hypotheses the data supports. Let's look at the classic analogy of the United States criminal justice system to help us further understand how to perform a hypothesis test. First, the starting assumption in any trial is that the defendant is not guilty. It is the defendant who is given the benefit of the doubt throughout the trial. Therefore, the null hypothesis is that the defendant is not guilty. The prosecutor, however, makes a claim that the defendant *is* guilty. It is then the prosecutor's responsibility to provide enough evidence *against* the assumption of innocence so that the jury will be persuaded beyond a reasonable doubt that the defendant is guilty of the crime. In just the same way, the null hypothesis is given the benefit of the doubt unless the evidence against it is overwhelming.

In a court of law, the jurors listen to the evidence and have to decide whether the evidence provides proof of guilt *beyond reasonable doubt*. Often this burden of proof depends on the expectation of the jurors themselves and is somewhat subjective. However, the *evidence* in a hypothesis test is provided by the data in the form of a statistic, which we call a **test statistic**. There is no subjective jury. Instead, a hypothesis test depends on the fact that test statistics are unlikely to take values far away from the value of the corresponding population parameter.

Recall from the earlier example about the chocolate bars that the null hypothesis is $\mu = 4$. What kind of data would convince us that the weight of the chocolate bars no longer averages 4 ounces? If the null hypothesis were true, then we would expect the mean of a sample of bars to be somewhere close to 4. If the sample mean were much larger or much smaller than 4, this evidence would certainly persuade us that there is a problem with the machinery. But how far away from four does the sample mean need to be in order for us to be persuaded? Would an average weight of 4.01 ounces be far enough away from the population mean of 4? How about 5 ounces? Where do we draw the line that distinguishes "far enough" from the population parameter?

To answer this question, we need to introduce the term *statistically significant*. The data is said to be **statistically significant** if it is unlikely that a sample similar to the one chosen would occur by chance if the null hypothesis is true. For example, if you toss a fair coin 100 times, it is very likely to land on heads 49 times. However, if you tossed a coin 100 times and it landed on heads only 20 times, you would probably say that this sample proportion was unlikely to occur by chance if the coin was fair, and that perhaps the coin is not fair after all. Thus we might say that the coin landing on heads 20 times, assuming that the coin is fair, is statistically significant.

To determine if the data is statistically significant, a confidence interval for the sample statistic can be calculated based on the assumptions of the null hypothesis, using similar methods to those we saw in Chapter 8. The level of confidence used for this confidence interval is analogous to the concept of *reasonable doubt*. The evidence is considered statistically significant when the test statistic is so inconsistent with the null hypothesis that it falls outside the confidence interval, i.e. *beyond reasonable doubt*.

Memory Booster!

c = level of confidence
$\alpha = 1 - c$
= level of significance

Recall that the level of confidence, c, is a measure of how certain we are that the calculated confidence interval contains the population parameter. If the level of confidence we choose for a particular hypothesis test is c, then the probability of committing an error in our testing is $1 - c$, which we denote as α. We call this value, α, the **level of significance**. Just as we stated the necessary level of confidence when constructing confidence intervals in the previous chapter, throughout this chapter we will state either the level of confidence or the level of significance when performing a hypothesis test. Recall that in the process of a statistical study, we decide the level of significance before collecting data.

In the following sections we will look at different ways to select and calculate the test statistic. For now, let's consider how the results of these calculations lead to a conclusion. If the data collected in a hypothesis test is found to be statistically significant, then we *reject the null hypothesis* in favor of the alternative hypothesis. However, if the evidence is not statistically significant, then the decision is to *fail to reject the null hypothesis*. Notice that in just the same sense that innocence is not proved in a trial, a hypothesis test never *proves* that either hypothesis is true. It can only say either that the data is strong enough to contradict the null hypothesis in favor of the alternative or else it is not strong enough to contradict the null hypothesis. The null hypothesis is presumed to be true unless there is statistically significant evidence to the contrary. Thus there are only two possible conclusions that can be drawn from a statistical hypothesis test, and these are restated in the box below.

Only Possible Conclusions in a Hypothesis Test:

- Reject the null hypothesis
- Fail to reject the null hypothesis

Once you have decided upon a conclusion for the hypothesis test, a discussion of the meaning of this conclusion in terms of the original claim is appropriate. Because it is not necessarily obvious what each conclusion implies, a well-worded statement of your findings that references the claim, and not just mathematical symbols, is a useful summary of your results. Let's look at two examples:

Example 4

It is generally accepted among leading educators that the average student studies no more than 15 hours per week. In their spring newsletter, a national student organization claims that its members study more per week than the national average. After performing a hypothesis test at the 95% level of confidence to evaluate the claim of the organization, the researchers' conclusion is to reject the null hypothesis. Do the researchers' findings support the organization's claim?

Solution:

We must first review the null and alternative hypotheses in order to evaluate the conclusion. Recall from Example 1 that the two hypotheses for this hypothesis test are

$$H_0: \mu \leq 15$$
$$H_a: \mu > 15.$$

Since the null hypothesis was rejected by the researchers, the evidence supports the alternative hypothesis, which is the organization's claim. Thus the organization can be 95% confident in its claim that its members study more per week than the national average.

Example 5

The Board of Education for one large school district uses at least 10% as the percentage of high school sophomores considering dropping out of school. A high school counselor in this district claims that this percentage is too high. A hypothesis test with $\alpha = 0.02$ is performed on the counselor's claim. The result is to fail to reject the null hypothesis. Do the findings support the counselor's claim?

Solution:

Once again we must write down the null and alternative hypotheses to evaluate the conclusion. The school counselor claims the percentage of high school sophomores in this district that are considering dropping out of school is less than the assumed rate of at least 10%. The counselor's claim is written mathematically as $p < 0.10$, and its logical opposite is $p \geq 0.10$. The established standard for the Board of Education is contained in the second of these hypotheses, so we write

$$H_0: p \geq 0.10$$
$$H_a: p < 0.10.$$

Since the test resulted in a failure to reject the null hypothesis, the evidence is not strong enough at this level of significance to support the counselor's claim that the dropout rate is too high. Remember that performing a hypothesis test does not prove the null hypothesis to be true. Therefore we can only say there is not sufficient evidence to reject it.

Now that we understand the basis for performing a hypothesis test, let's outline the general steps of any hypothesis test. They are given in the box below.

General Steps in Hypothesis Testing:

1. State the null and alternative hypotheses.
2. Set up the hypothesis test by choosing the test statistic and determine the values of the test statistic that would lead to rejecting the null hypothesis.
3. Gather data and calculate the necessary sample statistics.
4. Draw a conclusion using the process described in Step 2.

In this section we discussed in detail the process of setting up the null and alternative hypotheses, as well as the types of conclusions that can be made once the calculations in a hypothesis test have been completed. In the next several sections we will discuss the details of choosing the appropriate test statistic for a given population parameter and the various methods for determining the values of that test statistic that result in rejecting the null hypothesis.

Exercises

10.1

Directions: Decide whether the following statements are **True** or **False**. If the statement is false, tell why it is false.

1. The null hypothesis is always given the benefit of the doubt.

2. The first step in hypothesis testing is to gather data so that one knows how to set up the hypotheses.

3. When one rejects the null hypothesis, one has proven the alternative hypothesis to be true.

4. There are only two possible conclusions in a hypothesis test: reject or fail to reject the null hypothesis.

5. The alternative hypothesis is the starting assumption in a hypothesis test.

6. The null and alternative hypotheses form a pair of logically opposite statements.

Null and Alternative Hypotheses
Directions: State the null and alternative hypotheses for each of the following scenarios.

7. A sports analyst is testing his claim that this year's average NFL lineman is heavier than in past years. Over the past 5 years, the average lineman weighed 320 lbs. Set up the hypotheses needed to conduct a test with a 0.01 level of significance.

8. Based on past sales, a shoe manufacturer considers the average woman's shoe size to be 7.5. To test if this is still the case, set up the hypotheses needed to perform a hypothesis test with a 0.05 level of significance.

9. Austin's company manufactures test tubes. He needs his tubes to be exactly 4 mm in diameter. If they are too narrow or too wide, he must recalibrate his machine. Austin randomly measured 150 tubes off the production line in order to perform a hypothesis test with 99% confidence. Set up the hypotheses for Austin's test.

10. A nationwide study shows that the average child watches 3 hours of television per day. A mother in Louisiana believes that this is an underestimate for children in her area. She conducts a local survey to test her belief. Set up the hypotheses needed to conduct a test with 95% confidence.

11. While continuing to keep abreast of local trends in education, a school administrator read a journal article that reported only 42% of high school students study on a regular basis. He claims that this statistic is too low for his district. To test his claim, perform a hypothesis test with a 0.05 level of significance.

12. Guidelines for cereal manufacturers say that to claim a cereal is 'mostly fruit', at least 45% of the cereal must be fruit. A local consumer watch group tests 15 random boxes of a cereal claiming to be mostly fruit to see if it meets the guidelines. Set up the hypotheses for their test with a 0.02 level of significance.

13. As a general guideline, drinking water should contain 3 ppm (parts per million) chlorine throughout an automated watering system. A local council wants to test the chlorine levels on 15 randomly selected days over the next 2 months to check that the guidelines are being met. Set up the hypotheses needed to perform a hypothesis test with a 99% level of confidence.
Source: Edstrom Industries

14. Madison, an ambulance driver in a small town, believes that on a typical 24-hour shift, she receives less than 5 emergency calls on average. To test this belief, she records the number of calls she receives each shift for 10 shifts. Set up the hypotheses needed to conduct a hypothesis test with 95% confidence.

15. The ocean is believed to have an average fluoride concentration of 1.3 ppm (parts per million). A marine biologist is concerned that the level of fluoride is too high in a particular area and is killing the ocean life. Set up the hypotheses needed to perform a hypothesis test with a 95% level of confidence.
Source: en.wikipedia.org

16. The city council of Oxford is thinking of building a road around the city. They want to know if the majority of the residents are in favor of the new road before pursuing the issue further. Set up the hypotheses needed to perform a hypothesis test with a 90% level of confidence.

Interpreting Conclusions
Directions: Answer the following questions.

17. A television network has believed for many years that at least 40% of their viewers are below the age of 22. For marketing purposes, a potential advertiser wishes to test the claim that the percentage is actually less than 40%. After performing the test at the 0.05 level of significance, the advertiser decides to reject the null hypothesis. What does this conclusion lead us to believe about the potential advertiser's claim?

18. A pharmaceutical company has publicized that only 4% of people who take a particular drug experience significant side effects. A researcher is concerned that the percentage is more than 4%, and she decides to test her claim with a hypothesis test. Based on the sample she collects, she decides to fail to reject the null hypothesis at the 0.01 level of significance. What does this conclusion tell us about the researcher's claim?

19. A radio station has always believed that the average age of their listeners is 26. Because their ratings are slipping, they need to know if the average age has changed, so that they can alter their programs accordingly. After information is collected from 329 listeners and a hypothesis test is completed, the radio station decides that they should reject the null hypothesis with a 95% level of confidence. Based on this conclusion, should the radio station look into changing its programming?

20. A cellular phone company promotes the claim that its customers spend on average no more than \$60 a month on cell phone service. One skeptical college student is convinced that the average is higher than \$60 a month. He surveys his friends and performs a hypothesis test to test his claim. In the end, he decides to fail to reject the null hypothesis with $\alpha = 0.01$. What does this conclusion tell us about the college student's claim?

21. A prominent travel agency claims that the average 4-day theme park vacation will cost no more than \$600 per person. A researcher believes that the average is much higher, and decides to test her theory with a hypothesis test. The conclusion of the hypothesis test is to fail to reject the null hypothesis with $\alpha = 0.01$. What does this conclusion say about the researcher's claim?

Hypothesis Testing for Means (Small Samples)

10.2

Following the guidelines presented in the last section, we will now look at performing a complete hypothesis test for means. To begin, let's define some terminology. The **test statistic** is the value you use to test the null hypothesis. When we constructed confidence intervals in Chapter 8, we introduced all of the test statistics we will use for hypothesis testing now. They are the sample statistics used for the point estimates $\bar{x}$, $\hat{p}$, and s, which are then converted to scores of z, t, or χ^2.

To choose which distribution to use for testing means, we once again first consider whether σ is known or unknown. If sigma is known, which is rare in reality, we either use the z-test statistic or other more advanced methods. However, just as with confidence intervals, we will restrict our focus in this book to cases when σ is unknown.

If either the population is *normally distributed* or the sample size is *less than 30,* then the Student t-distribution should be used when testing the mean. In this section we will restrict our focus to these situations. Therefore, the test statistic for hypothesis testing involving the population mean with small samples is

$$t = \frac{\bar{x} - \mu}{\frac{s}{\sqrt{n}}}.$$

Recall from Section 6.5 that the Student t-distribution is completely defined by just one parameter, the number of degrees of freedom, which we abbreviate *d.f.* For the material covered in this section, the degrees of freedom will simply be $n - 1$. The larger the sample size, the closer the t-distribution resembles the normal distribution, and at $n = 30$ it is sufficiently close to use the normal distribution in its place.

The table that we will use when evaluating claims using the t-distribution is found in Table C of the appendix. An excerpt from that table is shown below. This table contains critical values of the t-distribution for certain degrees of freedom and commonly used levels of significance α. To find the desired critical t-score, t_α, simply scan through and find the cell in which the degrees of freedom row and α column intersect. Let's now look at an example of how to use the table to find a critical t-score.

d.f.	$t_{0.100}$	$t_{0.050}$	$t_{0.025}$	$t_{0.010}$	$t_{0.005}$
1	3.078	6.314	12.706	31.821	63.657
2	1.886	2.920	4.303	6.965	9.925
3	1.638	2.353	3.182	4.541	5.841
4	1.533	2.132	2.776	3.747	4.604

Example 6

Find the critical t-score that corresponds with 14 degrees of freedom at the 0.025 level of significance.

Solution:

Scanning through the table, we see that the row for 14 degrees of freedom and column for $\alpha = 0.025$ intersect at a critical t-score of 2.145. Hence $t_{0.025} = 2.145$.

d.f.	$t_{0.100}$	$t_{0.050}$	$t_{0.025}$	$t_{0.010}$	$t_{0.005}$
12	1.356	1.782	2.179	2.681	3.055
13	1.350	1.771	2.160	2.650	3.012
14	1.345	1.761	2.145	2.624	2.977
15	1.341	1.753	2.131	2.602	2.947

Let's next look at how to set up hypothesis tests using t-scores. Before the sample is drawn, there must be a standard in place for determining whether or not the test statistic calculated from the sample is statistically significant. The method we will use to evaluate t-scores is to use **rejection regions**. Rejection regions are determined by two things: (1) the type of hypothesis test and (2) the level of significance, α. To determine the type of test, simply look at the symbol contained in the alternative hypothesis. There are then three different types of tests: left-tailed tests (<), right-tailed tests (>), and two-tailed tests (≠).

Types of Hypothesis Tests

Alternative Hypothesis	Type of Test
$<$ Value	Left-tailed test
$>$ Value	Right-tailed test
$\neq$ Value	Two-tailed test

Before we look at each of these three cases individually, let's state the rule that will be used to determine the conclusion. The conclusion rule for rejection regions is that we reject the null hypothesis if the test statistic calculated from the sample data falls *within* the rejection region.

Conclusion Rule for Rejection Regions: Reject the null hypothesis if the test statistic calculated from the sample data falls within the rejection region.

Let's now look at how to construct rejection regions for each of the three types of hypothesis tests.

1. **Rejection Regions for Left-tailed Hypothesis Tests (H_a contains the symbol <)**
   As you would expect, the area shaded for the rejection region is in the left tail of the distribution. This area is equal to the level of significance, α. The boundary of the rejection region is marked by the critical *t*-score, t_α. Notice that because the critical *t*-score lies on the left side of the curve, it must be negative.

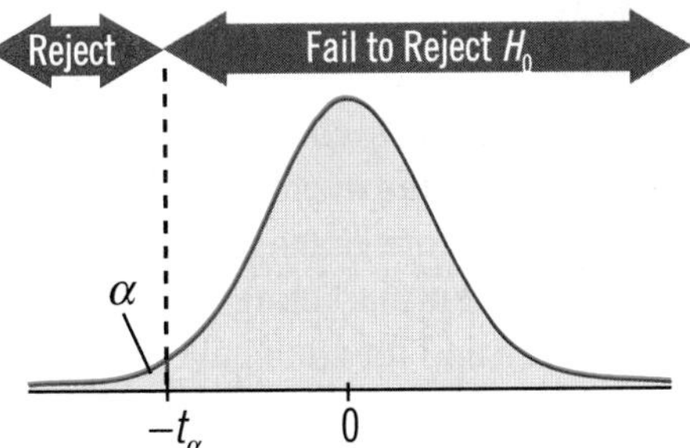

2. **Rejection Regions for Right-tailed Hypothesis Tests (H_a contains the symbol >)**
   The area α for a right-tailed test is shaded on the right side of the distribution. Again, the critical *t*-score marks the boundary of the rejection region. Because the critical *t*-score is on the right side of the curve, its value will be positive for these types of tests.

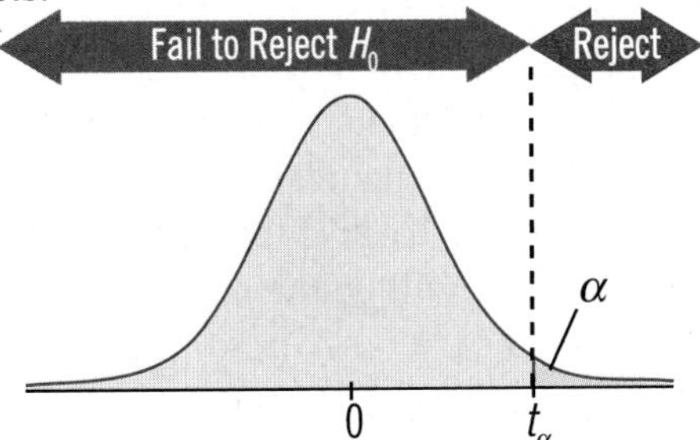

3. **Rejection Regions for Two-tailed Hypothesis Tests (H_a contains the symbol ≠)**
   For two-tailed tests, the area α is divided equally between the two tails of the distribution. Notice in the picture below that the area shaded in each tail is $\frac{\alpha}{2}$. It is important to be careful when using a table to find the appropriate critical *t*-score for a two-tailed test. Some tables give the critical values for two-tailed tests, and others give only values for one-tailed tests. If the table you are using gives only one-tailed values, then you must look up $\frac{\alpha}{2}$ rather than α.

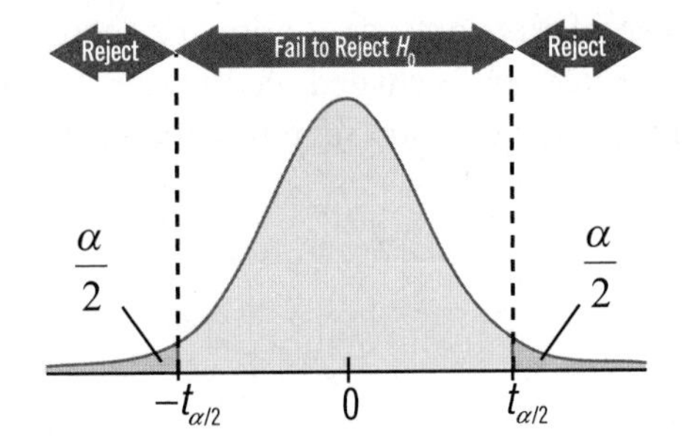

We can summarize these pictures with the following notation:

Reject H_0 if:
$t \le -t_\alpha$ for left-tailed
$t \ge t_\alpha$ for right-tailed
$|t| \ge t_{\alpha/2}$ for two-tailed

We now have all the tools we need to complete an entire hypothesis test. Let's take a look at a few full examples.

Example 7

A hometown department store has chosen its marketing strategies for many years under the assumption that the average shopper spends no more than $100 in their store. A newly hired store manager claims that the current average is higher, and wants to change their marketing scheme accordingly. A group of 27 shoppers is chosen at random and found to have spent on average $104.93 with a standard deviation of $9.07. Assume the distribution of spending is normal and test the store manager's claim at the 0.05 level of significance.

Solution:

We will follow the steps outlined in section 10.1 to complete the hypothesis test.

1. **State the null and alternative hypotheses.**
The store manager claims that the average amount spent is more than $100, which we write mathematically $\mu > 100$. The logical opposite for the claim is then $\mu \le 100$. Because the starting assumption is contained in this second hypothesis, it will be the null hypothesis for the test. We then have
$$H_0: \mu \le 100$$
$$H_a: \mu > 100.$$

2. **Set up the hypothesis test by choosing the test statistic and determining the values of the test statistic that would lead to rejecting the null hypothesis.**
The t-test statistic is appropriate to use in this case because the claim is about an average, and the sample size of 27 is less than 30. Next, notice from the symbol found in the alternative hypothesis that we are running a right-tailed test. We then need to obtain the critical t-score for the rejection region. From the table we see that a one-tailed distribution with $\alpha = 0.05$ and 26 degrees of freedom has a critical t-score of 1.706. Because the area in a right-tailed test is shaded on the right tail, the critical t-score will remain positive. The rejection region is shown on the following page.

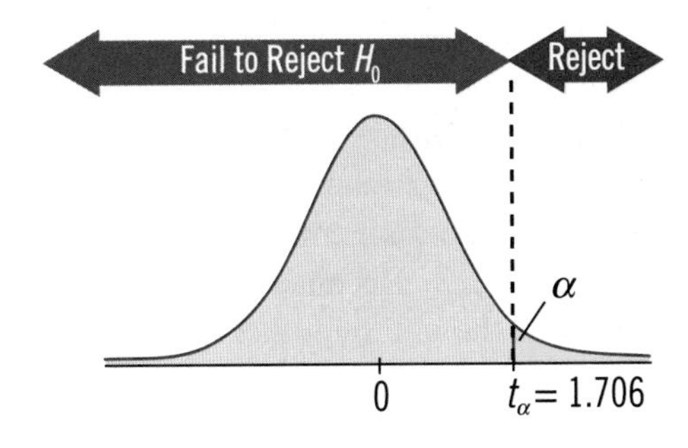

3. **Gather data and calculate the necessary sample statistics.**
   Substitute the information given in the scenario to obtain the t-test statistic.

$$t = \frac{\bar{x} - \mu}{\left(\frac{s}{\sqrt{n}}\right)} = \frac{104.93 - 100}{\left(\frac{9.07}{\sqrt{27}}\right)} \approx 2.824$$

4. **Draw a conclusion using the process described in Step 2.**
   Because the test statistic calculated from the sample data, $t = 2.824$, falls in the rejection region stated in Step 2, we must reject the null hypothesis.

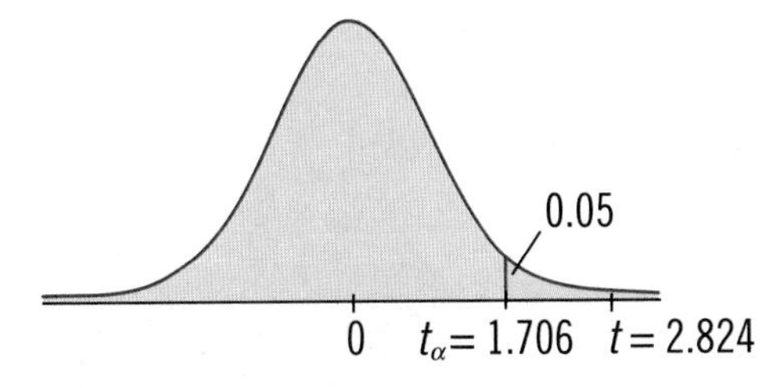

Based on the sample obtained, the evidence supports the manager's claim at the 0.05 level of significance that the average spent is more than \$100.

Example 8

The meat-packing department of a large grocery store packs ground beef in 2-pound portions. They are concerned that their machine is no longer packaging the beef to the required specifications. To test the claim that the average weight of ground beef packed is not 2 pounds, the store calculates the average weight for 20 packages of beef. The sample mean is 2.10 with a standard deviation of 0.33 pounds. Is there evidence at the 0.01 level of significance to show that the machine is not working properly? Assume that the weights are normally distributed.

cont'd on the next page

Solution:

1. **State the null and alternative hypotheses.**
   The store claims that the machine no longer averages 2 pounds per package, which we write mathematically as $\mu \neq 2$. The logical opposite for the claim is then $\mu = 2$. Because the established standard is the second hypothesis, it will be the null hypothesis for the test. We then have

$$H_0\text{: } \mu = 2$$
$$H_a\text{: } \mu \neq 2.$$

2. **Set up the hypothesis test by choosing the test statistic and determining the values of the test statistic that would lead to rejecting the null hypothesis.**
   The t-test statistic is appropriate to use in this case because the claim is about an average, the population is normally distributed, and the sample size of 20 is less than 30. Next, notice from the symbol found in the alternative hypothesis that we are running a two-tailed test. We then need to obtain the critical t-score for the rejection region. The table gives values for both one-tailed and two-tailed distributions, therefore we must look up α under the two-tailed section. We then look up the critical t-score for $\alpha = 0.01$ and 19 degrees of freedom. This gives a critical t-score of 2.861. The rejection region is shown below.

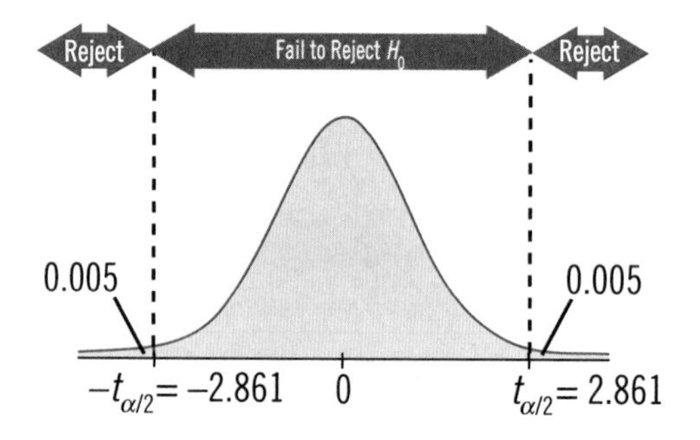

3. **Gather data and calculate the necessary sample statistics.**
   Substitute the information given in the scenario to obtain the t-test statistic.

$$\begin{aligned} t &= \frac{\bar{x} - \mu}{\left(\frac{s}{\sqrt{n}}\right)} \\ &= \frac{2.10 - 2}{\left(\frac{0.33}{\sqrt{20}}\right)} \\ &\approx 1.355 \end{aligned}$$

4. **Draw a conclusion using the process described in Step 2.**
Because the test statistic calculated from the sample, $t = 1.355$, does not fall in the rejection region stated in Step 2, we must fail to reject the null hypothesis.

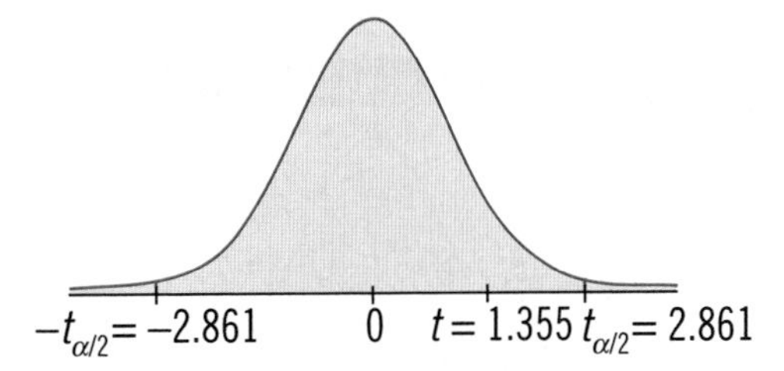

We interpret this conclusion to mean that the evidence collected is not strong enough at the 0.01 level of significance to say that the machine is working improperly.

Example 9

Nurses in a large teaching hospital have complained for many years that they are overworked and understaffed. The consensus among the nursing staff is that they average at least 8 patients per nurse each shift. The hospital administrators claim that the average is lower than 8. In order to prove their point to the nursing staff, the administrators gather information from 19 nurses. The sample mean is 7.5 with a standard deviation of 1.1 patients. Test the administrators' claim for $\alpha = 0.025$ and assume that the average number of patients per nurse follows a normal distribution.

Solution:

1. **State the null and alternative hypotheses.**
The hospital administrators' claim is that the average number of patients per nurse is less than 8, which we write mathematically $\mu < 8$. The logical opposite for the claim is then $\mu \geq 8$. Because the starting assumption is contained in the second hypothesis, it will be the null hypothesis for the test. We then have

$$H_0: \mu \geq 8$$
$$H_a: \mu < 8.$$

2. **Set up the hypothesis test by choosing the test statistic and determining the values of the test statistic that would lead to rejecting the null hypothesis.**
The t-test statistic is appropriate to use in this case because the claim is about an average, the population is normally distributed, and the sample size of 19 is less than 30. Next, notice from the symbol found in the alternative hypothesis that we are running a left-tailed test. We then need to obtain the critical t-score for the rejection region. From the table we see

cont'd on the next page

that a one-tailed distribution with $\alpha = 0.025$ and 18 degrees of freedom has a critical t-score of 2.101. Because the area in a left-tailed test is shaded on the left tail, the critical t-score will be negative, –2.101. The rejection region is shown below.

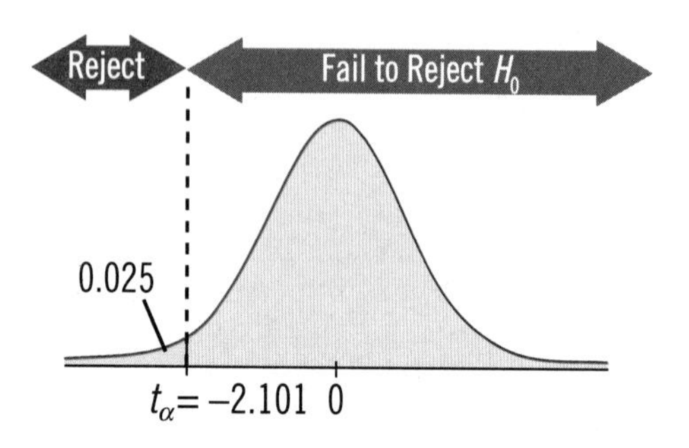

3. **Gather data and calculate the necessary sample statistics.**
   Substitute the information given in the scenario to obtain the t-test statistic.

$$t = \frac{\bar{x} - \mu}{\left(\frac{s}{\sqrt{n}}\right)} = \frac{7.5 - 8}{\left(\frac{1.1}{\sqrt{19}}\right)} \approx -1.981$$

4. **Draw a conclusion using the process described in Step 2.**
   Because the test statistic calculated from the sample, $t = -1.981$, does not fall in the rejection region given in Step 2, we must fail to reject the null hypothesis.

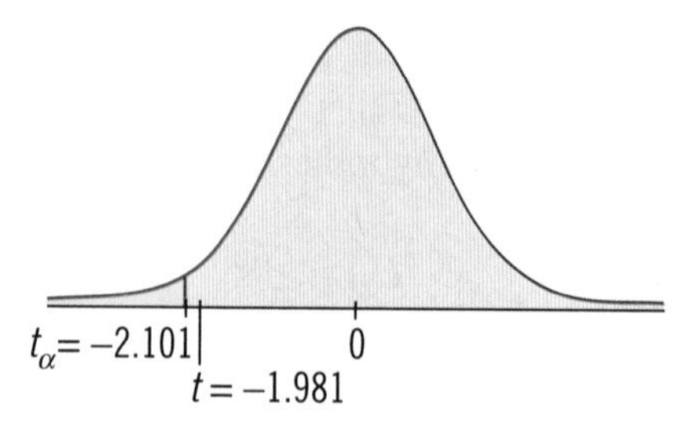

We can interpret this conclusion to mean that the evidence collected is not strong enough at the 0.025 level of significance to reject the null hypothesis in favor of the administrators' claim.

Exercises

10.2

Directions: Determine the rejection regions for the following hypothesis tests given the following information.

1. $\alpha = 0.05$; $d.f. = 14$, left-tailed test
2. $\alpha = 0.01$; $d.f. = 23$, right-tailed test
3. $\alpha = 0.05$; $d.f. = 10$, two-tailed test
4. $\alpha = 0.10$; $d.f. = 18$, H_a: $\mu \neq 18$
5. $\alpha = 0.025$; $d.f. = 17$, H_a: $\mu > 27$
6. $\alpha = 0.05$; $d.f. = 26$, H_a: $\mu < 0.18$
7. $\alpha = 0.10$; $d.f. = 9$, H_0: $\mu = 102$
8. $\alpha = 0.005$; $d.f. = 5$, H_0: $\mu \geq 203$

Directions: Draw the rejection region that corresponds with the given information.

9. A right-tailed hypothesis test with $\alpha = 0.05$ and $n = 11$
10. A left-tailed hypothesis test with $\alpha = 0.025$ and $n = 24$
11. A two-tailed hypothesis test with $\alpha = 0.05$ and $n = 19$
12. A right-tailed hypothesis test with $\alpha = 0.01$ and $n = 8$
13. A left-tailed hypothesis test with $\alpha = 0.005$ and $n = 11$
14. A two-tailed hypothesis test with $\alpha = 0.10$ and $n = 29$

Directions: Draw the rejection region and give the appropriate conclusion for the following information.

15. H_0: $\mu = 195$ and H_a: $\mu \neq 195$ with $t = -3.11$, $n = 20$, and $\alpha = 0.05$
16. H_0: $\mu \geq 17$ and H_a: $\mu < 17$ with $t = -2.05$, $n = 6$, and $\alpha = 0.05$
17. H_0: $\mu \leq 28$ and H_a: $\mu > 28$ with $t = 1.99$, $n = 29$, and $\alpha = 0.01$
18. H_0: $\mu = 7$ and H_a: $\mu \neq 7$ with $t = -3$, $n = 15$, and $\alpha = 0.01$

19. H_0: $\mu \geq 98$ and H_a: $\mu < 98$ with $t = -2.73$, $n = 10$, and $\alpha = 0.01$

20. H_0: $\mu \leq 39$ and H_a: $\mu > 39$ with $t = 2.01$, $n = 8$, and $\alpha = 0.05$

Directions: Complete the following hypothesis tests. For each question, complete the following steps. Assume that each population is approximately normal.

a. State the null and alternative hypotheses.

b. Set up the hypothesis test by choosing the test statistic and determine the values of the test statistic that would lead to rejecting the null hypothesis.

c. Calculate the test statistic from the sample data.

d. Draw a conclusion and interpret.

21. The faculty at a large university is irritated by students' cell phones. They have been complaining for the past few years that a cell phone rings in class at least 15 times a semester (which is about once a week.) A reporter for the school newspaper claims that students are more courteous with their cell phones now than in semesters past, and that the average is now lower than 15. The reporter asks select teachers to keep track of the number of times a cell phone rings in each of 12 different classes one semester. The sample mean is 14.8 with a standard deviation of 1.2 calls. Does the evidence support his claim at the 0.10 level of significance?

22. One cable company advertises that it has excellent customer service. In fact, the company advertises that a technician will be there within 30 minutes of when a service call is placed. One frustrated customer believes this is not accurate, claiming that it takes over 30 minutes for the cable repairman to arrive. The customer asks 9 of his neighbors how long it has taken for the cable repairman to arrive when they have called for him. The sample mean for this group is 33.2 minutes with a standard deviation of 3.4 minutes. Test the customer's claim at the 0.025 level of significance.

23. A parenting magazine reports that the average number of phone calls teenage girls make per night is at least 4. For her science fair project, Ella sets out to prove the magazine wrong. She claims that the average among teenage girls in her area is less than reported. Ella collects information from 25 teenage girls at her high school, and calculates an average of 3.4 calls with a standard deviation of 0.9 calls. Test Ella's claim at the 0.01 level of significance.

24. A children's clothing company sells hand-smocked dresses for girls. The length of one particular size of dress is designed to be 26 inches. The company regularly tests the length of the garments to insure quality control, and if the average length is found to be significantly longer or shorter than 26 inches the machines must be adjusted. The most recent sample of 28 dresses had an average length of 26.3 inches with a standard deviation of 0.77 inches. Test the claim at the 0.01 level of significance.

25. A pizza delivery chain advertises that they will deliver your pizza in no more than 20 minutes from when the order is placed. Being a skeptic, you decide to round up a few friends to test and see if the delivery time is actually more than 20 minutes on average. For the 7 friends who record the amount of time it takes for their pizzas to be delivered, the average is 22.7 minutes with a standard deviation of 4.3 minutes. Perform a hypothesis test using a 0.05 level of significance.

26. Teacher's salaries in one state are very low, so low that educators in that state regularly complain about their compensation. The state average is $33,600, but teachers in one district claim that the average in their district is significantly lower. They survey 22 teachers in the district and calculate an average salary of $32,400 with a standard deviation of $1520. Test the teachers' claim at the 0.01 level of significance.

27. It currently takes users 15 minutes on average to install the most popular computer program made by RodeTech, a software design company. After changes have been made to the program, the company wants to know if the new average is now different from 15 minutes so that they can change their advertising accordingly. A sample of 20 new customers are asked to time how long it takes for them to install the software. The sample mean is 14.1 minutes with a standard deviation of 1.9 minutes. Test the claim at the 0.05 level of significance.

Hypothesis Testing for Means (Large Samples)

10.3

The previous section dealt with choosing and calculating test statistics for hypothesis tests for means when the size of the sample is small. As we saw when we calculated confidence intervals, the *t*-distribution becomes very close to the normal distribution when the size of the sample grows larger than 30. In this section we will see how using the normal distribution changes the test statistic used in hypothesis testing. As always, we will assume that the population standard deviation is unknown and that all samples have an equal probability of being chosen.

Recall the general steps in a hypothesis test outlined below.

1. State the null and alternative hypotheses.
2. Set up the hypothesis test by choosing the test statistic and determine the values of the test statistic that would lead to rejecting the null hypothesis.
3. Gather data and calculate the necessary sample statistics.
4. Draw a conclusion using the process described in Step 2.

Let's look at an example of a hypothesis test for means using a large sample and rejection regions to draw the conclusion.

Example 10

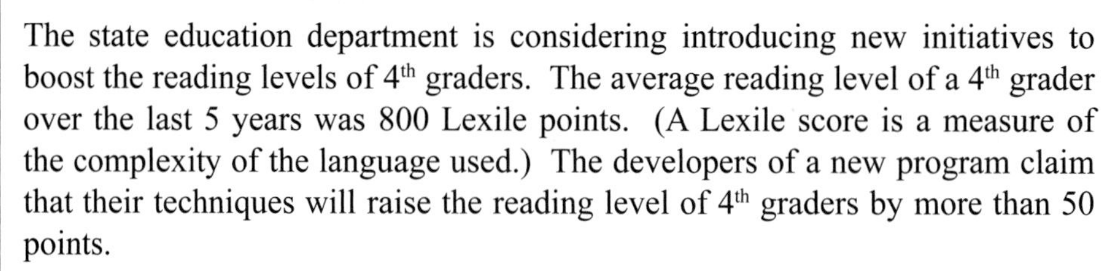

The state education department is considering introducing new initiatives to boost the reading levels of 4th graders. The average reading level of a 4th grader over the last 5 years was 800 Lexile points. (A Lexile score is a measure of the complexity of the language used.) The developers of a new program claim that their techniques will raise the reading level of 4th graders by more than 50 points.

To assess the impact of their initiative, the developers were given permission to pilot their ideas in the classrooms. At the end of the pilot a sample of 1000 4th graders produced an average reading level of 856 Lexile points with a sample standard deviation of 98. Using a 0.05 level of significance, should the findings of the study convince the education department of the validity of the developers' claim?

Solution:

1. **State the null and alternative hypotheses.**
   The developers' claim is that their classroom techniques will raise the average 4th grader's reading level to more than 850 Lexile points. This is written mathematically as $\mu > 850$, and since it is the research hypothesis it will be H_a. The mathematical opposite is $\mu \leq 850$. Thus we have

$$H_0: \mu \leq 850$$
$$H_a: \mu > 850.$$

2. **Set up the hypothesis test by choosing the test statistic and determining the values of the test statistic that would lead to rejecting the null hypothesis.**
   Since the hypotheses are statements about the population mean and $n = 1000$, which is greater than 30, we use the z-test statistic,

$$z = \frac{\bar{x} - \mu}{\left(\frac{s}{\sqrt{n}}\right)}.$$

   Recall from the previous section that we determine the type of test to use based on the alternative hypothesis. In this case, the alternative hypothesis contains '>' which indicates that this is a right-tailed test. To determine the rejection region, we need a z-value so that 0.05 of the area is to its right. If 0.05 of the area is to the right then $1 - 0.05 = 0.95$ is the area to the left. If we look up 0.95 in the cumulative z table, the corresponding critical z-value is 1.645. So the rejection region is $z \geq 1.645$.

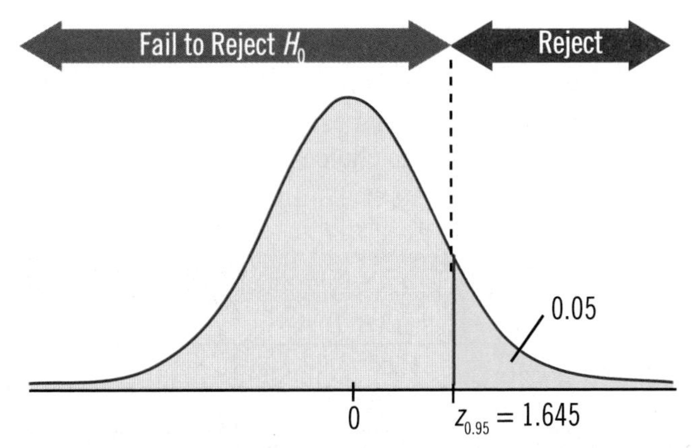

3. **Gather data and calculate the necessary sample statistics.**
   At the end of the pilot, a sample of 1000 4th graders produced an average reading level of 856 Lexile points with a sample standard deviation of 98. Thus the test statistic is

$$\begin{aligned} z &= \frac{\bar{x} - \mu}{\left(\frac{s}{\sqrt{n}}\right)} \\ &= \frac{856 - 850}{\left(\frac{98}{\sqrt{1000}}\right)} \\ &\approx 1.94. \end{aligned}$$

cont'd on the next page

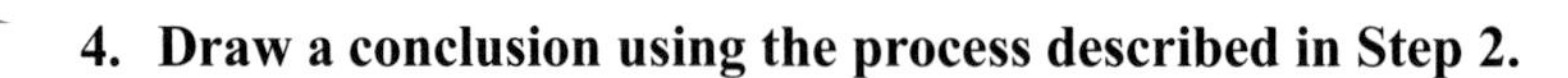

4. Draw a conclusion using the process described in Step 2.

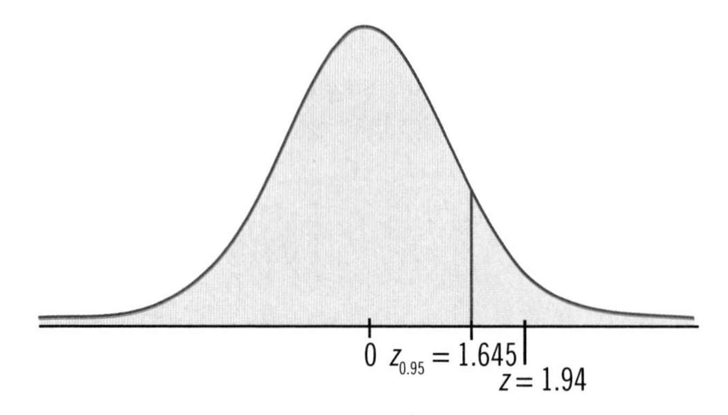

The z-value of 1.94 falls in the rejection region. So the conclusion is to reject the null hypothesis. Thus the evidence collected suggests that the education department can be 95% sure of the validity of the developers' claim, that the Lexile scores of 4th graders will increase by more than 50 points.

The choice for the level of significance is usually one of only a few values of alpha. In contrast to the t-distribution, we can easily create a table that allows us to look up critical z-values quickly for these choices of alpha for both one- and two-tailed tests.

Table 1 - Critical z-Values for Rejection Regions

c	One-Tailed Test	Two-Tailed Test
0.90	1.28	±1.645
0.95	1.645	±1.96
0.98	2.05	±2.33
0.99	2.33	±2.575

Note that the one-tailed column contains critical values for right-tailed tests. Since the standard normal distribution is symmetrical about $z = 0$, for left-tailed tests we simply make these values negative.

p-Values

Although using rejection regions to draw a conclusion in a hypothesis test is a perfectly viable method, it is not the method most commonly found in research articles or journals. Instead, one is more likely to encounter the use of a *p-value*.

A ***p*-value** is the probability of obtaining a sample more extreme than the one observed in your data, when H_0 is assumed to be true. The definition of "more extreme" is based on the type of hypothesis test. For example, if you were testing a left-tailed claim, then "more extreme" would indicate farther to the left.

For a left-tailed test, the probability of obtaining a sample more extreme than the one observed is the probability that the test statistic is *less than* the calculated value. We can calculate these probabilities using the cumulative normal curve table, or available technology, as we have done in previous chapters. Let's look at examples of calculating p-values for each of the three types of hypothesis tests.

Example 11

Calculate the p-value for a hypothesis test with the following hypotheses. Assume that data has been collected and the test statistic was calculated to be $z = -1.34$.

$$H_0: \mu \geq 0.15$$
$$H_a: \mu < 0.15$$

Solution:

The alternative hypothesis tells us that this is a left-tailed test. Therefore, the p-value for this situation is the probability that z is less than -1.34. Look up the area under the normal curve to the left of $z = -1.34$. Thus $p = 0.0901$.

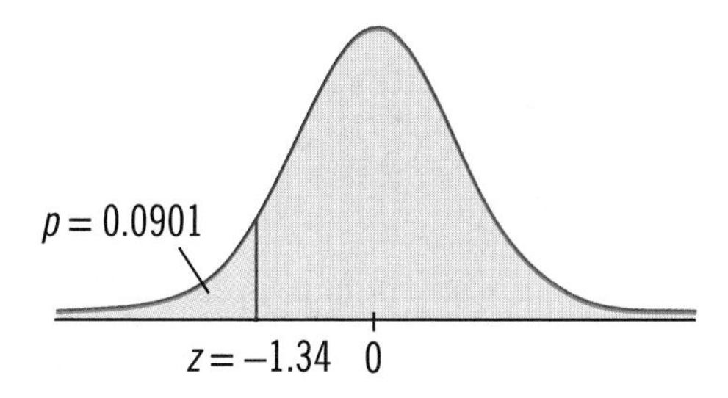

Example 12

Calculate the p-value for a hypothesis test with the following hypotheses. Assume that data has been collected and the test statistic was calculated to be $z = 2.78$.

$$H_0: \mu \leq 0.43$$
$$H_a: \mu > 0.43$$

Solution:

The alternative hypothesis tells us that this is a right-tailed test. Therefore, the p-value for this situation is the probability that z is greater than 2.78. Look up the area under the normal curve to the right of $z = 2.78$. Thus $p = 0.0027$.

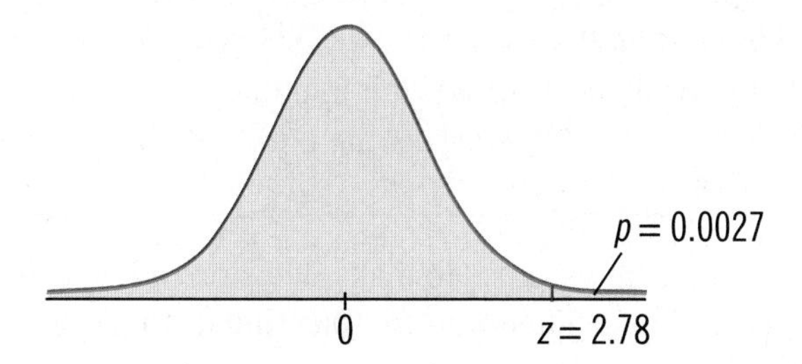

Example 13

Calculate the p-value for a hypothesis test with the following hypotheses. Assume that data has been collected and the test statistic was calculated to be $z = -2.15$.

$$H_0: \mu = 0.78$$
$$H_a: \mu \neq 0.78$$

Solution:

The alternative hypothesis tells us that this is a two-tailed test. Thus the p-value for this situation is the probability that z is either less than -2.15 or greater than 2.15. Look up the area under the normal curve to the left of $z = -2.15$. The area to the left of $z = -2.15$ is 0.0158. Since the normal curve is symmetric about the mean, the area to the right of $z = 2.15$ is also 0.0158. Thus

$$p = 0.0158(2)$$
$$p = 0.0316.$$

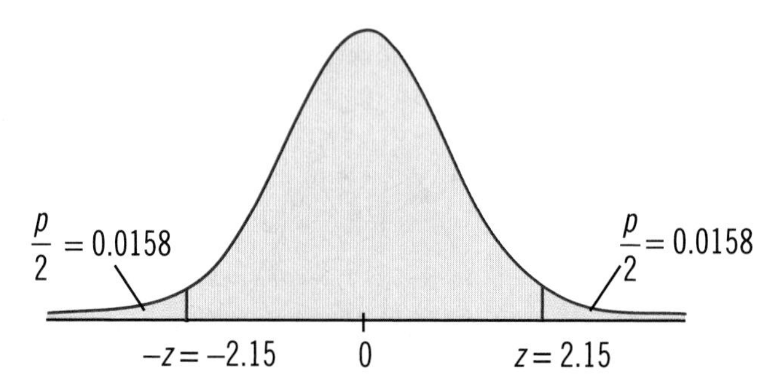

Recall from our discussion of statistical significance that if the data is found to be statistically significant, then we *reject the null hypothesis* in favor of the alternative hypothesis. We can determine if the difference between the claimed value of the parameter and the data collected (i.e. the sample statistic) is statistically significant by comparing the p-value of the test statistic to α, the level of significance. Throughout this text, we will state the necessary level of confidence or the level of significance when performing a hypothesis test. However, in practice it is common to simply quote the p-value and allow the reader to draw conclusions based on their own desired level of significance.

Once the test statistic and the p-value have been calculated, we compare the p-value to the stated value of α to draw a conclusion as described in the following box.

Conclusions Using *P*-Values

- If $p \leq \alpha$, then *reject* the null hypothesis.
- If $p > \alpha$, then *fail to reject* the null hypothesis.

Although we are using p-values to make decisions when using a z-test statistic, we can in fact use the same guidelines for rejecting the null hypothesis given a p-value and α for any test statistic.

Example 14

For a certain hypothesis test, the p-value is calculated to be $p = 0.0145$.

a. If the stated level of significance is 0.05, what is the conclusion to the hypothesis test?

b. If the level of confidence is 99%, what is the conclusion to the hypothesis test?

Solution:

a. The level of significance is $\alpha = 0.05$. Next, note that $0.0145 < 0.05$. Thus $p \le \alpha$, so we *reject* the null hypothesis.

b. The level of confidence is 99%, so

$$\begin{aligned} \alpha &= 1 - c \\ &= 1 - 0.99 \\ &= 0.01. \end{aligned}$$

Next, note that $0.0145 > 0.01$. Thus $p > \alpha$, so we *fail to reject* the null hypothesis.

Now that we are familiar with the necessary test statistic and method for determining a conclusion, let's perform complete hypothesis tests for population means with large samples.

Our general steps for hypothesis testing change slightly to reflect the use of p-values rather than rejection regions.

General Steps Using p-Values in Hypothesis Testing:

1. State the null and alternative hypotheses.
2. Set up the hypothesis test by choosing the test statistic and stating the level of significance.
3. Gather data and calculate the necessary sample statistics.
4. Draw a conclusion by comparing the p-value to the level of significance.

Example 15

Wikipedia, a free online encyclopedia, reports that in 2006, the average American woman is 25 years of age at her first marriage. A researcher claims that for women in California, this estimate is too low. Surveying 213 newlywed women in California, gave a mean of 25.4 years with a standard deviation of 2.3 years. Using a 95% level of confidence, determine if the data supports the researcher's claim.

Solution:

1. **State the null and alternative hypotheses.**
   The researcher's claim is that the average age at first marriage for women in California is more than 25. Written mathematically, this claim is $\mu > 25$. The logical opposite is $\mu \le 25$. Thus the null and alternative hypotheses are

$$H_0\colon \mu \le 25$$
$$H_a\colon \mu > 25.$$

2. **Set up the hypothesis test by choosing the test statistic and stating the level of significance.**
   Since the hypotheses are statements about the population mean and $n = 213$, which is greater than 30, we use the following test statistic:

$$z = \frac{\bar{x} - \mu}{\left(\frac{s}{\sqrt{n}}\right)}.$$

   We will draw a conclusion by computing the p-value for the calculated test statistic and comparing the value to α. For this hypothesis test,

$$\begin{aligned}\alpha &= 1 - c \\ &= 1 - 0.95 \\ &= 0.05.\end{aligned}$$

3. **Gather data and calculate the necessary sample statistics.**
   From the data given, we know that $\bar{x} = 25.4$, $s = 2.3$, and $n = 213$.

   Thus the test statistic is:

$$\begin{aligned}z &= \frac{\bar{x} - \mu}{\left(\frac{s}{\sqrt{n}}\right)} \\ &= \frac{25.4 - 25}{\left(\frac{2.3}{\sqrt{213}}\right)} \\ &\approx 2.54.\end{aligned}$$

4. **Draw a conclusion by comparing the *p*-value to the level of significance.**
   The alternative hypothesis tells us that we have a right-tailed test. Therefore, the p-value for this test statistic is the probability of obtaining a test statistic greater than $z = 2.54$. To find the p-value, we need to find the area under the normal curve to the right of $z = 2.54$.

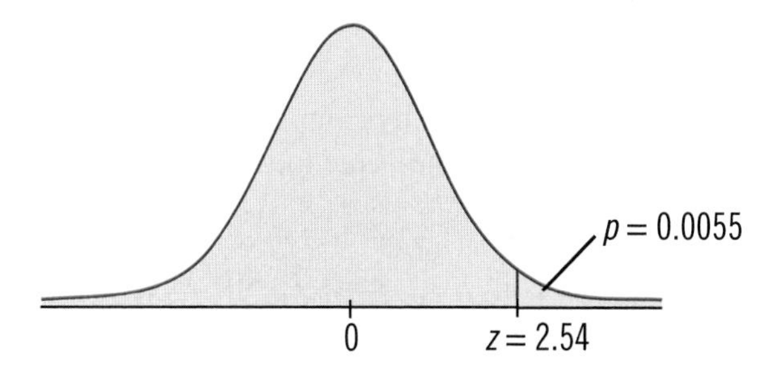

By looking up $z = -2.54$, we find the area $p = 0.0055$. Comparing this p-value to the level of significance, we see that $0.0055 < 0.05$, so $p < \alpha$. Thus the conclusion is to reject the null hypothesis. Therefore, we can say with 95% confidence that the data supports the researcher's claim that the average age at first marriage for women in California is more than 25.

Example 16

A recent study showed that the average number of children for women in Europe is 1.48. A global watch group claims that German women have an average fertility rate which is different from the rest of Europe. To test their claim, they surveyed 128 German women and found they had an average fertility rate of 1.39 children with a standard deviation of 0.84. Does this data support the claim made by the global watch group at the 90% level of confidence?

Solution:

1. **State the null and alternative hypotheses.**
   The group's claim is that the average number of children for women in Germany is *not equal to* the European average of 1.48. Written mathematically, this claim is $\mu \neq 1.48$. The logical opposite is $\mu = 1.48$. Thus the null and alternative hypotheses are

$$H_0\colon \mu = 1.48$$
$$H_a\colon \mu \neq 1.48.$$

cont'd on the next page

2. **Set up the hypothesis test by choosing the test statistic and stating the level of significance.**
   Since the hypotheses are statements about the population mean and $n = 128$, which is greater than 30, we use the following test statistic:

$$z = \frac{\bar{x} - \mu}{\left(\frac{s}{\sqrt{n}}\right)}.$$

   We will draw a conclusion by computing the p-value for the calculated test statistic and comparing the value to α. For this hypothesis test,

$$\begin{aligned} \alpha &= 1 - c \\ &= 1 - 0.90 \\ &= 0.10. \end{aligned}$$

3. **Gather data and calculate the necessary sample statistics.**
   From the data given, we know that $\bar{x} = 1.39$, $s = 0.84$, and $n = 128$.

   Thus the test statistic is:

$$\begin{aligned} z &= \frac{\bar{x} - \mu}{\left(\frac{s}{\sqrt{n}}\right)} \\ &= \frac{1.39 - 1.48}{\left(\frac{0.84}{\sqrt{128}}\right)} \\ &\approx -1.21. \end{aligned}$$

4. **Draw a conclusion by comparing the *p*-value to the level of significance.**
   The alternative hypothesis tells us that we have a two-tailed test. Therefore, the p-value for this test statistic is the probability of obtaining a test statistic which is either less than $z = -1.21$ or greater than $z = 1.21$. To find the p-value, we need to find the area under the normal curve to the left of $z = -1.21$ and to the right of $z = 1.21$.

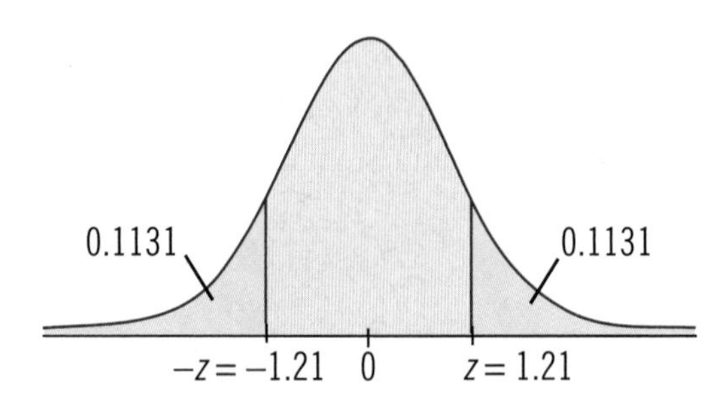

By looking up $z = -1.21$, we find the area to the left is equal to 0.1131. Since the normal curve is symmetric about the mean, the area to the right of $z = 1.21$ is also 0.1131. Thus

$$p = 0.1131(2)$$
$$p = 0.2262.$$

Comparing this p-value to the level of significance, we see that $0.2262 > 0.10$, so $p > \alpha$. Thus the conclusion is to fail to reject the null hypothesis. This means that the evidence does not support the watch group's claim with a 90% level of confidence that the fertility rate of German women is different from the rest of Europe.

Exercises

10.3

Directions: Draw the rejection region that corresponds to the given information.

1. H_0: $\mu \leq 65$ and H_a: $\mu > 65$ with $\alpha = 0.05$

2. H_0: $\mu = 5$ and H_a: $\mu \neq 5$ with $\alpha = 0.10$

3. H_0: $\mu \geq 3.2$ and H_a: $\mu < 3.2$ with $\alpha = 0.01$

4. H_0: $\mu \leq 700$ and H_a: $\mu > 700$ with $\alpha = 0.02$

5. H_0: $\mu = 0.19$ and H_a: $\mu \neq 0.19$ with $\alpha = 0.05$

6. H_0: $\mu \geq 0.166$ and H_a: $\mu < 0.166$ with $\alpha = 0.10$

Directions: Calculate the p-value for each of the following hypothesis tests.

7. H_0: $\mu \geq 60$ and H_a: $\mu < 60$ with $z = -1.48$

8. H_0: $\mu \leq 1.07$ and H_a: $\mu > 1.07$ with $z = 2.27$

9. H_0: $\mu = 2.89$ and H_a: $\mu \neq 2.89$ with $z = -2.21$

10. H_0: $\mu = 144$ and H_a: $\mu \neq 144$ with $z = 1.65$

11. H_0: $\mu \leq 0.56$ and H_a: $\mu > 0.56$ with $z = 1.61$

12. H_0: $\mu \geq 760$ and H_a: $\mu < 760$ with $z = -2$

13. H_0: $\mu = 3$ and H_a: $\mu \neq 3$ with $z = 2.04$

14. H_0: $\mu \leq 14.8$ and H_a: $\mu > 14.8$ with $z = 3.01$

Directions: Determine the appropriate conclusion for a hypothesis test that corresponds with the given information.

15. H_0: $\mu \geq 409$ and H_a: $\mu < 409$ with $\alpha = 0.05$ and $z = -1.87$

16. H_0: $\mu \leq 0.65$ and H_a: $\mu > 0.65$ with $\alpha = 0.10$ and $z = 1.22$

17. H_0: $\mu = 2$ and H_a: $\mu \neq 2$ with $\alpha = 0.02$ and $z = -2.28$

18. H_0: $\mu = 3010$ and H_a: $\mu \neq 3010$ with $\alpha = 0.01$ and $z = 2.69$

19. The calculated p-value = 0.0166 and $\alpha = 0.01$

20. The calculated p-value = 0.0197 and $\alpha = 0.02$

21. The calculated p-value = 0.0465 and $\alpha = 0.05$

22. The calculated p-value = 0.0488 and $\alpha = 0.02$

23. The calculated p-value = 0.1190 and $\alpha = 0.10$

24. The calculated p-value = 0.0094 and $\alpha = 0.01$

Directions: Complete the following hypothesis tests. For each question, complete the following steps:

a. State the null and alternative hypotheses.
b. Set up the hypothesis test by choosing the test statistic and stating the level of significance.
c. Calculate the test statistic from the sample data.
d. Draw a conclusion by comparing the p-value to the level of significance.

25. A manufacturer must test that his bolts are 2 cm long when they come off the assembly line. He must recalibrate his machines if the bolts are too long or too short. After sampling 100 bolts off the assembly line, he calculates the sample mean to be 1.9 cm and the standard deviation to be 0.5 cm. Assuming a level of significance of 0.05, is there evidence to show that the manufacturer needs to recalibrate the machines?

26. CNN Money reports that the average cost of a speeding ticket, including court costs, was $150 in 2002. A local police department claims that this amount has increased. To test their claim, they collect data from 160 drivers who have been fined in the last year and find that they paid an average of $154.00 per ticket with a standard deviation of $17.54. Is there evidence to support the police department's claim at the 0.01 level of significance?
Source: http://money.cnn.com/2002/05/22/news/q_speed_cost/

27. BeachCalifornia.com reports that theme park travelers spend on average \$839 per trip. A consumer watch group claims that this average is too low. After sampling 104 people leaving theme parks, they calculated a sample average of \$851.33 with a standard deviation of \$56.56. Is there evidence at the 0.01 level of significance to support the group's claim?
Source: http://www.beachcalifornia.com/california-theme-parks.html

28. A national business magazine reports that the average age of retirement for women executives is 61. A women's rights organization believes that this statistic does not accurately depict the current trend in retirement. To test this, they polled 95 recently retired women executives and found that they had an average age of retirement of 61.5 with a standard deviation of 2.5 years. Is there evidence to support the organization's claim at the 0.05 level?

29. According to the National Center for Health Statistics, the average weight for an adult female in the United States is 162.9 pounds. A group promoting healthier eating habits claims that they have made an impact in one community. They claim the average woman in this community weighs less than the national average. A sample of 39 women has an average weight of 160.1 pounds with a standard deviation of 5.6 pounds. Is there evidence at the 0.01 level of significance to say that the average woman weighs less than 162.9 pounds in this community?
Source: http://pediatrics.about.com/cs/growthcharts2/f/avg_wt_female.htm

30. The board of a major credit card company requires that the wait time for customers when they call customer service is at most 3 minutes. To make sure that the wait times are not exceeding the requirement, an assistant manager tracks the wait times of 45 calls. The average wait time was calculated to be 3.4 minutes with a standard deviation of 1.45 minutes. Is there evidence to say that the customers are waiting longer than 3 minutes with a 98% level of confidence?

Hypothesis Tests for Population Proportions

10.4

We have seen in the previous two sections how to apply the principles of hypothesis testing to questions regarding population means. In this section, we will turn our attention to questions of population proportions.

Recall the general steps in a hypothesis test for p-values is outlined below.

1. State the null and alternative hypotheses.
2. Set up the hypothesis test by choosing the test statistic and stating the level of significance.
3. Gather data and calculate the necessary sample statistics.
4. Draw a conclusion by comparing the p-value to the level of significance.

At this point you should be very familiar with stating the null and alternative hypotheses, so we can focus our attention on the test statistic for population proportions and the methods for determining when to reject the null hypothesis.

Recall from the discussion of confidence intervals in Chapter 8 that the best point estimate for a population proportion is the sample proportion, $\hat{p}$. Furthermore, if the conditions $np \geq 5$ and $n(1-p) \geq 5$ are both true, then the sampling distribution for the sample proportion is a normal distribution. Thus the test statistic for population proportions is a z-score. The formula for this test statistic for population proportions is

$$z = \frac{\hat{p} - p}{\sqrt{\dfrac{p(1-p)}{n}}}.$$

If the conditions for using the normal distribution are not met, then other methods beyond the scope of this text must be used to test the hypothesis, or the sample size must be increased to allow for the use of the normal distribution.

Example 17

For each scenario below, which test statistic should be used to test the claim?

a. At least 15% of listeners of non-profit radio stations generally favor commercials for other non-profit organizations on the station. A local non-profit radio station believes that less than 15% of their listeners favor such commercials. The station plans to survey 50 of its listeners to test their claim.

b. The college paper reports that at least 10% of students turn off their cell phone during their classes. Aggravated with the number of cell phones that ring during his classes, a professor believes that fewer than 10% of his students turn off their cell phones during the class period. He plans on testing his claim by asking each of his 20 students to show him their cell phones one day during class.

c. The average amount of rainfall in the southern part of Texas during the month of September is commonly believed to be no more than 3.02 inches. A meteorologist claims that the amount of rainfall this September has been higher than normal. He tests his claim by sampling the amount of rainfall in 12 random locations across the region. Assume the population standard deviation is unknown and the population is normally distributed.

Solution:

a. The claim refers to a population proportion, thus we must test the conditions necessary to use the z-test statistic.

$$np = 50(0.15) = 7.5 \geq 5$$
$$n(1-p) = 50(1-0.15) = 42.5 \geq 5$$

Both conditions are met, so the station should use the z-test statistic for population proportions to test their claim.

b. The claim refers to a population proportion, thus we must test the conditions necessary to use the z-test statistic.

$$np = 20(0.10) = 2 < 5$$

The first condition is not met; therefore, the z-test statistic cannot be used with the sample size given. If it is possible to increase the sample size to at least 50, then the z-test statistic for population proportions could be used. Otherwise, other methods are necessary.

c. The claim refers to a population mean, therefore we must use one of the test statistics discussed in the previous sections. Note that the population standard deviation is unknown and the population is assumed to be normally distributed. Since the sample size is less than 30, a t-test statistic should be used.

Once we have chosen the appropriate test statistic, we must decide under what circumstances to reject the null hypothesis. In the previous section, we introduced the method of using p-values to draw a conclusion for a hypothesis test. In this section, we will continue to use this method, because it is the most popular method in research journals. However, if we wished to use rejection regions, the method is identical to using rejection regions for population means since they both use a z-test statistic.

Example 18

The local school board has been advertising that at least 65% of voters favor a tax increase to pay for a new school. A local politician believes that less than 65% of his constituents favor this tax increase. To test his claim, he asked 50 of his constituents whether they favor the tax increase and 27 said that they would vote in favor of the tax increase. If the politician wishes to be 95% confident in his conclusion, does this information support his claim?

Memory Booster!

p can represent either *population proportion* or *p-value*. You must pay attention to the context in which the symbol is used in order to determine its meaning.

Solution:

1. **State the null and alternative hypotheses.**
   The claim is that less than 65% of the constituents favor a tax increase. Written mathematically, this claim is $p < 0.65$. The logical opposite of this claim is $p \geq 0.65$. Thus the null and alternative hypotheses are

$$H_0: p \geq 0.65$$
$$H_a: p < 0.65.$$

2. **Set up the hypothesis test by choosing the test statistic and stating the level of significance.**
   We are testing a population proportion, so we must check the necessary conditions to use the normal distribution and the z-test statistic.

$$np = 50(0.65) = 32.5 \geq 5$$
$$n(1-p) = 50(1-0.65) = 17.5 \geq 5$$

   Since both conditions are satisfied, we can use the z-test statistic for population proportions

$$z = \frac{\hat{p} - p}{\sqrt{\frac{p(1-p)}{n}}}.$$

   We will draw a conclusion by computing the p-value for the calculated test statistic and comparing that value to α. For this hypothesis test,

$$\begin{aligned}\alpha &= 1 - c \\ &= 1 - 0.95 \\ &= 0.05.\end{aligned}$$

3. **Gather data and calculate the necessary sample statistics.**
   The sample data shows that 27 out of 50 constituents favor the tax increase. Thus

$$\hat{p} = \frac{27}{50} = 0.54.$$

Substituting the necessary values into the formula for the test statistic for population proportions, we obtain

$$\begin{aligned} z &= \frac{\hat{p} - p}{\sqrt{\frac{p(1-p)}{n}}} \\ &= \frac{0.54 - 0.65}{\sqrt{\frac{0.65(1-0.65)}{50}}} \\ &\approx -1.63. \end{aligned}$$

4. **Draw a conclusion by comparing the *p*-value to the level of significance.**
The alternative hypothesis tells us that we have a left-tailed test. Therefore, the p-value for this test statistic is the probability of obtaining a test statistic less than $z = -1.63$. To find the p-value, we need to find the area under the normal curve less than $z = -1.63$.

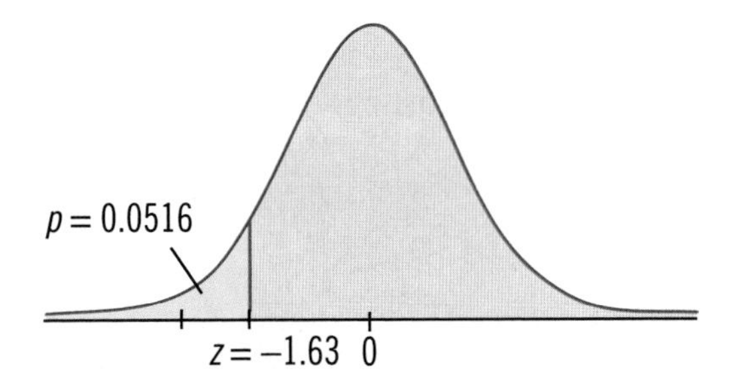

The p-value is 0.0516. Comparing this p-value to the level of significance, we see that 0.0516 > 0.05, so $p > \alpha$. Thus we fail to reject the null hypothesis. This means that at the 0.05 level of significance this evidence does not sufficiently support the politician's claim that less than 65% of his constituents favor a tax increase to pay for a new school.

Example 19

According to a report by the American Mathematical Association, 1% of bachelor's degrees in the U.S. are awarded in the field of mathematics. One recent mathematics graduate believes that this number is incorrect. She has no indication if the actual percentage is more or less than 1%. She decides to test her claim that the percentage of bachelor's degrees awarded in the field of mathematics is not 1% by using the 12,317 recent college graduates from public universities in her home state for her sample. According to the graduation programs from the 3 public universities in her home state, she notes that bachelor's degrees in mathematics were awarded to 148 graduates. Does this evidence support this graduate's claim that the percentage of bachelor's degrees awarded in the field of mathematics is not 1%? Use a 0.10 level of significance.

Source:
http://www.ams.org/cbms

cont'd on the next page

1. **State the null and alternative hypotheses.**
   The claim is that the percentage of bachelor's degrees awarded in the field of mathematics is not 1%. Written mathematically, this claim is that $p \neq 0.01$. The logical opposite of this claim is that $p = 0.01$. Thus, the null and alternative hypotheses are

$$H_0: p = 0.01$$
$$H_a: p \neq 0.01.$$

2. **Set up the hypothesis test by choosing the test statistic and stating the level of significance.**
   We are testing a population proportion, so we must check the necessary conditions to use the normal distribution and the z-test statistic.

$$np = 12{,}317(0.01) = 123.17 \geq 5$$
$$n(1-p) = 12{,}317(1-0.01) = 12{,}193.83 \geq 5$$

   Due to the large sample size, it should have been obvious that both of these conditions would be easily satisfied. Since both conditions are satisfied, we can use the z-test statistic for population proportions

$$z = \frac{\hat{p} - p}{\sqrt{\frac{p(1-p)}{n}}}.$$

   We will draw a conclusion by computing the p-value for the calculated test statistic and comparing that value to α. For this hypothesis test, we were told to use a level of significance of $\alpha = 0.10$.

3. **Gather data and calculate the necessary sample statistics.**
   The sample data shows that 148 out of 12,317 graduates obtained degrees in mathematics. Thus

$$\hat{p} = \frac{148}{12{,}317} \approx 0.012016.$$

   Substituting the necessary values into the formula for the test statistic for population proportions, we obtain

$$\begin{aligned} z &= \frac{\hat{p} - p}{\sqrt{\frac{p(1-p)}{n}}} \\ &= \frac{0.012016 - 0.01}{\sqrt{\frac{0.01(1-0.01)}{12{,}317}}} \\ &\approx 2.25. \end{aligned}$$

4. Draw a conclusion by comparing the *p*-value to the level of significance.

The alternative hypothesis tells us that we have a two-tailed test. Therefore, the *p*-value for this test statistic is the probability of obtaining a test statistic which is either less than $z = -2.25$ or greater than $z = 2.25$. To find the *p*-value, we need to find the area under the normal curve less than $z = -2.25$. This area is 0.0122. This area is just the area in the left tail. The total area is then twice this amount, or $0.0122(2) = 0.0244$.

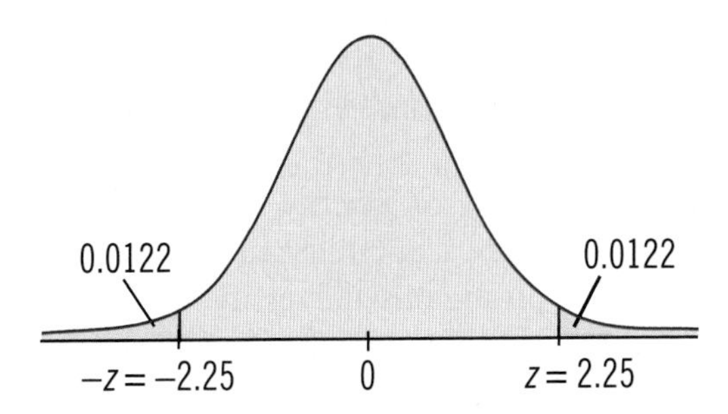

The *p*-value is 0.0244. Comparing this *p*-value to the level of significance, we see that $0.0244 < 0.10$, so $p < \alpha$. Thus we reject the null hypothesis. This means that, at the 0.10 level of significance, this evidence supports the graduate's claim that the percentage of bachelor's degrees awarded in the field of mathematics is not 1%. Notice that despite the sample proportion being larger than the assumed population proportion, our conclusion should reference only the original null and alternative hypotheses.

Exercises

10.4

Directions: Determine whether the *z*-test statistic for population proportions can be used to test a hypothesis under the given conditions.

1. $n = 43, p = 0.05$
2. $n = 21, p = 0.83$
3. $n = 54, p = 0.10$
4. $n = 97, p = 0.12$
5. $n = 52, p = 0.95$
6. $n = 39, p = 0.04$

Directions: Answer each of the following questions by performing a hypothesis test.

7. According to a report sponsored by the National Center for Health Statistics, 75% of American women have been married by the age of 30. Patrice, a single woman of 32 who has never been married, claims that this percentage is too high. To test her claim, she surveys 125 women between the ages of 18 and 30 and finds that 92 of them have been married. Does this evidence support Patrice's claim that less than 75% of women 30 years old or younger have been married? Use a 0.10 level of significance.
Source: http://www.cdc.gov/nchs/pressroom/02news/div_mar_cohab.htm

8. The National Academy of Science reported in a 1996 study that 40% of research in mathematics is published by U.S. authors. The mathematics chairperson of a prestigious university wishes to test the claim that this percentage is no longer 40%. He has no indication of whether the percentage has increased or decreased since that time. He surveys 130 of the latest articles published by reputable mathematics research journals and finds that 62 of these articles have U.S. authors. Does this evidence support the mathematics chairperson's claim that the percentage is no longer 40%? Use a 0.10 level of significance.
Source: http://www.nap.edu/openbook/NI000990/html

9. Sleep apnea is a condition in which the sufferers stop breathing momentarily while they are asleep. This condition results in lack of sleep and extreme fatigue during waking hours. A current estimate is that 22 million out of the 281 million Americans suffer from sleep apnea, or approximately 8%. A safety commission is concerned about the percentage of commercial truck drivers who suffer from sleep apnea. They do not have any reason to believe that it would be higher or lower than the population's percentage. To test the claim that the percentage of commercial truck drivers who suffer from sleep apnea is not 8%, 350 commercial truck drivers are interviewed by a medical expert, who concludes that 39 suffer from sleep apnea. Does this evidence support the claim that the percentage of commercial truck drivers who suffer from sleep apnea is not 8%? Use a 0.01 level of significance.
Source: http://www.sleepapnea.org, http://www.factfinder.census.gov

10. Sleep experts believe that sleep apnea is more likely to occur in men than in the general population. In other words, they claim that the percentage of men who suffer from sleep apnea is greater than 8%. To test this claim, one sleep expert interviews 75 men and determines that 10 of these men suffer from sleep apnea. Does this evidence support the claim that the percentage of men who suffer from sleep apnea is greater than 8%? Use a 0.05 level of significance.
Source: http://www.sleepapnea.org

11. A direct mail appeal for contributions from a university's alumni and supporters is considered cost effective if more than 15% of the alumni and supporters provide monetary contributions. To determine if a direct mail appeal is cost effective, the fundraising director sends the direct mail brochures to a random sample of 250 people on the alumni and supporters mailing lists. He receives monetary contributions from 40 people. Does this evidence support the cost effectiveness of a direct mail appeal? Use a 0.05 level of significance.

12. A study of HIV infection rates among intravenous drug users by a medical researcher reported that the rate of new infections has steadily dropped between the 1970's and 1990's. The latest report was that 2% of IV drug users are HIV positive. A drug counselor believes that this trend is continuing and that the percentage of IV drug users who are HIV positive is now less than 2%. The counselor surveys 415 IV drug users in 10 clinics across the U.S. and finds that 8 are HIV positive. Does this evidence support the counselor's claim? Use a 0.01 level of significance. (Round $\hat{p}$ to three decimal places.)
Source: http://www.thebody.com/schoofs/ivdrug.html from 1998

Types of Errors

10.5

Thus far in this chapter, we have learned how to complete a hypothesis test for population means and population proportions. Recall that there are only two possible conclusions to any hypothesis test: to reject the null hypothesis or to fail to reject the null hypothesis. When wording this conclusion you must be careful to state only the findings and not be tempted to say more than the evidence supports. Remember the analogy of the United States judicial system. In a U.S. trial, a finding of not guilty does not indicate that the defendant is innocent. It only means that the evidence presented in the trial did not support a finding of guilty. Similarly, if we fail to reject the null hypothesis, then we cannot say that the null hypothesis is supported. We can only say that the evidence does not support the alternative hypothesis. It is important to remember that no matter what the conclusion is, neither hypothesis has been proven.

Let's look at an example of a student scheduling classes for the next semester. Suppose this student is currently enrolled in English Composition I (Comp I). Next semester, the student will need to take English Composition II (Comp II), unless he fails Comp I. A potential problem arises during pre-registration. The student has not finished Comp I; therefore, he needs to use the evidence he has, i.e. his grades thus far, to draw a conclusion about whether or not he will pass Comp I. Let's give the student the benefit of the doubt, and assume that he will pass Comp I. Thus the null and alternative hypotheses can be written as follows:

H_0: The student passes Comp I.
H_a: The student does not pass Comp I.

If the student's grades are mostly F's, then we might conclude that the evidence is sufficient to reject the null hypothesis, supporting the alternative hypothesis that the student will not pass Comp I and therefore must repeat Comp I the following semester. If the final grade in the class is an F, then a correct decision was made. If the final grade is a C, then an error was made, and the student would need to correct his pre-registration schedule.

On the other hand, the student's grades could be mostly passing grades. In this case, we might conclude that the evidence is not sufficient to reject the null hypothesis. Thus we would fail to reject the null hypothesis and the student would schedule Comp II for the following semester. If the final grade in the class turns out to be a passing grade, then a correct decision was made. If the student's grade is an F, then an error was made, and the schedule would need to be corrected. The following table summarizes these four possible situations.

		Reality	
		The student passes Comp I.	**The student does not pass Comp I.**
Decision	**Comp I is scheduled for next semester.**	Wrong Schedule!	Schedule correct. Student retakes Comp I.
	Comp II is scheduled for next semester.	Schedule correct. Student takes Comp II.	Wrong Schedule!

Notice that there are two types of errors possible: 1) rejecting a true null hypothesis or 2) failing to reject a false null hypothesis. We call rejecting a true null hypothesis a **Type I error** and failing to reject a false null hypothesis is called a **Type II error**.

Replacing the specific statements in the table with general statements about the null hypothesis, the table becomes

		Reality	
		H_0 is true.	**H_0 is false.**
Decision	**H_0 is rejected.**	Type I error	Correct Decision
	H_0 is not rejected.	Correct Decision	Type II error

The decision criteria used in a hypothesis test are based on the idea of minimizing the chance of committing a Type I error. The probability of committing a Type I error, in other words the probability of rejecting a true null hypothesis, is called the **level of significance**, which we denote with the Greek symbol α. We have already seen this value in this chapter. This is the same level of significance used to draw the conclusions in hypothesis tests in the previous sections.

The level of significance is how willing we are to make a Type I error. In the case of creating a class schedule, how willing would you be to create an incorrect class schedule? Would you be willing to accept a 5% chance of being wrong? In this situation, it might depend on the amount of time and effort required to change a class schedule at your institution. If a computerized system allows changes to be made with minimal effort, you might be willing to make an error 10% of the time. If changing your schedule requires trips to several professors' offices and the registrar, then you may only be willing to accept a 5%, or even a 1% chance of making an error. The smaller the level of significance, the more evidence that will be required to reject the null hypothesis. The most common practice in academic research is to use a level of significance of 0.05, if one is stated at all. Many times, as we noted earlier, research does not provide a value for the level of significance, but simply states the *p*-value, thus allowing readers to determine for themselves whether the null hypothesis should be rejected.

Memory Booster!

α = probability of a Type I error

β = probability of a Type II error

Technically, you can choose the level of significance, α, to be any value you wish. So why wouldn't we choose the level of significance to be as low as possible? The answer lies in the inverse relationship between the probability of making a Type I error and the probability of making a Type II error. The probability of making a Type II error is represented by the Greek symbol β. The smaller the probability of making a Type I error, the larger the probability of making a Type II error will be. Thus you must choose a level of significance small enough to control the risk of a Type I error, while not choosing it too small, which would increase the chance of making a Type II error.

Example 20

A television executive believes that at least 99% of households own television sets. An intern at her company is given the task of testing the claim that the percentage is actually less than 99%. The hypothesis test is completed, and based on the sample collected the intern decides to fail to reject the null hypothesis. If, in reality, 97.5% of households own a television set, was an error made? If so, what type?

Solution:

Let's begin by writing the null and alternative hypotheses. The claim is that less than 99% of households own a television. Mathematically, this is written as $p < 0.99$. The opposite statement would be $p \geq 0.99$. The second statement describes the starting assumption, thus making it the null hypothesis for the test.

$$H_0: p \geq 0.99$$
$$H_a: p < 0.99$$

The hypothesis test fails to reject the null hypothesis. However, since the reality is that the null hypothesis is false, the intern failed to reject a false null hypothesis. This is a Type II error.

Example 21

Insurance companies commonly use 2000 miles as the average number of miles a car is driven per month. One insurance company claims that due to our more mobile society, the average is more than 2000 miles per month. The insurance company tests their claim with a hypothesis test and decides to reject the null hypothesis. In reality, the average number of miles a car is driven per month is 2250 miles. Was an error made? If so, what type?

Solution:

Begin by writing the null and alternative hypotheses. The claim is that the average is more than 2000, which is $\mu > 2000$. The opposite is $\mu \leq 2000$. Thus the null and alternative hypotheses are

$$H_0: \mu \leq 2000$$
$$H_a: \mu > 2000.$$

The decision was to reject the null hypothesis. The null hypothesis is false, so the decision was to reject a false null hypothesis, which is a correct decision.

Example 22

A study on the effects of television viewing on children reports that children watch an average of 4 hours of television per night. Kiko believes that the average number of hours children in her neighborhood watch television per night is not 4. She performs a hypothesis test and rejects the null hypothesis. In reality, children in her neighborhood do watch an average of 4 hours of television per night. Did she make an error? If so, what type?

Solution:

Begin by writing the null and alternative hypotheses. Kiko's claim is that the average is not 4, written mathematically as $\mu \neq 4$. The opposite is $\mu = 4$. Thus the null and alternative hypotheses are

$$H_0: \mu = 4$$
$$H_a: \mu \neq 4.$$

The decision was to reject the null hypothesis, when in reality it was true. This is a Type I error.

Exercises

10.5

Directions: Decide whether the following statements are **true** or **false**. If the statement is false, tell why it is false.

1. A Type I error is made when a true null hypothesis is rejected.

2. A Type II error is made when you fail to reject a true null hypothesis.

3. The level of significance is the probability of making a Type II error.

4. The probability of making a Type I error is inversely related to the probability of making a Type II error.

Directions: For each of the following scenarios, determine the type of error that was made, if any. (Hint: Begin by determining the null and alternative hypotheses.)

5. The percentage of children that leave foster care due to adoption is generally accepted to be 32%. A social worker claims that this percentage is incorrect. After performing a hypothesis test on his claim, he fails to reject the null hypothesis. In reality, the percentage of children leaving foster care due to adoption is 33%. Was an error made? If so, what type?

6. A software company advertises that its software can improve students' grades by 15%. One student conducts a hypothesis test to see if the percentage increase is less than 15%. The conclusion of the hypothesis test is to reject the null hypothesis. If the true percentage increase in grades for all students using the software is 13%, was an error committed?

7. A new whitening toothpaste advertises that it whitens teeth up to 3 shades whiter. The product has been so successful that the company wants to change their slogan to be more than 3 shades whiter. Before they do so, they want to test this new claim with a hypothesis test. According to the sample obtained, the decision is to fail to reject the null hypothesis. If the toothpaste in reality whitens teeth on average just 3 shades whiter, was an error made with the hypothesis test?

8. The national average for a gallon of unleaded gasoline is considered to be \$2.87 in August 2006. One consumer advocacy group claims that the average is more than \$2.87. A hypothesis test of the advocacy group's claim results in rejecting the null hypothesis. If the true average for a gallon of gasoline is \$2.91, was an error made? If so, what type?

9. The math department at one university reports that the failure rate for college algebra is no more than 35% in any given semester. A group of students believes the rate is higher, and they decided to use a hypothesis test to see if they are correct. Based on the sample they obtain, they decided to reject the null hypothesis. If, in reality, the failure rate is 34%, did the students make an error?

10. Dreamfilms Studios boasts that their summer features are so good that over half of all tickets sold one summer are for their films. A competing studio claims that less than half of all tickets sold are for Dreamfilms movies. After the hypothesis test is completed, the conclusion is to fail to reject the null hypothesis. If the true percentage of tickets sold for Dreamfilms Studios is only 48%, was an error committed?

11. An ice cream company must keep the temperature inside their delivery trucks at 30 degrees Fahrenheit. If the temperature is too warm, the ice cream will melt. The company also does not want to allow the temperature to be too low, because they do not want to waste the cost and energy it takes to keep the truck cool. Periodically the company runs a hypothesis test to ensure that the temperature is as claimed. Suppose that after one hypothesis test, the company fails to reject the null hypothesis. In reality, the average temperature of their trucks is 29 degrees Fahrenheit. Was an error committed? If so, what type?

12. In 2003, the National Assessment of Adult Literacy (NAAL), conducted by the U.S. Department of Education, found that fourteen percent of American adults are functionally illiterate. A local organization promoting literacy claims that they have impacted their community and the illiteracy rate in their community is lower than the national average. A hypothesis test of their claim results in rejecting the null hypothesis. If, in reality, the illiteracy rate in their community is 13%, was an error made? If so, what type?
Source: http://en.wikipedia.org/wiki/Literacy#United_States

Hypothesis Testing for Population Variance

10.6

In the previous sections, we looked at testing claims involving either the population mean or the population proportion of certain sets of data. However, just as we stated in the beginning of the chapter, hypothesis testing is a procedure for evaluating claims about population parameters in general, not just the mean and proportion.

Quite often, the variance of a population needs to be evaluated. Suppose for a moment you are a prominent artist. Your work becomes more valuable if you produce one-of-a-kind pieces of art. In other words, you want each piece to vary from the others in some way, and variance is a good thing. If however, you are a pharmaceutical company executive monitoring the quality of each drug you produce, you want to reproduce the same product over and over with the smallest possible variance between the pills. In this case the variance of the population is of particular interest and worthy of being evaluated.

In the same manner that we tested population means and proportions, we can look at hypothesis testing for population variance. Since the steps in hypothesis testing remain the same, let's look at what is different in hypothesis testing for the population variance. First, the following criteria must be met before a hypothesis test for variance can be performed.

- All possible samples of a given size must have an equal probability of being chosen, and
- The population must be normally distributed.

Now let's review the basic steps in a hypothesis test. Notice that only the first two steps have differences specific to hypothesis testing for population variance or standard deviation.

Hypothesis Testing for Population Variance:

1. **State the null and alternative hypotheses.**
   This will now require using the symbol for population variance, σ^2, or the symbol for population standard deviation, σ.

2. **Set up the hypothesis test by choosing the test statistic and determining the values of the test statistic that would lead to rejecting the null hypothesis.**
   The test statistic for population variance, or population standard deviation, is

$$\chi^2 = \frac{(n-1)s^2}{\sigma^2}.$$

This test statistic follows a chi-square distribution, and with it we will use rejection regions to draw a conclusion. We can compare the value of the test statistic to the critical value of chi-square found in Table G in Appendix A. Critical values for the chi-square distribution are found in much the same way as the critical value for the Student t-distribution. To find the critical value, we will need to find the number of degrees of freedom, which is $d.f. = n - 1$. To determine the rejection region associated with a specific critical value, we need to consider the type of test we are performing: left-tailed, right-tailed, or two-tailed.

- For a left-tailed test, reject H_0 when $\chi^2 \le \chi^2_{1-\alpha}$.
- For a right-tailed test, reject H_0 when $\chi^2 \ge \chi^2_{\alpha}$.
- For a two-tailed test, reject H_0 when $\chi^2 \le \chi^2_{1-\frac{\alpha}{2}}$ or $\chi^2 \ge \chi^2_{\frac{\alpha}{2}}$.

3. **Gather data and calculate the necessary sample statistics.**

4. **Draw a conclusion using the process described in Step 2.**

Let's look at a few examples of hypothesis tests for population variance.

Example 23

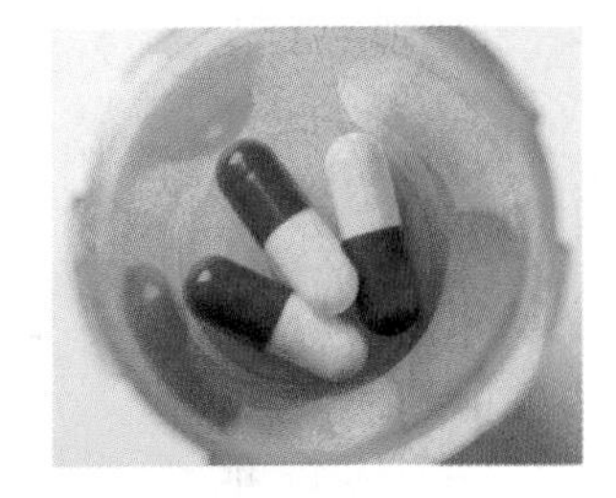

Suppose your pharmacy currently buys Drug A, a medication for high blood pressure. However, a new company says it has a "better" drug for blood pressure than Drug A. Both drugs have exactly the same active ingredient in them. The new company claims its pill is better because the variance in the amount of the active ingredient between its pills is smaller than in Drug A, which is known to have a variance of 0.0009. To test the new company's claim, 100 of the new pills are randomly selected and the amount of the active ingredient is found to have a mean of 2.47 mg and a standard deviation of 0.026 mg. Is this enough evidence, at the 0.01 level of significance, to support the new company's claim that its drug has a smaller variance in the amount of the active ingredient? Assume the amount of the active ingredient is normally distributed.

Solution:

1. **State the null and alternative hypotheses.**
   The new drug company's claim is that its drug's variance is smaller than 0.0009, which is written as $\sigma^2 < 0.0009$. The opposite of this claim is $\sigma^2 \ge 0.0009$. Thus the null and alternative hypotheses are

$$H_0: \sigma^2 \ge 0.0009$$
$$H_a: \sigma^2 < 0.0009.$$

cont'd on the next page

2. **Set up the hypothesis test by choosing the test statistic and determining the values of the test statistic that would lead to rejecting the null hypothesis.**

   Since we are testing a population variance, we must use the test statistic χ^2. This is a left-tailed test, so the rejection region is $\chi^2 \le \chi^2_{1-\alpha}$. The level of significance is $\alpha = 0.01$ and the number of degrees of freedom is $d.f. = n - 1 = 100 - 1 = 99$. Since the table only gives multiples of 10 after 30 degrees of freedom, we will approximate the critical value to the closest number of degrees of freedom given. The closest value to 99 is 100, so we will use $d.f. = 100$. Thus the critical value is $\chi^2_{0.990} = 70.065$. Therefore, reject H_0 if $\chi^2 \le 70.065$.

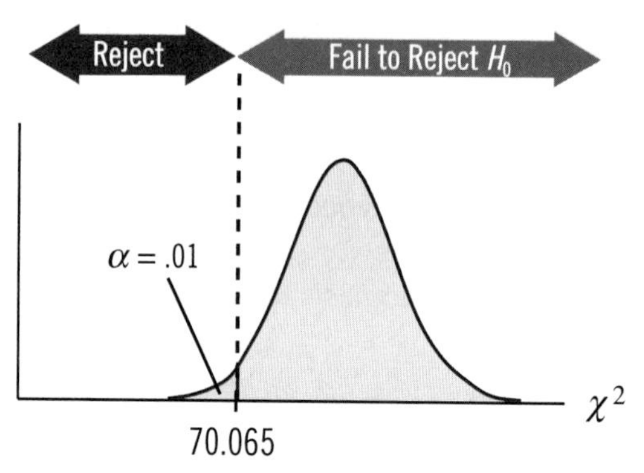

3. **Gather data and calculate the necessary sample statistics.**

   Using the information provided in the problem, we know that $n = 100$, $s^2 = (0.026)^2 = 0.000676$, and $\sigma^2 = 0.0009$. Substituting these values into the formula for our test statistic for population variance, we obtain

$$\chi^2 = \frac{(n-1)s^2}{\sigma^2}$$
$$= \frac{(100-1)(0.000676)}{(0.0009)}$$
$$= 74.360.$$

4. **Draw a conclusion using the process described in Step 2.**

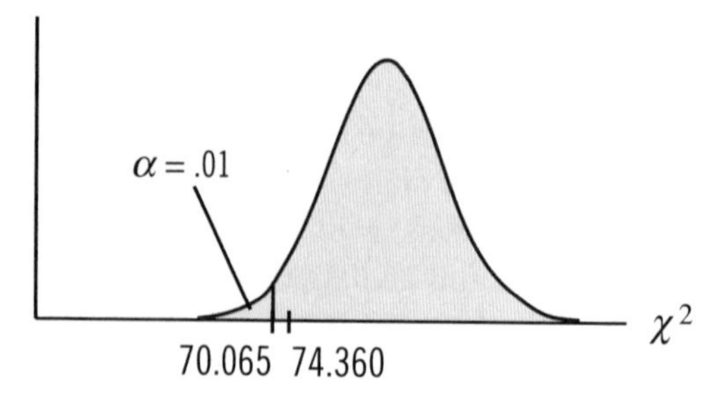

   Since $\chi^2 = 74.360$, which is not in the rejection region, we fail to reject the null hypothesis. Thus the evidence does not sufficiently support the new drug company's claim that the variance in the amount of the active ingredient in their drug is less than 0.0009 at the 0.01 level of significance.

Example 24

A manufacturer of golf balls requires that the weights of their golf balls have a standard deviation that will not exceed 0.08 ounces. One of the quality control inspectors claims that the machines need to be recalibrated because he believes the standard deviation is more than 0.08 ounces. To test this claim, he randomly selects 30 golf balls off the assembly line and finds that they have a mean weight of 1.62 ounces and a standard deviation of 0.0804. Does this evidence support the inspector's claim at the 0.05 level of significance?

Solution:

1. **State the null and alternative hypotheses.**
   The inspector's claim is that the standard deviation is more than 0.08 ounces, which is written as $\sigma > 0.08$. The opposite of this claim is $\sigma \leq 0.08$. Because the established standard is contained in the second hypothesis, it is the null hypothesis. Thus the hypotheses are

$$H_0\text{: } \sigma \leq 0.08$$
$$H_a\text{: } \sigma > 0.08.$$

2. **Set up the hypothesis test by choosing the test statistic and determining the values of the test statistic that would lead to rejecting the null hypothesis.**
   Since we are testing a population standard deviation, we must use the test statistic χ^2. This is a right-tailed test, so the rejection region is $\chi^2 \geq \chi^2_\alpha$. The level of significance is $\alpha = 0.05$ and the number of degrees of freedom is $d.f. = n - 1 = 30 - 1 = 29$. Thus the critical value is $\chi^2_{0.050} = 42.557$. Therefore, reject H_0 if $\chi^2 \geq 42.557$.

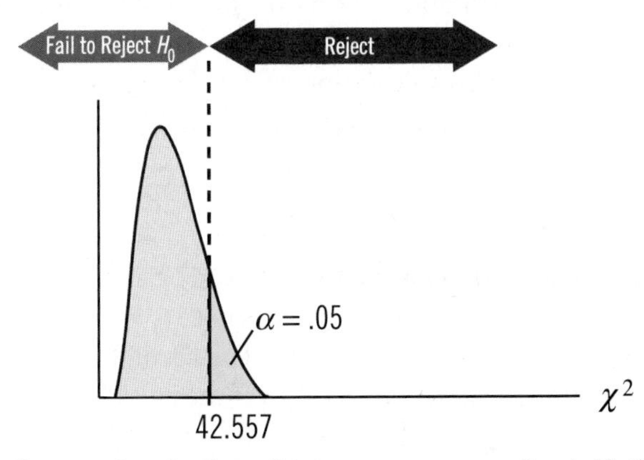

3. **Gather data and calculate the necessary sample statistics.**
   Using the information provided in the problem, we know that $n = 30$, $s = 0.0804$, and $\sigma = 0.08$. Substituting these values into the formula for our test statistic, we obtain

cont'd on the next page

$$\chi^2 = \frac{(n-1)s^2}{\sigma^2}$$

$$= \frac{(30-1)(0.0804)^2}{(0.08)^2}$$

$$\approx 29.291.$$

4. **Draw a conclusion using the process described in Step 2.**

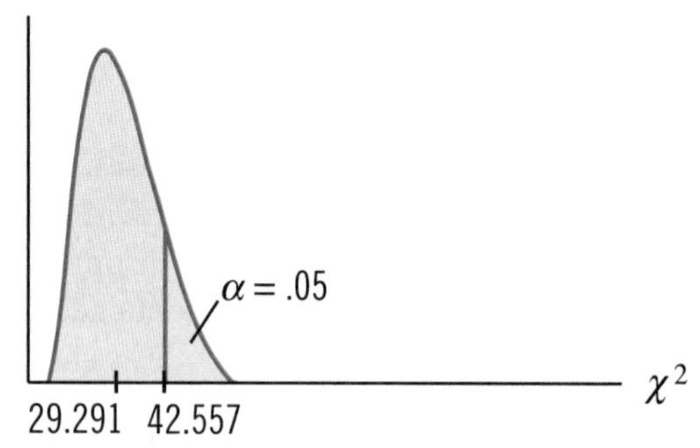

Since $\chi^2 = 29.291$, which is not in the rejection region, we fail to reject the null hypothesis. Thus the evidence does not sufficiently support the inspector's claim that the standard deviation is more than 0.08 ounces at the 0.05 level of significance.

Example 25

A tree farmer would like to know how much variance he gets in the heights of trees 2 years of age on his farm. It is generally accepted by botanists that after two years of growth the variance in the height of this variety of tree should be 16 feet. The farmer claims that after 2 years of growth, the variance in the heights of his trees is not 16 feet. He randomly selects forty 2-year-old trees on his farm and finds they have a standard deviation of 3.2 feet. Does this evidence, at the 0.10 level of significance, support the farmer's claim?

Solution:

1. **State the null and alternative hypotheses.**
   The farmer's claim is that the variance is not 16, which is written as $\sigma^2 \neq 16$. The opposite of this claim is $\sigma^2 = 16$. The second hypothesis contains the starting assumption, so it is the null hypothesis. Thus the hypotheses are

$$H_0: \sigma^2 = 16$$
$$H_a: \sigma^2 \neq 16.$$

2. **Set up the hypothesis test by choosing the test statistic and determining the values of the test statistic that would lead to rejecting the null hypothesis.**

   Since we are testing a population variance, we must use the test statistic χ^2.

   This is a two-tailed test, so the rejection region is $\chi^2 \le \chi^2_{1-\frac{\alpha}{2}}$ or $\chi^2 \ge \chi^2_{\frac{\alpha}{2}}$.

   The level of significance is $\alpha = 0.10$. The number of degrees of freedom is $d.f. = n - 1 = 40 - 1 = 39$. Since the table only gives multiples of 10 after 30 degrees of freedom, we will approximate the critical value to the closest number of degrees of freedom given. The closest value to 39 is 40, so we will use $d.f. = 40$. The left-hand critical value is $\chi^2_{0.950} = 26.509$ and the right-hand critical value is $\chi^2_{0.050} = 55.758$. Thus reject H_0 if $\chi^2 \le 26.509$ or $\chi^2 \ge 55.758$.

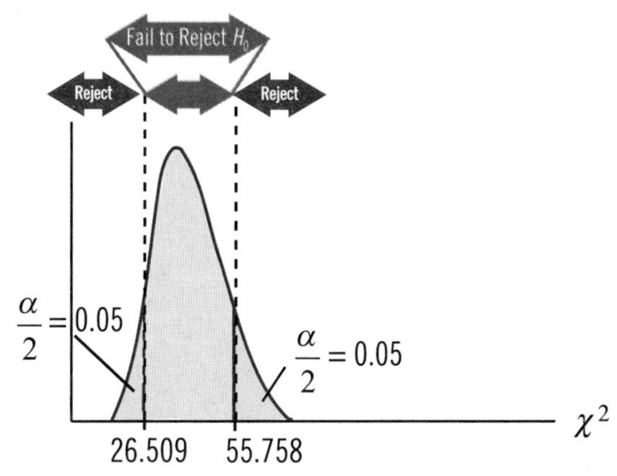

3. **Gather data and calculate the necessary sample statistics.**

   Using the information provided in the problem, we know that $n = 40$, $s = 3.2$, and $\sigma^2 = 16$. Substituting these values into the formula for our test statistic for population variance, we obtain

$$\chi^2 = \frac{(n-1)s^2}{\sigma^2} = \frac{(40-1)(3.2)^2}{(16)} = 24.960.$$

4. **Draw a conclusion using the process described in Step 2.**

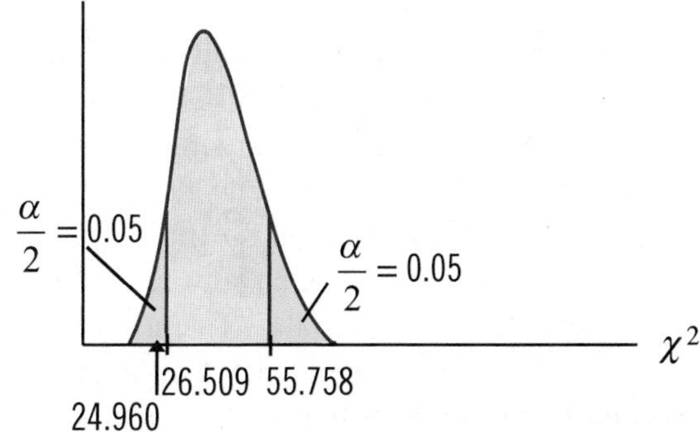

cont'd on the next page

Since $\chi^2 = 24.960$, which is in the rejection region, we reject the null hypothesis. Thus the evidence sufficiently supports the farmer's claim that the variance is not 16 at the 0.10 level of significance.

Exercises

10.6

Directions: Find the critical value(s) of the chi-square distribution for the following situations. Use the critical value(s) to determine the rejection region for each test.

1. $n = 27$, $\alpha = 0.05$, for a left-tailed test
2. $n = 16$, $\alpha = 0.10$, for a left-tailed test
3. $n = 42$, $\alpha = 0.01$, for a left-tailed test
4. $n = 18$, $\alpha = 0.05$, for a right-tailed test
5. $n = 31$, $\alpha = 0.10$, for a right-tailed test
6. $n = 58$, $\alpha = 0.01$, for a right-tailed test
7. $n = 25$, $\alpha = 0.05$, for a two-tailed test
8. $n = 30$, $\alpha = 0.01$, for a two-tailed test
9. $n = 67$, $\alpha = 0.10$, for a two-tailed test

Directions: Perform hypothesis tests, completing each of the following steps. Assume that all populations are normally distributed and that all possible samples of a given size have an equal probability of being chosen.

a. State the null and alternative hypotheses.
b. Determine the rejection region.
c. Calculate the test statistic, χ^2.
d. Draw a conclusion and interpret your results.

10. A dairy supplier fills hundreds of cartons a day with milk. He is contracted to fill each carton with exactly 1 gallon of milk. Because of the moving parts on the machinery that fills the cartons, the amount of milk dispensed begins to vary slightly over time. When this happens, the machinery must be serviced to realign it correctly. A variance of no more than 4 mL is acceptable. Servicing must occur when the variance between the cartons is more than 4 mL with a level of significance of 0.01. Twenty-five cartons are randomly chosen to be tested. If they are found to have a standard deviation of 2.5 mL, perform a hypothesis test to determine if the machine needs servicing.

11. A potato chip manufacturer produces bags of potato chips that have a net weight of 326 grams. Because the chips vary in size, it is difficult to fill the bags to the exact weight desired. However, the bags pass inspection so long as the standard deviation of their weights is no more than 3 grams. A quality control inspector wished to test the claim that one batch of bags has a standard deviation of more than 3 grams, and thus does not pass inspection. If a sample of 25 bags of potato chips is taken and the standard deviation is found to be 3.4 grams, does this evidence, at the 0.05 level of significance, support the claim that the bags should fail inspection?

12. At a shooting range, instructors can determine if a shooter is consistently missing the target because of his gun sight or because of his ability. If a gun's sight is off, the variance between shots will be small (even if the shots are not in the center of the target). A student claims that it is the sight that is off, not his aim, and wants the instructor to confirm his claim. If a skilled shooter fires a gun with a properly adjusted sight, the shot pattern will have a variance of no more than 0.33 mm. After the instructor shoots 20 shots at the target, the instructor calculates the variance between his shots to be 0.48 mm. Does this evidence support the student's claim that the gun's sight is off? Use a 0.10 level of significance.

13. A school district wants to make sure that the variance of the standard test scores at district schools is less than 0.05. To test this claim they looked at comprehensive test scores from the standardized testing done at each of the 18 schools in the district. The results from the survey found that the overall mean is a score of 192.56 with a standard deviation of 0.162. With $\alpha = 0.10$, perform a hypothesis test to determine if the variance is less than 0.05.

14. The manufacturer of a vaccine must ensure that the shots it sells contain 0.001 viral particles per vaccine. It is also essential that the variance of the active agent be less than 0.00002 viral particles per vaccine. For testing purposes, a random sample of 100 vaccines is taken, and the variance of the sample is found to be 0.0000175. Does the evidence support the claim that the variance is within the necessary bounds, at the 0.01 level of significance?

15. Claudia's bakery business has taken off. Her goods are so popular that she has decided to invest in kitchen equipment that will do some of the work for her. One machine in particular prepares the bread dough, and it is set to measure 8 grams of yeast per batch. Because baking requires precise measurement, she wants to test the new machine to make sure that the variance between batches is less than 0.3, at the 0.05 level of significance. A sample of 8 batches has a variance of 0.25. Does this evidence support the claim that the new machine produces dough with a variance between batches of less than 0.3?

16. The variance of temperatures in chicken incubators on a chicken farm is generally believed to be 0.5 degrees. The manager of the chicken farm claims that the variance has changed. He randomly tests 25 of his incubators and finds they have a standard deviation of 0.55 degrees. Does this evidence support the manager's claim that the variance is not 0.5 degrees, at the 0.05 level of significance?

17. A grocery store needs the refrigeration section to have its coolers stay at the same temperature on a daily basis with little variance to help ensure quality. The manager of the store assumes that the variance in temperature is 3.8 degrees. The assistant manager claims that the variance is not 3.8 degrees and decides to test his claim using a hypothesis test. Over the course of 30 days, he monitors the temperature and finds that the standard deviation is 2.9 degrees. At the 0.01 level of significance, does this evidence support the claim that the variance in the temperature is not 3.8 degrees?

18. A health club needs to ensure that the temperature in their heated pool stays constant throughout the winter months. Otherwise they need to invest in a new heater for the pool. It is assumed that the water temperature has a variance of 2.25 degrees, which is considered to be within normal limits for a properly operating heater. The pool manager needs to determine if the variance is still 2.25 degrees, so she performs a hypothesis test to test the claim that the variance in the temperature is not 2.25 degrees. After testing the pool water for 15 days, the pool manager finds a daily average temperature of 78.6 degrees with a variance of 3.81 degrees. At the 0.10 level of significance, does this evidence support the manager's claim that the variance in the water temperature is no longer 2.25 degrees?

Chi-Square Test for Goodness of Fit

10.7

In the previous section we used chi-square as a test statistic when presented with a hypothesis test involving population variance. These same critical values can also be used when we are trying to compare results that an experiment produces with what theoretically should be expected.

Consider the following scenario. Suppose you want to test if a die is actually fair (i.e. not tampered with in any way). If we record the outcomes of rolling the die one hundred times, we could compare them to what the expected outcomes should be. In other words, we would expect each number on the die to be rolled about $\frac{1}{6}$ of the time, or the expected value, $E(X)=\frac{100}{6}\approx 17$ out of 100 in our experiment. Let's say the following is what our experiment produced:

Table 2 – Results from Die Experiment

Number on Die	Number of Times it Appears
1	17
2	18
3	8
4	14
5	25
6	18
Total	100

Now, although each number did not get rolled exactly 17 times, are the results close enough for the die to be considered fair, or were the results far enough "off" to say the die is not fair? For testing purposes, it seems reasonable to look at the differences between the expected values and the actual values. However, because of the negatives and positives canceling each other out, we cannot simply add them up. In addition, we must also take into account that, unlike our example, in general not all of the possible outcomes may have the same probability. To adjust for both of these factors when calculating the test statistic, we will square the differences and divide each by its expected value as shown below:

$$\chi^2=\sum\frac{\left[N_i-E(N_i)\right]^2}{E(N_i)}\text{, where } i = i^{\text{th}} \text{ data value.}$$

For this example, the formula would have six terms as shown in the equation on the following page.

$$\chi^2 = \frac{[N_1 - E(N_1)]^2}{E(N_1)} + \frac{[N_2 - E(N_2)]^2}{E(N_2)} + \frac{[N_3 - E(N_3)]^2}{E(N_3)} + \frac{[N_4 - E(N_4)]^2}{E(N_4)} + \frac{[N_5 - E(N_5)]^2}{E(N_5)} + \frac{[N_6 - E(N_6)]^2}{E(N_6)}$$

As the formula suggests, this unlikely looking quantity is also actually distributed according to the chi-square distribution.

Let's calculate χ^2 for the previous example:

$$\chi^2 = \frac{\left[17 - \frac{100}{6}\right]^2}{\frac{100}{6}} + \frac{\left[18 - \frac{100}{6}\right]^2}{\frac{100}{6}} + \frac{\left[8 - \frac{100}{6}\right]^2}{\frac{100}{6}} + \frac{\left[14 - \frac{100}{6}\right]^2}{\frac{100}{6}} + \frac{\left[25 - \frac{100}{6}\right]^2}{\frac{100}{6}} + \frac{\left[18 - \frac{100}{6}\right]^2}{\frac{100}{6}}$$
$$= 9.32$$

As you can see from the formula, if the observed values are consistently far enough from their expected values, the result will be a large value of χ^2. The question is, "How large is too large?" We can use the critical values for the chi-square distribution to help us here. Recall that the chi-square distribution only has one parameter, the number of degrees of freedom. For this section, the number of degrees of freedom will be one fewer than the number of possible outcomes. In the previous example, there are six possible outcomes when we roll the die, so the degrees of freedom equate to $d.f. = 6 - 1 = 5$.

As before, we will use the table of critical values for the chi-square distribution. If our test statistic is larger than the critical value, reject the hypothesis that our outcomes are close enough to the expected outcome.

Rejection Rule:

If $\chi^2 \geq \chi^2_\alpha$, then reject H_0.

Let's use a level of significance of 0.10 and test the die.

d.f.	$\chi^2_{0.975}$	$\chi^2_{0.950}$	$\chi^2_{0.900}$	$\chi^2_{0.100}$	$\chi^2_{0.050}$
1	0.001	0.004	0.016	2.706	3.841
2	0.051	0.103	0.211	4.605	5.991
3	0.216	0.352	0.584	6.251	7.815
4	0.484	0.711	1.064	7.779	9.488
5	0.831	1.145	1.610	9.236	11.071
6	1.237	1.635	2.204	10.645	12.592
7	1.690	2.167	2.833	12.017	14.067

As shown in the previous table, the critical value is $\chi^2_{0.100} = 9.236$. Since the test statistic value is greater than this value, we must conclude that our throws were significantly "off" and therefore, at a confidence level of 90%, our die is not fair.

Before working some examples of goodness of fit, let's look at how to write the null and alternative hypotheses for these problem types. Recall that the null hypothesis contains the established standard. Therefore, when stating the null, the assumption is that the observed data does follow the theoretical distribution. In words we state the hypotheses as:

H_0: The data matches the expected values.
H_a: The data does not match the expected values.

Mathematically, when the expected probabilities are all the same, we can write the hypotheses as:

H_0: $p_1 = p_2 = \ldots = p_n$
H_a: any difference amongst the probabilities.

If the expected probabilities differ from one another, we write the hypotheses as:

H_0: p_1 = probability of N_1, p_2 = probability of N_2, ..., p_n = probability of N_n
H_a: any difference in the stated probabilities.

Now that we've discussed the concepts behind a Goodness of Fit test, let's work an example.

Example 26

The local bank wants to evaluate the usage of its ATM machine. Currently they assume that the ATM is used consistently throughout the week, including weekends. They decide to use statistical inference with a 0.05 level of significance to test their procedures and find out if they are meeting the needs of their customers. During a 2-week period, the bank recorded the customer usage on each day. The results are listed in the table below:

Table 3 – Results from ATM Experiment	
Monday	38
Tuesday	33
Wednesday	41
Thursday	25
Friday	22
Saturday	38
Sunday	45

cont'd on the next page

Solution:

For stating the hypotheses to be tested, take the null hypothesis to be that the proportion of customers is the same for each day.

H_0: The proportion of customers does not vary by the day of the week.
H_a: The proportion of customers does vary by the day of the week.

Mathematically, we can write the null and alternative hypotheses as

H_0: $p_1 = p_2 = p_3 = p_4 = p_5 = p_6 = p_7$
H_a: any difference amongst the probabilities.

Before we begin to calculate the test statistic, let's calculate the expected value for each day of the week. Since we are assuming that the number of customers does not vary for each day, then their probabilities will be the same, i.e.

$$E(X) = np$$
$$E(X) = 242 \cdot \frac{1}{7}$$
$$E(X) = \frac{242}{7}.$$

Where n = 242 (the total number of customers who used the ATM)

Let's calculate χ^2.

$$\chi^2 = \frac{\left[38 - \frac{242}{7}\right]^2}{\frac{242}{7}} + \frac{\left[33 - \frac{242}{7}\right]^2}{\frac{242}{7}} + \frac{\left[41 - \frac{242}{7}\right]^2}{\frac{242}{7}} + \frac{\left[25 - \frac{242}{7}\right]^2}{\frac{242}{7}}$$
$$+ \frac{\left[22 - \frac{242}{7}\right]^2}{\frac{242}{7}} + \frac{\left[38 - \frac{242}{7}\right]^2}{\frac{242}{7}} + \frac{\left[45 - \frac{242}{7}\right]^2}{\frac{242}{7}}$$
$$= 12.3140$$

Our degrees of freedom calculation is $d.f. = 7 - 1 = 6$, and $\alpha = 0.05$. Using the table, the critical value is $\chi^2_{0.050} = 12.592$. Comparing the two, $\chi^2 < \chi^2_{0.050}$, so we must fail to reject our null hypothesis. In other words, the bank cannot say that the ATM machine is used significantly more on the weekends than any other day of the week.

Example 27

A local fast food restaurant serves buffalo wings. They notice that they normally sell the following proportions of spices for their wings: 20% Spicy Garlic, 50% Classic Medium, 10% Teriyaki, 10% Hot BBQ, and 10% Asian Zing. After running a campaign to promote their non-traditional specialty wings, they want to know if the campaign has made an impact. The results after 10 days are listed in the table below:

Table 4 – Results from Wing Spice Experiment

Wing Spice	Number Sold
Spicy Garlic	251
Classic Medium	630
Teriyaki	115
Hot BBQ	141
Asian Zing	121

Is there evidence at the 0.05 level of significance to say that the campaign has made a difference in the proportion of types of wings sold?

Solution:

The null hypothesis here is that the proportions are the same as they were before the campaign and the alternative is that they are different. To write this mathematically, we need to state what the expected value is for each of the five types of orders. Let p_1, p_2, p_3, p_4, and p_5 be the proportions for Spicy Garlic, Classic Medium, Teriyaki, Hot BBQ, and Asian Zing respectively. Then we have the following:

$$p_1 = 0.20, p_2 = 0.50, p_3 = 0.10, p_4 = 0.10, p_5 = 0.10.$$

Therefore, the null and alternative hypotheses are:

H_0: $p_1 = 0.20, p_2 = 0.50, p_3 = 0.10, p_4 = 0.10, p_5 = 0.10$
H_a: at least one proportion differs from these values.

Next, to calculate χ^2 for the data given, let's begin by calculating the expected value for each of the spices since they are not all the same. Here, $n = 1258$ (the total number of wings sold).

$$E(\text{Spicy Garlic}) = (0.20)(1258) = 251.6$$
$$E(\text{Classic Medium}) = (0.50)(1258) = 629$$
$$E(\text{Teriyaki}) = (0.10)(1258) = 125.8$$
$$E(\text{Hot BBQ}) = (0.10)(1258) = 125.8$$
$$E(\text{Asian Zing}) = (0.10)(1258) = 125.8$$

cont'd on the next page

Now we can use these in the χ^2 formula as follows:

$$\chi^2 = \frac{(251-251.6)^2}{251.6} + \frac{(630-629)^2}{629} + \frac{(115-125.8)^2}{125.8} + \frac{(141-125.8)^2}{125.8} + \frac{(121-125.8)^2}{125.8}$$
$$= 2.950.$$

The number of degrees of freedom is $d.f. = 5 - 1 = 4$. Using the table, the critical value is $\chi^2_{0.050} = 9.488$. Comparing the two, $\chi^2 < \chi^2_{0.050}$, so we must fail to reject the null hypothesis. In other words, the evidence does not support the claim. The restaurant cannot say that the campaign has made any difference based on this evidence.

Exercises 10.7

Directions: State the null and alternative hypotheses in words for each scenario.

1. A pharmacist believes that more people get their prescriptions filled on Fridays than on any other day of the week.

2. A game warden thinks there are 3 times as many ducks as geese in the national park.

3. A salon thinks that they are twice as busy on Thursdays and Fridays than the rest of the week.

4. A music producer believes that Southerners prefer country music to pop music.

5. A cola company believes that the taste of their cola is preferred to that of their competitors'.

Directions: Calculate the test statistic, χ^2 with the following data.

6. Number of drive-through customers

Table for Exercise 6

	8:00	**12:00**	**3:00**	**5:00**
Observed Values	35	54	11	21
Expected Values, $E(X)$	$\frac{121}{4}$	$\frac{121}{4}$	$\frac{121}{4}$	$\frac{121}{4}$

7. Number of candy bars sold at the concession stand

Table for Exercise 7				
	A	**B**	**C**	**D**
Observed Values	11	12	5	15
Expected Values, $E(X)$	$\frac{43}{4}$	$\frac{43}{4}$	$\frac{43}{4}$	$\frac{43}{4}$

8. Pass rates on entrance tests for a company

Table for Exercise 8			
	Level 1	**Level 2**	**Level 3**
Observed Values	29	201	169
Expected Values, $E(X)$	20.5	218.8	159.7

9. Number of hamburgers sold at three competing restaurants

Table for Exercise 9			
	Handy Andy	**Philip's Grocery**	**Proud Larry**
Observed Values	37	26	24
Expected Values, $E(X)$	29	29	29

10. Number of ladies' jeans sold in various sizes

Table for Exercise 10				
	Size 6	**Size 8**	**Size 10**	**Size 12**
Observed Values	4	7	5	5
Expected Values, $E(X)$	5.25	5.25	5.25	5.25

11. Number of children at the local library

Table for Exercise 11				
	9:00 AM	**11:00 AM**	**1:00 PM**	**3:00 PM**
Observed Values	6	10	9	8
Expected Values, $E(X)$	8.25	8.25	8.25	8.25

Directions: Find the critical value for χ^2 for each of the following:

12. $\alpha = 0.10$, $n = 47$

13. $\alpha = 0.05$, $n = 22$

14. $\alpha = 0.005$, $n = 13$

15. $\alpha = 0.01$, $n = 9$

16. $\alpha = 0.05$, $n = 31$

Directions: Determine the conclusion.

17. $\alpha = 0.005$, $n = 15$, $\chi^2 = 29.674$

18. $\alpha = 0.10$, $n = 8$, $\chi^2 = 13.219$

19. $\alpha = 0.05$, $n = 29$, $\chi^2 = 43.222$

20. $\alpha = 0.025$, $n = 17$, $\chi^2 = 27.008$

21. $\alpha = 0.10$, $n = 42$, $\chi^2 = 49.305$

22. $\alpha = 0.01$, $n = 33$, $\chi^2 = 53.886$

Directions: Complete the following tests for goodness of fit.

23. A manufacturer of children's vitamins claims that their vitamins are mixed so that each batch of vitamins has exactly the following percentages of each color of vitamin: 20% green, 40% yellow, 10% red and 30% orange. To test this claim, 100 bottles of vitamins were pulled and the colors of vitamins were counted. The results are listed in the following table:

Table for Exercise 23

Color	Green	Yellow	Red	Orange
Observed Values	1149	1948	552	1401

a. State the null and alternative hypotheses.
b. Find the expected number for each color of vitamin.
c. Calculate χ^2.
d. At $\alpha = 0.05$, determine the conclusion.

24. A school principal claims that the number of students who are tardy in her school does not vary from month to month. A survey over the school year produced the following results:

Table for Exercise 24

Month	Aug.	Sept.	Oct.	Nov.	Dec.	Jan.	Feb.	Mar.	Apr.	May
Number	7	18	16	5	8	12	15	18	11	15

a. State the null and alternative hypotheses.
b. Find the expected number of students tardy for each month.
c. Calculate χ^2.
d. At $\alpha = 0.05$, determine the conclusion.

25. A service station owner believes that customers do not have a preference for the day of the week they buy gasoline. He surveyed 739 customers over a period of time to record their preferred day of the week. Here's what he found:

Table for Exercise 25							
Day	Mon.	Tues.	Wed.	Thurs.	Fri.	Sat.	Sun.
Preference	103	103	126	103	111	96	97

a. State the null and alternative hypotheses.
b. Find the expected number of customers who prefer each day.
c. Calculate χ^2.
d. At $\alpha = 0.10$, determine the conclusion.

26. At the emergency room at one hospital, the nurses are convinced that the number of patients that they see during the midnight shift is affected by the phases of the moon. The doctors think this is an old wives' tale. The nurses decide to test their theory by recording the number of patients they see during the midnight shift for each moon phase over the course of a month. The results are summarized in the table below.

Table for Exercise 26				
Phase of Moon	New Moon	1st Quarter	Full Moon	Last Quarter
Number of Patients	85	66	97	68

a. State the null and alternative hypotheses.
b. Find the expected number of patients for each moon phase.
c. Calculate χ^2.
d. At $\alpha = 0.10$, determine the conclusion.

27. The local zoo wants to know if all of its animal exhibits are equally popular. If there is significant evidence that some of the exhibits are not being visited frequently enough, then changes may need to take place within the zoo. A tally of visitors is taken for each of the following animals throughout the course of a week, and the results are contained in the table below.

Table for Exercise 27							
Exhibit	Elephants	Lions / tigers	Giraffes	Zebras	Monkeys	Birds	Reptiles
Number of Visits	157	154	168	162	185	129	133

a. State the null and alternative hypotheses.
b. Find the expected number of visitors for each animal exhibit.
c. Calculate χ^2.
d. At $\alpha = 0.05$, determine the conclusion.

28. The manager of the city pool has scheduled extra lifeguards to be on staff for Saturdays. However, he suspects that Fridays may be more popular than the other weekdays as well. If so, he will hire extra lifeguards for Fridays, too. In order to test his theory, he records the number of swimmers each day for the first week of summer.

Table for Exercise 28

Day	Monday	Tuesday	Wednesday	Thursday	Friday
Number	46	47	43	53	54

a. State the null and alternative hypotheses.
b. Find the expected number of swimmers for each day.
c. Calculate χ^2.
d. At $\alpha = 0.01$, determine the conclusion.

Chi-Square Test for Association

10.8

Up until this point in the book, we have only considered one variable in a given population. Let's consider whether two variables in the population are related. In assessing whether the two traits or characteristics are associated in some way, we will once again use the test statistic χ^2, so long as the variables satisfy the properties of a multinomial distribution, which are given below.

Properties of a Multinomial Distribution:

1. The experiment consists of n independent, identical trials.
2. There are k possible outcomes for each trial.
3. The probabilities of the k outcomes, p, are constant from trial to trial.
4. The random variables of interest are the counts in each of the k possible outcomes, $n_1, n_2, \ldots, n_k$.

When using the chi-square test for association, the data need not be numerical, nor does the relationship need to be linear. For instance, we could examine whether there is a dependent relationship between gender and choice of car color. Suppose we took a random sample of 1000 cars parked on the university campus. We have summarized the data in the table below, which is called a **contingency table**.

Table 5 – Observed Sample of 1000 Cars

	White	Black	Red	Other	Total
Male	190	117	215	53	575
Female	115	142	113	55	425
Total	305	259	328	108	1000

To test whether there is dependence, we set up a hypothesis test. Let's start with the null hypothesis assumption that they are not related, i.e. independent.

H_0: Gender and car color are independent.
H_a: Gender and car color are not independent.

The next thing we need to do is to calculate the expected value of each cell. The formula we will use to calculate the expected values is

$$E(X) = \frac{(\text{row total})(\text{column total})}{\text{grand total}}.$$

For example, let's begin by calculating the expected value for white cars owned by men. The row total for males is 575, the column total for white cars is 305, and the grand total for the contingency table is 1000. Substituting these values into the formula gives us an expected value of

$$E\left(N_{male,white}\right)=\frac{(575)(305)}{1000}=175.375.$$

We continue using this formula for each cell in the table until we have completed all of the required expected values. The table below shows the completed expected value contingency table.

Table 6 – Expected Values

	White	Black	Red	Other	Total
Male	175.375	148.925	188.6	62.1	575
Female	129.625	110.075	139.4	45.9	425
Total	305	259	328	108	1000

In the last section, we developed the χ^2 statistic to compare how data fits with theoretically expected values. We can use the same test statistic here.

$$\chi^2=\frac{\left[N_{male,white}-E\left(N_{male,white}\right)\right]^2}{E\left(N_{male,white}\right)}+\cdots+\frac{\left[N_{female,other}-E\left(N_{female,other}\right)\right]^2}{E\left(N_{female,other}\right)}$$

This quantity is distributed according to the chi-square distribution as we saw before. Let's calculate χ^2 for our particular example:

$$\chi^2=\frac{[190-175.375]^2}{175.375}+\frac{[117-148.925]^2}{148.925}+\cdots+\frac{[55-45.9]^2}{45.9}\approx 30.8054216$$

To decide whether this is significant or not, we need to know the degrees of freedom. For this setting, the degrees of freedom will be determined from the number of rows and columns in the contingency table.

$$d.f.=(\#\text{ of rows}-1)\times(\#\text{ of columns}-1)$$

Therefore, our example has $(2-1)\times(4-1)=3$ degrees of freedom.

As before, we will use the table of critical values for the chi-square distribution. If our test statistic is larger than the critical value, we must reject the idea that the traits are independent.

Rejection Rule:

If $\chi^2 \geq \chi^2_\alpha$, then reject H_0 of independence.

Let's use a level of significance of 0.01 and test our data.

d.f.	$\chi^2_{0.100}$	$\chi^2_{0.050}$	$\chi^2_{0.025}$	$\chi^2_{0.010}$	$\chi^2_{0.005}$
1	2.706	3.841	5.024	6.635	7.879
2	4.605	5.991	7.378	9.210	10.597
3	6.251	7.815	9.348	11.345	12.838
4	7.779	9.488	11.143	13.277	14.860
5	9.236	11.071	12.833	15.086	16.750

As shown in the table above, the critical value is $\chi^2_{0.01} = 11.345$. Since $\chi^2 > \chi^2_{0.01}$, the test statistic value is greater and we must conclude that the data does not fit with the assumption that gender and car color are independent. Therefore, with 99% confidence, we can say that gender does influence choice of car color.

Now that we've discussed the concepts behind a chi-square test for association, let's work another example.

Example 28

In a poll of 13,660 voters during the 2004 presidential election campaign, the following data were collected.

Table 7 – Observed Sample of 13,660 Voters

	Bush	Kerry	Other	Total
Male	3455	2764	65	6284
Female	3541	3762	73	7376
Total	6996	6526	138	13660

Is there evidence at the 0.05 level to say that gender influenced voting choice?

Solution:

We let the null hypothesis be that gender and voting preference are independent of one another.

H_0: Gender and voting preference are independent.
H_a: Gender and voting preference are not independent.

cont'd on the next page

Before we begin to calculate the test statistic, we must calculate the expected value for each cell in the contingency table.

Table 7 – Expected Values				
	Bush	**Kerry**	**Other**	**Total**
Male	3218.36	3002.15	63.48	6284
Female	3777.64	3523.85	74.52	7376
Total	6996	6526	138	13660

Let's calculate χ^2.

$$\chi^2 = \frac{[3455 - 3218.36]^2}{3218.36} + \frac{[2764 - 3002.15]^2}{3002.15} + \cdots + \frac{[73 - 74.52]^2}{74.52} = 67.2757$$

Our degrees of freedom calculation is $d.f. = (2 - 1) \times (3 - 1) = 2$, and $\alpha = 0.05$. Using the table, the critical value is $\chi^2_{0.050} = 5.991$. Comparing the two, $67.2757 > 5.991$, so we must reject the null hypothesis. In other words, gender played a part in the voting habits of the electorate.

Exercises

10.8

Directions: Use the observed contingency table to find the expected value contingency table for each of the following.

1. Observed values: Cereal Brands by Gender

Table for Exercise 1					
	Brand A	**Brand B**	**Brand C**	**Brand D**	**Total**
Male	28	14	11	27	80
Female	9	16	8	33	66
Total	37	30	19	60	146

2. Observed values: Preferred Writing Hand by Age

Table for Exercise 2				
	20 Year Olds	**30 Year Olds**	**40 Year Olds**	**Total**
Left hand	15	42	31	88
Right hand	35	41	50	126
Ambidextrous	10	5	8	23
Total	60	88	89	237

3. Observed values: Run Time of Four Different Computer Algorithms

Table for Exercise 3

	A	B	C	D	Total
Model 1	8	12	4	9	33
Model 2	11	15	6	3	35
Model 3	4	10	3	11	28
Model 4	6	11	4	2	23
Total	29	48	17	25	119

4. Observed values: Favorite Sports by Region

Table for Exercise 4

	Football	Basketball	Baseball	Soccer	Total
Northeast	27	25	23	20	95
Southeast	26	21	19	17	83
Midwest	24	25	18	11	78
West	20	22	17	21	80
Total	97	93	77	69	336

5. Observed values: Marriage Status and Hair Color

Table for Exercise 5

	Blonde	Brown	Red	Black	Total
Single Women	18	19	8	14	59
Married Women	20	18	9	17	64
Single Men	13	22	4	16	55
Married Men	12	24	3	15	54
Total	63	83	24	62	232

6. Observed values: College Classification and Spring Break Destination

Table for Exercise 6

	Beach	Mountains	City	Home	Total
Freshmen	19	2	7	24	52
Sophomores	15	4	3	20	42
Juniors	18	1	9	19	47
Seniors	21	6	4	20	51
Total	73	13	23	83	192

Directions: Use the 2 contingency tables to calculate χ^2.

7. Observed values: Eating Habits of Different Athletes

Table for Exercise 7			
	Prefer to Eat Before a Workout	**Prefer to Eat After a Workout**	**Total**
Runners	68	121	189
Swimmers	73	128	201
Total	141	249	390

Expected Values:

Table for Exercise 7			
	Prefer to Eat Before a Workout	**Prefer to Eat After a Workout**	**Total**
Runners	68.33	120.67	189
Swimmers	72.67	128.33	201
Total	141	249	390

8. Observed values: Age and Music Preference

Table for Exercise 8					
	Song 1	**Song 2**	**Song 3**	**Song 4**	**Total**
Male 18-22	34	3	46	20	103
Male 23-27	33	25	48	17	123
Female 18-22	28	18	17	9	72
Female 23-27	3	15	20	7	45
Total	98	61	131	53	343

Expected Values:

Table for Exercise 8					
	Song 1	**Song 2**	**Song 3**	**Song 4**	**Total**
Male 18-22	29.429	18.318	39.338	15.915	103
Male 23-27	35.143	21.875	46.977	19.006	123
Female 18-22	20.571	12.805	27.499	11.125	72
Female 23-27	12.857	8.003	17.187	6.953	45
Total	98	61	131	53	343

9. Observed values: Microwavable Meals and Nutrition

Table for Exercise 9				
	Less than 5 gm of Fat	**5 – 10 gm of Fat**	**More than 10 gm of Fat**	**Total**
Brand A	160	27	1380	1567
Brand B	86	9	789	884
Brand C	55	7	770	832
Total	301	43	2939	3283

Expected Values:

Table for Exercise 9				
	Less than 5 gm of Fat	**5 – 10 gm of Fat**	**More than 10 gm of Fat**	**Total**
Brand A	143.67	20.52	1402.81	1567
Brand B	81.05	11.58	791.37	884
Brand C	76.28	10.90	744.82	832
Total	301	43	2939	3283

10. Observed values: Movie Preferences by Class

Table for Exercise 10					
	Suspense	**Drama**	**Comedy**	**Horror**	**Total**
Freshmen	24	28	37	18	107
Sophomores	19	25	35	15	94
Juniors	31	33	30	12	106
Seniors	26	29	34	13	102
Total	100	115	136	58	409

Expected Values:

Table for Exercise 10					
	Suspense	**Drama**	**Comedy**	**Horror**	**Total**
Freshmen	26.16	30.09	35.58	15.17	107
Sophomores	22.98	26.43	31.26	13.33	94
Juniors	25.92	29.80	35.25	15.03	106
Seniors	24.94	28.68	33.92	14.46	102
Total	100	115	136	58	409

11. Observed values: Preferred Commercial by Age and Gender

Table for Exercise 11			
	Commercial 1	**Commercial 2**	**Total**
Males 18-30	14	6	20
Females 18-30	9	11	20
Males 31-45	12	8	20
Females 31-45	6	14	20
Total	41	39	80

Expected Values:

Table for Exercise 11			
	Commercial 1	**Commercial 2**	**Total**
Males 18-30	10.25	9.75	20
Females 18-30	10.25	9.75	20
Males 31-45	10.25	9.75	20
Females 31-45	10.25	9.75	20
Total	41	39	80

12. Observed values: Voting Preference and Profession

Table for Exercise 12			
	Democrat	**Republican**	**Total**
Doctor	13	35	48
Lawyer	33	19	52
Teacher	23	25	48
Farmer	39	11	50
Laborer	28	10	38
Total	136	100	236

Expected Values:

Table for Exercise 12			
	Democrat	**Republican**	**Total**
Doctor	27.66	20.34	48
Lawyer	29.97	22.03	52
Teacher	27.66	20.34	48
Farmer	28.81	21.19	50
Laborer	21.90	16.10	38
Total	136	100	236

Directions: Determine if the results of χ^2 indicate independence or dependence, i.e. to reject H_0 or fail to reject H_0.

13. $\chi^2 = 24.128$, $\alpha = 0.01$, Number of rows = 3, Number of columns = 4

14. $\chi^2 = 5.13$, $\alpha = 0.10$, Number of rows = 2, Number of columns = 3

15. $\chi^2 = 36.1$, $\alpha = 0.05$, Number of rows = 4, Number of columns = 6

16. $\chi^2 = 31.1$, $\alpha = 0.025$, Number of rows = 5, Number of columns = 5

17. $\chi^2 = 40.8$, $\alpha = 0.005$, Number of rows = 5, Number of columns = 7

18. $\chi^2 = 40.3$, $\alpha = 0.10$, Number of rows = 7, Number of columns = 6

Directions: Complete each test for association.

19. The Department of Education wants to know if there is a relationship between grades and particular subject and gender. A random cluster sample of schools in the district produce the following results. Is there evidence to show there is a relationship between grades in subjects and gender?

Observed values: Grades vs. Gender and Subject

Table for Exercise 19

	A	B	C	D	F	Total
Math, Female	12	30	44	15	9	110
Math, Male	5	24	37	36	8	110
English, Female	16	41	39	10	4	110
English, Male	10	40	35	19	6	110
Total	43	135	155	80	27	440

a. Write the null and alternative hypotheses.
b. Find the expected value contingency table.
c. Calculate χ^2.
d. At $\alpha = 0.005$, determine the conclusion.

20. An insurance company wants to know if the color of an automobile has a relationship to moving violations. The following contingency table gives the results of data they collected from police reports across the nation. The columns list the number of reported moving violations in a year.

Observed values: Moving Violations by Car Color

Table for Exercise 20				
	0-1	**2-3**	**More than 3**	**Total**
White	76	33	13	122
Black	44	21	8	73
Red	50	46	12	108
Silver	33	27	7	67
Other	28	34	11	73
Total	231	161	51	443

a. Write the null and alternative hypotheses.
b. Find the expected value contingency table.
c. Calculate χ^2.
d. At $\alpha = 0.01$, determine the conclusion.

21. A soft drink company is interested in knowing whether there is a relationship between cola preference and age. A random sample of 800 people is chosen for a taste test. The results of the study are found in the table below. Does the evidence gathered lead you to believe that there is an association between cola preference and age?

Observed values: Cola Preference by Age

Table for Exercise 21				
	Cola A	**Cola B**	**Cola C**	**Total**
15 - 29	94	102	105	301
30 - 44	99	97	86	282
45 - 59	68	73	76	217
Total	261	272	267	800

a. Write the null and alternative hypotheses.
b. Find the expected value contingency table.
c. Calculate χ^2.
d. At $\alpha = 0.005$, determine the conclusion.

22. An obstetrician has noticed that there seems to be a pattern to the sex of the babies that he delivers. That is, he tends to deliver boys and girls in clusters. In order to determine whether this pattern is true outside of his hospital, he calls 4 other hospitals in the region and obtains the following information regarding births from the previous six months. Does the evidence gathered support the doctor's claim?

Observed values: Pattern of Babies

Table for Exercise 22

	Nov/Dec	Jan/Feb	Mar/Apr	Total
Males	158	165	153	476
Females	149	147	154	450
Total	307	312	307	926

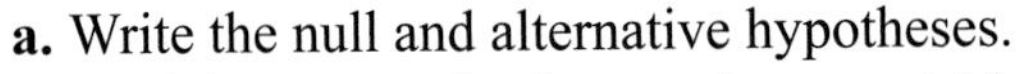

a. Write the null and alternative hypotheses.
b. Find the expected value contingency table.
c. Calculate χ^2.
d. At $\alpha = 0.01$, determine the conclusion.

23. Suppose that a bookseller wants to study the relationship between book preference and gender. A random sample of readers is chosen for the study, and each participant is asked to choose their favorite genre out of the following choices: mysteries, fiction, non-fiction, and self-help. The results are detailed below. Does the evidence gathered show a relationship between book preference and gender?

Observed values: Gender vs. Book Preference

Table for Exercise 23

	Mysteries	Fiction	Non-Fiction	Self-Help	Total
Males	28	30	41	22	121
Females	32	59	28	26	145
Total	60	89	69	48	266

a. Write the null and alternative hypotheses.
b. Find the expected value contingency table.
c. Calculate χ^2.
d. At $\alpha = 0.005$, determine the conclusion.

Chapter Review

Section 10.1 - Fundamentals of Hypothesis Testing — Pages

- **Definitions** — p. 468 – 471

  Null Hypothesis – written as H_0, specifies the currently accepted value for the population parameter.

  Alternative Hypothesis – written as H_a, is the logical opposite of the null hypothesis and the claim leading to the hypothesis test. Also referred to as the research hypothesis.

  Statistically Significant – Data is said to be statistically significant if it is unlikely that a sample similar to the one chosen would occur by chance if the null hypothesis is true.

  Level of Significance, $\alpha = 1 - c$

- **Only Possible Conclusions in a Hypothesis Test:** — p. 472
  - Reject the null hypothesis
  - Fail to reject the null hypothesis

- **General Steps in Hypothesis Testing:** — p. 473
  1. State the null and alternative hypotheses.
  2. Set up the hypothesis test by choosing the test statistic and determine the values of the test statistic that would lead to rejecting the null hypothesis.
  3. Gather data and calculate the necessary sample statistics.
  4. Draw a conclusion using the process described in Step 2.

Section 10.2 - Hypothesis Testing for Means (Small Samples) — Pages

- **Definitions** — p. 477

  Test Statistic – the value you use to test the null hypothesis

- **Test Statistic for Population Means (Small Samples)** — p. 477

$$t = \frac{\bar{x} - \mu}{\left(\frac{s}{\sqrt{n}}\right)} \text{ with } d.f. = n - 1$$

- **Types of Hypothesis Tests** — p. 478

Alternative Hypothesis	Type of Test
< Value	Left-tailed test
> Value	Right-tailed test
≠ Value	Two-tailed test

Section 10.2 - Hypothesis Testing for Means (Small Samples) (cont.)	Pages
• **Rejection regions are determined by two things:** **1.** The type of hypothesis test being run, **2.** The level of significance, α, that is required	p. 478
• **Conclusion Rule for Rejection Regions:** Reject the null hypothesis if the test statistic calculated from the sample data falls within the rejection region.	p. 478
• **Rejection Regions for Population Means (Small Samples)** Left-tailed: Reject H_0 if $t \leq -t_{\alpha}$. Right-tailed: Reject H_0 if $t \geq t_{\alpha}$. Two-tailed: Reject H_0 if $t \leq -t_{\frac{\alpha}{2}}$ or $t \geq t_{\frac{\alpha}{2}}$.	p. 480

Section 10.3 - Hypothesis Testing for Means (Large Samples)	Pages
• **Test Statistic for Population Means (Large Samples)** $z = \dfrac{\bar{x} - \mu}{\left(\dfrac{s}{\sqrt{n}}\right)}$	p. 489
• **Rejection Regions for Population Means (Large Samples)** Left-tailed: Reject H_0 if $z \leq -z_c$. Right-tailed: Reject H_0 if $z \geq z_c$. Two-tailed: Reject H_0 if $z \leq -z_c$ or $z \geq z_c$.	p. 489 – 492

Critical z-Values for Rejection Regions

c	One-Tailed Test	Two-Tailed Test
0.90	1.28	±1.645
0.95	1.645	±1.96
0.98	2.05	±2.33
0.99	2.33	±2.575

• A ***p*-value** is the probability of obtaining a sample more extreme than the one observed in your data, when H_0 is presumed to be true.	p. 490
• **Conclusions using *p*-values** • If $p \leq \alpha$, then *reject* the null hypothesis. • If $p > \alpha$, then *fail to reject* the null hypothesis.	p. 492

Section 10.3 - Hypothesis Testing for Means (Large Samples) (cont.)	Pages
• **General Steps Using *p*-Values in Hypothesis Testing:** **1.** State the null and alternative hypotheses. **2.** Set up the hypothesis test by choosing the test statistic and stating the level of significance. **3.** Gather data and calculate the necessary sample statistics. **4.** Draw a conclusion by comparing the *p*-value to the level of significance.	p. 493

Section 10.4 - Hypothesis Tests for Population Proportions	Pages
• **Test Statistic for Population Proportions** (for $np \geq 5$ and $n(1-p) \geq 5$) $z = \frac{\hat{p} - p}{\sqrt{\frac{p(1-p)}{n}}}$	p. 500

Section 10.5 - Types of Errors	Pages
• **Type I Error** – rejecting a true null hypothesis	p. 509
• **Type II Error** – failing to reject a false null hypothesis	
• The level of significance, α, is the probability of committing a Type I error.	
• The probability of committing a Type II error is denoted as β.	p. 510
• The values of α and β are inversely related.	

Reality

Decision	**H_0 is true.**	**H_0 is false.**
H_0 is rejected.	Type I error	Correct Decision
H_0 is not rejected.	Correct Decision	Type II error

Section 10.6 - Hypothesis Testing for Population Variance	Pages
• The following criteria must be met before a hypothesis test for variance can be performed. • All possible samples of a given size must have an equal chance of being chosen. • The population must be normally distributed.	p. 514
• **Test Statistic for Population Variance (or Standard Deviation)** $\chi^2 = \frac{(n-1)s^2}{\sigma^2}$	

Section 10.6 - Hypothesis Testing for Population Variance (cont.)	Pages
• **Rejection Regions for Population Variance** Left-tailed: Reject H_0 if $\chi^2 \le \chi^2_{1-\alpha}$. Right-tailed: Reject H_0 if $\chi^2 \ge \chi^2_{\alpha}$. Two-tailed: Reject H_0 if $\chi^2 \le \chi^2_{1-\frac{\alpha}{2}}$ or $\chi^2 \ge \chi^2_{\frac{\alpha}{2}}$.	p. 515

Section 10.7 - Chi-Square Test for Goodness of Fit	Pages
• **Null and Alternative Hypotheses for Goodness of Fit Test** **When expected probabilities are the same:** $H_0: p_1 = p_2 = ... = p_n$ H_a: any difference amongst the probabilities **When expected probabilities are different:** $H_0: p_1$ = probability of N_1; p_2 = probability of N_2; ...; p_n = probability of N_n H_a: any difference in the stated probabilities	p. 525
• **Test Statistic for Goodness of Fit** $\chi^2 = \sum \frac{[N_i - E(N_i)]^2}{E(N_i)}$ with $d.f. = n - 1$	p. 523 – 524
• **Rejection Rule for Goodness of Fit Test** Reject H_0 if $\chi^2 \ge \chi^2_{\alpha}$.	

Section 10.8 - Chi-Square Test for Association	Pages
• **Null and Alternative Hypotheses for Association Test** H_0: Variables are independent. H_a: Variables are not independent.	p. 533
• **Test Statistic for Association** $\chi^2 = \sum \frac{[N_i - E(N_i)]^2}{E(N_i)}$ with $d.f.$ = (# rows – 1)(# columns – 1)	p. 534
• **Rejection Rule for Association** Reject H_0 if $\chi^2 \ge \chi^2_{\alpha}$.	p. 535

Chapter Exercises

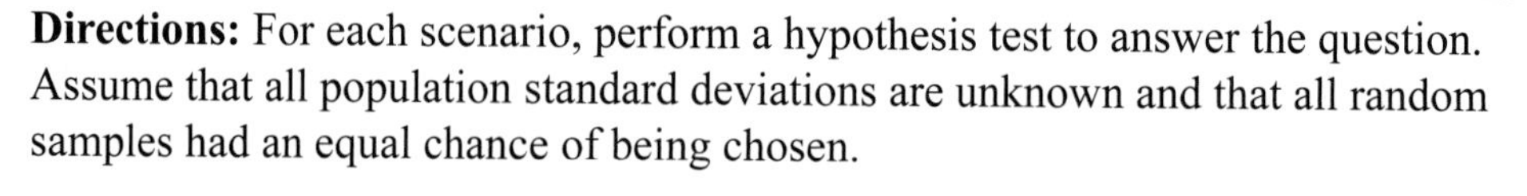

Directions: For each scenario, perform a hypothesis test to answer the question. Assume that all population standard deviations are unknown and that all random samples had an equal chance of being chosen.

1. A manufacturer is responsible for making the barrels to store crude oil for the United States, which federal requirements state must hold exactly 42 gallons of oil. As part of a routine quality check, you randomly test 27 barrels off the assembly line and find that their mean capacity is 41.7 gallons, with a standard deviation of 0.8 gallons. Can you, as the quality control administrator, say with 99% confidence that the requirements are being met? Assume that the population's distribution is approximately normal.

2. Sarah, a university junior, chose to live on the honors floor of her dorm because she was told that the girls on that floor would be quiet and studious. In fact, Sarah was told that students on the honors floor study at least 11 hours per week. After living on the honors floor for a few months, Sarah is convinced that the students around her study less than 11 hours a week. To test her claim, Sarah polls 10 girls on her floor regarding the number of hours per week they study. Use the results listed in the table below to test Sarah's claim at the 0.10 level of significance. Assume that the population is approximately normal.

Table for Exercise 2

Hours of Study	12	4	14	9	8	6	10	12	11	9

3. TransAir, an international airline company, has a new jumbo jet that they believe will decrease the average trans-Atlantic flight time from Newark to London Heathrow. In order to be faster than their competitors, the average must be less than 6 hr 50 min. Because of costs, the agency monitoring the flight times only requires that 8 trials be run. The results are given in the table below. If the FAA requires that the agency be 90% confident of its findings, what should the agency report? Assume that the population is approximately normal.

Table for Exercise 3

Newark to London	Flying Times
Flight 1	6 hr 32 min
Flight 2	6 hr 28 min
Flight 3	6 hr 51 min
Flight 4	7 hr 2 min
Flight 5	6 hr 51 min
Flight 6	7 hr 2 min
Flight 7	6 hr 36 min
Flight 8	6 hr 42 min

4. A computer program designer has developed an online questionnaire she believes will take an average time of 15 minutes to complete. Suppose she samples 35 volunteers and finds that the mean time to take the test is 14 minutes with a standard deviation of 2.5 minutes. Determine if there is sufficient evidence to conclude that the completion time of the online questionnaire differs from its intended duration. Use a 0.05 level of significance.

5. A manufacturer of hearing aid batteries has developed a new process which it believes will increase the usable life of the battery. Currently the usable life for this battery is 264 hours. Test the hypothesis that the process increases the usable life of a battery, at the 0.01 level of significance. A sample of 50 batteries is tested and found to have a mean of 271 hours with a standard deviation of 16.9 hours.

6. If it is true that less than 4 people per day use a particular public restroom at the state park, the park services will close the restroom. In order to determine the action the park should take, the number of uses is recorded each day for a 35-day period. Here are the results.

Table for Exercise 6

3	4	0	1	2	6	9
0	2	4	3	3	5	3
7	3	0	2	1	4	1
0	0	6	3	5	10	5
3	4	0	5	2	4	6

a. At the 5% level of significance, should the park shut the restroom down?
b. Because of the fact that the monitoring took place in an off season for the park, they ran the test again and determined that the actual usage amount was, on average, 7 times a day. Did the park make the right decision?

7. While trying to convince her friends that off campus living is preferred by more than half of all sophomores, Bren took a random survey and collected the following data:

Table for Exercise 7

	On Campus Living	Off Campus Living
Male	25	47
Female	38	19

a. Does this evidence support Bren's claim that more than half of all sophomores preferred off campus housing at the 0.10 level of significance?

b. A friend of Bren's went to the registrar's office and found out that actually 975 out of 3788 sophomores lived off campus this year. Based on this information, did Bren make an error in her conclusion? If so, what type of error did she make?

8. Don believes that less than one-third of the students on campus hold a part-time job while in school. To test his claim, he randomly selects 116 students from whom he collects the following data.

Table for Exercise 8	
	Part Time Jobs Held
Yes	32
No	84

a. Does this evidence support Don's claim at the 0.05 level of significance?
b. In a campus wide end-of-year survey, the office of research was able to report that 34% of students say they held a part-time job. Did Don draw the correct conclusion? If not, what type of error did he make?

9. A student organization that promotes diversity believes that the percentage of minority students is no longer 35%. From a random sample of 250 students, 97 of these students are minorities.
a. Does this evidence support the organization's claim that the percentage of minority students is not 35%, at the 0.01 level of significance?
b. The registrar reports that 38% of currently enrolled students are minorities according to their admissions records. Based on this new information, did the organization draw the correct conclusion? If not, what type of error was made?

10. A clothing designer produces thousands of pairs of blue jeans each week. The denim for each pair of jeans is cut by machine, and if the cuts are not made properly, then the jeans will not be properly sized once sewn together. To ensure that the jeans are properly sized, the variance in the lengths of the cuts must be less than 0.25 mm. To test the accuracy of the cuts, 20 lengths of denim are pulled from the line and measured. The variance in these lengths is found to be 0.1369 mm. Does this evidence support the claim that the variance in the lengths of the cuts is less than 0.25 mm, at the 0.10 level of significance?

11. From micro-minis to floor length skirts, a historian believes that the variance in the length of ladies' hems on dresses and skirts is more now than at any other point in recorded history. Specifically, she claims that the variance in the length of skirts' hems is more than 15.5. (Length measured in inches.) To test her claim, she randomly selects 25 dresses and skirts from a department store and measures the hem length of each. She finds that the variance in the length of the hem is 17.3. Does this evidence support the historian's claim, at the the 0.05 level of significance?

12. A statistics professor is concerned that the variance in the grades on the final exam will not be 138, as it has been in previous semesters. She randomly selects 35 of her many exams and finds that the variance in the grades is 142. Does this evidence support the professor's claim, at the 0.01 level of significance?

13. The local police department has always had an equal number of officers on patrol in each beat. However, each month they re-evaluate to make sure that they are utilizing their resources and manpower as they should. To do this, they record the number of calls they receive in each beat over the course of the month. The data is summarized below.

Table for Exercise 13

	Beat 1	Beat 2	Beat 3	Beat 4
Number of Calls	28	19	31	38

Does this evidence support the claim that there is a difference in the number of calls between the different areas of the city, at the 0.05 level of significance?

14. A workers' union is appealing to the management for more vacation time for its employees. The union claims that workers are more productive when they come back from vacation than before. The following contingency table contains the data that they collected from the employees.

Observed values: Production Rates per Hour

Table for Exercise 14

	100-125	156-150	More than 150	Total
Before Vacation	210 workers	319 workers	58 workers	587
After Vacation	178 workers	387 workers	60 workers	625
Total	388	706	118	1212

The union feels that it needs to be 90% confident of its claim before going into talks with the employers. Does the union have strong enough evidence to support its claim?

15. A sociologist wants to study the education levels of people living in various regions of the country. He randomly surveys 100 people in each of the following regions: Northeast, Southeast, Midwest, and West. The results obtained are found in the table on the following page.

Table for Exercise 15					
	Some High School	High School	College	Post College	Total
Northeast	11	29	45	15	100
Southeast	19	28	44	9	100
Midwest	15	25	47	13	100
West	13	31	42	14	100
Total	58	113	178	51	400

Does this evidence support the sociologist's claim that the level of education differs by region, at the 0.05 level of significance?

Chapter Project A: Hypothesis Testing for Means

Directions: Choose one of the three claims, collect data from friends in the desired population, and perform a hypothesis test on the claim to determine if the evidence supports the claim or does not support the claim. After you have written your conclusion, look at the "real" value and determine if your hypothesis test produced a correct decision, a Type I error, or a Type II error.

Pick one of the following claims to test:

- Parents of university freshmen believe that the average freshman spends at most $50 per week on eating out.
- It is believed that university sophomores see at least 4 movies at the theater per month.
- The librarian claims that university juniors visit the library twice a week.

Step 1 – State the null and alternative hypotheses.

Based on the claim you choose, what are the null and alternative hypotheses?

Step 2 – Determine your test statistic and level of significance.

You will be using a sample size of 10 as you will be asked to collect data from 10 friends that are in the desired population. Using this information, what is the test statistic? Also, choose a level of significance, either 0.10, 0.05, or 0.01.

Test Statistic: ______________________________ (formula only)

Step 3 – Collect data and calculate the test statistic.

Collect data on the claim from 10 of your classmates or friends who are in the desired population. Discuss which method of data collection you used. List any potential for bias. Calculate the sample mean and sample standard deviation. Round your answers to two decimal places.

$\bar{x} =$ ____________________

$s =$ ____________________

Calculate the test statistic using the values you just calculated from your sample.

Test Statistic = ____________________

Step 4 – Draw a conclusion using rejection regions.

This is a (left / right / two)-tailed test. (Circle your answer)

Draw a picture of your rejection region.

What is your conclusion?
In conclusion, we (reject / fail to reject) the null hypothesis. (Circle your answer.)
The original claim (is / is not) supported. (Circle your answer.)

Types of Error
Let's assume we find out that the truth is as follows:

- The average freshman spends $40 per week eating out.
- The average sophomore sees more than 4 movies a month.
- The average junior visits the library 4 times per week.

Based on your conclusion, did you make a Type I error, Type II error, or a correct decision? Explain.

Chapter Project B: Chi-Square Test for Goodness of Fit

**Note to instructors: This project requires that information be obtained from the registrar at your institution. If this information is not easily obtainable, i.e. public record on the web, it might be advisable to collect the data yourself rather than bombard the registrar with an influx of curious students.*

In this project, we will look at whether the makeup of your institution has significantly changed over the years. We are going to compare the proportion of freshmen, sophomores, juniors, and seniors of one year to that of another. Let's begin by collecting some data.

1. Choose 2 years that are not consecutive from which to collect data. (For instance, this might be the current academic year and four years before.) Find out the number of students enrolled at your institution for each classification and enter them in a table like the one below.

	Year 1	**Year 2**
Number of Freshmen Enrolled		
Number of Sophomores Enrolled		
Number of Juniors Enrolled		
Number of Seniors Enrolled		

2. Now calculate the percentage of students in each category in a new table.

	Year 1	**Year 2**
Percentage of Freshmen Enrolled		
Percentage of Sophomores Enrolled		
Percentage of Juniors Enrolled		
Percentage of Seniors Enrolled		

3. State the null and alternative hypotheses in words.

4. Specify the null and alternative hypotheses with mathematical symbols.

5. Calculate the expected value for each classification.

6. Calculate the chi-square test statistic.

7. Specify the critical value for the test if the level of significance is 0.10.

8. Determine your conclusion.

9. Is there sufficient evidence to conclude at $\alpha = 0.10$ that the makeup of your institution has significantly changed from the years __________ to __________? (Fill in the appropriate years in the blanks.)

Chapter Technology – Performing a Hypothesis Test

TI-84 PLUS

We can use a TI-84 Plus calculator to perform a hypothesis test by directly entering data or by entering the sample statistics. The process for performing any hypothesis test on a TI-84 Plus is basically the same:

Determine the type of test you need to perform and the null and alternative hypotheses.

1. Either enter your raw data in the List or Matrix Environments (whichever is appropriate) or calculate the necessary sample statistics for the type of test you are going to perform.
2. Press STAT.
3. Scroll over to the TESTS menu.
4. Scroll down the list and choose the type of test you need to perform.
5. Enter the appropriate information.
6. Choose Calculate.

Hypothesis Test for Means (Large Sample)

Example T.1

The average reading score for a class of third graders is believed to be 82. An educational researcher claims this figure is wrong. A sample of 45 third graders is chosen and the average reading score is calculated to be 81 with a standard deviation believed to be 5. Does the data gathered support the claim? Use a 0.05 level of significance to draw your conclusion. Assume the population is approximately normal.

Solution:

First, we notice that this will be a test for means with a large sample size. This is a z-test. The z-test is option 1 under the TESTS menu. We also need to write out our null and alternative hypotheses.

$$H_0: \mu = 82$$
$$H_a: \mu \neq 82$$

Next, note that we have the following sample statistics:

$\mu_0 = 82$
$\bar{x} = 81$
$s = 5$
$n = 45$
$\alpha = 0.05$

The calculator asks for σ. However, since our sample size is greater than 30, we can approximate σ with s. Since we have the sample statistics, we want to enter them in directly. Press STAT, then scroll to the TESTS menu and choose option 1. Enter your statistics in as shown in the following screen:

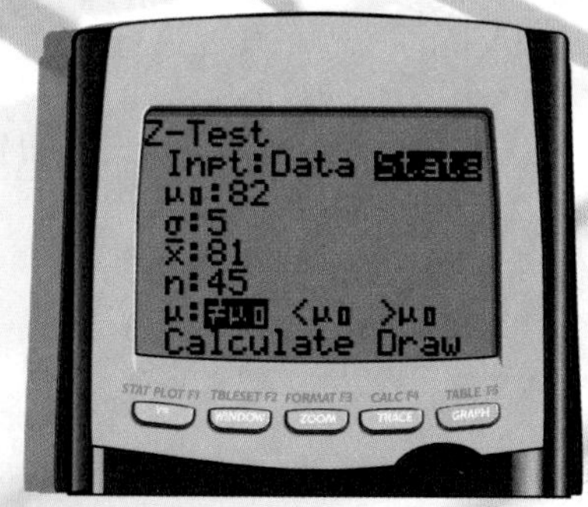

After choosing Calculate, the following screen appears.

This tells us that our alternative hypothesis is $\mu \neq 82$, our test statistic is $z = -1.34$, our p-value is 0.1797 and then repeats the sample mean and sample size we entered. Since our p-value is 0.1791 and it is greater than $\alpha = 0.05$, we would fail to reject the null hypothesis. Therefore, the evidence does not support the educational researcher's claim that the average reading score for a class of third graders is not 82.

Hypothesis Test for Means (Small Sample)

Example T.2

The average number of phone calls teenage girls make per night is believed to be more than 4. A concerned parent claims that the average is really more than 4. Ten teenage girls in a local neighborhood were asked how many phone calls they made last night. The average number of phone calls they made was 4.5 with a standard deviation of 0.85. Does this information support the claim? Use a 0.10 level of significance to draw your conclusion. Assume the population is approximately normal.

Solution:

The sample size for this hypothesis test is small, i.e. less than 30, so we would use a t-test. The t-test is option 2 under the TESTS menu. Next, we write our null and alternative hypotheses.

$$H_0: \mu \leq 4$$
$$H_a: \mu > 4$$

So we have a right-tailed t-test. Our sample statistics are given to us as follows:

$$\mu_0 = 4$$
$$\bar{x} = 4.5$$
$$s = 0.85$$
$$n = 10$$
$$\alpha = 0.10$$

cont'd on the next page

Press STAT, then scroll to the TESTS menu and choose option 2. Enter your statistics in as shown in the following screen:

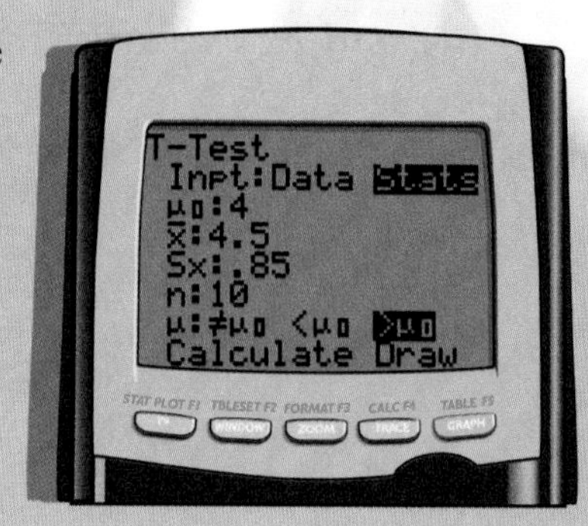

After choosing Calculate, the following screen appears.

This tells us that our alternative hypothesis is $\mu > 4$, our test statistic is $t = 1.86$, our p-value is 0.0479 and then repeats the sample mean, sample standard deviation, and sample size we entered. Since our p-value is 0.0479, and that is less than $\alpha = 0.10$, we reject the null hypothesis. Therefore, the evidence does support the parent's claim that the average number if phone calls teenage girls make per night is more than 4.

Hypothesis Test for Proportions

Example T.3

Let's consider Example 18 from the text. A local politician believes that less than 65% of his constituents will vote in favor of a tax increase to pay for a new school. He asks 50 of his constituents whether they favor the tax increase: 27 said that they would vote in favor of the tax increase. Does this information support the claim? Use a 0.01 level of significance to draw your conclusion.

Solution:

This example deals with proportions, so we would use a z-test for proportions. This particular z-test is option 5 under the TESTS menu. Next, we write our null and alternative hypotheses.

$$H_0: p \geq 0.65$$
$$H_a: p < 0.65$$

So we have a left-tailed 1-proportion z-test. Our sample statistics are given to us as follows:

$p_0 = 0.65$

$x = 27$ which is the number in the sample with the characteristic we are studying.

$n = 50$

$\alpha = 0.01$

Press STAT, then scroll to the TESTS menu and choose option 5. Enter your statistics in as shown in the following screen:

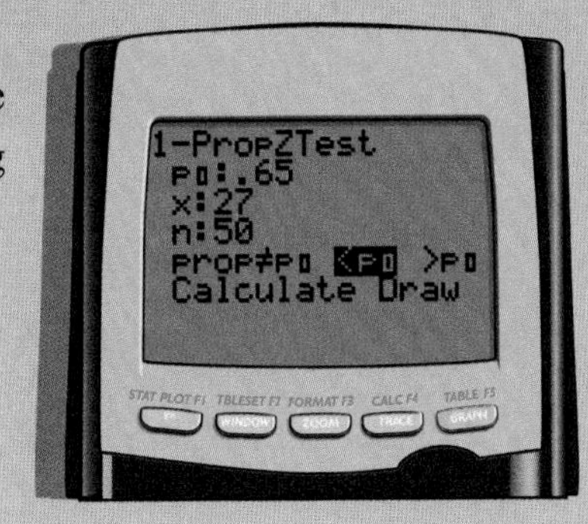

After choosing Calculate, the following screen appears.

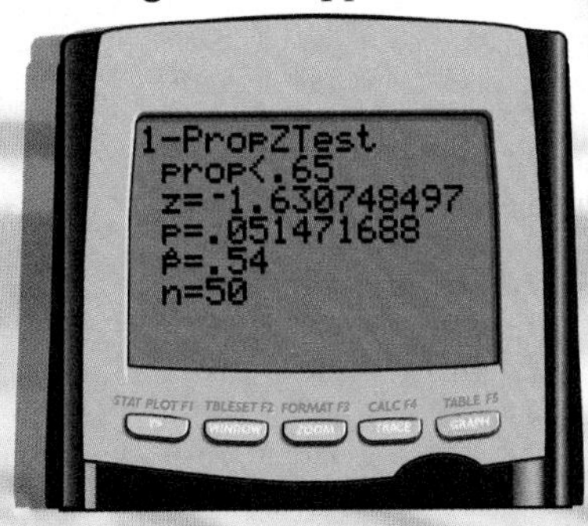

This tells us that our alternative hypothesis is $p < 0.65$, our test statistic is $z = -1.63$, our p-value is 0.0515, our sample proportion is $\hat{p} = 0.54$, and then repeats the sample size we entered. Since our p-value is 0.0515, which is greater than $\alpha = 0.01$, we fail to reject the null hypothesis. This means that at the 0.05 level of significance this evidence does not sufficiently support the politician's claim that less than 65% of his constituents favor a tax increase to pay for a new school.

χ^2 Test for Goodness of Fit

A TI-84 Plus cannot calculate a Goodness of Fit test directly, but we can use the List Environment to help us calculate χ^2. We would then use the χ^2 Critical Value Table to determine our conclusion.

Example T.4

Let's consider Example 26 from the text. The local bank wants to evaluate the usage of its ATM machine. Currently they assume that the ATM is used more frequently on the weekend. They decide to use statistical inference with a 0.05 level of confidence, to test their procedures and find out if they are meeting the needs of their customers. During a 2 week period, the bank recorded the customer usage on each day. The results are listed in the table below:

Monday	Tuesday	Wednesday	Thursday	Friday	Saturday	Sunday
38	33	41	25	22	38	45

Use your TI-84 Plus to calculate χ^2.

Solution:

We begin by calculating the expected value of each day. Since all of the days are the same, the expected value is just $\frac{242}{7}$. To calculate χ^2 on our TI-84 Plus, we need to enter the observed values in L_1 and the expected values in L_2. Your screen should appear as follows:

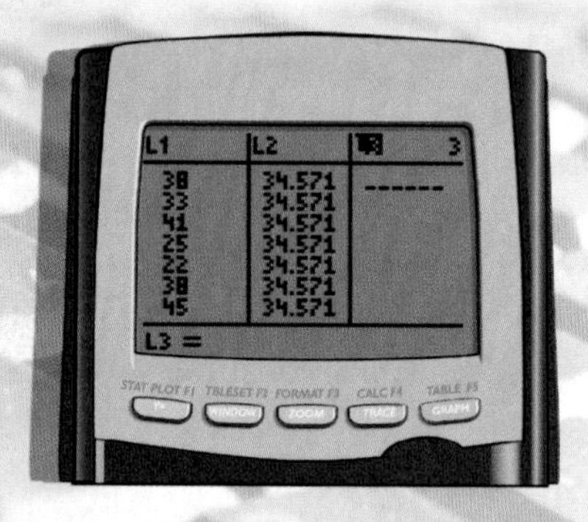

Recall that χ^2 is the difference of the observed value and the expected value squared divided by the expected value. To calculate this value for each row quickly, enter the following formula for L_3.

$$L_3 = \frac{(L_1 - L_2)^2}{L_2}$$

Entering this formula in for L_3 gives you the following values:

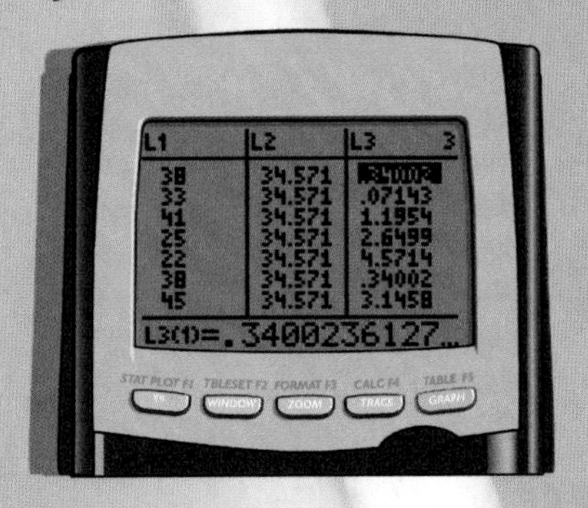

To calculate χ^2, we need to sum the values in L_3. Press 2ND, then STAT and scroll over to the MATH menu. Option 5 is 5: sum(. Choose option 5 and then press 2ND, 3 . This gives you the sum of the values in list 3.

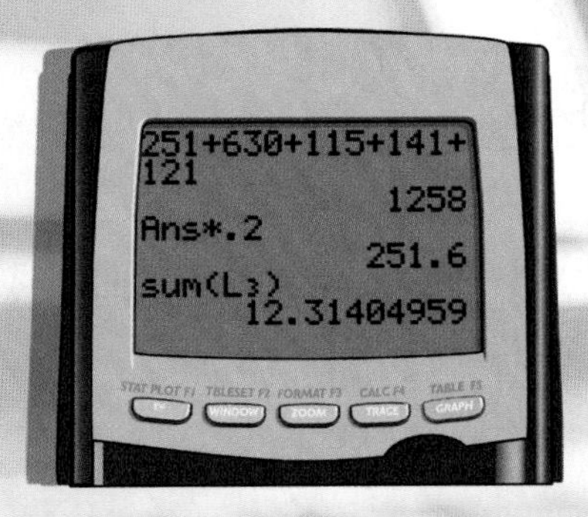

Thus, $\chi^2 = 12.3140$.

Using the table, the critical value is $\chi^2_{0.050} = 12.592$. Comparing the two, $\chi^2 < \chi^2_{0.050}$, so we must fail to reject our null hypothesis. In other words, the bank cannot say that the ATM machine is used significantly more on the weekends than any other day of the week.

χ^2 Test for Association

Example T.5

Let's consider Example 28 from the text. In a poll of 13,660 voters during the 2004 presidential election campaign, the following data was collected.

Observed Values				
	Bush	**Kerry**	**Other**	**Total**
Male	3455	2764	65	6284
Female	3541	3762	73	7376
Total	6996	6526	138	13660

Is there evidence at the 0.05 level to say that gender influenced voting choice?

Solution:

For stating our hypothesis to be tested, we let the null hypothesis be that gender and voting preference are independent of one another.

H_0: Gender and voting preference are independent.
H_a: Gender and voting preference are not independent.

Before we begin to calculate the test statistic, we must calculate the expected value for each cell in the contingency table.

Expected Values				
	Bush	**Kerry**	**Other**	**Total**
Male	3218.36	3002.15	63.48	6284
Female	3777.64	3523.85	74.52	7376
Total	6996	6526	138	13660

To calculate χ^2 on our TI-84 Plus, we need to enter the observed values in a matrix and the expected values in another matrix. We do not need to enter the totals. To enter values in a matrix, press 2ND , then x^{-1} to access the MATRIX menu. Scroll over to Edit and choose the matrix you want to edit (enter values into.) The first thing you must enter is the number of rows, then the number of columns. After that, you can begin entering the observed values in one matrix and then create another matrix in the same way to enter the expected values. The two matrices you create for this example are shown on the next page:

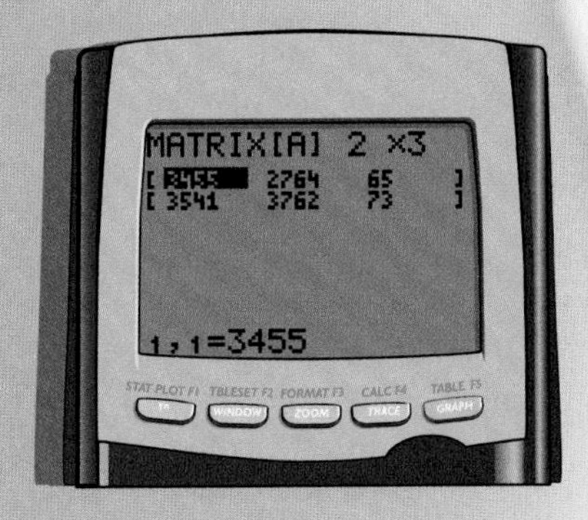

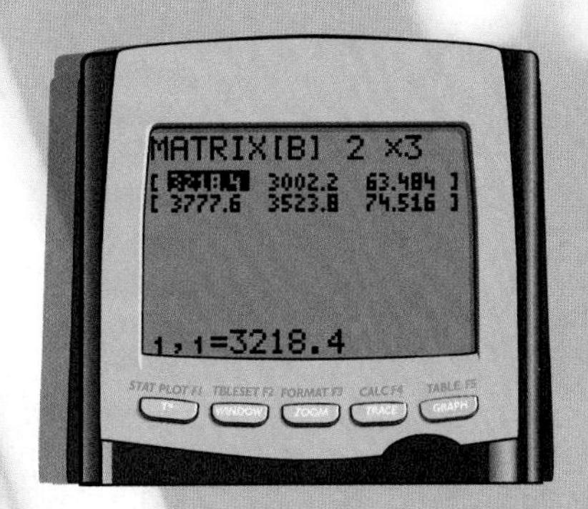

Once the matrices have been entered, press STAT and scroll over to the TESTS menu. The χ^2 test is option C. Since the observed values are in matrix A and the expected values are in matrix B, we do not need to change the default settings. Your screen should appear as follows:

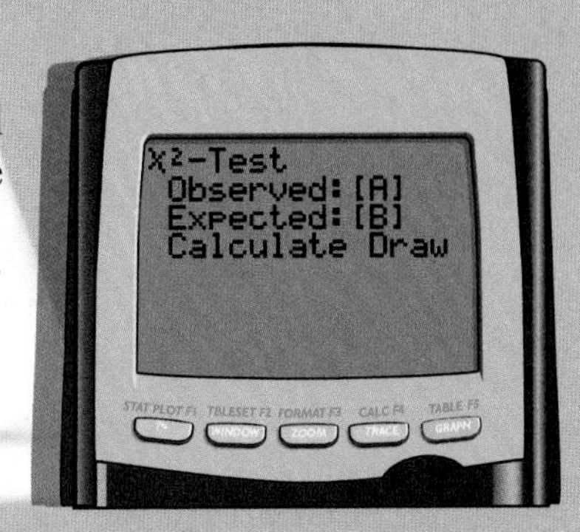

After choosing Calculate, the following screen appears.

This tells us the value of $\chi^2 = 67.2757$. It also tells us the p-value. For this example, the p-value is approximately 0 which is less than $\alpha = 0.05$, so we would reject the null hypothesis that these characteristics are independent.

Hypothesis Test for Means (Large Sample)

Excel can simplify hypothesis testing. This is especially useful when the process needs to be repeated several times.

Example T.6

Fred claims he has an 18 stroke golf handicap. This means he averages 18 strokes over par on a par 72 course. Ted thinks he is not telling the truth and finds 33 recent scorecards. From these he calculates the sample mean to be 88.1 with a standard deviation of 3.5. Can he claim with a 99% certainty that Fred has a handicap below 18?

Solution:

First, we recognize that the desired method is a *z*-test for means. Furthermore, we state the null and alternative hypotheses.

$$H_0: \mu \geq 90$$
$$H_a: \mu < 90$$

So we have a left-tailed *z*-test with the following sample statistics:

$$\mu_0 = 90$$
$$\bar{x} = 88.1$$
$$s = 3.5$$
$$n = 33$$
$$\alpha = 0.01$$

Now use Excel to calculate the test statistic and critical value. The test statistic, *z*, is produced by entering =(B1-B2)/(B3/SQRT(B4)). The critical value is produced by =NORMSINV(0.99). Since the $z < -z_\alpha$ the null hypothesis is rejected. With 99% certainty, Fred's handicap is below 18 strokes. The Excel spreadsheet is shown below.

	A	B
1	sample mean	88.1
2	population mean	90
3	s	3.5
4	n	33
5		
6	z	-3.118
7	critical z	2.33

Hypothesis Test for Means, Small Sample

Example T.7

A sanitation engineer believes one waste management district is producing more garbage than normal. He chooses a random sample of 10 trucks that will be serving the district and weighs them before and after pickup. He calculates the sample mean to be 10.6 tons with a standard deviation of 1.2 tons. He believes the population mean is 9.8 tons. Can his claim be substantiated at a confidence level of 95%?

Solution:

First, recognize that the desired test is a *t*-test for means and state the null and alternative hypotheses.

$$H_0: \mu \le 9.8$$
$$H_a: \mu > 9.8$$

So we have a left-tailed *t*-test with the following sample statistics:

$$\mu_0 = 9.8$$
$$\bar{x} = 10.6$$
$$s = 1.2$$
$$n = 10$$
$$\alpha = 0.05$$

Next, use Excel to calculate the test statistic and critical value. The test statistic, t, is produced by entering =(B1-B2)/(B3/SQRT(B4)). Since Excel evaluates critical values for a two-tailed test, we use probability = 2α and enter =TINV(0.1,9). Note $n - 1$ degrees of freedom are used. Since $t > t_\alpha$, we reject the null hypothesis and conclude that the district produces more trash than is anticipated. The Excel spreadsheet is shown below.

	A	B
1	sample mean	10.6
2	population mean	9.8
3	s	1.2
4	n	10
5		
6	t	2.108
7	critical t	1.83

Hypothesis Test for Proportions

Example T.8

Randy has heard that 10% of the population is left-handed. To test this proposition, he takes a sample and records the results. Out of the 65 people surveyed, 10 are left-handed. Using $\alpha = 0.10$, try to reject the claim that 10% of the population is left handed.

Solution:

Immediately, we recognize that we need to perform a z-test for population proportions. Since $np = 65(.10) = 6.5 > 5$ and $n(1-p) = 65(.90) = 58.5 > 5$, the conditions are met for utilizing this test. The null and alternative hypotheses are as follows:

$$H_0: p = 0.10$$
$$H_a: p \neq 0.10$$

Next, use Excel to calculate the test statistic and critical value. The test statistic is produced by entering =(10/65-0.10)/SQRT(0.10*(1-0.10)/65). The result of this calculation is 1.447. Since the test is two-tailed, we use the probability $1 - \frac{\alpha}{2}$ in the NORMSINV function. Enter =NORMSINV(1-0.1/2) to produce the critical value 1.64. Since the absolute value of the test statistic is less than the critical value, we fail to reject the null hypothesis. This means at the 0.10 level of significance this evidence does support the claim that 10% of the population is left-handed.

χ^2 Test for Goodness of Fit

Excel can save a good deal of time and effort in performing some hypothesis tests. The χ^2 test for goodness of fit certainly falls into this category.

Example T.9

A manager at a clothing store believes that pink shirts do not sell as well as other colors. Despite this observation, her buyer continues to supply the store with a uniform proportion of shirt colors. The manager tallies up last month's sales by shirt color. The results are shown below.

Color	Red	Green	Blue	Pink
Sales	27	25	26	17
Expected Sales	23.75	23.75	23.75	23.75

Can the manager prove with a confidence level of 90% that shirts do not sell in equal proportions?

Solution:

First, state the null and alternative hypotheses. We have

$$H_0: p_1 = p_2 = p_3 = p_4$$
$$H_a: \text{any difference amongst the probabilities}$$

Next, enter the data into an Excel spreadsheet and calculate the test statistic. The latter is actually a two part task. The first part requires calculating the difference from expected and uses the calculation $\frac{(\text{sales} - \text{expected sales})^2}{\text{expected sales}}$ for each shirt color. The second part is the sum of these four calculations. This sum, 2.642, is the test statistic. The critical value is calculated by entering =CHIINV(0.1, 3) where $\alpha = 0.10$ and there are $n - 1$ degrees of freedom. This critical value is 6.25. Then, since $2.642 < 6.25$, we fail reject the null hypothesis. Therefore at the 0.90 level of significance, the evidence does not support the manager's claim.

	A	B	C	D	E
1	Color	Red	Green	Blue	Pink
2	Sales	27	25	26	17
3	Expected Sales	23.75	23.75	23.75	23.75
4	(sales - expected sales)^2/expected sales	0.44474	0.0658	0.21316	1.91842
5					
6	Test Statistic (Sum)	2.642			
7	Critical Value	6.25			

χ^2 Test for Association

Example T.10

The U.S. Census reported that Charleston County, S.C. had the following population distribution in 2000:

Observed	Under 18	18 to 44	45 to 64	65 and over	Total
Male	37,462	65,186	32,173	14,966	149,787
Female	36,112	66,131	36,047	21,892	160,182
Total	73,574	131,317	68,220	36,858	309,969

Is there evidence at a 1% level to say that age is affected by gender?

Solution:

Our hypothesis is that age and gender are independent of one another. Symbolically,

H_0: Gender and age are independent.
H_a: Gender and age are not independent.

After entering the data into a table, the next step is calculating the expected value for each cell in the contingency table. These are created by multiplying the row probability by the column probability for each cell. Enter =B$4*$F2/$F$4 into cell B7 and copy this formula (Ctrl + v) into B7:E8. The only remaining step is calculating the *p*-value for the test. This is accomplished by entering =CHITEST(B2:E3,B7:E8). Since this value is below 0.01, we reject the null hypothesis. Therefore, at the 0.01 level of significance the evidence does support the claim that age is affected by gender.

	A	B	C	D	E	F
1		Under 18	18 to 44	45 to 64	65 and Over	Total
2	Male	37,462	65,186	32,173	14,966	149,787
3	Female	36,112	66,131	36,047	21,892	160,182
4	Total	73,574	131,317	68,220	36,858	309,969
5						
6						
7		35553.33	63456.60	32966.10	17810.97	
8		38020.67	67860.40	35253.90	19047.03	
9	p value	0.00				

MINITAB

Hypothesis Test for Proportions

Example T.11

Alfredo thinks he is a better golfer than Tiger Woods. He backs this up with the fact that he hit 133 out of 180 greens in regulation. Tiger is averaging a tour leading 72.5% of greens in regulation. Using $\alpha = 0.10$, test Alfredo's claim of golfing superiority.

Solution:

First, we identify the test as a hypothesis test for proportions. We have the following null and alternative hypotheses:

$$H_0: p \le 0.725$$
$$H_a: p > 0.725$$

This test will use the following parameters:

$$x = 133$$
$$n = 180$$
$$p = 0.725$$

Next, select STAT, BASIC STATISTICS, and *1 Proportion*. Choose **Summarized data** and enter 180 trials and 133 successes. Then select Options and set the **Confidence level** to 90, the **Test proportion** to .725, the select "greater than" for **Alternative**, and choose "Use test and interval based on normal distribution". Selecting OK for the options and main dialog produces the answer. With a p-value of 0.338, Alfredo cannot claim to be the better golfer.

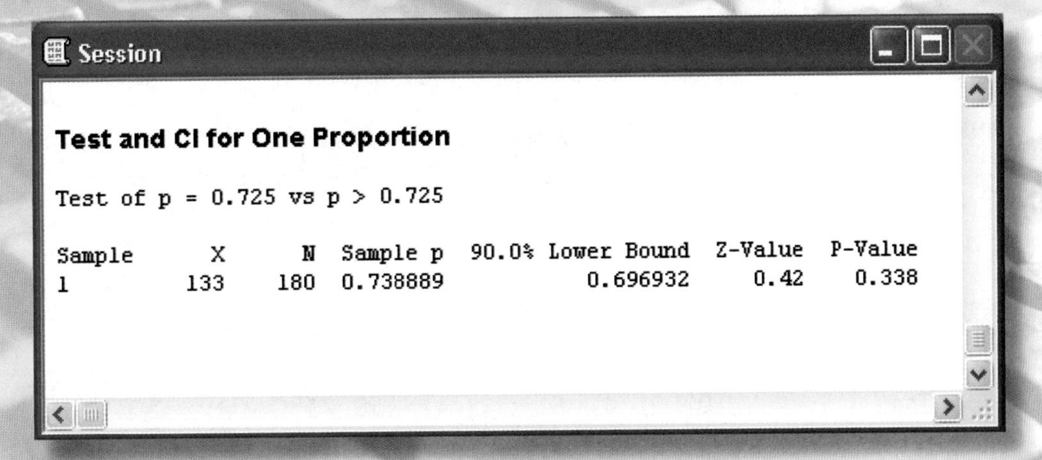

χ^2 Test for Goodness of Fit

Example T.12

A local library is open 20 hours per week. The librarian wants to know whether the number of books checked out each day follows the same distribution as hours of operation. She records the number of books checked out on each day. The results are as follows:

Day	Mon.	Tues.	Wed.	Thurs.	Fri.	Sat.	Sun.
Frequency	101	165	162	180	117	82	closed
Hours Open	2	4	4	5	3	2	closed

Calculate χ^2 with Minitab.

Solution:

First, enter the data into columns in the worksheet. Next, we need to calculate the expected number of checkouts on each day. Select CALC and then choose CALCULATOR. Enter the expression C3/SUM(C3)*SUM(C2) and store result in variable C4. Selecting OK finishes Column C4.

The next step is calculating the addends that compose the χ^2 statistic. This column is also produced with the calculator. Again, select CALC and then choose CALCULATOR. Enter the expression (C2-C4)**2/C4. This time you want to store result in variable C5. Choose OK and the column is produced.

Finally, select CALC and COLUMN STATISTICS. Select Sum with C5 as the input variable. The result is $\chi^2 = 7.6902$. The completed worksheet is shown below.

Worksheet 1 ***

↓	C1-T	C2	C3	C4	C5
	Day	Frequency	Hours Open	expected frequency	partial chi-squared
1	Mon	101	2	80.70	5.10644
2	Tues	165	4	161.40	0.08030
3	Wed	162	4	161.40	0.00223
4	Thurs	180	5	201.75	2.34480
5	Fri	117	3	121.05	0.13550
6	Sat	82	2	80.70	0.02094
7					

χ^2 Test for Association

Minitab can perform this otherwise time consuming process very quickly.

Example T.13

Gretchen wants to know whether gender has a relationship to beverage choice. She defines beverage choice as the first beverage consumed upon arrival at work and uses her workplace to take a sample. She records her coworkers' consumption habits one morning and the results are shown in a table below. Is there evidence at the 0.05 level to say that gender influenced beverage choice.

Gender	Coffee	Tea	Water	Other	Total
Male	10	7	18	9	44
Female	8	12	11	15	46
Total	18	19	29	24	90

Solution:

First, we state our null hypothesis–gender and beverage choice are independent. Symbolically,

H_0: Gender and beverage choice are independent.
H_a: Gender and beverage choice are not independent.

Now enter the data into the worksheet. Do not enter the total column or the total row into the spreadsheet. Next, select STAT, TABLES, and *Chi-Square Test*. Enter C2-C5 as the columns containing the table and select OK. Expected values, χ^2 value, degrees of freedom, and p-value are all displayed. Since $p = .196$, we cannot reject the null hypothesis at $\alpha = 0.05$.

Worksheet 1 ***

↓	C1-T	C2	C3	C4	C5
	Gender	Coffee	Tea	Water	Other
1	Male	10	7	18	9
2	Female	8	12	11	15

Hypothesis Testing (Two or More Populations)

Sections

Objectives

The objectives of this chapter are as follows:

1. Perform a hypothesis test comparing two population means using large independent samples, small independent samples, or dependent samples
2. Perform a hypothesis test comparing two population proportions using large independent samples
3. Perform a hypothesis test comparing two population variances
4. Learn the basics of an ANOVA table

Chapter 11

Introduction

The Battle of the Sexes ... the age old dispute about which gender is better. If we truly wanted to test the claim that one gender is better than another, we would need to do more than simply create a confidence interval; we would need to perform a hypothesis test. Take for example IQ scores. Using our current knowledge of hypothesis testing, we could begin by testing the claim that the average IQ of a man is greater than 100. Let's even assume that to be true! We could then test the claim that the average IQ of a woman is greater than 100. Assuming this is true as well, we could raise the value of the claim. We could next test to see if men, and then women, had a mean IQ greater than 105, and then 110, then 115, etc. We could continue in this manner until one gender's average IQ was greater than our value and the other's was not. This tedious process would be similar to the high jump; the bar keeps being raised until only one person can successfully jump over the bar.

Wouldn't it be simpler to perform only one test? We are not interested in the particular value for men or women's IQ, only in whether one of these values is larger than the other. Thus what we really want is a test that compares, not one that identifies the value.

Just as we discussed creating a confidence interval for one population parameter in Chapter 8 and then extended that process to creating confidence intervals for two population parameters in Chapter 9, we will extend the hypothesis testing procedure we learned in the previous chapter so that it is possible to test claims involving two or more populations. Then maybe we can settle the question of which gender is better once and for all.

Hypothesis Testing - Two Means (Large Independent Samples)

11.1

Recall the steps to a hypothesis test:

1. State the null and alternative hypotheses.

2. Set up the hypothesis test by choosing the test statistic and stating the level of significance.

3. Gather data and calculate the necessary sample statistics.

4. Draw a conclusion by comparing the p-value to the level of significance.

In this chapter we will be comparing two or more population parameters instead of testing a claim about a single parameter. However, the steps for a hypothesis test are the same with two or more parameters as they were before. All we need are the test statistics used for two populations and a method of writing the null and alternative hypotheses. The way that rejection regions are constructed, *p*-values calculated, and the conclusion determined is the same as in the previous chapter.

Memory Booster!

The null hypothesis contains the starting assumption.

First, let's look at how to write the null and alternative hypotheses for two population parameters. There are two different acceptable ways to write the null and alternative hypotheses. The first method directly compares the two populations. We can state that one mean is greater than or less than the other mean. For example,

$$H_0: \mu_1 \le \mu_2$$
$$H_a: \mu_1 > \mu_2.$$

The second way looks at the difference in the two means and compares it to zero. For example,

$$H_0: \mu_1 - \mu_2 \le 0$$
$$H_a: \mu_1 - \mu_2 > 0.$$

This second method is the way we will write the null and alternative hypotheses for population means since it directly relates to the test statistic we will use. The following example illustrates how to determine the hypotheses.

Example 1

Write the null and alternative hypotheses for the claim that population 1's mean is less than population 2's mean.

Solution:

We can write the phrase "population 1's mean is less than population 2's mean" symbolically as $\mu_1 < \mu_2$. By subtracting μ_2 from both sides, we have $\mu_1 - \mu_2 < 0$. Thus the claim is $\mu_1 - \mu_2 < 0$.

The mathematical opposite of the claim is $\mu_1 - \mu_2 \ge 0$. The alternative hypothesis contains the claim being tested. Thus

$$H_0: \mu_1 - \mu_2 \ge 0$$
$$H_a: \mu_1 - \mu_2 < 0.$$

Example 2

Write the null and alternative hypotheses for the claim that population 1's mean is greater than population 2's mean.

Solution:

We can write the phrase "population 1's mean is greater than population 2's mean" symbolically as $\mu_1 > \mu_2$. By subtracting μ_2 from both sides, we have $\mu_1 - \mu_2 > 0$. Thus the claim is $\mu_1 - \mu_2 > 0$.

The mathematical opposite of the claim is $\mu_1 - \mu_2 \leq 0$. Thus

$$H_0: \mu_1 - \mu_2 \leq 0$$
$$H_a: \mu_1 - \mu_2 > 0.$$

Example 3

Write the null and alternative hypotheses for the claim that population 1's mean is not equal to population 2's mean.

Solution:

We can write the phrase "population 1's mean is not equal to population 2's mean" symbolically as $\mu_1 \neq \mu_2$. By subtracting μ_2 from both sides, we have $\mu_1 - \mu_2 \neq 0$. Thus the claim is $\mu_1 - \mu_2 \neq 0$.

The mathematical opposite of the claim is $\mu_1 - \mu_2 = 0$. Thus

$$H_0: \mu_1 - \mu_2 = 0$$
$$H_a: \mu_1 - \mu_2 \neq 0.$$

Example 4

Write the null and alternative hypotheses for the claim that population 1's mean is more than 20 units above population 2's mean.

Solution:

We can write the phrase "population 1's mean is more than 20 units above population 2's mean" symbolically as $\mu_1 > \mu_2 + 20$. This may be rewritten as $\mu_1 - \mu_2 > 20$. The mathematical opposite of the claim is $\mu_1 - \mu_2 \le 20$. Thus

$$H_0: \mu_1 - \mu_2 \le 20$$
$$H_a: \mu_1 - \mu_2 > 20.$$

The test statistic for this type of hypothesis test is associated with the difference in the two means. When the following conditions are true, the z-test statistic should be used.

- each sample must contain at least 30 values ($n_1, n_2 \ge 30$),
- the data sets must be independent of each other,
- each population from which the samples have been chosen is normally distributed, and
- σ is unknown for both populations.

The formula for the z-test statistic for testing the difference between two population means with large independent samples is:

$$z = \frac{(\bar{x}_1 - \bar{x}_2) - (\mu_1 - \mu_2)}{\sqrt{\frac{s_1^2}{n_1} + \frac{s_2^2}{n_2}}}.$$

The following example shows how to use this test statistic to perform hypothesis tests when using large samples. The next two examples will only consider if there is a difference in the population means, not by how much they differ.

Example 5

A researcher wants to test the claim that grocery shoppers spend more when shopping on an empty stomach than when shopping on a full stomach. One morning 93 shoppers are surveyed upon leaving one grocery store. Out of the group, 41 of the shoppers did not eat breakfast that morning. This group spent an average of \$72.27 with a standard deviation of \$8.05. The other 52 shoppers did eat breakfast that morning. This group spent an average of \$69.43 with a standard deviation of \$9.22. Test the claim with a 0.05 level of significance.

cont'd on the next page

Solution:

First, let the empty stomach shoppers be population 1 and the full stomach shoppers be population 2.

1. **State the null and alternative hypotheses.**
Claim: Hungry shoppers spend more than shoppers who are not hungry. Written mathematically, the claim is $\mu_1 > \mu_2$. By subtracting μ_2 from both sides, we have $\mu_1 - \mu_2 > 0$. Thus

$$H_0: \mu_1 - \mu_2 \leq 0$$
$$H_a: \mu_1 - \mu_2 > 0.$$

2. **Set up the hypothesis test by choosing the test statistic and stating the level of significance.**
We are looking at the difference in population means with large independent samples, so we use the z-test statistic

$$z = \frac{(\bar{x}_1 - \bar{x}_2) - (\mu_1 - \mu_2)}{\sqrt{\frac{s_1^2}{n_1} + \frac{s_2^2}{n_2}}}.$$

We will draw a conclusion by computing the p-value for the calculated test statistic and comparing that value to α. For this hypothesis test, $\alpha = 0.05$.

3. **Gather data and calculate the necessary sample statistics.**
The following statistics were given in the problem

$$\bar{x}_1 = 72.27 \qquad \bar{x}_2 = 69.43$$
$$s_1 = 8.05 \qquad s_2 = 9.22$$
$$n_1 = 41 \qquad n_2 = 52$$

Substituting these values into the formula for the test statistic gives us:

$$z = \frac{(\bar{x}_1 - \bar{x}_2) - (\mu_1 - \mu_2)}{\sqrt{\frac{s_1^2}{n_1} + \frac{s_2^2}{n_2}}} = \frac{(72.27 - 69.43) - 0}{\sqrt{\frac{8.05^2}{41} + \frac{9.22^2}{52}}} \approx 1.58$$

*Note that in the formula above, $\mu_1 - \mu_2 = 0$ because H_0 contains that assertion.

Memory Booster!

If $p \leq \alpha$ – Reject H_0
If $p > \alpha$ – Fail to reject H_0

4. **Draw a conclusion by comparing the p-value to the level of significance.**
Because this is a right-tailed test, to find the p-value, we need the area under the normal curve to the right of $z = 1.58$. Using the normal distribution table, we find a p-value of 0.0571. This value is greater than α, which means that we do not have enough evidence to reject the null hypothesis. Therefore, there is not sufficient evidence at the 0.05 level of significance to say that hungry shoppers spend more on groceries than those shopping on a full stomach.

Example 6

Two universities in the same state are bitter rivals. Each university believes that their students are more physically fit than the other. To test the claim that there is a difference in the fitness of the students at each university, 36 students at the first university were surveyed and found to exercise on average 2.9 hours a week with a standard deviation of 1.1 hours. Thirty-eight students at the second university were also surveyed and found to have an average of 2.7 exercise hours a week with a standard deviation of 1.0 hour. Use a 0.05 level of significance to perform a hypothesis test with the given data.

Solution:

1. **State the null and alternative hypotheses.**
Claim: There is a difference in the physical fitness of students at the two schools. Written mathematically, this is $\mu_1 \neq \mu_2$. By subtracting μ_2 from both sides, we have $\mu_1 - \mu_2 \neq 0$. Thus,

$$H_0: \mu_1 - \mu_2 = 0$$
$$H_a: \mu_1 - \mu_2 \neq 0.$$

2. **Set up the hypothesis test by choosing the test statistic and stating the level of significance.**
We are looking at the difference in population means with large independent samples, so we use the z-test statistic

$$z = \frac{(\bar{x}_1 - \bar{x}_2) - (\mu_1 - \mu_2)}{\sqrt{\frac{s_1^2}{n_1} + \frac{s_2^2}{n_2}}}.$$

We will draw a conclusion by computing the p-value for the calculated test statistic and comparing that value to α. For this hypothesis test, $\alpha = 0.05$.

cont'd on the next page

3. **Gather data and calculate the necessary sample statistics.**
The following statistics were given in the problem:

$$\bar{x}_1 = 2.9 \qquad \bar{x}_2 = 2.7$$
$$s_1 = 1.1 \qquad s_2 = 1.0$$
$$n_1 = 36 \qquad n_2 = 38$$

Substituting these values into the formula for the test statistic gives us:

$$\begin{aligned} z &= \frac{(\bar{x}_1 - \bar{x}_2) - (\mu_1 - \mu_2)}{\sqrt{\frac{s_1^2}{n_1} + \frac{s_2^2}{n_2}}} \\ &= \frac{(2.9 - 2.7) - 0}{\sqrt{\frac{1.1^2}{36} + \frac{1.0^2}{38}}} \\ &\approx 0.82 \end{aligned}$$

*Once again note that in the formula above, $\mu_1 - \mu_2 = 0$ because H_0 contains that assertion.

4. **Draw a conclusion by comparing the *p*-value to the level of significance.**
Because this is a two-tailed test, to find the p-value, we need the area under the normal curve to the left of $z = -0.82$ and to the right of $z = 0.82$. Using the normal distribution table, we find a p-value of 0.4122. This value is greater than α, which means that we do not have enough evidence to reject the null hypothesis. Therefore, there is not sufficient evidence at the 0.05 level of significance to say that there is a difference in the fitness of the students at each university.

Now let's consider the case when the amount by which the populations differ is important to the claim.

Example 7

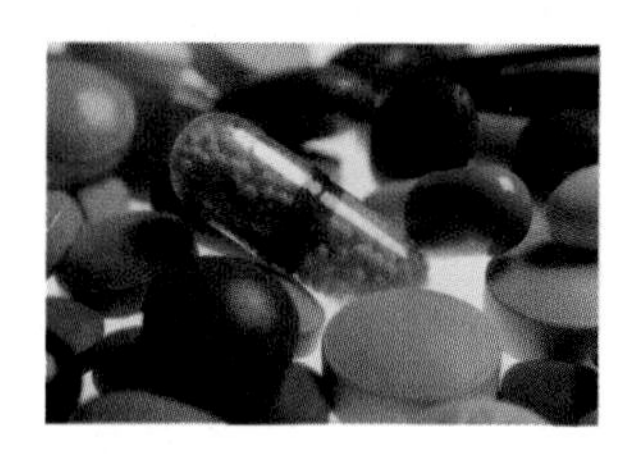

A drug manufacturer claims that its new cholesterol drug, when used together with a healthy diet and exercise plan, lowers cholesterol by over 20 points more than simply changing a patient's diet and exercise regimen. To test the claim, a group of 55 patients with high cholesterol is chosen to take the drug along with a change in diet and exercise. Over the course of 3 months, this group lowers their cholesterol by an average of 44.7 points with a standard deviation of 6.8 points. Another 55 patients change their diet and exercise regimen, but do not take the drug. This group lowers their cholesterol by an average of 23.1 points, with a standard deviation of 5.3 points. Test the claim using a 0.01 level of significance.

Solution:

First, let the group taking the drug be population 1 and the group not taking the drug be population 2.

1. **State the null and alternative hypotheses.**
   Claim: The drug lowers cholesterol by more than 20 points over diet and exercise alone. Notice that the average being tested is how much the cholesterol is lowered. Hence μ refers to the number of cholesterol points lost. Written mathematically, the claim is $\mu_1 > \mu_2 + 20$. By subtracting μ_2 from both sides, we have $\mu_1 - \mu_2 > 20$. Thus

$$H_0: \mu_1 - \mu_2 \le 20$$
$$H_a: \mu_1 - \mu_2 > 20.$$

2. **Set up the hypothesis test by choosing the test statistic and stating the level of significance.**
   We are looking at the difference in population means with large independent samples, so we use the z-test statistic

$$z = \frac{(\bar{x}_1 - \bar{x}_2) - (\mu_1 - \mu_2)}{\sqrt{\frac{s_1^2}{n_1} + \frac{s_2^2}{n_2}}}.$$

   We will draw a conclusion by computing the p-value for the calculated test statistic and comparing that value to α. For this hypothesis test, $\alpha = 0.01$.

3. **Gather data and calculate the necessary sample statistics.**
   The following statistics were given in the problem:

$$\bar{x}_1 = 44.7 \qquad \bar{x}_2 = 23.1$$
$$s_1 = 6.8 \qquad s_2 = 5.3$$
$$n_1 = 55 \qquad n_2 = 55$$

   Substituting these values into the formula for the test statistic gives us:

$$\begin{aligned} z &= \frac{(\bar{x}_1 - \bar{x}_2) - (\mu_1 - \mu_2)}{\sqrt{\frac{s_1^2}{n_1} + \frac{s_2^2}{n_2}}} \\ &= \frac{(44.7 - 23.1) - 20}{\sqrt{\frac{6.8^2}{55} + \frac{5.3^2}{55}}} \\ &\approx 1.38. \end{aligned}$$

cont'd on the next page

*Note that in the previous formula, $\mu_1 - \mu_2 = 20$ because H_0 contains that assertion.

4. **Draw a conclusion by comparing the *p*-value to the level of significance.**
Because this is a right-tailed test, to find the *p*-value, we need the area under the normal curve to the right of $z = 1.38$. Using the normal distribution table, we find a *p*-value of 0.0838. This value is greater than α, which means that we do not have enough evidence to reject the null hypothesis. Therefore, there is not sufficient evidence at the 0.01 level of significance to support the drug manufacturer's claim that the new drug, when used together with a healthy diet and exercise plan, lowers cholesterol by more than 20 points.

Exercises 11.1

Directions: Given the following statistics, calculate the test statistic, *z*.

1. $\bar{x}_1 = 9.21,\ s_1 = 2.01,\ n_1 = 45,\ \bar{x}_2 = 8.76,\ s_2 = 1.77,\ n_2 = 51,\ \mu_1 - \mu_2 \geq 0$

2. $\bar{x}_1 = 72.82,\ s_1 = 7.90,\ n_1 = 31,\ \bar{x}_2 = 75.11,\ s_2 = 6.54,\ n_2 = 39,\ \mu_1 - \mu_2 \leq 0$

3. $\bar{x}_1 = 118.4,\ s_1 = 5.93,\ n_1 = 64,\ \bar{x}_2 = 104.3,\ s_2 = 5.74,\ n_2 = 65,\ \mu_1 - \mu_2 \geq 15$

4. $\bar{x}_1 = 43.1,\ s_1 = 2.33,\ n_1 = 71,\ \bar{x}_2 = 34.3,\ s_2 = 2.96,\ n_2 = 70,\ \mu_1 - \mu_2 = 8$

Directions: Write the null and alternative hypotheses for the following scenarios.

5. The claim is that population 1's mean is less than 30 units less than population 2's mean.

6. Ann claims that the drive time home on a Friday (population 1) is less than the drive time home on a Thursday (population 2).

7. Carly claims that 6 months ago her average time to walk a mile was more than 4 minutes longer than it is currently (population 2).

8. A newspaper claims that the average age of its current readers (population 1) has dropped 3.6 years over the last 10 years (population 2).

Directions: Answer the following questions. Assume all exercises meet the following conditions:

- σ is unknown, and
- The distributions are independent and normally distributed.

9. A car company claims that its new SUV has better gas mileage than its competitor. A sample of 35 of their SUV's shows an average mileage of 12.6 miles per gallon with a standard deviation of 0.4 mpg. A sample of 31 competitor's SUV's shows an average mileage of 12.4 miles per gallon with a standard deviation of 0.3 mpg. Test the company's claim at the 0.05 level of significance.

10. A college student is interested in investigating the claim that students who graduate with a master's degree earn higher salaries than those who finish with a bachelor's degree. She surveys at random 42 recent graduates who completed their master's degree, and finds that they earn on average $38,400 with a standard deviation of $3100. She also surveys at random 45 recent graduates who completed their bachelor's degree, and finds that they earn on average $36,750 with a standard deviation of $3700. Test the claim at the 0.05 level of significance.

11. Rob and Phil are both internal medicine residents in the Southeast, but work at different hospitals. When they compare their schedules, Rob is convinced that on average, residents at his hospital work more than 3 hours more a week than those at Phil's. Rob asks 30 residents at his hospital to record their hours for the week. He finds that they average 74.3 hours worked with a standard deviation of 2.6 hours. Phil also asks 30 residents at his hospital to record their hours for the week. He calculates an average of 70.1 hours with a standard deviation of 2.9 hours. Test Rob's claim at the 0.05 level of significance. Let Rob's hours be population 1 and Phil's hours be population 2.

12. Fran is training for her first marathon, and she wonders if she will receive better training if she runs with a group rather than on her own. She wants to know if there is a significant difference in the number of miles run each week between group runners and individual runners. She interviews 32 people that she knows train in groups, and finds that they average 49 miles per week with a standard deviation of 4.2 miles. She also interviews 30 people who she knows that train on their own and finds that they run on average 47.2 miles per week with a standard deviation of 4.8 miles. Test the claim at the 0.05 level of significance.

13. A weight-loss company wants to make sure that their clients lose more weight than they would without the company's help. An independent researcher collects data on the amount of weight lost from 45 of the company's clients and finds an average weight loss of 12 pounds with a standard deviation of 7 pounds per month. Data from 50 dieters not using the company's services reported an average weight loss of 10 pounds with a standard deviation of 6 pounds. Test the company's claim that using their services results in greater weight loss at the 0.05 level of significance.

14. Two friends, Karen and Jodi, work different shifts for the same ambulance service. They wonder if the different shifts average the same number of calls. Looking at past records, Karen determines from 35 shifts that she averages 5.2 calls per shift with a standard deviation of 1.3 calls. Jodi calculates from 34 shifts that her average number of calls is 4.8 calls per shift with a standard deviation of 1.2 calls per shift. Test the claim that the two shifts average different numbers of calls at the 0.05 level of significance.

15. A professor believes that evening classes average more than 5 points less than morning classes on art history tests. He collects data from a random sample of 250 students from evening classes and finds they have an average grade of 80.2 with a standard deviation of 11.9. A random sample of 300 students from morning classes results in an average grade of 86.8 with a standard deviation of 10.2. Let the evening classes be population 1 and the morning classes be population 2. Test his claim with a 95% level of confidence.

16. A state board of directors for higher education is looking at the salaries of entry-level Ph.D. positions at the state's two major universities to make sure there is not a difference in entry-level pay. Is there a significant difference at the 0.05 level between the salaries if the following is what they found?

Table for Exercise 16 - Salaries of Entry-Level Ph.D. Positions

	University A	University B
Sample Size	42	51
Average Starting Salary	\$58,500	\$60,200
Sample Standard Deviation	\$3200	\$11,700

17. A parent interest group is looking at whether birth order affects scores on the ACT test. It was suggested that the first-born child in a family did more poorly on the ACT test than a second-born child. After surveying 100 first-born students, they found they had an average of 20.9 on the ACT with a standard deviation of 1.8. Surveying 175 second-born students, found an average of 21.1 with a standard deviation of 2.3. Is there enough evidence at the 10% level of significance to say that first-born students do worse on the ACT?

18. Joe wants to get the best possible price on a used luxury car. Because he lives on the border of two states, he believes that prices across the state line are better. Joe surveys 31 car dealerships in his state and finds that used luxury cars have a mean selling price of \$62,065 with a standard deviation of \$1625. In the other state, he surveys 40 dealerships and finds the cars have a mean price of \$61,300 with a standard deviation of \$1475. Is there evidence at the 0.05 level to suggest that car prices are lower in the other state?

19. Lauren and Keri live in different states and disagree about who has higher electric bills. To settle their disagreement, the girls decide to sample electric bills in their area and perform a hypothesis test. Lauren sampled 35 of her friends and found they have an average electric bill of \$104.53 with a standard deviation of \$17.81. Keri surveyed 51 of her friends and found that they have an average bill of \$101.48 with a standard deviation of \$25.30. Is there evidence at the 0.01 level to say that Lauren's area has higher electric bills?

20. A car servicing shop wants to use the best windshield wiper blades for its customers. They have kept track of the average number of sets of blades needed per year for two different brands of blades. 35 customers using Brand A, needed on average 1.2 sets of blades per year with a standard deviation of 0.3. At the same time 30 customers using Brand B needed 1.3 sets on average per year with a standard deviation of 0.7. Is there evidence at the 0.15 level to say that customers using Brand A need less sets of blades per year than those using Brand B?

21. Two college friends are big sports fans. While they are watching baseball one season, they think that there is probably a difference in batting averages between the SEC East and the SEC West. To test their theory they find the mean average of 40 SEC West players to be 0.260 with a standard deviation of 0.026. They also found the mean average of 50 SEC East players to be 0.249 with a standard deviation of 0.051. Is there evidence at the 0.10 level of significance to say that there is a difference in the batting averages?

Hypothesis Testing - Two Means (Small Independent Samples)

11.2

As we have seen in previous chapters, it is not always possible to collect data from large samples. If we wish to perform a hypothesis test comparing the means of two populations where one or both of the sample sizes is less than 30, then we must use the t-distribution.

We will restrict our focus in this section to cases that have the following characteristics:

- The samples are independent
- Each population from which the samples were drawn has a normal distribution
- $n < 30$ for one or both samples
- σ is unknown for both populations

When the above conditions are met, we then must consider whether or not the population variances are likely to be equal. Once again, although neither population variance is actually known, in some cases it is reasonable to assume the variances are the same.

Let's first consider the case when the population variances are assumed to be the same. When this is true, the data from the samples can be **pooled**, or combined. The test statistic is as follows:

$$t = \frac{(\bar{x}_1 - \bar{x}_2) - (\mu_1 - \mu_2)}{\sqrt{\frac{(n_1 - 1)s_1^2 + (n_2 - 1)s_2^2}{n_1 + n_2 - 2}}\sqrt{\frac{1}{n_1} + \frac{1}{n_2}}},$$

where the number of degrees of freedom is $d.f. = n_1 + n_2 - 2$.

On the other hand, it may not be appropriate to assume that the two samples are drawn from populations with the same variance. If the variances are *not assumed to be equal* then the data **cannot be pooled** and the formula for the test statistic becomes

$$t = \frac{(\bar{x}_1 - \bar{x}_2) - (\mu_1 - \mu_2)}{\sqrt{\frac{s_1^2}{n_1} + \frac{s_2^2}{n_2}}},$$

where the number of degrees of freedom is the smaller of the values $n_1 - 1$ and $n_2 - 1$.

Since we are using the t-distribution in either case, we will draw our conclusion based on rejection regions as discussed in Chapter 10. Recall that the rejection region is based on the type of hypothesis test we are performing.

Rejection Regions

- For a left-tailed test, reject H_0 if $t \le -t_\alpha$.
- For a right-tailed test, reject H_0 if $t \ge t_\alpha$.
- For a two-tailed test, reject H_0 if $|t| \ge t_{\alpha/2}$.

Because setting up the null and alternative hypotheses is done in the same manner as when the samples are large, we now have all of the information that we need to perform a hypothesis test comparing the means of two populations using small independent samples. Let's look at an example using each type of test statistic—one where the population variances are assumed to be the same and one where they are not assumed to be the same.

Example 8

A home improvement warehouse claims that people interested in home improvement projects can save time by attending their workshops. Specifically, the warehouse claims that people who attend their tiling workshops take less time to complete comparable projects than those who do not. To test the claim, a group of 10 people who attended their workshops is later surveyed about the time it took to finish the project. This group spent an average of 14.1 hours to complete their projects with a standard deviation of 2.3 hours. Another 10 people chosen for the study did not attend the workshops, and they spent on average 15.0 hours with a standard deviation of 2.4 hours. Test the claim at the 0.01 level of significance. Assume that the populations are normally distributed and that the variances of the populations are the same.

Solution:

First, let population 1 be those people who attended the workshops and population 2 be those who did not.

1. **State the null and alternative hypotheses.**
   Claim: People who attend the workshops spend less time on their tiling projects. Written mathematically, the claim is $\mu_1 < \mu_2$. By subtracting μ_2 from both sides, we have $\mu_1 - \mu_2 < 0$. Thus

$$H_0: \mu_1 - \mu_2 \ge 0$$
$$H_a: \mu_1 - \mu_2 < 0.$$

cont'd on the next page

2. **Set up the hypothesis test by choosing the test statistic and determining the values of the test statistic that would lead to rejecting the null hypothesis.**
   We are looking at the difference in two population means with small independent samples where the population variances are assumed to be the same, so use the *t*-test statistic for pooled data. We next need to find the rejection region. To do this, we first need to know the degrees of freedom for the problem. Using the formula, we get $n_1 + n_2 - 2 = 10 + 10 - 2 = 18$. Thus we have a left-tailed test with 18 degrees of freedom and $\alpha = 0.01$. Therefore, we will reject the null hypothesis if $t < -2.552$.

3. **Gather data and calculate the necessary sample statistics.**
   The following statistics were given in the problem:

$$\bar{x}_1 = 14.1 \qquad \bar{x}_2 = 15.0$$
$$s_1 = 2.3 \qquad s_2 = 2.4$$
$$n_1 = 10 \qquad n_2 = 10$$

   We are told that the population variances are the same, so we will use the pooled test statistic. Substitute the statistics into the formula to obtain the following:

$$\begin{aligned} t &= \frac{(\bar{x}_1 - \bar{x}_2) - (\mu_1 - \mu_2)}{\sqrt{\frac{(n_1-1)s_1^2 + (n_2-1)s_2^2}{n_1+n_2-2}}\sqrt{\frac{1}{n_1}+\frac{1}{n_2}}} \\ &= \frac{(14.1 - 15.0) - 0}{\sqrt{\frac{(10-1)2.3^2 + (10-1)2.4^2}{10+10-2}}\sqrt{\frac{1}{10}+\frac{1}{10}}} \\ &\approx -0.856 \end{aligned}$$

4. **Draw a conclusion using the process described in Step 2.**
   Because $-0.856 > -2.552$, we must fail to reject the null hypothesis. We interpret this conclusion to mean that there is not sufficient evidence at the 0.01 level to say that people who complete the tiling workshops take less time to complete their projects than those who do not.

Example 9

The Smith CPA Firm claims that their clients obtain larger tax refunds than clients of their competitor, Jones and Company CPA. To test the claim, 15 clients from the Smith firm are selected and found to have an average tax refund of $942 with a standard deviation of $103. At Jones and Company, 18 clients are surveyed and found to have an average refund of $898 with a standard deviation of $95. Test the claim at the 0.05 level of significance. Assume that both populations are normally distributed. Because the two firms are located in different parts of a large city, assume that the variances for the two populations are not equal.

Solution:

First, let Smith clients be population 1 and Jones and Company clients be population 2.

1. **State the null and alternative hypotheses.**
   Claim: Smith clients obtain larger tax refunds than Jones and Company clients. Written mathematically, the claim is $\mu_1 > \mu_2$. By subtracting μ_2 from both sides, we have $\mu_1 - \mu_2 > 0$. Thus

$$H_0: \mu_1 - \mu_2 \le 0$$
$$H_a: \mu_1 - \mu_2 > 0.$$

2. **Set up the hypothesis test by choosing the test statistic and determining the values of the test statistic that would lead to rejecting the null hypothesis.**
   We are looking at the difference in two population means with small independent samples where the population variances are assumed to be different, so use the t-test statistic for un-pooled data. We next need to find the rejection region. To do this, we first need to know the degrees of freedom. The number of degrees of freedom for populations with different variances is the smaller of $n_1 - 1 = 15 - 1 = 14$ and $n_2 - 1 = 18 - 1 = 17$. Thus the number of degrees of freedom for this problem is 14. For a right-tailed test with 14 degrees of freedom and $\alpha = 0.05$, we reject the null hypothesis if $t \ge 1.761$.

3. **Gather data and calculate the necessary sample statistics.**
   The following statistics were given in the problem:

$$\bar{x}_1 = 942 \qquad \bar{x}_2 = 898$$
$$s_1 = 103 \qquad s_2 = 95$$
$$n_1 = 15 \qquad n_2 = 18$$

   We are told that the population variances are different, so we will use the un-pooled test statistic. Substitute the statistics into the formula to obtain the following:

cont'd on the next page

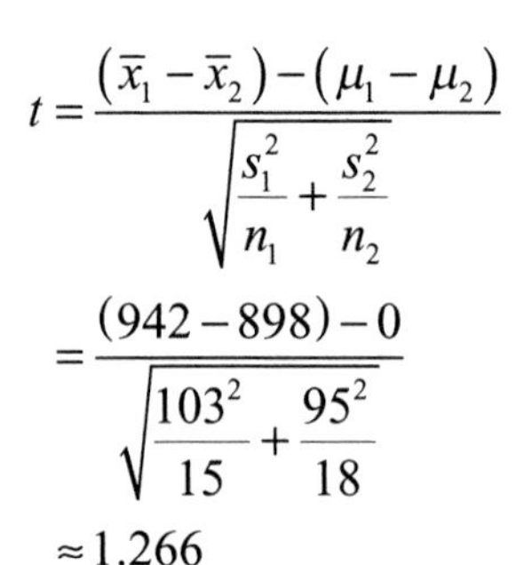

$$t = \frac{(\bar{x}_1 - \bar{x}_2) - (\mu_1 - \mu_2)}{\sqrt{\frac{s_1^2}{n_1} + \frac{s_2^2}{n_2}}} = \frac{(942 - 898) - 0}{\sqrt{\frac{103^2}{15} + \frac{95^2}{18}}} \approx 1.266$$

4. **Draw a conclusion using the process described in Step 2.**
   Because 1.266 < 1.761, we must fail to reject the null hypothesis. We interpret this conclusion to mean that the evidence does not sufficiently support Smith CPA's claim that their clients receive larger tax refunds than clients of Jones and Company.

Exercises

11.2

Directions: Calculate the t-test statistic and determine the number of degrees of freedom for the following.

1. $\bar{x}_1 = 93$, $s_1 = 10.4$, $n_1 = 21$, $\bar{x}_2 = 89$, $s_2 = 9.5$, $n_2 = 18$, $\mu_1 - \mu_2 = 0$
   Assume the variances are not the same.

2. $\bar{x}_1 = 3.4$, $s_1 = 0.3$, $n_1 = 5$, $\bar{x}_2 = 3.7$, $s_2 = 0.5$, $n_2 = 7$, $\mu_1 - \mu_2 \le 0$
   Assume the variances are not the same.

3. $\bar{x}_1 = 33$, $s_1 = 2.1$, $n_1 = 14$, $\bar{x}_2 = 31$, $s_2 = 2.8$, $n_2 = 11$, $\mu_1 - \mu_2 > 0$
   Assume the variances are the same.

4. $\bar{x}_1 = 24$, $s_1 = 1.3$, $n_1 = 19$, $\bar{x}_2 = 23$, $s_2 = 1.1$, $n_2 = 22$, $\mu_1 - \mu_2 = 0$
   Assume the variances are the same.

Directions: Assume all exercises meet the following conditions and perform hypothesis tests to test the claims:

- The samples are independent
- Each population from which the samples were drawn has a normal distribution
- $n < 30$ for one or both samples
- σ is unknown for both populations

5. A professor is concerned that the two sections of college algebra that he teaches are not performing at the same level. To test his claim, he looks at the exam average from a random sample of each of his classes. In Class

1, the exam average of 12 students is 79 with a standard deviation of 6. In Class 2, the exam average of 15 students is 81 with a standard deviation of 7. Test the professor's claim at the 0.05 level of significance. Assume that the population variances are the same.

6. A manufacturer fills soda bottles. Periodically they test to see if there is a difference in the amount of soda put in cola and diet cola bottles. A random sample of 14 cola bottles contains an average of 502 mL of cola with a standard deviation of 4 mL. A random sample of 16 diet cola bottles contains an average of 499 mL of cola with a standard deviation of 5 mL. Test the claim that there is a difference in the fill levels of the two types of soda using a 0.01 level of significance. Assume that the population variances are different since different machines are used for the filling process.

7. While shopping for a cookout, Ian noticed that a bag of charcoal briquets claimed to burn longer because they were 60% larger than the competitors. Feeling tired of being taken for a shopper who just believed what the manufacturer wanted him to believe, Ian decided to test this claim that charcoal from bag A is larger than that of bag B. Ian believed that there was probably not any real difference overall in the two charcoals. He bought a bag of each charcoal and measured some of the briquets. Here's what he found:

Table for Exercise 7 - Size of Charcoal Briquets		
	Charcoal from Bag A	**Charcoal from Bag B**
Sample Size	8	6
Average Size	3.2	2.1
Standard Deviation	0.5	0.9

Test Ian's claim that the charcoals are the same at the 0.02 level. Assume the variances are different.

8. A new small business wants to know if their current advertising is effective. They decide to look at the average number of customers who make a purchase in the store on days following advertising as compared to those days following no advertisements. They found that on 11 days following no advertisements, they averaged 18 purchasing customers with a standard deviation of 3.5 customers. On 6 days following advertising, they averaged 23 customers with a standard deviation of 2.8 customers. Test the claim at the 0.01 level, that the average number of customers who make a purchase in the store is less on days following no advertising compared to days following advertising. Assume the variances are the same.

9. Gary has discovered a new painting tool to help him in his work. If he can prove to himself that the painting tool helps reduce the amount of time it takes to paint a room, he has decided to invest in a tool for each of his helpers as well. Gary knows from his 6 painting jobs in the last week that on average it takes him 4.2 hours to paint a medium sized room without the tool with a standard deviation of 0.5 hours. After purchasing the tool, he used it to paint 4 medium sized rooms and found that it took him on average 3.9 hours to paint with a standard deviation of 0.7 hours. At the 0.10 level, can Gary say that without the tool, his average time for painting was greater than with the tool? Assume the variances are the same.

10. Joyce is trying to convince her husband that she spends more at the grocery store when she takes their children with her to shop, in hopes that she can make the trips alone in the future. She shows him that on 4 trips with the children she spent on average \$122.56 with a standard deviation of \$13.12. But, on 5 trips without the children, she only spent \$108.31 on average with a standard deviation of \$17.06. Assume the variances are different. At the 0.10 level, can Joyce convince her husband that it's cheaper to not take her children with her to the grocery store?

11. While running some quality control tests, a manager at a chip factory noticed a difference in the mean weight of chips coming off two different production lines. To see if the bags in line A did actually have a mean weight lower than those in line B, he randomly tested some of the bags off each line. Here's what he found:

Table for Exercise 11 - Mean Weight of Chips		
	Line A	**Line B**
Sample Size	20	15
Mean Weight	309.6 g	311.8 g
Standard Deviation	15.9 g	13.2 g

If he assumes the variances in the two lines to be different, can he conclude at the 0.10 level that the bags from line A weigh less than the bags from line B?

12. A physician wants to test the claim that the adult height of premature babies is different from that of full term babies. To do this, he finds 18 adults that were born prematurely and calculates that their average height is 68.1 inches with a standard deviation of 2.3 inches. He also finds 20 adults who were carried full term and calculates their average height to be 68.9 inches with a standard deviation of 2.0 inches. Assume that the population standard deviations are different, and test the claim at the 0.05 level of significance.

13. A pharmaceutical company needs to know if its new cholesterol drug, Praxor, is effective at lowering cholesterol. They believe that the 25 participants that take Praxor will average a greater loss in cholesterol than the 25 participants taking a placebo. After the experiment is complete, the researchers find that the treatment group lowered their cholesterol an average of 23.5 points with a standard deviation of 5.8 points. The participants in the control group lowered their cholesterol an average of 18.9 points with a standard deviation of 4.1 points. Assume that the population variances are different and test the company's claim at the 0.01 level.

14. A speech pathology professor believes from experience that boys begin talking at a later age than do girls. To test her theory, she gathers information from the parents of 12 boys and 14 girls that she knows. The boys began talking at 1.3 years on average with a standard deviation of 0.15 years. The girls began talking at 1.2 years on average with a standard deviation of 0.12 years. Assume that the population standard deviations are equal, and test the claim at the 0.05 level of significance.

15. Insurance Company A claims that their customers pay less for insurance than customers at their competitor, Company B. You wonder if this is true, so you take an informal survey among your friends. Of the 9 people you know who use Company A, they pay an average of $152 a month with a standard deviation of $17. Of the 11 people you know who use Company B, they pay an average of $155 a month with a standard deviation of $14. Assume that the population variances are the same, and test Company A's claim at the 0.01 level of significance.

16. A women's group believes that women are offered lower starting salaries than men applying for the same jobs. To test this claim, the group sends 10 women and 10 men to various job interviews around town. The women were offered on average a starting salary of $29,500 with a standard deviation of $1100. The men were offered on average a starting salary of $30,500 with a standard deviation of $950. Assume that the population standard deviations are different, and test the group's claim at the 0.05 level of significance.

17. A psychologist wants to test the claim that the time it takes to complete a work of art is different between male and female artists. He chooses for his sample 12 men and 11 women and asks each to estimate the amount of time it takes for them to complete a work. The men took on average 4.6 hours to complete a work with a standard deviation of 1.3 hours. The women took on average 5.4 hours with a standard deviation of 1.1 hours. Assume that the population variances are equal and test the claim at the 0.05 level of significance.

Hypothesis Testing - Two Means (Dependent Samples)

11.3

So far in this chapter we have looked at hypothesis testing using independent samples. We now turn our attention to situations when the samples are dependent. Recall from Chapter 9 that dependent data are connected in a specific manner, or **paired**. Further, we must be sure that all of the data are drawn from normal distributions.

When working with paired data, begin by calculating the differences between each pair of values. The hypotheses for dependent samples are based on the **mean** of the **paired differences**, μ_d, instead of the difference in the two populations' means like we used for independent samples. Let's look at an example of how to write the null and alternative hypotheses.

> **Memory Booster!**
>
> If subtracting the before treatment value from the after treatment value, a *reduction* in the value would be a negative number. The inequality signs will also be reversed: a "reduction of more than" would be <, and a "reduction of less than" would be >.

When calculating the differences between population 1's values and population 2's values, you will get opposite values depending on which population is first. For example, if you subtract population 1's value of 5 from population 2's value of 8, then you get the number 3. However, if you reverse the order, you get –3. The difference is the sign. So, for consistency, we need to have a standard way of deciding which population will be first. Let's impose the rule that in situations where there is a measurement before and after some treatment, the measurement after the treatment is population 2, thus it will always be the first value in the subtraction. Therefore, if we expect the treatment to reduce the value of the measurement, then we would expect to get negative differences. On the other hand, if we expect the treatment to increase the value of the measurement, we would expect to get positive values. This procedure for deciding on population 1 and 2 should help us correctly set up our hypotheses. Let's look at an example.

Example 10

A new diet pill claims to help a person lose more than 7 pounds in the first week. To test this claim, 15 people volunteer to take the diet pill for one week. Their weight is measured at the beginning and end of the week. State the null and alternative hypotheses for this hypothesis test.

Solution:

For this example, population 1 will be the starting weight and population 2 will be the ending weights. To consider the difference between population 1 and population 2 we want to subtract the person's starting weight from their final weight. The claim is that the person will lose more than 7 pounds, or that the difference in their starting weight and their ending weight will be more than 7 pounds. Since the claim is that the weight will be reduced, we should expect negative values. Thus if we are expecting people to lose more than 7 pounds, we should expect negative differences *less than* −7. (This is due to the nature of negative numbers.) Mathematically, this claim is written as $\mu_d < -7$. The opposite of this claim is $\mu_d \geq -7$. Thus the null and alternative hypotheses are

$$H_0: \mu_d \geq -7$$
$$H_a: \mu_d < -7.$$

The test statistic used to analyze $\bar{d}$ when there are n sets of paired data drawn from normal distributions is

$$t = \frac{\bar{d} - \mu_d}{\left(\frac{s_d}{\sqrt{n}}\right)}$$

with $n - 1$ degrees of freedom.

Memory Booster!

Recall that $\bar{d}$ is the mean of the paired differences.

$$\bar{d} = \frac{\sum d}{n}$$

Since we are using the t-distribution once again, we will draw our conclusion based on rejection regions as discussed in Chapter 10. Recall that the rejection region is based on the type of hypothesis test we are performing.

Rounding Rule

Round $\bar{d}$ to one more decimal place than the original data, when given.

Rejection Regions

- For a left-tailed test, reject H_0 if $t \leq -t_\alpha$.
- For a right-tailed test, reject H_0 if $t \geq t_\alpha$.
- For a two-tailed test, reject H_0 if $|t| \geq t_{\alpha/2}$.

The following example illustrates a complete hypothesis test for dependent samples.

Example 11

The standard course at a local defensive driving school often shows films depicting violent car crashes and graphic pictures of injuries sustained in these crashes. The driving school has shown these videos for many years, believing that they reduce their students' average speed on the highway. A group of concerned citizens, who feel that these videos are very disturbing, is not convinced that the videos reduce highway speed at all. In fact, they claim that these videos reduce a person's speed on the highway by less than 5 miles per hour. To test their claim, they install electronic data recorders (EDRs) on the vehicles of 10 volunteers, who agree to drive as they normally would for two weeks while the EDR records their vehicles' speed. After the initial driving period, each volunteer attends the defensive driving school. After finishing the course, including viewing the videos, the EDRs again record their vehicles' speed for another two weeks. The data on the following page

cont'd on the next page

represent the average highway speed for each volunteer before and after attending the defensive driving school. Use these data to test the citizens' claim that these videos reduce a person's speed on the highway by less than 5 miles per hour. Use a 0.05 level of significance.

Table 1 - Average Highway Speed (in Miles per Hour)										
Before	75.83	80.12	65.41	70.03	73.91	76.02	75.10	67.89	81.12	77.67
After	72.13	73.87	66.09	68.43	71.45	73.67	70.19	65.34	75.31	70.92

Solution:

1. **State the null and alternative hypotheses.**
We want to subtract the average speed before viewing the videos from their average speed after viewing the videos. The citizens' claim is that these videos reduce a person's speed on the highway by less than 5 miles per hour. Thus we should expect these values to be greater than –5. (A difference of –5 would be a reduction of 5 mph.) Therefore, the citizens' claim, written symbolically, is $\mu_d > -5$. The opposite of this claim is $\mu_d \leq -5$. Thus the hypotheses are

$$H_0\text{: } \mu_d \leq -5$$
$$H_a\text{: } \mu_d > -5.$$

2. **Set up the hypothesis test by choosing the test statistic and determining the values of the test statistic that would lead to rejecting the null hypothesis.**
Since we are talking about the population mean for paired data, we will use the test statistic

$$t = \frac{\bar{d} - \mu_d}{\left(\frac{s_d}{\sqrt{n}}\right)}.$$

The level of significance is $\alpha = 0.05$. For the rejection region, we have $d.f. = 10 - 1 = 9$, which gives a critical value of 1.833 from the table. Since this is a right-tailed test, we will reject the null hypothesis if $t \geq 1.833$.

3. **Gather data and calculate the necessary sample statistic.**
Since we were given raw data, we need to begin by calculating the differences, as well as the mean and standard deviation of these differences. The differences are listed in the table below.

Table 2 - Average Highway Speed (in Miles per Hour)										
Before	75.83	80.12	65.41	70.03	73.91	76.02	75.10	67.89	81.12	77.67
After	72.13	73.87	66.09	68.43	71.45	73.67	70.19	65.34	75.31	70.92
Difference, *d*	–3.70	–6.25	0.68	–1.60	–2.46	–2.35	–4.91	–2.55	–5.81	–6.75

The mean of these differences is −3.570 and the standard deviation is 2.353. Substituting these values into the formula for our test statistic, we obtain

$$t = \frac{\bar{d} - \mu_d}{\left(\frac{s_d}{\sqrt{n}}\right)} = \frac{-3.570 - (-5)}{\left(\frac{2.353}{\sqrt{10}}\right)} \approx 1.922.$$

4. **Draw a conclusion using the process described in Step 2.**
   Since $t = 1.922$, which is in the rejection region, we reject the null hypothesis. Thus there is enough evidence, at the 0.05 level of significance, to support the citizens' claim that these videos reduce a person's speed on the highway by less than 5 miles per hour.

Example 12

Dr. Xiong, a clinical psychologist, wishes to test the claim that there is a significant difference in a person's weight if their father instead of their mother raises them. Luckily, Dr. Xiong knows of five sets of identical twins that were raised separately, one twin by the mother and one twin by the father, and who are willing to participant in a study to help her test her claim. Each twin is weighed and identified as having been raised by their mother or their father. The table below lists the results. Does this evidence support Dr. Xiong's claim at the 0.01 level of significance?

Table 3 - Weight of Twin (in Pounds)					
Twin raised by father	143.67	235.91	156.34	187.21	129.81
Twin raised by mother	134.81	221.37	163.92	193.45	131.38

Solution:

1. **State the null and alternative hypotheses.**
   It does not matter how we label the groups, so let's label the twins raised by their father population 1 and the twins raised by their mother population 2. Dr. Xiong claims that there is a significant difference in a person's weight if their father instead of their mother raises them. If there is a significant difference in weight, then the difference is not zero. Written symbolically, the claim is $\mu_d \neq 0$. The opposite of this claim is $\mu_d = 0$. Thus the hypotheses are

cont'd on the next page

$$H_0: \mu_d = 0$$
$$H_a: \mu_d \neq 0.$$

2. **Set up the hypothesis test by choosing the test statistic and determining the values of the test statistic that would lead to rejecting the null hypothesis.**
   Since we are talking about the population mean for paired data, we will use the test statistic

$$t = \frac{\bar{d} - \mu_d}{\left(\frac{s_d}{\sqrt{n}}\right)}.$$

   The level of significance is $\alpha = 0.01$. For the rejection region, we have $d.f. = 5 - 1 = 4$, which gives a critical value of 4.604 from the table. Since this is a two-tailed test, we will reject the null hypothesis if $|t| \geq 4.604$.

3. **Gather data and calculate the necessary sample statistic.**
   Since we were given raw data, we need to begin by calculating the differences, as well as the mean and standard deviation of these differences. The differences are listed in the table below.

Table 4 - Weight of Twin (in Pounds)					
Twin raised by father	143.67	235.91	156.34	187.21	129.81
Twin raised by mother	134.81	221.37	163.92	193.45	131.38
Difference, *d*	–8.86	–14.54	7.58	6.24	1.57

   The mean of these differences is –1.602 and the standard deviation is approximately 9.694561. Substituting these values into the formula for our test statistic, we obtain

$$\begin{aligned} t &= \frac{\bar{d} - \mu_d}{\left(\frac{s_d}{\sqrt{n}}\right)} \\ &= \frac{-1.602 - 0}{\left(\frac{9.694561}{\sqrt{5}}\right)} \\ &\approx -0.370. \end{aligned}$$

4. **Draw a conclusion using the process described in Step 2.**
   Since $t = -0.370$, which is not in the rejection region, we fail to reject the null hypothesis. Thus there is not enough evidence, at the 0.01 level of significance, to support Dr. Xiong's claim that there is a significant difference in a person's weight if their father instead of their mother raises them.

Exercises

11.3

Directions: Answer each of the following questions.

1. An anger management course claims that, after completing their seminar, participants will lose their temper less often. Always a skeptic, you decide to test their claim. A group of 12 seminar participants is chosen and asked to record the number of times they each lost their temper in the two weeks prior to the course. After the course is over, the same participants are asked to record the number of times they lost their temper in the following two weeks. The table below lists the results of the survey. Test the claim at the 0.05 level of significance.

Table for Exercise 1 - Number of Times Temper was Lost

Participant	1	2	3	4	5	6	7	8	9	10	11	12
Before	8	10	6	7	4	11	12	5	6	3	6	4
After	6	5	6	6	5	9	4	5	4	4	5	4

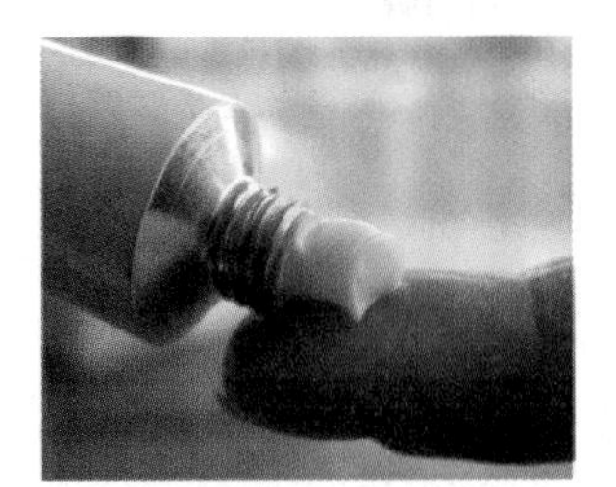

2. A new eye cream claims to reduce the appearance of fine lines and wrinkles after just 14 days of application. To test the claim, a group of 10 women are chosen to participate in the study. The number of fine lines and wrinkles visible around each of their eyes is recorded before and after the 14 days of treatment. The table below displays the results. Test the claim at the 0.01 level of significance.

Table for Exercise 2 - Number of Fine Lines and Wrinkles

Participant	1	2	3	4	5	6	7	8	9	10
Before	8	14	13	15	10	16	9	10	11	10
After	6	14	11	14	10	14	9	9	11	8

3. An SAT prep course claims to increase student scores by more than 60 points. A group of 9 students who have previously taken the SAT are chosen to take the prep course. Their SAT scores before and after completing the prep course are listed below. Test the claim at the 0.01 level of significance.

Table for Exercise 3 - SAT Scores

Participant	1	2	3	4	5	6	7	8	9
Pre-course SAT	1010	980	1170	1200	1040	1280	1450	1470	1500
Post-course SAT	1100	1260	1190	1280	1170	1370	1440	1500	1520

4. Sarah believes that cutting caffeine out of a person's diet completely will allow them more restful sleep at night. In fact, she believes that her friends will have over 2 more nights of restful sleep after cutting caffeine from their diets. She enlists 8 friends to help her test this theory. Each person is asked to consume 2 caffeinated beverages a day for 28 days, then cut back to no caffeinated beverages for the following 28 days. During each period, her friends are asked to record the number of nights that they had a restful night of sleep. The table below gives the results. Test Sarah's claim at the 0.05 level of significance.

Table for Exercise 4 - Number of Nights of Good Sleep								
Participant	1	2	3	4	5	6	7	8
Nights of good sleep (with caffeine)	21	20	26	20	24	21	18	15
Nights of good sleep (without caffeine)	22	24	27	23	21	26	22	23

5. To test the claim that children from the same parents do not necessarily have the same weight, some students in an upper level statistics class surveyed 9 families with at least 2 boys in the family having the same parents. Their weights, taken at the same age, are listed below. The data collected is listed in the table below. Is there enough evidence, at the 0.10 level of significance, to support the claim that children from the same parents do not necessarily have the same weight?

Table for Exercise 5 - Boy's Weight (in Pounds)									
Family	1	2	3	4	5	6	7	8	9
Boy A's Weight	94	138	171	131	159	110	148	90	170
Boy B's Weight	93	140	176	130	173	112	145	90	178

6. One of the top golf camps in the country advertises that a week with their coaches will lower your average golf score by two points. One disgruntled customer claims that the course does not live up to its claim. To test the customer's claim, eight men attending the camp agreed to participate in the study, and their pre-camp and post-camp average scores are listed in the table below. (These are average scores based on a par 72 course.) Test the customer's claim at the 0.10 level of significance.

Table for Exercise 6 - Average Golf Scores								
Participant	A	B	C	D	E	F	G	H
Pre-Camp	75	74	76	75	76	76	74	77
Post-Camp	73	72	73	74	74	72	74	75

7. A violin teacher wants to encourage parents of five year olds that after a year of lessons, the students will increase their stamina for standing in the correct position by more than 30 minutes. She recorded the times that the students could hold a correct position for practicing at the beginning of the year and then again at the end. Here's what she found:

Table for Exercise 7 - Stamina for Correct Position (in Minutes)								
Student	1	2	3	4	5	6	7	8
Start of Lessons	5	7	5	4	10	6	8	5
After a Year of Lessons	36	39	40	35	41	37	38	37

Is there evidence at the 0.10 level that the stamina times increase by more than 30 minutes in 5 year olds at the end of one year of violin lessons?

8. A psychology graduate student wants to test the claim that there is a significant difference in the IQ of husbands and their wives. To test this claim, she measures the IQ of 9 married couples using a standard IQ test. The results of the IQ tests are listed in the table below. Using a 0.05 level of significance, test the claim that there is a significant difference in the IQ of husbands and their wives.

Table for Exercise 8 - Husband and Wife's IQs									
Husband's IQ	100	110	132	120	90	115	124	121	107
Wife's IQ	98	111	134	119	95	116	122	118	110

9. A local school district is looking at implementing a new textbook that, according to the publishers, will increase standardized test scores by more than 10 points. Never willing to believe a publisher's claim without evidence to support it, the school board decides to test the publisher's claim. The school chooses two second grade classrooms for the study. One classroom was assigned the new textbook and the other classroom the traditional textbook. Eight children from each classroom are then paired based on demographics and ability. The following table lists the test scores for the pairs. Does this evidence support the company's claim at the 0.05 level of significance?

Table for Exercise 9 - Test Scores								
New Book	78	82	90	67	79	83	89	93
Old Book	67	70	79	54	68	71	78	82

10. An economist studying inflation in gas prices in 1996 and 1997 believes that the average price of gasoline, even after adjusting for inflation, has changed from year to year. To test his claim, he samples 10 different gas stations from each year and records the average price of gasoline at each station. He then adjusts the gas prices for inflation. His results are given in the table below. Test the economist's claim at the 0.01 level of significance.

Table for Exercise 10 - Gasoline Prices

Gas Station	1996	1997
A	\$1.29	\$1.31
B	\$1.25	\$1.26
C	\$1.29	\$1.29
D	\$1.19	\$1.21
E	\$1.23	\$1.26
F	\$1.25	\$1.24
G	\$1.30	\$1.31
H	\$1.29	\$1.29
I	\$1.28	\$1.30
J	\$1.30	\$1.31

Hypothesis Testing - Two Proportions (Large Independent Samples)

11.4

Let's now turn our attention to hypothesis tests to examine whether two population proportions differ. The steps in the hypothesis test are the same as in the previous sections, the only new piece of information that we need is the test statistic.

We will continue to write our null and alternative hypotheses in the form $p_1 - p_2 < 0$. Since H_0 contains an assertion of equality, we can pool the data from both samples to give an estimate of the common proportion, which we denote as $\bar{p}$. We will need this estimate to calculate the test statistic in these hypothesis tests.

> **Memory Booster!**
>
> x_i is the number of successes that occur in the ith sample.

The test statistic is

$$z = \frac{(\hat{p}_1 - \hat{p}_2) - (p_1 - p_2)}{\sqrt{\bar{p}(1-\bar{p})\left(\frac{1}{n_1} + \frac{1}{n_2}\right)}}$$

where $\bar{p} = \dfrac{x_1 + x_2}{n_1 + n_2}$.

We will restrict our discussion to comparing population proportions using large, independent samples. To determine if the samples are large enough for z to be normally distributed, make sure that the following inequalities all hold:

$$n_1\hat{p}_1 \geq 5 \qquad n_2\hat{p}_2 \geq 5$$

$$n_1(1 - \hat{p}_1) \geq 5 \qquad n_2(1 - \hat{p}_2) \geq 5$$

For all exercises in this section, you can safely assume that the sample sizes given are large enough to justify using the normal distribution to compare the population proportions. Since we are using a normal distribution, we can use either rejection regions or p-value methods discussed earlier to determine the conclusion. The following example illustrates how to use this new test statistic.

Example 13

The mayor's chief of staff thinks that a local newspaper article has changed the way that the community thinks about the mayor. To test his theory, he used a poll of the mayor's approval rating taken before the article came out and compared it to the mayor's approval rating after the article. Before the article ran in the paper, 480 out of 1200 voters thought the mayor was trustworthy. After the article, 550 out of 1180 thought he was trustworthy. Based on these statistics, use a 5% level of significance to test the claim that the mayor's approval rating has gone up in the area of trust.

cont'd on the next page

Solution:

Let the approval rating before the article be population 1 and after the article be population 2.

1. **State the null and alternative hypotheses.**
   Claim: The approval proportion before the newspaper article, p_1, is less than the approval proportion after the article, p_2. Written mathematically, the claim is $p_1 < p_2$ which can also be written as $p_1 - p_2 < 0$. Thus

$$H_0: p_1 - p_2 \geq 0$$
$$H_a: p_1 - p_2 < 0.$$

2. **Set up the hypothesis test by choosing the test statistic and stating the level of significance.**
   We are looking at the difference in population proportions with large independent samples, so we use the z-test statistic

$$z = \frac{(\hat{p}_1 - \hat{p}_2) - (p_1 - p_2)}{\sqrt{\bar{p}(1-\bar{p})\left(\frac{1}{n_1} + \frac{1}{n_2}\right)}}.$$

   We will draw a conclusion by computing the p-value for the calculated test statistic and comparing that value to α. For this hypothesis test, $\alpha = 0.05$.

3. **Gather data and calculate the necessary sample statistic.**
   Begin by calculating the sample proportions. The sample proportion for sample 1, before the article, is $\hat{p}_1 = \frac{480}{1200} = 0.400$. The sample proportion for sample 2, after the article, is $\hat{p}_2 = \frac{550}{1180} \approx 0.466102 \approx 0.466$.
   Lastly, we need $\bar{p}$ before substituting these values into the test statistic.

$$\begin{aligned} \bar{p} &= \frac{x_1 + x_2}{n_1 + n_2} \\ &= \frac{480 + 550}{1200 + 1180} \\ &\approx 0.432773 \\ &\approx 0.433 \end{aligned}$$

   Substituting these into the test statistic, z, we obtain the following:

$$z = \frac{(\hat{p}_1 - \hat{p}_2) - (p_1 - p_2)}{\sqrt{\bar{p}(1-\bar{p})\left(\frac{1}{n_1} + \frac{1}{n_2}\right)}}$$

Rounding Rule

Round $\hat{p}$ to 3 decimal places. If the first two decimal places are both zero, round to the first non-zero digit.

Rounding Rule

When dealing with intermediate calculation, do not round the intermediate steps. For example, although $\hat{p}$ is normally rounded to three decimal places, for intermediate calculation it is best to keep $\hat{p}$ out to six decimal places in order to not affect the remaining calculations with a rounding error.

$$=\frac{(0.40-0.466102)-0}{\sqrt{0.432773(1-0.432773)\left(\frac{1}{1200}+\frac{1}{1180}\right)}}$$

$$\approx -3.25$$

*Note that in the formula above, $p_1 - p_2 = 0$ because H_0 contains that assertion.

4. **Draw a conclusion by comparing the p-value to the level of significance.**
Because this is a left-tailed test, to find the p-value, we need the area under the normal curve to the left of $z = -3.25$. Using the normal distribution table, we find a p-value of 0.0006. Since the p-value is smaller than α, we reject the null hypothesis. Therefore, there is sufficient evidence at the 0.05 level of significance to support the chief of staff's claim that the mayor's approval rating has gone up.

Example 14

Last month's pass/fail record at two driving schools in town, Flynt's School of Safe Driving and Pass with Pops, are listed below.

Table 5 - Pass/Fail Record

	Pass	Fail
Flynt's School of Safe Driving	17	8
Pass with Pops	23	6

Based on these statistics, use a 10% level of significance to test whether the pass-fail proportions at the two centers are different.

Solution:

1. **State the null and alternative hypotheses.**
Claim: The pass proportion at Flynt's School, p_1, is different from the pass proportion at Pops', p_2. Written mathematically, the claim is $p_1 \neq p_2$ which can also be written as $p_1 - p_2 \neq 0$. Thus

$$H_0: p_1 - p_2 = 0$$
$$H_a: p_1 - p_2 \neq 0.$$

2. **Set up the hypothesis test by choosing the test statistic and stating the level of significance.**
We are looking at the difference in population proportions with large independent samples, so we use the z-test statistic

cont'd on the next page

$$z=\frac{(\hat{p}_1-\hat{p}_2)-(p_1-p_2)}{\sqrt{\bar{p}(1-\bar{p})\left(\frac{1}{n_1}+\frac{1}{n_2}\right)}}$$

We will draw a conclusion by computing the p-value for the calculated test statistic and comparing that value to α. For this hypothesis test, $\alpha = 0.10$.

3. **Gather data and calculate the necessary sample statistic.**
Begin by calculating the sample proportions.

$$\text{Flynt's: } \hat{p}_1=\frac{17}{25}=0.680. \qquad \text{Pops': } \hat{p}_2=\frac{23}{29}\approx 0.793103\approx 0.793.$$

Lastly, we need $\bar{p}$ before substituting into the test statistic.

$$\begin{aligned}\bar{p}&=\frac{x_1+x_2}{n_1+n_2}\\&=\frac{17+23}{25+29}\\&\approx 0.740741\\&\approx 0.741\end{aligned}$$

Substituting these into the test statistic, z, we obtain the following:

$$\begin{aligned}z&=\frac{(\hat{p}_1-\hat{p}_2)-(p_1-p_2)}{\sqrt{\bar{p}(1-\bar{p})\left(\frac{1}{n_1}+\frac{1}{n_2}\right)}}\\&=\frac{(0.68-0.793103)-0}{\sqrt{0.740741(1-0.740741)\left(\frac{1}{25}+\frac{1}{29}\right)}}\\&\approx -0.95\end{aligned}$$

*Note that in the formula above, $p_1 - p_2 = 0$ because H_0 contains that assertion.

4. **Draw a conclusion by comparing the p-value to the level of significance.**
Because this is a two-tailed test, to find the p-value, we need the area under the normal curve to the left of $z = -0.95$ and the area to the right of $z = 0.95$. Using the normal distribution table, we find a p-value of (0.1711)(2)=0.3422. This value is greater than α, which means that we do not have enough evidence to reject the null hypothesis. Therefore, there is not sufficient evidence at the 0.10 level of significance to say the pass-fail proportions at the two centers are different.

Exercises

11.4

Directions: Calculate the z-test statistic given the following information:

1. $x_1 = 15, n_1 = 28, x_2 = 21, n_2 = 33, p_1 - p_2 = 0$

2. $x_1 = 139, n_1 = 180, x_2 = 178, n_2 = 208, p_1 - p_2 = 0$

3. $\hat{p}_1 = \dfrac{18}{67}, \hat{p}_2 = \dfrac{20}{59}, p_1 - p_2 = 0$

4. $\hat{p}_1 = \dfrac{62}{101}, \hat{p}_2 = \dfrac{71}{130}, p_1 - p_2 = 0$

Directions: Answer each of the following questions.

5. A newspaper story claims that more houses are purchased by single women than single men. In a study of single women's buying habits, 100 out of 500 house purchases were made by single women. A similar study for men provided a sample of 500 house purchases in which 72 homes were bought by single men. Test the newspaper's claim using p-values with a 0.01 level of significance. Is there evidence to support the newspaper's claim?

6. Researchers claim that the birth rate in Bonn, Germany is more than the national average. A sample of 1200 Bonn residents produced 12 births in 2006, whereas a sample of 1000 people from all over Germany had 8 births the same year. Test this claim using p-values and a 0.05 level of significance

7. University officials hope that changes they have made have improved retention rates. In 2004, 1400 of 1926 freshmen returned as sophomores. In 2005, 1508 of 2011 freshmen returned as sophomores. Using p-values, determine if there is evidence at the 0.05 level to say that retention rates have improved.

8. Adrian hopes that his new training methods have improved his batting technique. Before starting his new regimen he was batting .220 in 50 at bats. Since changing his techniques he has had 24 at bats with a batting average of .375. Determine if there is sufficient evidence to say that his batting average has improved at the 0.10 level of significance.

9. There is an old wives' tale that women who eat chocolate during pregnancy are more likely to have happy babies. A pregnancy magazine wants to test this claim, and they gather 100 pregnant women for their study. Half of the group agrees to eat chocolate at least once a day, while the other group agrees to forego chocolate for the duration of their pregnancy. A year later, the ladies complete a survey regarding the overall happiness and contentedness of their babies. The results are given in the table below. At the 0.01 level of significance, test the claim of the old wives' tale.

Table for Exercise 9 - Number of Babies		
	Happy Babies	**Unhappy Babies**
With Chocolate	24	26
Without Chocolate	22	28

10. A new government program claims to lower high school dropout rates. In one school district of 3400 students, the current dropout rate is 4.5%. Two years after the start of the new program, the dropout rate has been lowered to 3.8% out of 1450 students. Is there enough evidence to say that the government program is effective at lowering the high school dropout rate? Test the government's claim at the 0.10 level of significance.

11. To test the fairness of law enforcement in their area, a local citizens' group wants to know whether women and men are unequally likely to get speeding tickets. Two hundred adults are randomly phoned and asked whether or not they had been cited for speeding in the last year. Using the results below and a 0.05 level of significance, test the claim of the citizens' group.

Table for Exercise 11 - Number of Speeding Tickets		
	Ticketed	**Not Ticketed**
Men	11	75
Women	12	102

Hypothesis Testing - Two Population Variances

11.5

The process for comparing two population variances using a hypothesis test is basically the same as we have used throughout this chapter. The main difference is the distribution and the formulas used to conduct the hypothesis test.

When comparing population means and proportions, we used either the normal distribution or the Student *t*-distribution. When comparing population variances, we will use the *F*-distribution, as we did when constructing confidence intervals for the ratio of population variances. Let's begin by reviewing the conditions necessary to use the *F*-distribution to compare population variances. In order to compare two population variances, we must ensure that the following two conditions are met:

- The two populations must be independent, not matched or paired in any way, and
- The two populations must be normally distributed.

Let's also review the basic properties of the *F*-distribution introduced in Section 9.5.

- The *F*-distribution is skewed to the right,
- The values of *F* are always greater than 0, and
- The shape of the *F*-distribution is completely determined by its two parameters, the degrees of freedom of the numerator and the degrees of freedom of the denominator of the ratio.

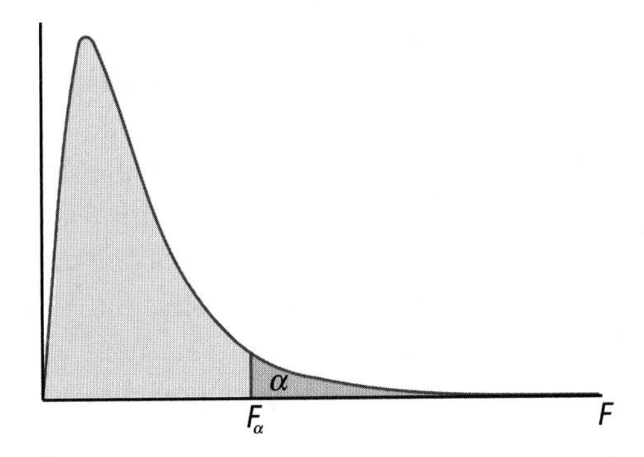

Figure 1 – *F*-Distribution

Let's walk through the basic steps of a hypothesis test, focusing specifically on comparing two population variances.

Step 1 – State the null and alternative hypotheses.
As always, the null and alternative hypotheses are logically opposite statements. Remember that the null hypothesis always contains the starting assumption regarding the population parameters. In this section, the null and alternative hypotheses for comparing two population variances will take on one of the following three forms.

$H_0: \sigma_1^2 \geq \sigma_2^2$	$H_0: \sigma_1^2 \leq \sigma_2^2$	$H_0: \sigma_1^2 = \sigma_2^2$
$H_a: \sigma_1^2 < \sigma_2^2$	$H_a: \sigma_1^2 > \sigma_2^2$	$H_a: \sigma_1^2 \neq \sigma_2^2$

Rounding Rule

In order to be consistent with the F-distribution tables, we will round calculations to four decimal places.

Step 2 – Set up the hypothesis test by choosing the test statistic and determining the values of the test statistic that would lead to rejecting the null hypothesis.
To compare population variances, we will use the ratio of the sample variances for the test statistic,

$$F = \frac{s_1^2}{s_2^2}.$$

We will no longer impose the rule that the larger of the two sample variances is s_1^2 and the smaller variance is s_2^2. As discussed in Section 9.5, this rule was not mathematically necessary and was only used as a means of restricting our discussion to right-tailed situations. We do not want to impose this rule if we wish to consider cases where $\sigma_1^2 < \sigma_2^2$, such as in a left-tailed hypothesis test.

Next, we must determine the values of the test statistic that would lead to a rejection of the null hypothesis. The method we will use in this section is that of rejection regions (first introduced in section 10.2). Recall that the size and location of the rejection region is determined by two things: (1) the type of hypothesis test being run and (2) the level of significance, α, that is desired.

We will use the F-distribution tables to determine the critical values for these rejection regions. The F-test statistic is the ratio of the variances being tested. If the population variances are equal, then their ratio is one and we would expect the ratio of the sample variances to be close to one. Thus the farther the F-test statistic is from one, the less likely that the population variances are equal. Let's look at how to find the critical F-value and rejection region for each of the three types of hypothesis tests.

1. Left-tailed test:

The critical value is $F_{1-\alpha}$. Thus the rejection region is $F \leq F_{1-\alpha}$.

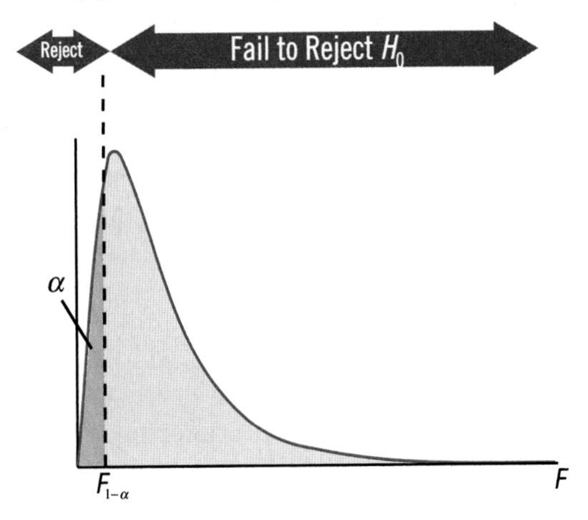

2. Right-tailed test:

The critical value is F_{α}. Thus the rejection region is $F \geq F_{\alpha}$.

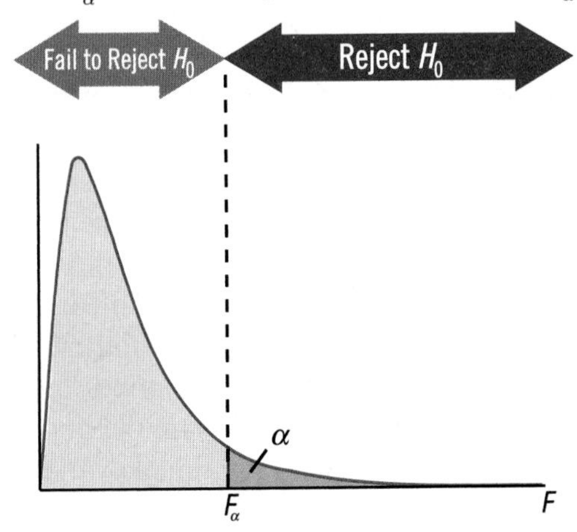

3. Two-tailed test:

The critical values are $F_{1-\frac{\alpha}{2}}$ and $F_{\frac{\alpha}{2}}$. Thus the rejection region is $F \leq F_{1-\frac{\alpha}{2}}$ or $F \geq F_{\frac{\alpha}{2}}$.

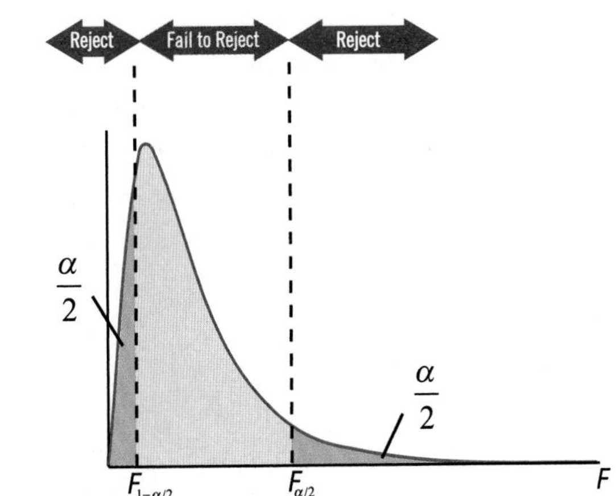

Note that the level of significance is usually stated in this step as well.

Step 3 – Gather data and calculate the necessary sample statistic.
At this point, we would collect sample data from two independent populations. Each sample should be chosen in such a way that all samples of the same size from the population have an equal chance of being chosen. Once the data is collected, calculate the sample variance for each sample. The test statistic is then calculated by dividing the first sample variance by the second sample variance. Again, the test statistic is

$$F = \frac{s_1^2}{s_2^2}.$$

Step 4 – Draw a conclusion using the process described in Step 2.
Remember that the conclusion rule for rejection regions is to *reject the null hypothesis if the test statistic calculated from the sample data falls within the rejection region.* Otherwise, you must fail to reject the null hypothesis. It is also important at this step to interpret the meaning of the conclusion in reference to the question you are trying to answer.

Example 15

A graduate student walking around a large college campus notices that the prices of professors' cars have a much smaller variance than the price variance of the cars owned by students. (Students, he notes, own vehicles ranging from small, broken-down trucks to Porsche convertibles while the cars owned by faculty are more similar in style and price.) To test his hypothesis that the prices of professors' cars have a smaller variance than the prices of students' cars, he collects data from 15 professors and 14 students on the price of their cars and calculates the variance from each sample. The sample variance for the sample of professors' cars is 34,057 and the sample variance for the sample of students' cars is 45,923. Conduct a hypothesis test using a 0.10 level of significance to test the graduate student's claim. Does the evidence support the student's claim?

Solution:

1. **State the null and alternative hypotheses.**
   Let's represent the population variance of the prices of professors' cars as σ_1^2 and the population variance of the prices of students' cars as σ_2^2. Then, the claim is that $\sigma_1^2 < \sigma_2^2$. The mathematical opposite of this claim is $\sigma_1^2 \geq \sigma_2^2$. The null and alternative hypotheses are

$$H_0\text{: } \sigma_1^2 \geq \sigma_2^2$$
$$H_a\text{: } \sigma_1^2 < \sigma_2^2.$$

2. **Set up the hypothesis test by choosing the test statistic and determining the values of the test statistic that would lead to rejecting the null hypothesis.**
   The test statistic for comparing population variances is $F = \frac{s_1^2}{s_2^2}$. In this problem, the level of significance is 0.10. Next, we must determine the rejection region. Looking at the alternative hypothesis, we see that this is a left-tailed test. The critical F-value for a left-tailed test is $F_{1-\alpha}$. Using 14 degrees of freedom for the numerator, 13 degrees of freedom for the denominator, and $\alpha = 0.10$ in Table H, we find that $F_{1-\alpha} = 0.4909$. Therefore, we will reject the null hypothesis if $F \le 0.4909$.

3. **Gather data and calculate the necessary sample statistics.**
   We have the sample statistics, so let's calculate the test statistic.

$$\begin{aligned} F &= \frac{s_1^2}{s_2^2} \\ &= \frac{34{,}057}{45{,}923} \\ &\approx 0.7416 \end{aligned}$$

4. **Draw a conclusion using the process described in Step 2.**
   Because $0.7416 > 0.4909$, we must fail to reject the null hypothesis. This means that at the 0.10 level of significance the graduate student's claim is not sufficiently supported.

Example 16

Suppose that a quality control inspector believes that the machines on Assembly Line A are not adjusted properly and that the variance in the size of the candy produced is greater than the variance in the size of the candy produced by Assembly Line B. A sample of 20 pieces of candy is taken from each assembly line and measured. The size of the candy from Assembly Line A has a sample variance of 1.45, while the size of the candy from Assembly Line B has a sample variance of 0.47. Using a 0.01 level of significance, perform a hypothesis test to test the quality control inspector's claim.

Solution:

1. **State the null and alternative hypotheses.**
   Let's represent the population variance of Assembly Line A as σ_1^2 and the population variance of Assembly Line B as σ_2^2. Then the quality control inspector's claim is that $\sigma_1^2 > \sigma_2^2$. The mathematical opposite of this claim is $\sigma_1^2 \le \sigma_2^2$. Thus

$$\begin{aligned} H_0&: \sigma_1^2 \le \sigma_2^2 \\ H_a&: \sigma_1^2 > \sigma_2^2. \end{aligned}$$

cont'd on the next page

2. **Set up the hypothesis test by choosing the test statistic and determining the values of the test statistic that would lead to rejecting the null hypothesis.**
   The test statistic for comparing population variances is $F=\frac{s_1^2}{s_2^2}$. In this problem, the level of significance is 0.01. Next, we must determine the rejection region. Looking at the alternative hypothesis, we see that this is a right-tailed test. The critical F-value for a right-tailed test is F_α. Using 19 degrees of freedom for the numerator, 19 degrees of freedom for the denominator, and $\alpha = 0.01$ in Table H, we find that $F_\alpha = 3.0274$. Therefore, we will reject the null hypothesis if $F \geq 3.0274$.

3. **Gather data and calculate the necessary sample statistics.**
   We have the sample statistics, so let's calculate the test statistic.

$$\begin{aligned} F &= \frac{s_1^2}{s_2^2} \\ &= \frac{1.45}{0.47} \\ &\approx 3.0851 \end{aligned}$$

4. **Draw a conclusion using the process described in Step 2.**
   Because 3.0851 > 3.0274, we must reject the null hypothesis. This means that at the 0.01 level of significance there is sufficient evidence to support the quality control inspector's claim.

Example 17

A professor claims that the variances in the test scores on two different versions of an exam are not equal. To test his claim, he chooses a sample from each version. A sample of 20 scores on version A has a sample variance of 3.815, while a sample of 20 scores from version B has a sample variance of 3.391. Conduct a hypothesis test to test the professor's claim that the variances of the test scores are not equal. Use a 0.05 level of significance.

Solution:

1. **State the null and alternative hypotheses.**
   Let the population variance for version A be represented by σ_1^2 and the population variance for version B be σ_2^2. Then the claim is that $\sigma_1^2 \neq \sigma_2^2$. The mathematical opposite of this claim is $\sigma_1^2 = \sigma_2^2$. The null and alternative hypotheses are then

$$\begin{aligned} &H_0\colon \sigma_1^2 = \sigma_2^2 \\ &H_a\colon \sigma_1^2 \neq \sigma_2^2. \end{aligned}$$

2. **Set up the hypothesis test by choosing the test statistic and determining the values of the test statistic that would lead to rejecting the null hypothesis.**
The test statistic for comparing population variances is $F = \frac{s_1^2}{s_2^2}$. In this problem, the level of significance is 0.05. Next, we must determine the rejection region. Looking at the alternative hypothesis, we see that this is a two-tailed test. The critical F-values for two-tailed tests are $F_{1-\frac{\alpha}{2}}$ and $F_{\frac{\alpha}{2}}$. Using 19 degrees of freedom for the numerator, 19 degrees of freedom for the denominator, and $\alpha = 0.05$ in Table H, we find that $F_{1-\frac{\alpha}{2}} = 0.3958$ and $F_{\frac{\alpha}{2}} = 2.5265$. Therefore, we will reject the null hypothesis if either $F \le 0.3958$ or $F \ge 2.5265$.

3. **Gather data and calculate the necessary sample statistics.**
We have the sample statistics, so let's calculate the test statistic.

$$\begin{aligned} F &= \frac{s_1^2}{s_2^2} \\ &= \frac{3.815}{3.391} \\ &\approx 1.1250 \end{aligned}$$

4. **Draw a conclusion using the process described in Step 2.**
Because $F = 1.1250$ is not in the rejection region, we must fail to reject the null hypothesis. This means that at the 0.05 level of significance there is not sufficient evidence to say that the two test versions have different variances.

Exercises

11.5

Directions: For each claim, write the null and alternative hypotheses.

1. A medical researcher believes that the variance in the lung capacity of smokers is less than that of non-smokers. Let σ_1^2 represent the population variance for smokers.

2. A professor believes that the variance of SAT scores of honor students is less than that of students in general. Let σ_1^2 represent the population variance for honor students.

3. A quality control inspector believes that variance in the diameter of soda cans produced by machine 1 is greater than the variance in the diameter of soda cans produced by machine 2. Let σ_1^2 represent the population variance for machine 1.

4. A consumer protection agency believes that the price of SUVs has a greater variance than the price of luxury cars. Let σ_1^2 represent the population variance for SUVs.

5. A golf pro believes that the variance in the distance hit with different brands of golf balls is not the same. He is especially interested in comparing Titleist golf balls to a generic store brand. Let σ_1^2 represent the population variance for Titleist golf balls.

6. A paint technician believes that the variance in the depth of the special coating in one tank is not the same as the variance of the depth of the coating in another tank. Let σ_1^2 represent the population variance for tank 1.

Directions: For each claim, calculate the F-test statistic.

7. $n_1 = 15,\ s_1^2 = 3.007,\ n_2 = 16,\ s_2^2 = 2.897$

8. $n_1 = 4,\ s_1^2 = 0.961,\ n_2 = 6,\ s_2^2 = 0.899$

9. $n_1 = 31,\ s_1^2 = 46.821,\ n_2 = 28,\ s_2^2 = 57.024$

10. $n_1 = 23,\ s_1^2 = 35{,}679,\ n_2 = 24,\ s_2^2 = 39{,}018$

Directions: For the information given, determine the rejection region and draw a conclusion.

11. $n_1 = 19,\ s_1^2 = 0.3891, n_2 = 24,\ s_2^2 = 0.9579, H_a\colon\ \sigma_1^2 < \sigma_2^2, \alpha = 0.05$

12. $n_1 = 14,\ s_1^2 = 3.152, n_2 = 11,\ s_2^2 = 9.300, H_a\colon\ \sigma_1^2 < \sigma_2^2, \alpha = 0.05$

13. $n_1 = 11,\ s_1^2 = 31{,}207\ n_2 = 11,\ s_2^2 = 38{,}916, H_a\colon\ \sigma_1^2 < \sigma_2^2, \alpha = 0.01$

14. $n_1 = 11,\ s_1^2 = 3.007, n_2 = 25,\ s_2^2 = 2.897, H_a\colon\ \sigma_1^2 > \sigma_2^2, \alpha = 0.05$

15. $n_1 = 20,\ s_1^2 = 10.453, n_2 = 23,\ s_2^2 = 3.199, H_a\colon\ \sigma_1^2 > \sigma_2^2, \alpha = 0.10$

16. $n_1 = 12,\ s_1^2 = 1893, n_2 = 26,\ s_2^2 = 1066, H_a\colon\ \sigma_1^2 > \sigma_2^2, \alpha = 0.01$

17. $n_1 = 16,\ s_1^2 = 18.01, n_2 = 21,\ s_2^2 = 17.07, H_a\colon\ \sigma_1^2 \neq \sigma_2^2, \alpha = 0.05$

18. $n_1 = 20,\ s_1^2 = 27.08, n_2 = 29,\ s_2^2 = 11.77, H_a\colon\ \sigma_1^2 \neq \sigma_2^2, \alpha = 0.05$

19. $n_1 = 20,\ s_1^2 = 8.12, n_2 = 18,\ s_2^2 = 16.78, H_a\colon\ \sigma_1^2 \neq \sigma_2^2, \alpha = 0.10$

20. $n_1 = 11,\ s_1^2 = 12{,}047, n_2 = 12,\ s_2^2 = 18{,}019, H_a\colon\ \sigma_1^2 \neq \sigma_2^2, \alpha = 0.01$

Directions: Answer the following questions.

21. A golf pro believes that the variance in the distance hit with different brands of golf balls is different. In particular, he believes that distances hit using Titleist golf balls have a smaller variance than distances hit using a generic store brand. He hits 10 Titleist golf balls and records a sample variance of 201.65. He hits 10 generic golf balls and records a sample variance of 364.57. Test the golf pro's claim using a 0.05 level of significance. Does the evidence support the golf pro's claim?

22. A quality control inspector believes that the variance in the diameter of soda cans produced by machine A is greater than the variance in the diameter of soda cans produced by machine B. The sample variance of a sample of 15 soda cans from machine A is 2.788. The sample variance from a sample of 17 soda cans from machine B is 1.982. Test the inspector's claim using a 0.10 level of significance. Does the evidence support the inspector's claim?

23. A medical researcher believes that the variance of cholesterol levels in men is greater than the variance of cholesterol levels in women. The sample variance for a sample of 8 men is 277. The sample variance for a sample of 7 women is 89. Test the researcher's claim using a 0.10 level of significance. Does the evidence support the researcher's claim?

24. A basketball coach believes that the variance in the height of basketball players is different from the variance of height in the general population. The sample variance for a sample of 12 basketball players is 24.76. The sample variance for a sample of 13 other men is 25.87. Test the coach's claim using a 0.01 level of significance. Does the evidence support the coach's claim?

25. One study claims that the variance in the heart rate of smokers is different than the variance in the heart rate of non-smokers. A medical student decides to test this claim. The sample variance for a sample of 5 smokers is 545. The sample variance for a sample of 5 non-smokers is 103. Test the student's claim using a 0.01 level of significance. Does the evidence support the student's claim?

26. A professor believes that the variance of ACT scores of honor students is less than that of students in general. The sample variance in ACT scores for a sample of 18 honors students is 12.14. The sample variance in ACT scores for a sample of 20 other students is 28.91. Test the professor's claim using a 0.05 level of significance. Does the evidence support the professor's claim?

ANOVA

11.6

Our discussion thus far has shown us how to compare population parameters from as many as two populations. But, what if you have four, eight, or even ninety-two? Consider the chair of an academic department who wants to know if the average final grade is the same for each of the 12 sections of an introductory statistics class. If she has nothing better to do, she could compare the first class to each of the other 11, then the second class to each of the other 10, then the third class, etc. using techniques we have seen previously. However, this would result in her having to calculate 66 different hypothesis tests! (Verify this yourself.) It would also reduce the level of confidence in the final result, because as we increase the number of tests performed, we increase the likelihood of finding a difference by chance. One such method that compares the means of multiple populations all at once is an ANOVA.

ANOVA stands for **AN**alysis **O**f **VA**riance and is used to compare the means of three or more populations. An ANOVA is a type of hypothesis test where the null hypothesis states that all of the population means are equal. Thus

$$H_0\colon \mu_1 = \mu_2 = \cdots = \mu_k$$
H_a: At least one mean differs from the others.

where k = the number of populations being studied.

From your study of statistics so far, it should come as no great surprise that if you take samples from various populations, the sample means will not necessarily be equal, even if the population means are equal. Thus an ANOVA seeks to determine if the variation in the sample means from the different populations is due to random sampling or to a true difference in the population means. So, the reason for the test being called an analysis of *variance* instead of an analysis of means is that the test is analyzing the variation between the sample means, not the means themselves.

In order for an ANOVA test to be used, the following assumptions must be met:

ANOVA Test

1. The distributions of all of the populations are approximately normal.
2. The variances of the populations are the same. (If the sample sizes are nearly equal, this assumption in not essential.)
3. Independent, simple random samples were taken from each population.

Before we look at an actual ANOVA table and learn how to determine if the means of the populations are equal, we will begin by briefly introducing a few formulas. The purpose of showing you these formulas is so that you can better understand how the numbers in an ANOVA table are related; however, using these formulas to calculate an ANOVA by hand would be tedious at best. Therefore, we will use a TI-84 Plus calculator or Excel to generate an ANOVA table, when necessary.

The first calculation is the **grand mean**, $\bar{\bar{x}}$, which is the weighted mean of the sample means from each of the populations. The formula for the grand mean,

$$\bar{\bar{x}} = \frac{\sum_{i=1}^{m}\left[\sum_{j=1}^{k} x_{ij}\right]}{\sum n_i}$$

where x_{ij} = the jth data value from population i,
n_i = the sample size from population i.
$\sum x_{ij}$ = the sum of the data values in the ith population.

The next two calculations that we need to look at are the values of two different types of variation, SST and SSE. First, SST stands for the Sum of Squares *among* Treatments, and is the variation resulting from the differences in the population means. This value is sometimes referred to as SS(Between groups), SS(Factor), or SS(Treatment), where the treatments or factors are what distinguish one population from another. The formula for SST is

$$\text{SST} = \sum_{i=1}^{k} n_i \left(\bar{x}_i - \bar{\bar{x}}\right)^2$$

where k = the number of populations,
n_i = the sample size from the ith population,
$\bar{x}_i$ = the sample mean from the ith population.

It is SST that most influences the results of our hypothesis test. As the difference in the populations' means gets larger, the between groups' variance grows larger as well. With a "large enough" SST value, we reject the null hypothesis.

SSE stands for the Sum of Squares of Error, and is the variation resulting from the variance *within* the populations. The formula is

$$\text{SSE} = \sum_{j=1}^{n_1}\left(x_{1j} - \bar{x}_1\right)^2 + \sum_{j=1}^{n_2}\left(x_{2j} - \bar{x}_2\right)^2 + \cdots + \sum_{j=1}^{n_k}\left(x_{kj} - \bar{x}_k\right)^2.$$

The **total variation** is the sum of the variations contributed by each sample. The formula for the total variation is

$$\text{Total Variation} = \sum\left(x_{1j} - \bar{\bar{x}}\right)^2 + \sum\left(x_{2j} - \bar{\bar{x}}\right)^2 + \cdots + \sum\left(x_{kj} - \bar{\bar{x}}\right)^2$$

for k populations. Each summation is the amount of variation contributed by that sample, thus $\left(x_{1j} - \bar{\bar{x}}\right)^2$ is the amount of variation contributed by sample 1. The total variation contains the two types of variation, SST and SSE. Thus we can say that

$$\text{Total Variation} = \text{SST} + \text{SSE}.$$

Two other calculations which are important to introduce at this point are MST and MSE, which stand for Mean Square for Treatment and Mean Square for Error respectively. Their formulas are given below:

$$\text{MST} = \frac{\text{SST}}{k-1}$$

$$\text{MSE} = \frac{\text{SSE}}{n-k}$$

where k is the number of populations and $n = \sum_{i=1}^{k} n_i$ is the sample size. If SST is very large, then dividing that number by $k - 1$ will result in a large number as well.

The last value that needs explanation is the F-statistic, named for R. A. Fisher who first worked out the F-distribution. An ANOVA test uses this statistic instead of z, t, or χ^2 as some other tests do. We have seen the F-distribution when we constructed confidence intervals for the ratio of two population variances and performed hypothesis tests comparing population variances. There are a few differences in how we will use this distribution for performing ANOVA tests, so let's look at this distribution again. For an ANOVA test, the F-test statistic is the quotient of the Mean Squares. Thus

$$F = \frac{\text{MST}}{\text{MSE}}.$$

The F-distribution is skewed to the right and is defined by only two parameters, v_1 and v_2. The first parameter, v_1, is the degrees of freedom of the numerator of F, which is $k - 1$. The second parameter is v_2, which is the degrees of freedom of the denominator of F, defined as $n - k$. A picture of the F-distribution is shown in Figure 2. Notice that the critical F-value, denoted F_α, is the value of F such that the area to the right of F is equal to α.

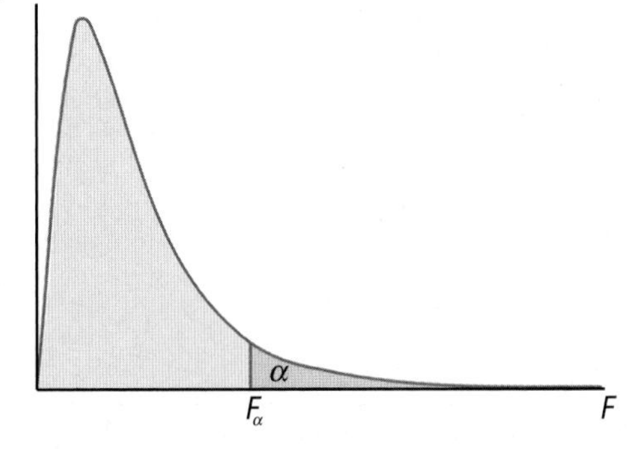

Figure 2 – F-Distribution

The larger the difference among the population means, the larger F will be. Thus a large F-value indicates that at least one of the population means is different from the others. It does NOT indicate which of the population means is different from the others, or how many are different. Therefore, a large F-value allows us to reject the null hypothesis. The question naturally arises at this point, "How large is large enough?" F is considered "large enough" to reject the null hypothesis if $F \geq F_{\alpha}$.

Now that we have discussed all of the elements to an ANOVA test, we can look at an ANOVA table, which is the summary of all of the calculations we discussed so far. An example of an ANOVA table is shown below.

	SS	DF	MS	F
Treatments	4035.51	9	448.39	0.96
Error	5112.14	11	464.74	
Total	9147.65	20		

Table 6 – Example of an ANOVA Table

Notice that SST is the value in the SS column and the Treatments (T) row. Similarly, MSE is the value in the MS column and the Error (E) row. Notice that the last row is the total (or sum) of the previous two rows. Also notice that the formulas we discussed so far are evident in the table. $\text{MST} = \frac{\text{SST}}{\text{DFT}}$, $\text{MSE} = \frac{\text{SSE}}{\text{DFE}}$, and $F = \frac{\text{MST}}{\text{MSE}}$. (Verify that these formulas are correct in this table.) The following table is a summary of the formulas we have discussed in this chapter.

	SS	DF	MS	F
Treatments	SST	$k-1$	$\frac{\text{SST}}{\text{DFT}}$	$\frac{\text{MST}}{\text{MSE}}$
Error	SSE	$n-k$	$\frac{\text{SSE}}{\text{DFE}}$	
Total	SST + SSE	DFT + DFE		

Table 7 – Calculations in the ANOVA Table

Example 18

Medical researchers studying the effect of Vitamin C on the common cold grouped study participants into three groups: 1) took 1000 mg of Vitamin C daily, 2) took 500 mg of Vitamin C daily, and 3) took no Vitamin C. The participants were followed for a year, during which time the number of colds each participant suffered was recorded. Using these data, researchers performed an ANOVA test, which resulted in the following ANOVA table. Unfortunately, a few of the cells in the table have been lost.

a. Complete the ANOVA table using the known values in the table.

b. Based on this completed ANOVA table, do the researchers have enough evidence, at the 0.05 level of significance, to support their claim that there is a significant difference in the number of colds suffered per year between patients who take some amount of Vitamin C and patients who do not?

c. From the ANOVA table, can you conclude if one group had fewer colds than the others?

	SS	DF	MS	*F*	*P*-value	*F* crit.
Treatments		2	1.4288		0.4288	3.3158
Error	49.2030					
Total	52.0606	32				

Solution:

a. To begin filling in the ANOVA table, we can use the fact that the Total Variation = SST + SSE. Thus

$$\text{Total} - \text{SSE} = \text{SST}$$

$$52.0606 - 49.2030 = 2.8576.$$

Similarly, DF = DFT + DFE. Thus

$$\text{DF} - \text{DFT} = \text{DFE}$$

$$32 - 2 = 30.$$

To find the next missing value, we must divide.

$$\text{MSE} = \frac{\text{SSE}}{\text{DFE}} = \frac{49.2030}{30} = 1.6401$$

Lastly, we must divide again to find the value of our *F*-test statistic.

$$\text{F} = \frac{\text{MST}}{\text{MSE}} = \frac{1.4288}{1.6401} \approx 0.8712$$

We can now fill in all of the missing values.

cont'd on the next page

	SS	DF	MS	F	P-value	F crit.
Treatments	2.8576	2	1.4288	0.8712	0.4288	3.3158
Error	49.2030	30	1.6401			
Total	52.0606	32				

b. First, let's state the null and alternative hypotheses for this test.

$$H_0: \mu_1 = \mu_2 = \cdots = \mu_k$$
H_a: At least one mean differs from the others.

We have a level of significance of 0.05, thus $\alpha = 0.05$. We can draw our conclusion to an ANOVA test by using p-values, as we have in previous sections. Notice that the ANOVA table stated that the p-value is 0.4288. Since $0.4288 > 0.05$, we fail to reject the null hypothesis. Thus there is not sufficient evidence at the 0.05 level of significance to support the researcher's claim that there is a significant difference in the number of colds suffered per year between patients who take some amount of Vitamin C and patients who do not.

c. Remember that an ANOVA test indicates if at least one of the population means is different, but it DOES NOT indicate which one differs, or by how much. This test did not provide support for the claim that one group had fewer (or more) colds than another group; however, even if it had, we would not be able to tell from the ANOVA table which group had fewer colds.

Now that we are comfortable interpreting an ANOVA table, let's see how to use technology to generate the values in the table from raw data. We will use Microsoft Excel to generate the values in the next example, as the output generated by that program is in the familiar format. A TI-84 Plus calculator can be used just as easily, but the output is in the form of a list, not a table. Directions for using a TI-84 Plus calculator are provided for you in the technology section at the end of this chapter.

Example 19

Researchers at a drug manufacturing company wish to test whether their drug is the best on the market at lowering cholesterol levels in middle-aged women. Thirty women with high-cholesterol volunteer to take part in the 6-month study. The women are divided into three groups and each group is given a different cholesterol-lowering drug, one of which is the drug made by the company. The women's cholesterol levels are measured at the beginning of the study, and again after six months. The difference in the before treatment and after treatment cholesterol levels for each woman is recorded in the table below. Use these data to perform an ANOVA test, at the 0.10 level of significance. What conclusion can you draw from the ANOVA test?

Table 8 - Cholesterol Levels

Drug 1	Drug 2	Drug 3
18	18	21
19	20	22
20	16	17
21	20	18
22	21	22
23	20	19
18	18	21
19	19	20
20	17	18
21	13	23

Solution:

To begin, we need to enter our data into Microsoft Excel. Enter the data, just as shown above, into columns A, B, and C. Click on TOOLS, then DATA ANALYSIS. In the Data Analysis menu, choose ANOVA: Single Factor. Next, fill out the ANOVA menu as shown below.

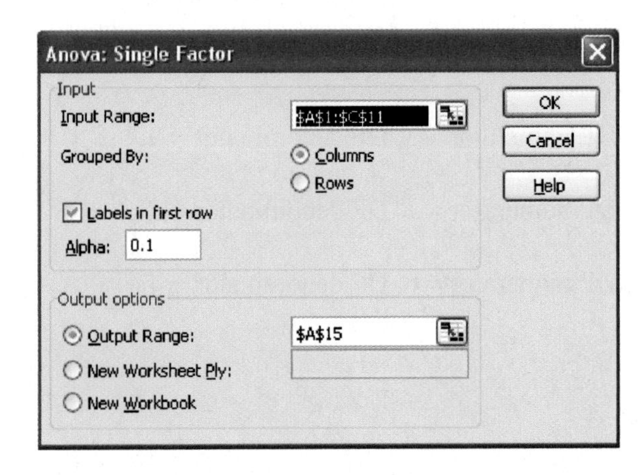

cont'd on the next page

After choosing OK, the ANOVA table will be displayed, along with descriptive statistics for each sample. Also, the p-value and the critical F-value are included in the ANOVA table.

SUMMARY

Groups	Count	Sum	Average	Variance
Drug 1	10	201	20.1	2.766667
Drug 2	10	182	18.2	5.733333
Drug 3	10	201	20.1	4.1

ANOVA

Source of Variation	SS	df	MS	F	P-value	F crit
Between Groups	24.06667	2	12.03333	2.865079	0.074407	2.510609
Within Groups	113.4	27	4.2			
Total	137.4667	29				

Notice that $p = 0.0744$, which is less than our 0.10 level of significance, so we reject the null hypothesis. Thus the researchers can conclude that there is enough evidence, at the 0.10 level of significance, to support the claim that at least one of the population means is different. Therefore, the average for at least one of the drugs is different from the others. However, we cannot conclude, from the ANOVA test alone how the averages differ. Looking at the descriptive statistics suggests that Drug 2 lowers cholesterol by fewer points than Drugs 1 or 3, but we cannot conclude that from the ANOVA test alone.

Exercises 11.6

Directions: Find the critical F-value for the given information:

1. $\alpha = 0.05$, DF numerator = 10, DF denominator = 12
2. $\alpha = 0.01$, DF numerator = 9, DF denominator = 15
3. $\alpha = 0.10$, DF numerator = 7, DF denominator = 8
4. $\alpha = 0.05$, DF numerator = 3, DF denominator = 9
5. $\alpha = 0.10$, DF numerator = 1, DF denominator = 5
6. $\alpha = 0.01$, DF numerator = 6, DF denominator = 4

Directions: Complete the following ANOVA tables.

7.

	SS	DF	MS	F
Treatments	4	5		
Error	11	6		
Total				

8.

	SS	DF	MS	F
Treatments	49.3	4		
Error		3		
Total	142.5			

9.

	SS	DF	MS	F
Treatments	50		12.5	
Error		3		
Total	74			

10.

	SS	DF	MS	F
Treatments		9		1.6330
Error	21.8		5.45	
Total	101.9			

11.

	SS	DF	MS	F
Treatments	313	6		
Error		8		
Total	649			

12. A regional manager wants to know if there is a difference in the time customers wait in line at the drive-through window between the three stores in her region. She samples the wait time at each store. Her data is given in the table below. Use an ANOVA test to determine if there is a difference in the wait time, at the 0.05 level of significance.

Table for Exercise 12 - Wait Time (in Minutes)		
Store 1	**Store 2**	**Store 3**
2.34	2.87	1.32
1.23	1.94	1.45
1.89	2.36	1.78
2.31	1.85	2.01
3.02	1.75	2.45
1.95	2.82	1.92
2.45	3.32	1.83

13. A manager is concerned that one of his workers is producing more defective parts than the other similarly skilled workers. He records the number of defective parts made by this worker and two others with similar experience during their shift each day for one week. The results are provided in the table below. At the 0.10 level of significance, can you conclude that there is a difference in the number of defective parts made between the three workers?

Table for Exercise 13 - Number of Defective Parts		
Worker A	**Worker B**	**Worker C**
3	2	1
2	2	2
2	1	1
4	0	1
2	1	2

14. The panhellenic council is comparing the GPA between the four sororities on campus. The GPA for 10 girls in each sorority is given in the table below. Based on this data, can you conclude that there is a difference in the average GPA between the four sororities on campus? Use a 0.05 level of significance.

Table for Exercise 14 - Sorority GPAs			
Tri-Delta	**Phi Mu**	**Chi Omega**	**Beta Phi**
2.5	3.1	2.7	3.1
3.4	3.4	2.9	3.2
4.0	3.6	2.8	3.1
3.8	3.6	2.8	2.8
2.7	3.7	2.9	2.9
2.8	3.0	3.1	3.1
2.8	4.0	2.9	3.0
3.2	3.9	2.7	3.3
3.1	4.0	2.6	2.7
3.9	3.4	3.3	2.6

15. A group of paramedics does not believe that the number of calls is the same for each shift. To test their claim, they record the number of calls received during each shift for seven days. Based on this evidence, can the paramedics conclude that the average number of calls is different between shifts? Use a 0.01 level of significance.

Table for Exercise 15 - Number of Calls		
Morning	**Afternoon**	**Night**
2	3	5
3	4	4
2	4	5
4	5	3
3	3	2
1	2	5
3	5	6

Chapter Review

Section 11.1 - Hypothesis Testing - Two Means (Large Independent Samples)	Pages
• **Steps to a Hypothesis Test** **1.** State the null and alternative hypotheses. **2.** Set up the hypothesis test by choosing the test statistic and stating the level of significance. **3.** Gather data and calculate the necessary sample statistics. **4.** Draw a conclusion by comparing the *p*-value to the level of significance.	p. 574
• **Test Statistic** $z=\dfrac{(\bar{x}_1-\bar{x}_2)-(\mu_1-\mu_2)}{\sqrt{\dfrac{s_1^2}{n_1}+\dfrac{s_2^2}{n_2}}}$	p. 577

Section 11.2 - Hypothesis Testing - Two Means (Small Independent Samples)	Pages
• **If the Population Variances are Equal** Degrees of Freedom: n_1+n_2-2 Test Statistic: $t=\dfrac{(\bar{x}_1-\bar{x}_2)-(\mu_1-\mu_2)}{\sqrt{\dfrac{(n_1-1)s_1^2+(n_2-1)s_2^2}{n_1+n_2-2}}\sqrt{\dfrac{1}{n_1}+\dfrac{1}{n_2}}}$	p. 586
• **If the Population Variances are Not Equal** Degrees of Freedom: smaller of n_1-1 or n_2-1 Test Statistic: $t=\dfrac{(\bar{x}_1-\bar{x}_2)-(\mu_1-\mu_2)}{\sqrt{\dfrac{s_1^2}{n_1}+\dfrac{s_2^2}{n_2}}}$	p. 586
• **Rejection Regions** • For a left-tailed test, reject H_0 if $t\le -t_\alpha$. • For a right-tailed test, reject H_0 if $t\ge t_\alpha$. • For a two-tailed test, reject H_0 if $\lvert t\rvert \ge t_{\alpha/2}$.	p. 587

Section 11.3 - Hypothesis Testing - Two Means (Dependent Samples)	Pages
• **Test Statistic** $$t=\frac{\overline{d}-\mu_d}{\left(\frac{s_d}{\sqrt{n}}\right)}$$	p. 595
• **Rejection Regions** • For a left-tailed test, reject H_0 if $t \le -t_\alpha$. • For a right-tailed test, reject H_0 if $t \ge t_\alpha$. • For a two-tailed test, reject H_0 if $\lvert t\rvert \ge t_{\alpha/2}$.	p. 595

Section 11.4 - Hypothesis Testing - Two Proportions (Large Independent Samples)	Pages
• **Test Statistic** $$z=\frac{\left(\hat{p}_1-\hat{p}_2\right)-\left(p_1-p_2\right)}{\sqrt{\overline{p}\left(1-\overline{p}\right)\left(\frac{1}{n_1}+\frac{1}{n_2}\right)}}$$	p. 603
• **Weighted Estimates** $$\overline{p}=\frac{x_1+x_2}{n_1+n_2}$$	p. 603

Section 11.5 - Hypothesis Testing - Two Population Variances	Pages
• **Test Statistic** $$F=\frac{s_1^2}{s_2^2}$$	p. 610
• **Critical *F*-values** • For a left-tailed test, the rejection region is $F \le F_{1-\alpha}$. • For a right-tailed test, the rejection region is $F \ge F_\alpha$. • For a two-tailed test, the rejection region is $F \le F_{1-\frac{\alpha}{2}}$ or $F \ge F_{\frac{\alpha}{2}}$.	p. 611

Section 11.6 - ANOVA	Pages
• **Definitions** **ANOVA** – Analysis of Variance, used to compare the means of three or more populations **Grand Mean** – the weighted mean of the sample means from each of the populations **Total Variation** – sum of the variations contributed by each sample	p. 619 – 620
• **For ANOVA Test:** **1.** The distributions of all of the populations are approximately normal. **2.** The variances of the populations are the same. (If the sample sizes are nearly equal, this assumption is not essential.) **3.** Independent, simple random samples were taken from each population.	p. 619
• **Null and Alternative Hypothesis** H_0: $\mu_1 = \mu_2 = \cdots = \mu_k$ H_a: At least one mean differs from the others	p. 619
• **ANOVA Table:**	p. 622

	SS	DF	MS	F
Treatments	SST	$k - 1$	$\frac{\text{SST}}{\text{DFT}}$	$\frac{\text{MST}}{\text{MSE}}$
Error	SSE	$n - k$	$\frac{\text{SSE}}{\text{DFE}}$	
Total	SST + SSE	DFT + DFE		

Chapter Exercises

Directions: Answer each of the following questions. Assume the necessary criteria are met for each scenario.

1. Psychologists believe that test anxiety can reduce a student's score on an exam. An educational psychologist administers a test to two groups of students, one group that has test anxiety and one group that does not. The following statistics were calculated on the test scores:

Table for Exercise 1 - Test Scores			
	n	$\bar{x}$	s
With Test Anxiety	35	65	8.9
Without Test Anxiety	38	73	11.5

   Test the claim that test anxiety reduces a student's exam score at the 0.05 level of significance.

2. A company that manufactures baseball bats believes that their new bat will allow you to hit the ball 30 feet farther than their current model. They hire a professional baseball player known for hitting home runs to hit 10 balls with each bat and they measure the distance each ball is hit to test their claim. The results of the batting experiment are shown below. Test the company's claim using a 0.05 level of significance. Assume the variances of the two populations are the same.

Table for Exercise 2 - Distance of Hitting a Baseball (in Feet)	
New Model	**Old Model**
235	200
240	210
253	231
267	218
243	210
237	209
250	210
241	229
251	234
248	231

3. Do husbands gain more weight after marriage than their wives? Answer this question by performing a hypothesis test using a 0.05 level of significance. Data on the amount of weight gained the first year of marriage was collected from 10 couples and is given on the following page. Negative values indicate weight loss.

Table for Exercise 3 - Weight Gain (in Pounds)										
Couple	1	2	3	4	5	6	7	8	9	10
Husband's Gain	10	15	8	7	3	10	12	20	–5	4
Wife's Gain	11	8	4	2	–1	8	15	15	–10	3

4. A blackjack pit boss at a casino is concerned that one of the tables is not generating the same revenue as the other 3 tables. The amount of revenue (in whole dollars) received during one particular 3-hour period is recorded at each table each night for 5 nights. The results are given in the table below. Based on this data, can the pit boss conclude that one of the tables is not generating the same revenue as the other tables? Use a 0.05 level of significance.

Table for Exercise 4 - Amount of Revenue			
Table 1	**Table 2**	**Table 3**	**Table 4**
8534	9821	10542	7367
7845	8997	9982	8021
8901	7905	8934	9034
9371	8923	9344	6703
7782	6675	8745	8432

5. The company manufacturing a new antibiotic ointment claims that it speeds healing time for minor cuts and scrapes. To test this claim, a researcher conducts an experiment involving 30 participants who have recently suffered a minor cut or scrape. Half of the group is asked to apply the new ointment twice daily, and the other group is not allowed to treat the cuts with any medication. Over the course of several weeks, the researchers track the length of time it takes for each cut to heal. The cuts treated with the new ointment take on average 4.6 days to heal, with a standard deviation of 1.3 days. The cuts that did not receive treatment took on average 6.1 days to heal, with a standard deviation of 1.6 days. Test the company's claim at the 0.01 level of significance. Assume that the population variances are not the same.

6. Are people with low incomes more likely to cheat on their taxes than people with high incomes? To test this claim, a researcher anonymously surveys 50 taxpayers with low annual incomes and 50 taxpayers with high annual incomes. Of those with low incomes, 7 admit that they were not completely honest the last time they paid their taxes. Of those with high incomes, 5 admit that they were not entirely honest the last time they paid their taxes. Evaluate the claim by performing a hypothesis test at the 0.05 level of significance.

7. An education professor believes single-gender classrooms do not have the same number of discipline problems as mixed-gender classrooms. For her study, the professor randomly selects 10 students from one elementary school. For one month, the students are separated into boys' and girls' classes. The next month, the students are assigned to classrooms with boys and girls. For each month, the number of times each student was disciplined is recorded. Test the professor's claim at the 0.01 level of significance.

Table for Exercise 7 - Number of Times Disciplined										
Participant	A	B	C	D	E	F	G	H	I	J
Single Gender	1	2	0	4	3	6	1	2	0	0
Mixed Gender	2	2	1	6	3	9	0	4	0	1

8. Do women read to their children more than men? A survey of 50 mothers found that 37 of them read to their children. A survey of 55 fathers found that 35 of them read to their children. Answer the question by performing a hypothesis test using a 0.05 level of significance.

9. A psychologist is convinced that attitude plays a central role in health. Along those lines, he believes that people who attend church regularly are healthier. To test this claim, he randomly selects 75 adults who regularly attend church and finds that in the past year they averaged 1.4 doctor's visits with a standard deviation of 0.4 visits. He also randomly selects 75 adults who do not regularly attend church, and he finds that in the past year they averaged 1.6 doctor's visits with a standard deviation of 0.5 visits. Use the 0.05 level of significance to test this claim.

10. A nutritionist believes that the variance in the weight of women on a diet is less than the variance in the weight of women in the general population. The sample variance for a sample of 15 women on a diet is 256.02. The sample variance for a sample of 15 other women is 367.91. Test the nutritionist's claim using a 0.05 level of significance. Does the evidence support the nutritionist's claim?

Chapter Project A: Hypothesis Tests Comparing Two Populations

Directions: Choose one of the three questions to answer, collect data from friends in the desired population, and perform the appropriate hypothesis test to answer the question. After you have written your conclusion, look at the "truth" and determine if your hypothesis test produced a correct decision, a Type I error, or a Type II error.

Pick one of the following questions to test:

- Do women in college spend more time studying than men in college? Ask at least 30 women and at least 30 men to estimate the amount of time they spend studying each week. Use a 0.10 level of significance.
- Do college freshmen spend less money a week eating out than seniors? Ask between 10 and 20 freshmen and between 10 and 20 seniors to estimate the amount of money they spend each week eating out. Assume the population variances are different. Use a 0.05 level of significance.
- Is the percentage of men and the percentage of women who exercise regularly the same? Ask at least 30 men and at least 30 women if they exercise at least 3 times per week. (Note: The participants do not need to be in college.) Record the number of men and the number of women who say "yes." Use a 0.01 level of significance.

Step 1 – State the null and alternative hypotheses.

What are the null and alternative hypotheses?

H_0 ______________________ H_a ______________________

Step 2 – Determine your test statistic and level of significance.

Test Statistic: ______________________ (formula only)

Level of Significance: ______________

Step 3 – Collect data and calculate the test statistic.

Collect data on the claim from the desired population. Discuss which method of data collection you used. List any potential for bias. Calculate the sample statistics needed for your test statistic formula. Round your answers to two decimal places.

Calculate the test statistic using your sample statistics.

Test Statistic = ______________

Step 4 – Draw a conclusion using rejection regions.

This is a (left / right / two)-tailed test. (Circle your answer.)

State the rejection region: ______________________

What is your conclusion? Be sure to answer the original question.

Types of Errors

Suppose the "truth" is:

- Women in college spend more time studying than men in college.
- Freshmen spend more money eating out than do seniors.
- The percentage of women who exercise regularly is the same as the percentage of men.

Based on your conclusion, did you make a Type I error, Type II error, or a correct decision? Explain.

Chapter Project B: ANOVA

Parking is one of the most common complaints of students on any college campus. Is there a difference in the average number of parking tickets received each semester by commuters, students who live in residential housing on campus, students who live in a fraternity house, or students who live in a sorority house? Let's perform an ANOVA test to help us answer this question.

To begin, label the populations as follows:

Population 1 - Commuters
Population 2 - Students living in residential housing on campus
Population 3 - Students living in a fraternity house
Population 4 - Students living in a sorority house

Step 1 – State the null and alternative hypotheses.

Based on the claim you choose, what are the null and alternative hypotheses?

H_0 ____________________ H_a ____________________

Step 2 – Determine your test statistic and level of significance.

Use the F-test statistic and a level of significance of 0.05.

$$F = \frac{\text{MST}}{\text{MSE}}$$

$$\alpha = 0.05$$

Step 3 – Collect data and calculate the test statistic.

Collect data from 5 students from each population. Record the average number of parking tickets they received each semester in a table similar to the one below.

Commuters	Residence Hall	Fraternity House	Sorority House

Use the formulas from the ANOVA section, or available technology (as described in the Using Technology section at the end of this chapter), to complete an ANOVA table.

Step 4 – Draw a conclusion.

a. Find the critical F-value for $\alpha = 0.05$ and the appropriate degrees of freedom.

b. Based on the calculated value of F from the ANOVA table, should you reject the null hypothesis?

c. Is there a difference in the average number of parking tickets received by commuters, students who live in residential housing on campus, students who live in a fraternity house, or students who live in a sorority house?

d. Does our hypothesis test provide any indication if one of the populations receives more parking tickets than another?

Chapter Technology – Hypothesis Tests with Two or More Samples

TI-84 PLUS

If we have the raw data, we can use either a TI-84 Plus, Microsoft Excel, or Minitab to perform a hypothesis test to compare two populations. If we only have the sample statistics, the TI-84 Plus is much easier to use.

Comparing Two Means (Large Independent Samples)

Example T.1

A researcher is looking at the study habits of college students. She believes that freshmen study less than seniors. A group of 42 freshmen reported an average study time per week of 19 hours with a standard deviation of 4.7 hours. A group of 39 seniors reported an average study time per week of 21 hours with a standard deviation of 5.2 hours. Test the researcher's claim with a 0.05 level of significance. Use a TI-84 Plus calculator.

Solution:

1. **State the null and alternative hypotheses.**
   Let freshmen be population 1 and seniors be population 2. Claim: Freshmen study less than seniors. Thus

$$H_0: \mu_1 - \mu_2 \geq 0$$
$$H_a: \mu_1 - \mu_2 < 0.$$

2. **Set up the hypothesis test by choosing the test statistic and stating the level of significance.**
   We are looking at the difference in population means with large independent samples, so we use z as the test statistic. We will draw a conclusion by computing the p-value for the calculated test statistic and comparing that value to α. For this hypothesis test, $\alpha = 0.05$.

3. **Gather data and calculate the necessary sample statistics.**
   The following statistics were given in the problem

$$\bar{x}_1 = 19 \qquad \bar{x}_2 = 21$$
$$s_1 = 4.7 \qquad s_2 = 5.2$$
$$n_1 = 42 \qquad n_2 = 39$$

cont'd on the next page

To use a TI-84 Plus to perform this test, press STAT, scroll over to TESTS and choose option 3: 2-SampZTest. Choose Stats, since we know the sample statistics. Fill in the sample values as shown in the screen shots below.

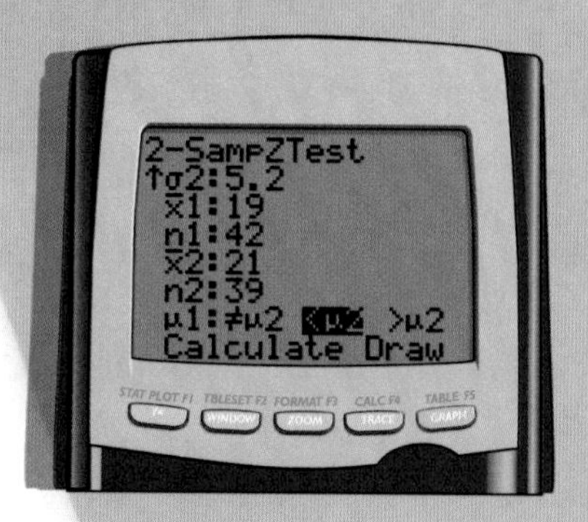

Choosing Calculate produces the following results.

First, the alternative hypothesis is stated. Note that the form of the alternative hypothesis is different, but equivalent to the one we used in this chapter. The z-score is -1.81 and the p-value is 0.0351. The sample statistics are then repeated.

4. **Draw a conclusion by comparing the p-value to the level of significance.**
   Since the p-value is 0.0351, which is smaller than the level of significance, we reject the null hypothesis. Therefore, there is sufficient evidence to support the researcher's claim that freshmen studied less than seniors, at $\alpha = 0.05$.

Comparing Two Means (Small Independent Samples)

Let's look at how to use a TI-84 Plus to perform a hypothesis test comparing population means using small independent samples. The process is the same for pooled data as for data that is not pooled. Whether or not the data is pooled is just one of the options in the menu.

Example T.2

Obstetricians are concerned that a certain pain reliever may be causing lower birth weights. A medical researcher collects data from 12 mothers who took the pain reliever while pregnant and calculates that the average birth weight of their babies was 5.6 pounds with a standard deviation of 1.8. She also collects data from 20 mothers who did not take the pain reliever and calculates that the average birth weight of their babies was 6.3 pounds with a standard deviation of 2.1. Assuming that the population variances are the same, test the hypothesis that using the pain reliever results in lower birth weights. Use a 0.01 level of significance.

Solution:

Let group 1 include the birth weights associated with taking the pain reliever and group 2 include the birth weights associated with not taking the pain reliever.

1. **State the null and alternative hypotheses.**
   Claim: Birth weights with the pain reliever are less than birth weights without the pain reliever, i.e. $\mu_1 < \mu_2$. Thus

$$H_0: \mu_1 - \mu_2 \ge 0$$
$$H_a: \mu_1 - \mu_2 < 0.$$

2. **Set up the hypothesis test by choosing the test statistic and determining the values of the test statistic that would lead to rejecting the null hypothesis.**
   We are looking at the difference in population means with small independent samples, so we use t as the test statistic. We will draw a conclusion by computing the p-value for the calculated test statistic and comparing that value to α. For this hypothesis test, $\alpha = 0.01$.

3. **Gather data and calculate the necessary sample statistics.**
   The following statistics were given in the problem:

$$\bar{x}_1 = 5.6 \qquad \bar{x}_2 = 6.3$$
$$s_1 = 1.8 \qquad s_2 = 2.1$$
$$n_1 = 12 \qquad n_2 = 20$$

cont'd on the next page

We are also told that the population variances are the same, so we will use the pooled test statistic. Now we are ready to enter the statistics in the calculator. Press STAT, scroll over to TESTS and choose option 4: 2-SampTTest. Enter the requested statistics. This is a left-tailed test, so choose $< \mu_2$. Since the variances are the same, choose YES next to 'Pooled:'.

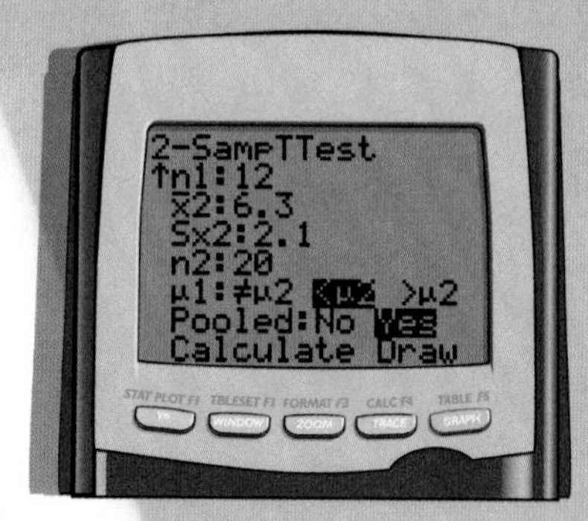

Choosing Calculate produces the following results.

4. **Draw a conclusion using the process described in Step 2.**
   $p = 0.1722$ is greater than $\alpha = 0.01$, so we fail to reject the null hypothesis. This means that the evidence does not support the claim that the pain reliever reduces birth weight.

Comparing Two Means (Dependent Samples)

The TI-84 Plus does not directly perform a hypothesis test for comparing the mean of two dependent samples, but we can perform a hypothesis test for a one sample t-test on the paired difference as seen in the following example.

Example T.3

In a CPR class students are given a pre-test at the beginning of the class to determine their initial knowledge of CPR and then a post-test after the class to determine their new knowledge of CPR. Educational researchers are interested if the class increases a student's knowledge of CPR. Test the claim that the class increases students' knowledge by more than 15 points using a 0.05 level of significance. Data from 10 students who took the CPR class are listed in the table below.

Student	1	2	3	4	5	6	7	8	9	10
Pre-Test Score	60	63	68	70	71	68	72	80	83	79
Post-Test Score	80	79	83	90	93	89	94	95	96	93

Solution:

1. **State the null and alternative hypotheses.**
   Claim: The post-test scores are more than 15 points higher than the pre-test scores. In other words, the difference in the mean scores is more than 15 points. Symbolically, this claim is $\mu_d > 15$. Thus

   $$H_0:\ \mu_d \le 15$$
   $$H_a:\ \mu_d > 15.$$

2. **Set up the hypothesis test by choosing the test statistic and determining the values of the test statistic that would lead to rejecting the null hypothesis.**
   We are looking at the difference in population means with dependent samples, so we use t as the test statistic. We will draw a conclusion by computing the p-value for the calculated test statistic and comparing that value to α. For this hypothesis test, $\alpha = 0.05$.

3. **Gather data and calculate the necessary sample statistics.**
   First, let's enter the pre-test scores in List 1 and Post-Test scores in List 2. Next, we want to calculate the paired difference in List 3. To do so, Highlight L3 and enter the formula to subtract the pre-test scores from the post-test scores by pressing 2ND 2 − 2ND 1 then ENTER.

cont'd on the next page

This will produce the formula L2 – L1 and calculate the paired difference for each pair of data.

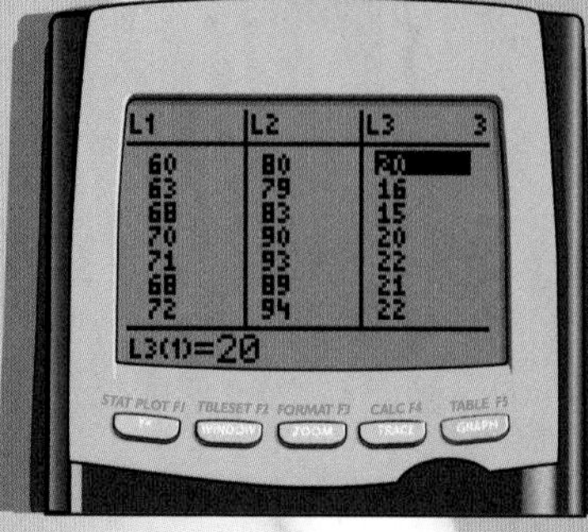

We now have the paired difference in List 3 and can create a one sample *t*-test from the raw data. Press STAT, scroll to TESTS and choose option 2: T-test. We want to perform the hypothesis test from raw data, so choose the Data option. $\mu_0 = 15$. The data is in List 3, so enter L3 by pressing 2ND then 3. The frequency of the data is the default, 1. We have a right-tailed test, so choose $> \mu_0$. The menu should then appear as it does in the screen shot shown at the right.

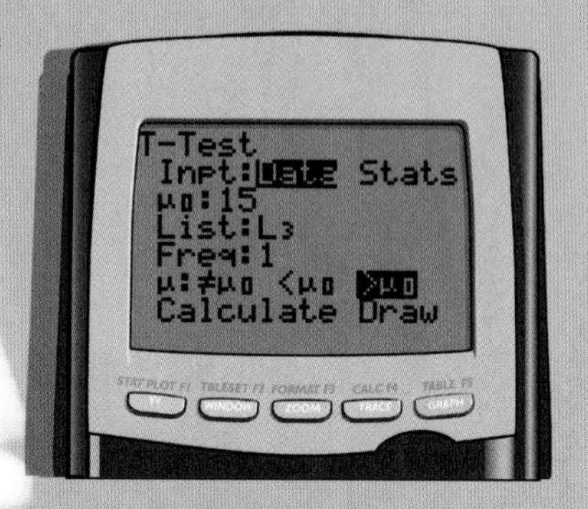

Choosing Calculate produces the following results.

4. **Draw a conclusion using the process described in Step 2.**
These results show $p = 0.0165$. This value is less than α, so we reject the null hypothesis. Thus there is enough evidence at the 0.05 level of significance to support the claim that the CPR class increases test scores by more than 15 points.

Comparing Two Proportions (Large Independent Samples)

A TI-84 Plus can perform a hypothesis for 2 proportions very simply. Let's look at an example.

Example T.4

School administrators believe that fewer students at a school in a lower economic district (School A) carry cell phones than students at a school in an upper economic district (School B). A survey of students is conducted at each school and the following results are tabulated:

	School A	School B
Yes	45	53
No	31	25

Test the administrators' claim using *p*-values and a 0.10 level of significance.

Solution:

1. **State the null and alternative hypotheses.**
Let School A be population 1 and School B be population 2. The claim is that $p_1 < p_2$. Thus

$$H_0: p_1 - p_2 \geq 0$$
$$H_a: p_1 - p_2 < 0.$$

2. **Set up the hypothesis test by choosing the test statistic and determining the values of the test statistic that would lead to rejecting the null hypothesis.**
We are looking at the difference in population proportions using large independent samples, so we use *z* as the test statistic. We will draw a conclusion by computing the *p*-value for the calculated test statistic and comparing that value to α. For this hypothesis test, $\alpha = 0.10$.

3. **Gather data and calculate the necessary sample statistics.**
To use a TI-84 Plus to perform this test, press STAT, scroll over to TESTS and choose option 6: 2-PropZTest. Enter the values provided in the table.

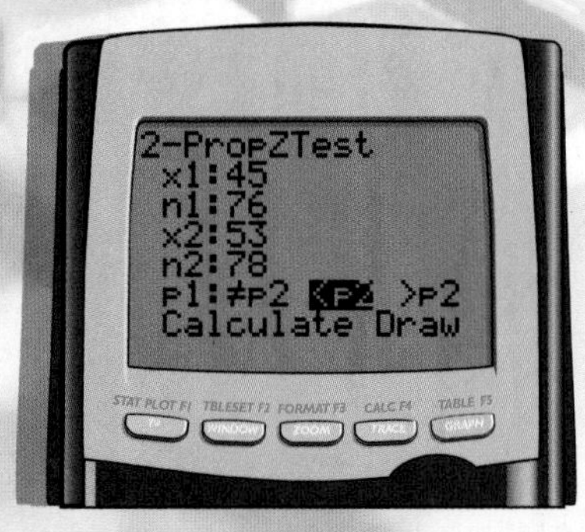

cont'd on the next page

Choosing Calculate produces the following results.

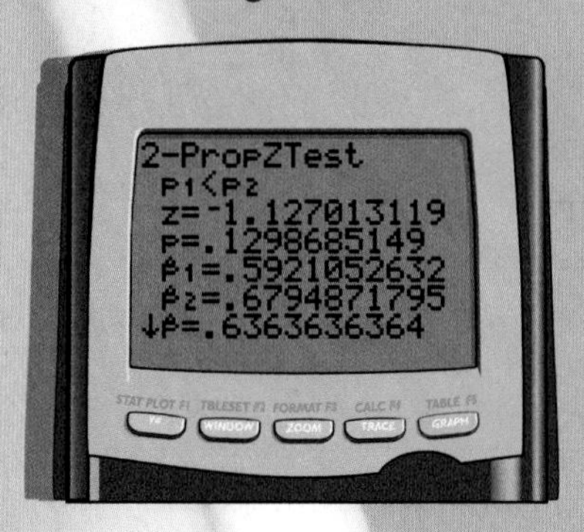

4. **Draw a conclusion by comparing the *p*-value to the level of significance.**
   Since the *p*-value is 0.1299, which is greater than the level of significance, we fail to reject the null hypothesis. Therefore, there is not sufficient evidence to say that fewer students carry a phone in the lower economic district at $\alpha = 0.10$.

Comparing Two Population Variances

The TI-84 Plus cannot construct a confidence interval for comparing population variances, but it can perform a hypothesis test for comparing two samples using the *F*-test statistic. Let's look at how to use the TI-84 Plus to solve the following problem.

Example T.5

A graduate student walking around a large college campus notices that the price of professors' cars seems to have a much smaller variance than the price variance of the cars owned by students. To test his hypothesis, he collects data from 15 professors and 14 students on the price of their cars and calculates the variance from each sample. The sample variance for the sample of professors' cars is 34,057 and the sample variance for the sample of students' cars is 45,923. Conduct a hypothesis test using a 0.10 level of significance to test the graduate student's claim. Does the evidence support the student's claim?

Solution:

1. **State the null and alternative hypotheses.**
   Let professors be population 1 and students be population 2. The claim is that $\sigma_1^2 < \sigma_2^2$. Thus

$$H_0: \sigma_1^2 \geq \sigma_2^2$$
$$H_a: \sigma_1^2 < \sigma_2^2.$$

   However, a TI-84 Plus calculator will only perform a hypothesis test on population standard deviation. Since variance is necessarily a positive number, we can convert these hypotheses regarding variance to hypotheses regarding standard deviation as follows.

$$H_0: \sigma_1 \geq \sigma_2$$
$$H_a: \sigma_1 < \sigma_2.$$

2. **Set up the hypothesis test by choosing the test statistic and determining the values of the test statistic that would lead to rejecting the null hypothesis.**
   We are looking at the difference in population standard deviations, so we use the ratio of the variances, *F*, as the test statistic. We will draw a conclusion by computing the *p*-value for the calculated test statistic and comparing that value to α. For this hypothesis test, $\alpha = 0.10$.

cont'd on the next page

3. **Gather data and calculate the necessary sample statistics.**
   The following statistics were given in the problem:

$$n_1 = 15 \qquad n_2 = 14$$
$$s_1^2 = 34{,}057 \quad s_2^2 = 45{,}923$$

In addition to these sample statistics, we need the sample standard deviations. Taking the square root of each sample variance gives us:

$$s_1 = 184.5454 \quad s_2 = 214.2965$$

Press STAT. Scroll over to TESTS and choose option D: 2-SampFTest. We want to enter the statistics, so choose STATS and enter the test statistics as shown in the screen shot to the right. This is a left-tailed test, so choose the "< σ2" option.

Choosing CALCULATE produces the results in the screen shown to the right. The first line is the alternative hypothesis. The second line is the F-test statistic. The third line is the p-value. The rest of the statistics are repeated from the problem.

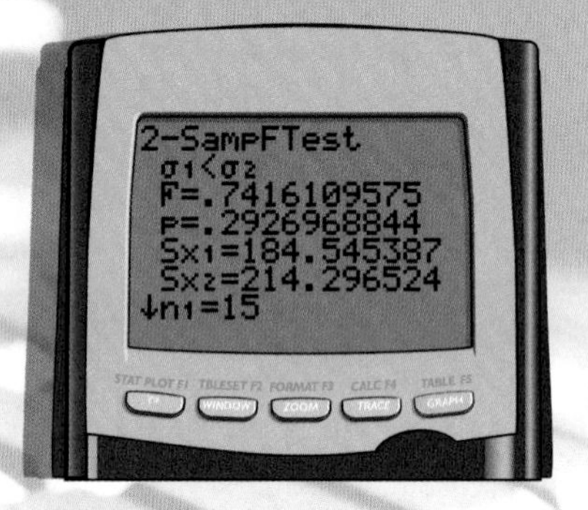

4. **Draw a conclusion using the process described in Step 2.**
   Since the p-value is 0.2927, which is greater than the level of significance, we fail to reject the null hypothesis. Therefore, there is not sufficient evidence at the 0.10 level of significance to support the claim that the standard deviation (or variance) in the price of professors' cars is less than the standard deviation (or variance) in the price of students' cars.

ANOVA

In order to use either a TI-84 Plus or Microsoft Excel to calculate an ANOVA test, you must have the raw data. Let's compare the mean of three populations. The sample data from each of these populations is listed below.

Sample 1	2	3	4	5	5	3	6
Sample 2	2	2	3	4	5	6	2
Sample 3	3	5	5	3	6	4	2

Example T.6

Use a TI-84 Plus to perform an ANOVA test on the above data. Use a 0.05 level of significance.

Solution:

To use a TI-84 Plus to perform an ANOVA test, begin by entering the sample data in the list environment. Next, press STAT, scroll to tests and choose option F: ANOVA. It will show ANOVA(on the screen. You need to enter the lists of data you want compared. In our case, we enter ANOVA(L1, L2, L3). Press ENTER. The following results will be displayed.

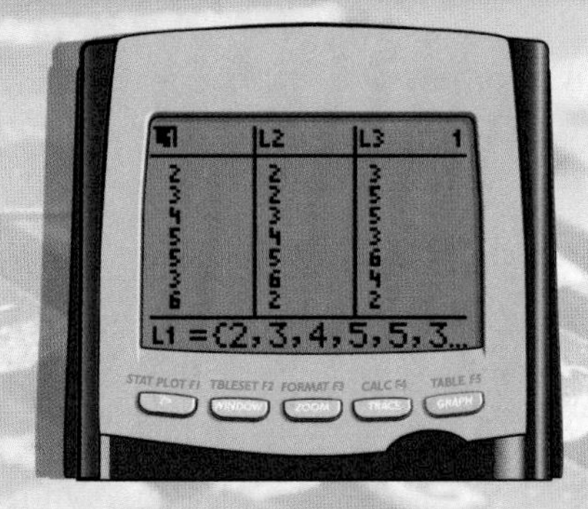

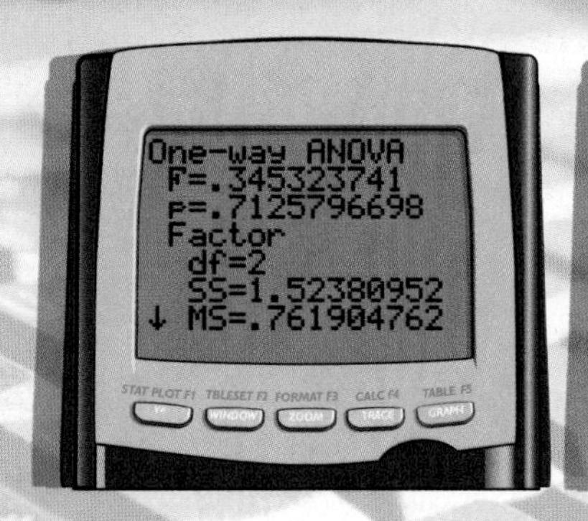

A traditional ANOVA table is not displayed, however, all of the values that would be in an ANOVA table are present. The first is the value of F, then the p-value. The values under Factor are the Treatment values and the values under Error are obviously the Error values. Since the p-value 0.7126 is much larger than the level of significance, we fail to reject the null hypothesis. Thus there is not enough evidence at the 0.05 level of significance to support the claim that the means are different.

MICROSOFT EXCEL

Comparing Two Means (Large Independent Samples)

Example T.7

The following data was collected on two different variables:

	1	2	3	4	5	6	7	8	9	10	11	12	13	14	15
Variable 1	35	32	36	39	34	31	37	39	40	41	42	45	38	31	46
Variable 2	2	3	6	5	4	9	8	2	3	6	5	7	8	9	4

	16	17	18	19	20	21	22	23	24	25	26	27	28	29	30
Variable 1	39	47	31	30	32	46	47	46	49	48	34	35	47	46	35
Variable 2	1	2	6	1	3	5	6	9	8	4	6	5	3	4	6

Determine if the difference in the two population means is greater than 34, with a level of significance of 0.05. Assume that the variance of the first population is known to be 37.93 and the variance of the second population is known to be 5.65. Use Microsoft Excel.

Solution:

To begin, we need to enter our data in the worksheet. Enter the above data in columns A and B. The hypotheses for this test are as follows:

$$H_0: \mu_1 - \mu_2 \leq 34$$
$$H_a: \mu_1 - \mu_2 > 34$$

Next, choose the DATA ANALYSIS option under the TOOLS menu. Then, choose z-Test: Two Sample for Means. The menu that appears should be filled in as follows:

z-Test: Two Sample for Means

Input

Variable 1 Range: $A$2:$A$31

Variable 2 Range: $B$2:$B$31

Hypothesized Mean Difference: 34

Variable 1 Variance (known): 37.93

Variable 2 Variance (known): 5.65

Labels

Alpha: 0.05

Output options

Output Range: $D$2

New Worksheet Ply:

New Workbook

OK

Cancel

Help

After pressing OK, the following results are displayed:

z-Test: Two Sample for Means		
	Variable 1	*Variable 2*
Mean	39.26666667	5
Known Variance	37.93	5.65
Observations	30	30
Hypothesized Mean Difference	34	
z	0.221251257	
P(Z<=z) one-tail	0.412448437	
z Critical one-tail	1.644853476	
P(Z<=z) two-tail	0.824896874	
z Critical two-tail	1.959962787	

This is a one-tailed test, so, we will use the p-value for the one-tailed test, labeled "P(Z<=z) one-tail." Since $p = 0.4124$, which is greater than $\alpha = 0.05$, we fail to reject the null hypothesis, so there is not sufficient evidence, at the 0.05 level of significance, to support the claim that the difference in the two means is greater than 34.

Comparing Two Means (Small Independent Samples)

Example T.8

A fleet supervisor tracks the mileage and fuel consumption of two types of vehicles. The first part of the fleet consists of 24 vehicles and had a sample mean of 22.1 miles per gallon with a standard deviation of 1.5. There are 36 vehicles of the second type. These vehicles averaged 19.5 miles per gallon with a standard deviation of 2.3. Can the fleet supervisor conclude with 99% certainty that there is a difference in fuel efficiency between the two vehicle types. Assume that both population variances are the same.

Solution:

Since one sample is less than 30, this will be a test for means with a small sample size. We will use a *t*-test with a pooled test statistic. We state the hypothesis.

$$H_0: \mu_1 - \mu_2 = 0$$
$$H_a: \mu_1 - \mu_2 \neq 0.$$

This is a two-tailed test with the following statistics:

$$\bar{x}_1 = 22.1 \qquad \bar{x}_2 = 19.5$$
$$s_1 = 1.5 \qquad s_2 = 2.3$$
$$n_1 = 24 \qquad n_2 = 36$$

Next, compute the test statistic by entering the following computation into cell B6
=(B3-C3)/(SQRT(((B2-1)*B4^2+(C2-1)*C4^2)/(B2+C2-2))*SQRT(1/B2+1/C2)).

Since this is a two-tailed test with pooled variance, the probability used in the TINV function is $\alpha = 0.01$ and the degrees of freedom are $n_1 + n_2 - 2$. Enter =TINV(0.01, B2+C2-2) and we see that the test statistic, 4.882, is greater than the critical value, 2.66. Therefore, the null hypothesis is rejected.

	A	B	C
1		Fleet 1	Fleet 2
2	n	24	36
3	x	22.1	19.5
4	s	1.5	2.3
5			
6	Test Statistic	4.882	
7	Critical Value	2.66	

Comparing Two Means (Small Independent Samples)

Excel can perform a paired, two sample test for means using the Data Analysis package.

Example T.9

Juan believes physical exercise leads to an average of 15 beats per minute increase in pulse rate. He measures the pulse rate of 15 volunteers before they run a mile and after they run a mile. The results are as follows:

Participant	1	2	3	4	5	6	7	8	9	10	11	12	13	14	15
Pulse Rate Before	65	75	73	78	62	58	76	79	64	73	78	55	63	69	72
Pulse Rate After	78	105	92	96	79	65	91	106	82	89	99	70	79	92	93

Is there evidence at the 0.10 level that pulse rates increase by more than 15 beats per minute with exercise?

Solution:

Since the data is in pairs, we will perform a paired two sample test for means. We state the hypotheses:

$$H_0\colon\ \mu_d \leq 15$$
$$H_a\colon\ \mu_d > 15.$$

Select TOOLS and DATA ANALYSIS. Next, choose t-Test: Paired Two Sample for Means and select OK. You are prompted to enter two ranges. The first (Variable 1 Range) corresponds to the pulse rates after exercise and the second (Variable 2 Range) corresponds to pulse rates before exercise. Enter 15 as the Hypothesized Mean Difference, 0.1 for Alpha, and D1 for the Output Range. This dialog box is shown below. Selecting OK produces the completed test. Since the p-value is less than 0.1 for the one-tailed test, we reject the null hypothesis.

	A	B	C	D	E	F
1	Before	After		t-Test: Paired Two Sample for Means		
2	65	78				
3	75	105			*Variable 1*	*Variable 2*
4	73	92		Mean	87.73333333	69.3333333
5	78	96		Variance	143.9238095	59.2380952
6	62	79		Observations	15	15
7	58	65		Pearson Correlation	0.930100732	
8	76	91		Hypothesized Mean Difference	15	
9	79	106		df	14	
10	64	82		t Stat	2.349955956	
11	73	89		P(T<=t) one-tail	0.016983142	
12	78	99		t Critical one-tail	1.345030375	
13	55	70		P(T<=t) two-tail	0.033966283	
14	63	79		t Critical two-tail	1.761310115	
15	69	92				
16	72	93				

Comparing Two Means (Small Independent Samples)

Example T.10

Renaldo thinks support for a referendum is stronger in one district than in a neighboring district. He takes a random sample from each district and records the results. The sample size was 85 from the first district with 36 votes in favor of the referendum. Out of 72 voters surveyed in district two, only 25 supported the referendum. Using $\alpha = 0.05$, try to reject the claim that support for the referendum is equal in the two districts.

Solution:

Immediately, we recognize that we need to perform a z-test for two population proportions. Since

$$n_1 \hat{p}_1 = 36 > 5$$
$$n_1 \left(1 - \hat{p}_1\right) = 49 > 5$$
$$n_2 \hat{p}_2 = 25 > 5$$
$$n_2 \left(1 - \hat{p}_2\right) = 47 > 5$$

the conditions are met for utilizing this test. The null and alternative hypotheses are as follows:

$$H_0: p_1 - p_2 = 0$$
$$H_a: p_1 - p_2 \neq 0$$

We are given the following statistics:

$$n_1 = 85 \quad n_2 = 72$$
$$x_1 = 36 \quad x_2 = 25$$

Next, enter this data into the spreadsheet and use Excel to calculate the test statistic and critical value. The completed spreadsheet is shown below. The test statistic is produced by entering

=(B4-C4)/SQRT(B6*B7*(1/B2+1/C2)).

The result of this calculation is 0.98. Since the test is two-tailed, we use the probability $1 - \frac{\alpha}{2}$ in the NORMSINV function. Enter =NORMSINV(1-0.05/2) to produce the critical value 1.96. Since the absolute value of the test statistic is less than the critical value, we cannot reject the null hypothesis. Therefore there is not sufficient evidence to reject the claim that support for the referendum is equal in the two districts.

	A	B	C
1		District 1	District 2
2	n	85	72
3	x	36	25
4	phat	0.42353	0.34722
5			
6	pbar	0.388535	
7	1-pbar	0.611465	
8	z	0.98	
9	Critical z	1.96	

Comparing Two Population Variances

Example T.11

A commuter believes the variance for her commute is higher in the afternoon. She decides to test this claim. The sample variance for the morning commute is 32.4 while the sample variance for the afternoon commute is 38.7. The sample size is 15 for both morning and afternoon travel times. Test the commuter's claim using a 0.05 level of significance. Does the evidence support her claim?

Solution:

We wish to compare two population variances. First, state the hypotheses. Let σ_1^2 = morning. Thus

$$H_0: \sigma_1^2 \geq \sigma_2^2$$
$$H_a: \sigma_1^2 < \sigma_2^2.$$

The test statistic for comparing variances is $F = \frac{s_1^2}{s_2^2}$. Enter the sample sizes and sample variances into your spreadsheet and use the calculations =B3/C3 and =FINV(0.95, 14, 14) to compute the test statistic and critical value, respectively. The first parameter in the FINV function is α for a right-tailed test, since this is a left-tailed test we use $1 - \alpha$ and the latter two are degrees of freedom for the numerator and denominator. The complete spreadsheet is shown below. Since the test statistic is greater than the critical value, we do not reject the null hypothesis.

	A	B	C
1		Morning	Afternoon
2	n	15	15
3	variance	32.4	38.7
4			
5	F	0.84	
6	Critical F	0.4	

MINITAB

Comparing Two Means (Dependent Samples)

Minitab can perform a hypothesis test for comparing the mean of two dependent samples in a single operation.

Example T.12

Cindy and her roommate both believe they have the longest commute. To settle this matter, both she and her roommate time their trips to and from work for 15 days. The total minutes each spent commuting is shown below. Use a 90% level of confidence to show there is a significant difference between their commute times.

Day	1	2	3	4	5	6	7	8	9	10	11	12	13	14	15
Cindy	24	25	28	32	19	22	27	36	32	28	22	24	21	29	31
Roommate	22	26	29	36	22	18	29	38	32	30	24	23	21	30	30

Solution:

First, we identify the necessary procedure. We will perform a hypothesis test for comparing the mean of two dependent samples. The hypotheses are as follows:

$$H_0: \mu_d = 0$$
$$H_a: \mu_d \neq 0.$$

This is a two sample *t*-test. Next, enter the data into columns in the worksheet. Now select STAT, select BASIC STATISTICS, and choose *Paired t*. The first sample is the column C1 and the second sample is column C2. Select Options in order to specify a confidence level of 90. Choose OK for both the options and main dialog windows and the *p*-value for the hypothesis test is produced. Since 0.230 > 0.10, we cannot reject the null hypothesis. The dialog box is shown below.

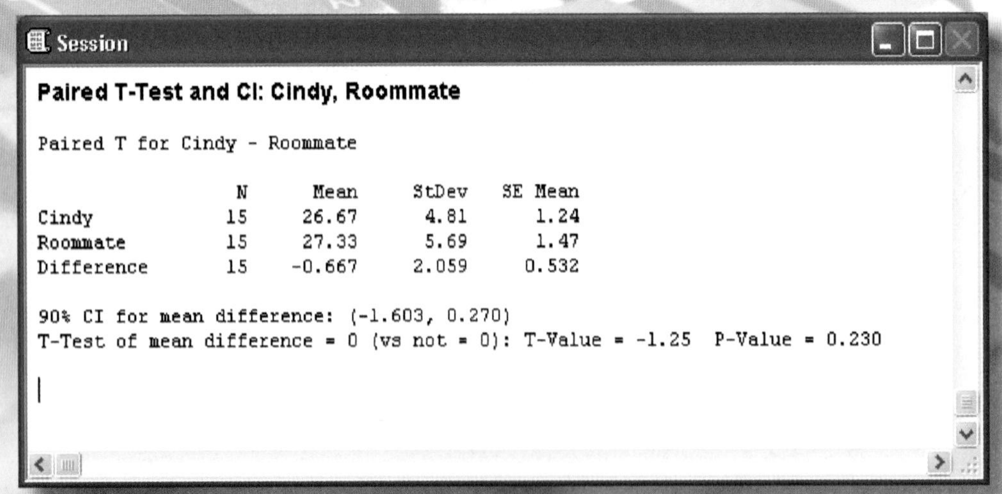

Comparing Two Proportions (Large Independent Samples)

Minitab can compare two proportions and conduct a hypothesis test with a single operation.

Example T.13

Willis and Annie play 72 holes of golf on their vacation. Annie collects 10 birdies and Willis shoots 8. Can Annie claim she is the better golfer? Use a 90% level of confidence.

Solution:

We want to compare the difference in their ratio of birdies to total holes played. Since we have

$$n_1\hat{p}_1 = 10 > 5$$
$$n_1\left(1-\hat{p}_1\right) = 62 > 5$$
$$n_2\hat{p}_2 = 8 > 5$$
$$n_2\left(1-\hat{p}_2\right) = 64 > 5$$

we can use a two sample z-test with proportions. The null and alternative hypotheses

$$H_0: p_1 - p_2 \le 0$$
$$H_a: p_1 - p_2 > 0$$

indicate that a right-tailed test will be utilized. Select STAT, select BASIC STATISTICS, and *2 Proportions….* Choose **Summarized data** and enter 72 trials for both samples, 10 successes for sample 1, and 8 successes for sample 2. Next, choose Options and enter 90 for **Confidence level** and select "greater than" for **Alternative**. Choose OK for the Options and main dialog and the results are displayed in the Session window. The dialog is shown below.

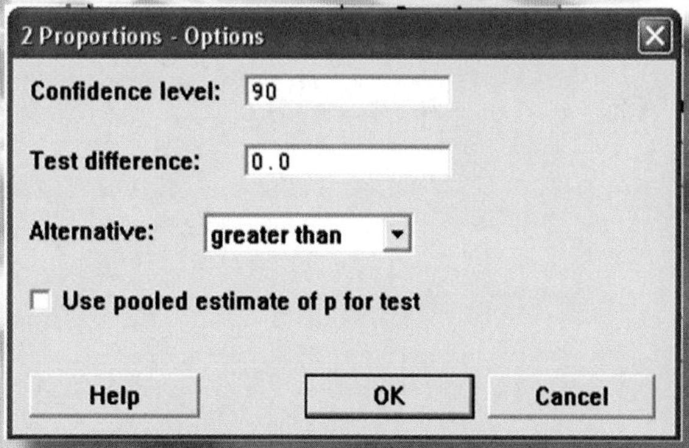

2 Proportions - Options

Confidence level: 90

Test difference: 0.0

Alternative: greater than

☐ Use pooled estimate of p for test

Help OK Cancel

cont'd on the next page

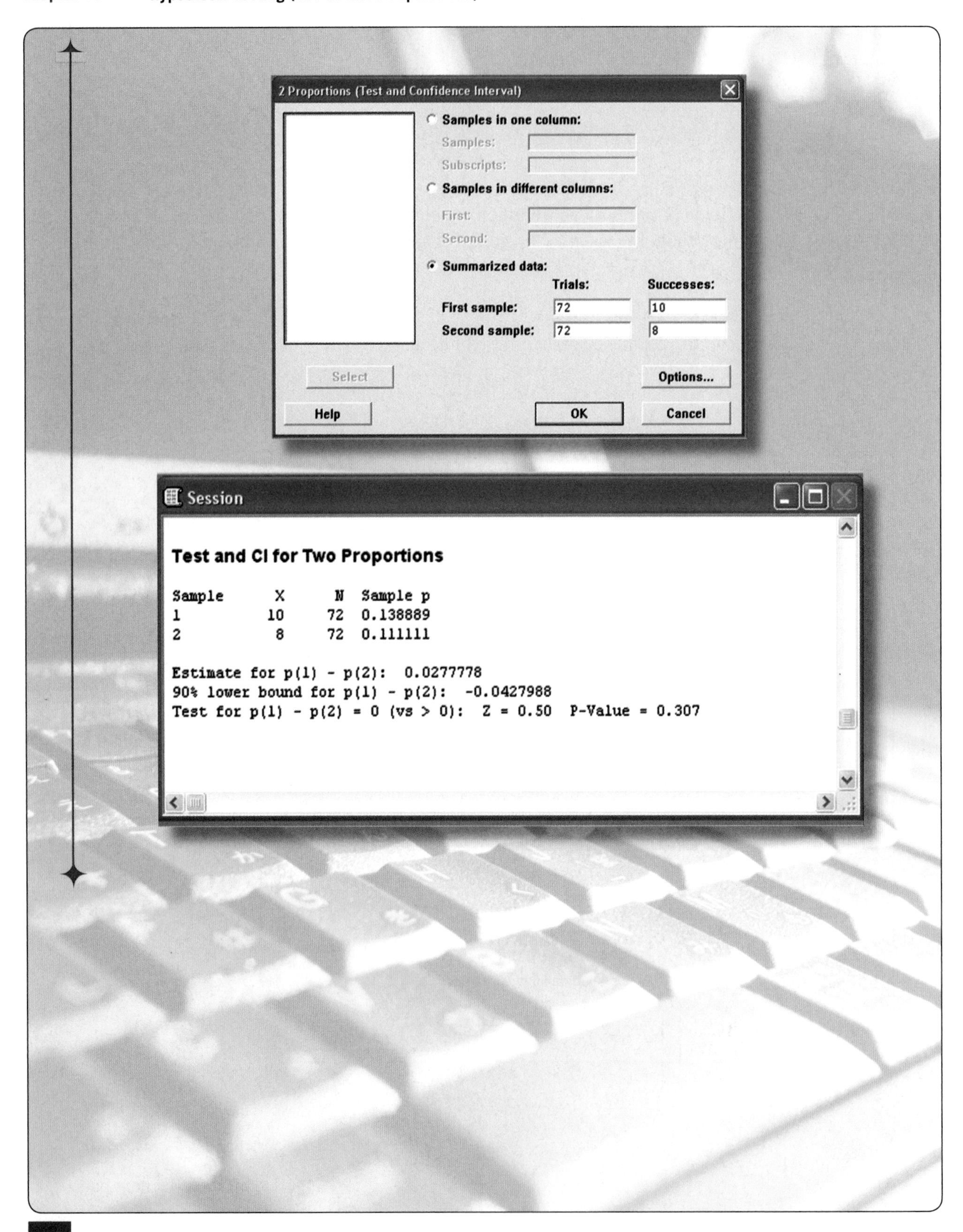
2 Proportions (Test and Confidence Interval)
Samples in one column:
Samples:
Subscripts:
Samples in different columns:
First:
Second:
Summarized data:
Trials:
Successes:
First sample:
72
10
Second sample:
72
8
Select
Options...
Help
OK
Cancel
Session
Test and CI for Two Proportions
Sample X N Sample p
1 10 72 0.138889
2 8 72 0.111111
Estimate for p(1) - p(2): 0.0277778
90% lower bound for p(1) - p(2): -0.0427988
Test for p(1) - p(2) = 0 (vs > 0): Z = 0.50 P-Value = 0.307

Comparing Two Proportions Variances

Example T.14

Ralph believes that the variance in batting average of catchers is greater than that of outfielders. He decides to test this claim and takes a sample of 20 catchers and finds their batting average has a sample variance of 0.0062. A similar sample of 23 outfielders yields a sample variance of 0.0043. Test his claim at a 0.10 level of significance. Does the evidence support Ralph's claim?

Solution:

We wish to compare two population variances. First, state the hypothesis.

$$H_0: \sigma_1^2 \le \sigma_2^2$$
$$H_a: \sigma_1^2 > \sigma_2^2.$$

The test statistic for comparing variances is $F = \frac{s_1^2}{s_2^2}$. Minitab can calculate this value by selecting CALC and then CALCULATOR. Enter 0.0062/0.0043 as the **Expression** and choose C1 under Store result in variable. This value is 1.44816. Next, we need to find the critical value for F. Choose CALC, PROBABILITY DISTRIBUTIONS, and F. On the dialog screen that appears, choose **Inverse cumulative probability**, set **Noncentrality parameter** to 0.0, set **Numerator degrees of freedom** to 19, set **Denominator degrees of freedom** to 22, set **Input constant** to 0.9, and select OK. The critical value, 1.7675, is displayed. The dialog box is shown below. Since the test statistic is less than the critical value, the null hypothesis is not rejected.

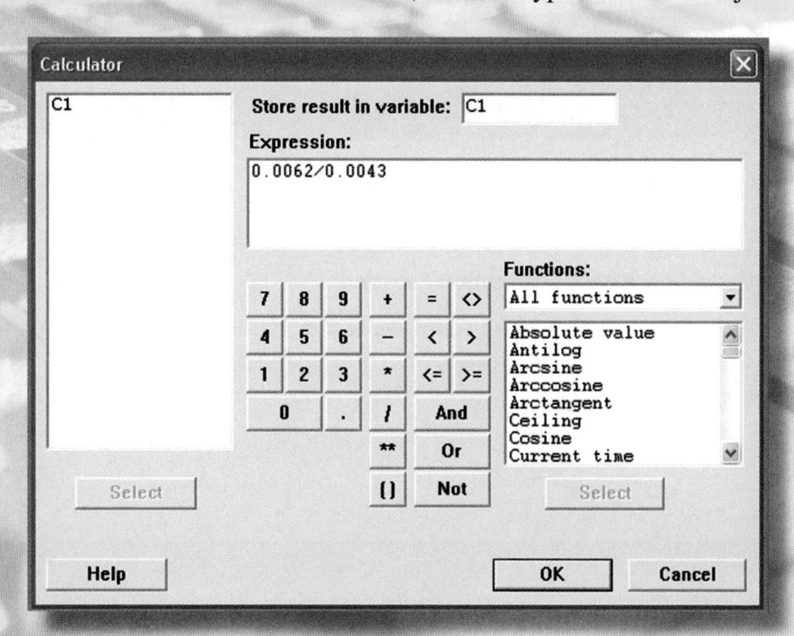

cont'd on the next page

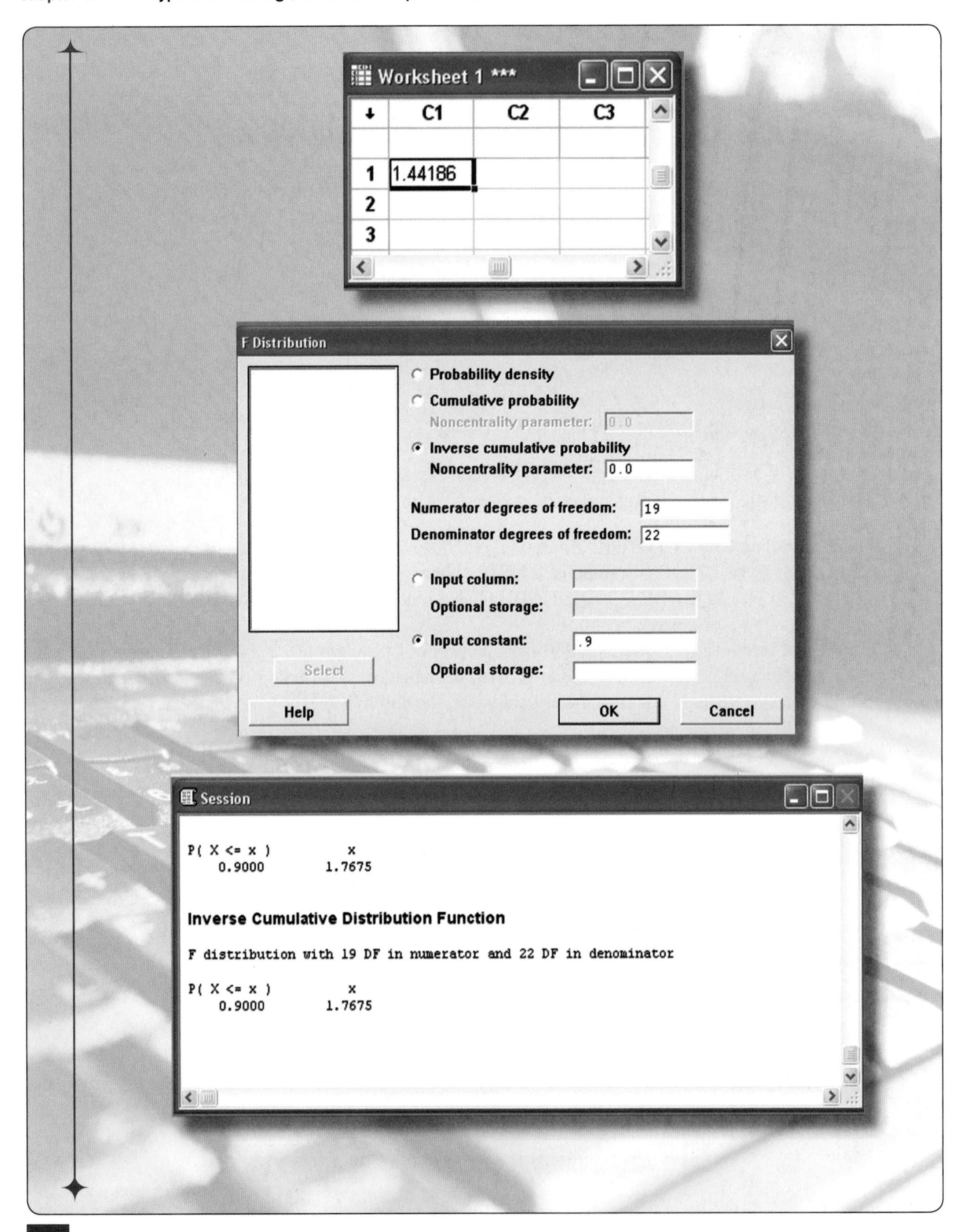

Worksheet 1 ***
C1
C2
C3
1
1.44186
2
3
F Distribution
Probability density
Cumulative probability
Noncentrality parameter:
0.0
Inverse cumulative probability
Noncentrality parameter:
0.0
Numerator degrees of freedom:
19
Denominator degrees of freedom:
22
Input column:
Optional storage:
Input constant:
.9
Optional storage:
Select
Help
OK
Cancel
Session
P(X <= x) x
0.9000 1.7675
Inverse Cumulative Distribution Function
F distribution with 19 DF in numerator and 22 DF in denominator
P(X <= x) x
0.9000 1.7675

ANOVA

To use Minitab to perform an ANOVA test, begin by entering your data into the columns, and then choose ANOVA, One-way (Unstacked). Let's look at an example.

Example T.15

You have just been promoted to sales manager of a company manufacturing robots used to assemble automobiles. Although your sales force is given a suggested price at which to sell the robots, they have considerable leeway in negotiating the final price. Past sales records indicate that sometimes there is a large difference in the sales price which different sales reps are able to negotiate. You are interested in knowing if this difference is significant, possibly because of a more effective negotiating strategy or exceptional interpersonal skills, or whether this observed difference in sales price is just due to random variation. You decide to randomly select four sales over the last year for each of your three sales representatives and observe the actual selling price of the robot. The following table shows the amounts at which the robots sold in thousands of dollars. Use $\alpha = 0.05$.

	Salesperson # 1	**Salesperson # 2**	**Salesperson # 3**	
	10	11	11	
	14	16	13	
	13	14	12	
	12	15	15	
Totals	49	56	51	156

Solution:

Enter the four sales for the three sales reps into columns C1, C2, and C3. Then, under STAT choose ANOVA, then *One-way (Unstacked)*. In the dialog box, use Select to input C1, C2, and C3 as the Responses. Select OK. Observe the Output Screen for the regression analysis.

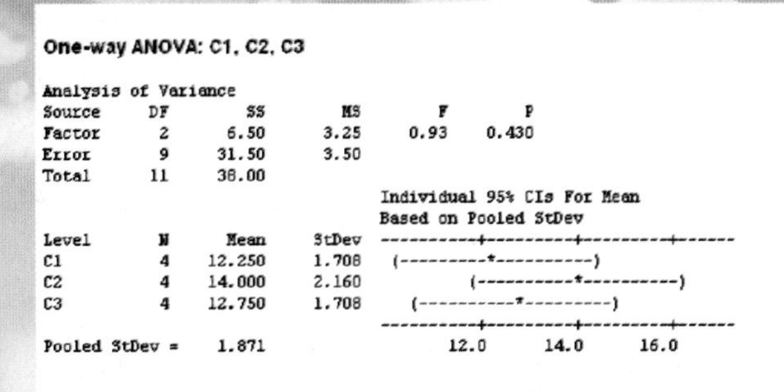

```
One-way ANOVA: C1, C2, C3

Analysis of Variance
Source     DF        SS        MS        F        P
Factor      2      6.50      3.25     0.93    0.430
Error       9     31.50      3.50
Total      11     38.00
                                  Individual 95% CIs For Mean
                                  Based on Pooled StDev
Level       N      Mean     StDev  ---------+---------+---------+------
C1          4    12.250     1.708   (---------*----------)
C2          4    14.000     2.160          (----------*----------)
C3          4    12.750     1.708     (----------*---------)
                                   ---------+---------+---------+------
Pooled StDev =    1.871                   12.0      14.0      16.0
```

Regression, Inference and Model Building

Objectives

The objectives of this chapter are as follows:

1. Construct and interpret scatter plots
2. Calculate and interpret the correlation between 2 variables
3. Determine if a correlation coefficient is statistically significant
4. Calculate the regression line for a given set of data
5. Use a regression line when appropriate to make predictions
6. Calculate residuals
7. Calculate the sum of squared errors
8. Calculate the standard error of estimate
9. Construct prediction intervals for an individual *y*
10. Construct confidence intervals for the slope of the regression line
11. Construct confidence intervals for the *y*-intercept of the regression line
12. Calculate a multiple regression equation

Sections

Chapter 12

Introduction

Relationships can be complicated. "Did he notice me?" "Should I call her?" "Would we make a good couple?" These are just a few of the many questions that you might ask yourself at the beginning of a relationship. Dating is fun, but often stressful and uncertain as well—but why? Dating can be nerve-wracking because it is hard sometimes to determine the connection you have with the other person. Do they like you? Do they like you enough to want to go out with you? Could this be the person you are going to marry?

Like dating, linear regression is all about relationships. Specifically, linear regression looks at the relationship between different sets of data. By using statistics, we can mathematically determine if two sets of data are related and, if so, how strong that relationship is. By knowing how the data are related, we can make a model that allows us to make predictions regarding the data sets. Too bad we don't have a social predictor that is this accurate!

Scatter Plots and Correlation

12.1

To begin analyzing the relationship between two characteristics of a population, it is helpful to look at the data visually first. The most common graph used to show the relationship between pairs of data is a scatter plot.

A **scatter plot** is a graph on the coordinate plane which contains one point for each pair of data. Unlike a line graph, these points are not connected with a line. The horizontal axis (x-axis) represents one characteristic and the vertical axis (y-axis) represents the other. The choice of which characteristic to put on which axis depends on whether or not a causal relationship *might* exist between the two variables. If a causal relationship exists, we say that the change in the values in one variable, called the **independent variable**, causes the change in the values of the other variable, called the **dependant variable**. Other names for the independent variable are the *predictor variable* or the *explanatory variable*. Another name for the dependant variable is the *response variable*. For example, the number of hours a student studies for an exam and their grade on an exam possibly has a causal relationship. In such a case, you should put the independent variable (the number of hours spent studying) on the x axis and the dependent variable (grade on the exam) on the y axis. If no causal relationship is assumed to exist, such as with GPA and ACT scores, it does not matter which variable is placed on which axis. The scatter plot will appear different, but the important aspects of the graphs will be the same.

Let's look at an example where creating a scatter plot would be useful in determining a relationship between two variables. A child development class wants to learn about the growth rate of children. They visit a local preschool and measure the heights of the children. The mean height of the whole school may be interesting, but their aim is to go one step deeper and to analyze the way in which children grow as they get older. The table below shows the heights of children from various ages.

Table 1 – Heights of Children														
Age (years)	1.0	1.5	1.8	2.0	2.2	2.5	3.2	3.3	3.5	3.9	4.2	4.5	4.8	5.1
Height (inches)	29.4	28.2	30.1	33.7	33.5	34.3	37.5	36.0	37.5	39.0	41.4	43.9	44.5	45.8

In order to produce a scatter plot displaying the above data, the first step would be to determine if a causal relationship exists between these two variables. Certainly, we would expect that as a child gets older, they would grow taller, thus the independent variable would be the age (in years) and should be placed on the x-axis. The dependent variable is then the height (in inches) and should be placed on the y-axis. The next step is to think of each age-height pair as a point on the coordinate plane. For instance, the height of 29.4″ for the 1 year old gives the point (1, 29.4), the height of 33.7″ for the two year old gives the point (2, 33.7), and so on. When we plot these points, we obtain the following graph.

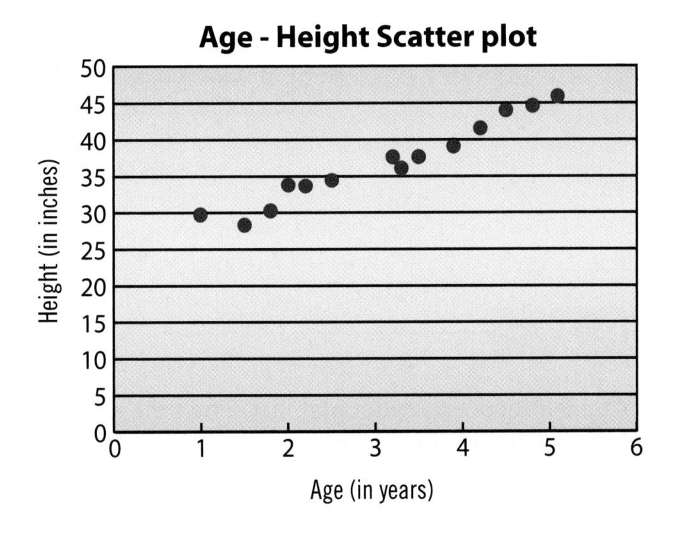

The importance of a scatter plot is to identify trends in the data. For example, do the points fall in a linear pattern, a curved pattern, or no pattern at all? If the points have a linear shape, how close to a straight line are they? Does the linear pattern have a positive slope (the points rise from left to right) or a negative slope (the points fall from left to right)? Notice that the points in our age-height scatter plot have a linear pattern that is very close to a straight line that has a positive slope. We will talk about the significance of the trend a little later in this section.

While at the preschool, the child development class also records the number of children in each age group and the average number of diaper changes for each child per day in each age group. That data is given in the table on the following page.

Table 2 – Class Size and Diaper Changes					
Age (years)	1	2	3	4	5
Class Size	7	15	8	14	6
Diaper Changes	3.5	2.3	1.8	0.65	0

Example 1

Use the data from the preschool to produce a scatter plot that shows the relationship between the age of a child and the number of children in the class. Discuss any noticeable trends.

Solution:

Looking at this scatter plot, we do not see a linear pattern. Actually, no pattern is evident. This probably indicates that these two variables do not have a relationship.

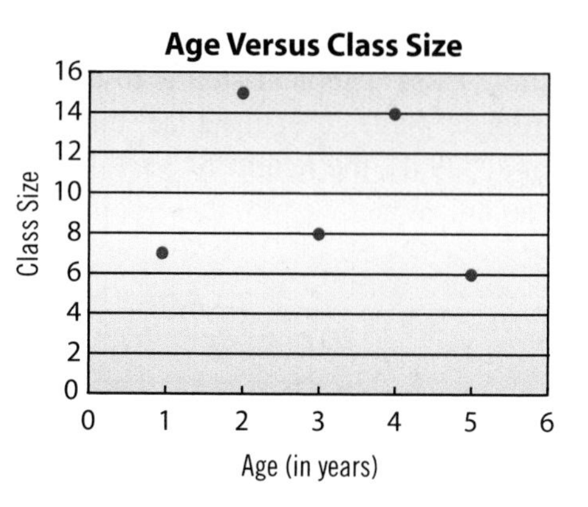

Example 2

Use the above data to produce a scatter plot that shows the relationship between the age of a child and the number of diaper changes per day at the preschool. Discuss any noticeable trends.

Solution:

Notice that the points tend to go down from left to right, and fall close to a straight line. Thus there is a linear pattern with a negative slope.

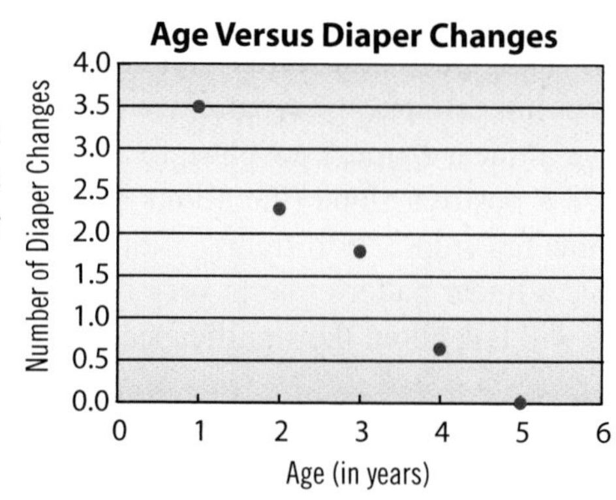

Let's compare the scatter plots from our examples. The age versus mean height scatter plot shows a predictable pattern. That is, it shows that height seems to increase as age increases (not a very deep observation about child development). In fact, the relationship seems to roughly follow a straight line, or a **linear relationship**. The third scatter plot, age versus the number of diaper changes, also has a linear pattern, but with a negative slope. This indicates, as you would expect, that as children get older, they require fewer diaper changes.

When the scatter plots do roughly follow a straight line, the direction of the pattern tells how the variables respond to each other. A **positive slope** indicates that as the values of one variable increase, so do the values of the other variable. A **negative slope** indicates that as the values of one variable increase, the values of the other variable decrease. The figure below demonstrates the difference between a positive and negative pattern.

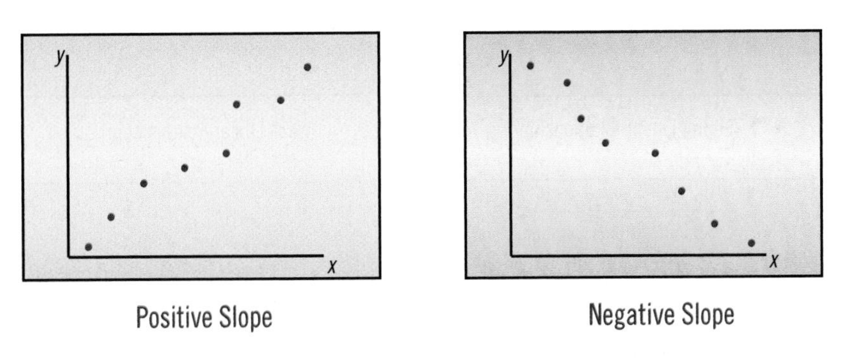

Positive Slope

Negative Slope

Figure 1 – Direction of a Linear Pattern

Example 3

Determine whether the pattern of a scatter plot between the two variables would have a positive slope, negative slope, or not follow a straight line pattern.

a. The number of hours you study for an exam and the grade you make on that exam

b. The price of a used car and the number of miles it was driven

c. The pressure on a gas pedal and the speed of the car

d. Shoe size and IQ for adults

Solution:

a. As the number of hours you study goes up, usually your grade goes up. Thus it would have a positive slope pattern.

b. As the number of miles a car was driven goes up, the price usually goes down. Thus it would have a negative slope pattern.

c. The more you push on the gas pedal, the faster the car will go. Thus the pattern would have a positive slope.

d. There is no apparent linear relationship between shoe size and IQ.

Let's consider a new parameter that measures how strongly one of the variables is linearly dependent upon the other, i.e. the linear relationship between the two variables. The strength of the linear relationship is determined by how close the points in the scatter plot lie to a straight line. Thus the stronger the relationship, the more the diagram looks like a straight line. The weaker the relationship, the more scattered the points are in the diagram. The figure below demonstrates the difference between a strong linear relationship and a weak linear relationship.

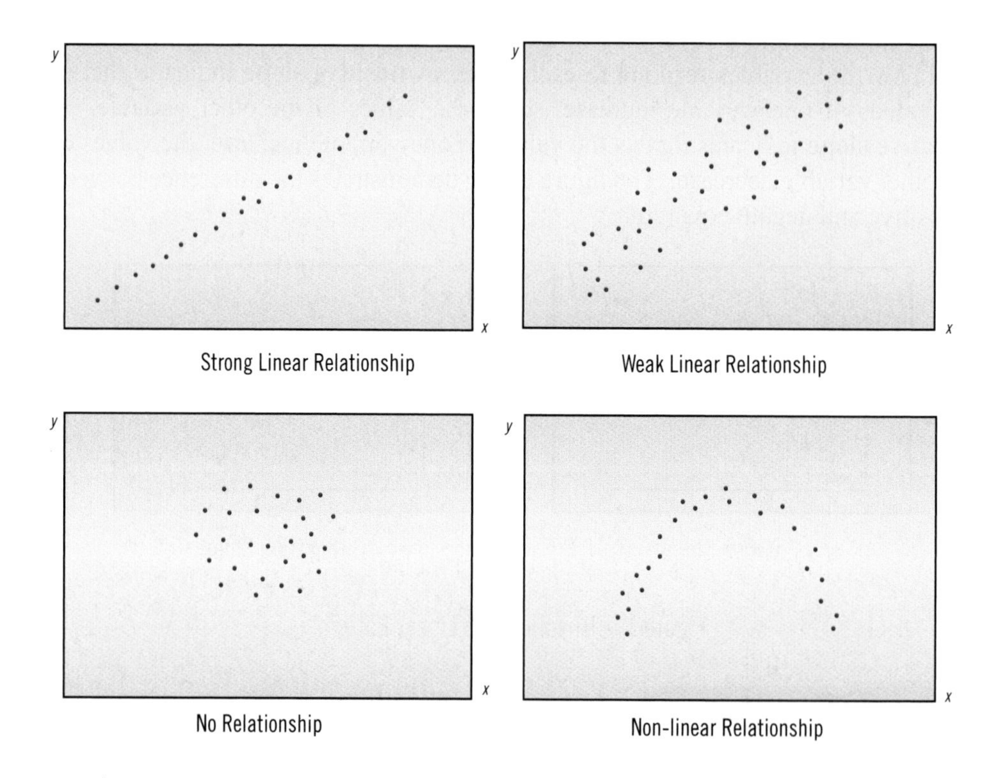

Figure 2 – Strength of a Linear Relationship

This parameter, which measures the strength of a linear relationship, is called the **Pearson correlation coefficient**, ρ, pronounced "rho". The correlation coefficient for a sample is denoted r. It always takes a value between −1 and 1, inclusive.

Correlation coefficient, r

$$-1 \leq r \leq 1$$

If r is positive, then the scatter plot has a positive slope pattern and the variables are said to have a positive relationship. Similarly, a negative value of r indicates a negative relationship between the variables. If $r = 0$, then no linear relationship exists between the two variables. If $r = 1$ then the data falls in a perfect line with a positive slope. Similarly, if $r = -1$, the data falls in a perfect line with a negative slope. The strength of the correlation is expressed by $|r|$. The larger $|r|$ is, the stronger the correlation.

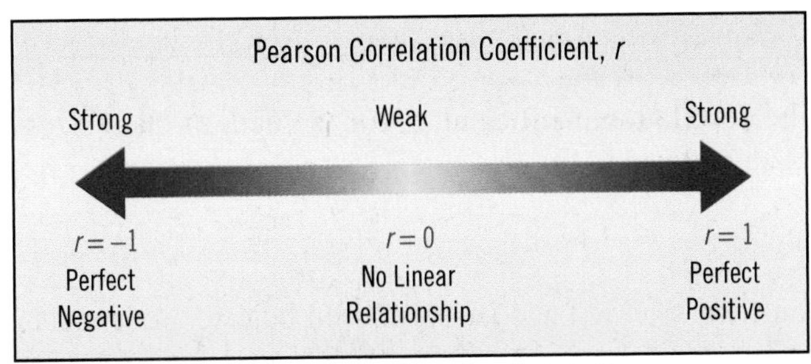

Figure 3 – Properties of a Correlation

To calculate the correlation coefficient, r, we could use the following formula.

$$r = \frac{n\sum xy - (\sum x)(\sum y)}{\sqrt{n\sum x^2 - (\sum x)^2}\sqrt{n\sum y^2 - (\sum y)^2}}$$

Using this formula to compute the correlation coefficient by hand is unnecessary with all of the technology available. We will use a TI-84 Plus calculator to find the correlation coefficient for the following examples.

A TI-84 Plus calculates the correlation coefficient simultaneously with the coefficient of determination, as well as the slope and y-intercept of the regression line, which we will also discuss in this chapter*. To begin, press STAT and then 1:EDIT and enter the values for the x-variable in List 1 and the values for the y variable in List 2. Then press STAT, choose CALC and option 4:LinReg(ax+b). Press ENTER twice. The output will include the correlation coefficient, r.

Let's use a TI-84 Plus calculator to calculate the correlation for the data of age versus height from Table 1. Begin by pressing STAT and then 1:EDIT. Enter the values for age (x) in List 1 and the values for height (y) in List 2. Then press STAT and choose CALC and option 4:LinReg(ax+b). Press ENTER twice. The output should appear as shown in the screen shot to the right.

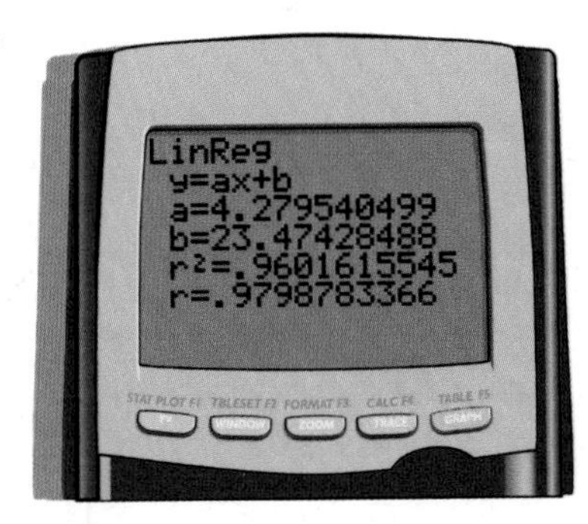

The output screen provides many values, one of which is the correlation. We can see that the correlation is $r \approx 0.980$, rounded to three decimal places. Notice that this value is positive, which matches with the positive slope of the scatter plot. Also, note that the points in the scatter plot closely follow a straight line, which corresponds to the r value being very close to 1.

**Footnote:* *In order to make sure that the output on a TI-84 Plus will contain all four of the values, begin by pressing 2ND and 0, which will bring up the Catalog. Scroll down to DiagnosticOn. Press ENTER twice. You need to perform this step only once.*

Rounding Rule

Round the correlation coefficient, r, to three decimal places.

Example 4

Calculate the correlation coefficient, r, for the data of class sizes by age, from the earlier data collected.

Solution:

The data for class size and age is reproduced below. Let's enter this data into our calculator.

Age	Class Size
1	7
2	15
3	8
4	14
5	6

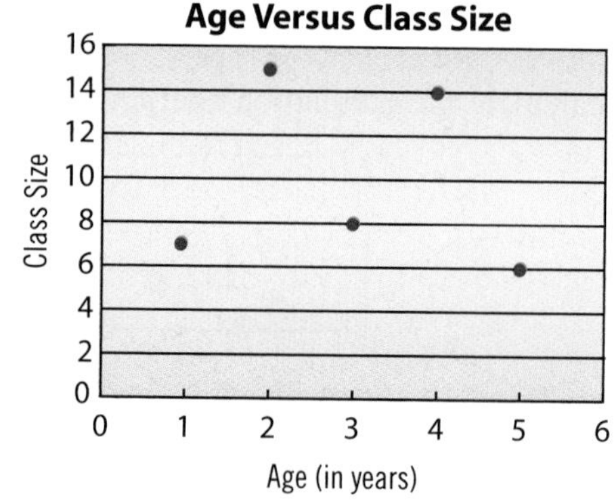

From the scatter plot, we would expect $|r|$ to be close to 0. And indeed, the calculator confirms that the correlation coefficient is $r = -0.113$. This indicates a weak negative relationship, if any relationship at all.

Example 5

Calculate the correlation coefficient, r, for the data of diaper changes by age, from the earlier data collected.

Solution:

The data for diaper changes and age is reproduced below. Let's enter this data into our calculator.

Age	Diaper Changes
1	3.5
2	2.3
3	1.8
4	0.65
5	0

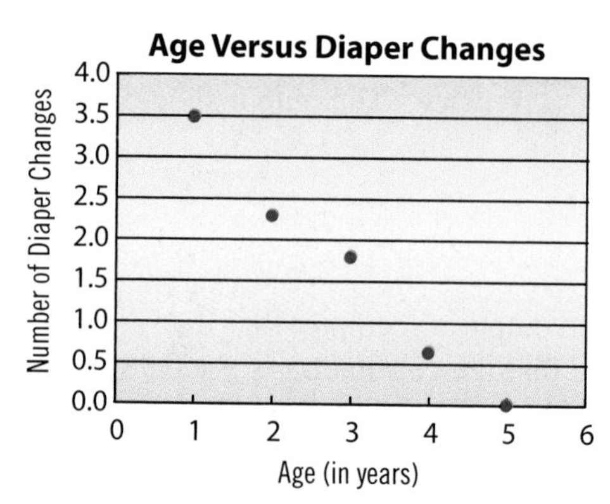

From the scatter plot, we would expect $|r|$ to be close to 1. And indeed, the calculator confirms that the correlation coefficient is $r = -0.993$. This indicates a very strong negative relationship.

✦ Significance

For each of our examples, we have discussed the linear relationship between two variables. However, we did not determine if the relationship occurred by chance, or if the relationship is statistically significant. The term *statistically significant* for the Pearson correlation coefficient indicates that there is a significant linear relationship between the two variables. In other words, if the linear relationship between the variables in the population is statistically significant, then the population's correlation, ρ, would not equal zero, i.e. $\rho \neq 0$. We can use the principles of hypothesis testing to determine if the linear relationship is significant.

First, the null and alternative hypotheses for testing the statistical significance of a linear relationship are

H_0: $\rho = 0$ (Implies there is no significant linear relationship.)
H_a: $\rho \neq 0$ (Implies there is a significant linear relationship.)

If you wish to test if the linear relationship not only exists, but specifically if it is positive or negative, the null and alternative hypotheses can be adjusted as follows:

Null and Alternative Hypotheses Testing for Linear Relationships:

Significant Linear Relationship (Two-tailed test):

H_0: $\rho = 0$ (Implies there is no significant linear relationship.)
H_a: $\rho \neq 0$ (Implies there is a significant linear relationship.)

Negative Linear Relationship (Left-tailed test):

$$H_0: \rho \geq 0$$
$$H_a: \rho < 0$$

Positive Linear Relationship (Right-tailed test):

$$H_0: \rho \leq 0$$
$$H_a: \rho > 0$$

Second, the test statistic for this hypothesis test is the following t-test statistic with $n - 2$ degrees of freedom:

$$t = \frac{r}{\sqrt{\frac{1-r^2}{n-2}}}$$

We can use rejection regions to draw conclusions in the same manner as we have previously with t-test statistics. Alternatively, we can use technology, such as a TI-84 Plus calculator, to find the p-value, which can be compared to the level of significance to draw a conclusion.

Rejection Rule for Testing Linear Relationships:

Significant Linear Relationship (Two-tailed test):
Reject H_0 if $|t| \geq t_{\frac{\alpha}{2}}$.

Negative Linear Relationship (Left-tailed test):
Reject H_0 if $t \leq -t_{\alpha}$.

Positive Linear Relationship (Right-tailed test):
Reject H_0 if $t \geq t_{\alpha}$.

The only sample statistics we need to calculate the test statistic are the correlation coefficient and the sample size. After we have calculated the test statistic, we can use rejections regions or p-values to draw a conclusion, whichever we prefer.

Though the one-sided forms of the hypotheses were presented, we will only focus on testing if there is a significant linear relationship (two-tailed test), as this is the most useful test. Let's look at an example of using a hypothesis test to determine if the linear relationship between two variables is significant.

Example 6

Use a hypothesis test to determine if the linear relationship between the number of parking tickets a student receives during a semester and their GPA during the same semester is statistically significant at the 0.05 level of significance. Refer to the data presented below.

Table 3 - GPA and Number of Speeding Tickets

Number of Tickets	0	0	0	0	1	1	1	2
GPA	3.6	3.9	2.4	3.1	3.5	4.0	3.6	2.8
Number of Tickets	2	2	3	3	5	7	8	
GPA	3.0	2.2	3.9	3.1	2.1	2.8	1.7	

Solution:

1. **State the null and alternative hypotheses.**
We wish to test the claim that there exists a significant linear relationship between the number of parking tickets a student receives during a semester and their GPA during the same semester.

$$H_0: \rho = 0$$
$$H_a: \rho \neq 0$$

2. **Set up the hypothesis test by choosing the test statistic and determining the values of the test statistic that would lead to rejecting the null hypothesis.**
   We will use the t-test statistic, with $\alpha = 0.05$. We will use rejection regions in this example to draw the conclusion. (Example 16 in Section 12.3 briefly shows how to use technology to calculate the p-value for the hypothesis test.) Since the sample size for this example is 15, the number of degrees of freedom is $n - 2 = 15 - 2 = 13$. Thus $t_{\frac{\alpha}{2}} = 2.160$ with 13 degrees of freedom. So we will reject H_0 if $|t| \geq 2.160$.

3. **Gather data and calculate the necessary sample statistics.**
   We need to begin by calculating the correlation coefficient. We get $r = -0.587$ from the calculator and we know that $n = 15$. Substituting these values into our test statistic formula yields

$$\begin{aligned} t &= \frac{r}{\sqrt{\dfrac{1-r^2}{n-2}}} \\ &= \frac{-0.587}{\sqrt{\dfrac{1-(-0.587)^2}{15-2}}} \\ &= -2.614. \end{aligned}$$

4. **Draw a conclusion using the process described in Step 2.**
   $|t| = 2.614 \geq 2.160$, so we reject the null hypothesis. Therefore, there is sufficient evidence at the 0.05 level of significance to support the claim that there is a significant linear relationship between the number of parking tickets a student receives during a semester and their GPA during the same semester.

Something To Ponder...

Do you think it is always true that there is a relationship between a student's GPA and the number of parking tickets that he or she has? Would you ever say that a cause and effect relationship exists between the variables? Or, is it possible that a third factor is influencing both GPA and the number of parking tickets obtained? Is this sample necessarily representative of the entire population?

Instead of using a formal hypothesis test, there is an alternate method that does not require the calculation of the test statistic. We know that the closer the correlation coefficient is to 1 or −1, the stronger the linear relationship between the two variables, but how close is close enough to imply statistical significance? Table I lists critical values for the Pearson correlation coefficient at particular levels of significance. To use this table, you need to know the level of significance, α, and the sample size, n, to look up the critical value for the correlation coefficient, r_α. Once we know the critical value, r_α, we can compare the correlation coefficient to the critical value to draw a conclusion about statistical significance.

Using Critical Values to Determine Statistical Significance

The correlation coefficient, r, is statistically significant if the absolute value of the correlation is greater than the critical value in the table.

$$|r| > r_\alpha$$

Example 7

Use the critical values in Table I to determine if the correlation between age and class size is statistically significant. Use a 0.05 level of significance.

Solution:

Begin by finding the critical value for $\alpha = 0.05$ with $n = 5$. Thus $r_\alpha = 0.878$. Comparing this critical value to the correlation coefficient we found for this data in Example 4, we have that $|r| = 0.113 < 0.878$. So the linear relationship between the variables is not statistically significant at the 0.05 level of significance. Therefore, we do not have sufficient evidence, at the 0.05 level of significance, to conclude that there exists a linear relationship between age and class size.

Suppose that for a given set of data r is not statistically significant. Does this mean that there is definitely not a correlation between the variables? No—there is just not enough evidence based on the sample taken. Another sample may yield different results. On the other hand, it is possible that regardless of the number of samples taken, there still may not be a significant correlation between the variables. However, the more trials that are completed, the better the prediction regarding the correlation will likely be.

Before leaving the significance of the correlation coefficient, let's take a moment to look at what statistical significance does NOT tell us. Many times we are tempted to say that because two things are statistically correlated, one must cause the other, or that we have proven such a thing. However, the world is much more complicated than that. When we do have correlation, one of at least four different things may actually be going on, as are illustrated on the following page.

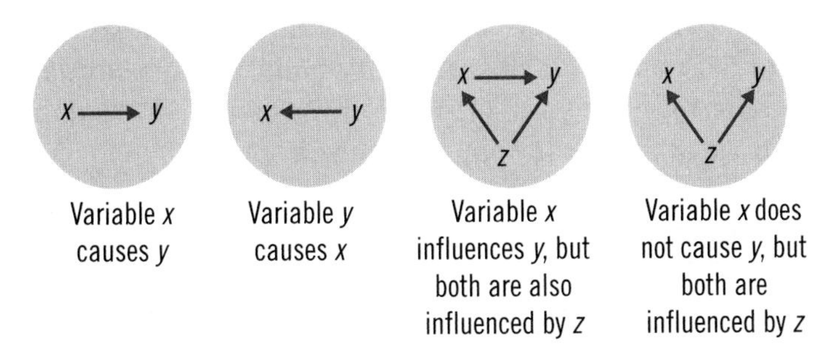

Figure 4 – Types of Relationships Between Two Correlated Variables

Suppose there is a correlation between two variables, x and y. The natural assumption is that x causes y. However, if we reversed the variables, in reality, y may be causing x to change. On the other hand, we might have a combination of factors playing a part. Variable x may indeed be influencing y, but a third variable, z, might also be influencing them both. Take for instance the type of anesthesia used in surgery and death rates associated with various surgeries. It is possible that the type of anesthesia might influence the number of deaths. However, it is the type of surgery being performed that affects both the type of anesthesia and the death rate.

It might be the case that even though a correlation exists between two variables, one is not influencing the other at all. Take for instance the correlation between watermelon consumption and the number of drownings per month. There is a positive correlation between the two, but an outside variable, temperature, influences both. As the temperature goes up, so does watermelon consumption and the number of people swimming who might drown. Eating watermelon certainly does not increase your chances of drowning! Think for yourself about ice cream sales and flu cases, which have a negative correlation. Should we all eat loads of ice cream during flu season as a preventative measure, or does another factor play into their negative correlation?

✦ Coefficient of Determination

We can actually say more than just whether or not a correlation is significant. If we would like to say how much of the variation in a child's height is associated with the relationship between their age and height, we can calculate the **coefficient of determination**. The coefficient of determination is a measure of the amount of variance in the dependent variable that is associated with the relationship between the two variables.

Coefficient of Determination ✦

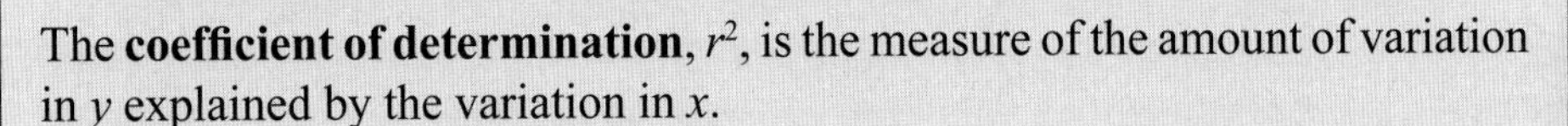

The **coefficient of determination**, r^2, is the measure of the amount of variation in y explained by the variation in x.

In our height versus age example, $r = 0.97988$ (unrounded), so $r^2 = 0.96016$. Thus 96.016% of the variation in the height of a child, y, can be explained by the relationship between the child's age and height. The other 3.984% of the variation is associated with other factors or even sampling error.

Example 8

If the correlation between the number of rooms in a house and its price is $r = 0.65$, how much of the variation in price can be explained by the relationship between the two variables?

Solution:

If $r = 0.65$, then $r^2 = 0.4225$, so about 42% of the variation in the price of a house can be explained by the relationship between the two variables.

Exercises 12.1

Directions: Determine the following correlations in words, i.e. strong or weak, positive or negative, or none.

1. Height and shoe size
2. Cholesterol level and IQ
3. Hours of physical activity per week and percentage body fat
4. Number of fruits and vegetables eaten a day and risk of heart disease
5. Number of pets owned and household income

Directions: For each of the following data sets, complete the following:

a. Draw a scatter plot

b. Estimate the correlation in words (i.e. positive, negative, or no correlation.)

c. Calculate the correlation coefficient, r.

6. Consider the relationship between the number of bids an item on eBay receives and the item's selling price. The following is a sample of 10 items sold through auction on eBay.

Table for Exercise 6 - Number of Bids vs. Price in Dollars

Number of Bids	7	18	32	11	20	1	5	4	2	1
Price in Dollars	106.00	100.00	129.99	176.00	200.00	25.00	20.50	36.00	36.50	49.99

7. The following data give the number of hours 10 students spent studying and their corresponding grades on their midterm exams.

Table for Exercise 7 - Hours Studying vs. Midterm Grade										
Hours Studying	0	0	0.5	1.5	2	3	3.5	3.5	4.5	6
Midterm Grades	64	83	72	74	85	89	79	93	98	95

8. The following table gives the average number of hours 12 junior high students were left unsupervised each day and their corresponding overall grade averages.

Table for Exercise 8 - Hours Unsupervised vs. Overall Average												
Hours Unsupervised	0	0	0.5	1	1	1.5	2	3	3	4	4.5	5.5
Overall Average	96	91	88	92	94	91	87	85	81	80	77	72

9. The following list compares the completion percentage and interception percentage of 8 NFL quarterbacks.

Table for Exercise 9 - Completion Percentage vs. Interception Percentage								
Completion %	62.5	67	60.2	65.4	57.8	57.5	59.1	59.2
Interception %	1.8	1.8	2.3	4.5	4.2	2.3	1.5	3.6

10. The following table gives the weeks of gestation and corresponding birth weights for a sample of 14 babies.

Table for Exercise 10 - Weeks in Gestation vs. Weight (in lbs)														
Weeks	34	35	37	37	38	38	39	39	39	40	40	40	41	41
Weight (in lbs)	4.8	5.9	5.3	6.7	5.6	7.3	7.7	8.4	8.2	9.5	7.6	8.6	8.9	9.4

11. A study on bone density at the women's hospital produced the following results.

Table for Exercise 11 - Age vs. Bone Density (mg/cm^2)					
Age	34	43	49	51	65
Bone Density (mg/cm^2)	946	875	804	723	691

Directions: Determine whether the following correlation coefficients are statistically significant at the specified level of significance and sample size.

12. $r = 0.731$, $\alpha = 0.01$, $n = 11$

13. $r = 0.638$, $\alpha = 0.05$, $n = 11$

14. $r = -0.499$, $\alpha = 0.01$, $n = 26$

15. $r = -0.443$, $\alpha = 0.01$, $n = 30$

16. $r = -0.462$, $\alpha = 0.05$, $n = 18$

17. $r = 0.305$, $\alpha = 0.05$, $n = 45$

Directions: For each of the following data sets, complete the following:

a. Calculate the correlation coefficient, r.
b. Determine if r is statistically significant at the 0.05 level.
c. Calculate the coefficient of determination, r^2.
d. Interpret the meaning of r^2 for the given set of data.

18. Listed below is data from 16 former Elementary Statistics students, where x is their number of absences for the semester and y is their final class average.

Table for Exercise 18 - Number of Absences vs. Grade

Absences	2	2	3	10	3	7	9	1	12	9	1	1	13	1	10	3
Grade	86	83	81	53	92	71	68	79	53	78	77	85	62	97	54	79

19. The table below gives data collected on the birth weight of mothers versus the birth weight of their babies. (Hint: To do calculations on this data, first convert the data to a single unit of measurement.)

Table for Exercise 19 - Mother's Birth Weight (in lbs) vs. Baby's Birth Weight (in lbs)

Mother	lbs	6	6	7	8	7	7	7	6	9	6	9	8	6	5	6	5
	oz	6	2	9	10	6	15	7	5	3	10	3	9	1	0	1	0
Baby	lbs	5	6	7	7	10	5	9	5	8	7	9	8	5	7	7	8
	oz	11	14	10	2	0	13	0	9	13	11	10	6	8	15	2	5

20. Many universities use a student's ACT score to help with admission decisions as well as placement decisions. From a sample of 20 students, the following data is obtained:

Table for Exercise 20 - ACT score vs. GPA

ACT	GPA
16	1.85
18	2.2
24	2.8
25	3.5
34	4.0
27	3.18
29	3.9
25	2.9
30	4.0
21	2.6
17	2.5
21	3.65
28	3.1
31	3.72
35	3.24
18	2.3
17	1.7
26	3.1
28	3.5
23	2.76

Linear Regression

12.2

If two variables x and y have a significant linear correlation, then we can predict what value y might take, given a specific value of x. Take, for instance, the relationship between class size and achievement test scores in one school district. A sample of 8 classrooms from the school district is given in the table below.

Table 4 - Class Size versus Test Scores								
Class Size	15	17	18	20	21	24	26	29
Average Score	85.3	86.2	85.0	82.7	81.9	78.8	75.3	72.1

Let's go ahead and assume that there is, indeed, a significant correlation between these variables. Now, it would be helpful to be able to make *predictions* regarding the data, such as the average test score for a class of 25 students. To do this, we will need to use a **regression line**, or line of best fit, for the data. As you might imagine, the regression line is the one line that fits the data set best. Look now at the scatter plot below depicting the school data. There are several lines shown that fall among all of the points on the scatter plot, but which one is the best fit?

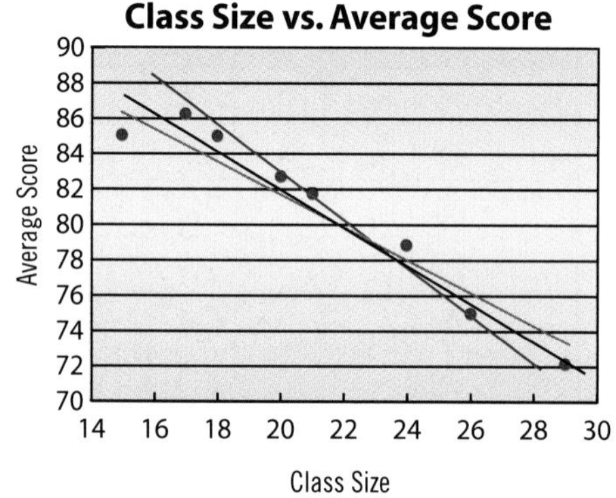

Memory Booster!

Commonly used forms for the regression line:

$\hat{y} = b_0 + b_1x$

$\hat{y} = mx + b$

$y = b + mx$

$y = ax + b$

$y = a + bx$

There is a particular line called the Least-Squares Regression line, or "Line of Best Fit", and it is the line for which the average variation from the data is the smallest. We can find the equation of this line by using some algebra. The regression line $\hat{y} = b_0 + b_1x$ is determined by its slope b_1 and its y-intercept b_0. Alternate forms for the regression line include: $\hat{y} = mx + b$, $y = b + mx$, $y = ax + b$, or $y = a + bx$.

The formula to find the slope is:

$$b_1 = \frac{n\sum xy - \sum x \sum y}{n\sum x^2 - \left(\sum x\right)^2}$$

and the formula for the y-intercept is:

$$b_0 = \frac{\sum y}{n} - b_1 \frac{\sum x}{n}.$$

We mentioned in the previous section that when calculating the correlation coefficient using a TI-84 Plus, it produces the slope and y-intercept of the regression line. Thus we do not need to do any additional work to find the regression line once we have calculated the correlation on a TI-84 Plus.

Rounding Rule

Round the slope and y-intercept for the regression line to three decimal places.

We have discussed the applications of r and r^2 for data whose correlation is statistically significant. What happens if the correlation shown by the data is not statistically significant? If r fails to be statistically significant, then the results are not meaningful enough to use. In other words, there is not a strong enough relationship shown by the data to determine that there is a correlation between the two variables. A linear regression model should not be used if r is not statistically significant.

Example 9

The local school board wants to evaluate the relationship between class size and performance on the state achievement test. They decide to collect data from various schools in the county, and a portion of the data collected is listed below. Each pair represents the class size and corresponding average score on the achievement test for 8 schools.

Table 5 - Class Size versus Test Scores

Class Size	15	17	18	20	21	24	26	29
Average Score	85.3	86.2	85.0	82.7	81.9	78.8	75.3	72.1

Solution:

First, we must decide which variable should be the x variable and which variable should be y. Consider whether there is a *possibility* that one of these variables influences the values of the other. In this case, we are interested in the *possibility* that class size influences the average test score. Thus class size is the independent variable, x, and average test score is the dependent variable, y.

Next, press STAT and then 1:EDIT and enter the values for class size (x) in List 1 and the values for average test score (y) in List 2. Then press STAT and choose CALC and option 4:LinReg(ax+b). Press ENTER twice. The output will include the correlation coefficient, r, and the coefficient of determination, r^2, that we discussed in the previous section. In addition, we see the slope, a, and the y-intercept, b, of the regression line.

cont'd on the next page

Memory Booster!

Remember, to ensure that the output on a TI-84 Plus will contain all four of the values, begin by pressing 2ND and 0, which will bring up the Catalog. Scroll down to DiagnosticOn. Press ENTER twice. **You need to perform this step only once.**

Before we write out the formula for the regression line, we must be sure that there is sufficient evidence to claim that a linear relationship exists between these two variables. Based on the critical value for a 0.05 level of significance, the absolute value of the correlation must be greater than 0.707. The absolute value of the correlation, $r \approx -0.978$ is certainly greater than 0.707, so there is enough evidence at the 0.05 level of significance to conclude that a linear relationship exists between these two variables.

Now it is appropriate to consider the linear regression model. The slope of the regression line is $a = b_1 \approx -1.043$ and the y-intercept of the regression line is $b = b_0 \approx 103.085$. Thus the regression line, in the form $\hat{y} = b_0 + b_1x$, is

$$\hat{y} = 103.085 - 1.043x.$$

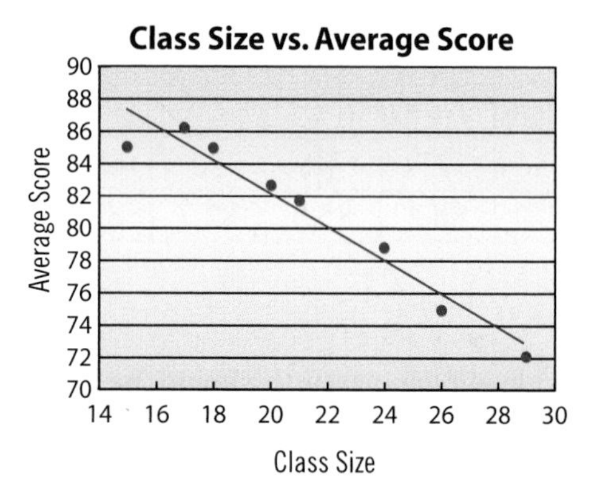

A few words of caution when using a regression model to make predictions. First, as stated previously, it is not appropriate to make a prediction if r is not statistically significant. Also, you should only make predictions for values that are *within the range of the sample data*. The regression model should not be applied outside of the known range. Finally, predictions should only be made *for the population from which data was drawn*. For example, if a regression model was created using data from Caucasian women ages 25 to 40, then it would not be appropriate to use this model to make a prediction about Asian women or even Caucasian women in another age bracket. This particular regression model should be used to make predictions for Caucasian women ages 25 to 40 only.

A prediction should *not* be made with a regression model if...

1. the correlation is not statistically significant,
2. you are using a value outside of the range of the sample data, or
3. the population is different than that of the sample data.

Example 10

Use the equation of the regression line from the previous example,

$$\hat{y} = 103.085 - 1.043x,$$

to predict what the average achievement test score will be for the following class sizes:

a. 16 **b.** 19 **c.** 25 **d.** 45

Solution:

a. $\hat{y} = 103.085 - 1.043(16) = 86.397$
b. $\hat{y} = 103.085 - 1.043(19) = 83.268$
c. $\hat{y} = 103.085 - 1.043(25) = 77.010$
d. It is not meaningful to predict the value of y for this class size because the variable $x = 45$ is outside the range of the original data. The original data only considered class sizes between 15 and 29, so we should only predict the average achievement test score for classes within this range.

Example 11

The table below lists the data collected on the price of used Land Rover Freelanders and their age in years. Create a linear regression model for predicting the price of a used Land Rover Freelander based on its age in years, if appropriate.

Table 6 - Land Rover Price versus Age

Age	3	4	1	2	2
Price	15,500	14,995	30,795	28,995	23,995
Age	3	4	3	4	1
Price	20,900	20,500	19,995	19,888	29,995

cont'd on the next page

Solution:

To begin, we must determine which variable is the independent variable (x) and which is the dependent variable (y). We want to use the age of a used Freelander to predict its price, thus age (x) is the independent variable and price (y) is the dependent variable.

Next, press STAT and then EDIT and enter the values for age (x) in List 1 and the values for price (y) in List 2. Then press STAT and choose CALC and option 4:LinReg(ax+b). Press ENTER twice. The output will include the correlation coefficient, r, and the coefficient of determination, r^2, the slope, a, and the y-intercept, b, of the regression line.

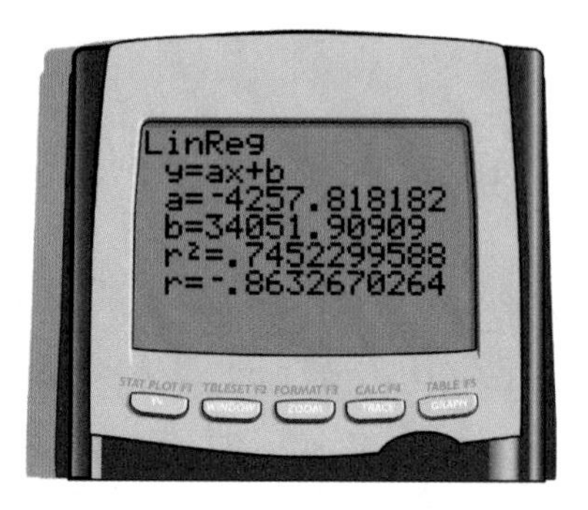

Before we write out the formula for the regression line, we must be sure that there is sufficient evidence to claim that a linear relationship exists between these two variables. Based on the critical value for a 0.05 level of significance, the absolute value of the correlation must be greater than 0.632. The absolute value of the correlation, $r \approx -0.863$ is greater than 0.632, so there is enough evidence at the 0.05 level of significance to conclude that a linear relationship exists between these two variables.

It is appropriate to consider the linear regression model. The slope of the regression line is $a = b_1 \approx -4257.818$ and the y-intercept of the regression line is $b = b_0 \approx 34{,}051.909$. Thus the regression line, in the form $\hat{y} = b_0 + b_1x$, is

$$\hat{y} = 34{,}051.909 - 4257.818x.$$

Example 12

Use the linear regression model from the previous example,

$$\hat{y} = 34{,}051.909 - 4257.818x,$$

to predict the following:

a. The price of a Land Rover Freelander that is 2.5 years old.
b. The price of a Land Rover Freelander that is 10 years old.
c. The price of a Land Rover Range Rover that is 3 years old.

Solution:

a. Substitute the value $x = 2.5$ into the regression line and solve for $\hat{y}$.

$$\hat{y} = 34{,}051.909 - 4257.818x$$
$$\hat{y} = 34{,}051.909 - 4257.818(2.5)$$
$$\hat{y} = 23{,}407.36$$

Thus a Freelander that is 2.5 years old would sell for approximately \$23,407.36.

b. The original data only contains ages from 1 to 5 years old; therefore, it is inappropriate to use this model to predict the price of a Freelander that is more than 5 years old.

c. The population is Freelanders, not Range Rovers. Thus it is inappropriate to use this model to predict the price of a Range Rover, no matter how old it is.

Now, let's put all of the information we have learned about linear regression together for the last example.

Example 13

The table below gives the average monthly temperature and corresponding monthly precipitation totals for one year in Key West, Florida.

Table 7 - Temperatures (in °F) and Precipitation in Key West, Florida						
Average Temperature (in °F)	75	76	79	82	85	88
Inches of Precipitation	2.22	1.51	1.86	2.06	3.48	4.57
Average Temperature (in °F)	89	90	88	85	81	77
Inches of Precipitation	3.27	5.40	5.45	4.34	2.64	2.14

cont'd on the next page

a. Create a scatter plot for the data. Does there appear to be a linear relationship between x and y?

b. Calculate the correlation coefficient, r.

c. Verify that the correlation coefficient is statistically significant at the 0.05 level of significance.

d. Calculate the equation of the line of best fit.

e. Calculate and interpret the coefficient of determination, r^2.

f. If appropriate, make a prediction for the monthly precipitation for a month in which the average temperature is 80 degrees.

g. If appropriate, make a prediction for the monthly precipitation in Destin, Florida, for a month in which the average temperature is 83 degrees.

Solution:

a. In order to create a scatter plot, we must first decide which variable will be x and which will be y. Let's choose the average temperatures for the x-values and the inches of precipitation for the y-values. Below you will see a scatter plot of the data.

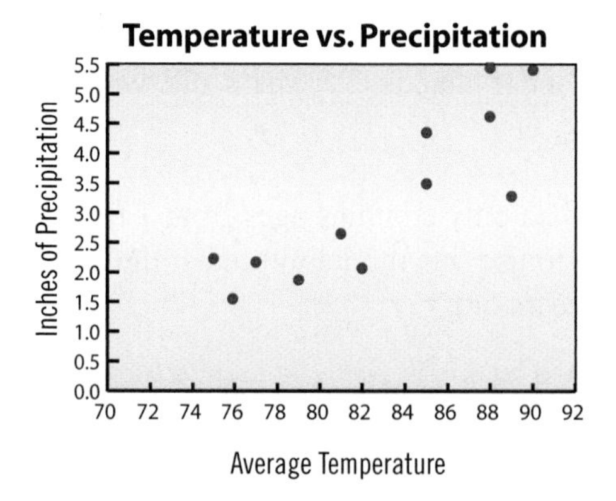

The scatter plot shows a positive trend in the data. Because the points are pretty close together, we estimate a strong positive correlation.

b. Using a calculator, we calculate $r = 0.859$

c. Using the table to obtain the critical value, $|0.859| > 0.576$, which tells us that r is indeed statistically significant at the 0.05 level.

d. From the calculator, we see that the equation of the line of best fit is

$$\hat{y} = -15.424 + 0.225x.$$

e. The coefficient of determination is 0.738. This tells us that approximately 73.8% of the variation in precipitation can be attributed to the linear relationship between temperature and precipitation. The remaining 26.2% of the variation is from unknown sources.

f. Because r is statistically significant, we can use the best fit line to make predictions regarding the data. Because we designated the temperatures as the x-values, substitute $x = 80$ into the regression equation to obtain an estimate for the precipitation when the temperature is 80 degrees. Then $\hat{y} = -15.424 + 0.225(80) = 2.58$. Thus a reasonable estimate for the precipitation for a month in which the average temperature is 80 degrees is 2.58 inches.

g. The data was collected in Key West—not Destin, Florida. Therefore, it is not appropriate to use the linear regression line to make predictions regarding the precipitation in Destin.

Exercises

12.2

Directions: For each of the following data sets

a. determine the regression line. Remember to round the slope and y-intercept to three decimal places.

b. use the regression line to make a prediction for y given a value for x.

1. a.

Table for Exercise 1

x	55	40	71	82	90	50	83	75	65	52	77	93
y	58	43	68	86	87	51	87	70	67	55	77	90

b. If $x = 70$, make a prediction for y.

2. a.

Table for Exercise 2

x	5	3	8	3	9	4	7	2	6	3	2	5	6
y	18	15	18	13	19	11	12	13	15	14	13	17	19

b. If $x = 3.2$, make a prediction for y.

3. a.

Table for Exercise 3

x	121	156	145	109	111	137	128	142	149	152	155	128
y	99	123	119	87	90	101	98	121	131	134	135	97

b. If $x = 130$, make a prediction for y.

4. a.

Table for Exercise 4								
x	12	15	14	10	11	13	12	14
y	99	63	79	115	108	82	98	73

b. If $x = 11.5$, make a prediction for y.

5. The following data shows the value of the dollar to the British pound (£) through the course of 8 days.

a.

Table for Exercise 5 - Value of the Dollar to Pound per Date								
Date	12th	13th	14th	15th	16th	17th	18th	19th
$ to £	1.7448	1.7453	1.7449	1.7461	1.7488	1.7513	1.7534	1.7542

b. Make a prediction for the value of the dollar against the pound on the 25th.

6. The following data shows the value of a certain stock price through 10 days.

a.

Table for Exercise 6 - Stock Price (in Dollars) per Date										
Date	2/9	2/10	2/13	2/14	2/15	2/16	2/17	2/21	2/22	2/23
Price	25.01	24.65	25.46	26	24.45	24.88	25.09	25.2	25.75	26.28

b. Make a prediction for the stock price for February 28th based on this data.

7. The following data shows the average mathematics and science achievement scores of eighth-grade students from 12 countries around the world.

a.

Table for Exercise 7 - Mathematics vs. Science Achievement Scores												
Math	525	558	511	531	392	585	476	496	520	582	532	502
Science	540	535	518	533	420	569	450	538	535	530	552	515

b. Make a prediction for the average science score for a math score of 570.

8. The following table shows the average amount of time wasted at work per day by year of birth.

a.

Table for Exercise 8 - Year of Birth vs. Time Wasted per Day (in Hours)					
Year of Birth	1940	1955	1965	1975	1985
Time Wasted (in Hours)	0.5	0.68	1.19	1.61	1.95

b. Make a prediction for the amount of time that would be wasted from someone born in 1980.

Regression Analysis

12.3

In the first section we looked at how to evaluate whether or not a linear model is appropriate for a given set of data by using statistical significance. We also learned how to quantify the amount of variation in one variable that is associated with the amount of variation in the other variable by finding the coefficient of determination. In this section we will analyze the linear regression model in more detail.

One important point that we have yet to discuss is that the regression line, or line of best fit, that we calculated in the previous section is a sample statistic, because it results from calculations performed on a sample of the population's data. The true regression line would only come from a census of the entire population, and would thus be a population parameter. We represent the true regression line using Greek letters, as we have with most population parameters.

$$y = \beta_0 + \beta_1 x \text{ (Population parameter)}$$

$$\hat{y} = b_0 + b_1 x \text{ (Sample statistic)}$$

Thus y represents the actual value occurring in the population and $\hat{y}$ represents the predicted value based on sample data. As you might expect, these two values are often not the same. The difference in the actual value and the predicated value is called a **residual**.

Residual

A **residual** is the difference in the actual value and the predicted value.

$$\text{Residual} = y - \hat{y}$$

Example 14

The table below gives data from a local school district on a child's age and their reading level. For this data, a reading level of 4.3 would indicate 3/10 of the year through the fourth grade. Children's ages are given in years. Using a TI-84 Plus calculator as an efficient method for quickly determining the values in the linear model, we calculate the regression line to be $\hat{y} = -3.811 + 0.865x$. Use this equation to calculate an estimate, $\hat{y}$, for each value of the independent variable, x, and then use the estimate to calculate the residual for each value of y.

Table 8 - Age versus Reading Level

Age	6	7	8	9	10	11	12	13	14	15
Reading Level	1.3	2.2	3.7	4.1	4.9	5.2	6.0	7.1	8.5	9.7

cont'd on the next page

Solution:

We can use a TI-84 Plus to perform all of the necessary calculations at once. Age is the independent variable, x. Reading level is the dependent variable, y. Next we need to enter this data into our calculator. Begin by pressing STAT and entering the ages in List 1 and the reading levels in List 2. Then highlight L3 and enter the formula –3.811 + 0.865*L1. This will quickly calculate the predicated y-value for each x-value. Lastly, highlight L4 and enter the formula L2-L3. This formula will calculate each of the residuals.

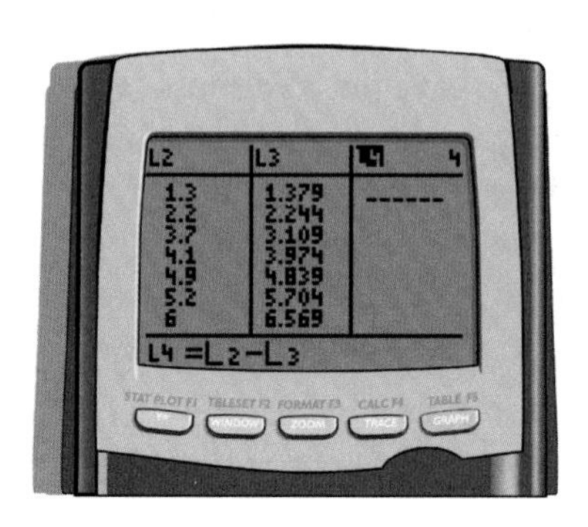

The results will be as follows:

Table 9 - Age versus Reading Level			
Age x	**Reading Level** y	**Predicted Value** $\hat{y}$	**Residual** $y-\hat{y}$
6	1.3	1.379	–0.079
7	2.2	2.244	–0.044
8	3.7	3.109	0.591
9	4.1	3.974	0.126
10	4.9	4.839	0.061
11	5.2	5.704	–0.504
12	6.0	6.569	–0.569
13	7.1	7.434	–0.334
14	8.5	8.299	0.201
15	9.7	9.164	0.536

The residual of each value reflects how far the original data point is from the point on the regression line. Graphically, the residual is the vertical distance from the original data point to the point on the regression line. If we graph the regression line on a scatter plot of the data, we can draw vertical bars from the regression line to the original points to indicate the residual for each point.

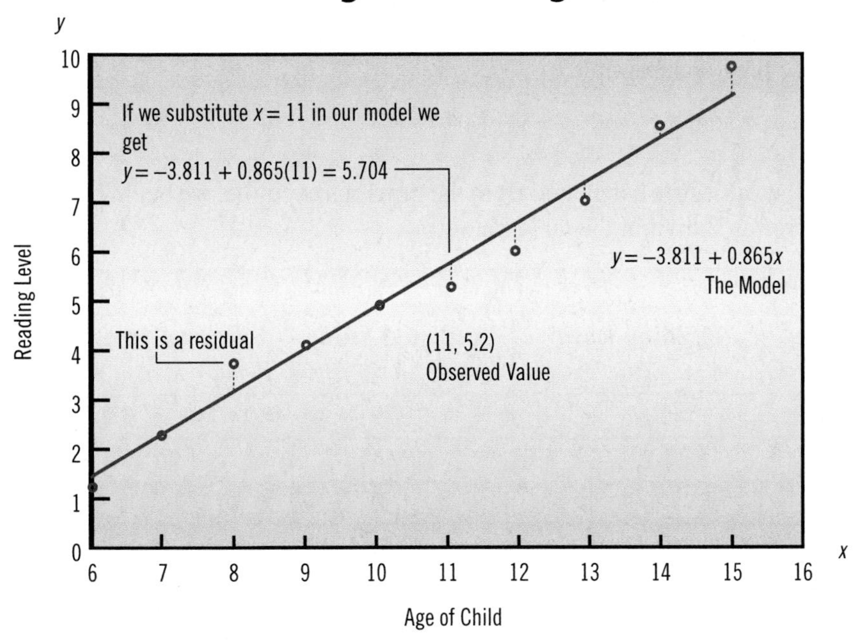

Figure 5 – Errors Shown Graphically

Another name for a residual is an *error in the predicted value*. We could find the "total error" for the regression line by summing all of the errors; however, as was the case with standard deviation, simply summing the errors would result in negative errors canceling out positive errors. To eliminate this situation, we square the errors before adding. The value calculated by summing the square of the errors is called the **sum of squared errors**, which we abbreviate SSE.

Sum of Squared Errors 

The value calculated by summing the square of the errors is called the **sum of squared errors**, abbreviated SSE.

$$\text{SSE} = \sum\left(y - \hat{y}\right)^2$$

To interpret the value for the sum of squared errors, consider the size of the value. If the data points are very far from the regression line, then the sum of squared errors will be large. If the data points are close to the regression line, then the sum of squared errors will be small. Thus the larger the value of SSE, the worse the linear model will be at predicting the value of y. The smaller the value of SSE, the better the linear model will be at predicting the value of y. Therefore, the line that fits the data "best" would be the one with the smallest value of SSE.

Example 15

Calculate the sum of squared errors, SSE, for the data on age and reading level from the previous example.

Solution:

Using the values we calculated from the previous example, we begin by squaring each error as shown in the table below.

Table 10- Age versus Reading Level				
Age x	**Reading Level** y	**Predicted Value** $\hat{y}$	**Error** $y-\hat{y}$	**Squared Error** $(y-\hat{y})^2$
6	1.3	1.379	−0.079	0.00624
7	2.2	2.244	−0.044	0.00194
8	3.7	3.109	0.591	0.34928
9	4.1	3.974	0.126	0.01588
10	4.9	4.839	0.061	0.00372
11	5.2	5.704	−0.504	0.25402
12	6.0	6.569	−0.569	0.32376
13	7.1	7.434	−0.334	0.11156
14	8.5	8.299	0.201	0.04040
15	9.7	9.164	0.536	0.28730

The last column lists the errors squared. The sum of the squared errors is the sum of this last column. Thus SSE = 1.394.

We can compare the calculation of the sum of squared errors to the calculation of the standard deviation of a set of data. This comparison is not a coincidence. We will use this value to calculate the **standard error of estimate**, which is a measure of how much the data points deviate from the regression line. This is analogous to how the standard deviation measures how much the data deviates from the sample mean. Thus the smaller the value of the standard error of estimate is, the closer the data points are to the regression line. The formula is also similar to the standard deviation formula. The formula for the standard error of estimate, denoted S_e, is

$$S_e = \sqrt{\frac{\sum\left(y-\hat{y}\right)^2}{n-2}}.$$

Note that we are dividing the sum of the squared errors by the sample size *minus 2*. This sample statistic is easily calculated using a TI-84 Plus calculator.

Example 16

Calculate the standard error of estimate for the data given for age and reading level in Example 14 (repeated below).

Table 11 - Age versus Reading Level										
Age	6	7	8	9	10	11	12	13	14	15
Reading Level	1.3	2.2	3.7	4.1	4.9	5.2	6.0	7.1	8.5	9.7

Solution:

Begin by pressing STAT and choosing 1:EDIT. Enter the age data into List 1 and reading level data in List 2. Next, press STAT and choose TESTS and option E:LinRegTTest. Enter L1 for the Xlist and L2 for the Ylist. The option "Freq:" should be 1. Choose $\rho \neq 0$ to test the significance of the linear relationship. Enter the regression equation into RegEQ if you have already calculated it. If not, you may leave this blank. Choose CALCULATE and press ENTER.

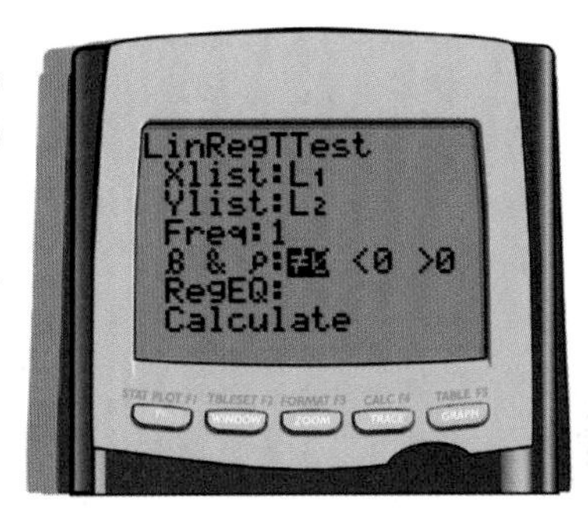

The results show us the *t*-test statistic for testing the significance of the linear relationship. It also gives us the *p*-value for that hypothesis test and the degrees of freedom. The slope and *y*-intercept of the regression line are also given. The last two values given are the coefficient of determination and the correlation. The value we are interested in is *s*, the third to last value given.

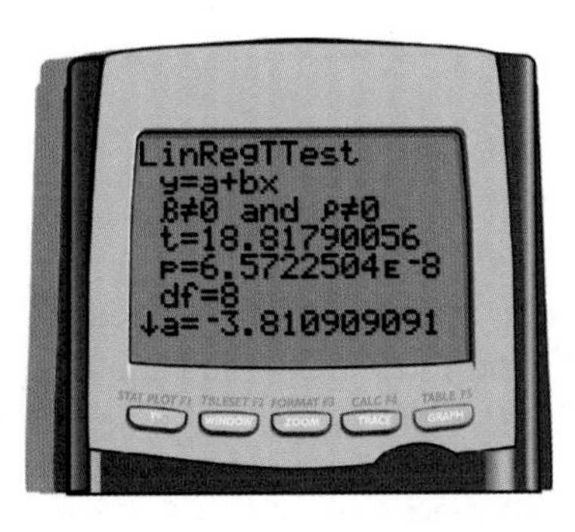

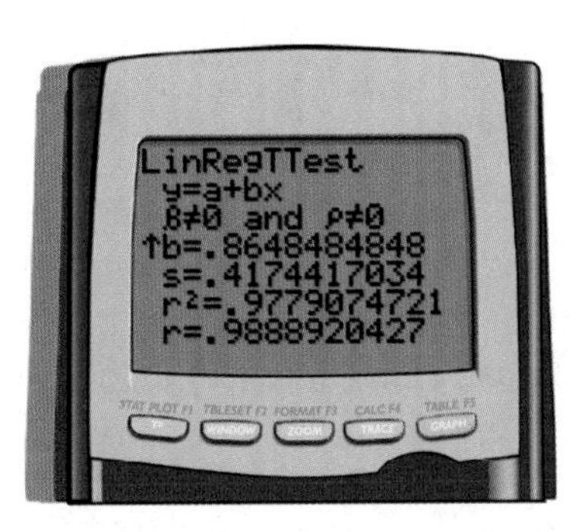

Thus the standard error of estimate for the data on age and reading level is $S_e = 0.417$. Since this value is close to zero, this value tells us that the data points do not deviate very much from the regression line.

✦ Prediction Intervals for an Individual y value

In Chapter 8, we learned that it is preferable to use an interval estimate instead of a point estimate for a population parameter. So far in this chapter, we have only presented point estimates for the various parameters discussed. For example, the predicted value, $\hat{y}$, is a point estimate for the population parameter, y. We can use the standard error of estimate to create a confidence interval for y at a given fixed value of the independent variable, x. We call this confidence interval for the predicted dependent variable a **prediction interval**.

Assume that for any given fixed value of the independent variable, x_0, the possible sample values of the dependent variable, y, are normally distributed about the regression line, with the mean of the normal distribution equal to $\hat{y}$ and the standard deviation of the normal distribution the same for each value of x_0. This type of distribution with two variables is called a **bivariate normal distribution**. Figure 6 below illustrates this distribution.

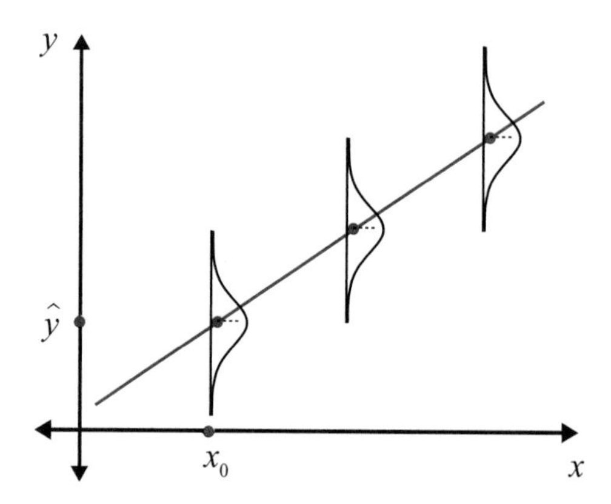

Figure 6 – Bivariate Normal Distribution

Thus for a given fixed value, x_0, the prediction interval for an individual y can be found by calculating the margin of error, E, and then adding and subtracting that margin of error to the point estimate, $\hat{y}$. The formula for the margin of error is

$$E = t_{\frac{\alpha}{2}} S_e \sqrt{1 + \frac{1}{n} + \frac{n\left(x_0 - \bar{x}\right)^2}{n\left(\sum x^2\right) - \left(\sum x\right)^2}}.$$

The value $t_{\frac{\alpha}{2}}$ has $n - 2$ degrees of freedom for an $\alpha = 1 - c$ level of significance. The standard error of estimate, S_e is found as discussed previously. Lastly, x_0 is the fixed value of x, and n is the sample size.

Prediction Interval for an Individual y

$$\hat{y} - E < y < \hat{y} + E$$

or

$$(\hat{y} - E,\ \hat{y} + E)$$

Example 17

Construct a 95% prediction interval for the reading level of a child who is 8 years old. Use the data from Example 14 on age and reading levels as your sample data (repeated below).

Table 12 - Age versus Reading Level

Age	6	7	8	9	10	11	12	13	14	15
Reading Level	1.3	2.2	3.7	4.1	4.9	5.2	6.0	7.1	8.5	9.7

Solution:

Neither a TI-84 Plus nor Microsoft Excel will directly calculate a prediction interval*, so we must calculate the margin of error by hand, and use this value to calculate the prediction interval.

Step 1 – Find the regression equation for the sample data.
We know from previous examples that the regression equation is

$$\hat{y} = -3.811 + 0.865x.$$

Step 2 – Use the regression equation to calculate the point estimate, $\hat{y}$, for the given value of x.
In this example, $x = 8$. Thus

$$\hat{y} = -3.811 + 0.865x$$
$$\hat{y} = -3.811 + 0.865(8)$$
$$\hat{y} = 3.109$$

Step 3 – Calculate the sample statistics necessary to calculate the margin of error.
Using a TI-84 Plus calculator, we can enter the values for age in List 1 and the values for reading level in List 2. Next, press STAT, then choose CALC and option 2: 2-Var Stats. This will give us many of the statistics we need:

$$n = 10,\quad \bar{x} = 10.5,\quad \sum x = 105,\quad \sum x^2 = 1185.$$

Next, recall that we found that $S_e = 0.4174417$ (unrounded) in the previous example. This value was also found using a TI-84 Plus. Lastly, using Table C we find that $t_{\frac{\alpha}{2}} = 2.306$ for $10 - 2 = 8$ degrees of freedom and $\alpha = 1 - 0.95 = 0.05$.

cont'd on the next page

**Footnote: However, many statistical packages, such as Minitab, will calculate a prediction interval directly.*

Step 4 – Calculate the margin of error.
Substituting the necessary statistics into the margin of error formula, we obtain

$$E = t_{\frac{\alpha}{2}} S_e \sqrt{1 + \frac{1}{n} + \frac{n\left(x_0 - \bar{x}\right)^2}{n\left(\sum x^2\right) - \left(\sum x\right)^2}}$$

$$= (2.306)(0.4174417)\sqrt{1 + \frac{1}{10} + \frac{10(8 - 10.5)^2}{10(1185) - (105)^2}}$$

$$= 1.044$$

Step 5 – Construct the prediction interval.
Subtracting the margin of error from the point estimate of $\hat{y}$ = 3.109 gives us the lower endpoint for the prediction interval of

$$3.109 - 1.044 = 2.065.$$

By adding the margin of error to the point estimate, we obtain the upper endpoint for the prediction interval as follows

$$3.109 + 1.044 = 4.153.$$

Thus the confidence interval for the population proportion is

$$2.065 < y < 4.153, \text{ or}$$
$$(2.065, 4.153).$$

Thus for an 8-year old child, we can be 95% confident that they would have a reading level between 2.065 and 4.153, or be reading between the second and fourth grade level.

✦ Confidence Intervals for the Slope and the y-Intercept of the Regression Equation

We can now construct a prediction interval of an individual value of y given a specific value of x, but $\hat{y}$ is not the only sample statistic in the regression equation. The values for the slope, b_1, and the y-intercept, b_0, are also sample statistics and thus, point estimates for the population parameters β_1 and β_0 respectively. We can construct confidence intervals for these parameters as well. Actually, Microsoft Excel easily calculates these confidence intervals from the raw sample data.

Example 18

Construct 95% confidence intervals for the slope, β_1, and the y-intercept, β_0, of the regression equation for age and reading level. Use the sample data from Example 14 (repeated below).

Table 13 - Age versus Reading Level

Age	6	7	8	9	10	11	12	13	14	15
Reading Level	1.3	2.2	3.7	4.1	4.9	5.2	6.0	7.1	8.5	9.7

Solution:

Begin by entering the sample data into Microsoft Excel as shown in the screen shot to the right.

	A	B
1	Age	Reading Level
2	6	1.3
3	7	2.2
4	8	3.7
5	9	4.1
6	10	4.9
7	11	5.2
8	12	6
9	13	7.1
10	14	8.5
11	15	9.7

Next, choose DATA ANALYSIS from the TOOLS menu. Choose REGRESSION from the options listed. Enter the necessary information into the Regression menu as shown on the following page. Click on OK.

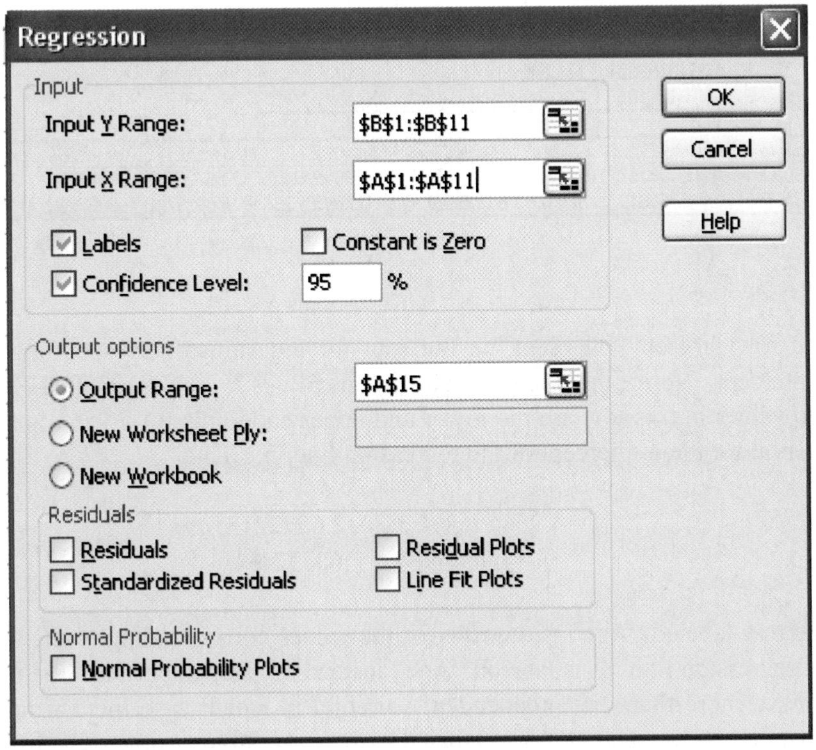

Regression

Input

Input Y Range: $B$1:$B$11

Input X Range: $A$1:$A$11

☑ Labels ☐ Constant is Zero

☑ Confidence Level: 95 %

OK

Cancel

Help

Output options

◉ Output Range: $A$15

○ New Worksheet Ply:

○ New Workbook

Residuals

☐ Residuals ☐ Residual Plots

☐ Standardized Residuals ☐ Line Fit Plots

Normal Probability

☐ Normal Probability Plots

cont'd on the next page

The results provide an abundance of information, much of which we have discussed throughout this chapter.

1. "Multiple R" is the correlation coefficient for this data.
2. "R Square" is just that, r^2.
3. Standard Error is the standard error of estimate, S_e.

The ANOVA table will be discussed in the next section, since it is more meaningful when discussing more than one independent variable. However, it does contain a few of the important values we discussed so far in this section.

4. The intersection of the Residual row and the SS column is the Sum of Squared Errors, SSE.
5. The upper and lower endpoints of the confidence intervals for the y-intercept and slope.
6. 7. The values for the coefficients in the regression line.

These upper and lower endpoints are the values we are interested in this example.

SUMMARY OUTPUT

Regression Statistics	
Multiple R (1)	0.988892043
R Square (2)	0.977907472
Adjusted R Square	0.975145906
Standard Error (3)	0.417441703
Observations	10

ANOVA

	df	*SS*	*MS*	*F*	*Significance F*
Regression	1	61.70693939	61.70694	354.1134	6.57225E-08
Residual	8	1.394060606 (4)	0.174258		
Total	9	63.101			

	Coefficients	*Standard Error*	*t Stat*	*P-value*
Intercept	-3.810909091 (6)	0.500297157	-7.61729	6.2E-05
Age	0.864848485 (7)	0.045958819	18.8179	6.57E-08

	Lower 95%	*Upper 95%*	*Lower 95.0%*	*Upper 95.0%*
Intercept	-4.964596403	-2.657222	-4.9645964 (5)	-2.6572218 (5)
Age	0.758867258	0.9708297	0.758867258 (5)	0.97082971 (5)

The row labeled "Intercept" is the row for the values corresponding to the y-intercept. Notice that the first value in this row is $b_0 = -3.81090909$. The last two values in this row are the lower and upper endpoints for a 95% confidence interval for the y-intercept of the regression line, β_0. Thus

$$-4.9645964 < \beta_0 < -2.6572218, \text{ or}$$
$$(-4.9645964, -2.6572218).$$

The row labeled "Age" is the row for the values corresponding to the slope of the regression line. It is labeled "Age" instead of "slope" because it is possible to have more than one independent variable, in which case there would be a separate row for each variable, labeled with the variable's name. The first value in this row is $b_1 = 0.864848485$. The last two values in this row are the lower and upper endpoints for a 95% confidence interval for the slope of the regression line, β_1. Thus

$$0.758867258 < \beta_1 < 0.97082971, \text{ or}$$
$$(0.758867258, 0.97082971).$$

Exercises

12.3

Directions: For each of the following data sets:

- **a.** Calculate the Sum of Squared Errors, SSE.
- **b.** Calculate the Standard Error of Estimate, S_e.
- **c.** Construct a 95% prediction interval for the given value of the independent variable.
- **d.** Construct a 95% confidence interval for the y-intercept of the regression line.
- **e.** Construct a 95% confidence interval for the slope of the regression line.

Note: Microsoft Excel can be used to calculate **a.**, **b.**, **d.**, and **e.** simultaneously.

1. The following data is a sample of 10 women's weights and their times to run/walk 1 mile. The times are in minutes and the weights are in pounds. Construct a 95% prediction interval for time given $x = 175$.

Table for Exercise 1 - Weight vs. Time

Weight	178	182	180	165	159	170	189	193	195	203
Time	13.24	15.32	15.21	12.04	12.21	13.10	16.75	16.98	17.02	17.19

2. The following data is a sample of 10 students' grades on their final exam in the first semester of English Comprehension and their corresponding grades on their final exam in their second semester of English Comprehension. Construct a 95% prediction interval for the final exam in the second semester given $x = 70$.

Table for Exercise 2 - Exam 1 vs. Exam 2

Final Exam 1	87	65	99	78	81	86	61	63	90	100
Final Exam 2	89	72	93	81	75	87	72	69	88	89

3. The table below is a sample of 15 couples and the corresponding weight gain of each partner during the first year after marriage.
Construct a 95% prediction interval for the husband's weight gain given $x = 20$.

Table for Exercise 3 - Wife's Weight Gain vs. Husband's Weight Gain															
Weight Gain (Wife)	15	5	32	24	33	20	19	8	12	40	21	15	24	3	10
Weight Gain (Husband)	20	9	12	25	38	12	23	10	9	33	22	17	18	5	11

4. The following data is a sample of golf scores and the average length of the golfer's drive.
Construct a 95% prediction interval for the length of the golfer's drive given $x = 72$.

Table for Exercise 4 - Golf Score vs. Length of Drive					
Golf Score	68	72	69	70	75
Length of Drive	275	236	289	245	225

5. The following data is a sample of the average number of grams of fat a person has per day in their diet and the amount of weight lost (in pounds) in a 6-month period.
Construct a 95% prediction interval for the amount of weight loss given $x = 55$.

Table for Exercise 5 - Grams of Fat vs. Weight Loss								
Grams of Fat	40	20	25	50	65	75	78	80
Weight Loss	23	30	38	15	10	8	7	5

6. The following data is a sample of measurements of a person's physical fitness (measured on a scale of 1 to 10 with 10 being perfectly fit) and the number of days it took for the person to recover from gall bladder surgery.
Construct a 95% prediction interval for the length of the recovery time given $x = 5$.

Table for Exercise 6 - Physical Fitness vs. Recovery Time								
Physical Fitness	8	7	6	5	4	3	2	1
Recovery Time	5	5	6	7	7	9	10	12

7. The following data is a sample of the number of bowls of chicken soup a person ate while having cold symptoms and the number of days the cold symptoms persisted.
Construct a 95% prediction interval for the duration of the cold given $x = 3$.

Table for Exercise 7 - Bowls of Soup vs. Duration of Cold								
Bowls of Soup	5	5	4	4	3	2	1	0
Duration of Cold	3	4	3	5	4	5	6	7

8. The following data is a sample of the weights of pairs of husbands and wives.
Construct a 95% prediction interval for the weight of the wife given $x = 225$.

Table for Exercise 8 - Weight of Husband vs. Weight of Wife					
Weight of Husband	220	230	267	200	195
Weight of Wife	154	167	186	132	125

9. The following data shows the number of parking tickets during one semester from a sample of students and their corresponding monthly incomes (counting allowances from parents as well as paychecks from employment as income).
Construct a 95% prediction interval for the monthly income given $x = 6$.

Table for Exercise 9 - Number of Tickets vs. Monthly Income										
Number of Tickets	10	8	3	2	0	5	4	2	1	0
Monthly Income	4000	3800	1500	2000	870	2500	1800	1000	1200	1400

10. The following data was collected from a sample of fathers and sons. The heights are given in inches.
Construct a 95% prediction interval for the height of the son given $x = 70$.

Table for Exercise 10 - Height of Father vs. Height of Son									
Height of Father	72	70	73	68	74	73	69	70	75
Height of Son	73	69	73	70	73	70	68	71	76

11. The following data was collected from a sample of students on the number of tardies in their World History course and their final exam grade in the course.
Construct a 95% prediction interval for the final grade given $x = 6$.

Table for Exercise 11 - Number of Tardies vs. Final Grade							
Number of Tardies	9	8	7	6	5	2	0
Final Exam Grade	57	64	69	70	72	84	91

Multiple Regression Equations

12.4

Thus far, we have only looked at linear regression models that had one independent variable. One example that we considered in detail was using a child's age to predict their reading level. However, as you might expect, there are many more factors that influence a child's reading level than just their age. Two other possible factors are the experience of their current reading teacher and the amount of education of their parents. Could we create a model that used all three factors (age, teacher's experience, and parents' education) to predict a child's reading level? The answer is yes! Actually, the techniques we have discussed so far can easily be expanded to include any number of independent variables.

A linear regression model using two or more independent variables to predict a dependent variable is called a **multiple regression model**. The equation for a multiple regression model is

$$\hat{y} = b_0 + b_1x_1 + b_2x_2 + \cdots + b_kx_k$$

where $x_1, x_2, \cdots, x_k$ are the k independent variables in the model and $b_1, b_2, \cdots, b_k$ are the corresponding coefficients of the independent variables. These values, $b_1, b_2, \cdots, b_k$, are the sample estimates of the corresponding population parameters, $\beta_1, \beta_2, \cdots, \beta_k$. The y-intercept of the multiple regression equation is b_0, which is the sample estimate of the population parameter, β_0.

Constructing a multiple regression model from sample data, and analyzing the validity of the model, is relatively easy with the use of statistical tools, such as those in Microsoft Excel and TI-84 Plus calculators. In the previous section when we constructed the confidence intervals for the slope and y-intercept, we briefly mentioned that the ANOVA table produced by Microsoft Excel was more useful when we had multiple independent variables. This is the type of ANOVA table that we will use to construct and analyze multiple regression equations.

Let's consider the following data collected from 10 children: the child's age, the teacher's experience, the parent's education, and the child's reading level. The child's age is measured to the nearest year. The teacher's experience is the number of years the child's current reading teacher has taught reading to children of this age. The parents' education is the average number of years of schooling obtained by parents with whom the child lives. Thus 12 years indicates that the average education of the parents is a high school diploma. Lastly, the child's reading level is the grade level at which the child is reading. As before, a reading level of 4.3 indicates that the child is reading at a level equivalent to 3/10 of the way through fourth grade.

Table 14 – Children's Reading Level			
Child's Age (in years)	**Teacher's Experience (in years)**	**Parents' Education**	**Child's Reading Level**
6	5	13.1	1.3
7	10	14.5	2.2
8	7	16.1	3.7
9	3	12.5	4.1
10	2	11.8	4.9
11	4	11.1	5.2
12	1	10.2	6
13	3	12.6	7.1
14	6	12.1	8.5
15	5	14.3	9.7

Using this data, the following ANOVA table is produced using the Regression Analysis Tool in Microsoft Excel. (The directions for creating this table will be presented in the next example. For now, we want to focus on interpreting the output.)

SUMMARY OUTPUT

Regression Statistics	
Multiple R	0.996706497
R Square	0.993423841
Adjusted R Square	0.990135761
Standard Error	0.262983593
Observations	10

ANOVA

	df	*SS*	*MS*	*F*	*Significance F*
Regression	3	62.68603778	20.89535	302.1289	6.20571E-07
Residual	6	0.41496222	0.06916		
Total	9	63.101			

	Coefficients	*Standard Error*	*t Stat*	*P-value*	*Lower 95%*	*Upper 95%*	*Lower 95.0%*	*Upper 95.0%*
Intercept	-6.983348818	0.932463961	-7.48914	0.000293	-9.26500593	-4.701692	-9.2650059	-4.7016917
Child's Age	0.89810517	0.031202521	28.7831	1.16E-07	0.821755352	0.974455	0.82175535	0.97445499
Teacher's Experience	-0.033200086	0.050469878	-0.65782	0.535062	-0.15669543	0.0902953	-0.1566954	0.09029526
Parents' Education	0.231953619	0.075147789	3.086633	0.02148	0.048073603	0.4158336	0.0480736	0.41583364

Figure 7 – Microsoft Excel Output for Regression

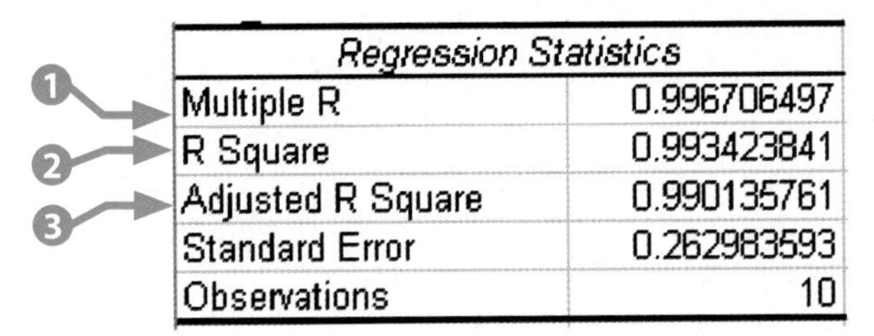

Regression Statistics	
Multiple R	0.996706497
R Square	0.993423841
Adjusted R Square	0.990135761
Standard Error	0.262983593
Observations	10

Figure 8 – Regression Statistics

To begin, the Regression Statistics are calculated, as shown in Figure 8. These values refer to the multiple regression equation in general. ❶ Notice that the first value given is Multiple R, which is the correlation coefficient of the multiple regression model. We use a capital R to represent multiple regression correlations. So, $R = 0.997$. Thus there is a strong positive correlation between the values in this model.

❷ The next value is R Square. This is the **multiple coefficient of determination**, R^2. This value is analogous to the coefficient of determination of a single independent variable, r^2. Just as with r^2, the closer the value of R^2 is to 1, the better the multiple regression model fits the sample data. A value of 1 indicates that the sample data falls on a perfectly straight line. For this example, $R^2 = 0.993$, signifying that the multiple regression model fits the data very well.

Unfortunately, as more independent variables are added to the multiple regression model, the value of R^2 increases as a general rule. So should we continue to add independent variables until we obtain a very large value of R^2? Of course not! Not all of the possible independent variables affect the dependent variable equally, and not all of the possible independent variables are useful in predicting the value of the dependent variable. ❸ So the **adjusted coefficient of determination** is the value of the multiple coefficient of determination adjusted for the number of independent variables and the sample size. This value appears as Adjusted R Square. For this example, adjusted $R^2 = 0.990$. Though this value is less than $R^2 = 0.993$, it still indicates that the multiple regression model fits the data well.

ANOVA					
	df	*SS*	*MS*	*F*	*Significance F*
Regression	3	62.68603778	20.89535	302.1289	6.20571E-07
Residual	6	0.41496222	0.06916		
Total	9	63.101			

❹

Figure 9 – ANOVA Table for Regression

The next block of data is the ANOVA table, as shown in Figure 9. This ANOVA table further analyzes how well the regression model fits the sample data. For a single regression model, we ran a t-test to determine if the linear relationship between the two variables was statistically significant. We were testing the claim that the population's correlation did not equal 0, i.e. $\rho \neq 0$. Equivalently, we could have used the same t-test to test the claim that the coefficient of the independent variable was not equal to 0, i.e. $\beta_1 \neq 0$. You may have even noticed that the output for the t-test on the TI-84 Plus listed *both* of these alternative hypotheses. When using the ANOVA table to analyze the statistical significance of the linear relationship between the variables in a multiple regression, we test the claim that at least one of the independent variables' coefficients is not equal to 0. Thus the null and alternative hypotheses for this ANOVA test are

H_0: $\beta_1 = \beta_2 = \cdots = \beta_k = 0$
H_a: At least one coefficient does not equal 0.

If all of the coefficients of the independent variables equal zero, then this sample data implies that there is not a significant linear relationship between these variables, and therefore, the regression model should not be used for predictions. However, if the null hypothesis is rejected, then we can conclude that there is enough evidence to support the claim that a significant linear relationship exists between these values. Therefore, the multiple regression equation can be used to predict values of the dependent variable.

❹ To test the null hypothesis, consider the p-value given under "Significance F." For our example, the p-value is so small, it is given in scientific notation. In standard notation $p = 0.00000062$, which is extremely small. Thus we reject the null hypothesis and conclude that there is sufficient evidence at the 0.05 level of significance to support the claim that this multiple regression model fits the data well and can be used for predictions.

	Coefficients	Standard Error	t Stat	P-value	Lower 95%	Upper 95%	Lower 95.0%	Upper 95.0%
Intercept	-6.983348818	0.932463961	-7.48914	0.000293	-9.26500593	-4.701692	-9.2650059	-4.7016917
Child's Age	0.89810517	0.031202521	28.7831	1.16E-07	0.821755352	0.974455	[illegible]175535	0.97445499
Teacher's Experience	-0.033200086	0.050469878	-0.65782	0.53[illegible]	-0.15669543	0.0902953	-0.1566954	0.09029526
Parents' Education	0.231953619	0.075147789	3.086633	0.02148	0.048073603	0.4158336	0.0480736	0.41583364

Figure 10 – Multiple Regression Model

The last block of information, shown in Figure 10, gives the coefficients of the multiple regression equation and the confidence intervals for the coefficients of the independent variables and y-intercept. The first row is the value ❺ and confidence interval ❻ for the y-intercept, β_0. The second row is the value ❼ and confidence interval ❽ for the first independent variable, child's age, β_1. The third row is the value ❾ and confidence ❿ interval for the second independent variable, teacher's experience, β_2. The fourth row is the value ⓫ and confidence interval ⓬ for the third independent variable, parents' education, β_3. Putting the values together, we can construct the multiple regression model for predicting a child's reading level.

$$\hat{y} = -6.983 + 0.898x_1 - 0.033x_2 + 0.232x_3$$

Let's use this regression model to predict the reading level of a child who is 10 years old with a teacher who has 8 years of experience and parents with an average of 17.2 years of education. Thus $x_1 = 10$, $x_2 = 8$, and $x_3 = 17.2$. Substituting these values into the multiple regression equation yields

$$\begin{aligned} \hat{y} &= -6.983 + 0.898x_1 - 0.033x_2 + 0.232x_3 \\ \hat{y} &= -6.983 + 0.898(10) - 0.033(8) + 0.232(17.2) \\ \hat{y} &= 5.723 \end{aligned}$$

Based on this regression equation, we would predict that a child with these characteristics would be reading at a fifth grade level.

Now let's consider the individual independent variables more closely. Look at the ⓭ *p*-values for the independent variables.

	Coefficients	*Standard Error*	*t Stat*	*P-value*	*Lower 95%*	*Upper 95%*	*Lower 95.0%*	*Upper 95.0%*
Intercept	-6.983348818	0.932463961	-7.48914	0.000293	-9.26500593	-4.701692	-9.2650059	-4.7016917
Child's Age	0.89810517	0.031202521	28.7831	1.16E-07	0.821755352	0.974455	0.82175535	0.97445499
Teacher's Experience	-0.033200086	0.050469878	-0.65782	0.535062	-0.15669543	0.0902953	-0.1566954	0.09029526
Parents' Education	0.231953619	0.075147789	3.086633	0.02148	0.048073603	0.4158336	0.0480736	0.41583364

Figure 11 – *P*-value for Independent Variables

These *p*-values test the null hypothesis that the coefficient of a particular independent variable equals 0. Mathematically, the null and alternative hypotheses for the first independent variable would be

$$H_0: \beta_1 = 0$$
$$H_a: \beta_1 \neq 0$$

The other null and alternative hypotheses are similar. A small *p*-value (such as one less than 0.05) indicates that there is sufficient evidence to support the claim that the coefficient is not 0, and therefore this particular variable has a statistically significant effect upon the dependent variable. If the *p*-value is not small (for instance, those greater than 0.05) then this particular variable may not be useful in predicting the value of the dependent variable.

In our example of predicting a child's reading level, notice that the *p*-values for the independent variables of child's age and parents' education both are small. This indicates that these independent variables are very likely to be effective in predicting the dependent variable. However, the *p*-value for teacher's experience is large. Thus including teacher's experience in the multiple regression model may not be useful. We could recalculate the multiple regression model without this variable. If the adjusted value of R^2 is larger for the model without this variable, then omitting this variable from the regression model would be advisable.

Example 19

Construct and analyze a multiple regression equation for predicting a child's reading level based on the following sample data, which omits the variable of teacher's experience from the multiple regression model we have been discussing. Use this new model to predict the reading level for a child who is 10 years old and has parents with an average of 17.2 years of education. Which of the two multiple regression models is better at predicting a child's reading level, based on the adjusted value of R^2?

Table 15 – Children's Reading Level

Child's Age (in years)	Parents' Education	Child's Reading Level
6	13.1	1.3
7	14.5	2.2
8	16.1	3.7
9	12.5	4.1
10	11.8	4.9
11	11.1	5.2
12	10.2	6
13	12.6	7.1
14	12.1	8.5
15	14.3	9.7

Solution:

We will use Microsoft Excel to construct and analyze a multiple regression model for this data. Begin by entering the data as it appears in columns A, B and C. Next, choose DATA ANALYSIS from the TOOLS menu. Choose REGRESSION from the options listed. Enter the necessary information into the Regression menu as shown on the following page. Click on OK.

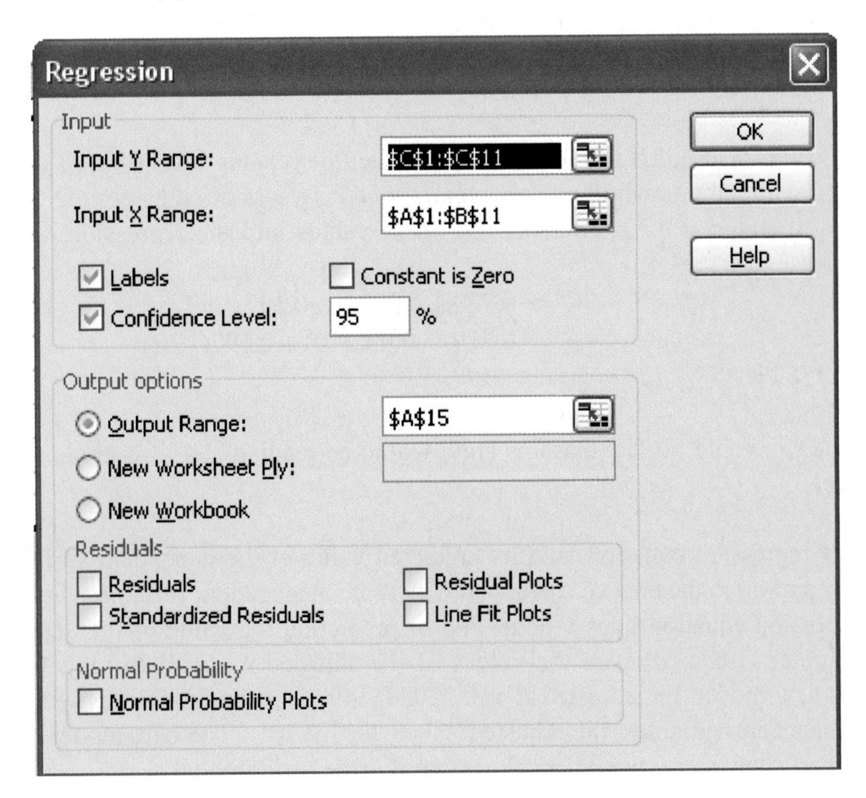

cont'd on the next page

The output is shown below.

SUMMARY OUTPUT

Regression Statistics	
Multiple R	0.996468545
R Square	0.992949561
Adjusted R Square	0.990935149
Standard Error	0.252102523
Observations	10

ANOVA

	df	SS	MS	F	Significance F
Regression	2	62.65611023	31.32806	492.9229633	2.94277E-08
Residual	7	0.444889774	0.063556		
Total	9	63.101			

	Coefficients	Standard Error	t Stat	P-value	Lower 95%	Upper 95%	Lower 95.0%	Upper 95.0%
Intercept	-6.728743363	0.813242326	-8.27397	7.34694E-05	-8.651755889	-4.8057308	-8.6517559	-4.80573084
Child's Age	0.901985643	0.029372093	30.70893	1.00214E-08	0.832531679	0.97143961	0.83253168	0.97143961
Parents' Education	0.197029939	0.050984389	3.864515	0.006175322	0.076471015	0.31758886	0.07647102	0.31758886

To begin, notice that the p-value for the regression equation is $p = 2.94277E{-}08$ which is extremely small. Thus this multiple regression equation fits the sample data extremely well. This can also be seen in the adjusted value of $R^2 = 0.991$, which is close to 1. Notice that the p-values for the coefficients of the individual independent variables have changed. These p-values are both small, which indicates that both of these variables are useful in predicting the value of the dependent variable.

The multiple regression equation is

$$\hat{y} = -6.729 + 0.902x_1 + 0.197x_2.$$

We can now use this new equation to predict the reading level of a 10 year old child with parents who have an average of 17.2 years of education. Note that $x_1 = 10$ and $x_2 = 17.2$. Substituting these values into the regression equation yields

$$\begin{aligned}\hat{y} &= -6.729 + 0.902x_1 + 0.197x_2 \\ \hat{y} &= -6.729 + 0.902(10) + 0.197(17.2) \\ \hat{y} &= 5.679\end{aligned}$$

Thus we would predict that this child would be reading at a fifth-grade reading level.

This regression equation, and its predicted value of the dependent variable, is very similar to the one we calculated with three independent variables. So which regression equation does a better job of predicting the value of the dependent variable? Let's compare the values of the adjusted R^2 for both models. For this new model, the adjusted $R^2 = 0.990935149$. For the first model with three independent variables, the adjusted $R^2 = 0.990135761$. The new model has the higher value, so we would conclude that it is the better model.

In this section we extended the concepts of linear regression to include multiple independent variables. We revisited the ANOVA tables to determine if the regression equation was statistically significant. The ANOVA tables also provided information on how closely each independent variable was related to the dependent variable. Lastly, we discussed possible ways to change the regression equation to eliminate unnecessary variables, and how to compare the new regression equation to the old equation to determine which equation is the best model for the given data. Most importantly, however, you should remember that a multiple regression model does not imply causation any more than a single regression model does. Just as we know that an increase in a child's age does not *cause* their reading level to increase, we should not presume that adding more variables to a regression equation accounts for all the possible influences on the dependent variable, or that the variables included are even *causing* the changes in the dependent variable at all.

Exercises

12.4

Directions: The following ANOVA tables were generated using Microsoft Excel. Based on the ANOVA table given, is there enough evidence at the 0.05 level of significance to conclude that the linear relationship between the independent variables and the dependent variable is statistically significant?

1.

	d.f.	*SS*	*MS*	*F*	*Significance F*
Regression	2	2677.934566	1338.967283	167.6684672	5.21957E-10
Residual	13	103.8154339	7.985802609		
Total	15	2781.75			

2.

	d.f.	*SS*	*MS*	*F*	*Significance F*
Regression	2	4.272742323	2.136371162	8.88737913	0.004285466
Residual	12	2.88459101	0.240382584		
Total	14	7.157333333			

3.

	d.f.	*SS*	*MS*	*F*	*Significance F*
Regression	3	1691672.173	563890.7244	11.50954064	0.080992754
Residual	2	97986.66027	48993.33013		
Total	5	1789658.833			

4.

	d.f.	*SS*	*MS*	*F*	*Significance F*
Regression	2	8.553919903	4.276959952	3.555389644	0.061293877
Residual	12	14.43541343	1.202951119		
Total	14	22.98933333			

Directions: For each of the following computer outputs, write the multiple regression equation. Define each of the variables used in the multiple regression equation. Could one of the independent variables be eliminated from the model? If so, which one?

5. The following table was generated from sample data from 16 statistics students regarding the number of absences in the class, the average number of hours they studied for the class per week, and their final grade in the class.

	Coefficients	*Standard Error*	*t-Stat*	*P-Value*
Intercept	63.88047809	4.914092338	12.99944602	7.95988E-09
Absences	−1.416204746	0.422223932	−3.354155552	0.005179159
Hours Studied per Week	4.168586523	0.73626183	5.661826206	7.77404E-05

6. The following table was generated from sample data from 16 babies regarding the weight of the mother at birth, the weight of the father at birth, and the weight of the baby at birth. All weights are measured in pounds.

	Coefficients	*Standard Error*	*t-Stat*	*P-Value*
Intercept	−2.882473037	0.9737576	−2.960154597	0.01105403
Mother's Weight	0.356055126	0.090343376	3.941131517	0.001689591
Father's Weight	1.02230869	0.099308161	10.29430688	1.28376E-07

7. The following table was generated from a sample of 15 babies regarding the number of weeks of gestation, the number of prenatal doctor visits the mother had, and the birth weight of the baby in pounds.

	Coefficients	*Standard Error*	*t-Stat*	*P-Value*
Intercept	−19.56998222	5.716106062	−3.423656245	0.005044399
Weeks Gestation	0.684805526	0.12588231	5.44004576	0.000150053
Number of Prenatal Visits	0.063077042	0.126763644	0.497595684	0.627761713

8. The following table was generated from a sample of 20 college freshman regarding their high school GPA, their ACT score, and their GPA after their first year at college.

	Coefficients	*Standard Error*	*t-Stat*	*P-Value*
Intercept	−0.616112354	0.345507101	−1.783211841	0.092415337
High School GPA	0.975068267	0.186279465	5.234437779	6.7297E-05
ACT	0.015000337	0.019048101	0.787497806	0.441832554

Directions: For each of the data sets below, use available technology, such as Microsoft Excel, to determine if there exists a statistically significant linear relationship between the independent variables and the dependent variable. If the linear relationship is significant, calculate the multiple regression equation that best fits the data.

9. The following data was collected to explore how tuition costs, student to faculty ratios, and female to male ratios affect freshmen enrollment. The student to faculty ratio is expressed as the number of students per faculty member. The female to male ratio is expressed as the number of female students per male student.

Table for Exercise 9 - Affects on Freshman Enrollment

Tuition Cost	Student to Faculty Ratio	Female to Male Ratio	Freshman Enrollment
5495	18	1.5	3219
7912	18	2	2781
5812	6	1.7	3556
7180	4	2.1	2987
8883	15	3	1794
6495	19	1.8	3014

10. The following data was collected to explore how the average number of hours a junior high student is unsupervised per night and the average number of hours per night the student watches television affect their Overall Average.

Table for Exercise 10 - Affects on Junior High Students' Overall Average		
Hours Unsupervised	**Hours Watching TV**	**Overall Average**
0	2	96
0	3	91
0.5	3	88
1	2	92
1	1	94
1.5	2	91
2	4	87
3	3	85
3	4	81
4	5	80
4.5	4	77
5.5	5	72

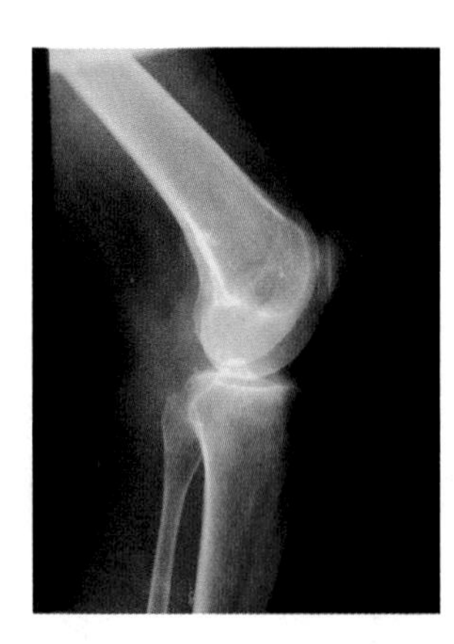

11. The following data was collected to explore how women's age and their average daily calcium intake (measured in mg per day) affect their bone mineral density.

Table for Exercise 11 - Affects on Bone Mineral Density (mg/cm^2)		
Age	**Daily Calcium Intake**	**Bone Density (mg/cm^2)**
34	791	950
43	832	875
47	865	804
52	901	723
58	923	691

12. Common predictors for freshman grade point average are ACT scores and high school GPA. Let's explore how well two unusual factors, a student's age when entering college and the number of parking tickets they receive during their first semester at school, predict a student's GPA after one semester at college.

Table for Exercise 12 - Affects on GPA

Student's Age	Number of Parking Tickets	GPA
20	0	3.6
21	0	3.9
17	0	2.4
18	0	3.1
19	1	3.5
35	1	4
23	1	3.6
19	2	2.8
18	2	3
18	2	2.2
24	3	3.9
19	3	3.1
18	5	2.1
17	7	2.8
18	8	1.7

Challenge Questions:
Directions: Answer each of the following questions.

13. Which value in the ANOVA table tells you the number of independent variables in the multiple regression model?

14. Should you always use as many independent variables as possible?

15. Which regression statistic should you use to compare one multiple regression model to another to determine which one better fits the sample data?

16. What is the disadvantage of the multiple coefficient of determination?

Chapter Review

Section 12.1 -Scatter Plots and Correlation	Pages
• **Definitions** **Scatter Plots** – a graph on the coordinate plane which contains one point for each pair of data **Correlation Coefficient** – measures how strongly one variable is linearly dependent upon the other **Statistically Significant** – indicates that the results are meaningful and likely to be representative of the entire population **Coefficient of Determination** – The amount of variation in y explained by the variation in x	p. 664 – 675
• **Correlation Coefficient, r** $-1 \le r \le 1$ $r = \frac{n\sum xy - (\sum x)(\sum y)}{\sqrt{n\sum x^2 - (\sum x)^2}\sqrt{n\sum y^2 - (\sum y)^2}}$	p. 668 – 669

Section 12.2 - Linear Regression	Pages
• **Definitions** **Regression Line (Line of Best Fit)** – the line for which the average variation from the data is the smallest, $\hat{y} = b_0 + b_1 x$	p. 680
• **Slope** $b_1 = \frac{n\sum xy - \sum x \sum y}{n\sum x^2 - \left(\sum x\right)^2}$	p. 680
• ***y*-Intercept** $b_0 = \frac{\sum y}{n} - b_1 \frac{\sum x}{n}.$	p. 681

Section 12.3 - Regression Analysis	Pages

- **Definitions** — p. 689 – 691

  Residual – the difference between the actual value of y from the original data and the predicted value of $\hat{y}$ found using the regression line

  Sum of Squared Errors – the value calculated by summing the square of the residuals

  Prediction Interval – confidence interval for the predicted dependent variable

- **Standard Error of Estimate, S_e** — p. 692

$$S_e = \sqrt{\frac{\sum (y - \hat{y})^2}{n-2}}.$$

- **Margin of Error formula for Prediction Interval** — p. 694

$$E = t_{\frac{\alpha}{2}} S_e \sqrt{1 + \frac{1}{n} + \frac{n(x_0 - \bar{x})^2}{n(\sum x^2) - (\sum x)^2}}$$

- **Prediction Interval** — p. 695

$$\hat{y} - E < y < \hat{y} + E$$

or

$$(\hat{y} - E, \hat{y} + E)$$

Section 12.4 - Multiple Regression Equations	Pages

- **Multiple Regression Model** – A linear regression model using two or more independent variables to predict a dependent variable. — p. 702

$$\hat{y} = b_0 + b_1x_1 + b_2x_2 + \cdots + b_kx_k$$

- **Multiple coefficient of determination,** R^2, is analogous to the coefficient of determination of a single independent variable, r^2. — p. 704

- **Adjusted coefficient of determination** - the value of the multiple coefficient of determination adjusted for the number of independent variables and the sample size. — p. 704

When using the ANOVA table to analyze the statistical significance of the linear relationship between the variables in a multiple regression, the null and alternative hypotheses are — p. 704

H_0: $\beta_1 = \beta_2 = \cdots = \beta_k = 0$

H_a: At least one coefficent does not equal 0.

Chapter Exercises

Directions: Answer each of the following questions.

1. The following tables gives the heights and corresponding batting averages of a sample of 12 professional baseball players.

Table for Exercise 1 - Height (in Inches) vs. Batting Average												
Height (in inches)	70	73	73	71	68	76	71	75	73	73	75	73
Batting Average	.248	.273	.245	.202	.304	.349	.226	.250	.233	.273	.237	.229

a. Calculate the correlation coefficient.

b. Is the correlation coefficient statistically significant at either the 0.05 or 0.01 level?

c. Should a regression line be used to model the data and make predictions? Why or why not?

2. The following table gives the college GPA and corresponding annual income for a sample of 9 young adults.

Table for Exercise 2 - GPA vs. Annual Income									
GPA	3.1	3.8	2.9	1.8	2.7	3.5	3.2	2.2	4.0
Annual Income	$38000	$36500	$39500	$29000	$33000	$40500	$26000	$30000	$61000

a. Before calculating the correlation coefficient, what kind of relationship would you expect to see between the given variables?

b. Calculate the correlation coefficient and coefficient of determination.

c. Is r statistically significant at the 0.05 level? At the 0.01 level?

d. Suppose that your college GPA is a 3.3. Using the regression model for this data, make a prediction for your future annual income.

3. The average monthly interest rate for a 30-year mortgage and corresponding number of new loans taken out at one bank are given for the following 12 months.

Table for Exercise 3 - Interest Rate vs. New Loans												
Interest Rate	7.12	7.23	7.25	7.04	6.99	6.87	6.68	6.91	7.03	7.09	7.15	7.11
New Loans	5	6	4	5	7	9	9	10	8	7	4	4

a. Before calculating the correlation coefficient, what kind of relationship would you expect to see between the given variables?

b. Calculate the correlation coefficient and coefficient of determination.

c. Is r statistically significant at the 0.05 level? At the 0.01 level?

d. Suppose that the average interest rate for a given month is 6.75%. Using the regression model for this data, make a prediction for the number of new loans taken out.

e. Suppose that the first column of data in the table represents the month of January, the second February, etc. What other factor may have influenced the rise in home purchases?

4. The following table gives the average price of gas and the corresponding SUV sales at a local car dealership for 10 randomly selected months, in 2003.

Table for Exercise 4 - Price of Gas vs. Number of SUVs Sold										
Price of Gas	$1.26	$1.24	$1.51	$1.37	$1.53	$1.43	$1.35	$1.49	$1.42	$1.58
SUVs Sold	45	44	36	45	40	43	47	39	47	37

a. Before calculating the correlation coefficient, what kind of relationship would you expect to see between the given variables?

b. Calculate the correlation coefficient and coefficient of determination.

c. Is r statistically significant at the 0.05 level? At the 0.01 level?

d. Suppose that the average gas price for one month is $1.40. How many SUVs should the car dealership expect to sell?

5. The following data was collected over the course of a year:

Table for Exercise 5 - Number of Swimsuits vs. Number of Seagulls												
# of Swimsuits	0	8	11	28	77	120	150	250	100	44	12	5
# of Seagulls	2	78	92	106	102	200	300	376	298	55	15	3

a. From analyzing the data above, can you conclude that there is a significant linear relationship between the two variables? At what level?

b. Could you predict how many seagulls would be visible if you can see 200 swimsuits? What about 500 swimsuits?

c. What reasonable conclusions can you make regarding your findings? Are these the type of results you would publish? Why or why not?

6. Consider the following data from a recent study:

Table for Exercise 6 - Cancer Risk vs. Number of Years Using a Cell Phone														
Cancer Risk	0.35	0.11	0.45	0.28	0.55	0.78	0.55	0.38	0.74	0.81	0.15	0.09	0.66	0.2
# of Years Using a Cell Phone	2	3.1	1.5	0.9	4	1.5	3.9	2	5	4.4	2.1	4.9	8	6.2

a. From analyzing the data above, can you conclude that there is a significant linear relationship between the two variables? At what level?

b. Is it reasonable to make predictions of cancer risk based on cell phone usage?

c. If you were a cell phone company, what would you do with the results of this study?

7. The following data was collected over the course of a year:

Table for Exercise 7 - Mileage vs. Price of Used Cars	
Mileage	**Price**
23,547	$21,803
48,710	$17,995
33,952	$18,995
14,876	$20,863
133,691	$4970
30,417	$28,500
29,104	$22,597
54,000	$17,600
63,079	$14,999
133,172	$10,995
110,000	$8000
63,263	$15,995

a. Can you conclude that there is a significant linear relationship between the two variables? At what level?

b. Predict the price of a vehicle which has 42,000 miles on it.

c. What amount of the change in price has to do with the change in the number of miles on the car?

8. Consider the following data collected on airline tickets. The following lists the distance from Dallas, Texas to the destination compared to the price of the ticket.

Table for Exercise 8 - Price of Ticket vs. Distance Traveling		
Distance (in miles)	**Destination**	**Price**
1573.61	New York	$212
1329.96	Washington DC	$234
1316.38	Miami	$275
782.50	Atlanta	$231
2203.26	Seattle	$312
967.73	Chicago	$185
1368.18	Baltimore	$267
453.18	Memphis	$229
521.11	New Orleans	$345

a. Is there a significant linear relationship at the 0.05 level between distance traveled and price of ticket from Dallas, Texas?

b. Predict the price of a ticket to San Francisco from Dallas if it's 1465 miles.

9. The following data represents the thread count of various bed sheets and their corresponding price. Calculate the following based on this data:

Table for Exercise 9 - Thread Count vs. Price								
Thread Count	150	200	225	250	275	300	350	400
Price	$18	$21	$25	$28	$30	$31	$35	$45

a. Calculate the Sum of Squared Errors (SSE).

b. Calculate the Standard Error of Estimate, S_e.

c. Construct a 95% prediction interval for the price of sheets with a 350 thread count.

d. Construct a 95% confidence interval for the *y*-intercept of the regression line.

e. Construct a 95% confidence interval for the slope of the regression line.

10. The following data was collected on the number of years of post high-school education and the income of eight people ten years after graduation. Calculate the following based on this data:

Table for Exercise 10 - Education vs. Income								
Education (in years)	2	3	3	4	5	6	6	8
Income (in $10,000)	35	41	38	45	65	89	95	115

a. Calculate the Sum of Squared Errors (SSE).

b. Calculate the Standard Error of Estimate, S_e.

c. Construct a 95% prediction interval for the income of someone ten years after graduation who had 7 years of post high-school education.

d. Construct a 95% confidence interval for the *y*-intercept of the regression line.

e. Construct a 95% confidence interval for the slope of the regression line.

11. In addition to the thread count of bed sheets, let's explore if the number of colors used is linearly related to the price of the sheets. Determine if there exists a statistically significant linear relationship between these independent variables and the dependent variable. If the linear relationship is significant, calculate the multiple regression equation that best fits the data.

Table for Exercise 11 - Affects on Price of Sheets		
Thread Count	**Number of Colors**	**Price (in Dollars)**
150	2	18
200	1	21
225	2	25
250	4	28
275	3	30
300	2	31
350	1	35
400	4	45

12. In addition to the number of years of post high-school education, the number of years of work experience also influences a person's salary. Determine if there exists a statistically significant linear relationship between these independent variables and the dependent variable. If the linear relationship is significant, calculate the multiple regression equation that best fits the data.

Table for Exercise 12 - Affects on Income		
Education	**Experience**	**Income (in Thousands of Dollars)**
2	8	35
3	7	41
3	5	38
4	6	45
5	6	65
6	5	89
6	4	95
8	4	115

Chapter Project A: Arm Span versus Height

Is it true that a person's arm span is equal to their height? If so, then there should be a perfect linear relationship between arm span and height. Let's find out if this is true by collecting data and creating a linear regression model.

Step 1 – Collect data from 10 people. Measure each person's height (in inches) and then their arm span (in inches), which would be from finger tip to finger tip as they hold their arms outstretched. Record your results in a table similar to the one below. Your results will be more generalizable if you collect data from people of many different heights, from children to tall men.

Height										
Arm Span										

Step 2 – Create a scatter plot of the data. Use Height as the x variable and Arm Span as the y variable.

a. Does there appear to be a linear relationship between x and y?

b. Is the relationship positive or negative?

c. Is the relationship strong or weak?

d. Does the graph seem to support the claim that a person's height is equal to their arm span?

Step 3 – Calculate the correlation coefficient, r.

a. Is the correlation statistically significant at the 0.05 level of significance?

b. What about the 0.01 level of significance?

c. Interpret what it means for the correlation to be statistically significant.

Step 4 – Calculate the equation of the regression line, $y = b_0 + b_1x$.

a. Draw the regression line on your scatter plot.

b. If it is true that a person's arm span is the same as their height, what would you expect the slope of the regression line to be?

c. How close is the actual slope of the regression line to the expected value?

d. Calculate a 95% confidence interval for the slope of the regression line.

e. Based on your confidence interval, could you conclude that it is likely that a person's arm span is equal to their height.

Step 5 – Calculate the coefficient of determination, r^2.

a. Interpret what r^2 means for this scenario.

b. Does the value of r^2 support the claim that arm span is equal to height?

Step 6 – Based on all of the above information, formulate a conclusion as to whether a person's arm span is equal to their height.

Chapter Project B: Attractiveness versus Parental Attentiveness

Have you ever commented that someone is a "perfect 10," meaning that they were extremely attractive? A Canadian researcher believes that a person's attractiveness can predict how people respond to them. Specifically, he hypothesizes that a child's attractiveness is a predictor of parental attentiveness. He thinks that the cuter the child, the more attention the parents pay to the child and the child thus receives better care. In a study performed in Canada, researchers rated a child's attractiveness on a scale of 1 to 10 and then counted the number of times the child was allowed to wander away from the parent in a 10 minute interval. (Source: http://www.futurepundit.com/archives/002711.html)

In order to test this hypothesis, we will perform a modified version of this research study.

Step 1 – Collect data on 6 children in a busy public setting. Begin by rating the child's overall attractiveness (how "cute" they are) on a scale of 1 to 10, where 1 is not attractive and 10 is extremely cute. Second, count the number of times in a 10 minute time period that their parent allows the child to wander more than 10 feet away. Record your results in a table similar to the one below.

Attractiveness						
Attention						

Step 2 – Create a scatter plot of the data. Use Attractiveness as the x variable and Attention as the y variable.

a. Does there appear to be a linear relationship between x and y?
b. Is the relationship positive or negative?
c. Is the relationship strong or weak?
d. Does the graph seem to support the claim that a child's attractiveness is a predictor of their parents attentiveness?

Step 3 – Calculate the correlation coefficient, r.

a. Is the correlation statistically significant at the 0.05 level of significance?
b. What about the 0.01 level of significance?
c. Interpret what it means for the correlation to be statistically significant.

Step 4 – Calculate the equation of the regression line, $y = b_0 + b_1x$. Draw the regression line on your scatterplot.

Step 5 – Calculate the coefficient of determination, r^2.

a. Interpret what r^2 means for this scenario.
b. Does the value of r^2 support the claim that a child's attractiveness is a predictor of their parents attentiveness?

Step 6 – How did any of the following factors affect your results, if at all?

a. The way in which you chose which children to watch.
b. The different ages of the children.
c. The location of your data collecting.
d. The way you rated the children's attractiveness or the parental attentiveness.

Step 7 – Based on all of the above information, formulate a conclusion as to whether a child's attractiveness is a predictor of their parents' attentiveness.

Chapter Technology – Scatter Plots and Linear Regression

TI-84 PLUS

Scatter Plots

The first step to creating any type of graph using technology is to enter your data. On a TI-84 Plus, enter the values of the x variable in L1 and the values of the y variable in L2. Recall that to enter data in the List environment, press STAT and then choose EDIT and option 1: Edit.

After the data has been entered in your TI-84 Plus, to create a scatter plot we must set up the STAT PLOT menu. Press 2ND and Y= to open the STAT PLOT menu. Highlight Plot 1 and then press ENTER.

Turn Plot1 on by selecting ON and then pressing ENTER. Under the type of graph, choose the first one, which is a small picture of a scatter plot. The defaults for the rest of the menu should be correct if your data are in lists.

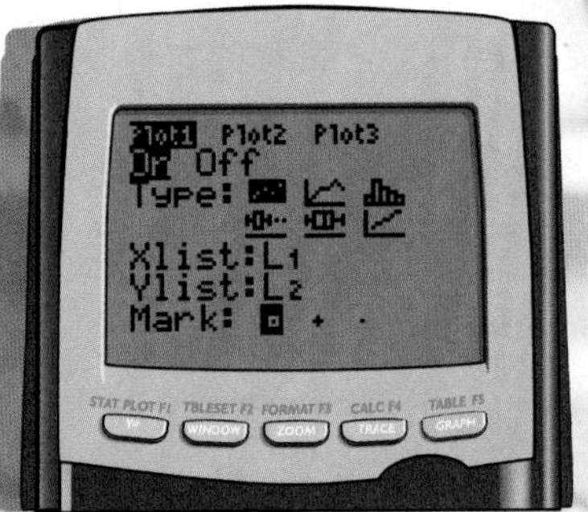

To create the scatter plot, press GRAPH. If a scatter plot does not immediately appear on the screen, which is likely, you may need to zoom in. The TI-84 Plus has a built-in zoom feature which will automatically zoom correctly for a statistical graph. Press ZOOM, then choose option 9: ZoomStat.

Example T.1

Use your TI-84 Plus to create a scatter plot for the data below:

Age (x)	1	2	3	4	5
Diaper Changes (y)	3.5	2.3	1.8	0.65	0

Solution:

We need to begin by entering our data into L1 and L2 as shown on the following page:

cont'd on the next page

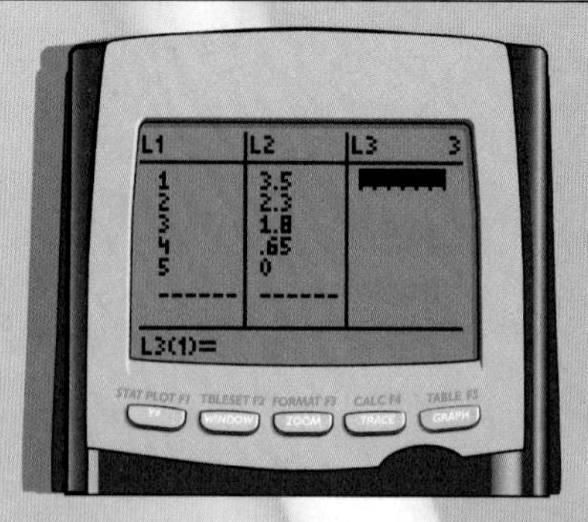

Once the data is entered, we can set up the STAT PLOT menu as described above. Next, press GRAPH. The following is what might appear on the screen, if anything:

In order to adjust the screen to better view the scatter plot, we can press ZOOM and then choose option 9. This will create the following scatter plot.

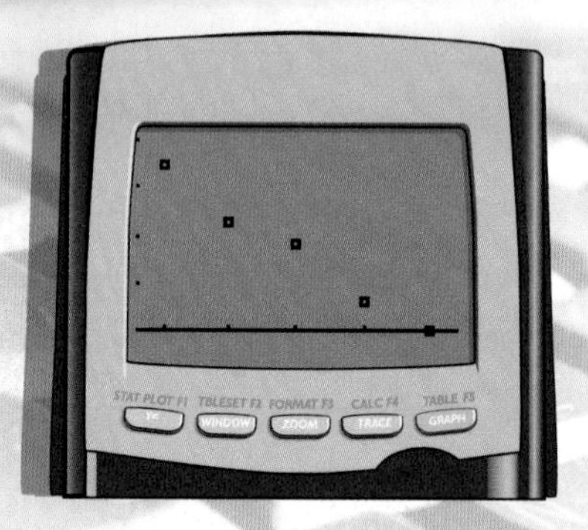

MICROSOFT EXCEL

To use Microsoft Excel to create a scatter plot, enter your data in columns A and B in the worksheet. Then, click on the Chart Wizard Button in the toolbar.

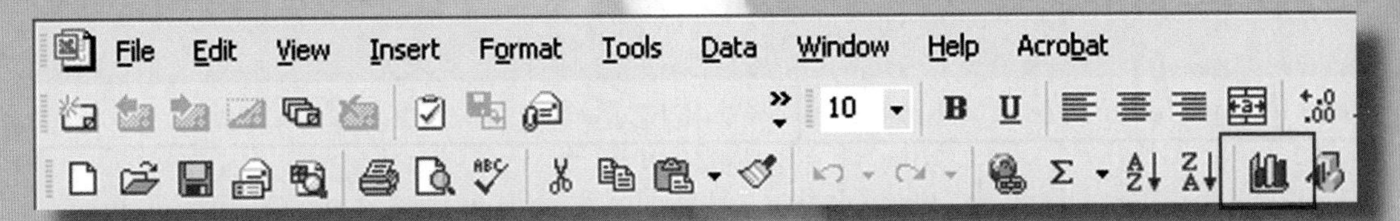

This wizard will walk you through the steps used to create a graph. Simply follow the steps in the wizard, clicking on the next button to move from one step to the next. The wizard takes you through the steps of choosing the type of chart, locating your source data, titling your chart, and creating the final output.

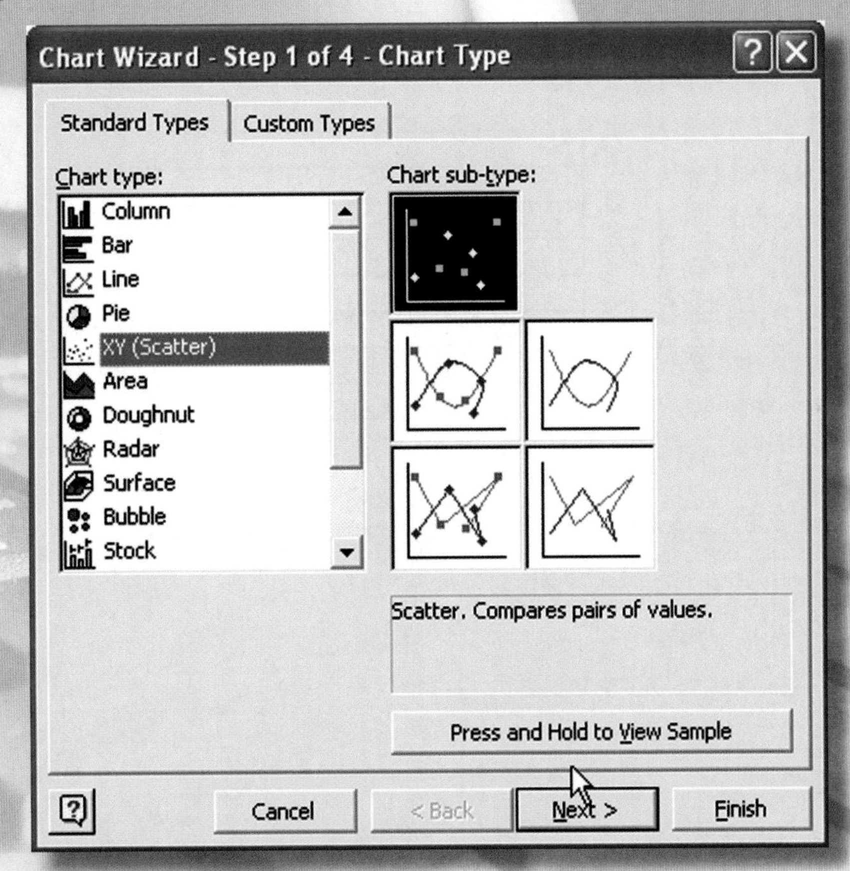

Example T.2

Use Microsoft Excel to create the same scatter plot of the age versus number of diaper changes as we created on our TI-84 Plus.

Solution:

Enter the data for age down column A and the data for the number of diaper changes down column B. In the chart wizard, choose the XY (Scatter) option for the type of graph. Click through the rest of the wizard, making any changes that you wish. After making various changes to the titles and colors, our scatter plot looks like the one below, which is the same one we generated in an example earlier in the chapter. Your scatter plot may be different, depending on the options you chose in the chart wizard.

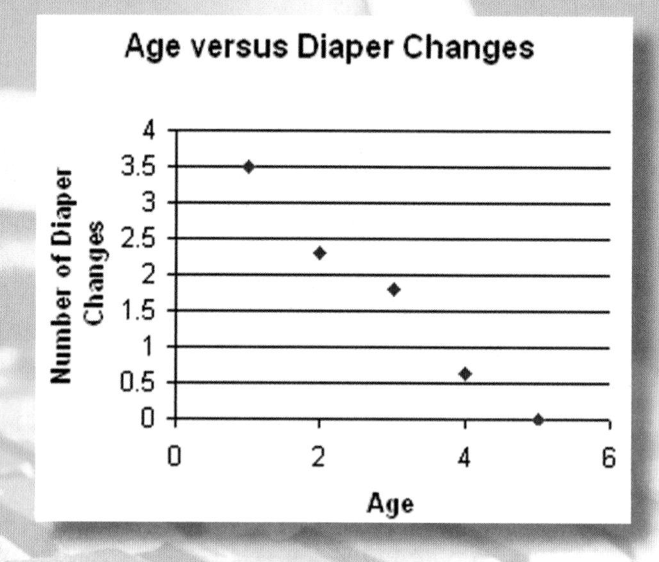

MINITAB

Minitab can create a scatter plot and create a linear regression model in a single operation.

Example T.3

The data below describes the population of a hypothetical small city. Create a scatter plot and linear regression model for the supplied data.

x	2000	2001	2002	2003	2004	2005
y	52,324	54,350	55,990	57,123	60,005	62,357

Solution:

First, enter the data into columns in the worksheet. Next, select STAT, REGRESSION, and *Fitted Line Plot*. In the dialog box that appears, enter C2 for Response(Y) and C1 for Predictor(X). Make sure **Linear** is the **Type of Regression Model** selected and select OK. The Dialog box is shown below. The linear regression model and scatter plot appear immediately. An ANOVA table is produced in the session window. Note that the p-value is produced which describes the fit of the model produced to the data.

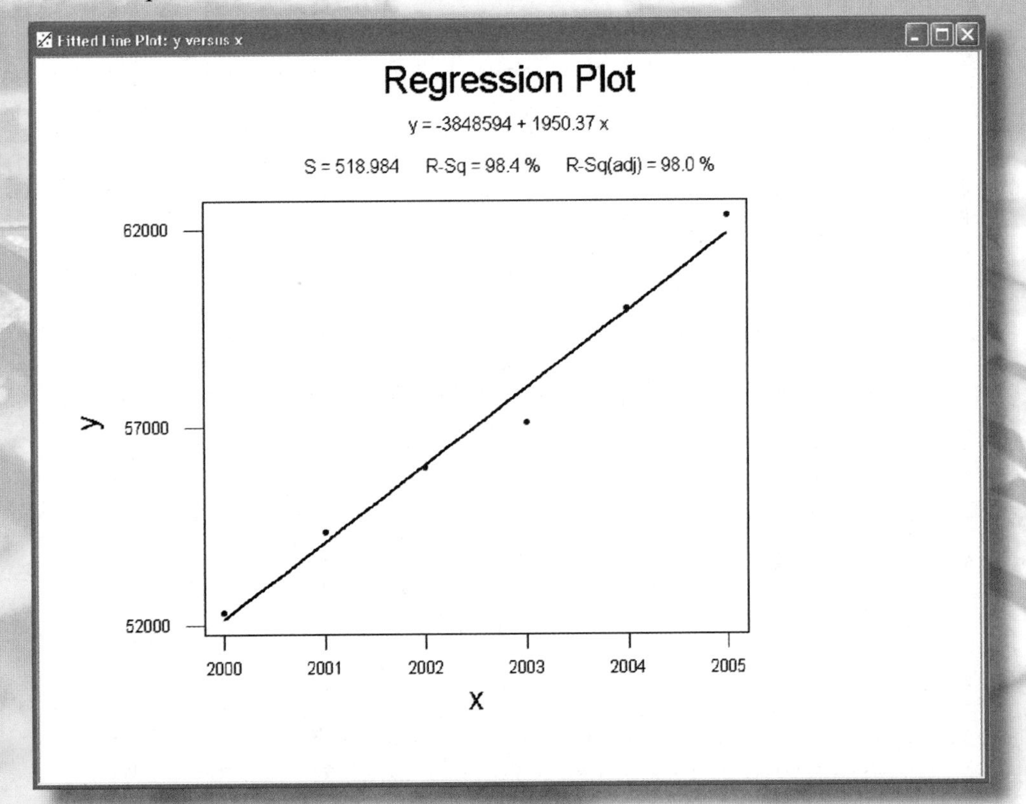

Appendix A

Statistical Tables

TABLE A

Standard Normal Distribution

Numerical entries represent the probability that a standard normal random variable is between $-\infty$ and z where $z = (x - \mu)/\sigma$.

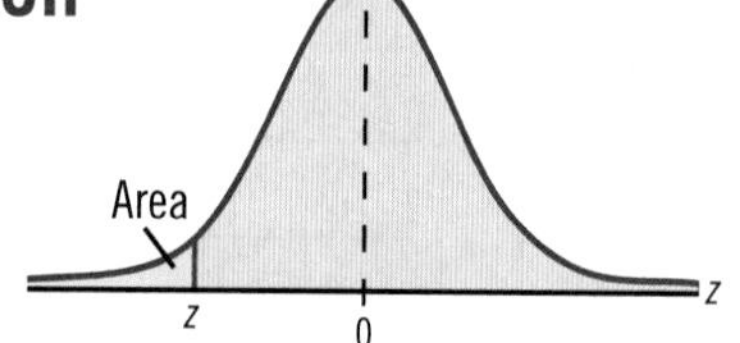

z	0.09	0.08	0.07	0.06	0.05	0.04	0.03	0.02	0.01	0.00
-3.4	0.0002	0.0003	0.0003	0.0003	0.0003	0.0003	0.0003	0.0003	0.0003	0.0003
-3.3	0.0003	0.0004	0.0004	0.0004	0.0004	0.0004	0.0004	0.0005	0.0005	0.0005
-3.2	0.0005	0.0005	0.0005	0.0006	0.0006	0.0006	0.0006	0.0006	0.0007	0.0007
-3.1	0.0007	0.0007	0.0008	0.0008	0.0008	0.0008	0.0009	0.0009	0.0009	0.0010
-3.0	0.0010	0.0010	0.0011	0.0011	0.0011	0.0012	0.0012	0.0013	0.0013	0.0013
-2.9	0.0014	0.0014	0.0015	0.0015	0.0016	0.0016	0.0017	0.0018	0.0018	0.0019
-2.8	0.0019	0.0020	0.0021	0.0021	0.0022	0.0023	0.0023	0.0024	0.0025	0.0026
-2.7	0.0026	0.0027	0.0028	0.0029	0.0030	0.0031	0.0032	0.0033	0.0034	0.0035
-2.6	0.0036	0.0037	0.0038	0.0039	0.0040	0.0041	0.0043	0.0044	0.0045	0.0047
-2.5	0.0048	0.0049	0.0051	0.0052	0.0054	0.0055	0.0057	0.0059	0.0060	0.0062
-2.4	0.0064	0.0066	0.0068	0.0069	0.0071	0.0073	0.0075	0.0078	0.0080	0.0082
-2.3	0.0084	0.0087	0.0089	0.0091	0.0094	0.0096	0.0099	0.0102	0.0104	0.0107
-2.2	0.0110	0.0113	0.0116	0.0119	0.0122	0.0125	0.0129	0.0132	0.0136	0.0139
-2.1	0.0143	0.0146	0.0150	0.0154	0.0158	0.0162	0.0166	0.0170	0.0174	0.0179
-2.0	0.0183	0.0188	0.0192	0.0197	0.0202	0.0207	0.0212	0.0217	0.0222	0.0228
-1.9	0.0233	0.0239	0.0244	0.0250	0.0256	0.0262	0.0268	0.0274	0.0281	0.0287
-1.8	0.0294	0.0301	0.0307	0.0314	0.0322	0.0329	0.0336	0.0344	0.0351	0.0359
-1.7	0.0367	0.0375	0.0384	0.0392	0.0401	0.0409	0.0418	0.0427	0.0436	0.0446
-1.6	0.0455	0.0465	0.0475	0.0485	0.0495	0.0505	0.0516	0.0526	0.0537	0.0548
-1.5	0.0559	0.0571	0.0582	0.0594	0.0606	0.0618	0.0630	0.0643	0.0655	0.0668
-1.4	0.0681	0.0694	0.0708	0.0721	0.0735	0.0749	0.0764	0.0778	0.0793	0.0808
-1.3	0.0823	0.0838	0.0853	0.0869	0.0885	0.0901	0.0918	0.0934	0.0951	0.0968
-1.2	0.0985	0.1003	0.1020	0.1038	0.1056	0.1075	0.1093	0.1112	0.1131	0.1151
-1.1	0.1170	0.1190	0.1210	0.1230	0.1251	0.1271	0.1292	0.1314	0.1335	0.1357
-1.0	0.1379	0.1401	0.1423	0.1446	0.1469	0.1492	0.1515	0.1539	0.1562	0.1587
-0.9	0.1611	0.1635	0.1660	0.1685	0.1711	0.1736	0.1762	0.1788	0.1814	0.1841
-0.8	0.1867	0.1894	0.1922	0.1949	0.1977	0.2005	0.2033	0.2061	0.2090	0.2119
-0.7	0.2148	0.2177	0.2206	0.2236	0.2266	0.2296	0.2327	0.2358	0.2389	0.2420
-0.6	0.2451	0.2483	0.2514	0.2546	0.2578	0.2611	0.2643	0.2676	0.2709	0.2743
-0.5	0.2776	0.2810	0.2843	0.2877	0.2912	0.2946	0.2981	0.3015	0.3050	0.3085
-0.4	0.3121	0.3156	0.3192	0.3228	0.3264	0.3300	0.3336	0.3372	0.3409	0.3446
-0.3	0.3483	0.3520	0.3557	0.3594	0.3632	0.3669	0.3707	0.3745	0.3783	0.3821
-0.2	0.3859	0.3897	0.3936	0.3974	0.4013	0.4052	0.4090	0.4129	0.4168	0.4207
-0.1	0.4247	0.4286	0.4325	0.4364	0.4404	0.4443	0.4483	0.4522	0.4562	0.4602
0.0	0.4641	0.4681	0.4721	0.4761	0.4801	0.4840	0.4880	0.4920	0.4960	0.5000

TABLE B
Standard Normal Distribution

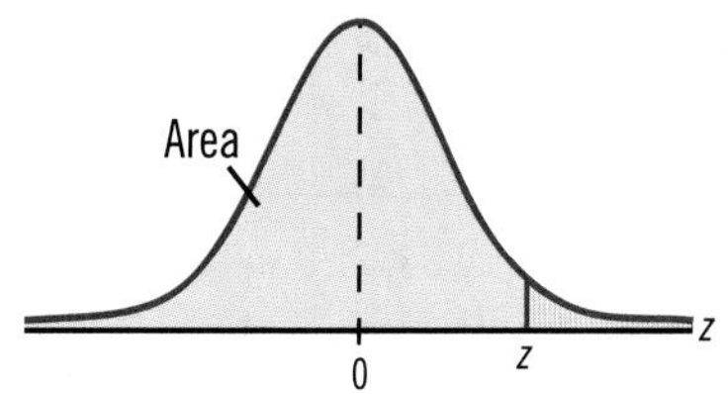

z	0.00	0.01	0.02	0.03	0.04	0.05	0.06	0.07	0.08	0.09
0.0	0.5000	0.5040	0.5080	0.5120	0.5160	0.5199	0.5239	0.5279	0.5319	0.5359
0.1	0.5398	0.5438	0.5478	0.5517	0.5557	0.5596	0.5636	0.5675	0.5714	0.5753
0.2	0.5793	0.5832	0.5871	0.5910	0.5948	0.5987	0.6026	0.6064	0.6103	0.6141
0.3	0.6179	0.6217	0.6255	0.6293	0.6331	0.6368	0.6406	0.6443	0.6480	0.6517
0.4	0.6554	0.6591	0.6628	0.6664	0.6700	0.6736	0.6772	0.6808	0.6844	0.6879
0.5	0.6915	0.6950	0.6985	0.7019	0.7054	0.7088	0.7123	0.7157	0.7190	0.7224
0.6	0.7257	0.7291	0.7324	0.7357	0.7389	0.7422	0.7454	0.7486	0.7517	0.7549
0.7	0.7580	0.7611	0.7642	0.7673	0.7704	0.7734	0.7764	0.7794	0.7823	0.7852
0.8	0.7881	0.7910	0.7939	0.7967	0.7995	0.8023	0.8051	0.8078	0.8106	0.8133
0.9	0.8159	0.8186	0.8212	0.8238	0.8264	0.8289	0.8315	0.8340	0.8365	0.8389
1.0	0.8413	0.8438	0.8461	0.8485	0.8508	0.8531	0.8554	0.8577	0.8599	0.8621
1.1	0.8643	0.8665	0.8686	0.8708	0.8729	0.8749	0.8770	0.8790	0.8810	0.8830
1.2	0.8849	0.8869	0.8888	0.8907	0.8925	0.8944	0.8962	0.8980	0.8997	0.9015
1.3	0.9032	0.9049	0.9066	0.9082	0.9099	0.9115	0.9131	0.9147	0.9162	0.9177
1.4	0.9192	0.9207	0.9222	0.9236	0.9251	0.9265	0.9279	0.9292	0.9306	0.9319
1.5	0.9332	0.9345	0.9357	0.9370	0.9382	0.9394	0.9406	0.9418	0.9429	0.9441
1.6	0.9452	0.9463	0.9474	0.9484	0.9495	0.9505	0.9515	0.9525	0.9535	0.9545
1.7	0.9554	0.9564	0.9573	0.9582	0.9591	0.9599	0.9608	0.9616	0.9625	0.9633
1.8	0.9641	0.9649	0.9656	0.9664	0.9671	0.9678	0.9686	0.9693	0.9699	0.9706
1.9	0.9713	0.9719	0.9726	0.9732	0.9738	0.9744	0.9750	0.9756	0.9761	0.9767
2.0	0.9772	0.9778	0.9783	0.9788	0.9793	0.9798	0.9803	0.9808	0.9812	0.9817
2.1	0.9821	0.9826	0.9830	0.9834	0.9838	0.9842	0.9846	0.9850	0.9854	0.9857
2.2	0.9861	0.9864	0.9868	0.9871	0.9875	0.9878	0.9881	0.9884	0.9887	0.9890
2.3	0.9893	0.9896	0.9898	0.9901	0.9904	0.9906	0.9909	0.9911	0.9913	0.9916
2.4	0.9918	0.9920	0.9922	0.9925	0.9927	0.9929	0.9931	0.9932	0.9934	0.9936
2.5	0.9938	0.9940	0.9941	0.9943	0.9945	0.9946	0.9948	0.9949	0.9951	0.9952
2.6	0.9953	0.9955	0.9956	0.9957	0.9959	0.9960	0.9961	0.9962	0.9963	0.9964
2.7	0.9965	0.9966	0.9967	0.9968	0.9969	0.9970	0.9971	0.9972	0.9973	0.9974
2.8	0.9974	0.9975	0.9976	0.9977	0.9977	0.9978	0.9979	0.9979	0.9980	0.9981
2.9	0.9981	0.9982	0.9982	0.9983	0.9984	0.9984	0.9985	0.9985	0.9986	0.9986
3.0	0.9987	0.9987	0.9987	0.9988	0.9988	0.9989	0.9989	0.9989	0.9990	0.9990
3.1	0.9990	0.9991	0.9991	0.9991	0.9992	0.9992	0.9992	0.9992	0.9993	0.9993
3.2	0.9993	0.9993	0.9994	0.9994	0.9994	0.9994	0.9994	0.9995	0.9995	0.9995
3.3	0.9995	0.9995	0.9995	0.9996	0.9996	0.9996	0.9996	0.9996	0.9996	0.9997
3.4	0.9997	0.9997	0.9997	0.9997	0.9997	0.9997	0.9997	0.9997	0.9997	0.9998

Critical Values

Level of Confidence	$z_{\alpha/2}$
0.80	1.28
0.85	1.44
0.90	1.645
0.95	1.96
0.98	2.33
0.99	2.575

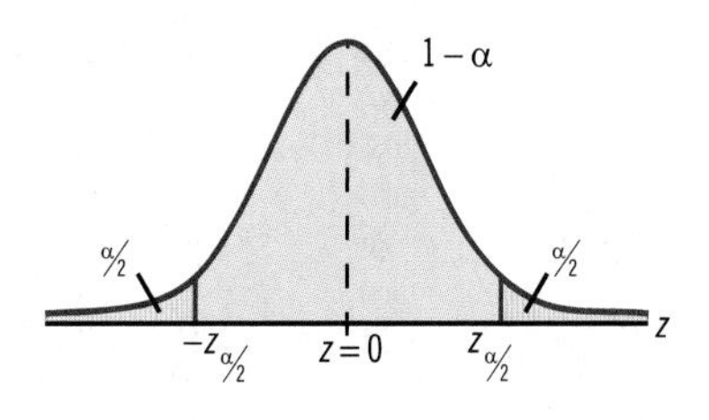

TABLE C

Critical Values of t

	Area in One Tail				
	$t_{.100}$	$t_{.050}$	$t_{.025}$	$t_{.010}$	$t_{.005}$
	Area in Two Tails				
d.f.	$t_{.200}$	$t_{.100}$	$t_{.050}$	$t_{.020}$	$t_{.010}$
1	3.078	6.314	12.706	31.821	63.657
2	1.886	2.920	4.303	6.965	9.925
3	1.638	2.353	3.182	4.541	5.841
4	1.533	2.132	2.776	3.747	4.604
5	1.476	2.015	2.571	3.365	4.032
6	1.440	1.943	2.447	3.143	3.707
7	1.415	1.895	2.365	2.998	3.499
8	1.397	1.860	2.306	2.896	3.355
9	1.383	1.833	2.262	2.821	3.250
10	1.372	1.812	2.228	2.764	3.169
11	1.363	1.796	2.201	2.718	3.106
12	1.356	1.782	2.179	2.681	3.055
13	1.350	1.771	2.160	2.650	3.012
14	1.345	1.761	2.145	2.624	2.977
15	1.341	1.753	2.131	2.602	2.947
16	1.337	1.746	2.120	2.583	2.921
17	1.333	1.740	2.110	2.567	2.898
18	1.330	1.734	2.101	2.552	2.878
19	1.328	1.729	2.093	2.539	2.861
20	1.325	1.725	2.086	2.528	2.845
21	1.323	1.721	2.080	2.518	2.831
22	1.321	1.717	2.074	2.508	2.819
23	1.319	1.714	2.069	2.500	2.807
24	1.318	1.711	2.064	2.492	2.797
25	1.316	1.708	2.060	2.485	2.787
26	1.315	1.706	2.056	2.479	2.779
27	1.314	1.703	2.052	2.473	2.771
28	1.313	1.701	2.048	2.467	2.763
29	1.311	1.699	2.045	2.462	2.756
30	1.310	1.697	2.042	2.457	2.750
31	1.309	1.696	2.040	2.453	2.744
32	1.309	1.694	2.037	2.449	2.738
34	1.307	1.691	2.032	2.441	2.728
36	1.306	1.688	2.028	2.434	2.719
38	1.304	1.686	2.024	2.429	2.712
40	1.303	1.684	2.021	2.423	2.704
45	1.301	1.679	2.014	2.412	2.690
50	1.299	1.676	2.009	2.403	2.678
55	1.297	1.673	2.004	2.396	2.668
60	1.296	1.671	2.000	2.390	2.660
65	1.295	1.669	1.997	2.385	2.654
70	1.294	1.667	1.994	2.381	2.648
75	1.293	1.665	1.992	2.377	2.643
80	1.292	1.664	1.990	2.374	2.639
90	1.291	1.662	1.987	2.368	2.632
100	1.290	1.660	1.984	2.364	2.656
120	1.289	1.658	1.980	2.358	2.617
200	1.286	1.653	1.972	2.345	2.601
300	1.284	1.650	1.968	2.339	2.592
400	1.284	1.649	1.966	2.336	2.588
500	1.283	1.648	1.965	2.334	2.586
750	1.283	1.647	1.963	2.331	2.582
1000	1.282	1.646	1.962	2.330	2.581
2000	1.282	1.646	1.961	2.328	2.578
∞	1.282	1.645	1.960	2.326	2.576

Left Tail

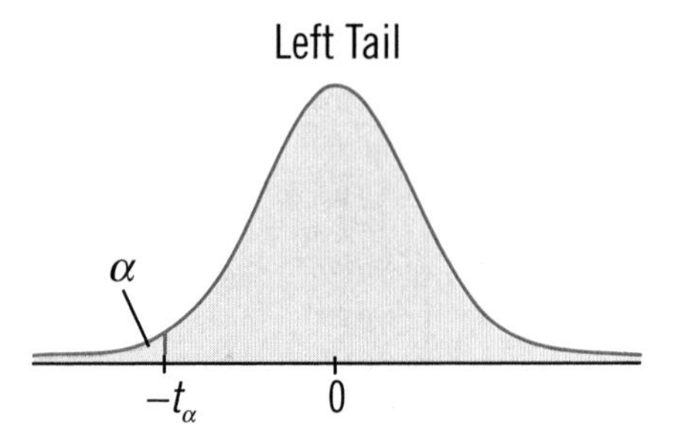

Right Tail

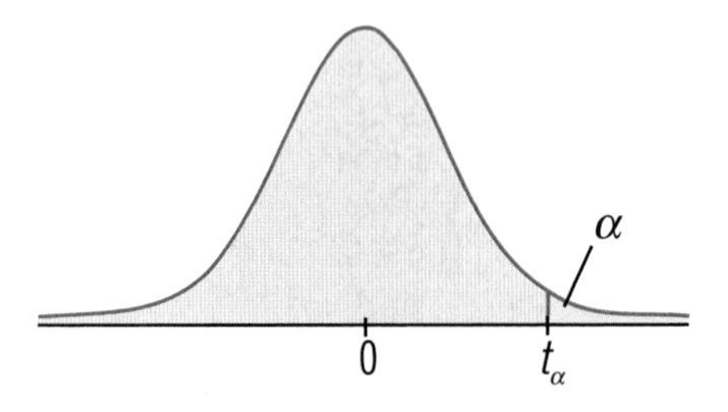

Two Tails

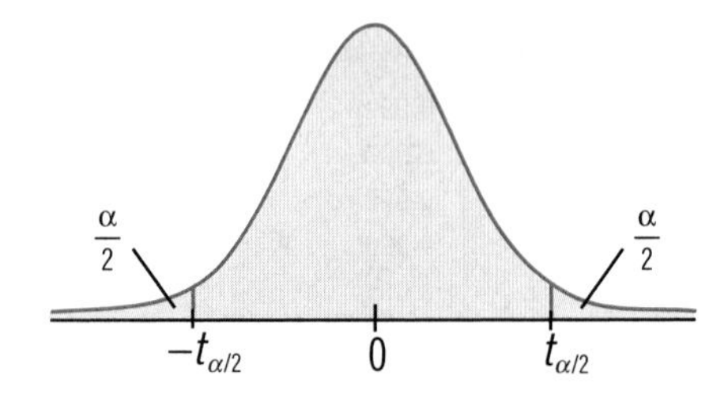

TABLE D
Binomial Probabilities

					p					
n	*x*	0.1	0.2	0.3	0.4	0.5	0.6	0.7	0.8	0.9
1	0	0.900	0.800	0.700	0.600	0.500	0.400	0.300	0.200	0.100
	1	0.100	0.200	0.300	0.400	0.500	0.600	0.700	0.800	0.900
2	0	0.810	0.640	0.490	0.360	0.250	0.160	0.090	0.040	0.010
	1	0.180	0.320	0.420	0.480	0.500	0.480	0.420	0.320	0.180
	2	0.010	0.040	0.090	0.160	0.250	0.360	0.490	0.640	0.810
3	0	0.729	0.512	0.343	0.216	0.125	0.064	0.027	0.008	0.001
	1	0.243	0.384	0.441	0.432	0.375	0.288	0.189	0.096	0.027
	2	0.027	0.096	0.189	0.288	0.375	0.432	0.441	0.384	0.243
	3	0.001	0.008	0.027	0.064	0.125	0.216	0.343	0.512	0.729
4	0	0.656	0.410	0.240	0.130	0.062	0.026	0.008	0.002	0.000
	1	0.292	0.410	0.412	0.346	0.250	0.154	0.076	0.026	0.004
	2	0.049	0.154	0.265	0.346	0.375	0.346	0.265	0.154	0.049
	3	0.004	0.026	0.076	0.154	0.250	0.346	0.412	0.410	0.292
	4	0.000	0.002	0.008	0.026	0.062	0.130	0.240	0.410	0.656
5	0	0.590	0.328	0.168	0.078	0.031	0.010	0.002	0.000	0.000
	1	0.328	0.410	0.360	0.259	0.156	0.077	0.028	0.006	0.000
	2	0.073	0.205	0.309	0.346	0.312	0.230	0.132	0.051	0.008
	3	0.008	0.051	0.132	0.230	0.312	0.346	0.309	0.205	0.073
	4	0.000	0.006	0.028	0.077	0.156	0.259	0.360	0.410	0.328
	5	0.000	0.000	0.002	0.010	0.031	0.078	0.168	0.328	0.590
6	0	0.531	0.262	0.118	0.047	0.016	0.004	0.001	0.000	0.000
	1	0.354	0.393	0.303	0.187	0.094	0.037	0.010	0.002	0.000
	2	0.098	0.246	0.324	0.311	0.234	0.138	0.060	0.015	0.001
	3	0.015	0.082	0.185	0.276	0.312	0.276	0.185	0.082	0.015
	4	0.001	0.015	0.060	0.138	0.234	0.311	0.324	0.246	0.098
	5	0.000	0.002	0.010	0.037	0.094	0.187	0.303	0.393	0.354
	6	0.000	0.000	0.001	0.004	0.016	0.047	0.118	0.262	0.531
7	0	0.478	0.210	0.082	0.028	0.008	0.002	0.000	0.000	0.000
	1	0.372	0.367	0.247	0.131	0.055	0.017	0.004	0.000	0.000
	2	0.124	0.275	0.318	0.261	0.164	0.077	0.025	0.004	0.000
	3	0.023	0.115	0.227	0.290	0.273	0.194	0.097	0.029	0.003
	4	0.003	0.029	0.097	0.194	0.273	0.290	0.227	0.115	0.023
	5	0.000	0.004	0.025	0.077	0.164	0.261	0.318	0.275	0.124
	6	0.000	0.000	0.004	0.017	0.055	0.131	0.247	0.367	0.372
	7	0.000	0.000	0.000	0.002	0.008	0.028	0.082	0.210	0.478
8	0	0.430	0.168	0.058	0.017	0.004	0.001	0.000	0.000	0.000
	1	0.383	0.336	0.198	0.090	0.031	0.008	0.001	0.000	0.000
	2	0.149	0.294	0.296	0.209	0.109	0.041	0.010	0.001	0.000
	3	0.033	0.147	0.254	0.279	0.219	0.124	0.047	0.009	0.000
	4	0.005	0.046	0.136	0.232	0.273	0.232	0.136	0.046	0.005
	5	0.000	0.009	0.047	0.124	0.219	0.279	0.254	0.147	0.033
	6	0.000	0.001	0.010	0.041	0.109	0.209	0.296	0.294	0.149
	7	0.000	0.000	0.001	0.008	0.031	0.090	0.198	0.336	0.383
	8	0.000	0.000	0.000	0.001	0.004	0.017	0.058	0.168	0.430
9	0	0.387	0.134	0.040	0.010	0.002	0.000	0.000	0.000	0.000
	1	0.387	0.302	0.156	0.060	0.018	0.004	0.000	0.000	0.000

TABLE D
Binomial Probabilities (continued)

n	x	p 0.1	0.2	0.3	0.4	0.5	0.6	0.7	0.8	0.9
9	2	0.172	0.302	0.267	0.161	0.070	0.021	0.004	0.000	0.000
	3	0.045	0.176	0.267	0.251	0.164	0.074	0.021	0.003	0.000
	4	0.007	0.066	0.172	0.251	0.246	0.167	0.074	0.017	0.001
	5	0.001	0.017	0.074	0.167	0.246	0.251	0.172	0.066	0.007
	6	0.000	0.003	0.021	0.074	0.164	0.251	0.267	0.176	0.045
	7	0.000	0.000	0.004	0.021	0.070	0.161	0.267	0.302	0.172
	8	0.000	0.000	0.000	0.004	0.018	0.060	0.156	0.302	0.387
	9	0.000	0.000	0.000	0.000	0.002	0.010	0.040	0.134	0.387
10	0	0.349	0.107	0.028	0.006	0.001	0.000	0.000	0.000	0.000
	1	0.387	0.268	0.121	0.040	0.010	0.002	0.000	0.000	0.000
	2	0.194	0.302	0.233	0.121	0.044	0.011	0.001	0.000	0.000
	3	0.057	0.201	0.267	0.215	0.117	0.042	0.009	0.001	0.000
	4	0.011	0.088	0.200	0.251	0.205	0.111	0.037	0.006	0.000
	5	0.001	0.026	0.103	0.201	0.246	0.201	0.103	0.026	0.001
	6	0.000	0.006	0.037	0.111	0.205	0.251	0.200	0.088	0.011
	7	0.000	0.001	0.009	0.042	0.117	0.215	0.267	0.201	0.057
	8	0.000	0.000	0.001	0.011	0.044	0.121	0.233	0.302	0.194
	9	0.000	0.000	0.000	0.002	0.010	0.040	0.121	0.268	0.387
	10	0.000	0.000	0.000	0.000	0.001	0.006	0.028	0.107	0.349
11	0	0.314	0.086	0.020	0.004	0.000	0.000	0.000	0.000	0.000
	1	0.384	0.236	0.093	0.027	0.005	0.001	0.000	0.000	0.000
	2	0.213	0.295	0.200	0.089	0.027	0.005	0.001	0.000	0.000
	3	0.071	0.221	0.257	0.177	0.081	0.023	0.004	0.000	0.000
	4	0.016	0.111	0.220	0.236	0.161	0.070	0.017	0.002	0.000
	5	0.002	0.039	0.132	0.221	0.226	0.147	0.057	0.010	0.000
	6	0.000	0.010	0.057	0.147	0.226	0.221	0.132	0.039	0.002
	7	0.000	0.002	0.017	0.070	0.161	0.236	0.220	0.111	0.016
	8	0.000	0.000	0.004	0.023	0.081	0.177	0.257	0.221	0.071
	9	0.000	0.000	0.001	0.005	0.027	0.089	0.200	0.295	0.213
	10	0.000	0.000	0.000	0.001	0.005	0.027	0.093	0.236	0.384
	11	0.000	0.000	0.000	0.000	0.000	0.004	0.020	0.086	0.314
12	0	0.282	0.069	0.014	0.002	0.000	0.000	0.000	0.000	0.000
	1	0.377	0.206	0.071	0.017	0.003	0.000	0.000	0.000	0.000
	2	0.230	0.283	0.168	0.064	0.016	0.002	0.000	0.000	0.000
	3	0.085	0.236	0.240	0.142	0.054	0.012	0.001	0.000	0.000
	4	0.021	0.133	0.231	0.213	0.121	0.042	0.008	0.001	0.000
	5	0.004	0.053	0.158	0.227	0.193	0.101	0.029	0.003	0.000
	6	0.000	0.016	0.079	0.177	0.226	0.177	0.079	0.016	0.000
	7	0.000	0.003	0.029	0.101	0.193	0.227	0.158	0.053	0.004
	8	0.000	0.001	0.008	0.042	0.121	0.213	0.231	0.133	0.021
	9	0.000	0.000	0.001	0.012	0.054	0.142	0.240	0.236	0.085
	10	0.000	0.000	0.000	0.002	0.016	0.064	0.168	0.283	0.230
	11	0.000	0.000	0.000	0.000	0.003	0.017	0.071	0.206	0.377
	12	0.000	0.000	0.000	0.000	0.000	0.002	0.014	0.069	0.282
13	0	0.254	0.055	0.001	0.001	0.000	0.000	0.000	0.000	0.000
	1	0.367	0.179	0.054	0.011	0.002	0.000	0.000	0.000	0.000

Table D

Binomial Probabilities (continued)

n	x	p = 0.1	0.2	0.3	0.4	0.5	0.6	0.7	0.8	0.9
13	2	0.245	0.268	0.139	0.045	0.010	0.001	0.000	0.000	0.000
	3	0.100	0.246	0.218	0.111	0.035	0.006	0.001	0.000	0.000
	4	0.028	0.154	0.234	0.184	0.087	0.024	0.003	0.000	0.000
	5	0.006	0.069	0.180	0.221	0.157	0.066	0.014	0.001	0.000
	6	0.001	0.023	0.103	0.197	0.209	0.131	0.044	0.006	0.000
	7	0.000	0.006	0.044	0.131	0.209	0.197	0.103	0.023	0.001
	8	0.000	0.001	0.014	0.066	0.157	0.221	0.180	0.069	0.006
	9	0.000	0.000	0.003	0.024	0.087	0.184	0.234	0.154	0.028
	10	0.000	0.000	0.001	0.006	0.035	0.111	0.218	0.246	0.100
	11	0.000	0.000	0.000	0.001	0.010	0.045	0.139	0.268	0.245
	12	0.000	0.000	0.000	0.000	0.002	0.011	0.054	0.179	0.367
	13	0.000	0.000	0.000	0.000	0.000	0.001	0.010	0.055	0.254
14	0	0.229	0.044	0.007	0.001	0.000	0.000	0.000	0.000	0.000
	1	0.356	0.154	0.041	0.007	0.001	0.000	0.000	0.000	0.000
	2	0.257	0.250	0.113	0.032	0.006	0.001	0.000	0.000	0.000
	3	0.114	0.250	0.194	0.085	0.022	0.003	0.000	0.000	0.000
	4	0.035	0.172	0.229	0.155	0.061	0.014	0.001	0.000	0.000
	5	0.008	0.086	0.196	0.207	0.122	0.041	0.007	0.000	0.000
	6	0.001	0.032	0.126	0.207	0.183	0.092	0.023	0.002	0.000
	7	0.000	0.009	0.062	0.157	0.209	0.157	0.062	0.009	0.000
	8	0.000	0.002	0.023	0.092	0.183	0.207	0.126	0.032	0.001
	9	0.000	0.000	0.007	0.041	0.122	0.207	0.196	0.086	0.008
	10	0.000	0.000	0.001	0.014	0.061	0.155	0.229	0.172	0.035
	11	0.000	0.000	0.000	0.003	0.022	0.085	0.194	0.250	0.114
	12	0.000	0.000	0.000	0.001	0.006	0.032	0.113	0.250	0.257
	13	0.000	0.000	0.000	0.000	0.001	0.007	0.041	0.154	0.356
	14	0.000	0.000	0.000	0.000	0.000	0.001	0.007	0.044	0.229
15	0	0.206	0.035	0.005	0.000	0.000	0.000	0.000	0.000	0.000
	1	0.343	0.132	0.031	0.005	0.000	0.000	0.000	0.000	0.000
	2	0.267	0.231	0.092	0.022	0.003	0.000	0.000	0.000	0.000
	3	0.129	0.250	0.170	0.063	0.014	0.002	0.000	0.000	0.000
	4	0.043	0.188	0.219	0.127	0.042	0.007	0.001	0.000	0.000
	5	0.010	0.103	0.206	0.186	0.092	0.024	0.003	0.000	0.000
	6	0.002	0.043	0.147	0.207	0.153	0.061	0.012	0.001	0.000
	7	0.000	0.014	0.081	0.177	0.196	0.118	0.035	0.003	0.000
	8	0.000	0.003	0.035	0.118	0.196	0.177	0.081	0.014	0.000
	9	0.000	0.001	0.017	0.061	0.153	0.207	0.147	0.043	0.002
	10	0.000	0.000	0.003	0.024	0.092	0.186	0.206	0.103	0.010
	11	0.000	0.000	0.001	0.007	0.042	0.127	0.219	0.188	0.043
	12	0.000	0.000	0.000	0.002	0.014	0.063	0.170	0.250	0.129
	13	0.000	0.000	0.000	0.000	0.003	0.022	0.092	0.231	0.267
	14	0.000	0.000	0.000	0.000	0.000	0.005	0.031	0.132	0.343
	15	0.000	0.000	0.000	0.000	0.000	0.000	0.005	0.035	0.206
16	0	0.185	0.028	0.003	0.000	0.000	0.000	0.000	0.000	0.000
	1	0.329	0.113	0.023	0.003	0.000	0.000	0.000	0.000	0.000
	2	0.275	0.211	0.073	0.015	0.002	0.000	0.000	0.000	0.000

TABLE D
Binomial Probabilities (continued)

n	x	0.1	0.2	0.3	0.4	0.5	0.6	0.7	0.8	0.9
16	3	0.142	0.246	0.146	0.047	0.009	0.001	0.000	0.000	0.000
	4	0.051	0.200	0.204	0.101	0.028	0.004	0.000	0.000	0.000
	5	0.014	0.120	0.210	0.162	0.067	0.014	0.001	0.000	0.000
	6	0.003	0.055	0.165	0.198	0.122	0.039	0.006	0.000	0.000
	7	0.000	0.020	0.101	0.189	0.175	0.084	0.019	0.001	0.000
	8	0.000	0.006	0.049	0.142	0.196	0.142	0.049	0.006	0.000
	9	0.000	0.001	0.019	0.084	0.175	0.189	0.101	0.020	0.000
	10	0.000	0.000	0.006	0.039	0.122	0.198	0.165	0.055	0.003
	11	0.000	0.000	0.001	0.014	0.067	0.162	0.210	0.120	0.014
	12	0.000	0.000	0.000	0.004	0.028	0.101	0.204	0.200	0.051
	13	0.000	0.000	0.000	0.001	0.009	0.047	0.146	0.246	0.142
	14	0.000	0.000	0.000	0.000	0.002	0.015	0.073	0.211	0.275
	15	0.000	0.000	0.000	0.000	0.000	0.003	0.023	0.113	0.329
	16	0.000	0.000	0.000	0.000	0.000	0.000	0.003	0.028	0.185
17	0	0.167	0.023	0.002	0.000	0.000	0.000	0.000	0.000	0.000
	1	0.315	0.096	0.017	0.002	0.000	0.000	0.000	0.000	0.000
	2	0.280	0.191	0.058	0.010	0.001	0.000	0.000	0.000	0.000
	3	0.156	0.239	0.125	0.034	0.005	0.000	0.000	0.000	0.000
	4	0.060	0.209	0.187	0.080	0.018	0.002	0.000	0.000	0.000
	5	0.017	0.136	0.208	0.138	0.047	0.008	0.001	0.000	0.000
	6	0.004	0.068	0.178	0.184	0.094	0.024	0.003	0.000	0.000
	7	0.001	0.027	0.120	0.193	0.148	0.057	0.009	0.000	0.000
	8	0.000	0.008	0.064	0.161	0.185	0.107	0.028	0.002	0.000
	9	0.000	0.002	0.028	0.107	0.185	0.161	0.064	0.008	0.000
	10	0.000	0.000	0.009	0.057	0.148	0.193	0.120	0.027	0.001
	11	0.000	0.000	0.003	0.024	0.094	0.184	0.178	0.068	0.004
	12	0.000	0.000	0.001	0.008	0.047	0.138	0.208	0.136	0.017
	13	0.000	0.000	0.000	0.002	0.018	0.080	0.187	0.209	0.060
	14	0.000	0.000	0.000	0.000	0.005	0.034	0.125	0.239	0.156
	15	0.000	0.000	0.000	0.000	0.001	0.010	0.058	0.191	0.280
	16	0.000	0.000	0.000	0.000	0.000	0.002	0.017	0.096	0.315
	17	0.000	0.000	0.000	0.000	0.000	0.000	0.002	0.023	0.167
18	0	0.150	0.018	0.002	0.000	0.000	0.000	0.000	0.000	0.000
	1	0.300	0.081	0.013	0.001	0.000	0.000	0.000	0.000	0.000
	2	0.284	0.172	0.046	0.007	0.001	0.000	0.000	0.000	0.000
	3	0.168	0.230	0.105	0.025	0.003	0.000	0.000	0.000	0.000
	4	0.070	0.215	0.168	0.061	0.012	0.001	0.000	0.000	0.000
	5	0.022	0.151	0.202	0.115	0.033	0.004	0.000	0.000	0.000
	6	0.005	0.082	0.187	0.166	0.071	0.015	0.001	0.000	0.000
	7	0.001	0.035	0.138	0.189	0.121	0.037	0.005	0.000	0.000
	8	0.000	0.012	0.081	0.173	0.167	0.077	0.015	0.001	0.000
	9	0.000	0.003	0.039	0.128	0.185	0.128	0.039	0.003	0.000
	10	0.000	0.001	0.015	0.077	0.167	0.173	0.081	0.012	0.000
	11	0.000	0.000	0.005	0.037	0.121	0.189	0.138	0.035	0.001
	12	0.000	0.000	0.001	0.015	0.071	0.166	0.187	0.082	0.005
	13	0.000	0.000	0.000	0.004	0.033	0.115	0.202	0.151	0.022

Table D
Binomial Probabilities (continued)

n	x	p = 0.1	0.2	0.3	0.4	0.5	0.6	0.7	0.8	0.9
18	14	0.000	0.000	0.000	0.001	0.012	0.061	0.168	0.215	0.070
	15	0.000	0.000	0.000	0.000	0.003	0.025	0.105	0.230	0.168
	16	0.000	0.000	0.000	0.000	0.001	0.007	0.046	0.172	0.284
	17	0.000	0.000	0.000	0.000	0.000	0.001	0.013	0.081	0.300
	18	0.000	0.000	0.000	0.000	0.000	0.000	0.002	0.018	0.150
19	0	0.135	0.014	0.001	0.000	0.000	0.000	0.000	0.000	0.000
	1	0.285	0.068	0.009	0.001	0.000	0.000	0.000	0.000	0.000
	2	0.285	0.154	0.036	0.005	0.000	0.000	0.000	0.000	0.000
	3	0.180	0.218	0.087	0.017	0.002	0.000	0.000	0.000	0.000
	4	0.080	0.218	0.149	0.047	0.007	0.001	0.000	0.000	0.000
	5	0.027	0.164	0.192	0.093	0.022	0.002	0.000	0.000	0.000
	6	0.007	0.095	0.192	0.145	0.052	0.008	0.001	0.000	0.000
	7	0.001	0.044	0.153	0.180	0.096	0.024	0.002	0.000	0.000
	8	0.000	0.017	0.098	0.180	0.144	0.053	0.008	0.000	0.000
	9	0.000	0.005	0.051	0.146	0.176	0.098	0.022	0.001	0.000
	10	0.000	0.001	0.022	0.098	0.176	0.146	0.051	0.005	0.000
	11	0.000	0.000	0.008	0.053	0.144	0.180	0.098	0.017	0.000
	12	0.000	0.000	0.002	0.024	0.096	0.180	0.153	0.044	0.001
	13	0.000	0.000	0.001	0.008	0.052	0.145	0.192	0.095	0.007
	14	0.000	0.000	0.000	0.002	0.022	0.093	0.192	0.164	0.027
	15	0.000	0.000	0.000	0.001	0.007	0.047	0.149	0.218	0.080
	16	0.000	0.000	0.000	0.000	0.002	0.017	0.087	0.218	0.180
	17	0.000	0.000	0.000	0.000	0.000	0.005	0.036	0.154	0.285
	18	0.000	0.000	0.000	0.000	0.000	0.001	0.009	0.068	0.285
	19	0.000	0.000	0.000	0.000	0.000	0.000	0.001	0.014	0.135
20	0	0.122	0.012	0.001	0.000	0.000	0.000	0.000	0.000	0.000
	1	0.270	0.058	0.007	0.000	0.000	0.000	0.000	0.000	0.000
	2	0.285	0.137	0.028	0.003	0.000	0.000	0.000	0.000	0.000
	3	0.190	0.205	0.072	0.012	0.001	0.000	0.000	0.000	0.000
	4	0.090	0.218	0.130	0.035	0.005	0.000	0.000	0.000	0.000
	5	0.032	0.175	0.179	0.075	0.015	0.001	0.000	0.000	0.000
	6	0.009	0.109	0.192	0.124	0.037	0.005	0.000	0.000	0.000
	7	0.002	0.055	0.164	0.166	0.074	0.015	0.001	0.000	0.000
	8	0.000	0.022	0.114	0.180	0.120	0.035	0.004	0.000	0.000
	9	0.000	0.007	0.065	0.160	0.160	0.071	0.012	0.000	0.000
	10	0.000	0.002	0.031	0.117	0.176	0.117	0.031	0.002	0.000
	11	0.000	0.000	0.012	0.071	0.160	0.160	0.065	0.007	0.000
	12	0.000	0.000	0.004	0.035	0.120	0.180	0.114	0.022	0.000
	13	0.000	0.000	0.001	0.015	0.074	0.166	0.164	0.055	0.002
	14	0.000	0.000	0.000	0.005	0.037	0.124	0.192	0.109	0.009
	15	0.000	0.000	0.000	0.001	0.015	0.075	0.179	0.175	0.032
	16	0.000	0.000	0.000	0.000	0.005	0.035	0.130	0.218	0.090
	17	0.000	0.000	0.000	0.000	0.001	0.012	0.072	0.205	0.190
	18	0.000	0.000	0.000	0.000	0.000	0.003	0.028	0.137	0.285
	19	0.000	0.000	0.000	0.000	0.000	0.000	0.007	0.058	0.270
	20	0.000	0.000	0.000	0.000	0.000	0.000	0.001	0.012	0.122

TABLE E

Cumulative Binomial Probabilities

		p								
n	x	0.1	0.2	0.3	0.4	0.5	0.6	0.7	0.8	0.9
1	0	0.900	0.800	0.700	0.600	0.500	0.400	0.300	0.200	0.100
1	1	1.000	1.000	1.000	1.000	1.000	1.000	1.000	1.000	1.000
2	0	0.810	0.640	0.490	0.360	0.250	0.160	0.090	0.040	0.010
2	1	0.990	0.960	0.910	0.840	0.750	0.640	0.510	0.360	0.190
2	2	1.000	1.000	1.000	1.000	1.000	1.000	1.000	1.000	1.000
3	0	0.729	0.512	0.343	0.216	0.125	0.064	0.027	0.008	0.001
3	1	0.972	0.896	0.784	0.648	0.500	0.352	0.216	0.104	0.028
3	2	0.999	0.992	0.973	0.936	0.875	0.784	0.657	0.488	0.271
3	3	1.000	1.000	1.000	1.000	1.000	1.000	1.000	1.000	1.000
4	0	0.656	0.410	0.240	0.130	0.063	0.026	0.008	0.002	0.000
4	1	0.948	0.819	0.652	0.475	0.313	0.179	0.084	0.027	0.004
4	2	0.996	0.973	0.916	0.821	0.688	0.525	0.348	0.181	0.052
4	3	1.000	0.998	0.992	0.974	0.938	0.870	0.760	0.590	0.344
4	4	1.000	1.000	1.000	1.000	1.000	1.000	1.000	1.000	1.000
5	0	0.591	0.328	0.168	0.078	0.031	0.010	0.002	0.000	0.000
5	1	0.919	0.737	0.528	0.337	0.188	0.087	0.031	0.007	0.000
5	2	0.991	0.942	0.837	0.683	0.500	0.317	0.163	0.058	0.009
5	3	1.000	0.993	0.969	0.913	0.813	0.663	0.472	0.263	0.081
5	4	1.000	1.000	0.998	0.990	0.969	0.922	0.832	0.672	0.410
5	5	1.000	1.000	1.000	1.000	1.000	1.000	1.000	1.000	1.000
6	0	0.531	0.262	0.118	0.047	0.016	0.004	0.001	0.000	0.000
6	1	0.886	0.655	0.420	0.233	0.109	0.041	0.011	0.002	0.000
6	2	0.984	0.901	0.744	0.544	0.344	0.179	0.070	0.017	0.001
6	3	0.999	0.983	0.930	0.821	0.656	0.456	0.256	0.099	0.016
6	4	1.000	0.998	0.989	0.959	0.891	0.767	0.580	0.345	0.114
6	5	1.000	1.000	0.999	0.996	0.984	0.953	0.882	0.738	0.469
6	6	1.000	1.000	1.000	1.000	1.000	1.000	1.000	1.000	1.000
7	0	0.478	0.210	0.082	0.028	0.008	0.002	0.000	0.000	0.000
7	1	0.850	0.577	0.329	0.159	0.063	0.019	0.004	0.000	0.000
7	2	0.974	0.852	0.647	0.420	0.227	0.096	0.029	0.005	0.000
7	3	0.997	0.967	0.874	0.710	0.500	0.290	0.126	0.033	0.003
7	4	1.000	0.995	0.971	0.904	0.773	0.580	0.353	0.148	0.026
7	5	1.000	1.000	0.996	0.981	0.938	0.841	0.671	0.423	0.150
7	6	1.000	1.000	1.000	0.998	0.992	0.972	0.918	0.790	0.522
7	7	1.000	1.000	1.000	1.000	1.000	1.000	1.000	1.000	1.000
8	0	0.430	0.168	0.058	0.017	0.004	0.001	0.000	0.000	0.000
8	1	0.813	0.503	0.255	0.106	0.035	0.009	0.001	0.000	0.000
8	2	0.962	0.797	0.552	0.315	0.145	0.050	0.011	0.001	0.000
8	3	0.995	0.944	0.806	0.594	0.363	0.174	0.058	0.010	0.000
8	4	1.000	0.990	0.942	0.826	0.637	0.406	0.194	0.056	0.005
8	5	1.000	0.999	0.989	0.950	0.855	0.685	0.448	0.203	0.038
8	6	1.000	1.000	0.999	0.991	0.965	0.894	0.745	0.497	0.187
8	7	1.000	1.000	1.000	0.999	0.996	0.983	0.942	0.832	0.570
8	8	1.000	1.000	1.000	1.000	1.000	1.000	1.000	1.000	1.000
9	0	0.387	0.134	0.040	0.010	0.002	0.000	0.000	0.000	0.000
9	1	0.775	0.436	0.196	0.071	0.020	0.004	0.000	0.000	0.000

TABLE E
Cumulative Binomial Probabilities (continued)

n	x	p = 0.1	0.2	0.3	0.4	0.5	0.6	0.7	0.8	0.9
9	2	0.947	0.738	0.463	0.232	0.090	0.025	0.004	0.000	0.000
9	3	0.992	0.914	0.730	0.483	0.254	0.099	0.025	0.003	0.000
9	4	0.999	0.980	0.901	0.733	0.500	0.267	0.099	0.020	0.001
9	5	1.000	0.997	0.975	0.901	0.746	0.517	0.270	0.086	0.008
9	6	1.000	1.000	0.996	0.975	0.910	0.768	0.537	0.262	0.053
9	7	1.000	1.000	1.000	0.996	0.980	0.929	0.804	0.564	0.225
9	8	1.000	1.000	1.000	1.000	0.998	0.990	0.960	0.866	0.613
9	9	1.000	1.000	1.000	1.000	1.000	1.000	1.000	1.000	1.000
10	0	0.349	0.107	0.028	0.006	0.001	0.000	0.000	0.000	0.000
10	1	0.736	0.376	0.149	0.046	0.011	0.002	0.000	0.000	0.000
10	2	0.930	0.678	0.383	0.167	0.055	0.012	0.002	0.000	0.000
10	3	0.987	0.879	0.650	0.382	0.172	0.055	0.011	0.001	0.000
10	4	0.998	0.967	0.850	0.633	0.377	0.166	0.047	0.006	0.000
10	5	1.000	0.994	0.953	0.834	0.623	0.367	0.150	0.033	0.002
10	6	1.000	0.999	0.989	0.945	0.828	0.618	0.350	0.121	0.013
10	7	1.000	1.000	0.998	0.988	0.945	0.833	0.617	0.322	0.070
10	8	1.000	1.000	1.000	0.998	0.989	0.954	0.851	0.624	0.264
10	9	1.000	1.000	1.000	1.000	0.999	0.994	0.972	0.893	0.651
10	10	1.000	1.000	1.000	1.000	1.000	1.000	1.000	1.000	1.000
11	0	0.314	0.086	0.020	0.004	0.000	0.000	0.000	0.000	0.000
11	1	0.697	0.322	0.113	0.030	0.006	0.001	0.000	0.000	0.000
11	2	0.910	0.617	0.313	0.119	0.033	0.006	0.001	0.000	0.000
11	3	0.981	0.839	0.570	0.296	0.113	0.029	0.004	0.000	0.000
11	4	0.997	0.950	0.790	0.533	0.274	0.099	0.022	0.002	0.000
11	5	1.000	0.988	0.922	0.754	0.500	0.247	0.078	0.012	0.000
11	6	1.000	0.998	0.978	0.901	0.726	0.467	0.210	0.050	0.003
11	7	1.000	1.000	0.996	0.971	0.887	0.704	0.430	0.161	0.019
11	8	1.000	1.000	0.999	0.994	0.967	0.881	0.687	0.383	0.090
11	9	1.000	1.000	1.000	0.999	0.994	0.970	0.887	0.678	0.303
11	10	1.000	1.000	1.000	1.000	1.000	0.996	0.980	0.914	0.686
11	11	1.000	1.000	1.000	1.000	1.000	1.000	1.000	1.000	1.000
12	0	0.282	0.069	0.014	0.002	0.000	0.000	0.000	0.000	0.000
12	1	0.659	0.275	0.085	0.020	0.003	0.000	0.000	0.000	0.000
12	2	0.889	0.558	0.253	0.083	0.019	0.003	0.000	0.000	0.000
12	3	0.974	0.795	0.493	0.225	0.073	0.015	0.002	0.000	0.000
12	4	0.996	0.927	0.724	0.438	0.194	0.057	0.009	0.001	0.000
12	5	0.999	0.981	0.882	0.665	0.387	0.158	0.039	0.004	0.000
12	6	1.000	0.996	0.961	0.842	0.613	0.335	0.118	0.019	0.001
12	7	1.000	0.999	0.991	0.943	0.806	0.562	0.276	0.073	0.004
12	8	1.000	1.000	0.998	0.985	0.927	0.775	0.507	0.205	0.026
12	9	1.000	1.000	1.000	0.997	0.981	0.917	0.747	0.442	0.111
12	10	1.000	1.000	1.000	1.000	0.997	0.980	0.915	0.725	0.341
12	11	1.000	1.000	1.000	1.000	1.000	0.998	0.986	0.931	0.718
12	12	1.000	1.000	1.000	1.000	1.000	1.000	1.000	1.000	1.000
13	0	0.254	0.055	0.010	0.001	0.000	0.000	0.000	0.000	0.000
13	1	0.621	0.234	0.064	0.013	0.002	0.000	0.000	0.000	0.000

Table E
Cumulative Binomial Probabilities (continued)

		p								
n	x	0.1	0.2	0.3	0.4	0.5	0.6	0.7	0.8	0.9
13	2	0.866	0.502	0.202	0.058	0.011	0.001	0.000	0.000	0.000
13	3	0.966	0.747	0.421	0.169	0.046	0.008	0.001	0.000	0.000
13	4	0.994	0.901	0.654	0.353	0.133	0.032	0.004	0.000	0.000
13	5	0.999	0.970	0.835	0.574	0.291	0.098	0.018	0.001	0.000
13	6	1.000	0.993	0.938	0.771	0.500	0.229	0.062	0.007	0.000
13	7	1.000	0.999	0.982	0.902	0.709	0.426	0.165	0.030	0.001
13	8	1.000	1.000	0.996	0.968	0.867	0.647	0.346	0.099	0.006
13	9	1.000	1.000	0.999	0.992	0.954	0.831	0.579	0.253	0.034
13	10	1.000	1.000	1.000	0.999	0.989	0.942	0.798	0.498	0.134
13	11	1.000	1.000	1.000	1.000	0.998	0.987	0.936	0.766	0.379
13	12	1.000	1.000	1.000	1.000	1.000	0.999	0.990	0.945	0.746
13	13	1.000	1.000	1.000	1.000	1.000	1.000	1.000	1.000	1.000
14	0	0.229	0.044	0.007	0.001	0.000	0.000	0.000	0.000	0.000
14	1	0.585	0.198	0.047	0.008	0.001	0.000	0.000	0.000	0.000
14	2	0.842	0.448	0.161	0.040	0.006	0.001	0.000	0.000	0.000
14	3	0.956	0.698	0.355	0.124	0.029	0.004	0.000	0.000	0.000
14	4	0.991	0.870	0.584	0.279	0.090	0.018	0.002	0.000	0.000
14	5	0.999	0.956	0.781	0.486	0.212	0.058	0.008	0.000	0.000
14	6	1.000	0.988	0.907	0.692	0.395	0.150	0.031	0.002	0.000
14	7	1.000	0.998	0.969	0.850	0.605	0.308	0.093	0.012	0.000
14	8	1.000	1.000	0.992	0.942	0.788	0.514	0.219	0.044	0.001
14	9	1.000	1.000	0.998	0.983	0.910	0.721	0.416	0.130	0.009
14	10	1.000	1.000	1.000	0.996	0.971	0.876	0.645	0.302	0.044
14	11	1.000	1.000	1.000	0.999	0.994	0.960	0.839	0.552	0.158
14	12	1.000	1.000	1.000	1.000	0.999	0.992	0.953	0.802	0.415
14	13	1.000	1.000	1.000	1.000	1.000	0.999	0.993	0.956	0.771
14	14	1.000	1.000	1.000	1.000	1.000	1.000	1.000	1.000	1.000
15	0	0.206	0.035	0.005	0.000	0.000	0.000	0.000	0.000	0.000
15	1	0.549	0.167	0.035	0.005	0.000	0.000	0.000	0.000	0.000
15	2	0.816	0.398	0.127	0.027	0.004	0.000	0.000	0.000	0.000
15	3	0.944	0.648	0.297	0.091	0.018	0.002	0.000	0.000	0.000
15	4	0.987	0.836	0.516	0.217	0.059	0.009	0.001	0.000	0.000
15	5	0.998	0.939	0.722	0.403	0.151	0.034	0.004	0.000	0.000
15	6	1.000	0.982	0.869	0.610	0.304	0.095	0.015	0.001	0.000
15	7	1.000	0.996	0.950	0.787	0.500	0.213	0.050	0.004	0.000
15	8	1.000	0.999	0.985	0.905	0.696	0.390	0.131	0.018	0.000
15	9	1.000	1.000	0.996	0.966	0.849	0.597	0.278	0.061	0.002
15	10	1.000	1.000	0.999	0.991	0.941	0.783	0.485	0.164	0.013
15	11	1.000	1.000	1.000	0.998	0.982	0.910	0.703	0.352	0.056
15	12	1.000	1.000	1.000	1.000	0.996	0.973	0.873	0.602	0.184
15	13	1.000	1.000	1.000	1.000	1.000	0.995	0.965	0.833	0.451
15	14	1.000	1.000	1.000	1.000	1.000	1.000	0.995	0.965	0.794
15	15	1.000	1.000	1.000	1.000	1.000	1.000	1.000	1.000	1.000
16	0	0.185	0.028	0.003	0.000	0.000	0.000	0.000	0.000	0.000
16	1	0.515	0.141	0.026	0.003	0.000	0.000	0.000	0.000	0.000
16	2	0.789	0.352	0.099	0.018	0.002	0.000	0.000	0.000	0.000

TABLE E
Cumulative Binomial Probabilities (continued)

		p								
n	*X*	0.1	0.2	0.3	0.4	0.5	0.6	0.7	0.8	0.9
16	3	0.932	0.598	0.246	0.065	0.011	0.001	0.000	0.000	0.000
16	4	0.983	0.798	0.450	0.167	0.038	0.005	0.000	0.000	0.000
16	5	0.997	0.918	0.660	0.329	0.105	0.019	0.002	0.000	0.000
16	6	1.000	0.973	0.825	0.527	0.227	0.058	0.007	0.000	0.000
16	7	1.000	0.993	0.926	0.716	0.402	0.142	0.026	0.001	0.000
16	8	1.000	0.999	0.974	0.858	0.598	0.284	0.074	0.007	0.000
16	9	1.000	1.000	0.993	0.942	0.773	0.473	0.175	0.027	0.001
16	10	1.000	1.000	0.998	0.981	0.895	0.671	0.340	0.082	0.003
16	11	1.000	1.000	1.000	0.995	0.962	0.833	0.550	0.202	0.017
16	12	1.000	1.000	1.000	0.999	0.989	0.935	0.754	0.402	0.068
16	13	1.000	1.000	1.000	1.000	0.998	0.982	0.901	0.648	0.211
16	14	1.000	1.000	1.000	1.000	1.000	0.997	0.974	0.859	0.485
16	15	1.000	1.000	1.000	1.000	1.000	1.000	0.997	0.972	0.815
16	16	1.000	1.000	1.000	1.000	1.000	1.000	1.000	1.000	1.000
17	0	0.167	0.023	0.002	0.000	0.000	0.000	0.000	0.000	0.000
17	1	0.482	0.118	0.019	0.002	0.000	0.000	0.000	0.000	0.000
17	2	0.762	0.310	0.077	0.012	0.001	0.000	0.000	0.000	0.000
17	3	0.917	0.549	0.202	0.046	0.006	0.000	0.000	0.000	0.000
17	4	0.978	0.758	0.389	0.126	0.025	0.003	0.000	0.000	0.000
17	5	0.995	0.894	0.597	0.264	0.072	0.011	0.001	0.000	0.000
17	6	0.999	0.962	0.775	0.448	0.166	0.035	0.003	0.000	0.000
17	7	1.000	0.989	0.895	0.641	0.315	0.092	0.013	0.001	0.000
17	8	1.000	0.997	0.960	0.801	0.500	0.199	0.040	0.003	0.000
17	9	1.000	1.000	0.987	0.908	0.685	0.360	0.105	0.011	0.000
17	10	1.000	1.000	0.997	0.965	0.834	0.552	0.225	0.038	0.001
17	11	1.000	1.000	0.999	0.989	0.928	0.736	0.403	0.106	0.005
17	12	1.000	1.000	1.000	0.997	0.975	0.874	0.611	0.242	0.022
17	13	1.000	1.000	1.000	1.000	0.994	0.954	0.798	0.451	0.083
17	14	1.000	1.000	1.000	1.000	0.999	0.988	0.923	0.690	0.238
17	15	1.000	1.000	1.000	1.000	1.000	0.998	0.981	0.882	0.518
17	16	1.000	1.000	1.000	1.000	1.000	1.000	0.998	0.977	0.833
17	17	1.000	1.000	1.000	1.000	1.000	1.000	1.000	1.000	1.000
18	0	0.150	0.018	0.002	0.000	0.000	0.000	0.000	0.000	0.000
18	1	0.450	0.099	0.014	0.001	0.000	0.000	0.000	0.000	0.000
18	2	0.734	0.271	0.060	0.008	0.001	0.000	0.000	0.000	0.000
18	3	0.902	0.501	0.165	0.033	0.004	0.000	0.000	0.000	0.000
18	4	0.972	0.716	0.333	0.094	0.015	0.001	0.000	0.000	0.000
18	5	0.994	0.867	0.534	0.209	0.048	0.006	0.000	0.000	0.000
18	6	0.999	0.949	0.722	0.374	0.119	0.020	0.001	0.000	0.000
18	7	1.000	0.984	0.859	0.563	0.240	0.058	0.006	0.000	0.000
18	8	1.000	0.996	0.940	0.737	0.407	0.135	0.021	0.001	0.000
18	9	1.000	0.999	0.979	0.865	0.593	0.263	0.060	0.004	0.000
18	10	1.000	1.000	0.994	0.942	0.760	0.437	0.141	0.016	0.000
18	11	1.000	1.000	0.999	0.980	0.881	0.626	0.278	0.051	0.001
18	12	1.000	1.000	1.000	0.994	0.952	0.791	0.466	0.133	0.006
18	13	1.000	1.000	1.000	0.999	0.985	0.906	0.667	0.284	0.028

TABLE E
Cumulative Binomial Probabilities (continued)

p

n	x	0.1	0.2	0.3	0.4	0.5	0.6	0.7	0.8	0.9
18	14	1.000	1.000	1.000	1.000	0.996	0.967	0.835	0.499	0.098
18	15	1.000	1.000	1.000	1.000	0.999	0.992	0.940	0.729	0.266
18	16	1.000	1.000	1.000	1.000	1.000	0.999	0.986	0.901	0.550
18	17	1.000	1.000	1.000	1.000	1.000	1.000	0.998	0.982	0.850
18	18	1.000	1.000	1.000	1.000	1.000	1.000	1.000	1.000	1.000
19	0	0.135	0.014	0.001	0.000	0.000	0.000	0.000	0.000	0.000
19	1	0.420	0.083	0.010	0.001	0.000	0.000	0.000	0.000	0.000
19	2	0.705	0.237	0.046	0.005	0.000	0.000	0.000	0.000	0.000
19	3	0.885	0.455	0.133	0.023	0.002	0.000	0.000	0.000	0.000
19	4	0.965	0.673	0.282	0.070	0.010	0.001	0.000	0.000	0.000
19	5	0.991	0.837	0.474	0.163	0.032	0.003	0.000	0.000	0.000
19	6	0.998	0.932	0.666	0.308	0.084	0.012	0.001	0.000	0.000
19	7	1.000	0.977	0.818	0.488	0.180	0.035	0.003	0.000	0.000
19	8	1.000	0.993	0.916	0.667	0.324	0.088	0.011	0.000	0.000
19	9	1.000	0.998	0.967	0.814	0.500	0.186	0.033	0.002	0.000
19	10	1.000	1.000	0.989	0.912	0.676	0.333	0.084	0.007	0.000
19	11	1.000	1.000	0.997	0.965	0.820	0.512	0.182	0.023	0.000
19	12	1.000	1.000	0.999	0.988	0.916	0.692	0.335	0.068	0.002
19	13	1.000	1.000	1.000	0.997	0.968	0.837	0.526	0.163	0.009
19	14	1.000	1.000	1.000	0.999	0.990	0.930	0.718	0.327	0.035
19	15	1.000	1.000	1.000	1.000	0.998	0.977	0.867	0.545	0.115
19	16	1.000	1.000	1.000	1.000	1.000	0.995	0.954	0.763	0.295
19	17	1.000	1.000	1.000	1.000	1.000	0.999	0.990	0.917	0.580
19	18	1.000	1.000	1.000	1.000	1.000	1.000	0.999	0.986	0.865
19	19	1.000	1.000	1.000	1.000	1.000	1.000	1.000	1.000	1.000
20	0	0.122	0.012	0.001	0.000	0.000	0.000	0.000	0.000	0.000
20	1	0.392	0.069	0.008	0.001	0.000	0.000	0.000	0.000	0.000
20	2	0.677	0.206	0.035	0.004	0.000	0.000	0.000	0.000	0.000
20	3	0.867	0.411	0.107	0.016	0.001	0.000	0.000	0.000	0.000
20	4	0.957	0.630	0.238	0.051	0.006	0.000	0.000	0.000	0.000
20	5	0.989	0.804	0.416	0.126	0.021	0.002	0.000	0.000	0.000
20	6	0.998	0.913	0.608	0.250	0.058	0.006	0.000	0.000	0.000
20	7	1.000	0.968	0.772	0.416	0.132	0.021	0.001	0.000	0.000
20	8	1.000	0.990	0.887	0.596	0.252	0.057	0.005	0.000	0.000
20	9	1.000	0.997	0.952	0.755	0.412	0.128	0.017	0.001	0.000
20	10	1.000	0.999	0.983	0.872	0.588	0.245	0.048	0.003	0.000
20	11	1.000	1.000	0.995	0.943	0.748	0.404	0.113	0.010	0.000
20	12	1.000	1.000	0.999	0.979	0.868	0.584	0.228	0.032	0.000
20	13	1.000	1.000	1.000	0.994	0.942	0.750	0.392	0.087	0.002
20	14	1.000	1.000	1.000	0.998	0.979	0.874	0.584	0.196	0.011
20	15	1.000	1.000	1.000	1.000	0.994	0.949	0.763	0.370	0.043
20	16	1.000	1.000	1.000	1.000	0.999	0.984	0.893	0.589	0.133
20	17	1.000	1.000	1.000	1.000	1.000	0.996	0.965	0.794	0.323
20	18	1.000	1.000	1.000	1.000	1.000	0.999	0.992	0.931	0.608
20	19	1.000	1.000	1.000	1.000	1.000	1.000	0.999	0.988	0.878
20	20	1.000	1.000	1.000	1.000	1.000	1.000	1.000	1.000	1.000

TABLE F
Poisson Probabilities

λ

X	0.02	0.03	0.04	0.05	0.06	0.07	0.08	0.09	0.10	0.20	0.30
0	0.9802	0.9704	0.9608	0.9512	0.9418	0.9324	0.9231	0.9139	0.9048	0.8187	0.7408
1	0.0196	0.0291	0.0384	0.0476	0.0565	0.0653	0.0739	0.0823	0.0905	0.1637	0.2222
2	0.0002	0.0004	0.0008	0.0012	0.0017	0.0023	0.0030	0.0037	0.0045	0.0164	0.0333
3	0.0000	0.0000	0.0000	0.0000	0.0000	0.0001	0.0001	0.0001	0.0002	0.0011	0.0033
4	0.0000	0.0000	0.0000	0.0000	0.0000	0.0000	0.0000	0.0000	0.0000	0.0001	0.0003

X	0.40	0.50	0.60	0.70	0.80	0.90	1.00	1.10	1.20	1.30	1.40
0	0.6703	0.6065	0.5488	0.4966	0.4493	0.4066	0.3679	0.3329	0.3012	0.2725	0.2466
1	0.2681	0.3033	0.3293	0.3476	0.3595	0.3659	0.3679	0.3662	0.3614	0.3543	0.3452
2	0.0536	0.0758	0.0988	0.1217	0.1438	0.1647	0.1839	0.2014	0.2169	0.2303	0.2417
3	0.0072	0.0126	0.0198	0.0284	0.0383	0.0494	0.0613	0.0738	0.0867	0.0998	0.1128
4	0.0007	0.0016	0.0030	0.0050	0.0077	0.0111	0.0153	0.0203	0.0260	0.0324	0.0395
5	0.0001	0.0002	0.0004	0.0007	0.0012	0.0020	0.0031	0.0045	0.0062	0.0084	0.0111
6	0.0000	0.0000	0.0000	0.0001	0.0002	0.0003	0.0005	0.0008	0.0012	0.0018	0.0026
7	0.0000	0.0000	0.0000	0.0000	0.0000	0.0000	0.0001	0.0001	0.0002	0.0003	0.0005
8	0.0000	0.0000	0.0000	0.0000	0.0000	0.0000	0.0000	0.0000	0.0000	0.0001	0.0001

X	1.50	1.60	1.70	1.80	1.90	2.00	2.10	2.20	2.30	2.40	2.50
0	0.2231	0.2019	0.1827	0.1653	0.1496	0.1353	0.1225	0.1108	0.1003	0.0907	0.0821
1	0.3347	0.3230	0.3106	0.2975	0.2842	0.2707	0.2572	0.2438	0.2306	0.2177	0.2052
2	0.2510	0.2584	0.2640	0.2678	0.2700	0.2707	0.2700	0.2681	0.2652	0.2613	0.2565
3	0.1255	0.1378	0.1496	0.1607	0.1710	0.1804	0.1890	0.1966	0.2033	0.2090	0.2138
4	0.0471	0.0551	0.0636	0.0723	0.0812	0.0902	0.0992	0.1082	0.1169	0.1254	0.1336
5	0.0141	0.0176	0.0216	0.0260	0.0309	0.0361	0.0417	0.0476	0.0538	0.0602	0.0668
6	0.0035	0.0047	0.0061	0.0078	0.0098	0.0120	0.0146	0.0174	0.0206	0.0241	0.0278
7	0.0008	0.0011	0.0015	0.0020	0.0027	0.0034	0.0044	0.0055	0.0068	0.0083	0.0099
8	0.0001	0.0002	0.0003	0.0005	0.0006	0.0009	0.0011	0.0015	0.0019	0.0025	0.0031
9	0.0000	0.0000	0.0001	0.0001	0.0001	0.0002	0.0003	0.0004	0.0005	0.0007	0.0009
10	0.0000	0.0000	0.0000	0.0000	0.0000	0.0000	0.0001	0.0001	0.0001	0.0002	0.0002

X	2.60	2.70	2.80	2.90	3.00	3.10	3.20	3.30	3.40	3.50	3.60
0	0.0743	0.0672	0.0608	0.0550	0.0498	0.0450	0.0408	0.0369	0.0334	0.0302	0.0273
1	0.1931	0.1815	0.1703	0.1596	0.1494	0.1397	0.1304	0.1217	0.1135	0.1057	0.0984
2	0.2510	0.2450	0.2384	0.2314	0.2240	0.2165	0.2087	0.2008	0.1929	0.1850	0.1771
3	0.2176	0.2205	0.2225	0.2237	0.2240	0.2237	0.2226	0.2209	0.2186	0.2158	0.2125
4	0.1414	0.1488	0.1557	0.1622	0.1680	0.1734	0.1781	0.1823	0.1858	0.1888	0.1912
5	0.0735	0.0804	0.0872	0.0940	0.1008	0.1075	0.1140	0.1203	0.1264	0.1322	0.1377
6	0.0319	0.0362	0.0407	0.0455	0.0504	0.0555	0.0608	0.0662	0.0716	0.0771	0.0826
7	0.0118	0.0139	0.0163	0.0188	0.0216	0.0246	0.0278	0.0312	0.0348	0.0385	0.0425
8	0.0038	0.0047	0.0057	0.0068	0.0081	0.0095	0.0111	0.0129	0.0148	0.0169	0.0191
9	0.0011	0.0014	0.0018	0.0022	0.0027	0.0033	0.0040	0.0047	0.0056	0.0066	0.0076
10	0.0003	0.0004	0.0005	0.0006	0.0008	0.0010	0.0013	0.0016	0.0019	0.0023	0.0028
11	0.0001	0.0001	0.0001	0.0002	0.0002	0.0003	0.0004	0.0005	0.0006	0.0007	0.0009
12	0.0000	0.0000	0.0000	0.0000	0.0001	0.0001	0.0001	0.0001	0.0002	0.0002	0.0003
13	0.0000	0.0000	0.0000	0.0000	0.0000	0.0000	0.0000	0.0000	0.0000	0.0001	0.0001

TABLE F
Poisson Probabilities (continued)

λ

X	3.70	3.80	3.90	4.00	4.10	4.20	4.30	4.40	4.50	4.60	4.70
0	0.0247	0.0224	0.0202	0.0183	0.0166	0.0150	0.0136	0.0123	0.0111	0.0101	0.0091
1	0.0915	0.0850	0.0789	0.0733	0.0679	0.0630	0.0583	0.0540	0.0500	0.0462	0.0427
2	0.1692	0.1615	0.1539	0.1465	0.1393	0.1323	0.1254	0.1188	0.1125	0.1063	0.1005
3	0.2087	0.2046	0.2001	0.1954	0.1904	0.1852	0.1798	0.1743	0.1687	0.1631	0.1574
4	0.1931	0.1944	0.1951	0.1954	0.1951	0.1944	0.1933	0.1917	0.1898	0.1875	0.1849
5	0.1429	0.1477	0.1522	0.1563	0.1600	0.1633	0.1662	0.1687	0.1708	0.1725	0.1738
6	0.0881	0.0936	0.0989	0.1042	0.1093	0.1143	0.1191	0.1237	0.1281	0.1323	0.1362
7	0.0466	0.0508	0.0551	0.0595	0.0640	0.0686	0.0732	0.0778	0.0824	0.0869	0.0914
8	0.0215	0.0241	0.0269	0.0298	0.0328	0.0360	0.0393	0.0428	0.0463	0.0500	0.0537
9	0.0089	0.0102	0.0116	0.0132	0.0150	0.0168	0.0188	0.0209	0.0232	0.0255	0.0281
10	0.0033	0.0039	0.0045	0.0053	0.0061	0.0071	0.0081	0.0092	0.0104	0.0118	0.0132
11	0.0011	0.0013	0.0016	0.0019	0.0023	0.0027	0.0032	0.0037	0.0043	0.0049	0.0056
12	0.0003	0.0004	0.0005	0.0006	0.0008	0.0009	0.0011	0.0014	0.0016	0.0019	0.0022
13	0.0001	0.0001	0.0002	0.0002	0.0002	0.0003	0.0004	0.0005	0.0006	0.0007	0.0008
14	0.0000	0.0000	0.0000	0.0001	0.0001	0.0001	0.0001	0.0001	0.0002	0.0002	0.0003
15	0.0000	0.0000	0.0000	0.0000	0.0000	0.0000	0.0000	0.0000	0.0001	0.0001	0.0001

X	4.80	4.90	5.00	5.10	5.20	5.30	5.40	5.50	5.60	5.70	5.80
0	0.0082	0.0074	0.0067	0.0061	0.0055	0.0050	0.0045	0.0041	0.0037	0.0033	0.0030
1	0.0395	0.0365	0.0337	0.0311	0.0287	0.0265	0.0244	0.0225	0.0207	0.0191	0.0176
2	0.0948	0.0894	0.0842	0.0793	0.0746	0.0701	0.0659	0.0618	0.0580	0.0544	0.0509
3	0.1517	0.1460	0.1404	0.1348	0.1293	0.1239	0.1185	0.1133	0.1082	0.1033	0.0985
4	0.1820	0.1789	0.1755	0.1719	0.1681	0.1641	0.1600	0.1558	0.1515	0.1472	0.1428
5	0.1747	0.1753	0.1755	0.1753	0.1748	0.1740	0.1728	0.1714	0.1697	0.1678	0.1656
6	0.1398	0.1432	0.1462	0.1490	0.1515	0.1537	0.1555	0.1571	0.1584	0.1594	0.1601
7	0.0959	0.1002	0.1044	0.1086	0.1125	0.1163	0.1200	0.1234	0.1267	0.1298	0.1326
8	0.0575	0.0614	0.0653	0.0692	0.0731	0.0771	0.0810	0.0849	0.0887	0.0925	0.0962
9	0.0307	0.0334	0.0363	0.0392	0.0423	0.0454	0.0486	0.0519	0.0552	0.0586	0.0620
10	0.0147	0.0164	0.0181	0.0200	0.0220	0.0241	0.0262	0.0285	0.0309	0.0334	0.0359
11	0.0064	0.0073	0.0082	0.0093	0.0104	0.0116	0.0129	0.0143	0.0157	0.0173	0.0190
12	0.0026	0.0030	0.0034	0.0039	0.0045	0.0051	0.0058	0.0065	0.0073	0.0082	0.0092
13	0.0009	0.0011	0.0013	0.0015	0.0018	0.0021	0.0024	0.0028	0.0032	0.0036	0.0041
14	0.0003	0.0004	0.0005	0.0006	0.0007	0.0008	0.0009	0.0011	0.0013	0.0015	0.0017
15	0.0001	0.0001	0.0002	0.0002	0.0002	0.0003	0.0003	0.0004	0.0005	0.0006	0.0007
16	0.0000	0.0000	0.0000	0.0001	0.0001	0.0001	0.0001	0.0001	0.0002	0.0002	0.0002
17	0.0000	0.0000	0.0000	0.0000	0.0000	0.0000	0.0000	0.0000	0.0001	0.0001	0.0001

X	5.90	6.00	6.10	6.20	6.30	6.40	6.50	6.60	6.70	6.80	6.90
0	0.0027	0.0025	0.0022	0.0020	0.0018	0.0017	0.0015	0.0014	0.0012	0.0011	0.0010
1	0.0162	0.0149	0.0137	0.0126	0.0116	0.0106	0.0098	0.0090	0.0082	0.0076	0.0070
2	0.0477	0.0446	0.0417	0.0390	0.0364	0.0340	0.0318	0.0296	0.0276	0.0258	0.0240
3	0.0938	0.0892	0.0848	0.0806	0.0765	0.0726	0.0688	0.0652	0.0617	0.0584	0.0552
4	0.1383	0.1339	0.1294	0.1249	0.1205	0.1162	0.1118	0.1076	0.1034	0.0992	0.0952
5	0.1632	0.1606	0.1579	0.1549	0.1519	0.1487	0.1454	0.1420	0.1385	0.1349	0.1314

TABLE F
Poisson Probabilities (continued)

λ

X	5.90	6.00	6.10	6.20	6.30	6.40	6.50	6.60	6.70	6.80	6.90
6	0.1605	0.1606	0.1605	0.1601	0.1595	0.1586	0.1575	0.1562	0.1546	0.1529	0.1511
7	0.1353	0.1377	0.1399	0.1418	0.1435	0.1450	0.1462	0.1472	0.1480	0.1486	0.1489
8	0.0998	0.1033	0.1066	0.1099	0.1130	0.1160	0.1188	0.1215	0.1240	0.1263	0.1284
9	0.0654	0.0688	0.0723	0.0757	0.0791	0.0825	0.0858	0.0891	0.0923	0.0954	0.0985
10	0.0386	0.0413	0.0441	0.0469	0.0498	0.0528	0.0558	0.0588	0.0618	0.0649	0.0679
11	0.0207	0.0225	0.0245	0.0265	0.0285	0.0307	0.0330	0.0353	0.0377	0.0401	0.0426
12	0.0102	0.0113	0.0124	0.0137	0.0150	0.0164	0.0179	0.0194	0.0210	0.0227	0.0245
13	0.0046	0.0052	0.0058	0.0065	0.0073	0.0081	0.0089	0.0099	0.0108	0.0119	0.0130
14	0.0019	0.0022	0.0025	0.0029	0.0033	0.0037	0.0041	0.0046	0.0052	0.0058	0.0064
15	0.0008	0.0009	0.0010	0.0012	0.0014	0.0016	0.0018	0.0020	0.0023	0.0026	0.0029
16	0.0003	0.0003	0.0004	0.0005	0.0005	0.0006	0.0007	0.0008	0.0010	0.0011	0.0013
17	0.0001	0.0001	0.0001	0.0002	0.0002	0.0002	0.0003	0.0003	0.0004	0.0004	0.0005
18	0.0000	0.0000	0.0000	0.0001	0.0001	0.0001	0.0001	0.0001	0.0001	0.0002	0.0002
19	0.0000	0.0000	0.0000	0.0000	0.0000	0.0000	0.0000	0.0000	0.0001	0.0001	0.0001

X	7.00	7.10	7.20	7.30	7.40	7.50	7.60	7.70	7.80	7.90	8.00
0	0.0009	0.0008	0.0007	0.0007	0.0006	0.0006	0.0005	0.0005	0.0004	0.0004	0.0003
1	0.0064	0.0059	0.0054	0.0049	0.0045	0.0041	0.0038	0.0035	0.0032	0.0029	0.0027
2	0.0223	0.0208	0.0194	0.0180	0.0167	0.0156	0.0145	0.0134	0.0125	0.0116	0.0107
3	0.0521	0.0492	0.0464	0.0438	0.0413	0.0389	0.0366	0.0345	0.0324	0.0305	0.0286
4	0.0912	0.0874	0.0836	0.0799	0.0764	0.0729	0.0696	0.0663	0.0632	0.0602	0.0573
5	0.1277	0.1241	0.1204	0.1167	0.1130	0.1094	0.1057	0.1021	0.0986	0.0951	0.0916
6	0.1490	0.1468	0.1445	0.1420	0.1394	0.1367	0.1339	0.1311	0.1282	0.1252	0.1221
7	0.1490	0.1489	0.1486	0.1481	0.1474	0.1465	0.1454	0.1442	0.1428	0.1413	0.1396
8	0.1304	0.1321	0.1337	0.1351	0.1363	0.1373	0.1382	0.1388	0.1392	0.1395	0.1396
9	0.1014	0.1042	0.1070	0.1096	0.1121	0.1144	0.1167	0.1187	0.1207	0.1224	0.1241
10	0.0710	0.0740	0.0770	0.0800	0.0829	0.0858	0.0887	0.0914	0.0941	0.0967	0.0993
11	0.0452	0.0478	0.0504	0.0531	0.0558	0.0585	0.0613	0.0640	0.0667	0.0695	0.0722
12	0.0264	0.0283	0.0303	0.0323	0.0344	0.0366	0.0388	0.0411	0.0434	0.0457	0.0481
13	0.0142	0.0154	0.0168	0.0181	0.0196	0.0211	0.0227	0.0243	0.0260	0.0278	0.0296
14	0.0071	0.0078	0.0086	0.0095	0.0104	0.0113	0.0123	0.0134	0.0145	0.0157	0.0169
15	0.0033	0.0037	0.0041	0.0046	0.0051	0.0057	0.0062	0.0069	0.0075	0.0083	0.0090
16	0.0014	0.0016	0.0019	0.0021	0.0024	0.0026	0.0030	0.0033	0.0037	0.0041	0.0045
17	0.0006	0.0007	0.0008	0.0009	0.0010	0.0012	0.0013	0.0015	0.0017	0.0019	0.0021
18	0.0002	0.0003	0.0003	0.0004	0.0004	0.0005	0.0006	0.0006	0.0007	0.0008	0.0009
19	0.0001	0.0001	0.0001	0.0001	0.0002	0.0002	0.0002	0.0003	0.0003	0.0003	0.0004
20	0.0000	0.0000	0.0000	0.0001	0.0001	0.0001	0.0001	0.0001	0.0001	0.0001	0.0002
21	0.0000	0.0000	0.0000	0.0000	0.0000	0.0000	0.0000	0.0000	0.0000	0.0001	0.0001

X	8.10	8.20	8.30	8.40	8.50	8.60	8.70	8.80	8.90	9.00	9.10
0	0.0003	0.0003	0.0002	0.0002	0.0002	0.0002	0.0002	0.0002	0.0001	0.0001	0.0001
1	0.0025	0.0023	0.0021	0.0019	0.0017	0.0016	0.0014	0.0013	0.0012	0.0011	0.0010
2	0.0100	0.0092	0.0086	0.0079	0.0074	0.0068	0.0063	0.0058	0.0054	0.0050	0.0046
3	0.0269	0.0252	0.0237	0.0222	0.0208	0.0195	0.0183	0.0171	0.0160	0.0150	0.0140
4	0.0544	0.0517	0.0491	0.0466	0.0443	0.0420	0.0398	0.0377	0.0357	0.0337	0.0319

TABLE F
Poisson Probabilities (continued)

λ

X	8.10	8.20	8.30	8.40	8.50	8.60	8.70	8.80	8.90	9.00	9.10
5	0.0882	0.0849	0.0816	0.0784	0.0752	0.0722	0.0692	0.0663	0.0635	0.0607	0.0581
6	0.1191	0.1160	0.1128	0.1097	0.1066	0.1034	0.1003	0.0972	0.0941	0.0911	0.0881
7	0.1378	0.1358	0.1338	0.1317	0.1294	0.1271	0.1247	0.1222	0.1197	0.1171	0.1145
8	0.1395	0.1392	0.1388	0.1382	0.1375	0.1366	0.1356	0.1344	0.1332	0.1318	0.1302
9	0.1256	0.1269	0.1280	0.1290	0.1299	0.1306	0.1311	0.1315	0.1317	0.1318	0.1317
10	0.1017	0.1040	0.1063	0.1084	0.1104	0.1123	0.1140	0.1157	0.1172	0.1186	0.1198
11	0.0749	0.0776	0.0802	0.0828	0.0853	0.0878	0.0902	0.0925	0.0948	0.0970	0.0991
12	0.0505	0.0530	0.0555	0.0579	0.0604	0.0629	0.0654	0.0679	0.0703	0.0728	0.0752
13	0.0315	0.0334	0.0354	0.0374	0.0395	0.0416	0.0438	0.0459	0.0481	0.0504	0.0526
14	0.0182	0.0196	0.0210	0.0225	0.0240	0.0256	0.0272	0.0289	0.0306	0.0324	0.0342
15	0.0098	0.0107	0.0116	0.0126	0.0136	0.0147	0.0158	0.0169	0.0182	0.0194	0.0208
16	0.0050	0.0055	0.0060	0.0066	0.0072	0.0079	0.0086	0.0093	0.0101	0.0109	0.0118
17	0.0024	0.0026	0.0029	0.0033	0.0036	0.0040	0.0044	0.0048	0.0053	0.0058	0.0063
18	0.0011	0.0012	0.0014	0.0015	0.0017	0.0019	0.0021	0.0024	0.0026	0.0029	0.0032
19	0.0005	0.0005	0.0006	0.0007	0.0008	0.0009	0.0010	0.0011	0.0012	0.0014	0.0015
20	0.0002	0.0002	0.0002	0.0003	0.0003	0.0004	0.0004	0.0005	0.0005	0.0006	0.0007
21	0.0001	0.0001	0.0001	0.0001	0.0001	0.0002	0.0002	0.0002	0.0002	0.0003	0.0003
22	0.0000	0.0000	0.0000	0.0000	0.0001	0.0001	0.0001	0.0001	0.0001	0.0001	0.0001

X	9.20	9.30	9.40	9.50	9.60	9.70	9.80	9.90	10.00	11.00	12.00
0	0.0001	0.0001	0.0001	0.0001	0.0001	0.0001	0.0001	0.0001	0.0000	0.0000	0.0000
1	0.0009	0.0009	0.0008	0.0007	0.0007	0.0006	0.0005	0.0005	0.0005	0.0002	0.0001
2	0.0043	0.0040	0.0037	0.0034	0.0031	0.0029	0.0027	0.0025	0.0023	0.0010	0.0004
3	0.0131	0.0123	0.0115	0.0107	0.0100	0.0093	0.0087	0.0081	0.0076	0.0037	0.0018
4	0.0302	0.0285	0.0269	0.0254	0.0240	0.0226	0.0213	0.0201	0.0189	0.0102	0.0053
5	0.0555	0.0530	0.0506	0.0483	0.0460	0.0439	0.0418	0.0398	0.0378	0.0224	0.0127
6	0.0851	0.0822	0.0793	0.0764	0.0736	0.0709	0.0682	0.0656	0.0631	0.0411	0.0255
7	0.1118	0.1091	0.1064	0.1037	0.1010	0.0982	0.0955	0.0928	0.0901	0.0646	0.0437
8	0.1286	0.1269	0.1251	0.1232	0.1212	0.1191	0.1170	0.1148	0.1126	0.0888	0.0655
9	0.1315	0.1311	0.1306	0.1300	0.1293	0.1284	0.1274	0.1263	0.1251	0.1085	0.0874
10	0.1210	0.1219	0.1228	0.1235	0.1241	0.1245	0.1249	0.1250	0.1251	0.1194	0.1048
11	0.1012	0.1031	0.1049	0.1067	0.1083	0.1098	0.1112	0.1125	0.1137	0.1194	0.1144
12	0.0776	0.0799	0.0822	0.0844	0.0866	0.0888	0.0908	0.0928	0.0948	0.1094	0.1144
13	0.0549	0.0572	0.0594	0.0617	0.0640	0.0662	0.0685	0.0707	0.0729	0.0926	0.1056
14	0.0361	0.0380	0.0399	0.0419	0.0439	0.0459	0.0479	0.0500	0.0521	0.0728	0.0905
15	0.0221	0.0235	0.0250	0.0265	0.0281	0.0297	0.0313	0.0330	0.0347	0.0534	0.0724
16	0.0127	0.0137	0.0147	0.0157	0.0168	0.0180	0.0192	0.0204	0.0217	0.0367	0.0543
17	0.0069	0.0075	0.0081	0.0088	0.0095	0.0103	0.0111	0.0119	0.0128	0.0237	0.0383
18	0.0035	0.0039	0.0042	0.0046	0.0051	0.0055	0.0060	0.0065	0.0071	0.0145	0.0256
19	0.0017	0.0019	0.0021	0.0023	0.0026	0.0028	0.0031	0.0034	0.0037	0.0084	0.0161
20	0.0008	0.0009	0.0010	0.0011	0.0012	0.0014	0.0015	0.0017	0.0019	0.0046	0.0097
21	0.0003	0.0004	0.0004	0.0005	0.0006	0.0006	0.0007	0.0008	0.0009	0.0024	0.0055
22	0.0001	0.0002	0.0002	0.0002	0.0002	0.0003	0.0003	0.0004	0.0004	0.0012	0.0030
23	0.0001	0.0001	0.0001	0.0001	0.0001	0.0001	0.0001	0.0002	0.0002	0.0006	0.0016
24	0.0000	0.0000	0.0000	0.0000	0.0000	0.0000	0.0001	0.0001	0.0001	0.0003	0.0008

Table F
Poisson Probabilities (continued)

λ

X	13.00	14.00	15.00	16.00	17.00	18.00	19.00	20.00	21.00	22.00	23.00
0	0.0000	0.0000	0.0000	0.0000	0.0000	0.0000	0.0000	0.0000	0.0000	0.0000	0.0000
1	0.0000	0.0000	0.0000	0.0000	0.0000	0.0000	0.0000	0.0000	0.0000	0.0000	0.0000
2	0.0002	0.0001	0.0000	0.0000	0.0000	0.0000	0.0000	0.0000	0.0000	0.0000	0.0000
3	0.0008	0.0004	0.0002	0.0001	0.0000	0.0000	0.0000	0.0000	0.0000	0.0000	0.0000
4	0.0027	0.0013	0.0006	0.0003	0.0001	0.0001	0.0000	0.0000	0.0000	0.0000	0.0000
5	0.0070	0.0037	0.0019	0.0010	0.0005	0.0002	0.0001	0.0001	0.0000	0.0000	0.0000
6	0.0152	0.0087	0.0048	0.0026	0.0014	0.0007	0.0004	0.0002	0.0001	0.0000	0.0000
7	0.0281	0.0174	0.0104	0.0060	0.0034	0.0019	0.0010	0.0005	0.0003	0.0001	0.0001
8	0.0457	0.0304	0.0194	0.0120	0.0072	0.0042	0.0024	0.0013	0.0007	0.0004	0.0002
9	0.0661	0.0473	0.0324	0.0213	0.0135	0.0083	0.0050	0.0029	0.0017	0.0009	0.0005
10	0.0859	0.0663	0.0486	0.0341	0.0230	0.0150	0.0095	0.0058	0.0035	0.0020	0.0012
11	0.1015	0.0844	0.0663	0.0496	0.0355	0.0245	0.0164	0.0106	0.0067	0.0041	0.0024
12	0.1099	0.0984	0.0829	0.0661	0.0504	0.0368	0.0259	0.0176	0.0116	0.0075	0.0047
13	0.1099	0.1060	0.0956	0.0814	0.0658	0.0509	0.0378	0.0271	0.0188	0.0127	0.0083
14	0.1021	0.1060	0.1024	0.0930	0.0800	0.0655	0.0514	0.0387	0.0282	0.0199	0.0136
15	0.0885	0.0989	0.1024	0.0992	0.0906	0.0786	0.0650	0.0516	0.0395	0.0292	0.0209
16	0.0719	0.0866	0.0960	0.0992	0.0963	0.0884	0.0772	0.0646	0.0518	0.0401	0.0301
17	0.0550	0.0713	0.0847	0.0934	0.0963	0.0936	0.0863	0.0760	0.0640	0.0520	0.0407
18	0.0397	0.0554	0.0706	0.0830	0.0909	0.0936	0.0911	0.0844	0.0747	0.0635	0.0520
19	0.0272	0.0409	0.0557	0.0699	0.0814	0.0887	0.0911	0.0888	0.0826	0.0735	0.0629
20	0.0177	0.0286	0.0418	0.0559	0.0692	0.0798	0.0866	0.0888	0.0867	0.0809	0.0724
21	0.0109	0.0191	0.0299	0.0426	0.0560	0.0684	0.0783	0.0846	0.0867	0.0847	0.0793
22	0.0065	0.0121	0.0204	0.0310	0.0433	0.0560	0.0676	0.0769	0.0828	0.0847	0.0829
23	0.0037	0.0074	0.0133	0.0216	0.0320	0.0438	0.0559	0.0669	0.0756	0.0810	0.0829
24	0.0020	0.0043	0.0083	0.0144	0.0226	0.0328	0.0442	0.0557	0.0661	0.0743	0.0794
25	0.0010	0.0024	0.0050	0.0092	0.0154	0.0237	0.0336	0.0446	0.0555	0.0654	0.0731
26	0.0005	0.0013	0.0029	0.0057	0.0101	0.0164	0.0246	0.0343	0.0449	0.0553	0.0646
27	0.0002	0.0007	0.0016	0.0034	0.0063	0.0109	0.0173	0.0254	0.0349	0.0451	0.0551
28	0.0001	0.0003	0.0009	0.0019	0.0038	0.0070	0.0117	0.0181	0.0262	0.0354	0.0452
29	0.0001	0.0002	0.0004	0.0011	0.0023	0.0044	0.0077	0.0125	0.0190	0.0269	0.0359
30	0.0000	0.0001	0.0002	0.0006	0.0013	0.0026	0.0049	0.0083	0.0133	0.0197	0.0275
31	0.0000	0.0000	0.0001	0.0003	0.0007	0.0015	0.0030	0.0054	0.0090	0.0140	0.0204
32	0.0000	0.0000	0.0001	0.0001	0.0004	0.0009	0.0018	0.0034	0.0059	0.0096	0.0147
33	0.0000	0.0000	0.0000	0.0001	0.0002	0.0005	0.0010	0.0020	0.0038	0.0064	0.0102
34	0.0000	0.0000	0.0000	0.0000	0.0001	0.0002	0.0006	0.0012	0.0023	0.0041	0.0069
35	0.0000	0.0000	0.0000	0.0000	0.0000	0.0001	0.0003	0.0007	0.0014	0.0026	0.0045
36	0.0000	0.0000	0.0000	0.0000	0.0000	0.0001	0.0002	0.0004	0.0008	0.0016	0.0029
37	0.0000	0.0000	0.0000	0.0000	0.0000	0.0000	0.0001	0.0002	0.0005	0.0009	0.0018
38	0.0000	0.0000	0.0000	0.0000	0.0000	0.0000	0.0000	0.0001	0.0003	0.0005	0.0011
39	0.0000	0.0000	0.0000	0.0000	0.0000	0.0000	0.0000	0.0001	0.0001	0.0003	0.0006
40	0.0000	0.0000	0.0000	0.0000	0.0000	0.0000	0.0000	0.0000	0.0001	0.0002	0.0004
41	0.0000	0.0000	0.0000	0.0000	0.0000	0.0000	0.0000	0.0000	0.0000	0.0001	0.0002
42	0.0000	0.0000	0.0000	0.0000	0.0000	0.0000	0.0000	0.0000	0.0000	0.0000	0.0001
43	0.0000	0.0000	0.0000	0.0000	0.0000	0.0000	0.0000	0.0000	0.0000	0.0000	0.0001

Table G

Critical Values of χ^2

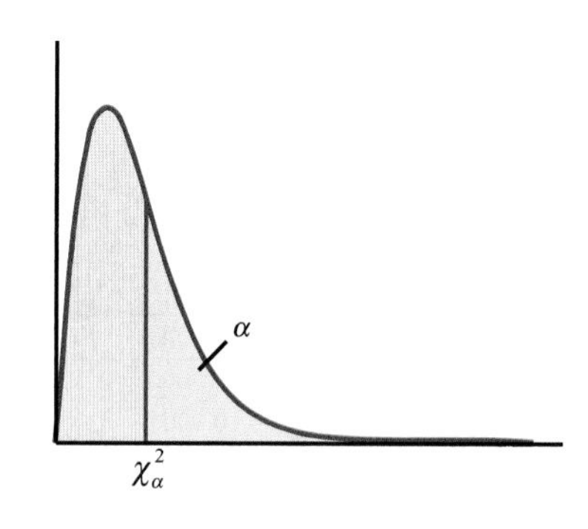

Area to the Right of the Critical Value

DF	$\chi^2_{.995}$	$\chi^2_{.990}$	$\chi^2_{.975}$	$\chi^2_{.950}$	$\chi^2_{.900}$	$\chi^2_{.100}$	$\chi^2_{.050}$	$\chi^2_{.025}$	$\chi^2_{.010}$	$\chi^2_{.005}$
1	–	–	0.001	0.004	0.016	2.706	3.841	5.024	6.635	7.879
2	0.010	0.020	0.051	0.103	0.211	4.605	5.991	7.378	9.210	10.597
3	0.072	0.115	0.216	0.352	0.584	6.251	7.815	9.348	11.345	12.838
4	0.207	0.297	0.484	0.711	1.064	7.779	9.488	11.143	13.277	14.860
5	0.412	0.554	0.831	1.145	1.610	9.236	11.071	12.833	15.086	16.750
6	0.676	0.872	1.237	1.635	2.204	10.645	12.592	14.449	16.812	18.548
7	0.989	1.239	1.690	2.167	2.833	12.017	14.067	16.013	18.475	20.278
8	1.344	1.646	2.180	2.733	3.490	13.362	15.507	17.535	20.090	21.955
9	1.735	2.088	2.700	3.325	4.168	14.684	16.919	19.023	21.666	23.589
10	2.156	2.558	3.247	3.940	4.865	15.987	18.307	20.483	23.209	25.188
11	2.603	3.053	3.816	4.575	5.578	17.275	19.675	21.920	24.725	26.757
12	3.074	3.571	4.404	5.226	6.304	18.549	21.026	23.337	26.217	28.299
13	3.565	4.107	5.009	5.892	7.042	19.812	22.362	24.736	27.688	29.819
14	4.075	4.660	5.629	6.571	7.790	21.064	23.685	26.119	29.141	31.319
15	4.601	5.229	6.262	7.261	8.547	22.307	24.996	27.488	30.578	32.801
16	5.142	5.812	6.908	7.962	9.312	23.542	26.296	28.845	32.000	34.267
17	5.697	6.408	7.564	8.672	10.085	24.769	27.587	30.191	33.409	35.718
18	6.265	7.015	8.231	9.390	10.865	25.989	28.869	31.526	34.805	37.156
19	6.844	7.633	8.907	10.117	11.651	27.204	30.144	32.852	36.191	38.582
20	7.434	8.260	9.591	10.851	12.443	28.412	31.410	34.170	37.566	39.997
21	8.034	8.897	10.283	11.591	13.240	29.615	32.671	35.479	38.932	41.401
22	8.643	9.542	10.982	12.338	14.042	30.813	33.924	36.781	40.289	42.796
23	9.260	10.196	11.689	13.091	14.848	32.007	35.172	38.076	41.638	44.181
24	9.886	10.856	12.401	13.848	15.659	33.196	36.415	39.364	42.980	45.559
25	10.520	11.524	13.120	14.611	16.473	34.382	37.652	40.646	44.314	46.928
26	11.160	12.198	13.844	15.379	17.292	35.563	38.885	41.923	45.642	48.290
27	11.808	12.879	14.573	16.151	18.114	36.741	40.113	43.194	46.963	49.645
28	12.461	13.565	15.308	16.928	18.939	37.916	41.337	44.461	48.278	50.993
29	13.121	14.257	16.047	17.708	19.768	39.087	42.557	45.722	49.588	52.336
30	13.787	14.954	16.791	18.493	20.599	40.256	43.773	46.979	50.892	53.672
40	20.707	22.164	24.433	26.509	29.051	51.805	55.758	59.342	63.691	66.766
50	27.991	29.707	32.357	34.764	37.689	63.167	67.505	71.420	76.154	79.490
60	35.534	37.485	40.482	43.188	46.459	74.397	79.082	83.298	88.379	91.952
70	43.275	45.442	48.758	51.739	55.329	85.527	90.531	95.023	100.425	104.215
80	51.172	53.540	57.153	60.391	64.278	96.578	101.879	106.629	112.329	116.321
90	59.196	61.754	65.647	69.126	73.291	107.565	113.145	118.136	124.116	128.299
100	67.328	70.065	74.222	77.929	82.358	118.498	124.342	129.561	135.807	140.169

TABLE H

Critical Values of the *F*-Distribution ($\alpha = 0.995$)

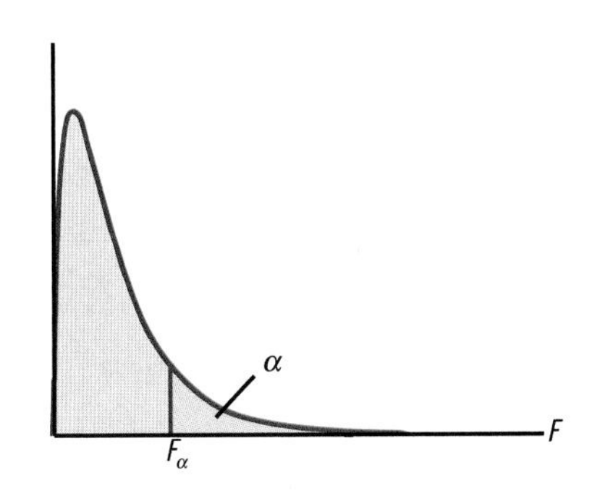

Numerator Degrees of Freedom

Denominator Degrees of Freedom

	1	2	3	4	5	6	7	8	9
1	0.0001	0.0050	0.0180	0.0319	0.0439	0.0537	0.0616	0.0681	0.0735
2	0.0001	0.0050	0.0201	0.0380	0.0546	0.0688	0.0806	0.0906	0.0989
3	0.0000	0.0050	0.0211	0.0412	0.0605	0.0774	0.0919	0.1042	0.1147
4	0.0000	0.0050	0.0216	0.0432	0.0643	0.0831	0.0995	0.1136	0.1257
5	0.0000	0.0050	0.0220	0.0445	0.0669	0.0872	0.1050	0.1205	0.1338
6	0.0000	0.0050	0.0223	0.0455	0.0689	0.0903	0.1092	0.1258	0.1402
7	0.0000	0.0050	0.0225	0.0462	0.0704	0.0927	0.1125	0.1300	0.1452
8	0.0000	0.0050	0.0227	0.0468	0.0716	0.0946	0.1152	0.1334	0.1494
9	0.0000	0.0050	0.0228	0.0473	0.0726	0.0962	0.1175	0.1363	0.1529
10	0.0000	0.0050	0.0229	0.0477	0.0734	0.0976	0.1193	0.1387	0.1558
11	0.0000	0.0050	0.0230	0.0480	0.0741	0.0987	0.1209	0.1408	0.1584
12	0.0000	0.0050	0.0230	0.0483	0.0747	0.0997	0.1223	0.1426	0.1606
13	0.0000	0.0050	0.0231	0.0485	0.0752	0.1005	0.1235	0.1441	0.1625
14	0.0000	0.0050	0.0232	0.0487	0.0757	0.1012	0.1246	0.1455	0.1642
15	0.0000	0.0050	0.0232	0.0489	0.0761	0.1019	0.1255	0.1468	0.1658
16	0.0000	0.0050	0.0233	0.0491	0.0764	0.1025	0.1263	0.1479	0.1671
17	0.0000	0.0050	0.0233	0.0492	0.0767	0.1030	0.1271	0.1489	0.1684
18	0.0000	0.0050	0.0233	0.0494	0.0770	0.1035	0.1278	0.1498	0.1695
19	0.0000	0.0050	0.0234	0.0495	0.0773	0.1039	0.1284	0.1506	0.1705
20	0.0000	0.0050	0.0234	0.0496	0.0775	0.1043	0.1290	0.1513	0.1715
21	0.0000	0.0050	0.0234	0.0497	0.0777	0.1046	0.1295	0.1520	0.1723
22	0.0000	0.0050	0.0234	0.0498	0.0779	0.1050	0.1300	0.1526	0.1731
23	0.0000	0.0050	0.0234	0.0499	0.0781	0.1053	0.1304	0.1532	0.1739
24	0.0000	0.0050	0.0235	0.0499	0.0782	0.1055	0.1308	0.1538	0.1745
25	0.0000	0.0050	0.0235	0.0500	0.0784	0.1058	0.1312	0.1543	0.1752
26	0.0000	0.0050	0.0235	0.0501	0.0785	0.1060	0.1315	0.1547	0.1758
27	0.0000	0.0050	0.0235	0.0501	0.0787	0.1063	0.1319	0.1552	0.1763
28	0.0000	0.0050	0.0235	0.0502	0.0788	0.1065	0.1322	0.1556	0.1768
29	0.0000	0.0050	0.0235	0.0502	0.0789	0.1067	0.1325	0.1560	0.1773
30	0.0000	0.0050	0.0235	0.0503	0.0790	0.1069	0.1327	0.1563	0.1778
40	0.0000	0.0050	0.0236	0.0506	0.0798	0.1082	0.1347	0.1590	0.1812
60	0.0000	0.0050	0.0237	0.0510	0.0806	0.1096	0.1368	0.1619	0.1848
120	0.0000	0.0050	0.0238	0.0514	0.0815	0.1111	0.1390	0.1649	0.1887
∞	0.0000	0.0050	0.0239	0.0517	0.0823	0.1126	0.1413	0.1681	0.1928

Table H

Critical Values of the *F*-Distribution ($\alpha = 0.995$)

(continued)

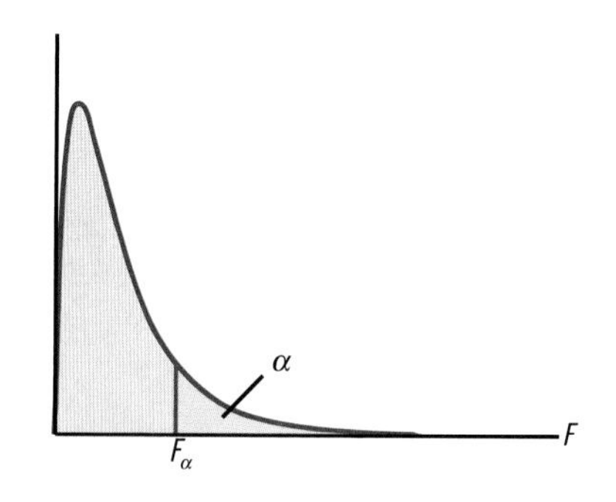

Numerator Degrees of Freedom

Denominator Degrees of Freedom

	10	11	12	13	14	15	16	17	18
1	0.0780	0.0818	0.0851	0.0879	0.0904	0.0926	0.0946	0.0963	0.0979
2	0.1061	0.1122	0.1175	0.1222	0.1262	0.1299	0.1331	0.1360	0.1386
3	0.1238	0.1316	0.1384	0.1444	0.1497	0.1544	0.1586	0.1625	0.1659
4	0.1362	0.1453	0.1533	0.1604	0.1667	0.1723	0.1774	0.1819	0.1861
5	0.1455	0.1557	0.1647	0.1727	0.1798	0.1861	0.1919	0.1971	0.2018
6	0.1528	0.1639	0.1737	0.1824	0.1902	0.1972	0.2035	0.2093	0.2145
7	0.1587	0.1705	0.1810	0.1904	0.1988	0.2063	0.2131	0.2193	0.2250
8	0.1635	0.1760	0.1871	0.1970	0.2059	0.2139	0.2212	0.2278	0.2339
9	0.1676	0.1806	0.1922	0.2026	0.2120	0.2204	0.2281	0.2351	0.2415
10	0.1710	0.1846	0.1966	0.2075	0.2172	0.2261	0.2341	0.2414	0.2481
11	0.1740	0.1880	0.2005	0.2117	0.2218	0.2310	0.2393	0.2469	0.2539
12	0.1766	0.1910	0.2038	0.2154	0.2258	0.2353	0.2439	0.2518	0.2591
13	0.1789	0.1936	0.2068	0.2187	0.2294	0.2392	0.2481	0.2562	0.2637
14	0.1810	0.1960	0.2094	0.2216	0.2326	0.2426	0.2517	0.2601	0.2678
15	0.1828	0.1981	0.2118	0.2242	0.2355	0.2457	0.2551	0.2636	0.2715
16	0.1844	0.2000	0.2139	0.2266	0.2381	0.2485	0.2581	0.2669	0.2749
17	0.1859	0.2017	0.2159	0.2287	0.2404	0.2511	0.2608	0.2698	0.2780
18	0.1873	0.2032	0.2177	0.2307	0.2426	0.2534	0.2634	0.2725	0.2809
19	0.1885	0.2047	0.2193	0.2325	0.2446	0.2556	0.2657	0.2749	0.2835
20	0.1896	0.2060	0.2208	0.2342	0.2464	0.2576	0.2678	0.2772	0.2859
21	0.1906	0.2072	0.2221	0.2357	0.2481	0.2594	0.2698	0.2793	0.2881
22	0.1916	0.2083	0.2234	0.2371	0.2496	0.2611	0.2716	0.2813	0.2902
23	0.1925	0.2093	0.2246	0.2384	0.2511	0.2627	0.2733	0.2831	0.2922
24	0.1933	0.2103	0.2257	0.2397	0.2524	0.2641	0.2749	0.2848	0.2940
25	0.1941	0.2112	0.2267	0.2408	0.2537	0.2655	0.2764	0.2864	0.2957
26	0.1948	0.2120	0.2276	0.2419	0.2549	0.2668	0.2778	0.2879	0.2973
27	0.1954	0.2128	0.2285	0.2429	0.2560	0.2680	0.2791	0.2893	0.2988
28	0.1961	0.2135	0.2294	0.2438	0.2570	0.2692	0.2803	0.2906	0.3002
29	0.1967	0.2142	0.2302	0.2447	0.2580	0.2702	0.2815	0.2919	0.3015
30	0.1972	0.2149	0.2309	0.2455	0.2589	0.2712	0.2826	0.2930	0.3028
40	0.2014	0.2197	0.2365	0.2519	0.2660	0.2789	0.2909	0.3020	0.3124
60	0.2058	0.2250	0.2425	0.2587	0.2736	0.2873	0.3001	0.3119	0.3230
120	0.2105	0.2306	0.2491	0.2661	0.2819	0.2965	0.3102	0.3229	0.3348
∞	0.2156	0.2367	0.2562	0.2742	0.2910	0.3067	0.3214	0.3351	0.3480

TABLE H

Critical Values of the F-Distribution ($\alpha = 0.995$)
(continued)

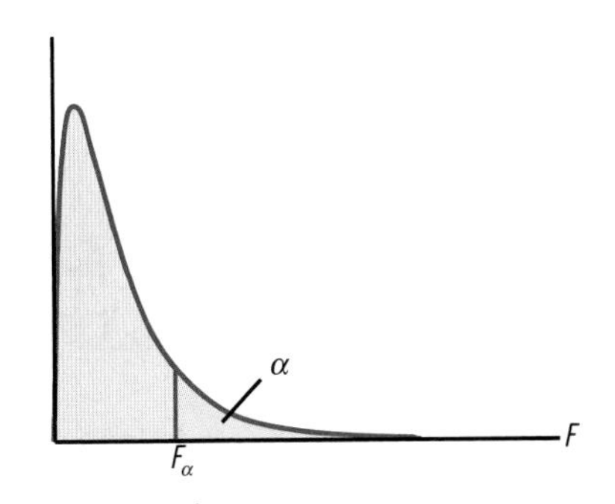

Numerator Degrees of Freedom

Denominator Degrees of Freedom	19	20	24	30	40	60	120
1	0.0993	0.1006	0.1047	0.1089	0.1133	0.1177	0.1223
2	0.1410	0.1431	0.1501	0.1574	0.1648	0.1726	0.1805
3	0.1690	0.1719	0.1812	0.1909	0.2010	0.2115	0.2224
4	0.1898	0.1933	0.2045	0.2163	0.2286	0.2416	0.2551
5	0.2061	0.2100	0.2229	0.2365	0.2509	0.2660	0.2818
6	0.2192	0.2236	0.2380	0.2532	0.2693	0.2864	0.3044
7	0.2302	0.2349	0.2506	0.2673	0.2850	0.3038	0.3239
8	0.2394	0.2445	0.2613	0.2793	0.2985	0.3190	0.3410
9	0.2474	0.2528	0.2706	0.2898	0.3104	0.3324	0.3561
10	0.2543	0.2599	0.2788	0.2990	0.3208	0.3443	0.3697
11	0.2603	0.2663	0.2860	0.3072	0.3302	0.3550	0.3819
12	0.2657	0.2719	0.2924	0.3146	0.3386	0.3647	0.3931
13	0.2706	0.2769	0.2982	0.3212	0.3463	0.3735	0.4033
14	0.2749	0.2815	0.3034	0.3272	0.3532	0.3816	0.4127
15	0.2788	0.2856	0.3081	0.3327	0.3596	0.3890	0.4215
16	0.2824	0.2893	0.3124	0.3377	0.3654	0.3959	0.4296
17	0.2856	0.2927	0.3164	0.3423	0.3708	0.4023	0.4371
18	0.2886	0.2958	0.3200	0.3466	0.3758	0.4082	0.4442
19	0.2914	0.2987	0.3234	0.3506	0.3805	0.4137	0.4508
20	0.2939	0.3014	0.3265	0.3542	0.3848	0.4189	0.4570
21	0.2963	0.3039	0.3295	0.3577	0.3889	0.4238	0.4629
22	0.2985	0.3062	0.3322	0.3609	0.3927	0.4283	0.4684
23	0.3006	0.3083	0.3347	0.3639	0.3963	0.4326	0.4737
24	0.3025	0.3104	0.3371	0.3667	0.3997	0.4367	0.4787
25	0.3043	0.3123	0.3393	0.3693	0.4029	0.4406	0.4834
26	0.3059	0.3140	0.3414	0.3718	0.4059	0.4442	0.4879
27	0.3075	0.3157	0.3434	0.3742	0.4087	0.4477	0.4922
28	0.3090	0.3173	0.3452	0.3764	0.4114	0.4510	0.4964
29	0.3104	0.3188	0.3470	0.3785	0.4140	0.4542	0.5003
30	0.3118	0.3202	0.3487	0.3805	0.4164	0.4572	0.5040
40	0.3220	0.3310	0.3616	0.3962	0.4356	0.4810	0.5345
60	0.3333	0.3429	0.3762	0.4141	0.4579	0.5096	0.5725
120	0.3459	0.3564	0.3927	0.4348	0.4846	0.5452	0.6229
∞	0.3602	0.3717	0.4119	0.4596	0.5177	0.5922	0.6988

Table H

Critical Values of the F-Distribution ($\alpha = 0.990$)

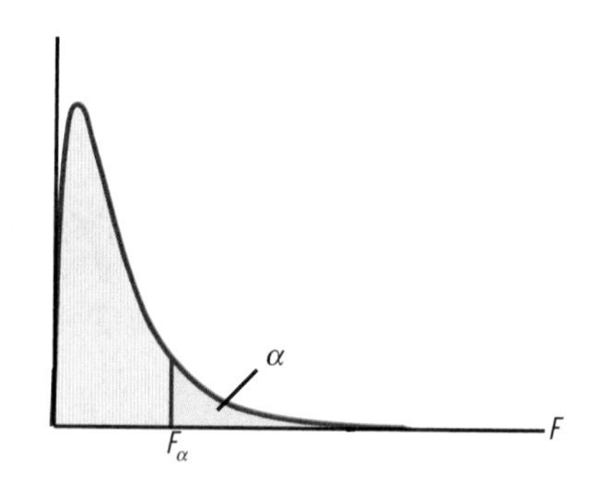

Numerator Degrees of Freedom

Denominator Degrees of Freedom	1	2	3	4	5	6	7	8	9
1	0.0002	0.0102	0.0293	0.0472	0.0615	0.0728	0.0817	0.0888	0.0947
2	0.0002	0.0101	0.0325	0.0556	0.0753	0.0915	0.1047	0.1156	0.1247
3	0.0002	0.0101	0.0339	0.0599	0.0829	0.1023	0.1183	0.1317	0.1430
4	0.0002	0.0101	0.0348	0.0626	0.0878	0.1093	0.1274	0.1427	0.1557
5	0.0002	0.0101	0.0354	0.0644	0.0912	0.1143	0.1340	0.1508	0.1651
6	0.0002	0.0101	0.0358	0.0658	0.0937	0.1181	0.1391	0.1570	0.1724
7	0.0002	0.0101	0.0361	0.0668	0.0956	0.1211	0.1430	0.1619	0.1782
8	0.0002	0.0101	0.0364	0.0676	0.0972	0.1234	0.1462	0.1659	0.1829
9	0.0002	0.0101	0.0366	0.0682	0.0984	0.1254	0.1488	0.1692	0.1869
10	0.0002	0.0101	0.0367	0.0687	0.0995	0.1270	0.1511	0.1720	0.1902
11	0.0002	0.0101	0.0369	0.0692	0.1004	0.1284	0.1529	0.1744	0.1931
12	0.0002	0.0101	0.0370	0.0696	0.1011	0.1296	0.1546	0.1765	0.1956
13	0.0002	0.0101	0.0371	0.0699	0.1018	0.1306	0.1560	0.1783	0.1978
14	0.0002	0.0101	0.0371	0.0702	0.1024	0.1315	0.1573	0.1799	0.1998
15	0.0002	0.0101	0.0372	0.0704	0.1029	0.1323	0.1584	0.1813	0.2015
16	0.0002	0.0101	0.0373	0.0707	0.1033	0.1330	0.1594	0.1826	0.2031
17	0.0002	0.0101	0.0373	0.0708	0.1037	0.1336	0.1603	0.1837	0.2045
18	0.0002	0.0101	0.0374	0.0710	0.1041	0.1342	0.1611	0.1848	0.2058
19	0.0002	0.0101	0.0374	0.0712	0.1044	0.1347	0.1618	0.1857	0.2069
20	0.0002	0.0101	0.0375	0.0713	0.1047	0.1352	0.1625	0.1866	0.2080
21	0.0002	0.0101	0.0375	0.0715	0.1050	0.1356	0.1631	0.1874	0.2090
22	0.0002	0.0101	0.0375	0.0716	0.1052	0.1360	0.1636	0.1881	0.2099
23	0.0002	0.0101	0.0376	0.0717	0.1054	0.1364	0.1641	0.1888	0.2107
24	0.0002	0.0101	0.0376	0.0718	0.1056	0.1367	0.1646	0.1894	0.2115
25	0.0002	0.0101	0.0376	0.0719	0.1058	0.1371	0.1651	0.1900	0.2122
26	0.0002	0.0101	0.0376	0.0720	0.1060	0.1374	0.1655	0.1905	0.2128
27	0.0002	0.0101	0.0377	0.0721	0.1062	0.1376	0.1659	0.1910	0.2135
28	0.0002	0.0101	0.0377	0.0721	0.1063	0.1379	0.1662	0.1915	0.2141
29	0.0002	0.0101	0.0377	0.0722	0.1065	0.1381	0.1666	0.1920	0.2146
30	0.0002	0.0101	0.0377	0.0723	0.1066	0.1383	0.1669	0.1924	0.2151
40	0.0002	0.0101	0.0379	0.0728	0.1076	0.1400	0.1692	0.1955	0.2190
60	0.0002	0.0101	0.0380	0.0732	0.1087	0.1417	0.1717	0.1987	0.2231
120	0.0002	0.0101	0.0381	0.0738	0.1097	0.1435	0.1743	0.2022	0.2274
∞	0.0002	0.0101	0.0383	0.0743	0.1109	0.1453	0.1770	0.2058	0.2320

TABLE H

Critical Values of the *F*-Distribution ($\alpha = 0.990$)
(continued)

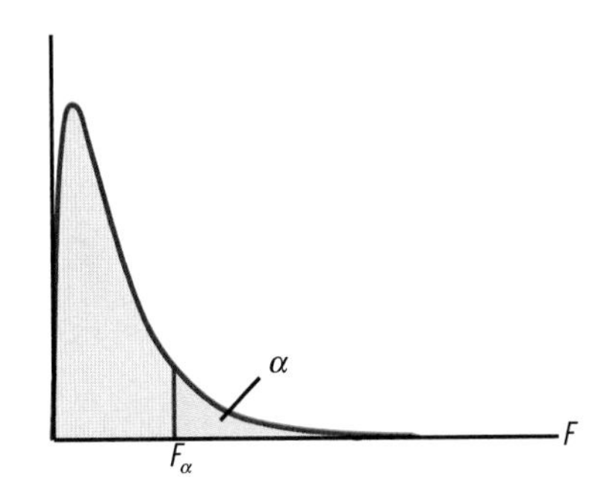

Numerator Degrees of Freedom

Denominator Degrees of Freedom	10	11	12	13	14	15	16	17	18
1	0.0996	0.1037	0.1072	0.1102	0.1128	0.1152	0.1172	0.1191	0.1207
2	0.1323	0.1388	0.1444	0.1492	0.1535	0.1573	0.1606	0.1636	0.1663
3	0.1526	0.1609	0.1680	0.1742	0.1797	0.1846	0.1890	0.1929	0.1964
4	0.1668	0.1764	0.1848	0.1921	0.1986	0.2044	0.2095	0.2142	0.2184
5	0.1774	0.1881	0.1975	0.2057	0.2130	0.2195	0.2254	0.2306	0.2354
6	0.1857	0.1973	0.2074	0.2164	0.2244	0.2316	0.2380	0.2438	0.2491
7	0.1923	0.2047	0.2155	0.2252	0.2338	0.2415	0.2484	0.2547	0.2604
8	0.1978	0.2108	0.2223	0.2324	0.2415	0.2497	0.2571	0.2638	0.2699
9	0.2023	0.2159	0.2279	0.2386	0.2482	0.2568	0.2645	0.2716	0.2780
10	0.2062	0.2203	0.2328	0.2439	0.2538	0.2628	0.2709	0.2783	0.2850
11	0.2096	0.2241	0.2370	0.2485	0.2588	0.2681	0.2765	0.2842	0.2912
12	0.2125	0.2274	0.2407	0.2525	0.2631	0.2728	0.2815	0.2894	0.2967
13	0.2151	0.2303	0.2439	0.2561	0.2670	0.2769	0.2859	0.2941	0.3015
14	0.2174	0.2329	0.2468	0.2592	0.2704	0.2806	0.2898	0.2982	0.3059
15	0.2194	0.2352	0.2494	0.2621	0.2735	0.2839	0.2933	0.3020	0.3099
16	0.2212	0.2373	0.2517	0.2647	0.2763	0.2869	0.2966	0.3054	0.3134
17	0.2229	0.2392	0.2539	0.2670	0.2789	0.2897	0.2995	0.3085	0.3167
18	0.2244	0.2409	0.2558	0.2691	0.2812	0.2922	0.3022	0.3113	0.3197
19	0.2257	0.2425	0.2576	0.2711	0.2833	0.2945	0.3046	0.3139	0.3224
20	0.2270	0.2440	0.2592	0.2729	0.2853	0.2966	0.3069	0.3163	0.3250
21	0.2281	0.2453	0.2607	0.2745	0.2871	0.2985	0.3090	0.3185	0.3273
22	0.2292	0.2465	0.2620	0.2761	0.2888	0.3003	0.3109	0.3206	0.3295
23	0.2302	0.2476	0.2633	0.2775	0.2903	0.3020	0.3127	0.3225	0.3315
24	0.2311	0.2487	0.2645	0.2788	0.2918	0.3036	0.3144	0.3243	0.3334
25	0.2320	0.2497	0.2656	0.2800	0.2931	0.3050	0.3160	0.3260	0.3352
26	0.2328	0.2506	0.2667	0.2812	0.2944	0.3064	0.3174	0.3276	0.3369
27	0.2335	0.2515	0.2676	0.2823	0.2956	0.3077	0.3188	0.3290	0.3385
28	0.2342	0.2523	0.2685	0.2833	0.2967	0.3089	0.3201	0.3304	0.3399
29	0.2348	0.2530	0.2694	0.2842	0.2977	0.3101	0.3213	0.3317	0.3413
30	0.2355	0.2537	0.2702	0.2851	0.2987	0.3111	0.3225	0.3330	0.3426
40	0.2401	0.2591	0.2763	0.2919	0.3062	0.3193	0.3313	0.3424	0.3527
60	0.2450	0.2648	0.2828	0.2993	0.3143	0.3282	0.3409	0.3528	0.3637
120	0.2502	0.2710	0.2899	0.3072	0.3232	0.3379	0.3515	0.3642	0.3760
∞	0.2558	0.2776	0.2975	0.3159	0.3329	0.3486	0.3633	0.3769	0.3897

Table H

Critical Values of the F-Distribution ($\alpha = 0.990$)

(continued)

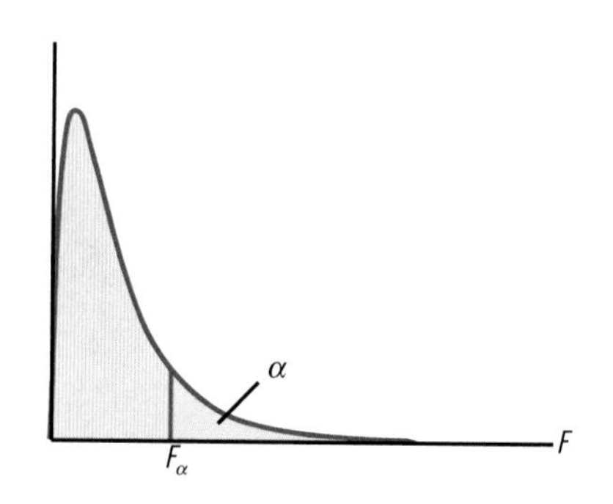

Numerator Degrees of Freedom

Denominator Degrees of Freedom

	19	20	24	30	40	60	120
1	0.1222	0.1235	0.1278	0.1322	0.1367	0.1413	0.1460
2	0.1688	0.1710	0.1781	0.1855	0.1931	0.2009	0.2089
3	0.1996	0.2025	0.2120	0.2217	0.2319	0.2424	0.2532
4	0.2222	0.2257	0.2371	0.2489	0.2612	0.2740	0.2874
5	0.2398	0.2437	0.2567	0.2703	0.2846	0.2995	0.3151
6	0.2539	0.2583	0.2727	0.2879	0.3039	0.3206	0.3383
7	0.2656	0.2704	0.2860	0.3026	0.3201	0.3386	0.3582
8	0.2754	0.2806	0.2974	0.3152	0.3341	0.3542	0.3755
9	0.2839	0.2893	0.3071	0.3261	0.3463	0.3679	0.3908
10	0.2912	0.2969	0.3156	0.3357	0.3571	0.3800	0.4045
11	0.2977	0.3036	0.3232	0.3442	0.3667	0.3908	0.4168
12	0.3033	0.3095	0.3299	0.3517	0.3753	0.4006	0.4280
13	0.3084	0.3148	0.3359	0.3585	0.3830	0.4095	0.4382
14	0.3130	0.3195	0.3413	0.3647	0.3901	0.4177	0.4476
15	0.3171	0.3238	0.3462	0.3703	0.3966	0.4251	0.4563
16	0.3209	0.3277	0.3506	0.3755	0.4025	0.4320	0.4643
17	0.3243	0.3313	0.3548	0.3802	0.4080	0.4384	0.4718
18	0.3274	0.3346	0.3585	0.3846	0.4131	0.4443	0.4788
19	0.3303	0.3376	0.3620	0.3886	0.4178	0.4498	0.4854
20	0.3330	0.3404	0.3652	0.3924	0.4221	0.4550	0.4915
21	0.3355	0.3430	0.3682	0.3959	0.4262	0.4598	0.4973
22	0.3378	0.3454	0.3710	0.3991	0.4301	0.4644	0.5028
23	0.3399	0.3476	0.3736	0.4022	0.4337	0.4687	0.5079
24	0.3419	0.3497	0.3761	0.4050	0.4371	0.4727	0.5128
25	0.3438	0.3517	0.3784	0.4077	0.4403	0.4766	0.5175
26	0.3455	0.3535	0.3805	0.4103	0.4433	0.4802	0.5219
27	0.3472	0.3553	0.3826	0.4127	0.4461	0.4836	0.5261
28	0.3487	0.3569	0.3845	0.4149	0.4488	0.4869	0.5301
29	0.3502	0.3584	0.3863	0.4171	0.4514	0.4900	0.5340
30	0.3516	0.3599	0.3880	0.4191	0.4538	0.4930	0.5376
40	0.3622	0.3711	0.4012	0.4349	0.4730	0.5165	0.5673
60	0.3739	0.3835	0.4161	0.4529	0.4952	0.5446	0.6040
120	0.3870	0.3973	0.4329	0.4738	0.5216	0.5793	0.6523
∞	0.4017	0.4130	0.4523	0.4984	0.5541	0.6247	0.7244

Table H

Critical Values of the *F*-Distribution ($\alpha = 0.975$)

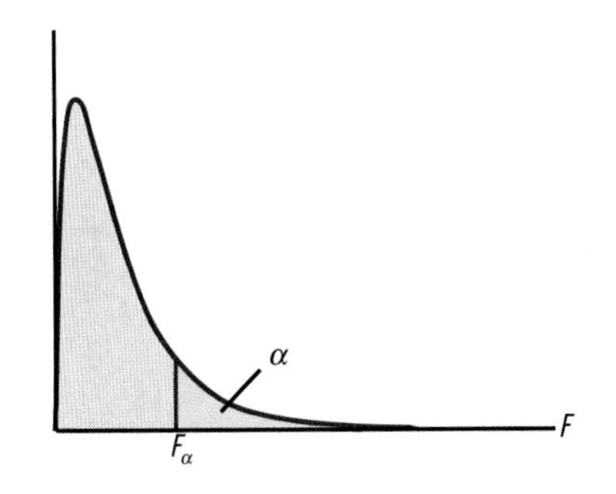

Numerator Degrees of Freedom (columns); Denominator Degrees of Freedom (rows)

	1	2	3	4	5	6	7	8	9
1	0.0015	0.0260	0.0573	0.0818	0.0999	0.1135	0.1239	0.1321	0.1387
2	0.0013	0.0256	0.0623	0.0939	0.1186	0.1377	0.1529	0.1650	0.1750
3	0.0012	0.0255	0.0648	0.1002	0.1288	0.1515	0.1698	0.1846	0.1969
4	0.0011	0.0255	0.0662	0.1041	0.1354	0.1606	0.1811	0.1979	0.2120
5	0.0011	0.0254	0.0672	0.1068	0.1399	0.1670	0.1892	0.2076	0.2230
6	0.0011	0.0254	0.0679	0.1087	0.1433	0.1718	0.1954	0.2150	0.2315
7	0.0011	0.0254	0.0684	0.1102	0.1459	0.1756	0.2002	0.2208	0.2383
8	0.0010	0.0254	0.0688	0.1114	0.1480	0.1786	0.2041	0.2256	0.2438
9	0.0010	0.0254	0.0691	0.1123	0.1497	0.1810	0.2073	0.2295	0.2484
10	0.0010	0.0254	0.0694	0.1131	0.1511	0.1831	0.2100	0.2328	0.2523
11	0.0010	0.0254	0.0696	0.1137	0.1523	0.1849	0.2123	0.2357	0.2556
12	0.0010	0.0254	0.0698	0.1143	0.1533	0.1864	0.2143	0.2381	0.2585
13	0.0010	0.0254	0.0699	0.1147	0.1541	0.1877	0.2161	0.2403	0.2611
14	0.0010	0.0254	0.0700	0.1152	0.1549	0.1888	0.2176	0.2422	0.2633
15	0.0010	0.0254	0.0702	0.1155	0.1556	0.1898	0.2189	0.2438	0.2653
16	0.0010	0.0254	0.0703	0.1158	0.1562	0.1907	0.2201	0.2453	0.2671
17	0.0010	0.0254	0.0704	0.1161	0.1567	0.1915	0.2212	0.2467	0.2687
18	0.0010	0.0254	0.0704	0.1164	0.1572	0.1922	0.2222	0.2479	0.2702
19	0.0010	0.0254	0.0705	0.1166	0.1576	0.1929	0.2231	0.2490	0.2715
20	0.0010	0.0253	0.0706	0.1168	0.1580	0.1935	0.2239	0.2500	0.2727
21	0.0010	0.0253	0.0706	0.1170	0.1584	0.1940	0.2246	0.2510	0.2738
22	0.0010	0.0253	0.0707	0.1172	0.1587	0.1945	0.2253	0.2518	0.2749
23	0.0010	0.0253	0.0708	0.1173	0.1590	0.1950	0.2259	0.2526	0.2758
24	0.0010	0.0253	0.0708	0.1175	0.1593	0.1954	0.2265	0.2533	0.2767
25	0.0010	0.0253	0.0708	0.1176	0.1595	0.1958	0.2270	0.2540	0.2775
26	0.0010	0.0253	0.0709	0.1178	0.1598	0.1962	0.2275	0.2547	0.2783
27	0.0010	0.0253	0.0709	0.1179	0.1600	0.1965	0.2280	0.2552	0.2790
28	0.0010	0.0253	0.0710	0.1180	0.1602	0.1968	0.2284	0.2558	0.2797
29	0.0010	0.0253	0.0710	0.1181	0.1604	0.1971	0.2288	0.2563	0.2803
30	0.0010	0.0253	0.0710	0.1182	0.1606	0.1974	0.2292	0.2568	0.2809
40	0.0010	0.0253	0.0712	0.1189	0.1619	0.1995	0.2321	0.2604	0.2853
60	0.0010	0.0253	0.0715	0.1196	0.1633	0.2017	0.2351	0.2642	0.2899
120	0.0010	0.0253	0.0717	0.1203	0.1648	0.2039	0.2382	0.2682	0.2948
∞	0.0010	0.0253	0.0719	0.1211	0.1662	0.2062	0.2414	0.2725	0.3000

TABLE H

Critical Values of the F-Distribution ($\alpha = 0.975$)

(continued)

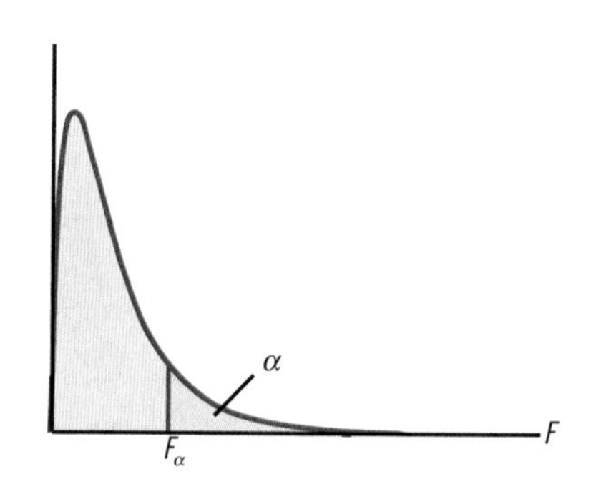

Numerator Degrees of Freedom

Denominator Degrees of Freedom

	10	11	12	13	14	15	16	17	18
1	0.1442	0.1487	0.1526	0.1559	0.1588	0.1613	0.1635	0.1655	0.1673
2	0.1833	0.1903	0.1962	0.2014	0.2059	0.2099	0.2134	0.2165	0.2193
3	0.2072	0.2160	0.2235	0.2300	0.2358	0.2408	0.2453	0.2493	0.2529
4	0.2238	0.2339	0.2426	0.2503	0.2569	0.2629	0.2681	0.2729	0.2771
5	0.2361	0.2473	0.2570	0.2655	0.2730	0.2796	0.2855	0.2909	0.2957
6	0.2456	0.2577	0.2682	0.2774	0.2856	0.2929	0.2993	0.3052	0.3105
7	0.2532	0.2661	0.2773	0.2871	0.2959	0.3036	0.3106	0.3169	0.3226
8	0.2594	0.2729	0.2848	0.2952	0.3044	0.3126	0.3200	0.3267	0.3327
9	0.2646	0.2787	0.2910	0.3019	0.3116	0.3202	0.3280	0.3350	0.3414
10	0.2690	0.2836	0.2964	0.3077	0.3178	0.3268	0.3349	0.3422	0.3489
11	0.2729	0.2879	0.3011	0.3127	0.3231	0.3325	0.3409	0.3485	0.3554
12	0.2762	0.2916	0.3051	0.3171	0.3279	0.3375	0.3461	0.3540	0.3612
13	0.2791	0.2948	0.3087	0.3210	0.3320	0.3419	0.3508	0.3589	0.3663
14	0.2817	0.2977	0.3119	0.3245	0.3357	0.3458	0.3550	0.3633	0.3709
15	0.2840	0.3003	0.3147	0.3276	0.3391	0.3494	0.3587	0.3672	0.3750
16	0.2860	0.3026	0.3173	0.3304	0.3421	0.3526	0.3621	0.3708	0.3787
17	0.2879	0.3047	0.3196	0.3329	0.3448	0.3555	0.3652	0.3741	0.3821
18	0.2896	0.3066	0.3217	0.3352	0.3473	0.3582	0.3681	0.3770	0.3853
19	0.2911	0.3084	0.3237	0.3373	0.3496	0.3606	0.3706	0.3798	0.3881
20	0.2925	0.3100	0.3254	0.3393	0.3517	0.3629	0.3730	0.3823	0.3908
21	0.2938	0.3114	0.3271	0.3410	0.3536	0.3649	0.3752	0.3846	0.3932
22	0.2950	0.3128	0.3286	0.3427	0.3554	0.3668	0.3773	0.3868	0.3955
23	0.2961	0.3140	0.3300	0.3442	0.3570	0.3686	0.3792	0.3888	0.3976
24	0.2971	0.3152	0.3313	0.3456	0.3586	0.3703	0.3809	0.3907	0.3996
25	0.2981	0.3163	0.3325	0.3470	0.3600	0.3718	0.3826	0.3924	0.4014
26	0.2990	0.3173	0.3336	0.3482	0.3614	0.3733	0.3841	0.3940	0.4031
27	0.2998	0.3183	0.3347	0.3494	0.3626	0.3746	0.3856	0.3956	0.4048
28	0.3006	0.3191	0.3357	0.3505	0.3638	0.3759	0.3869	0.3970	0.4063
29	0.3013	0.3200	0.3366	0.3515	0.3649	0.3771	0.3882	0.3984	0.4077
30	0.3020	0.3208	0.3375	0.3525	0.3660	0.3783	0.3894	0.3997	0.4091
40	0.3072	0.3267	0.3441	0.3598	0.3739	0.3868	0.3986	0.4095	0.4194
60	0.3127	0.3329	0.3512	0.3676	0.3825	0.3962	0.4087	0.4201	0.4308
120	0.3185	0.3397	0.3588	0.3761	0.3919	0.4063	0.4196	0.4319	0.4433
∞	0.3247	0.3469	0.3670	0.3853	0.4021	0.4175	0.4317	0.4450	0.4573

TABLE H

Critical Values of the F-Distribution ($\alpha = 0.975$)
(continued)

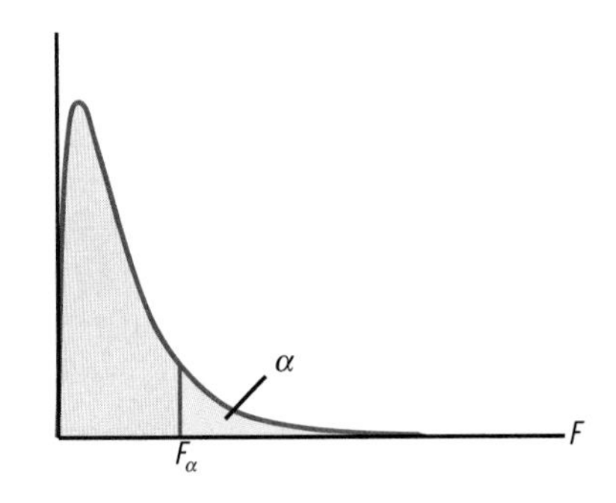

Numerator Degrees of Freedom

Denominator Degrees of Freedom	19	20	24	30	40	60	120
1	0.1689	0.1703	0.1749	0.1796	0.1844	0.1892	0.1941
2	0.2219	0.2242	0.2315	0.2391	0.2469	0.2548	0.2628
3	0.2562	0.2592	0.2687	0.2786	0.2887	0.2992	0.3099
4	0.2810	0.2845	0.2959	0.3077	0.3199	0.3325	0.3455
5	0.3001	0.3040	0.3170	0.3304	0.3444	0.3589	0.3740
6	0.3153	0.3197	0.3339	0.3488	0.3644	0.3806	0.3976
7	0.3278	0.3325	0.3480	0.3642	0.3811	0.3989	0.4176
8	0.3383	0.3433	0.3598	0.3772	0.3954	0.4147	0.4349
9	0.3472	0.3525	0.3700	0.3884	0.4078	0.4284	0.4501
10	0.3550	0.3605	0.3788	0.3982	0.4187	0.4405	0.4636
11	0.3617	0.3675	0.3866	0.4069	0.4284	0.4513	0.4757
12	0.3677	0.3737	0.3935	0.4146	0.4370	0.4610	0.4867
13	0.3730	0.3792	0.3997	0.4215	0.4448	0.4698	0.4966
14	0.3778	0.3842	0.4052	0.4278	0.4519	0.4778	0.5057
15	0.3821	0.3886	0.4103	0.4334	0.4583	0.4851	0.5141
16	0.3860	0.3927	0.4148	0.4386	0.4642	0.4919	0.5219
17	0.3896	0.3964	0.4190	0.4434	0.4696	0.4981	0.5291
18	0.3928	0.3998	0.4229	0.4477	0.4747	0.5039	0.5358
19	0.3958	0.4029	0.4264	0.4518	0.4793	0.5093	0.5421
20	0.3986	0.4058	0.4297	0.4555	0.4836	0.5143	0.5480
21	0.4011	0.4084	0.4327	0.4590	0.4877	0.5190	0.5535
22	0.4035	0.4109	0.4356	0.4623	0.4914	0.5234	0.5587
23	0.4057	0.4132	0.4382	0.4653	0.4950	0.5275	0.5636
24	0.4078	0.4154	0.4407	0.4682	0.4983	0.5314	0.5683
25	0.4097	0.4174	0.4430	0.4709	0.5014	0.5351	0.5727
26	0.4115	0.4193	0.4452	0.4734	0.5044	0.5386	0.5769
27	0.4132	0.4210	0.4472	0.4758	0.5072	0.5419	0.5809
28	0.4148	0.4227	0.4491	0.4780	0.5098	0.5451	0.5847
29	0.4163	0.4243	0.4510	0.4802	0.5123	0.5481	0.5883
30	0.4178	0.4258	0.4527	0.4822	0.5147	0.5509	0.5917
40	0.4286	0.4372	0.4660	0.4978	0.5333	0.5734	0.6195
60	0.4406	0.4498	0.4808	0.5155	0.5547	0.6000	0.6536
120	0.4539	0.4638	0.4975	0.5358	0.5800	0.6325	0.6980
∞	0.4688	0.4795	0.5167	0.5597	0.6108	0.6747	0.7631

TABLE H

Critical Values of the F-Distribution ($\alpha = 0.950$)

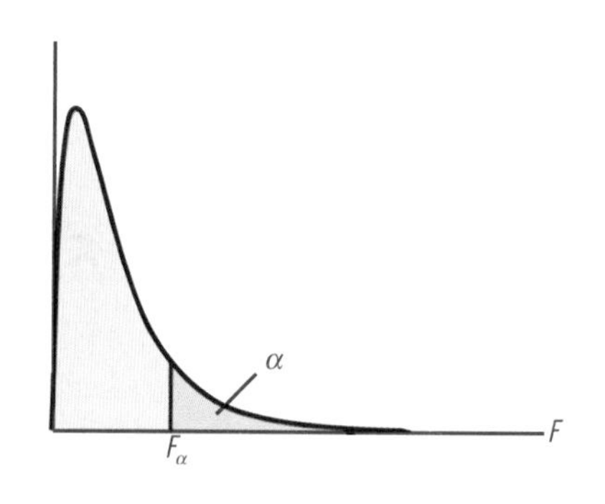

Numerator Degrees of Freedom

Denominator Degrees of Freedom	1	2	3	4	5	6	7	8	9
1	0.0062	0.0540	0.0987	0.1297	0.1513	0.1670	0.1788	0.1881	0.1954
2	0.0050	0.0526	0.1047	0.1440	0.1728	0.1944	0.2111	0.2243	0.2349
3	0.0046	0.0522	0.1078	0.1517	0.1849	0.2102	0.2301	0.2459	0.2589
4	0.0045	0.0520	0.1097	0.1565	0.1926	0.2206	0.2427	0.2606	0.2752
5	0.0043	0.0518	0.1109	0.1598	0.1980	0.2279	0.2518	0.2712	0.2872
6	0.0043	0.0517	0.1118	0.1623	0.2020	0.2334	0.2587	0.2793	0.2964
7	0.0042	0.0517	0.1125	0.1641	0.2051	0.2377	0.2641	0.2857	0.3037
8	0.0042	0.0516	0.1131	0.1655	0.2075	0.2411	0.2684	0.2909	0.3096
9	0.0042	0.0516	0.1135	0.1667	0.2095	0.2440	0.2720	0.2951	0.3146
10	0.0041	0.0516	0.1138	0.1677	0.2112	0.2463	0.2750	0.2988	0.3187
11	0.0041	0.0515	0.1141	0.1685	0.2126	0.2483	0.2775	0.3018	0.3223
12	0.0041	0.0515	0.1144	0.1692	0.2138	0.2500	0.2797	0.3045	0.3254
13	0.0041	0.0515	0.1146	0.1697	0.2148	0.2515	0.2817	0.3068	0.3281
14	0.0041	0.0515	0.1147	0.1703	0.2157	0.2528	0.2833	0.3089	0.3305
15	0.0041	0.0515	0.1149	0.1707	0.2165	0.2539	0.2848	0.3107	0.3327
16	0.0041	0.0515	0.1150	0.1711	0.2172	0.2550	0.2862	0.3123	0.3346
17	0.0040	0.0514	0.1152	0.1715	0.2178	0.2559	0.2874	0.3138	0.3363
18	0.0040	0.0514	0.1153	0.1718	0.2184	0.2567	0.2884	0.3151	0.3378
19	0.0040	0.0514	0.1154	0.1721	0.2189	0.2574	0.2894	0.3163	0.3393
20	0.0040	0.0514	0.1155	0.1723	0.2194	0.2581	0.2903	0.3174	0.3405
21	0.0040	0.0514	0.1156	0.1726	0.2198	0.2587	0.2911	0.3184	0.3417
22	0.0040	0.0514	0.1156	0.1728	0.2202	0.2593	0.2919	0.3194	0.3428
23	0.0040	0.0514	0.1157	0.1730	0.2206	0.2598	0.2926	0.3202	0.3438
24	0.0040	0.0514	0.1158	0.1732	0.2209	0.2603	0.2932	0.3210	0.3448
25	0.0040	0.0514	0.1158	0.1733	0.2212	0.2608	0.2938	0.3217	0.3456
26	0.0040	0.0514	0.1159	0.1735	0.2215	0.2612	0.2944	0.3224	0.3465
27	0.0040	0.0514	0.1159	0.1737	0.2217	0.2616	0.2949	0.3231	0.3472
28	0.0040	0.0514	0.1160	0.1738	0.2220	0.2619	0.2954	0.3237	0.3479
29	0.0040	0.0514	0.1160	0.1739	0.2222	0.2623	0.2958	0.3242	0.3486
30	0.0040	0.0514	0.1161	0.1740	0.2224	0.2626	0.2962	0.3247	0.3492
40	0.0040	0.0514	0.1164	0.1749	0.2240	0.2650	0.2994	0.3286	0.3539
60	0.0040	0.0513	0.1167	0.1758	0.2257	0.2674	0.3026	0.3327	0.3588
120	0.0039	0.0513	0.1170	0.1767	0.2274	0.2699	0.3060	0.3370	0.3640
∞	0.0039	0.0513	0.1173	0.1777	0.2291	0.2726	0.3096	0.3416	0.3695

TABLE H

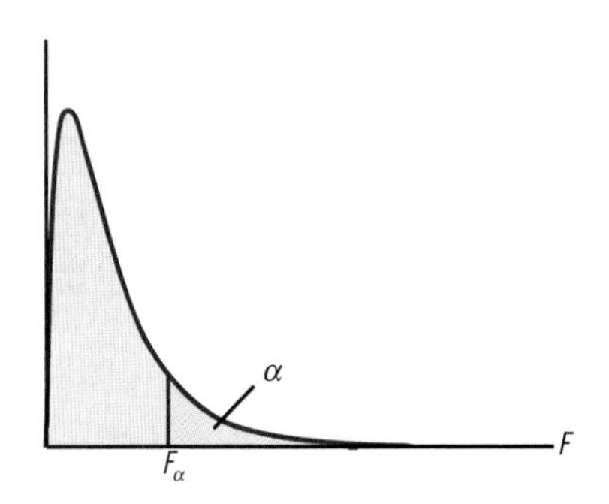

Critical Values of the *F*-Distribution ($\alpha = 0.950$)
(continued)

Numerator Degrees of Freedom

Denominator Degrees of Freedom	10	11	12	13	14	15	16	17	18
1	0.2014	0.2064	0.2106	0.2143	0.2174	0.2201	0.2225	0.2247	0.2266
2	0.2437	0.2511	0.2574	0.2628	0.2675	0.2716	0.2752	0.2784	0.2813
3	0.2697	0.2788	0.2865	0.2932	0.2991	0.3042	0.3087	0.3128	0.3165
4	0.2875	0.2979	0.3068	0.3146	0.3213	0.3273	0.3326	0.3373	0.3416
5	0.3007	0.3121	0.3220	0.3305	0.3380	0.3447	0.3506	0.3559	0.3606
6	0.3108	0.3231	0.3338	0.3430	0.3512	0.3584	0.3648	0.3706	0.3758
7	0.3189	0.3320	0.3432	0.3531	0.3618	0.3695	0.3763	0.3825	0.3881
8	0.3256	0.3392	0.3511	0.3614	0.3706	0.3787	0.3859	0.3925	0.3984
9	0.3311	0.3453	0.3576	0.3684	0.3780	0.3865	0.3941	0.4009	0.4071
10	0.3358	0.3504	0.3632	0.3744	0.3843	0.3931	0.4010	0.4082	0.4146
11	0.3398	0.3549	0.3680	0.3796	0.3898	0.3989	0.4071	0.4145	0.4212
12	0.3433	0.3587	0.3722	0.3841	0.3946	0.4040	0.4124	0.4201	0.4270
13	0.3464	0.3621	0.3759	0.3881	0.3988	0.4085	0.4171	0.4250	0.4321
14	0.3491	0.3651	0.3792	0.3916	0.4026	0.4125	0.4214	0.4294	0.4367
15	0.3515	0.3678	0.3821	0.3948	0.4060	0.4161	0.4251	0.4333	0.4408
16	0.3537	0.3702	0.3848	0.3976	0.4091	0.4193	0.4285	0.4369	0.4445
17	0.3556	0.3724	0.3872	0.4002	0.4118	0.4222	0.4316	0.4402	0.4479
18	0.3574	0.3744	0.3893	0.4026	0.4144	0.4249	0.4345	0.4431	0.4510
19	0.3590	0.3762	0.3913	0.4047	0.4167	0.4274	0.4371	0.4459	0.4539
20	0.3605	0.3779	0.3931	0.4067	0.4188	0.4296	0.4395	0.4484	0.4565
21	0.3618	0.3794	0.3948	0.4085	0.4207	0.4317	0.4417	0.4507	0.4589
22	0.3631	0.3808	0.3964	0.4102	0.4225	0.4336	0.4437	0.4528	0.4612
23	0.3643	0.3821	0.3978	0.4117	0.4242	0.4354	0.4456	0.4548	0.4632
24	0.3653	0.3833	0.3991	0.4132	0.4258	0.4371	0.4473	0.4567	0.4652
25	0.3663	0.3844	0.4004	0.4145	0.4272	0.4386	0.4490	0.4584	0.4670
26	0.3673	0.3855	0.4015	0.4158	0.4286	0.4401	0.4505	0.4600	0.4687
27	0.3681	0.3864	0.4026	0.4170	0.4299	0.4415	0.4520	0.4616	0.4703
28	0.3689	0.3874	0.4036	0.4181	0.4311	0.4427	0.4533	0.4630	0.4718
29	0.3697	0.3882	0.4046	0.4191	0.4322	0.4439	0.4546	0.4643	0.4732
30	0.3704	0.3890	0.4055	0.4201	0.4332	0.4451	0.4558	0.4656	0.4746
40	0.3758	0.3951	0.4122	0.4275	0.4412	0.4537	0.4650	0.4753	0.4848
60	0.3815	0.4016	0.4194	0.4354	0.4499	0.4629	0.4749	0.4858	0.4959
120	0.3876	0.4085	0.4272	0.4440	0.4592	0.4730	0.4857	0.4973	0.5081
∞	0.3940	0.4159	0.4355	0.4532	0.4693	0.4841	0.4976	0.5101	0.5217

Table H

Critical Values of the F-Distribution ($\alpha = 0.950$)

(continued)

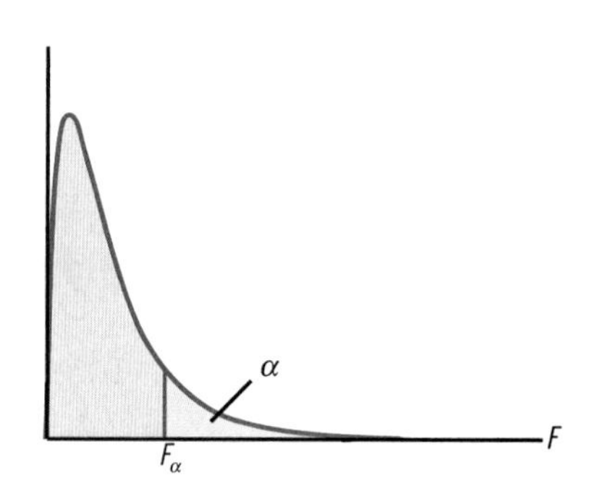

Numerator Degrees of Freedom

Denominator Degrees of Freedom

	19	20	24	30	40	60	120
1	0.2283	0.2298	0.2348	0.2398	0.2448	0.2499	0.2551
2	0.2839	0.2863	0.2939	0.3016	0.3094	0.3174	0.3255
3	0.3198	0.3227	0.3324	0.3422	0.3523	0.3626	0.3731
4	0.3454	0.3489	0.3602	0.3718	0.3837	0.3960	0.4086
5	0.3650	0.3689	0.3816	0.3947	0.4083	0.4222	0.4367
6	0.3805	0.3848	0.3987	0.4131	0.4281	0.4436	0.4598
7	0.3932	0.3978	0.4128	0.4284	0.4446	0.4616	0.4792
8	0.4038	0.4087	0.4246	0.4413	0.4587	0.4769	0.4959
9	0.4128	0.4179	0.4347	0.4523	0.4708	0.4902	0.5105
10	0.4205	0.4259	0.4435	0.4620	0.4814	0.5019	0.5234
11	0.4273	0.4329	0.4512	0.4705	0.4908	0.5122	0.5350
12	0.4333	0.4391	0.4580	0.4780	0.4991	0.5215	0.5453
13	0.4386	0.4445	0.4641	0.4847	0.5066	0.5299	0.5548
14	0.4433	0.4494	0.4695	0.4908	0.5134	0.5376	0.5634
15	0.4476	0.4539	0.4745	0.4963	0.5196	0.5445	0.5713
16	0.4515	0.4579	0.4789	0.5013	0.5253	0.5509	0.5785
17	0.4550	0.4615	0.4830	0.5059	0.5305	0.5568	0.5853
18	0.4582	0.4649	0.4868	0.5102	0.5353	0.5623	0.5916
19	0.4612	0.4679	0.4902	0.5141	0.5397	0.5674	0.5974
20	0.4639	0.4708	0.4934	0.5177	0.5438	0.5721	0.6029
21	0.4665	0.4734	0.4964	0.5210	0.5477	0.5765	0.6080
22	0.4688	0.4758	0.4991	0.5242	0.5512	0.5806	0.6129
23	0.4710	0.4781	0.5017	0.5271	0.5546	0.5845	0.6174
24	0.4730	0.4802	0.5041	0.5298	0.5577	0.5882	0.6217
25	0.4749	0.4822	0.5063	0.5324	0.5607	0.5916	0.6258
26	0.4767	0.4840	0.5084	0.5348	0.5635	0.5949	0.6297
27	0.4784	0.4858	0.5104	0.5371	0.5661	0.5980	0.6333
28	0.4799	0.4874	0.5123	0.5393	0.5686	0.6009	0.6368
29	0.4814	0.4890	0.5141	0.5413	0.5710	0.6037	0.6402
30	0.4828	0.4904	0.5157	0.5432	0.5733	0.6064	0.6434
40	0.4935	0.5016	0.5286	0.5581	0.5907	0.6272	0.6688
60	0.5052	0.5138	0.5428	0.5749	0.6108	0.6518	0.6998
120	0.5181	0.5273	0.5588	0.5940	0.6343	0.6815	0.7397
∞	0.5325	0.5425	0.5770	0.6164	0.6627	0.7198	0.7975

Table H

Critical Values of the *F*-Distribution ($\alpha = 0.900$)

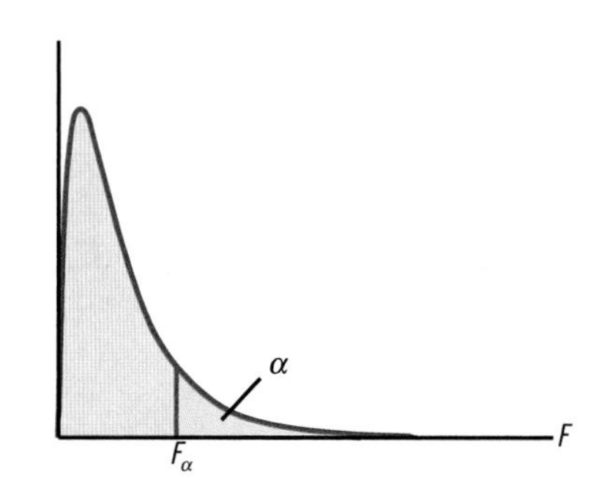

Numerator Degrees of Freedom

Denominator Degrees of Freedom

	1	2	3	4	5	6	7	8	9
1	0.0251	0.1173	0.1806	0.2200	0.2463	0.2648	0.2786	0.2892	0.2976
2	0.0202	0.1111	0.1831	0.2312	0.2646	0.2887	0.3070	0.3212	0.3326
3	0.0187	0.1091	0.1855	0.2386	0.2763	0.3041	0.3253	0.3420	0.3555
4	0.0179	0.1082	0.1872	0.2435	0.2841	0.3144	0.3378	0.3563	0.3714
5	0.0175	0.1076	0.1884	0.2469	0.2896	0.3218	0.3468	0.3668	0.3831
6	0.0172	0.1072	0.1892	0.2494	0.2937	0.3274	0.3537	0.3748	0.3920
7	0.0170	0.1070	0.1899	0.2513	0.2969	0.3317	0.3591	0.3811	0.3992
8	0.0168	0.1068	0.1904	0.2528	0.2995	0.3352	0.3634	0.3862	0.4050
9	0.0167	0.1066	0.1908	0.2541	0.3015	0.3381	0.3670	0.3904	0.4098
10	0.0166	0.1065	0.1912	0.2551	0.3033	0.3405	0.3700	0.3940	0.4139
11	0.0165	0.1064	0.1915	0.2560	0.3047	0.3425	0.3726	0.3971	0.4173
12	0.0165	0.1063	0.1917	0.2567	0.3060	0.3443	0.3748	0.3997	0.4204
13	0.0164	0.1062	0.1919	0.2573	0.3071	0.3458	0.3767	0.4020	0.4230
14	0.0164	0.1062	0.1921	0.2579	0.3080	0.3471	0.3784	0.4040	0.4253
15	0.0163	0.1061	0.1923	0.2584	0.3088	0.3483	0.3799	0.4058	0.4274
16	0.0163	0.1061	0.1924	0.2588	0.3096	0.3493	0.3812	0.4074	0.4293
17	0.0163	0.1060	0.1926	0.2592	0.3102	0.3503	0.3824	0.4089	0.4309
18	0.0162	0.1060	0.1927	0.2595	0.3108	0.3511	0.3835	0.4102	0.4325
19	0.0162	0.1059	0.1928	0.2598	0.3114	0.3519	0.3845	0.4114	0.4338
20	0.0162	0.1059	0.1929	0.2601	0.3119	0.3526	0.3854	0.4124	0.4351
21	0.0162	0.1059	0.1930	0.2604	0.3123	0.3532	0.3862	0.4134	0.4363
22	0.0162	0.1059	0.1930	0.2606	0.3127	0.3538	0.3870	0.4143	0.4373
23	0.0161	0.1058	0.1931	0.2608	0.3131	0.3543	0.3877	0.4152	0.4383
24	0.0161	0.1058	0.1932	0.2610	0.3134	0.3548	0.3883	0.4160	0.4392
25	0.0161	0.1058	0.1932	0.2612	0.3137	0.3553	0.3889	0.4167	0.4401
26	0.0161	0.1058	0.1933	0.2614	0.3140	0.3557	0.3894	0.4174	0.4408
27	0.0161	0.1058	0.1934	0.2615	0.3143	0.3561	0.3900	0.4180	0.4416
28	0.0161	0.1058	0.1934	0.2617	0.3146	0.3565	0.3904	0.4186	0.4423
29	0.0161	0.1057	0.1935	0.2618	0.3148	0.3568	0.3909	0.4191	0.4429
30	0.0161	0.1057	0.1935	0.2620	0.3151	0.3571	0.3913	0.4196	0.4435
40	0.0160	0.1056	0.1938	0.2629	0.3167	0.3596	0.3945	0.4235	0.4480
60	0.0159	0.1055	0.1941	0.2639	0.3184	0.3621	0.3977	0.4275	0.4528
120	0.0159	0.1055	0.1945	0.2649	0.3202	0.3647	0.4012	0.4317	0.4578
∞	0.0158	0.1054	0.1948	0.2659	0.3221	0.3674	0.4047	0.4362	0.4631

TABLE H

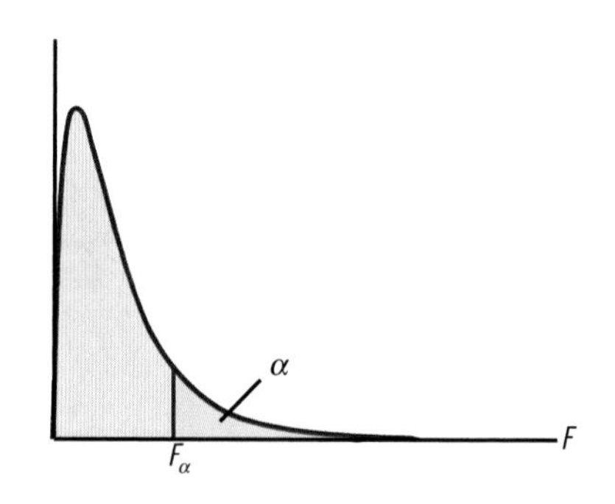

Critical Values of the F-Distribution ($\alpha = 0.900$)

(continued)

Numerator Degrees of Freedom

Denominator Degrees of Freedom

	10	11	12	13	14	15	16	17	18
1	0.3044	0.3101	0.3148	0.3189	0.3224	0.3254	0.3281	0.3304	0.3326
2	0.3419	0.3497	0.3563	0.3619	0.3668	0.3710	0.3748	0.3781	0.3811
3	0.3666	0.3759	0.3838	0.3906	0.3965	0.4016	0.4062	0.4103	0.4139
4	0.3838	0.3943	0.4032	0.4109	0.4176	0.4235	0.4287	0.4333	0.4375
5	0.3966	0.4080	0.4177	0.4261	0.4335	0.4399	0.4457	0.4508	0.4554
6	0.4064	0.4186	0.4290	0.4380	0.4459	0.4529	0.4591	0.4646	0.4696
7	0.4143	0.4271	0.4381	0.4476	0.4560	0.4634	0.4699	0.4758	0.4811
8	0.4207	0.4340	0.4455	0.4555	0.4643	0.4720	0.4789	0.4851	0.4907
9	0.4260	0.4399	0.4518	0.4621	0.4713	0.4793	0.4865	0.4930	0.4988
10	0.4306	0.4448	0.4571	0.4678	0.4772	0.4856	0.4931	0.4998	0.5058
11	0.4344	0.4490	0.4617	0.4727	0.4824	0.4910	0.4987	0.5056	0.5119
12	0.4378	0.4527	0.4657	0.4770	0.4869	0.4958	0.5037	0.5108	0.5172
13	0.4408	0.4560	0.4692	0.4807	0.4909	0.5000	0.5081	0.5154	0.5220
14	0.4434	0.4589	0.4723	0.4841	0.4945	0.5037	0.5120	0.5194	0.5262
15	0.4457	0.4614	0.4751	0.4871	0.4976	0.5070	0.5155	0.5231	0.5300
16	0.4478	0.4638	0.4776	0.4897	0.5005	0.5101	0.5187	0.5264	0.5334
17	0.4497	0.4658	0.4799	0.4922	0.5031	0.5128	0.5215	0.5294	0.5365
18	0.4514	0.4677	0.4819	0.4944	0.5054	0.5153	0.5241	0.5321	0.5394
19	0.4530	0.4694	0.4838	0.4964	0.5076	0.5176	0.5265	0.5346	0.5420
20	0.4544	0.4710	0.4855	0.4983	0.5096	0.5197	0.5287	0.5370	0.5444
21	0.4557	0.4725	0.4871	0.5000	0.5114	0.5216	0.5308	0.5391	0.5466
22	0.4569	0.4738	0.4886	0.5015	0.5131	0.5234	0.5327	0.5411	0.5487
23	0.4580	0.4750	0.4899	0.5030	0.5146	0.5250	0.5344	0.5429	0.5506
24	0.4590	0.4762	0.4912	0.5044	0.5161	0.5266	0.5360	0.5446	0.5524
25	0.4600	0.4773	0.4923	0.5056	0.5174	0.5280	0.5375	0.5462	0.5540
26	0.4609	0.4783	0.4934	0.5068	0.5187	0.5294	0.5390	0.5477	0.5556
27	0.4617	0.4792	0.4944	0.5079	0.5199	0.5306	0.5403	0.5491	0.5571
28	0.4625	0.4801	0.4954	0.5089	0.5210	0.5318	0.5415	0.5504	0.5584
29	0.4633	0.4809	0.4963	0.5099	0.5220	0.5329	0.5427	0.5516	0.5597
30	0.4639	0.4816	0.4971	0.5108	0.5230	0.5340	0.5438	0.5528	0.5610
40	0.4691	0.4874	0.5035	0.5177	0.5305	0.5419	0.5522	0.5616	0.5702
60	0.4746	0.4936	0.5103	0.5251	0.5384	0.5504	0.5613	0.5712	0.5803
120	0.4804	0.5001	0.5175	0.5331	0.5470	0.5597	0.5712	0.5817	0.5914
∞	0.4865	0.5071	0.5253	0.5417	0.5564	0.5698	0.5820	0.5932	0.6036

Table H

Critical Values of the F-Distribution ($\alpha = 0.900$)
(continued)

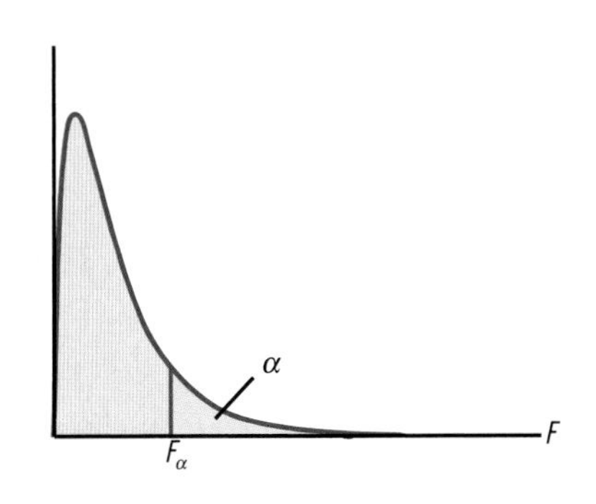

Numerator Degrees of Freedom

Denominator Degrees of Freedom

	19	20	24	30	40	60	120
1	0.3345	0.3362	0.3416	0.3471	0.3527	0.3583	0.3639
2	0.3838	0.3862	0.3940	0.4018	0.4098	0.4178	0.4260
3	0.4172	0.4202	0.4297	0.4394	0.4492	0.4593	0.4695
4	0.4412	0.4447	0.4556	0.4668	0.4783	0.4900	0.5019
5	0.4596	0.4633	0.4755	0.4880	0.5008	0.5140	0.5275
6	0.4741	0.4782	0.4914	0.5050	0.5190	0.5334	0.5483
7	0.4859	0.4903	0.5044	0.5190	0.5340	0.5496	0.5658
8	0.4958	0.5004	0.5153	0.5308	0.5468	0.5634	0.5807
9	0.5041	0.5089	0.5246	0.5408	0.5578	0.5754	0.5937
10	0.5113	0.5163	0.5326	0.5496	0.5673	0.5858	0.6052
11	0.5176	0.5228	0.5397	0.5573	0.5757	0.5951	0.6154
12	0.5231	0.5284	0.5459	0.5641	0.5832	0.6033	0.6245
13	0.5280	0.5335	0.5514	0.5702	0.5900	0.6108	0.6328
14	0.5323	0.5380	0.5564	0.5757	0.5960	0.6175	0.6403
15	0.5363	0.5420	0.5608	0.5806	0.6015	0.6237	0.6472
16	0.5398	0.5457	0.5649	0.5851	0.6066	0.6293	0.6536
17	0.5431	0.5490	0.5686	0.5893	0.6112	0.6345	0.6595
18	0.5460	0.5521	0.5720	0.5930	0.6154	0.6393	0.6649
19	0.5487	0.5549	0.5751	0.5965	0.6194	0.6437	0.6700
20	0.5512	0.5575	0.5780	0.5998	0.6230	0.6479	0.6747
21	0.5535	0.5598	0.5807	0.6028	0.6264	0.6517	0.6792
22	0.5557	0.5621	0.5831	0.6056	0.6295	0.6554	0.6833
23	0.5576	0.5641	0.5854	0.6082	0.6325	0.6587	0.6873
24	0.5595	0.5660	0.5876	0.6106	0.6353	0.6619	0.6910
25	0.5612	0.5678	0.5896	0.6129	0.6379	0.6649	0.6945
26	0.5629	0.5695	0.5915	0.6150	0.6403	0.6678	0.6978
27	0.5644	0.5711	0.5933	0.6171	0.6427	0.6705	0.7010
28	0.5658	0.5726	0.5950	0.6190	0.6449	0.6730	0.7040
29	0.5672	0.5740	0.5966	0.6208	0.6469	0.6754	0.7068
30	0.5684	0.5753	0.5980	0.6225	0.6489	0.6777	0.7095
40	0.5781	0.5854	0.6095	0.6356	0.6642	0.6957	0.7312
60	0.5887	0.5964	0.6222	0.6504	0.6816	0.7167	0.7574
120	0.6003	0.6085	0.6364	0.6672	0.7019	0.7421	0.7908
∞	0.6132	0.6221	0.6524	0.6866	0.7263	0.7743	0.8385

Table H

Critical Values of the F-Distribution ($\alpha = 0.100$)

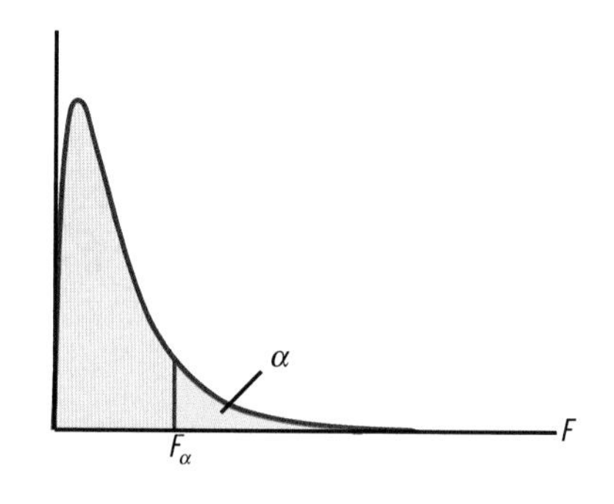

Numerator Degrees of Freedom

Denominator Degrees of Freedom

	1	2	3	4	5	6	7	8	9
1	39.8635	49.5000	53.5932	55.8330	57.2401	58.2044	58.9060	59.4390	59.8576
2	8.5263	9.0000	9.1618	9.2434	9.2926	9.3255	9.3491	9.3668	9.3805
3	5.5383	5.4624	5.3908	5.3426	5.3092	5.2847	5.2662	5.2517	5.2400
4	4.5448	4.3246	4.1909	4.1072	4.0506	4.0097	3.9790	3.9549	3.9357
5	4.0604	3.7797	3.6195	3.5202	3.4530	3.4045	3.3679	3.3393	3.3163
6	3.7759	3.4633	3.2888	3.1808	3.1075	3.0546	3.0145	2.9830	2.9577
7	3.5894	3.2574	3.0741	2.9605	2.8833	2.8274	2.7849	2.7516	2.7247
8	3.4579	3.1131	2.9238	2.8064	2.7264	2.6683	2.6241	2.5893	2.5612
9	3.3603	3.0065	2.8129	2.6927	2.6106	2.5509	2.5053	2.4694	2.4403
10	3.2850	2.9245	2.7277	2.6053	2.5216	2.4606	2.4140	2.3772	2.3473
11	3.2252	2.8595	2.6602	2.5362	2.4512	2.3891	2.3416	2.3040	2.2735
12	3.1765	2.8068	2.6055	2.4801	2.3940	2.3310	2.2828	2.2446	2.2135
13	3.1362	2.7632	2.5603	2.4337	2.3467	2.2830	2.2341	2.1953	2.1638
14	3.1022	2.7265	2.5222	2.3947	2.3069	2.2426	2.1931	2.1539	2.1220
15	3.0732	2.6952	2.4898	2.3614	2.2730	2.2081	2.1582	2.1185	2.0862
16	3.0481	2.6682	2.4618	2.3327	2.2438	2.1783	2.1280	2.0880	2.0553
17	3.0262	2.6446	2.4374	2.3077	2.2183	2.1524	2.1017	2.0613	2.0284
18	3.0070	2.6239	2.4160	2.2858	2.1958	2.1296	2.0785	2.0379	2.0047
19	2.9899	2.6056	2.3970	2.2663	2.1760	2.1094	2.0580	2.0171	1.9836
20	2.9747	2.5893	2.3801	2.2489	2.1582	2.0913	2.0397	1.9985	1.9649
21	2.9610	2.5746	2.3649	2.2333	2.1423	2.0751	2.0233	1.9819	1.9480
22	2.9486	2.5613	2.3512	2.2193	2.1279	2.0605	2.0084	1.9668	1.9327
23	2.9374	2.5493	2.3387	2.2065	2.1149	2.0472	1.9949	1.9531	1.9189
24	2.9271	2.5383	2.3274	2.1949	2.1030	2.0351	1.9826	1.9407	1.9063
25	2.9177	2.5283	2.3170	2.1842	2.0922	2.0241	1.9714	1.9292	1.8947
26	2.9091	2.5191	2.3075	2.1745	2.0822	2.0139	1.9610	1.9188	1.8841
27	2.9012	2.5106	2.2987	2.1655	2.0730	2.0045	1.9515	1.9091	1.8743
28	2.8938	2.5028	2.2906	2.1571	2.0645	1.9959	1.9427	1.9001	1.8652
29	2.8870	2.4955	2.2831	2.1494	2.0566	1.9878	1.9345	1.8918	1.8568
30	2.8807	2.4887	2.2761	2.1422	2.0492	1.9803	1.9269	1.8841	1.8490
40	2.8354	2.4404	2.2261	2.0909	1.9968	1.9269	1.8725	1.8289	1.7929
60	2.7911	2.3933	2.1774	2.0410	1.9457	1.8747	1.8194	1.7748	1.7380
120	2.7478	2.3473	2.1300	1.9923	1.8959	1.8238	1.7675	1.7220	1.6842
∞	2.7055	2.3026	2.0838	1.9449	1.8473	1.7741	1.7167	1.6702	1.6315

Table H

Critical Values of the *F*-Distribution ($\alpha = 0.100$) (continued)

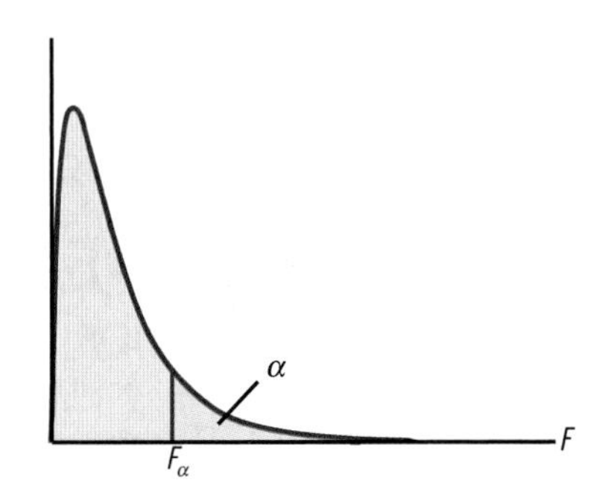

Numerator Degrees of Freedom (columns); **Denominator Degrees of Freedom** (rows)

	10	11	12	13	14	15	16	17	18
1	60.1950	60.4727	60.7052	60.9028	61.0727	61.2203	61.3499	61.4644	61.5664
2	9.3916	9.4006	9.4081	9.4145	9.4200	9.4247	9.4289	9.4325	9.4358
3	5.2304	5.2224	5.2156	5.2098	5.2047	5.2003	5.1964	5.1929	5.1898
4	3.9199	3.9067	3.8955	3.8859	3.8776	3.8704	3.8639	3.8582	3.8531
5	3.2974	3.2816	3.2682	3.2567	3.2468	3.2380	3.2303	3.2234	3.2172
6	2.9369	2.9195	2.9047	2.8920	2.8809	2.8712	2.8626	2.8550	2.8481
7	2.7025	2.6839	2.6681	2.6545	2.6426	2.6322	2.6230	2.6148	2.6074
8	2.5380	2.5186	2.5020	2.4876	2.4752	2.4642	2.4545	2.4458	2.4380
9	2.4163	2.3961	2.3789	2.3640	2.3510	2.3396	2.3295	2.3205	2.3123
10	2.3226	2.3018	2.2841	2.2687	2.2553	2.2435	2.2330	2.2237	2.2153
11	2.2482	2.2269	2.2087	2.1930	2.1792	2.1671	2.1563	2.1467	2.1380
12	2.1878	2.1660	2.1474	2.1313	2.1173	2.1049	2.0938	2.0839	2.0750
13	2.1376	2.1155	2.0966	2.0802	2.0658	2.0532	2.0419	2.0318	2.0227
14	2.0954	2.0729	2.0537	2.0370	2.0224	2.0095	1.9981	1.9878	1.9785
15	2.0593	2.0366	2.0171	2.0001	1.9853	1.9722	1.9605	1.9501	1.9407
16	2.0281	2.0051	1.9854	1.9682	1.9532	1.9399	1.9281	1.9175	1.9079
17	2.0009	1.9777	1.9577	1.9404	1.9252	1.9117	1.8997	1.8889	1.8792
18	1.9770	1.9535	1.9333	1.9158	1.9004	1.8868	1.8747	1.8638	1.8539
19	1.9557	1.9321	1.9117	1.8940	1.8785	1.8647	1.8524	1.8414	1.8314
20	1.9367	1.9129	1.8924	1.8745	1.8588	1.8449	1.8325	1.8214	1.8113
21	1.9197	1.8956	1.8750	1.8570	1.8412	1.8271	1.8146	1.8034	1.7932
22	1.9043	1.8801	1.8593	1.8411	1.8252	1.8111	1.7984	1.7871	1.7768
23	1.8903	1.8659	1.8450	1.8267	1.8107	1.7964	1.7837	1.7723	1.7619
24	1.8775	1.8530	1.8319	1.8136	1.7974	1.7831	1.7703	1.7587	1.7483
25	1.8658	1.8412	1.8200	1.8015	1.7853	1.7708	1.7579	1.7463	1.7358
26	1.8550	1.8303	1.8090	1.7904	1.7741	1.7596	1.7466	1.7349	1.7243
27	1.8451	1.8203	1.7989	1.7802	1.7638	1.7492	1.7361	1.7243	1.7137
28	1.8359	1.8110	1.7895	1.7708	1.7542	1.7395	1.7264	1.7146	1.7039
29	1.8274	1.8024	1.7808	1.7620	1.7454	1.7306	1.7174	1.7055	1.6947
30	1.8195	1.7944	1.7727	1.7538	1.7371	1.7223	1.7090	1.6970	1.6862
40	1.7627	1.7369	1.7146	1.6950	1.6778	1.6624	1.6486	1.6362	1.6249
60	1.7070	1.6805	1.6574	1.6372	1.6193	1.6034	1.5890	1.5760	1.5642
120	1.6524	1.6250	1.6012	1.5803	1.5617	1.5450	1.5300	1.5164	1.5039
∞	1.5987	1.5705	1.5458	1.5240	1.5046	1.4871	1.4714	1.4570	1.4439

TABLE H

Critical Values of the F-Distribution ($\alpha = 0.100$)
(continued)

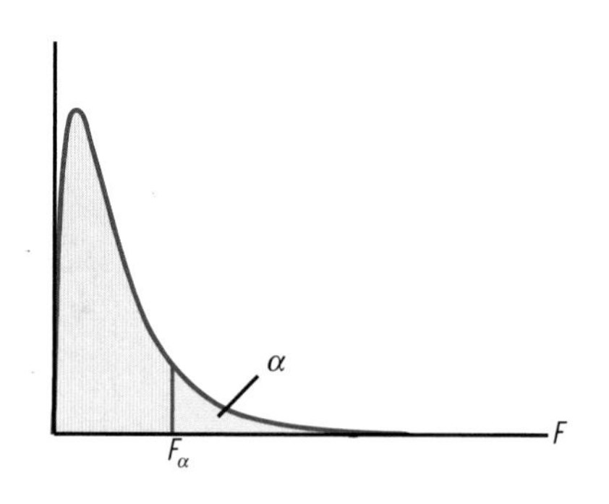

Numerator Degrees of Freedom

Denominator Degrees of Freedom

	19	20	24	30	40	60	120
1	61.6579	61.7403	62.0020	62.2650	62.5291	62.7943	63.0606
2	9.4387	9.4413	9.4496	9.4579	9.4662	9.4746	9.4829
3	5.1870	5.1845	5.1764	5.1681	5.1597	5.1512	5.1425
4	3.8485	3.8443	3.8310	3.8174	3.8036	3.7896	3.7753
5	3.2117	3.2067	3.1905	3.1741	3.1573	3.1402	3.1228
6	2.8419	2.8363	2.8183	2.8000	2.7812	2.7620	2.7423
7	2.6008	2.5947	2.5753	2.5555	2.5351	2.5142	2.4928
8	2.4310	2.4246	2.4041	2.3830	2.3614	2.3391	2.3162
9	2.3050	2.2983	2.2768	2.2547	2.2320	2.2085	2.1843
10	2.2077	2.2007	2.1784	2.1554	2.1317	2.1072	2.0818
11	2.1302	2.1230	2.1000	2.0762	2.0516	2.0261	1.9997
12	2.0670	2.0597	2.0360	2.0115	1.9861	1.9597	1.9323
13	2.0145	2.0070	1.9827	1.9576	1.9315	1.9043	1.8759
14	1.9701	1.9625	1.9377	1.9119	1.8852	1.8572	1.8280
15	1.9321	1.9243	1.8990	1.8728	1.8454	1.8168	1.7867
16	1.8992	1.8913	1.8656	1.8388	1.8108	1.7816	1.7507
17	1.8704	1.8624	1.8362	1.8090	1.7805	1.7506	1.7191
18	1.8450	1.8368	1.8103	1.7827	1.7537	1.7232	1.6910
19	1.8224	1.8142	1.7873	1.7592	1.7298	1.6988	1.6659
20	1.8022	1.7938	1.7667	1.7382	1.7083	1.6768	1.6433
21	1.7840	1.7756	1.7481	1.7193	1.6890	1.6569	1.6228
22	1.7675	1.7590	1.7312	1.7021	1.6714	1.6389	1.6041
23	1.7525	1.7439	1.7159	1.6864	1.6554	1.6224	1.5871
24	1.7388	1.7302	1.7019	1.6721	1.6407	1.6073	1.5715
25	1.7263	1.7175	1.6890	1.6589	1.6272	1.5934	1.5570
26	1.7147	1.7059	1.6771	1.6468	1.6147	1.5805	1.5437
27	1.7040	1.6951	1.6662	1.6356	1.6032	1.5686	1.5313
28	1.6941	1.6852	1.6560	1.6252	1.5925	1.5575	1.5198
29	1.6849	1.6759	1.6465	1.6155	1.5825	1.5472	1.5090
30	1.6763	1.6673	1.6377	1.6065	1.5732	1.5376	1.4989
40	1.6146	1.6052	1.5741	1.5411	1.5056	1.4672	1.4248
60	1.5534	1.5435	1.5107	1.4755	1.4373	1.3952	1.3476
120	1.4926	1.4821	1.4472	1.4094	1.3676	1.3203	1.2646
∞	1.4318	1.4206	1.3832	1.3419	1.2951	1.2400	1.1686

TABLE H

Critical Values of the *F*-Distribution ($\alpha = 0.050$)

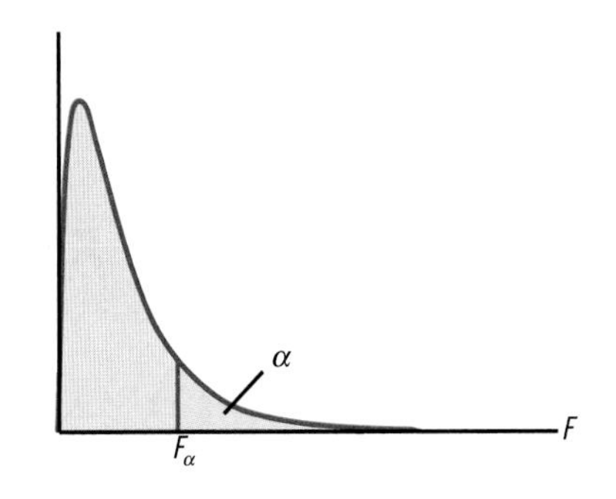

Numerator Degrees of Freedom

Denominator Degrees of Freedom

	1	2	3	4	5	6	7	8	9
1	161.4476	199.5000	215.7073	224.5832	230.1619	233.9860	236.7684	238.8827	240.5433
2	18.5128	19.0000	19.1643	19.2468	19.2964	19.3295	19.3532	19.3710	19.3848
3	10.1280	9.5521	9.2766	9.1172	9.0135	8.9406	8.8867	8.8452	8.8123
4	7.7086	6.9443	6.5914	6.3882	6.2561	6.1631	6.0942	6.0410	5.9988
5	6.6079	5.7861	5.4095	5.1922	5.0503	4.9503	4.8759	4.8183	4.7725
6	5.9874	5.1433	4.7571	4.5337	4.3874	4.2839	4.2067	4.1468	4.0990
7	5.5914	4.7374	4.3468	4.1203	3.9715	3.8660	3.7870	3.7257	3.6767
8	5.3177	4.4590	4.0662	3.8379	3.6875	3.5806	3.5005	3.4381	3.3881
9	5.1174	4.2565	3.8625	3.6331	3.4817	3.3738	3.2927	3.2296	3.1789
10	4.9646	4.1028	3.7083	3.4780	3.3258	3.2172	3.1355	3.0717	3.0204
11	4.8443	3.9823	3.5874	3.3567	3.2039	3.0946	3.0123	2.9480	2.8962
12	4.7472	3.8853	3.4903	3.2592	3.1059	2.9961	2.9134	2.8486	2.7964
13	4.6672	3.8056	3.4105	3.1791	3.0254	2.9153	2.8321	2.7669	2.7144
14	4.6001	3.7389	3.3439	3.1122	2.9582	2.8477	2.7642	2.6987	2.6458
15	4.5431	3.6823	3.2874	3.0556	2.9013	2.7905	2.7066	2.6408	2.5876
16	4.4940	3.6337	3.2389	3.0069	2.8524	2.7413	2.6572	2.5911	2.5377
17	4.4513	3.5915	3.1968	2.9647	2.8100	2.6987	2.6143	2.5480	2.4943
18	4.4139	3.5546	3.1599	2.9277	2.7729	2.6613	2.5767	2.5102	2.4563
19	4.3807	3.5219	3.1274	2.8951	2.7401	2.6283	2.5435	2.4768	2.4227
20	4.3512	3.4928	3.0984	2.8661	2.7109	2.5990	2.5140	2.4471	2.3928
21	4.3248	3.4668	3.0725	2.8401	2.6848	2.5727	2.4876	2.4205	2.3660
22	4.3009	3.4434	3.0491	2.8167	2.6613	2.5491	2.4638	2.3965	2.3419
23	4.2793	3.4221	3.0280	2.7955	2.6400	2.5277	2.4422	2.3748	2.3201
24	4.2597	3.4028	3.0088	2.7763	2.6207	2.5082	2.4226	2.3551	2.3002
25	4.2417	3.3852	2.9912	2.7587	2.6030	2.4904	2.4047	2.3371	2.2821
26	4.2252	3.3690	2.9752	2.7426	2.5868	2.4741	2.3883	2.3205	2.2655
27	4.2100	3.3541	2.9604	2.7278	2.5719	2.4591	2.3732	2.3053	2.2501
28	4.1960	3.3404	2.9467	2.7141	2.5581	2.4453	2.3593	2.2913	2.2360
29	4.1830	3.3277	2.9340	2.7014	2.5454	2.4324	2.3463	2.2783	2.2229
30	4.1709	3.3158	2.9223	2.6896	2.5336	2.4205	2.3343	2.2662	2.2107
40	4.0847	3.2317	2.8387	2.6060	2.4495	2.3359	2.2490	2.1802	2.1240
60	4.0012	3.1504	2.7581	2.5252	2.3683	2.2541	2.1665	2.0970	2.0401
120	3.9201	3.0718	2.6802	2.4472	2.2899	2.1750	2.0868	2.0164	1.9588
∞	3.8415	2.9957	2.6049	2.3719	2.2141	2.0986	2.0096	1.9384	1.8799

Table H

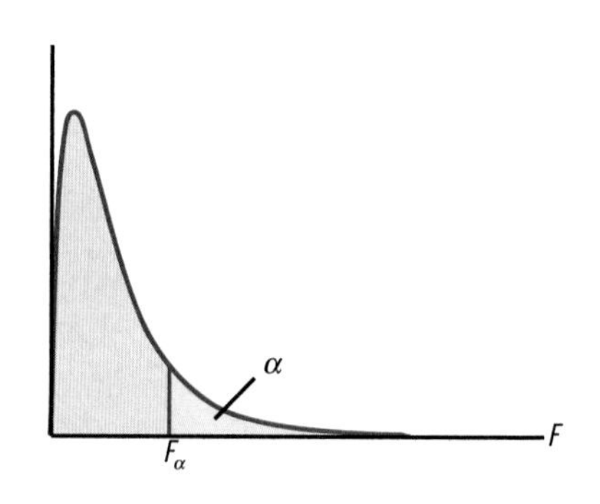

Critical Values of the F-Distribution ($\alpha = 0.050$)

(continued)

Numerator Degrees of Freedom

Denominator Degrees of Freedom

	10	11	12	13	14	15	16	17	18
1	241.8817	242.9835	243.9060	244.6898	245.3640	245.9499	246.4639	246.9184	247.3232
2	19.3959	19.4050	19.4125	19.4189	19.4244	19.4291	19.4333	19.4370	19.4402
3	8.7855	8.7633	8.7446	8.7287	8.7149	8.7029	8.6923	8.6829	8.6745
4	5.9644	5.9358	5.9117	5.8911	5.8733	5.8578	5.8441	5.8320	5.8211
5	4.7351	4.7040	4.6777	4.6552	4.6358	4.6188	4.6038	4.5904	4.5785
6	4.0600	4.0274	3.9999	3.9764	3.9559	3.9381	3.9223	3.9083	3.8957
7	3.6365	3.6030	3.5747	3.5503	3.5292	3.5107	3.4944	3.4799	3.4669
8	3.3472	3.3130	3.2839	3.2590	3.2374	3.2184	3.2016	3.1867	3.1733
9	3.1373	3.1025	3.0729	3.0475	3.0255	3.0061	2.9890	2.9737	2.9600
10	2.9782	2.9430	2.9130	2.8872	2.8647	2.8450	2.8276	2.8120	2.7980
11	2.8536	2.8179	2.7876	2.7614	2.7386	2.7186	2.7009	2.6851	2.6709
12	2.7534	2.7173	2.6866	2.6602	2.6371	2.6169	2.5989	2.5828	2.5684
13	2.6710	2.6347	2.6037	2.5769	2.5536	2.5331	2.5149	2.4987	2.4841
14	2.6022	2.5655	2.5342	2.5073	2.4837	2.4630	2.4446	2.4282	2.4134
15	2.5437	2.5068	2.4753	2.4481	2.4244	2.4034	2.3849	2.3683	2.3533
16	2.4935	2.4564	2.4247	2.3973	2.3733	2.3522	2.3335	2.3167	2.3016
17	2.4499	2.4126	2.3807	2.3531	2.3290	2.3077	2.2888	2.2719	2.2567
18	2.4117	2.3742	2.3421	2.3143	2.2900	2.2686	2.2496	2.2325	2.2172
19	2.3779	2.3402	2.3080	2.2800	2.2556	2.2341	2.2149	2.1977	2.1823
20	2.3479	2.3100	2.2776	2.2495	2.2250	2.2033	2.1840	2.1667	2.1511
21	2.3210	2.2829	2.2504	2.2222	2.1975	2.1757	2.1563	2.1389	2.1232
22	2.2967	2.2585	2.2258	2.1975	2.1727	2.1508	2.1313	2.1138	2.0980
23	2.2747	2.2364	2.2036	2.1752	2.1502	2.1282	2.1086	2.0910	2.0751
24	2.2547	2.2163	2.1834	2.1548	2.1298	2.1077	2.0880	2.0703	2.0543
25	2.2365	2.1979	2.1649	2.1362	2.1111	2.0889	2.0691	2.0513	2.0353
26	2.2197	2.1811	2.1479	2.1192	2.0939	2.0716	2.0518	2.0339	2.0178
27	2.2043	2.1655	2.1323	2.1035	2.0781	2.0558	2.0358	2.0179	2.0017
28	2.1900	2.1512	2.1179	2.0889	2.0635	2.0411	2.0210	2.0030	1.9868
29	2.1768	2.1379	2.1045	2.0755	2.0500	2.0275	2.0073	1.9893	1.9730
30	2.1646	2.1256	2.0921	2.0630	2.0374	2.0148	1.9946	1.9765	1.9601
40	2.0772	2.0376	2.0035	1.9738	1.9476	1.9245	1.9037	1.8851	1.8682
60	1.9926	1.9522	1.9174	1.8870	1.8602	1.8364	1.8151	1.7959	1.7784
120	1.9105	1.8693	1.8337	1.8026	1.7750	1.7505	1.7285	1.7085	1.6904
∞	1.8307	1.7886	1.7522	1.7202	1.6918	1.6664	1.6435	1.6228	1.6039

Table H

Critical Values of the F-Distribution ($\alpha = 0.050$)
(continued)

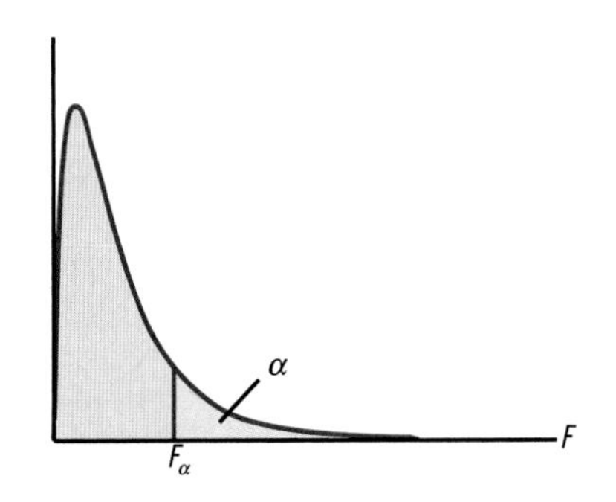

Numerator Degrees of Freedom

Denominator Degrees of Freedom	19	20	24	30	40	60	120
1	247.6861	248.0131	249.0518	250.0951	251.1432	252.1957	253.2529
2	19.4431	19.4458	19.4541	19.4624	19.4707	19.4791	19.4874
3	8.6670	8.6602	8.6385	8.6166	8.5944	8.5720	8.5494
4	5.8114	5.8025	5.7744	5.7459	5.7170	5.6877	5.6581
5	4.5678	4.5581	4.5272	4.4957	4.4638	4.4314	4.3985
6	3.8844	3.8742	3.8415	3.8082	3.7743	3.7398	3.7047
7	3.4551	3.4445	3.4105	3.3758	3.3404	3.3043	3.2674
8	3.1613	3.1503	3.1152	3.0794	3.0428	3.0053	2.9669
9	2.9477	2.9365	2.9005	2.8637	2.8259	2.7872	2.7475
10	2.7854	2.7740	2.7372	2.6996	2.6609	2.6211	2.5801
11	2.6581	2.6464	2.6090	2.5705	2.5309	2.4901	2.4480
12	2.5554	2.5436	2.5055	2.4663	2.4259	2.3842	2.3410
13	2.4709	2.4589	2.4202	2.3803	2.3392	2.2966	2.2524
14	2.4000	2.3879	2.3487	2.3082	2.2664	2.2229	2.1778
15	2.3398	2.3275	2.2878	2.2468	2.2043	2.1601	2.1141
16	2.2880	2.2756	2.2354	2.1938	2.1507	2.1058	2.0589
17	2.2429	2.2304	2.1898	2.1477	2.1040	2.0584	2.0107
18	2.2033	2.1906	2.1497	2.1071	2.0629	2.0166	1.9681
19	2.1683	2.1555	2.1141	2.0712	2.0264	1.9795	1.9302
20	2.1370	2.1242	2.0825	2.0391	1.9938	1.9464	1.8963
21	2.1090	2.0960	2.0540	2.0102	1.9645	1.9165	1.8657
22	2.0837	2.0707	2.0283	1.9842	1.9380	1.8894	1.8380
23	2.0608	2.0476	2.0050	1.9605	1.9139	1.8648	1.8128
24	2.0399	2.0267	1.9838	1.9390	1.8920	1.8424	1.7896
25	2.0207	2.0075	1.9643	1.9192	1.8718	1.8217	1.7684
26	2.0032	1.9898	1.9464	1.9010	1.8533	1.8027	1.7488
27	1.9870	1.9736	1.9299	1.8842	1.8361	1.7851	1.7306
28	1.9720	1.9586	1.9147	1.8687	1.8203	1.7689	1.7138
29	1.9581	1.9446	1.9005	1.8543	1.8055	1.7537	1.6981
30	1.9452	1.9317	1.8874	1.8409	1.7918	1.7396	1.6835
40	1.8529	1.8389	1.7929	1.7444	1.6928	1.6373	1.5766
60	1.7625	1.7480	1.7001	1.6491	1.5943	1.5343	1.4673
120	1.6739	1.6587	1.6084	1.5543	1.4952	1.4290	1.3519
∞	1.5865	1.5705	1.5173	1.4591	1.3940	1.3180	1.2214

Table H

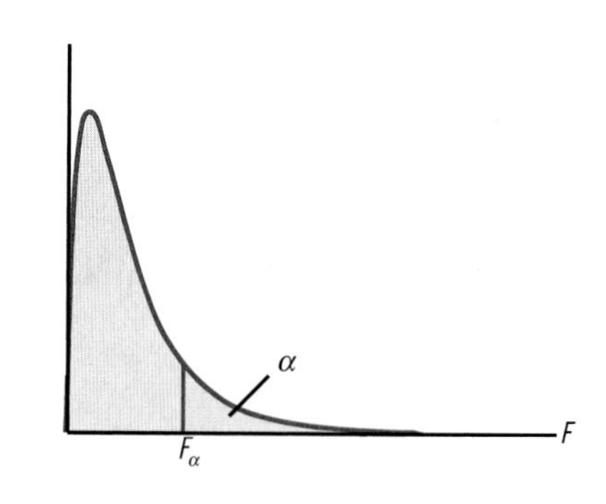

Critical Values of the *F*-Distribution ($\alpha = 0.025$)

Denominator Degrees of Freedom	Numerator Degrees of Freedom								
	1	2	3	4	5	6	7	8	9
1	647.7890	799.5000	864.1630	899.5833	921.8479	937.1111	948.2169	956.6562	963.2846
2	38.5063	39.0000	39.1655	39.2484	39.2982	39.3315	39.3552	39.3730	39.3869
3	17.4434	16.0441	15.4392	15.1010	14.8848	14.7347	14.6244	14.5399	14.4731
4	12.2179	10.6491	9.9792	9.6045	9.3645	9.1973	9.0741	8.9796	8.9047
5	10.0070	8.4336	7.7636	7.3879	7.1464	6.9777	6.8531	6.7572	6.6811
6	8.8131	7.2599	6.5988	6.2272	5.9876	5.8198	5.6955	5.5996	5.5234
7	8.0727	6.5415	5.8898	5.5226	5.2852	5.1186	4.9949	4.8993	4.8232
8	7.5709	6.0595	5.4160	5.0526	4.8173	4.6517	4.5286	4.4333	4.3572
9	7.2093	5.7147	5.0781	4.7181	4.4844	4.3197	4.1970	4.1020	4.0260
10	6.9367	5.4564	4.8256	4.4683	4.2361	4.0721	3.9498	3.8549	3.7790
11	6.7241	5.2559	4.6300	4.2751	4.0440	3.8807	3.7586	3.6638	3.5879
12	6.5538	5.0959	4.4742	4.1212	3.8911	3.7283	3.6065	3.5118	3.4358
13	6.4143	4.9653	4.3472	3.9959	3.7667	3.6043	3.4827	3.3880	3.3120
14	6.2979	4.8567	4.2417	3.8919	3.6634	3.5014	3.3799	3.2853	3.2093
15	6.1995	4.7650	4.1528	3.8043	3.5764	3.4147	3.2934	3.1987	3.1227
16	6.1151	4.6867	4.0768	3.7294	3.5021	3.3406	3.2194	3.1248	3.0488
17	6.0420	4.6189	4.0112	3.6648	3.4379	3.2767	3.1556	3.0610	2.9849
18	5.9781	4.5597	3.9539	3.6083	3.3820	3.2209	3.0999	3.0053	2.9291
19	5.9216	4.5075	3.9034	3.5587	3.3327	3.1718	3.0509	2.9563	2.8801
20	5.8715	4.4613	3.8587	3.5147	3.2891	3.1283	3.0074	2.9128	2.8365
21	5.8266	4.4199	3.8188	3.4754	3.2501	3.0895	2.9686	2.8740	2.7977
22	5.7863	4.3828	3.7829	3.4401	3.2151	3.0546	2.9338	2.8392	2.7628
23	5.7498	4.3492	3.7505	3.4083	3.1835	3.0232	2.9023	2.8077	2.7313
24	5.7166	4.3187	3.7211	3.3794	3.1548	2.9946	2.8738	2.7791	2.7027
25	5.6864	4.2909	3.6943	3.3530	3.1287	2.9685	2.8478	2.7531	2.6766
26	5.6586	4.2655	3.6697	3.3289	3.1048	2.9447	2.8240	2.7293	2.6528
27	5.6331	4.2421	3.6472	3.3067	3.0828	2.9228	2.8021	2.7074	2.6309
28	5.6096	4.2205	3.6264	3.2863	3.0626	2.9027	2.7820	2.6872	2.6106
29	5.5878	4.2006	3.6072	3.2674	3.0438	2.8840	2.7633	2.6686	2.5919
30	5.5675	4.1821	3.5894	3.2499	3.0265	2.8667	2.7460	2.6513	2.5746
40	5.4239	4.0510	3.4633	3.1261	2.9037	2.7444	2.6238	2.5289	2.4519
60	5.2856	3.9253	3.3425	3.0077	2.7863	2.6274	2.5068	2.4117	2.3344
120	5.1523	3.8046	3.2269	2.8943	2.6740	2.5154	2.3948	2.2994	2.2217
∞	5.0239	3.6889	3.1161	2.7858	2.5665	2.4082	2.2875	2.1918	2.1136

TABLE H

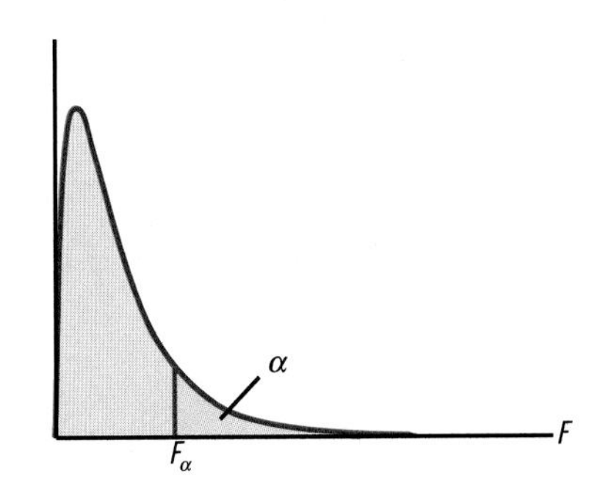

Critical Values of the F-Distribution ($\alpha = 0.025$)
(continued)

Numerator Degrees of Freedom

Denominator Degrees of Freedom	10	11	12	13	14	15	16	17	18
1	968.6274	973.0252	976.7079	979.8368	982.5278	984.8668	986.9187	988.7331	990.3490
2	39.3980	39.4071	39.4146	39.4210	39.4265	39.4313	39.4354	39.4391	39.4424
3	14.4189	14.3742	14.3366	14.3045	14.2768	14.2527	14.2315	14.2127	14.1960
4	8.8439	8.7935	8.7512	8.7150	8.6838	8.6565	8.6326	8.6113	8.5924
5	6.6192	6.5678	6.5245	6.4876	6.4556	6.4277	6.4032	6.3814	6.3619
6	5.4613	5.4098	5.3662	5.3290	5.2968	5.2687	5.2439	5.2218	5.2021
7	4.7611	4.7095	4.6658	4.6285	4.5961	4.5678	4.5428	4.5206	4.5008
8	4.2951	4.2434	4.1997	4.1622	4.1297	4.1012	4.0761	4.0538	4.0338
9	3.9639	3.9121	3.8682	3.8306	3.7980	3.7694	3.7441	3.7216	3.7015
10	3.7168	3.6649	3.6209	3.5832	3.5504	3.5217	3.4963	3.4737	3.4534
11	3.5257	3.4737	3.4296	3.3917	3.3588	3.3299	3.3044	3.2816	3.2612
12	3.3736	3.3215	3.2773	3.2393	3.2062	3.1772	3.1515	3.1286	3.1081
13	3.2497	3.1975	3.1532	3.1150	3.0819	3.0527	3.0269	3.0039	2.9832
14	3.1469	3.0946	3.0502	3.0119	2.9786	2.9493	2.9234	2.9003	2.8795
15	3.0602	3.0078	2.9633	2.9249	2.8915	2.8621	2.8360	2.8128	2.7919
16	2.9862	2.9337	2.8890	2.8506	2.8170	2.7875	2.7614	2.7380	2.7170
17	2.9222	2.8696	2.8249	2.7863	2.7526	2.7230	2.6968	2.6733	2.6522
18	2.8664	2.8137	2.7689	2.7302	2.6964	2.6667	2.6404	2.6168	2.5956
19	2.8172	2.7645	2.7196	2.6808	2.6469	2.6171	2.5907	2.5670	2.5457
20	2.7737	2.7209	2.6758	2.6369	2.6030	2.5731	2.5465	2.5228	2.5014
21	2.7348	2.6819	2.6368	2.5978	2.5638	2.5338	2.5071	2.4833	2.4618
22	2.6998	2.6469	2.6017	2.5626	2.5285	2.4984	2.4717	2.4478	2.4262
23	2.6682	2.6152	2.5699	2.5308	2.4966	2.4665	2.4396	2.4157	2.3940
24	2.6396	2.5865	2.5411	2.5019	2.4677	2.4374	2.4105	2.3865	2.3648
25	2.6135	2.5603	2.5149	2.4756	2.4413	2.4110	2.3840	2.3599	2.3381
26	2.5896	2.5363	2.4908	2.4515	2.4171	2.3867	2.3597	2.3355	2.3137
27	2.5676	2.5143	2.4688	2.4293	2.3949	2.3644	2.3373	2.3131	2.2912
28	2.5473	2.4940	2.4484	2.4089	2.3743	2.3438	2.3167	2.2924	2.2704
29	2.5286	2.4752	2.4295	2.3900	2.3554	2.3248	2.2976	2.2732	2.2512
30	2.5112	2.4577	2.4120	2.3724	2.3378	2.3072	2.2799	2.2554	2.2334
40	2.3882	2.3343	2.2882	2.2481	2.2130	2.1819	2.1542	2.1293	2.1068
60	2.2702	2.2159	2.1692	2.1286	2.0929	2.0613	2.0330	2.0076	1.9846
120	2.1570	2.1021	2.0548	2.0136	1.9773	1.9450	1.9161	1.8900	1.8663
∞	2.0483	1.9927	1.9447	1.9027	1.8656	1.8326	1.8028	1.7759	1.7515

TABLE H

Critical Values of the *F*-Distribution ($\alpha = 0.025$)
(continued)

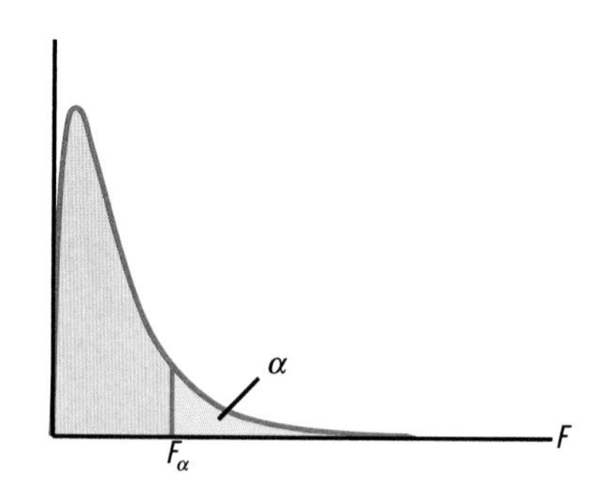

Numerator Degrees of Freedom

Denominator Degrees of Freedom	19	20	24	30	40	60	120
1	991.7973	993.1028	997.2492	1001.4144	1005.5981	1009.8001	1014.0202
2	39.4453	39.4479	39.4562	39.4646	39.4729	39.4812	39.4896
3	14.1810	14.1674	14.1241	14.0805	14.0365	13.9921	13.9473
4	8.5753	8.5599	8.5109	8.4613	8.4111	8.3604	8.3092
5	6.3444	6.3286	6.2780	6.2269	6.1750	6.1225	6.0693
6	5.1844	5.1684	5.1172	5.0652	5.0125	4.9589	4.9044
7	4.4829	4.4667	4.4150	4.3624	4.3089	4.2544	4.1989
8	4.0158	3.9995	3.9472	3.8940	3.8398	3.7844	3.7279
9	3.6833	3.6669	3.6142	3.5604	3.5055	3.4493	3.3918
10	3.4351	3.4185	3.3654	3.3110	3.2554	3.1984	3.1399
11	3.2428	3.2261	3.1725	3.1176	3.0613	3.0035	2.9441
12	3.0896	3.0728	3.0187	2.9633	2.9063	2.8478	2.7874
13	2.9646	2.9477	2.8932	2.8372	2.7797	2.7204	2.6590
14	2.8607	2.8437	2.7888	2.7324	2.6742	2.6142	2.5519
15	2.7730	2.7559	2.7006	2.6437	2.5850	2.5242	2.4611
16	2.6980	2.6808	2.6252	2.5678	2.5085	2.4471	2.3831
17	2.6331	2.6158	2.5598	2.5020	2.4422	2.3801	2.3153
18	2.5764	2.5590	2.5027	2.4445	2.3842	2.3214	2.2558
19	2.5265	2.5089	2.4523	2.3937	2.3329	2.2696	2.2032
20	2.4821	2.4645	2.4076	2.3486	2.2873	2.2234	2.1562
21	2.4424	2.4247	2.3675	2.3082	2.2465	2.1819	2.1141
22	2.4067	2.3890	2.3315	2.2718	2.2097	2.1446	2.0760
23	2.3745	2.3567	2.2989	2.2389	2.1763	2.1107	2.0415
24	2.3452	2.3273	2.2693	2.2090	2.1460	2.0799	2.0099
25	2.3184	2.3005	2.2422	2.1816	2.1183	2.0516	1.9811
26	2.2939	2.2759	2.2174	2.1565	2.0928	2.0257	1.9545
27	2.2713	2.2533	2.1946	2.1334	2.0693	2.0018	1.9299
28	2.2505	2.2324	2.1735	2.1121	2.0477	1.9797	1.9072
29	2.2313	2.2131	2.1540	2.0923	2.0276	1.9591	1.8861
30	2.2134	2.1952	2.1359	2.0739	2.0089	1.9400	1.8664
40	2.0864	2.0677	2.0069	1.9429	1.8752	1.8028	1.7242
60	1.9636	1.9445	1.8817	1.8152	1.7440	1.6668	1.5810
120	1.8447	1.8249	1.7597	1.6899	1.6141	1.5299	1.4327
∞	1.7291	1.7085	1.6402	1.5660	1.4835	1.3883	1.2684

TABLE H

Critical Values of the F-Distribution ($\alpha = 0.010$)

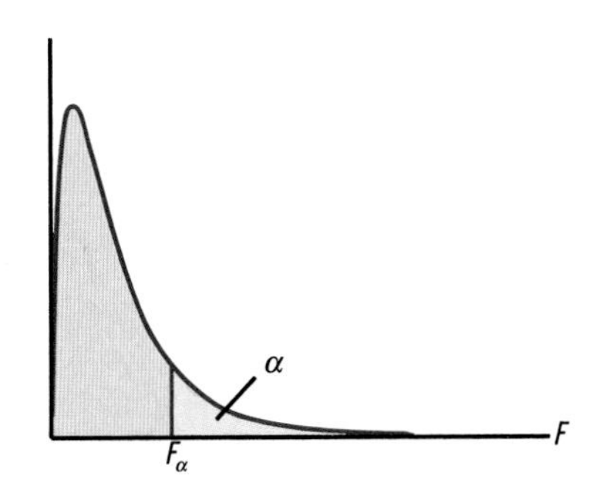

Numerator Degrees of Freedom

Denominator Degrees of Freedom

	1	2	3	4	5	6	7	8	9
1	4052.1807	4999.5000	5403.3520	5624.5833	5763.6496	5858.9861	5928.3557	5981.0703	6022.4732
2	98.5025	99.0000	99.1662	99.2494	99.2993	99.3326	99.3564	99.3742	99.3881
3	34.1162	30.8165	29.4567	28.7099	28.2371	27.9107	27.6717	27.4892	27.3452
4	21.1977	18.0000	16.6944	15.9770	15.5219	15.2069	14.9758	14.7989	14.6591
5	16.2582	13.2739	12.0600	11.3919	10.9670	10.6723	10.4555	10.2893	10.1578
6	13.7450	10.9248	9.7795	9.1483	8.7459	8.4661	8.2600	8.1017	7.9761
7	12.2464	9.5466	8.4513	7.8466	7.4604	7.1914	6.9928	6.8400	6.7188
8	11.2586	8.6491	7.5910	7.0061	6.6318	6.3707	6.1776	6.0289	5.9106
9	10.5614	8.0215	6.9919	6.4221	6.0569	5.8018	5.6129	5.4671	5.3511
10	10.0443	7.5594	6.5523	5.9943	5.6363	5.3858	5.2001	5.0567	4.9424
11	9.6460	7.2057	6.2167	5.6683	5.3160	5.0692	4.8861	4.7445	4.6315
12	9.3302	6.9266	5.9525	5.4120	5.0643	4.8206	4.6395	4.4994	4.3875
13	9.0738	6.7010	5.7394	5.2053	4.8616	4.6204	4.4410	4.3021	4.1911
14	8.8616	6.5149	5.5639	5.0354	4.6950	4.4558	4.2779	4.1399	4.0297
15	8.6831	6.3589	5.4170	4.8932	4.5556	4.3183	4.1415	4.0045	3.8948
16	8.5310	6.2262	5.2922	4.7726	4.4374	4.2016	4.0259	3.8896	3.7804
17	8.3997	6.1121	5.1850	4.6690	4.3359	4.1015	3.9267	3.7910	3.6822
18	8.2854	6.0129	5.0919	4.5790	4.2479	4.0146	3.8406	3.7054	3.5971
19	8.1849	5.9259	5.0103	4.5003	4.1708	3.9386	3.7653	3.6305	3.5225
20	8.0960	5.8489	4.9382	4.4307	4.1027	3.8714	3.6987	3.5644	3.4567
21	8.0166	5.7804	4.8740	4.3688	4.0421	3.8117	3.6396	3.5056	3.3981
22	7.9454	5.7190	4.8166	4.3134	3.9880	3.7583	3.5867	3.4530	3.3458
23	7.8811	5.6637	4.7649	4.2636	3.9392	3.7102	3.5390	3.4057	3.2986
24	7.8229	5.6136	4.7181	4.2184	3.8951	3.6667	3.4959	3.3629	3.2560
25	7.7698	5.5680	4.6755	4.1774	3.8550	3.6272	3.4568	3.3239	3.2172
26	7.7213	5.5263	4.6366	4.1400	3.8183	3.5911	3.4210	3.2884	3.1818
27	7.6767	5.4881	4.6009	4.1056	3.7848	3.5580	3.3882	3.2558	3.1494
28	7.6356	5.4529	4.5681	4.0740	3.7539	3.5276	3.3581	3.2259	3.1195
29	7.5977	5.4204	4.5378	4.0449	3.7254	3.4995	3.3303	3.1982	3.0920
30	7.5625	5.3903	4.5097	4.0179	3.6990	3.4735	3.3045	3.1726	3.0665
40	7.3141	5.1785	4.3126	3.8283	3.5138	3.2910	3.1238	2.9930	2.8876
60	7.0771	4.9774	4.1259	3.6490	3.3389	3.1187	2.9530	2.8233	2.7185
120	6.8509	4.7865	3.9491	3.4795	3.1735	2.9559	2.7918	2.6629	2.5586
∞	6.6349	4.6052	3.7816	3.3192	3.0173	2.8020	2.6393	2.5113	2.4073

TABLE H

Critical Values of the F-Distribution ($\alpha = 0.010$)

(continued)

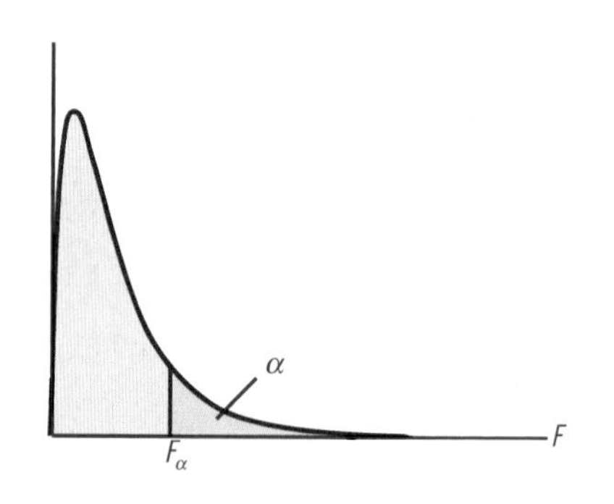

Numerator Degrees of Freedom

Denominator Degrees of Freedom

	10	11	12	13	14	15	16	17	18
1	6055.8467	6083.3168	6106.3207	6125.8647	6142.6740	6157.2846	6170.1012	6181.4348	6191.5287
2	99.3992	99.4083	99.4159	99.4223	99.4278	99.4325	99.4367	99.4404	99.4436
3	27.2287	27.1326	27.0518	26.9831	26.9238	26.8722	26.8269	26.7867	26.7509
4	14.5459	14.4523	14.3736	14.3065	14.2486	14.1982	14.1539	14.1146	14.0795
5	10.0510	9.9626	9.8883	9.8248	9.7700	9.7222	9.6802	9.6429	9.6096
6	7.8741	7.7896	7.7183	7.6575	7.6049	7.5590	7.5186	7.4827	7.4507
7	6.6201	6.5382	6.4691	6.4100	6.3590	6.3143	6.2750	6.2401	6.2089
8	5.8143	5.7343	5.6667	5.6089	5.5589	5.5151	5.4766	5.4423	5.4116
9	5.2565	5.1779	5.1114	5.0545	5.0052	4.9621	4.9240	4.8902	4.8599
10	4.8491	4.7715	4.7059	4.6496	4.6008	4.5581	4.5204	4.4869	4.4569
11	4.5393	4.4624	4.3974	4.3416	4.2932	4.2509	4.2134	4.1801	4.1503
12	4.2961	4.2198	4.1553	4.0999	4.0518	4.0096	3.9724	3.9392	3.9095
13	4.1003	4.0245	3.9603	3.9052	3.8573	3.8154	3.7783	3.7452	3.7156
14	3.9394	3.8640	3.8001	3.7452	3.6975	3.6557	3.6187	3.5857	3.5561
15	3.8049	3.7299	3.6662	3.6115	3.5639	3.5222	3.4852	3.4523	3.4228
16	3.6909	3.6162	3.5527	3.4981	3.4506	3.4089	3.3720	3.3391	3.3096
17	3.5931	3.5185	3.4552	3.4007	3.3533	3.3117	3.2748	3.2419	3.2124
18	3.5082	3.4338	3.3706	3.3162	3.2689	3.2273	3.1904	3.1575	3.1280
19	3.4338	3.3596	3.2965	3.2422	3.1949	3.1533	3.1165	3.0836	3.0541
20	3.3682	3.2941	3.2311	3.1769	3.1296	3.0880	3.0512	3.0183	2.9887
21	3.3098	3.2359	3.1730	3.1187	3.0715	3.0300	2.9931	2.9602	2.9306
22	3.2576	3.1837	3.1209	3.0667	3.0195	2.9779	2.9411	2.9082	2.8786
23	3.2106	3.1368	3.0740	3.0199	2.9727	2.9311	2.8943	2.8613	2.8317
24	3.1681	3.0944	3.0316	2.9775	2.9303	2.8887	2.8519	2.8189	2.7892
25	3.1294	3.0558	2.9931	2.9389	2.8917	2.8502	2.8133	2.7803	2.7506
26	3.0941	3.0205	2.9578	2.9038	2.8566	2.8150	2.7781	2.7451	2.7153
27	3.0618	2.9882	2.9256	2.8715	2.8243	2.7827	2.7458	2.7127	2.6830
28	3.0320	2.9585	2.8959	2.8418	2.7946	2.7530	2.7160	2.6830	2.6532
29	3.0045	2.9311	2.8685	2.8144	2.7672	2.7256	2.6886	2.6555	2.6257
30	2.9791	2.9057	2.8431	2.7890	2.7418	2.7002	2.6632	2.6301	2.6003
40	2.8005	2.7274	2.6648	2.6107	2.5634	2.5216	2.4844	2.4511	2.4210
60	2.6318	2.5587	2.4961	2.4419	2.3943	2.3523	2.3148	2.2811	2.2507
120	2.4721	2.3990	2.3363	2.2818	2.2339	2.1915	2.1536	2.1194	2.0885
∞	2.3209	2.2477	2.1847	2.1299	2.0815	2.0385	2.0000	1.9652	1.9336

TABLE H

Critical Values of the F-Distribution ($\alpha = 0.010$)
(continued)

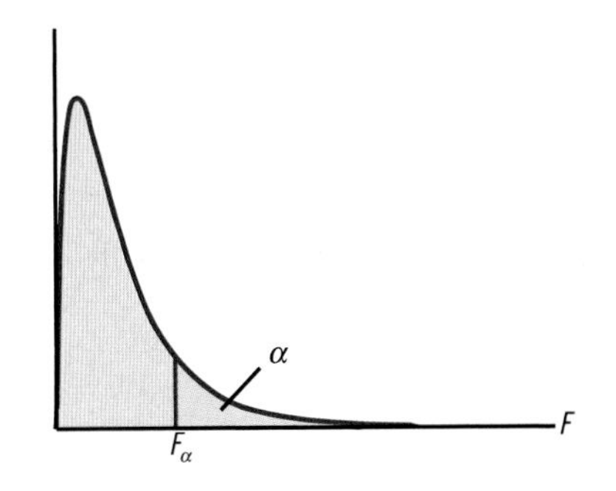

Numerator Degrees of Freedom

Denominator Degrees of Freedom

	19	20	24	30	40	60	120
1	6200.5756	6208.7302	6234.6309	6260.6486	6286.7821	6313.0301	6339.3913
2	99.4465	99.4492	99.4575	99.4658	99.4742	99.4825	99.4908
3	26.7188	26.6898	26.5975	26.5045	26.4108	26.3164	26.2211
4	14.0480	14.0196	13.9291	13.8377	13.7454	13.6522	13.5581
5	9.5797	9.5526	9.4665	9.3793	9.2912	9.2020	9.1118
6	7.4219	7.3958	7.3127	7.2285	7.1432	7.0567	6.9690
7	6.1808	6.1554	6.0743	5.9920	5.9084	5.8236	5.7373
8	5.3840	5.3591	5.2793	5.1981	5.1156	5.0316	4.9461
9	4.8327	4.8080	4.7290	4.6486	4.5666	4.4831	4.3978
10	4.4299	4.4054	4.3269	4.2469	4.1653	4.0819	3.9965
11	4.1234	4.0990	4.0209	3.9411	3.8596	3.7761	3.6904
12	3.8827	3.8584	3.7805	3.7008	3.6192	3.5355	3.4494
13	3.6888	3.6646	3.5868	3.5070	3.4253	3.3413	3.2548
14	3.5294	3.5052	3.4274	3.3476	3.2656	3.1813	3.0942
15	3.3961	3.3719	3.2940	3.2141	3.1319	3.0471	2.9595
16	3.2829	3.2587	3.1808	3.1007	3.0182	2.9330	2.8447
17	3.1857	3.1615	3.0835	3.0032	2.9205	2.8348	2.7459
18	3.1013	3.0771	2.9990	2.9185	2.8354	2.7493	2.6597
19	3.0274	3.0031	2.9249	2.8442	2.7608	2.6742	2.5839
20	2.9620	2.9377	2.8594	2.7785	2.6947	2.6077	2.5168
21	2.9039	2.8796	2.8010	2.7200	2.6359	2.5484	2.4568
22	2.8518	2.8274	2.7488	2.6675	2.5831	2.4951	2.4029
23	2.8049	2.7805	2.7017	2.6202	2.5355	2.4471	2.3542
24	2.7624	2.7380	2.6591	2.5773	2.4923	2.4035	2.3100
25	2.7238	2.6993	2.6203	2.5383	2.4530	2.3637	2.2696
26	2.6885	2.6640	2.5848	2.5026	2.4170	2.3273	2.2325
27	2.6561	2.6316	2.5522	2.4699	2.3840	2.2938	2.1985
28	2.6263	2.6017	2.5223	2.4397	2.3535	2.2629	2.1670
29	2.5987	2.5742	2.4946	2.4118	2.3253	2.2344	2.1379
30	2.5732	2.5487	2.4689	2.3860	2.2992	2.2079	2.1108
40	2.3937	2.3689	2.2880	2.2034	2.1142	2.0194	1.9172
60	2.2230	2.1978	2.1154	2.0285	1.9360	1.8363	1.7263
120	2.0604	2.0346	1.9500	1.8600	1.7628	1.6557	1.5330
∞	1.9048	1.8783	1.7908	1.6964	1.5923	1.4730	1.3246

Table H

Critical Values of the F-Distribution ($\alpha = 0.005$)

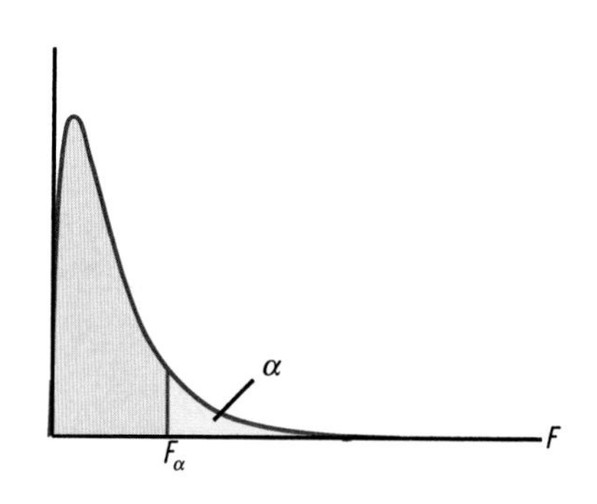

Numerator Degrees of Freedom

Denominator Degrees of Freedom	1	2	3	4	5	6	7	8	9
1	16210.7227	19999.5000	21614.7414	22499.5833	23055.7982	23437.1111	23714.5658	23925.4062	24091.0041
2	198.5013	199.0000	199.1664	199.2497	199.2996	199.3330	199.3568	199.3746	199.3885
3	55.5520	49.7993	47.4672	46.1946	45.3916	44.8385	44.4341	44.1256	43.8824
4	31.3328	26.2843	24.2591	23.1545	22.4564	21.9746	21.6217	21.3520	21.1391
5	22.7848	18.3138	16.5298	15.5561	14.9396	14.5133	14.2004	13.9610	13.7716
6	18.6350	14.5441	12.9166	12.0275	11.4637	11.0730	10.7859	10.5658	10.3915
7	16.2356	12.4040	10.8824	10.0505	9.5221	9.1553	8.8854	8.6781	8.5138
8	14.6882	11.0424	9.5965	8.8051	8.3018	7.9520	7.6941	7.4959	7.3386
9	13.6136	10.1067	8.7171	7.9559	7.4712	7.1339	6.8849	6.6933	6.5411
10	12.8265	9.4270	8.0807	7.3428	6.8724	6.5446	6.3025	6.1159	5.9676
11	12.2263	8.9122	7.6004	6.8809	6.4217	6.1016	5.8648	5.6821	5.5368
12	11.7542	8.5096	7.2258	6.5211	6.0711	5.7570	5.5245	5.3451	5.2021
13	11.3735	8.1865	6.9258	6.2335	5.7910	5.4819	5.2529	5.0761	4.9351
14	11.0603	7.9216	6.6804	5.9984	5.5623	5.2574	5.0313	4.8566	4.7173
15	10.7980	7.7008	6.4760	5.8029	5.3721	5.0708	4.8473	4.6744	4.5364
16	10.5755	7.5138	6.3034	5.6378	5.2117	4.9134	4.6920	4.5207	4.3838
17	10.3842	7.3536	6.1556	5.4967	5.0746	4.7789	4.5594	4.3894	4.2535
18	10.2181	7.2148	6.0278	5.3746	4.9560	4.6627	4.4448	4.2759	4.1410
19	10.0725	7.0935	5.9161	5.2681	4.8526	4.5614	4.3448	4.1770	4.0428
20	9.9439	6.9865	5.8177	5.1743	4.7616	4.4721	4.2569	4.0900	3.9564
21	9.8295	6.8914	5.7304	5.0911	4.6809	4.3931	4.1789	4.0128	3.8799
22	9.7271	6.8064	5.6524	5.0168	4.6088	4.3225	4.1094	3.9440	3.8116
23	9.6348	6.7300	5.5823	4.9500	4.5441	4.2591	4.0469	3.8822	3.7502
24	9.5513	6.6609	5.5190	4.8898	4.4857	4.2019	3.9905	3.8264	3.6949
25	9.4753	6.5982	5.4615	4.8351	4.4327	4.1500	3.9394	3.7758	3.6447
26	9.4059	6.5410	5.4091	4.7852	4.3844	4.1027	3.8928	3.7297	3.5989
27	9.3423	6.4885	5.3611	4.7396	4.3402	4.0594	3.8501	3.6875	3.5571
28	9.2838	6.4403	5.3170	4.6977	4.2996	4.0197	3.8110	3.6487	3.5186
29	9.2297	6.3958	5.2764	4.6591	4.2622	3.9831	3.7749	3.6131	3.4832
30	9.1797	6.3547	5.2388	4.6234	4.2276	3.9492	3.7416	3.5801	3.4505
40	8.8279	6.0664	4.9758	4.3738	3.9860	3.7129	3.5088	3.3498	3.2220
60	8.4946	5.7950	4.7290	4.1399	3.7599	3.4918	3.2911	3.1344	3.0083
120	8.1788	5.5393	4.4972	3.9207	3.5482	3.2849	3.0874	2.9330	2.8083
∞	7.8794	5.2983	4.2794	3.7151	3.3499	3.0913	2.8968	2.7444	2.6210

TABLE H

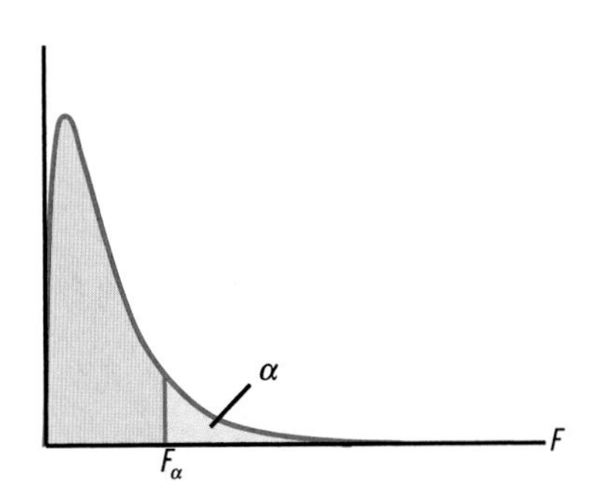

Critical Values of the *F*-Distribution ($\alpha = 0.005$)
(continued)

Numerator Degrees of Freedom

Denominator Degrees of Freedom	10	11	12	13	14	15	16	17	18
1	24224.4868	24334.3581	24426.3662	24504.5356	24571.7673	24630.2051	24681.4673	24726.7982	24767.1704
2	199.3996	199.4087	199.4163	199.4227	199.4282	199.4329	199.4371	199.4408	199.4440
3	43.6858	43.5236	43.3874	43.2715	43.1716	43.0847	43.0083	42.9407	42.8804
4	20.9667	20.8243	20.7047	20.6027	20.5148	20.4383	20.3710	20.3113	20.2581
5	13.6182	13.4912	13.3845	13.2934	13.2148	13.1463	13.0861	13.0327	12.9850
6	10.2500	10.1329	10.0343	9.9501	9.8774	9.8140	9.7582	9.7086	9.6644
7	8.3803	8.2697	8.1764	8.0967	8.0279	7.9678	7.9148	7.8678	7.8258
8	7.2106	7.1045	7.0149	6.9384	6.8721	6.8143	6.7633	6.7180	6.6775
9	6.4172	6.3142	6.2274	6.1530	6.0887	6.0325	5.9829	5.9388	5.8994
10	5.8467	5.7462	5.6613	5.5887	5.5257	5.4707	5.4221	5.3789	5.3403
11	5.4183	5.3197	5.2363	5.1649	5.1031	5.0489	5.0011	4.9586	4.9205
12	5.0855	4.9884	4.9062	4.8358	4.7748	4.7213	4.6741	4.6321	4.5945
13	4.8199	4.7240	4.6429	4.5733	4.5129	4.4600	4.4132	4.3716	4.3344
14	4.6034	4.5085	4.4281	4.3591	4.2993	4.2468	4.2005	4.1592	4.1221
15	4.4235	4.3295	4.2497	4.1813	4.1219	4.0698	4.0237	3.9827	3.9459
16	4.2719	4.1785	4.0994	4.0314	3.9723	3.9205	3.8747	3.8338	3.7972
17	4.1424	4.0496	3.9709	3.9033	3.8445	3.7929	3.7473	3.7066	3.6701
18	4.0305	3.9382	3.8599	3.7926	3.7341	3.6827	3.6373	3.5967	3.5603
19	3.9329	3.8410	3.7631	3.6961	3.6378	3.5866	3.5412	3.5008	3.4645
20	3.8470	3.7555	3.6779	3.6111	3.5530	3.5020	3.4568	3.4164	3.3802
21	3.7709	3.6798	3.6024	3.5358	3.4779	3.4270	3.3818	3.3416	3.3054
22	3.7030	3.6122	3.5350	3.4686	3.4108	3.3600	3.3150	3.2748	3.2387
23	3.6420	3.5515	3.4745	3.4083	3.3506	3.2999	3.2549	3.2148	3.1787
24	3.5870	3.4967	3.4199	3.3538	3.2962	3.2456	3.2007	3.1606	3.1246
25	3.5370	3.4470	3.3704	3.3044	3.2469	3.1963	3.1515	3.1114	3.0754
26	3.4916	3.4017	3.3252	3.2594	3.2020	3.1515	3.1067	3.0666	3.0306
27	3.4499	3.3602	3.2839	3.2182	3.1608	3.1104	3.0656	3.0256	2.9896
28	3.4117	3.3222	3.2460	3.1803	3.1231	3.0727	3.0279	2.9879	2.9520
29	3.3765	3.2871	3.2110	3.1454	3.0882	3.0379	2.9932	2.9532	2.9173
30	3.3440	3.2547	3.1787	3.1132	3.0560	3.0057	2.9611	2.9211	2.8852
40	3.1167	3.0284	2.9531	2.8880	2.8312	2.7811	2.7365	2.6966	2.6607
60	2.9042	2.8166	2.7419	2.6771	2.6205	2.5705	2.5259	2.4859	2.4498
120	2.7052	2.6183	2.5439	2.4794	2.4228	2.3727	2.3280	2.2878	2.2514
∞	2.5188	2.4324	2.3583	2.2938	2.2371	2.1868	2.1417	2.1011	2.0642

Table H

Critical Values of the F-Distribution ($\alpha = 0.005$)

(continued)

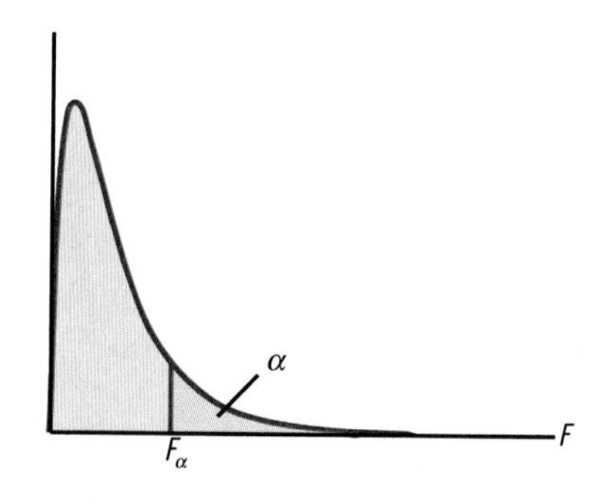

Numerator Degrees of Freedom

Denominator Degrees of Freedom

	19	20	24	30	40	60	120
1	24803.3549	24835.9709	24939.5653	25043.6277	25148.1532	25253.1369	25358.5734
2	199.4470	199.4496	199.4579	199.4663	199.4746	199.4829	199.4912
3	42.8263	42.7775	42.6222	42.4658	42.3082	42.1494	41.9895
4	20.2104	20.1673	20.0300	19.8915	19.7518	19.6107	19.4684
5	12.9422	12.9035	12.7802	12.6556	12.5297	12.4024	12.2737
6	9.6247	9.5888	9.4742	9.3582	9.2408	9.1219	9.0015
7	7.7881	7.7540	7.6450	7.5345	7.4224	7.3088	7.1933
8	6.6411	6.6082	6.5029	6.3961	6.2875	6.1772	6.0649
9	5.8639	5.8318	5.7292	5.6248	5.5186	5.4104	5.3001
10	5.3055	5.2740	5.1732	5.0706	4.9659	4.8592	4.7501
11	4.8863	4.8552	4.7557	4.6543	4.5508	4.4450	4.3367
12	4.5606	4.5299	4.4314	4.3309	4.2282	4.1229	4.0149
13	4.3008	4.2703	4.1726	4.0727	3.9704	3.8655	3.7577
14	4.0888	4.0585	3.9614	3.8619	3.7600	3.6552	3.5473
15	3.9127	3.8826	3.7859	3.6867	3.5850	3.4803	3.3722
16	3.7641	3.7342	3.6378	3.5389	3.4372	3.3324	3.2240
17	3.6372	3.6073	3.5112	3.4124	3.3108	3.2058	3.0971
18	3.5275	3.4977	3.4017	3.3030	3.2014	3.0962	2.9871
19	3.4318	3.4020	3.3062	3.2075	3.1058	3.0004	2.8908
20	3.3475	3.3178	3.2220	3.1234	3.0215	2.9159	2.8058
21	3.2728	3.2431	3.1474	3.0488	2.9467	2.8408	2.7302
22	3.2060	3.1764	3.0807	2.9821	2.8799	2.7736	2.6625
23	3.1461	3.1165	3.0208	2.9221	2.8197	2.7132	2.6015
24	3.0920	3.0624	2.9667	2.8679	2.7654	2.6585	2.5463
25	3.0429	3.0133	2.9176	2.8187	2.7160	2.6088	2.4961
26	2.9981	2.9685	2.8728	2.7738	2.6709	2.5633	2.4501
27	2.9571	2.9275	2.8318	2.7327	2.6296	2.5217	2.4079
28	2.9194	2.8899	2.7941	2.6949	2.5916	2.4834	2.3690
29	2.8847	2.8551	2.7594	2.6600	2.5565	2.4479	2.3331
30	2.8526	2.8230	2.7272	2.6278	2.5241	2.4151	2.2998
40	2.6281	2.5984	2.5020	2.4015	2.2958	2.1838	2.0636
60	2.4171	2.3872	2.2898	2.1874	2.0789	1.9622	1.8341
120	2.2183	2.1881	2.0890	1.9840	1.8709	1.7469	1.6055
∞	2.0306	1.9998	1.8983	1.7891	1.6691	1.5325	1.3637

TABLE I
Pearson Correlation Coefficient

n	$\alpha = .05$	$\alpha = .01$
4	0.950	0.990
5	0.878	0.959
6	0.811	0.917
7	0.754	0.875
8	0.707	0.834
9	0.666	0.798
10	0.632	0.765
11	0.602	0.735
12	0.576	0.708
13	0.553	0.684
14	0.532	0.661
15	0.514	0.641
16	0.497	0.623
17	0.482	0.606
18	0.468	0.590
19	0.456	0.575
20	0.444	0.561
21	0.433	0.549
22	0.423	0.537
23	0.413	0.526
24	0.404	0.515
25	0.396	0.505
26	0.388	0.496
27	0.381	0.487
28	0.374	0.479
29	0.367	0.471
30	0.361	0.463
35	0.334	0.430
40	0.312	0.403
45	0.294	0.380
50	0.279	0.361
55	0.266	0.345
60	0.254	0.330
65	0.244	0.317
70	0.235	0.306
75	0.227	0.296
80	0.220	0.286
85	0.213	0.278
90	0.207	0.270
95	0.202	0.268
100	0.197	0.256

Note:

r is statistically significant if $|r|$ is greater than the value given in the table.

Appendix B

Getting Started with Microsoft Excel

The Basics of Excel

When you open Excel, you see a large electronic spreadsheet program with 65,536 rows and 256 columns in the current worksheet. Columns run left to right and are labeled with letters. Rows run top to bottom and are labeled with numbers. In the Office 2003 version, there are 3 tabs at the bottom of the current worksheet that let you change to another clean worksheet with the same storage capacity without creating a new file. One of these worksheets on paper would produce a sheet more than 20 feet wide and longer than 4 football fields.

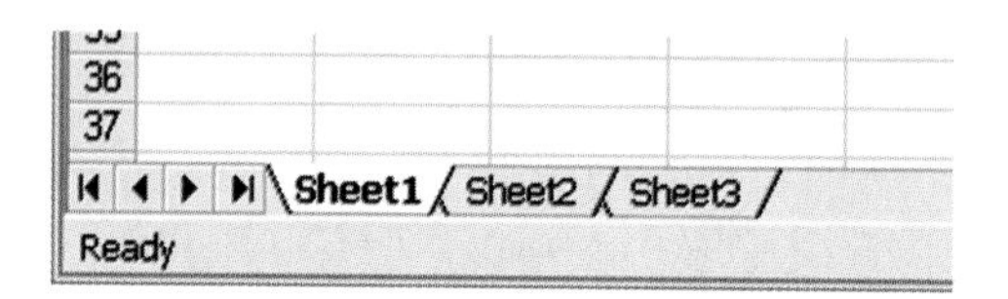

Cells

Excel uses cells to contain data and text. The cells' locations are described by their position in terms of rows and columns. Rows are listed along the left side of the worksheet and are labeled with numbers. Rows run up and down the spreadsheet. Columns are listed across the top of the worksheet and are labeled with letters. Columns run left and right in the spreadsheet.

A thick black border will outline the active cell. To help you identify the cell address, the column and row headers are also highlighted. Another way to identify the active cell's address is to look at the window just above the header for column A. A column letter followed by a row number describes the active cell position. For example "B2" would be referring to the cell found in Column B at Row 2.

To the right of the cell address window is the formula window. This displays the contents of the active cell. In the formula window type an "=" sign. This is the command to start a formula. Cells or data can be added, subtracted, multiplied, divided and manipulated in any combination through formulas.

To change the active cell, move the mouse to the desired cell and click. The border will now be surrounding the new cell and the address will have changed in the active cell address window. The arrow keys can also control navigation of the active cell.

As you enter data in the active cell, it is displayed in the cell and in the formula window at the top of the worksheet.

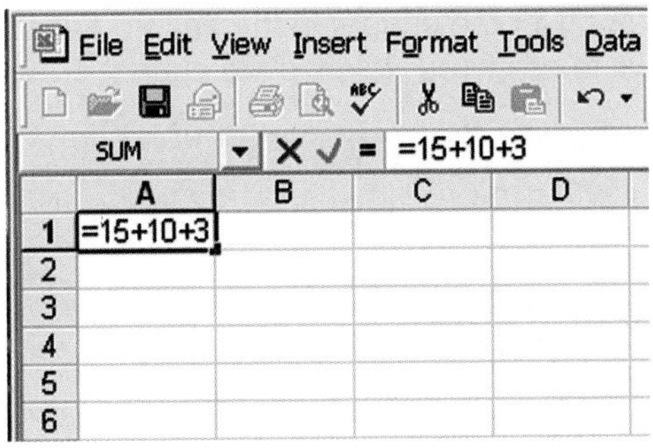

When you finish entering data in a cell, press Enter. The active cell will now move one row down. The answer to the formula, rather than the formula itself, will be displayed in the cell.

Excel will try to anticipate a reoccurring word or phrase within a column. When you type a word in a cell, you can then move to a lower cell in that column. If you type the first letter of the word again, Excel will fill in the rest of the word or phrase. To accept the automatic completion of the word just hit the Enter key or click in a new cell. If you do not want to use the automatic completion of the word, simply keep typing in the cell and the new word or phrase will appear.

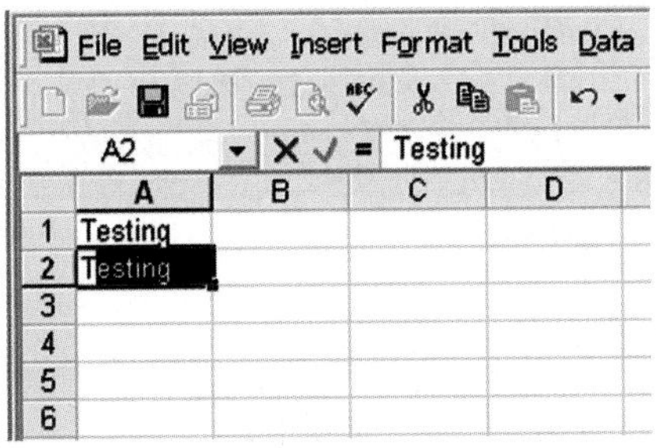

Filling Cells

Excel has a helpful tool when entering a series of data. We will refer to this as filling cells. Try entering "1" into cell A1 and "2" into cell A2. Then highlight both of the cells. You will see a small black square at the bottom right of the highlighted selection.

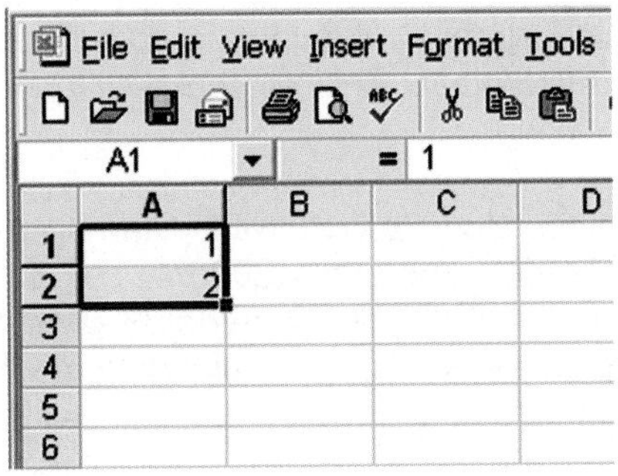

If you move your cursor over this it should change into a narrow plus sign. Now click your left mouse button and drag the mouse down to include cells A3 through A10. When you release the mouse button, the cells should now be filled with values from 1 to 10.

The fill cells tool will work for other values as well. Excel finds the relationship between two data points and replicates this in the fill cells. Now try this with multiples of 5. Enter 5 in cell A1 and 10 in A2. Highlight cells A1 and A2 again and put your cursor over the square in the bottom right of the highlighted selection. Click the left mouse button and pull down to cell A10. Release the button. Now the values should have changed from 5 through 50.

File Edit View Insert Format Tools Data

A1 = 5

	A	B	C	D
1	5			
2	10			
3	15			
4	20			
5	25			
6	30			
7	35			
8	40			
9	45			
10	50			
11				

You can also use this feature with sequential labels such as days of the week or months of the year.

File Edit View Insert Format Tools D

A1 = Sunday

	A	B	C	D
1	Sunday	Monday	Tuesday	Wednes
2				
3				

Now that you have the basic skills to work with Excel, start a new worksheet and we will try some examples.

Put the label "Checking balance" in cell A1, and "Savings balance" in cell B1. You will notice that the label you typed in cell A1 is cut off. We need to resize the columns to allow for the entire label. Move your cursor to the line that separates the column labels for A and B. Your cursor will change to an arrow pointing off to the left and right. You can click and drag the cursor to the right to increase the column width, or simply double click to have the width auto size to fit the label. Repeat this for Column B.

Next select the column labels A and B and you should see both columns highlighted. With the column highlighted, click on the $ button on the toolbar. This will make all the values we put in those two columns be formatted as dollars.

Next we will use the fill tool to get some values to work with. Start by entering "100" in cell A2 and "200" in cell A3. Now move to cell B2, enter "1000", and enter "2000" in cell B3. Now highlight cell A2 through B3 by clicking in A2 and dragging down and to the right to include B3. With those cells highlighted, click on the Fill tool at the bottom right of highlighted selection. Click and drag down to row 11. When you release the mouse button, the cells should fill with data and end with 10000 in cell B11. You should now have a screen that looks like this:

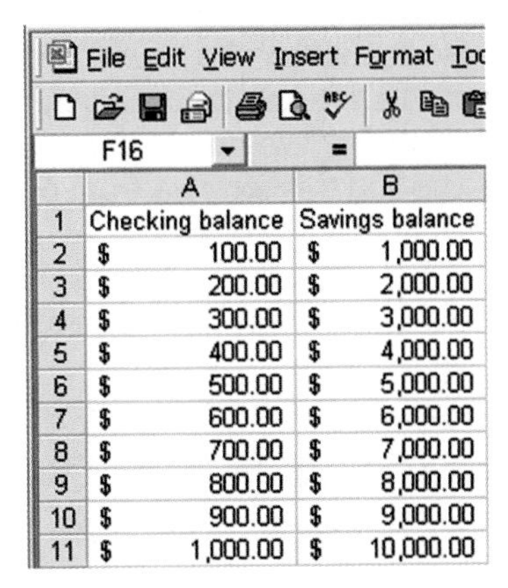

	A	B
1	Checking balance	Savings balance
2	$ 100.00	$ 1,000.00
3	$ 200.00	$ 2,000.00
4	$ 300.00	$ 3,000.00
5	$ 400.00	$ 4,000.00
6	$ 500.00	$ 5,000.00
7	$ 600.00	$ 6,000.00
8	$ 700.00	$ 7,000.00
9	$ 800.00	$ 8,000.00
10	$ 900.00	$ 9,000.00
11	$ 1,000.00	$ 10,000.00

Formulas can be applied to manipulate data between cells. We will see however, that special attention needs to be paid to the copying of formulas. We will continue working with the above worksheet with the Checking and Savings balances while we try some formulas.

Formulas and Addressing

We want to see what your total bank balance would be if you had the amounts listed in a particular row in your checking and savings accounts. Click in cell C2 type "=" to start a formula, then click cell A2 type "+" and click in cell B2 then press Enter. You have just added cells A2 and B2 together; the results should be shown in cell C2. Now click in cell C2. Select Copy from the toolbar and then highlight cells C3 through C11. Select Paste from the toolbar. The formula has now been copied to all of the rows for their respective balances. Notice that the value $1,100 that was displayed in cell C2 was not copied, but rather the formula that made up that value. The formula changed for each of the rows and substituted the new row value in the formula. This is called Relative Addressing. The formula changes relative to its address. The last cell should show the value $11,000.

Suppose we wanted to see what the annual interest would be on our savings account at the levels listed in Column B. Create a new label in cell A13 called "Interest rate". Now in cell B13 input the interest rate ".045". Format the cell for percentages. To do this highlight the cell B13 and click the % button on the toolbar. If the number changes to 5%, we can adjust the number of decimals displayed by clicking on the Increase Decimal button pictured here.

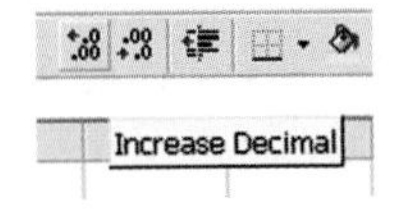

Now we want to use this value to compute interest for each level of savings. Label cell D1 "Savings interest". Resize the cell to fit the label. In cell D2 we will put our formula. Type "=" to start the formula, and then select cell B13 to get our percentage rate. Then type "*" and select cell B2. The resulting value will be the annual interest gained on $1000 in savings at 4.5% APR.

Since we have the formula we can copy it to the rest of the table and get the interest income for each of the levels of savings. So copy the formula from D2 to cell D3. Is this the result you expected? Probably not. Remember the rules of relative addressing—as you copy the formula down one row, the formula values change by one row. So in cell D3 the formula is pulling from cell B14 for the interest rate and cell D2 for the principle amount. You will notice that cell B14 is empty.

Absolute Addressing

We want to use cell B13 for each of these formulas. So how do we lock in the address of cell B13? We use "Absolute Addressing". The "$" symbol is the key to locking the position in the formula. In cell D2 we need to change the formula to read "=B$13*B2". We do not need to put a "$" in front of the "B" because it refers to the column, and we will not be moving this formula from column to column. We are only copying the formula between rows.

Go back and copy the new formula to cells D3 through D11. The results should now look like this:

File Edit View Insert Format Tools Data Window Help

F24 =

	A	B	C	D
1	Checking balance	Savings balance		Savings interest
2	$ 100.00	$ 1,000.00	$ 1,100.00	$ 45.00
3	$ 200.00	$ 2,000.00	$ 2,200.00	$ 90.00
4	$ 300.00	$ 3,000.00	$ 3,300.00	$ 135.00
5	$ 400.00	$ 4,000.00	$ 4,400.00	$ 180.00
6	$ 500.00	$ 5,000.00	$ 5,500.00	$ 225.00
7	$ 600.00	$ 6,000.00	$ 6,600.00	$ 270.00
8	$ 700.00	$ 7,000.00	$ 7,700.00	$ 315.00
9	$ 800.00	$ 8,000.00	$ 8,800.00	$ 360.00
10	$ 900.00	$ 9,000.00	$ 9,900.00	$ 405.00
11	$ 1,000.00	$ 10,000.00	$11,000.00	$ 450.00
12				
13	Interest rate	4.5%		
14				

Charts

Suppose you had data about ticket sales from your traveling circus. Adult tickets are $20, child tickets are $12. The sales for the last week are listed in the spreadsheet below.

	A	B	C
1		Adult	Child
2	Sunday	$10,120.00	$ 5,760.00
3	Monday	$ 9,040.00	$ 3,600.00
4	Tuesday	$ 8,380.00	$ 3,360.00
5	Wednesday	$ 7,620.00	$ 3,132.00
6	Thursday	$ 7,900.00	$ 2,520.00
7	Friday	$10,560.00	$ 4,944.00
8	Saturday	$13,260.00	$ 7,740.00
9			

From this data you can see how the ticket sales change daily and the comparison of child and adult ticket sales. This information might be easier to understand if it were graphically displayed in a chart. Enter the data as seen above in a new worksheet and use a bar chart for this particular example.

To create a chart we need to select the data to be graphed. Of course we want to chart the ticket sales for both the adults and children. In a bar chart the dollar amount will determine the height of the bar. What about the days of the week? The day of the week should be the labels for the bars but is not part of the charted data.

Highlight the data in column B and C from row 1 to row 8 (Excel can interpret the first row in a set of data as labels and not part of the data). Now click the Chart Wizard button on the toolbar, which looks like this:

Select the "Clustered Column chart". A column chart is a bar chart that displays the bars vertically.

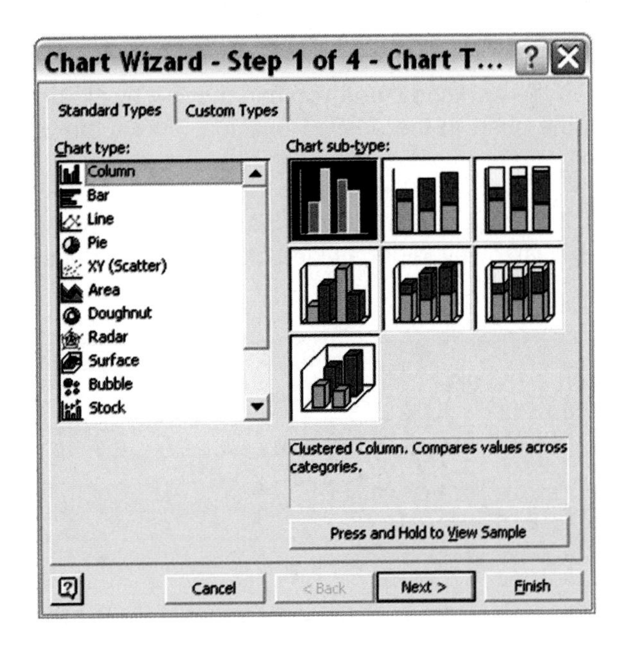

Then select the Next button.

Excel displays a sample view of the chart in this window with some formatting options. The data range is displayed in the "Data Range" window. Here we see that the data is coming from Sheet 2, cells B1 to C8.

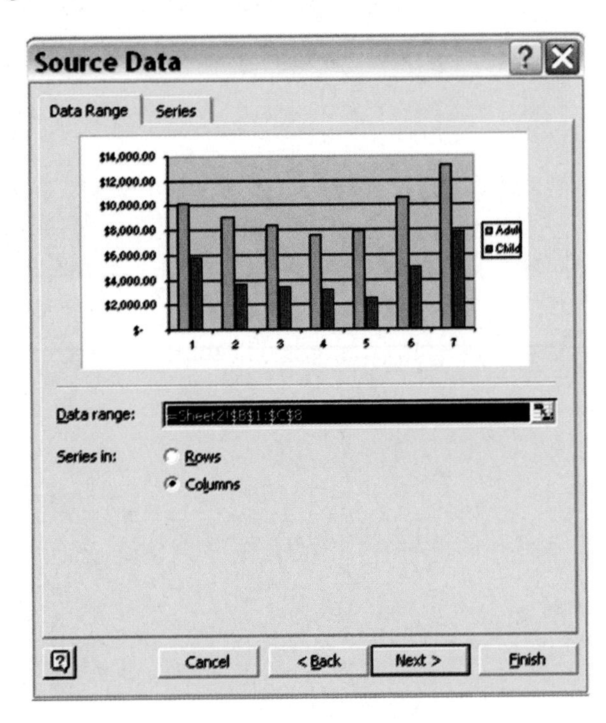

From the “Series” tab you can add or remove sets of data to be included in your chart.

Click Next again and you will be given the option to assign titles and various other graph options. Add the title “This Week” in the chart title field. Click Next, and you will be given the option to create this chart in a new worksheet or as an object in the existing sheet. Select an object in the existing sheet. Then click Finish.

Your chart will now be placed in the worksheet and you can move it by clicking on it and holding down the mouse button while dragging the chart to the desired position. Move the chart to the left margin of the worksheet.

You now have a good graphical representation of total sales, and sales by Adult and Child classifications. As you can see, this is a much easier way to get an idea of your sales trends versus looking at the data in the table constructed earlier.

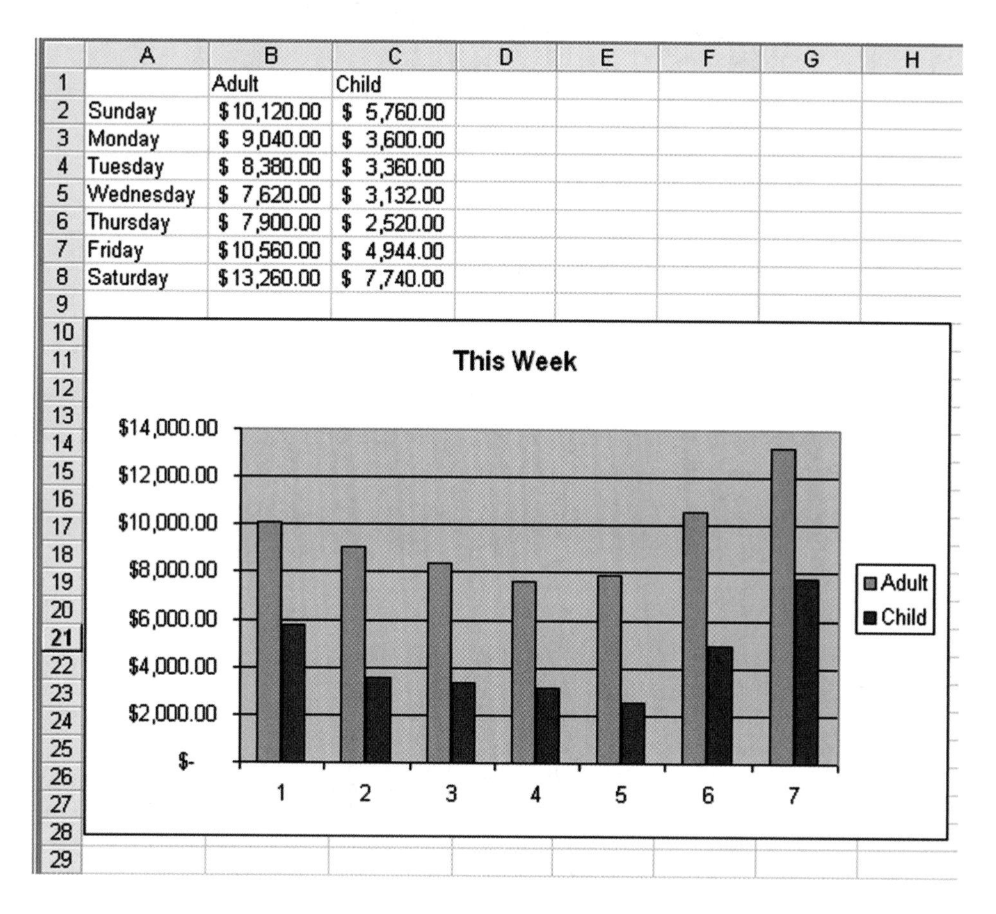

	A	B	C
1		Adult	Child
2	Sunday	$10,120.00	$ 5,760.00
3	Monday	$ 9,040.00	$ 3,600.00
4	Tuesday	$ 8,380.00	$ 3,360.00
5	Wednesday	$ 7,620.00	$ 3,132.00
6	Thursday	$ 7,900.00	$ 2,520.00
7	Friday	$10,560.00	$ 4,944.00
8	Saturday	$13,260.00	$ 7,740.00

Installing Data Analysis Tools

To get the most out of Excel as a statistical tool, you will want to use the Data Analysis Tools. If you did a default installation of Microsoft Office, you probably do not have these tools installed. First verify if the Data Analysis Tools are installed. At the top of the screen is a menu bar. On that menu bar you will see Tools. Clicking on Tools, you will look for Data Analysis. If you see the category Data Analysis listed you are ready to go.

If you do not have the statistical analysis tools installed you can install them by following these steps:

Click on the word Tools on the menu bar.

Select Add-Ins...

Check the boxes next to Analysis ToolPak and Analysis ToolPak – VBA.

Click OK.

You will be prompted for your original Microsoft Office Installation CD.

Follow the onscreen directions to complete the installation.

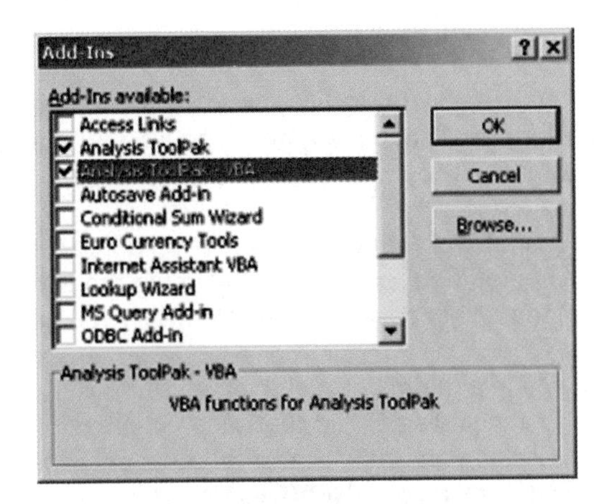

With these tools installed you can streamline statistical functions. Commonly used statistical formulas are already included. Generally all you need to do is select values you have set up in your worksheet to solve statistical functions. A fill in the blank wizard is available for most of the calculation formulas.

Appendix C

Getting Started with Minitab

The Basics of Minitab

By simply opening Minitab, you see a large electronic spreadsheet-type program with a large number of rows and up to 4,000 columns. The actual number of rows and cells depends on your computer's capabilities.

Managing files is easy with Minitab. Simply go to File, click Save Project As, and save your file under a new name and location. When you return to Minitab later, go to the file manager program in Windows and double-click your file. Doing this will automatically open Minitab and your file.

Opening a new Minitab file provides a blank worksheet. Enter data into a cell by typing into the dark-bordered cell. To move to another cell, move left/right or up/down by using the arrow keys. Copy a value in a cell by dragging the right-lower corner of the cell to cover all cells for which you wish to have that value.

When you open Minitab, you will notice a menu bar at the top of the screen. On that menu bar you will see Stat. Clicking on Stat you will notice a list including Basic Statistics, Regression, and ANOVA. Clicking on each of these reveals other possibilities. Clicking on Basic Statistics leads to Display Descriptive Statistics; clicking on Regression leads to Fitted Line Plot; clicking on ANOVA leads to One-way (Unstacked) or Two-way.

Also on the menu bar you will notice Calc. Clicking on Calc you will see a list including Random Data and Probability Distributions. Clicking on Random Data leads to many possibilities such as Integer or Chi-Square; clicking on Probability Distributions leads to choices such as Binomial or Poisson.

Both Stat and Calc require inputs into the cells before clicking on that menu item. End of chapter examples explain how to provide these inputs for the particular application and give further explanation as to how to use dialog boxes required along the way.

Above each worksheet there is a session window. The results of each session are displayed there. To print what is in the window, go to File, Print Session Window, and then press OK.

Cells

To create a new worksheet click File, New, and select Minitab Worksheet. Press OK. This opens a new worksheet to use in the current project. Columns run left to right and are labeled with the letter C and a number. Rows run top to bottom and are labeled with numbers. To refer to a specific cell, we will use the format "Column name, Row name". For example, if we were referring to column 2, row 3 we would use "C2,3" to reference the cell. Minitab provides a row that is not labeled for you to insert your column headings.

The cells in Minitab contain data and text. If you are inputting text in the cells other than the cell for the column heading, the column label will add "-T" to the column name. For example, if we typed the weekdays in the first five rows of column C1, the column name would change to "C1–T".

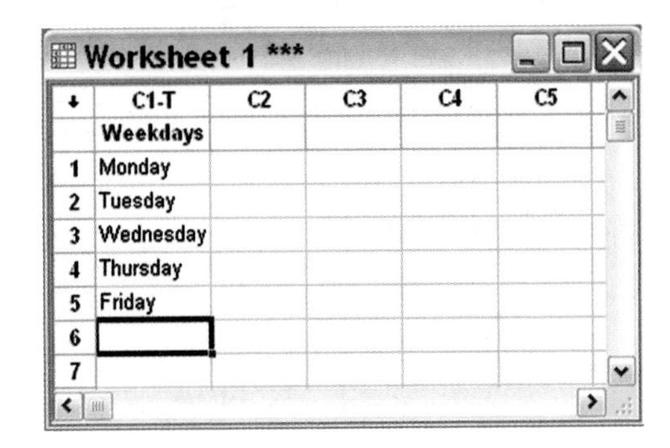

Filling Cells and Formulas

Now, let's input some data values into a new project to show some of the calculation features of Minitab. Choose File, New, and select Minitab Project. Press OK.

Minitab allows you to fill cells when entering a series of data. Enter "100" into cell C1,1 and "200" into cell C1,2. Then highlight both of the cells. Locate the small black square at the bottom right of the highlighted selection. When you move your cursor over the square, a plus sign should appear. Click the left mouse button and drag the mouse down to include C1,3 through C1,10. The cells should now be filled with values from 100 to 1000.

We can perform some basic functions on these values. From the menu bar, choose Calc and Column Statistics. This allows you to get the sum, the mean, the standard deviation, and many other values. Choose the mean and click in the Input variable text box. While your cursor is in this box, all the columns that contain data will be listed in the box to the left. You can click on the column and press Select to input C1 as the Input variable.

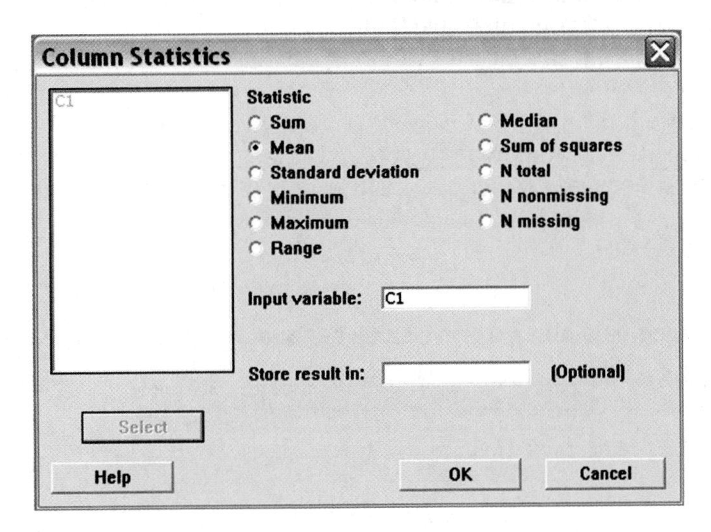

Press OK and the mean of C1 will appear in the session window.

To perform calculations on the data in the cells, select Calc and Calculator from the menu bar. In Store results in variable, enter C2. Under the Functions box, scroll down until you find the function Round. Select Round and it will appear in the Expression box along with the order of the parameters needed for the function. For the parameter "number" use "C1" and for the parameter "num_digits" enter 1. We can manipulate the data in C1. Press "* 7 / 3" after C1.

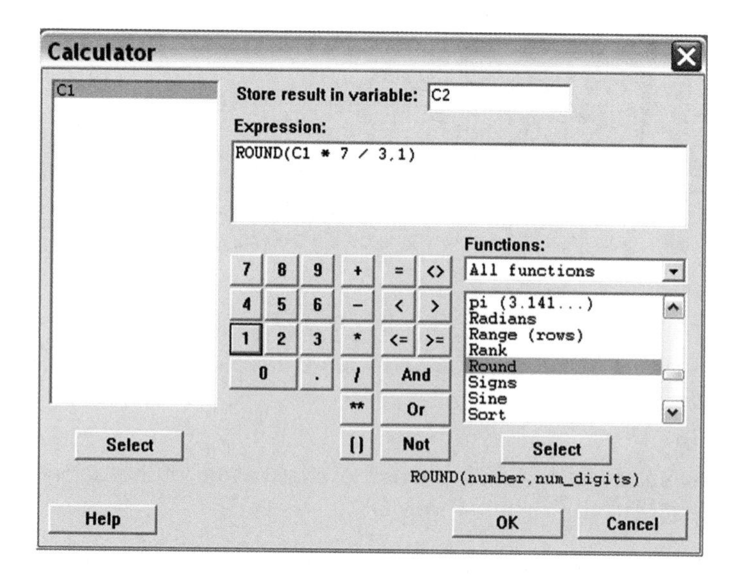

After pressing OK, you will see the values of this expression in column C2 rounded to one decimal place.

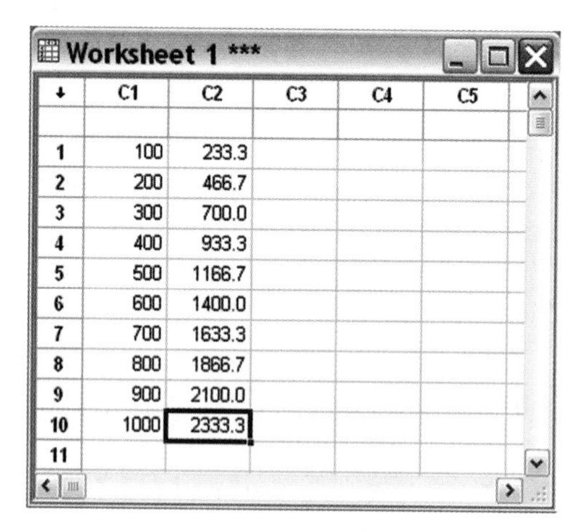

Worksheet 1 ***

↓	C1	C2	C3	C4	C5
1	100	233.3			
2	200	466.7			
3	300	700.0			
4	400	933.3			
5	500	1166.7			
6	600	1400.0			
7	700	1633.3			
8	800	1866.7			
9	900	2100.0			
10	1000	2333.3			
11					

Graphs

Suppose you had the following data about a firm's sales per quarter for a particular year.

Worksheet 1 ***

↓	C1-T	C2	C3	C4	C5
	Quarter	Sales			
1	First	356210			
2	Second	349800			
3	Third	370355			
4	Fourth	402775			
5					
6					
7					
8					
9					
10					

To graphically display this information, select Graph and Time Series Plot. Enter C2 in the Graph variables and press OK. A time series plot is shown in a separate window.

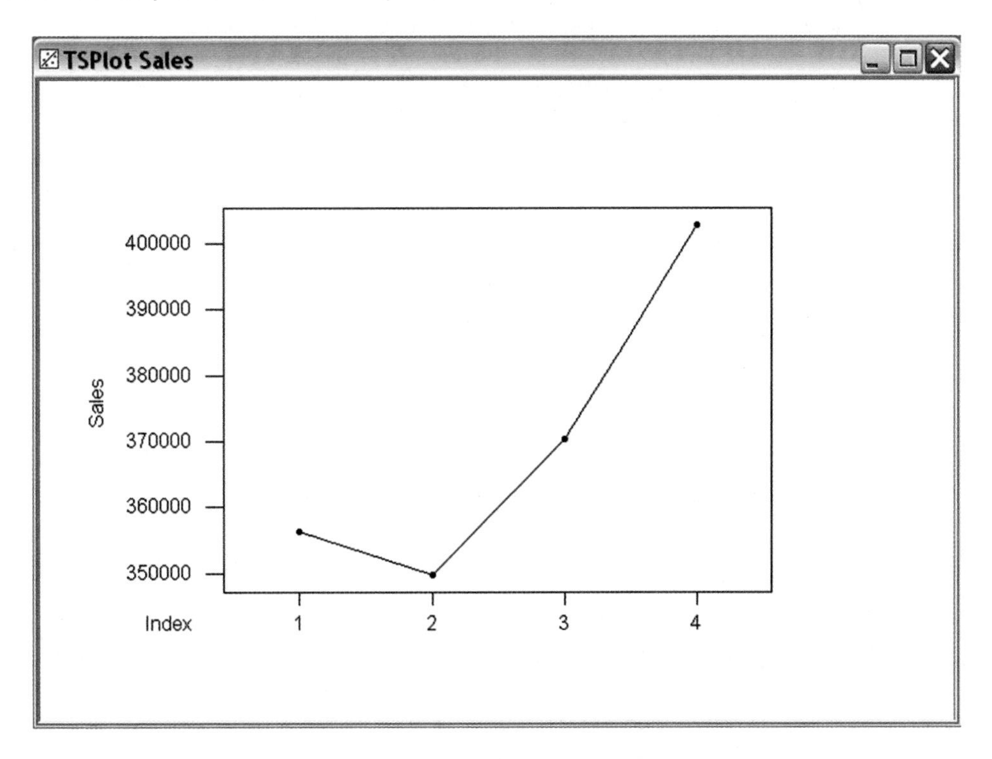

Data Analysis

One of the most important uses of Minitab is its statistical analysis. Start a new Minitab project and enter the following data for the number of phone calls a company receives each hour of the work day.

Worksheet 1 ***

↓	C1	C2	C3	C4	C5
	Phone Calls				
1	10				
2	27				
3	21				
4	24				
5	30				
6	23				
7	36				
8	30				
9	15				
10					

Under Stat, choose Basic Statistics and Display Descriptive Statistics. In the Variable input box, enter C1. The following descriptive statistics will be displayed in the session window of your project.

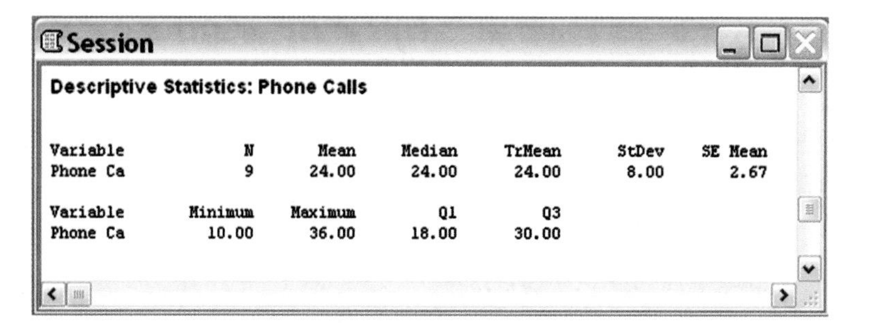
Session

Descriptive Statistics: Phone Calls

Variable	N	Mean	Median	TrMean	StDev	SE Mean
Phone Ca	9	24.00	24.00	24.00	8.00	2.67

Variable	Minimum	Maximum	Q1	Q3
Phone Ca	10.00	36.00	18.00	30.00

These are just a few examples of the analyses that may be performed using Minitab. There are many other statistical features of Minitab that you will encounter throughout the book.

Appendix D

Answer Key

✦ Chapter 1

Section 1.1 Exercises

1. Sample
3. Population
5. Population
7. Population
9. Sample
11. P - All Americans
S - Readers who mail in their ballots
13. P - All shoppers
S - 100 random shoppers between 1 & 5 p.m. on Thursday
15. P - All Christmas trees sold in your city
S - 45 random Christmas trees
17. Statistic
19. Statistic
21. Statistic
23. Parameter
25. Statistic
27. Inferential
29. Descriptive

Section 1.2 Exercises

1. **a.** Quantitative **b.** Discrete **c.** Ratio
3. **a.** Quantitative **b.** Discrete **c.** Interval
5. **a.** Quantitative **b.** Continuous **c.** Ratio
7. **a.** Quantitative **b.** Continuous **c.** Ratio
9. **a.** Quantitative **b.** Discrete **c.** Ratio
11. **a.** Quantitative **b.** Continuous **c.** Ratio
13. **a.** Quantitative **b.** Continuous **c.** Ratio
15. **a.** Qualitative **b.** N/A **c.** Nominal
17. **a.** Qualitative **b.** N/A **c.** Ordinal
19. **a.** Qualitative **b.** N/A **c.** Ordinal
21. **a.** Quantitative **b.** Discrete **c.** Ratio
23. **a.** Quantitative **b.** Discrete **c.** Interval

Section 1.3 Exercises

1. False
3. True
5. True
7. Double-masked
9. Control
11. Placebo
13. Placebo effect
15. Treatment
17. Experimental
None since it is an experiment
19. Observational
Lists of teenage girls at local high schools
21. Cluster
23. Stratified
25. Convenience
27. Census
29. Census
31. Random
33. Case Study
35. Meta-Analysis
37. Meta-Analysis

Section 1.4 Exercises

1. Informed consent
3. Researcher bias
5. Dropout
7. Non-adhere
9. Confounding variables

Chapter 1 Exercises

1. **a.** Statistic
   b. Inferential
3. **a.** Qualitative, Ordinal
   b. Qualitative, Ordinal
   c. Quantitative, Ratio
5. No
   Yes
7. **a.** All shoppers
   b. Convenience
   c. No
9. Answers may vary
11. **a.** Pop = All Adults
   Study = Experiment
   b. Answers may vary
13. **a.** Stratified
   b. Answers may vary
15. **a.** Adults who purchased vacation packages in the past year
   b. 724 adults
   c. 20% (one in five)
   d. Answers may vary

✦ Chapter 2

Section 2.1 Exercises

1.

Class	Bounds
15–19	14.5–19.5
20–24	19.5–24.5
25–29	24.5–29.5
30–34	29.5–34.5
35–39	34.5–39.5

3. Each class has a class width of 4

5.

Class	Midpoint
15–24	19.5
25–34	29.5
35–44	39.5
45–54	49.5
55–64	59.5

7.

Class	Relative f
15–18	2/31
19–22	7/31
23–26	4/31
27–30	15/31
31–34	3/31

9.

Class	Cumululative f
15–19	12
20–24	20
25–29	35
30–34	47
35–39	56

11.

Class	f
69.0–69.9	3
70.0–70.9	4
71.0–71.9	5
72.0–72.9	9
73.0–73.9	3
74.0–74.9	3
75.0–75.9	3

13.

Class	f
1100–1199	1
1200–1299	3
1300–1399	4
1400–1499	4
1500–1599	4
1600–1699	4

15.

Class	f	Class Boundary	Mid	Rel f	Cumul f
15–19	1	14.5–19.5	17	1/16	1
20–24	1	19.5–24.5	22	1/16	2
25–29	5	24.5–29.5	27	5/16	7
30–34	5	29.5–34.5	32	5/16	12
35–39	2	34.5–39.5	37	1/8	14
40–44	2	39.5–44.5	42	1/8	16

17.

Class	f	Class Boundary	Mid	Rel f	Cumul f
1800–2199	1	1799.5–2199.5	1999.5	1/15	1
2200–2599	4	2199.5–2599.5	2399.5	4/15	5
2600–2999	6	2599.5–2999.5	2799.5	2/5	11
3000–3399	3	2999.5–3399.5	3199.5	1/5	14
3400–3799	1	3399.5–3799.5	3599.5	1/15	15

Section 2.2 Exercises

1.

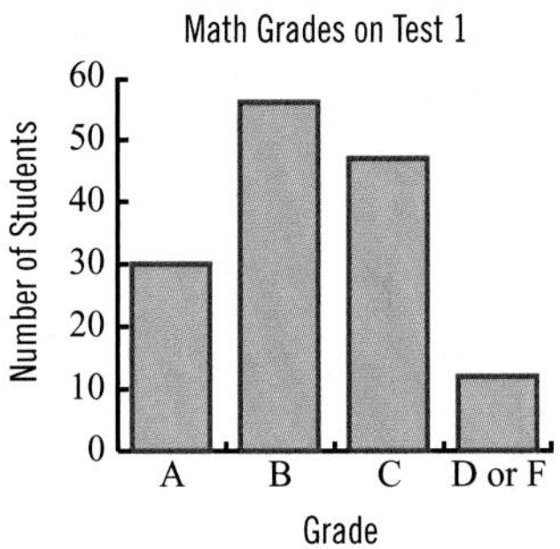

A pareto chart is not appopriate.

3. Bar Chart:

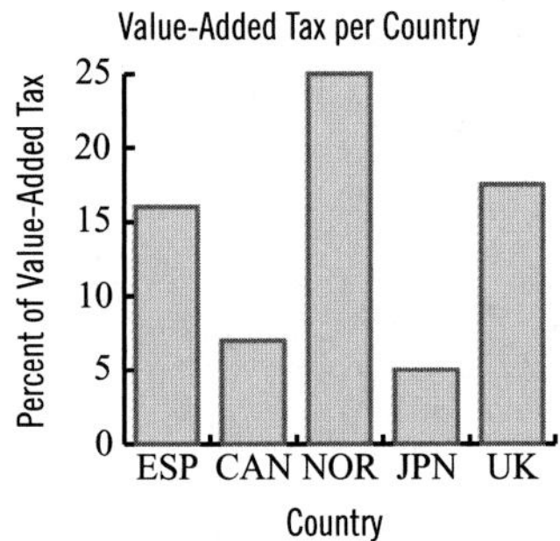

Pareto Chart:

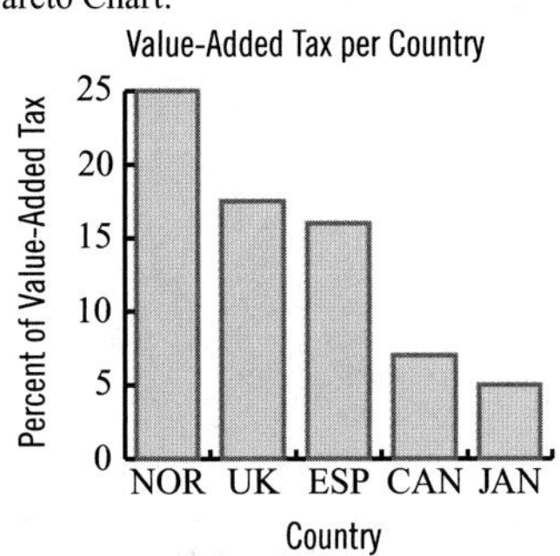

5.

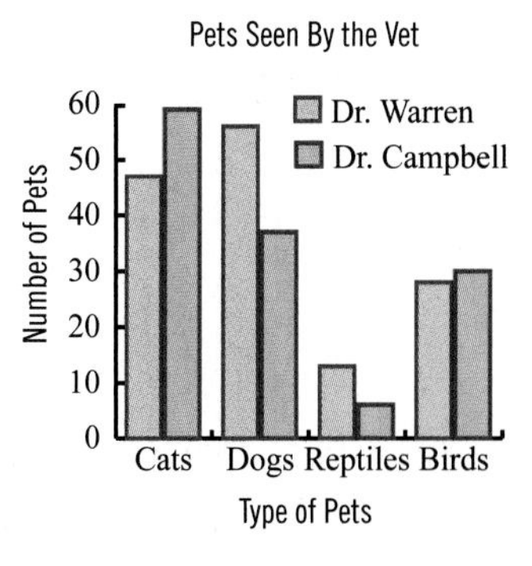

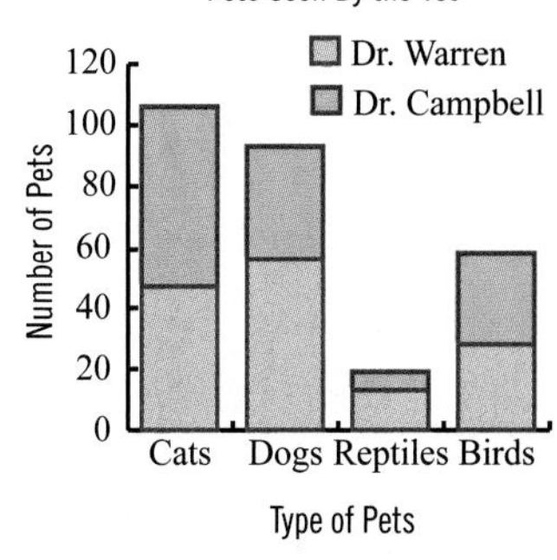

a. Dr. Warren, side-by-side bar
b. Reptiles, stacked bar
c. Dr. Campbell, side-by-side bar
d. Reptiles, side-by-side bar

7. English = 45°
Business = 149°
Education = 91°
Science = 75°

9.

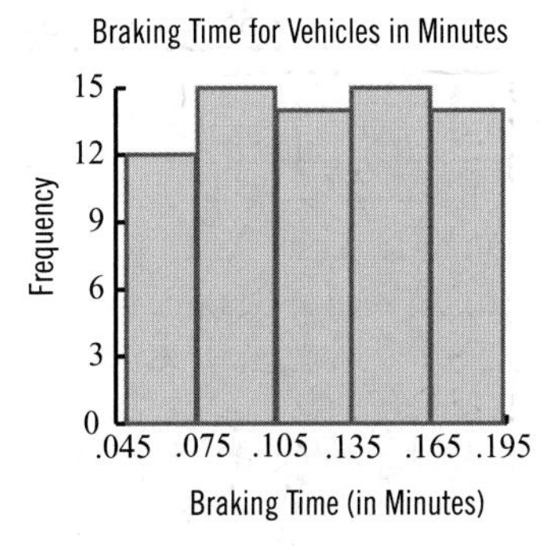

11.

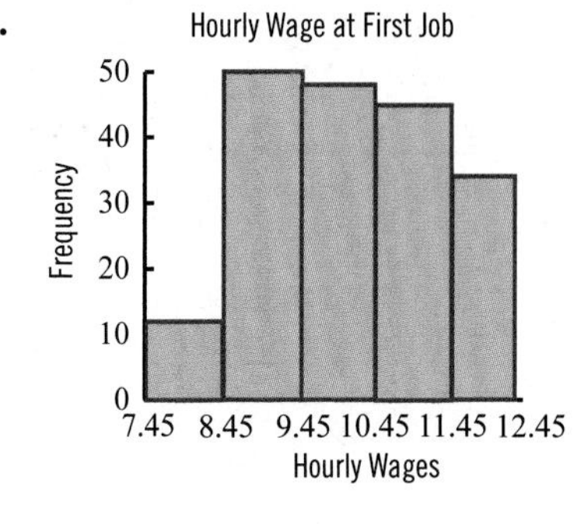

13.

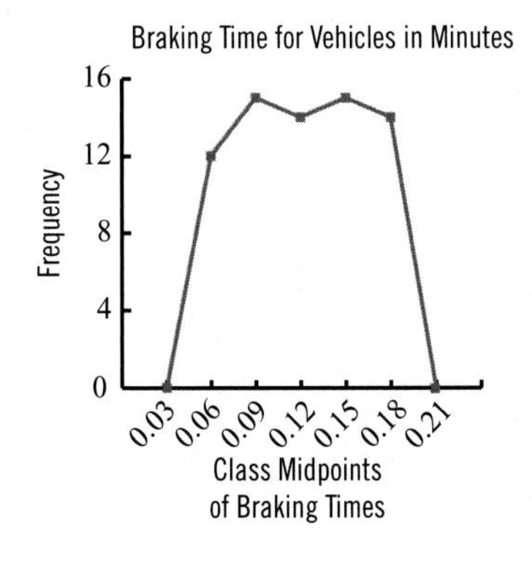

15.

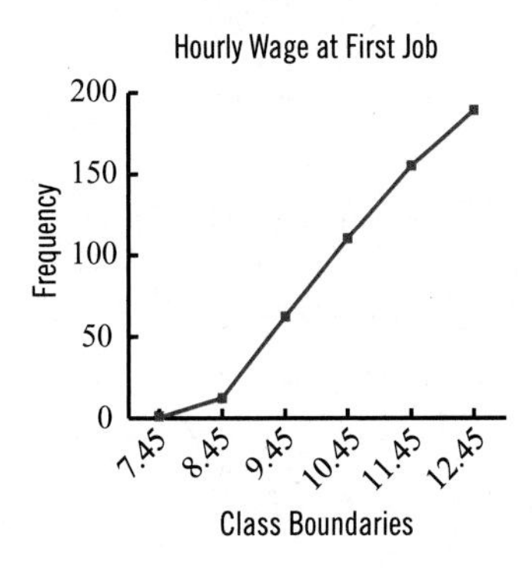

17. Caloric Intake for Men Ages 20 - 39

Stem	Leaves
1	8
2	2 2
2	5 5 6 6 7 8 8 9
3	0 0 1
3	5

Key: 1 | 8 = 1800

19. September Rainfall in Inches for Alaska

Stem	Leaves
2.5	0 4 5 6 9
2.6	1 2 4 4 7 8
2.7	0 3 4 7 8
2.8	1 1 6 8 9

Key: 2.6 | 1 = 2.61

21. **a.** about 290 cents
b. about 180 cents
c. 1/03/05
d. 9/12/05

23. **a.** 12
b. 34
c. 18
d. 29

Section 2.3 Exercises

1. Answers may vary

3. Answers may vary

5. Answers may vary

7. Skewed to the left

9. Uniform

11. Time-Series

13. Cross-Sectional

15. Answers may vary. A possible answer could be that the scale is not a good choice since it shows a dramatic change over a few cents.

17. The class widths are not equal.

Chapter 2 Exercises

1.

Class	f	Class Boundary	Mid	Rel f	Cumul f
0–4	7	–0.5–4.5	2	7/24	7
5–9	6	4.5–9.5	7	1/4	13
10–14	5	9.5–14.5	12	5/24	18
15–19	3	14.5–19.5	17	1/8	21
20–24	2	19.5–24.5	22	1/12	23
25–29	1	24.5–29.5	27	1/24	24

3. Saline Concentration

Stem	Leaves
101	7 8 9 9
102	0 1 1 1 2 2 2 2 2 3 3 3 4 5

Key: 101 | 7 = 1.017

5. **a.** Line graph
b. Bar graph
c. Stacked bar graph
d. Stem and leaf plot

7. **a.** 7
b. 1, Not very accurate
c. Wives who work less do more leisure activity
d. This graph is not easy to follow. A pair of side-by-side bar graphs, one for husbands and one for wives

9. **a.** 19 %
b. 53 %
c. 27 %
d. 47 %

11. **a.** Symmetric
b. Skewed to the left.
c. Answers may vary. One possibility is that students knew roughly the same amount of information. The scores were low because the material had not yet been taught.
d. Answers may vary. One possibility is that the majority of students learned most of the material.

✦ Chapter 3

Section 3.1 Exercises

1. Mean ≈ 7.6
Median = 8
Unimodal at 10

3. Mean ≈ 8.9
Median = 5.5
Unimodal at 5

5. Mean = 6.8
Median = 6
No mode

7. Mean ≈ 5.6
Median = 5
No mode

9. Mean = 5
Median = 5
Bimodal at 1 and 5

11. Mean ≈ 7.49
Median = 7.5
Bimodal at 7.3 and 7.5

13. Mean ≈ 16.56
Median = 14.9
Unimodal at 14.9

15. Mean ≈ 33.7
Median = 32
Multimodal at 29, 31, and 32

17. 2.78

19. $341.82

21. Mode

23. Mean

25. Mean = A
Median = B
Mode = C

Section 3.2 Exercises

1. Range = 13
Variance ≈ 14.5
Standard Deviation ≈ 3.8

3. Range = 22
Variance ≈ 42.2
Standard Deviation ≈ 6.5

5. Range = 9
Variance ≈ 9.1
Standard Deviation ≈ 3.0

7. Range = 0
Variance = 0
Standard Deviation = 0

9. Range = 1.5
Variance ≈ 0.22
Standard Deviation ≈ 0.47

11. 1.14

13. 12.52

15. 7.8

17. False

19. True

21. $s \approx 0.042$
$s^2 \approx 0.002$

23. $s \approx 1.20$
$s^2 \approx 1.44$

25. A ≈ 29.8%
B ≈ 11.5%
Larger Spread = A

27. A ≈ 16.8%
A ≈ 10.8%
Smaller Spread = B

29. 95%

31. 97.5%

33. 97.5%

35. At least 88.9%

Section 3.3 Exercises

1. 4, 8, 10, 12, 17

3. 13, 16.5, 18, 19.5, 21

5. **a.** B
b. A
c. A

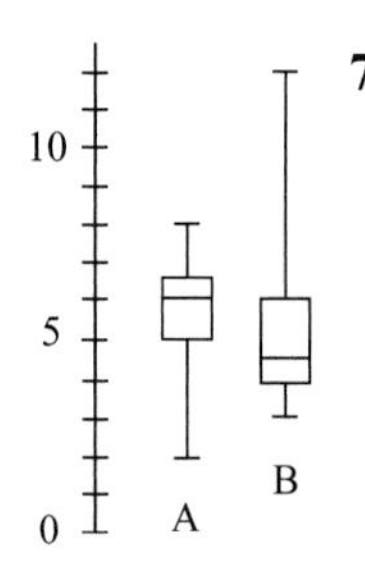

7. **a.** B
b. B
c. B

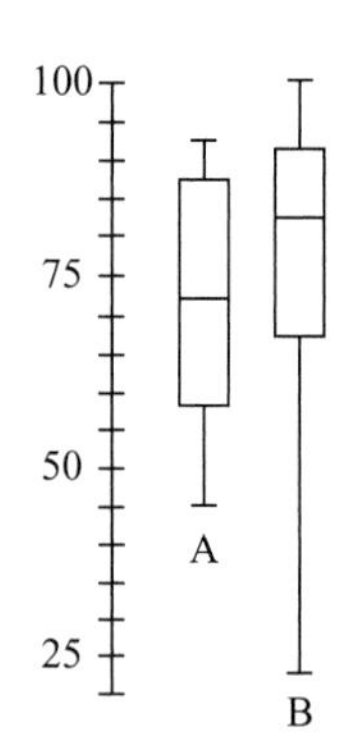

9. **a.** 7.4
b. 67th

11. **a.** 31.1
b. 38th

13. **a.** 29
b. 15th

15. −0.38

17. −0.4

19. ACT

21. Charity Tournament

23. a. −1.3

Chapter 3 Exercises

1. $10.92

3. **a.** False
b. True

5. 83

7. True

✦ Chapter 4

Section 4.1 Exercises

1. {HH, HT, TH, TT}

3. {BBBB, BBBG, BBGB, BGBB, GBBB, BBGG, BGBG, GBBG, BGGB, GBGB, GGBB, BGGG, GBGG, GGBG, GGGB, GGGG}

5. {RSunL, RSunC, RNoL, RNoC, BSunL, BSunC, BNoL, BNoC, SSunL, SSunC, SNoL, SNoC}

7. {1A, 1N, 2A, 2N, 3A, 3N}

9. Emperical

11. Classical

13. Subjective

15. **a.** $\frac{51}{175} \approx 0.291$
b. $\frac{40}{175} \approx 0.229$
c. $\frac{106}{175} \approx 0.606$

17. $\frac{16}{66} \approx 0.242$

19. $\frac{31}{365} \approx 0.085$

21. $\frac{150}{350} \approx 0.429$

23. $\frac{1}{45} \approx 0.022$

25. $\frac{10}{1000} = 0.01$

Section 4.2 Exercises

1. 211 apple trees that are not ready for harvesting.
3. The 17 players who are not left-handed.
5. The 30% of viewers who are 30 years old or younger.
7. 0.909
9. 0.969
11. Not Mutually Exclusive
13. Not Mutually Exclusive
15. Not Mutually Exclusive
17. Mutually Exclusive
19. Independent
21. Dependent
23. Independent
25. 0.019
27. 0.005
29. 0.444
31. 0.750
33. 0.68
35. 0.017
37. 0.059

Section 4.3 Exercises

1. 100,000
3. 216
5. 500,000
7. 120
9. 35
11. 40,320
13. 336
15. 56
17. 56
19. 5040
21. 60
23. 5040
25. 8
27. 6
29. 56
31. 1
33. 12
35. 2
37. 205
39. 132
41. 3060
43. 336
45. 3003
47. 495
49. 32,760
51. 50,400
53. $\frac{3125}{100{,}000} \approx 0.031$
55. $\frac{1}{276} \approx 0.004$
57. $\frac{1}{15600} \approx 0.00006$
59. 1
61. n
63. $n!$

Section 4.4 Exercises

1. 1,728,000
3. 1120
5. 22,308
7. 15,750
9. 106,400
11. 4845
13. 1.56×10^{22}

Chapter 4 Exercises

1. 0.37 or 37%
3. $\frac{1}{1319} \approx 0.0008$
5. 0.26 or 26%
7. $\frac{2}{252} \approx 0.008$
9. $\frac{1}{16} \approx 0.063$
11. $\frac{270724}{270725}$
13. $\frac{650}{1000} = 0.65$
15. 4.89×10^{34}
17. 32
19. 336
21. 0.006
23. 630,630

✦ Chapter 5

Section 5.1 Exercises

1.

x	$P(x)$
0	$\frac{1}{16}$
1	$\frac{4}{16}=\frac{1}{4}$
2	$\frac{6}{16}=\frac{3}{8}$
3	$\frac{4}{16}=\frac{1}{4}$
4	$\frac{1}{16}$

3.

x	$P(x)$
0	$\frac{6}{36}=\frac{1}{6}$
1	$\frac{10}{36}=\frac{5}{18}$
2	$\frac{8}{36}=\frac{2}{9}$
3	$\frac{6}{36}=\frac{1}{6}$
4	$\frac{4}{36}=\frac{1}{9}$
5	$\frac{2}{36}=\frac{1}{18}$

5. 17.8

7. 12.2

9. **a.** –$3.24
b. –$324

11. **a.** –$10
b. –$150
c. $2,105,900

13. **a.** $43.20
b. $432,000

15. **a.** –$0.50, –$1, –$1.50
b. 1

17. **a.** –$53.33
b. $4800

Section 5.2 Exercises

1. No; More than 2 outcomes. No fixed number of trials.

3. No; More than 2 outcomes

5. Yes

7. 0.200

9. 0.271

11. 0.500

13. 0.850

15. 0.787

17. **a.** 0.253
b. 0.350

19. **a.** 0.420
b. 0.901

21. **a.** 0.902
b. 0.550
c. 0.150

23. **a.** 0.023
b. 0.00002
c. 0.055

25. **a.** 0.782
b. 0.218
c. 0.00003

Section 5.3 Exercises

1. 12

3. 62.5

5. $\frac{35}{3}$

7. 0.0743

9. 0.1269

11. 0.2881

13. 0.9901

15. 0.2588

17. **a.** 0.0067
b. 0.5438

19. **a.** 0.2381
b. 0.7851

21. **a.** 0.2510
b. 0.5578

23. **a.** 0.0307
b. 0.4335

Section 5.4 Exercises

1. **a.** 0.2743
   b. 0.0005
3. **a.** 0.2054
   b. 0.0036
5. **a.** 0.2308
   b. 0.5000
7. **a.** 0.0511
   b. 0.3576
9. **a.** 0.2962
   b. 0.2176

Chapter 5 Exercises

1. **a.** 0.336
   b. 0.428
3. 0.2084
5. 0.0353
7. 0.015
9. 0.2696
11. 0.0483

✦ Chapter 6

Section 6.1 Exercises

1. False. There are an **unlimited** number of normal distributions.
3. False. The mean of the **standard** normal distribution is always 0.
5. False. The standard deviation of the standard normal distribution is always **1**.
7. True
9. True
11. Normal
13. Not normal, it is skewed to the left since it is a relatively easy exam.
15. Not normal because the outcomes are discrete, not continuous, and the distribution would be uniform.

17.

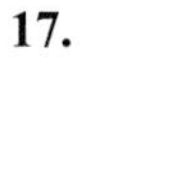

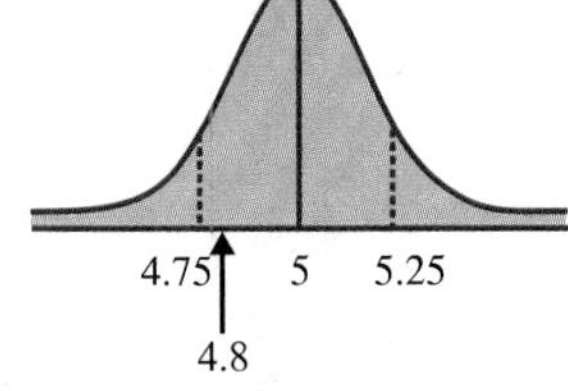

19.

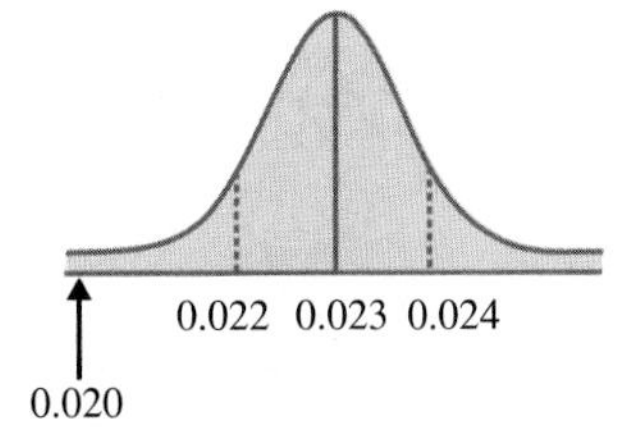

21.

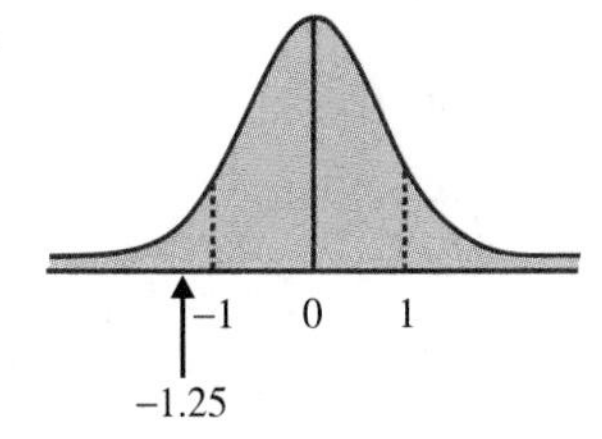

23.

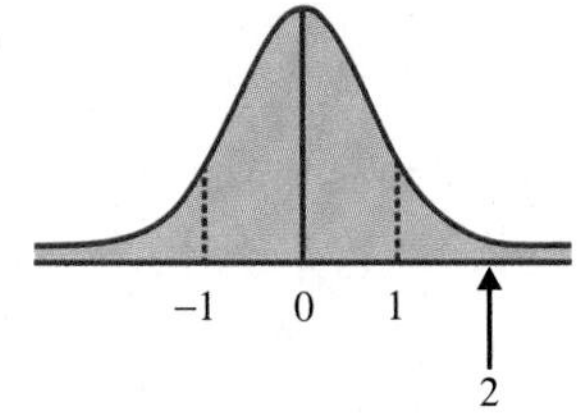

25.

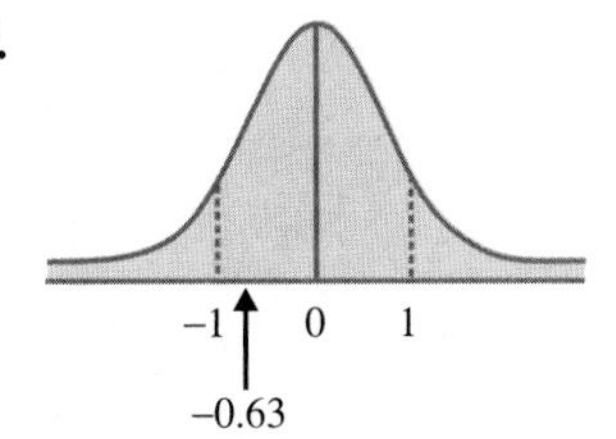

Section 6.2 Exercises

Note: Answers found using technology will vary slightly.

1. 0.9906 **3.** 0.1056 **5.** 0.9772 **7.** ≈ 1 **9.** 0.0885

11. 0.9940 **13.** 0.8413 **15.** 0.3310 **17.** 0.1336 **19.** 0.9678

21. 0.0348 **23.** 0.0579 **25.** 0.1815 **27.** 0.9317 **29.** 0.9629

31. 0.0008 **33.** 0.0033 **35.** 0.9678 **37.** 0.4973 **39.** 0.2076

Section 6.3 Exercises

Note: Answers found using technology will vary slightly.

1. **a.** 15.87%
b. 2.28%
c. 49.72%
d. 50.28%

3. **a.** 0.1587
b. 0.0526
c. 0.3933
d. 0.8104

5. **a.** 0.0228
b. 0.0139
c. 0.8644
d. 0.1096

7. **a.** 0.5832
b. 0.0985
c. 0.2615
d. 0.3174

9. **a.** 84.13%
b. Answers will vary. For example, lowering the number of years in the warranty would decrease the company's expense on returned batteries.

11. Both a. and b. are possible answers

Section 6.4 Exercises

Note: Answers found using technology will vary slightly.

1. −2.67 **3.** −2.03 **5.** −1.89 **7.** 2.575 **9.** 1.96

11. 2.19 **13.** 1.645 **15.** 0.67 **17.** −0.67 **19.** $34,680

21. 136.6 pounds **23.** 132.11 minutes **25.** $x = 51.24$, therefore prepare for 52 calls

Section 6.5 Exercises

1. 1.753 **3.** 2.779 **5.** 2.462 **7.** 22 **9.** 1.833

11. 3.365 **13.** −1.812 **15.** −3.055 **17.** 3.012 **19.** 2.681

21. 2.064 **23.** 1.708

Chapter 6 Exercises

1. 1 **3.** median, mode **5.** mean, standard deviation **7.** degrees of freedom **9.** 47.72%

11. 0.1574 **13.** 25% **15.** c. 1.04 **17.** −2.583 **19.** 1.333

✦ Chapter 7

Section 7.1 Exercises

1. False. A sampling distribution refers to **groups** rather than **individuals**.
3. True
5. $u_{\bar{x}} = 35, \sigma_{\bar{x}} = 1.1$
7. $u_{\bar{x}} = 12, \sigma_{\bar{x}} = 0.38$
9. $u_{\bar{x}} = 9.5, \sigma_{\bar{x}} = 1.60$
11. $u_{\bar{x}} = \$1.87$
13. $u_{\bar{x}} = 7$ days
15. $\sigma_{\bar{x}} = 0.02$
17. $\sigma_{\bar{x}} = 0.58$
19. $\sigma = \$191.00$

Section 7.2 Exercises

Note: Answers found using technology will vary slightly.

1. 0.44
3. −2.06
5. −1.33
7. **a.** 0.1587
   b. 0
   c. 0.0262
   d. 0.2420
9. **a.** 0.3707
   b. 0.0091
   c. 0.9909
   d. 0.0091
   e. 0.0182
11. **a.** 0.2881
    b. 0.0057
    c. 0.0089
    d. 0.0836
13. 0.9500
15. ≈ 0
17. 0.7258
19. 0.0376
21. 0.0401
23. 0.4744
25. 0.0040
27. 0.9490
29. 0.9886

Section 7.3 Exercises

1. $\hat{p} = 0.35$
   $z = 0.21$
3. $\hat{p} = 0.83$
   $z = 0.40$
5. $\hat{p} = 0.09$
   $z = -18.79$
7. 0.1357
9. 0.6879
11. 0.0436
13. 0.6744
15. 0.2206

Section 7.4 Exercises

1. Area to the right of 39.5
3. Area to the right of 500.5
5. Area to the left of 30.5
7. Area to the left of 11.5
9. Area between 34.5 and 35.5
11. Yes
13. No
15. 0.2236
17. ≈ 1
19. 0.9938
21. 0.9976
23. 0.8790
25. 0.0735
27. 0.0244
29. 0.0453

Chapter 7 Exercises

1. 0.1483 **3.** 0.7088 **5.** 0.1388 **7.** 0.0016 **9.** 0.9909

11. 0.0749 **13.** 0.1788 **15.** 0.7839 **17.** 0.9250

✦ Chapter 8

Section 8.1 Exercises

1. \$18 **3.** 177 yards **5.** (16.30, 19.70) **7.** (9.2, 11.8) **9.** Use t

11. Use z **13.** Use z **15.** Use more advanced methods **17.** Use z

19. Use more advanced methods **21.** Use t

Section 8.2 Exercises

1. 0.69 **3.** 0.25 **5.** (44.58, 45.42) **7.** (18.25, 18.65) **9.** 424

11. 67 **13.** 0.4 hours **15.** 0.6 hours **17.** (\$63, \$71)

19. (6.7, 7.3) **21.** 58 students **23.** 97 students

Section 8.3 Exercises

1. 2.179 **3.** 2.898 **5.** 0.02 **7.** 0.49

9. (92.2, 97.8) **11.** (5.5, 8.5) **13.** 3.2 miles **15.** 45 pounds

17. (5, 93) **19.** (3838, 4412)

Section 8.4 Exercises

1. $\frac{41}{50} \approx 0.82 = 82\%$ **3.** $\frac{48}{112} \approx 0.429 = 42.9\%$ **5.** (80.0%, 84.0%) **7.** $0.039 = 3.9\%$

9. $0.058 = 5.8\%$ **11.** (33.3%, 46.7%) **13.** (30.5%, 44.2%) **15.** 373 college students

17. 1637 women

Section 8.5 Exercises

1. 4.84

3. 0.76

5. 5

7. 25

9. $\chi^2_{0.025} = 39.364$
$\chi^2_{0.975} = 12.401$

11. $\chi^2_{0.050} = 37.671$
$\chi^2_{0.950} = 11.591$

13. (15.0, 106.0)

15. (6.1, 30.3)

17. (1.9, 3.1)

19. (6.1, 10.4)

21. 212

23. 337

25. (0.02, 0.15)

27. (1.37, 1.84)

Chapter 8 Exercises

1. ($43.00, $47.00)

3. 0.016 = 1.6%

5. 0.35

7. (38.2, 43.8)

9. (0.001, 0.003)

11. 2653 people

13. The width will increase.

15. 99%

17. The width will decrease.

✦ Chapter 9

Section 9.1 Exercises

1. 3

3. 0.2

5. 1.17

7. 0.25

9. (–5, 1)

11. (–6, 0)

13. (–$13720, $7120)

Section 9.2 Exercises

1. 2.045

3. 1.796

5. 3.182

7. 2.539

9. 2

11. 5

13. (–2.9, –1.1)

15. (0.4, 2.0)

17. (–1.05, 14.07)

19. (–2.5, –0.1)

21. (0.7, 3.1)

23. (–$30, $12)

Section 9.3 Exercises

1. $\bar{d} = 1.9, s_d = 1.6$

3. $\bar{d} = 1.3, s_d = 6.0$

5. (1.99, 2.47)

7. (–0.10, 2.38)

9. (0.1, 2.2)

11. (–12.2, –7.4)

13. (7.2, 10.3)

15. (–9.6, 3.3)

17. (–8.2, 0.4)

19. (5.3, 16.2)

21. (–3.0, 19.6)

Section 9.4 Exercises

1. Conditions are met

3. Conditions are not met $n_2\left(1-\hat{p}_2\right)=3.92<5$

5. −0.176

7. 0.019

9. 0.085

11. 0.199

13. 0.107

15. (−0.185, 0.266)

17. (−0.576, −0.275)

19. (−0.022, 0.295)

21. (−0.098, 0.405)

23. (−0.067, 0.107)

25. (0.010, 0.355)

Section 9.5 Exercises

1. 1.194

3. 1.113

5. $F_{\frac{\alpha}{2}}=2.4478$
$F_{1-\frac{\alpha}{2}}=0.3868$

7. $F_{\frac{\alpha}{2}}=2.1555$
$F_{1-\frac{\alpha}{2}}=0.4679$

9. $F_{\frac{\alpha}{2}}=4.9884$
$F_{1-\frac{\alpha}{2}}=0.1910$

11. (0.5021, 3.4589)

13. (0.3324, 3.7861)

15. (0.2165, 8.7358)

17. (0.3590, 3.0610)

19. (0.4836, 2.1496)

21. (0.1824, 6.9829)

23. (0.4867, 2.2882)

25. (0.5793, 3.7090)

27. (0.2512, 4.6427)

Chapter 9 Exercises

1. (0.32, 5.68)

3. (−3.65,−2.01)

5. (−0.2, 1.4)

7. (0.2, 0.6)

9. (4, 6)

11. (1.0, 7.6)

13. (0.3701, 3.2841)

15. (0.4407, 2.8219)

✦ Chapter 10

Section 10.1 Exercises

1. True

3. False, we **never** prove either hypothesis to be true or false.

5. False, the **null** hypothesis is the starting assumption in a hypothesis test.

7. $H_0\colon \mu\le 320$
$H_a\colon \mu>320$

9. $H_0\colon \mu=4$
$H_a\colon \mu\ne 4$

11. $H_0\colon p\le 0.42$
$H_a\colon p>0.42$

13. $H_0\colon \mu=3$
$H_a\colon \mu\ne 3$

15. $H_0\colon \mu\le 1.3$
$H_a\colon \mu>1.3$

17. The evidence supports the claim at the 0.05 level of significance that the percentage is less than 40%.

19. Yes. With a 95% level of confidence, the evidence suggests that the average age of their listeners is no longer 26.

21. With α= 0.01, the evidence does not support the claim that the average is higher than $600.

Section 10.2 Exercises

1. $t \le -1.761$

3. $|t| \ge 2.228$

5. $t \ge 2.110$

7. $|t| \ge 1.833$

9.

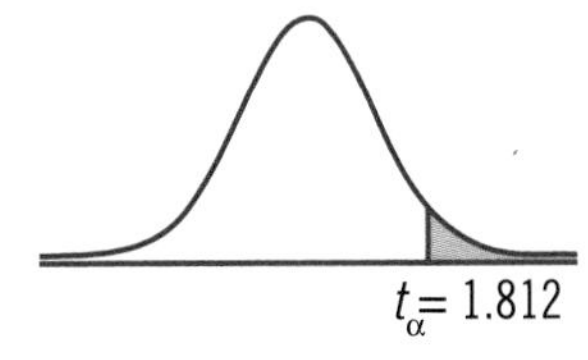

11.

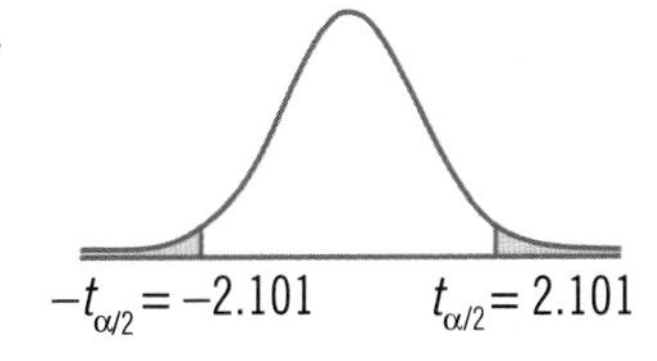

13.

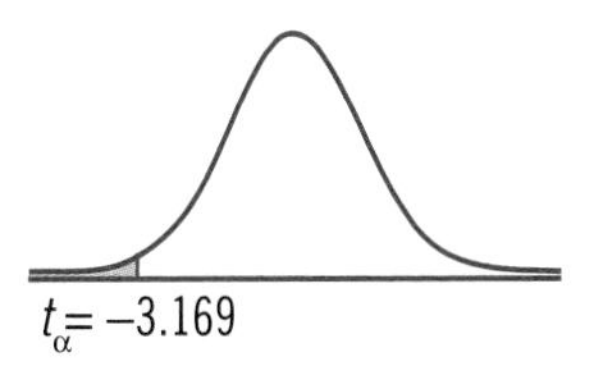

15.

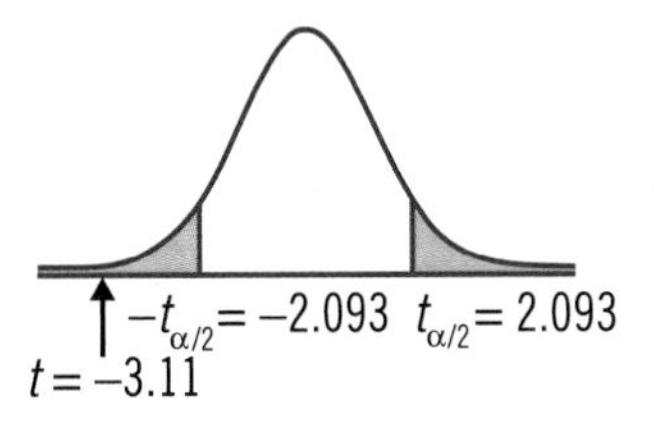

Reject H_0

17.

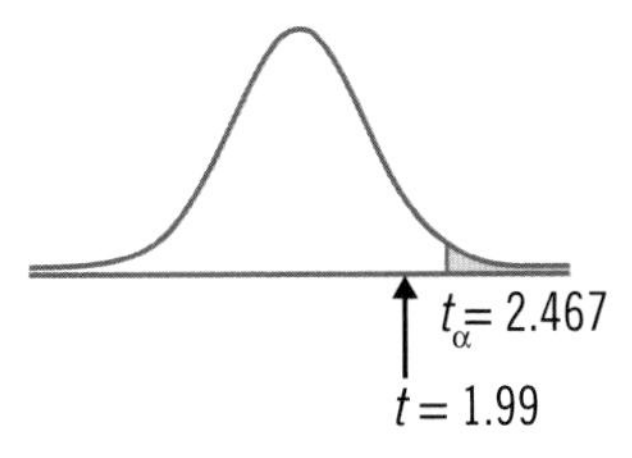

Fail to reject H_0

19.

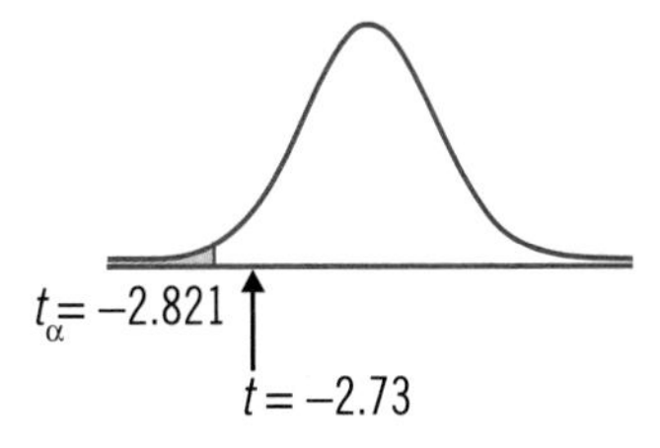

Fail to reject H_0

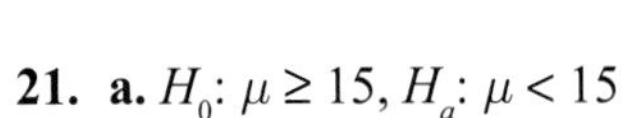

21. **a.** H_0: $\mu \ge 15$, H_a: $\mu < 15$
b. t-distribution, $\alpha = 0.10$

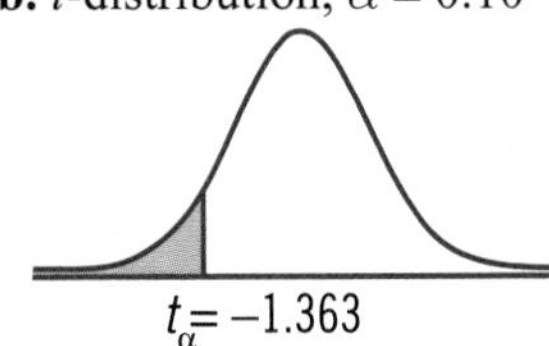

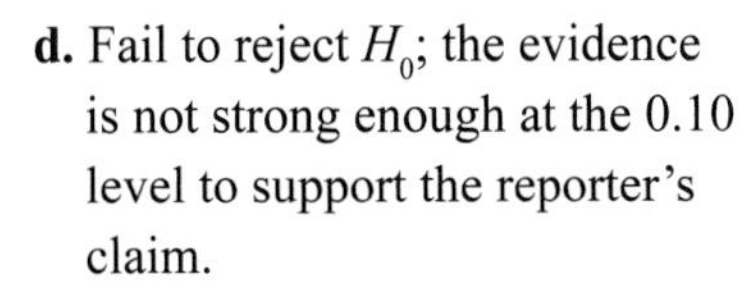

c. -0.577
d. Fail to reject H_0; the evidence is not strong enough at the 0.10 level to support the reporter's claim.

23. **a.** H_0: $\mu \ge 4$, H_a: $\mu < 4$
b. t-distribution, $\alpha = 0.01$

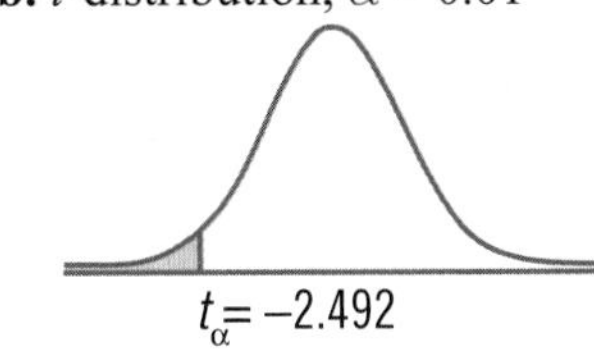

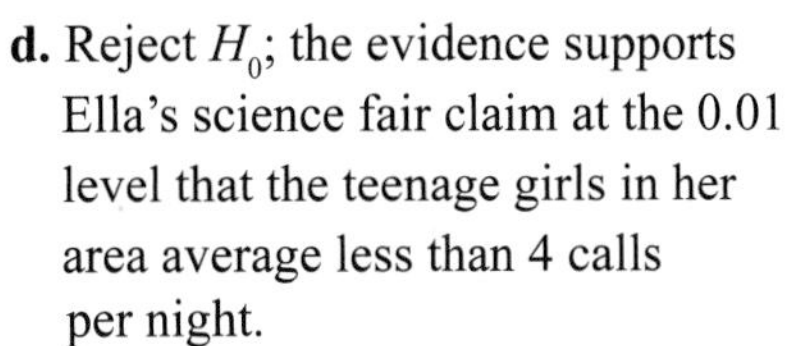

c. -3.33
d. Reject H_0; the evidence supports Ella's science fair claim at the 0.01 level that the teenage girls in her area average less than 4 calls per night.

25. **a.** H_0: $\mu \le 20$, H_a: $\mu > 20$
b. t-distribution, $\alpha = 0.05$

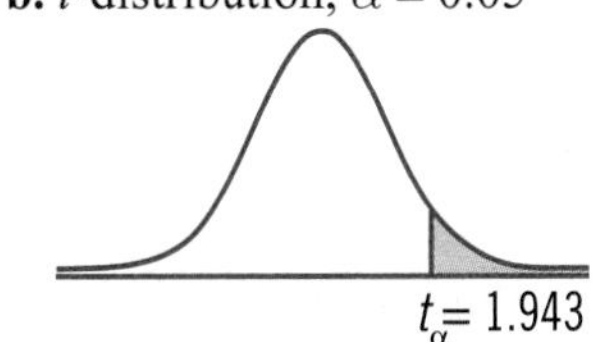

c. 1.661
d. Fail to reject H_0; the evidence is not strong enough at the 0.05 level of significance to say that the average delivery time is more than 20 minutes.

27. **a.** H_0: $\mu = 15$, H_a: $\mu \ne 15$
b. t-distribution, $\alpha = 0.05$

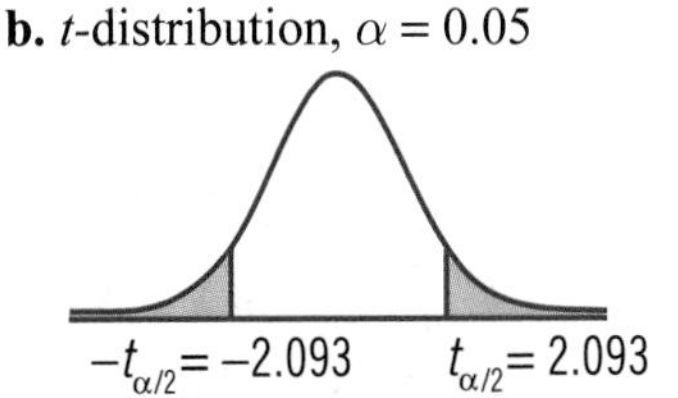

c. -2.118
d. Reject H_0; the evidence collected supports the claim at the 0.05 level that the average is no longer 15. The company may need to change their advertising.

Section 10.3 Exercises

1.

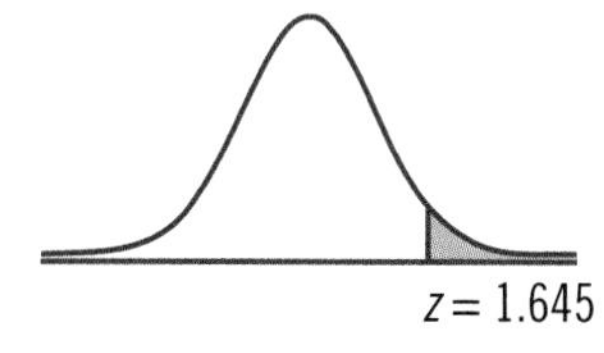

3.

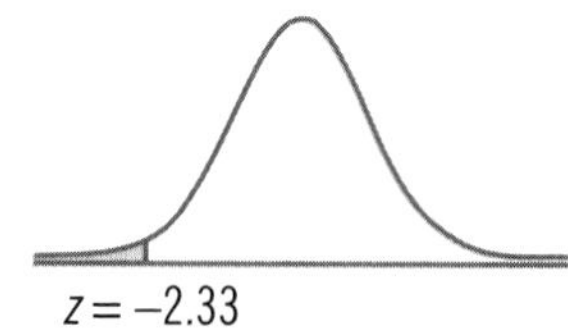

5.

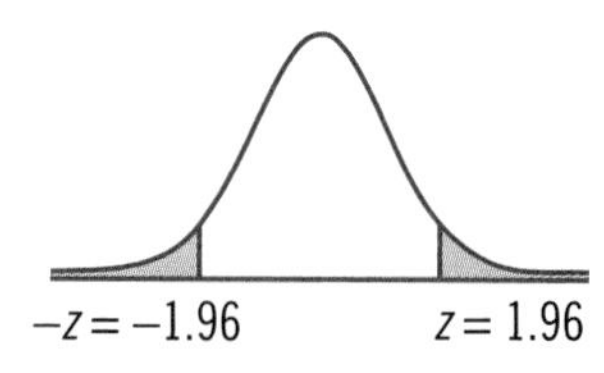

7. 0.0694

9. 0.0272

11. 0.0537

13. 0.0414

15. Reject H_0

17. Fail to reject H_0

19. Fail to reject H_0

21. Reject H_0

23. Fail to reject H_0

25. **a.** H_0: $\mu = 2$, H_a: $\mu \neq 2$
b. z-score for population means, $\alpha = 0.05$
c. $z = -2$
d. p-value = 0.0456, therefore reject H_0. The evidence supports the claim at the 0.05 level of significance that the bolts are not 2 cm and the manufacturer needs to recalibrate the machines.

27. **a.** H_0: $\mu \leq 839$, H_a: $\mu > 839$
b. z-score for population means, $\alpha = 0.01$
c. $z = 2.22$
d. p-value = 0.0132, therefore fail to reject H_0. At the 0.01 level of significance, the evidence does not support the group's claim that the average theme park traveler spends more than $839 per trip.

29. **a.** H_0: $\mu \geq 162.9$, H_a: $\mu < 162.9$
b. z-score for population means, $\alpha = 0.01$
c. $z = -3.12$
d. p-value = 0.0009, therefore reject H_0. The evidence does not support the group's claim at the 0.01 level of significance that the average woman in this community weighs less than 162.9 pounds.

Section 10.4 Exercises

1. No, $np < 5$

3. Yes

5. No, $n(1-p) < 5$

7. H_0: $p \geq 0.75$, H_a: $p < 0.75$
$z = -0.36$
$p = 0.3594$
Fail to reject H_0; the evidence does not support her claim at the 0.10 level of significance.

9. H_0: $p = 0.08$, H_a: $p \neq 0.08$
$z = 2.17$
$p = 0.0300$
Fail to reject H_0; the evidence does not support the claim at the 0.01 level of significance that the percentage of truck drivers with sleep apnea is different from 8%.

11. H_0: $p \leq 0.15$, H_a: $p > 0.15$
$z = 0.44$
$p = 0.3300$
Fail to reject H_0; the evidence does not support the cost effectiveness claim of the direct mail campaign at the 0.05 level of significance.

Section 10.5 Exercises

1. True

3. False. The level of significance is the probability of making a **Type I** error.

5. Yes; Type II error

7. No; Correct decision

9. Yes; Type I error

11. Yes; Type II error

Section 10.6 Exercises

1. $\chi^2_{0.950} = 15.379$, $\chi^2 \leq 15.379$

3. $\chi^2_{0.990} = 22.164$, $\chi^2 \leq 22.164$

5. $\chi^2_{0.100} = 40.256$, $\chi^2 \geq 40.256$

7. $\chi^2_{0.975} = 12.401$, $\chi^2_{0.025} = 39.364$
$\chi^2 \leq 12.401$, $\chi^2 \geq 39.364$

9. $\chi^2_{0.950} = 51.739$, $\chi^2_{0.050} = 90.531$
$\chi^2 \leq 51.739$, $\chi^2 \geq 90.531$

11. **a.** H_0: $\sigma \leq 3$, H_a: $\sigma > 3$
b. Reject H_0 if $\chi^2 \geq 36.415$
c. $\chi^2 = 30.827$
d. Fail to reject H_0; thus the evidence does not support the claim that the bags should fail inspection at the 0.05 level of significance.

13. **a.** H_0: $\sigma^2 \geq 0.05$, H_a: $\sigma^2 < 0.05$
b. Reject H_0 if $\chi^2 \leq 10.085$
c. $\chi^2 = 8.923$
d. Reject H_0; thus the evidence supports the claim that the variance of the standard test scores at district schools is less than 0.05 at the 0.10 level of significance.

15. **a.** H_0: $\sigma^2 \geq 0.3$, H_a: $\sigma < 0.3$
b. Reject H_0 if $\chi^2 \leq 2.167$
c. $\chi^2 = 5.833$
d. Fail to reject H_0; thus the evidence does not support the claim that the new machine produces dough with a variance between batches of less than 0.3, at the 0.05 level of significance.

17. **a.** H_0: $\sigma^2 = 3.8$, H_a: $\sigma^2 \neq 3.8$
b. Reject H_0 if $\chi^2 \leq 13.121$ or $\chi^2 \geq 52.336$
c. $\chi^2 = 64.182$
d. Reject H_0; thus the evidence supports the assistant manager's claim that the variance in the water temperature is not 3.8, at the 0.01 level of significance.

Section 10.7 Exercises

1. H_0: The proportion of customers does not vary by day of the week
H_a: The proportion of customers does vary by day of the week

3. H_0: The proportion of customers does not vary by day of the week
H_a: The proportion of customers does vary by day of the week

5. H_0: The proportion of customers' cola tastes does not vary
H_a: The proportion of customers' cola tastes does vary

7. 4.907

9. 3.379

11. 1.061

13. 32.671

15. 20.090

17. 31.319, Fail to reject H_0

19. 41.337, Reject H_0

21. 51.805, Fail to reject H_0

23. **a.** H_0: $p_1 = 0.2, p_2 = 0.4, p_3 = 0.1, p_4 = 0.3$
H_a: At least one proportion differs from these values.
b. 1010, 2020, 505, 1515
c. $\chi^2 = 34.649$
d. $\chi^2_\alpha = 7.815$. Reject H_0; the evidence supports the alternative hypothesis that the proportion of colors is different from what the manufacturer claims.

25. **a.** H_0: The proportion of customers does not vary by day of the week
H_a: The proportion of customers does vary by day of the week
b. 105.571 for each day
c. $\chi^2 = 5.984$
d. $\chi^2_\alpha = 10.645$
Fail to reject H_0; the evidence does not support the claim that the proportion of customers does vary by day of the week.

27. **a.** H_0: The proportion of visitors to each animal does not vary
H_a: The proportion of visitors to each animal does vary
b. 155.429 for each exhibit
c. $\chi^2 = 14.680$
d. $\chi^2_\alpha = 12.592$
Reject H_0; the evidence supports the claim that the proportion of visitors is not the same for all animals. Some changes may need to take place.

Section 10.8 Exercises

1.

20.27	16.44	10.41	32.88
16.73	13.56	8.59	27.12

3.

8.04	13.31	4.71	6.93
8.53	14.12	5	7.35
6.82	11.29	4	5.88
5.61	9.28	3.29	4.83

5.

16.02	21.10	6.10	15.77
17.38	22.90	6.62	17.10
14.94	19.68	5.69	14.69
14.66	19.32	5.59	14.43

7. 0.005

9. 13.336

11. 7.355

13. Dependent

15. Dependent

17. Independent

19. **a.** H_0: The grades, particular subjects and gender are independent
H_a: The grades, particular subjects and gender are dependent

b.

10.75	33.75	38.75	20	6.75
10.75	33.75	38.75	20	6.75
10.75	33.75	38.75	20	6.75
10.75	33.75	38.75	20	6.75

c. $\chi^2 = 34.225$
d. $\chi^2_\alpha = 28.299$, The grade and subject and gender are dependent

21. **a.** H_0: Cola preference and age are independent
H_a: Cola preference and age are dependent

b.

98.20	102.34	100.46
92.00	95.88	94.12
70.80	73.78	72.42

c. $\chi^2 = 1.927$
d. $\chi^2_\alpha = 14.860$,
Cola preference and age are independent.

23. **a.** H_0: Book preference and gender are independent
H_a: Book preference and gender are dependent

b.

27.29	40.48	31.39	21.83
32.71	48.52	37.61	26.17

c. $\chi^2 = 10.418$
d. $\chi^2_\alpha = 12.838$, Book preference and gender are independent

Chapter 10 Exercises

1. H_0: $\mu = 42$; H_a: $\mu \neq 42$
Use a t-test statistic;
Reject H_0 if $|t| \geq 2.779$
$t = -1.949$
Fail to reject H_0; based on the sample chosen, the manufacturer can be 99% confident that the barrels meet the necessary standards.

3. H_0: $\mu \geq 6.83$; H_a: $\mu < 6.83$
Use a t-test and reject H_0 if $t \leq -1.415$
$t = -0.974$
Fail to reject H_0; based on the sample chosen, the agency does not have enough evidence to be 90% confident that the new jumbo jet will decrease the average trans-Atlantic flight time from Newark to London Heathrow.

5. H_0: $\mu \leq 264$; H_a: $\mu > 264$
Use a z-test and reject H_0 if $p \leq 0.01$
$z = 2.93$, $p = 0.0017$
Reject H_0; this evidence supports the claim, at the 0.01 level of significance, that the usable life of the batteries is increased with the new process.

7. H_0: $p \leq 0.50$; H_a: $p > 0.50$
Use a z-test statistic;
Reject H_0 if $p \leq 0.10$
$z = 0.26$, $p = 0.3974$
a. Fail to reject H_0; thus this evidence does not support Bren's claim at the 0.10 level of significance that more than half of all sophomores prefer off-campus housing.
b. Bren failed to reject a true null hypothesis, so she made a correct decision.

9. H_0: $p = 0.35$; H_a: $p \neq 0.35$
Use a z-test statistic;
Reject H_0 if $p \leq 0.01$
$z = 1.26$, $p = 0.2076$
a. Fail to reject H_0; thus this evidence does not support the organization's claim at the 0.01 level of significance that the percentage is different from 35%.
b. The organization failed to reject a false null hypothesis, so they committed a Type II error.

11. H_0: $\sigma^2 \leq 15.5$; H_a: $\sigma^2 > 15.5$
Use a χ^2 test statistic;
Reject H_0 if $\chi^2 \geq 36.415$
$\chi^2 = 26.787$
Fail to reject H_0; thus this evidence does not support the historian's claim at the 0.05 level of significance.

13. H_0: There is no difference in the number of calls per beat.
H_a: There is a difference in the number of calls per beat.
Use a χ^2 test statistic; Reject H_0 if $\chi^2 > 7.815$
$\chi^2 = 6.414$
Fail to reject H_0; the evidence does not support the claim that there is a difference in the number of calls per beat, at the 0.05 level of significance.

15. H_0: Education levels and regions in the U.S. are independent
H_a: Education levels and regions in the U.S. are not independent
Use a χ^2 test statistic; Reject H_0 if $\chi^2 > 16.919$
$\chi^2 = 4.997$
Fail to reject H_0; the evidence does not support the claim that education levels and regions in the U.S. are not independent.

✦ Chapter 11

Section 11.1 Exercises

1. 1.16

3. –0.88

5. H_0: $\mu_1 - \mu_2 \geq -30$
H_a: $\mu_1 - \mu_2 < -30$

7. H_0: $\mu_1 - \mu_2 \leq 4$
H_a: $\mu_1 - \mu_2 > 4$

9. H_0: $\mu_1 - \mu_2 \leq 0$; H_a: $\mu_1 - \mu_2 > 0$
$z = 2.31, p = 0.0104$
Reject H_0; there is sufficient evidence at the 0.05 level to support the car company's claim that its new SUV has better gas mileage than its competitor.

11. H_0: $\mu_1 - \mu_2 \leq 3$; H_a: $\mu_1 - \mu_2 > 3$
$z = 1.69, p = 0.0455$
Reject H_0; there is sufficient evidence at the 0.05 level to support the claim that residents at Rob's hospital work more than 3 hours a week more than those at Phil's.

13. H_0: $\mu_1 - \mu_2 \leq 0$; H_a: $\mu_1 - \mu_2 > 0$
$z = 1.49, p = 0.0681$
Fail to reject H_0; there is not sufficient evidence at the 0.05 level to say that clients lose more weight with the company's help than without it.

15. H_0: $\mu_1 - \mu_2 \geq -5$;
H_a: $\mu_1 - \mu_2 < -5$
$z = -1.67, p = 0.0475$
Reject H_0; there is sufficient evidence at the 0.05 level of significance to say that evening classes average more than 5 points less than morning classes on art history tests.

17. H_0: $\mu_1 - \mu_2 \geq 0$, H_a: $\mu_1 - \mu_2 < 0$
$z = -0.80, p = 0.2119$
Fail to reject H_0; there is not sufficient evidence at the 10% level of significance to say that first-born children did more poorly on the ACT than second-born children.

19. H_0: $\mu_1 - \mu_2 \leq 0$; H_a: $\mu_1 - \mu_2 > 0$
$z = 0.66, p = 0.2546$
Fail to reject H_0; there is not enough evidence at the 0.01 level to say that Lauren's area has higher electric bills.

21. H_0: $\mu_1 - \mu_2 = 0$, H_a: $\mu_1 - \mu_2 \neq 0$
$z = 1.33, p = 0.1836$
Fail to reject H_0; there is not sufficient evidence at the 0.10 level to say there is a difference in the batting averages of the SEC East and SEC West baseball players.

Section 11.2 Exercises

1. $t = 1.255$, $d.f. = 17$

3. $t = 2.043$, $d.f. = 23$

5. H_0: $\mu_1 - \mu_2 = 0$; H_a: $\mu_1 - \mu_2 \neq 0$
Reject H_0 if $|t| \geq 2.060$
$t = -0.785$
Fail to reject H_0; at the 0.05 level, there is not sufficient evidence to say that the classes are performing differently.

7. H_0: $\mu_1 - \mu_2 = 0$; H_a: $\mu_1 - \mu_2 \neq 0$
Reject H_0 if $|t| \geq 3.365$
$t = 2.698$
Fail to reject H_0; at the 0.02 level, there is not sufficient evidence to say that briquets from Bag A are larger than those in Bag B.

9. H_0: $\mu_1 - \mu_2 \leq 0$; H_a: $\mu_1 - \mu_2 > 0$
Reject H_0 if $t \geq 1.397$
$t = 0.797$
Fail to reject H_0; at the 0.10 level, there is not sufficient evidence to say say that there is a reduction in time to paint a room using the new tool.

11. H_0: $\mu_1 - \mu_2 \geq 0$; H_a: $\mu_1 - \mu_2 < 0$
Reject H_0 if $t \leq -1.345$
$t = -0.447$
Fail to reject H_0; at the 0.10 level, there is not sufficient evidence to that the bags from line A weigh less than bags in line B.

13. H_0: $\mu_1 - \mu_2 \leq 0$; H_a: $\mu_1 - \mu_2 > 0$
Reject H_0 if $|t| \geq 2.492$
$t = 3.238$
Reject H_0; at the 0.01 level, there is sufficient evidence to say that Praxor lowers cholesterol more efficiently than a placebo.

15. H_0: $\mu_1 - \mu_2 \geq 0$; H_a: $\mu_1 - \mu_2 < 0$
Reject H_0 if $t \leq -2.552$
$t = -0.433$
Fail to reject H_0; at the 0.01 level, there is not sufficient evidence to say that customers at company A pay less for their insurance than those at company B.

17. H_0: $\mu_1 - \mu_2 = 0$; H_a: $\mu_1 - \mu_2 \neq 0$
Reject H_0 if $|t| \geq 2.080$
$t = -1.585$
Fail to reject H_0; at the 0.05 level, there is not sufficient evidence to say that completion times are different between male and female artists.

Section 11.3 Exercises

1. H_0: $\mu_d \geq 0$, H_a: $\mu_d < 0$
Reject H_0 if $t \leq -1.796$
$\overline{d} = -1.583$, $s_d = 2.610$, $t = -2.102$
Reject H_0; there is sufficient evidence, at the 0.05 level of significance, to support the claim that participants in the anger management course will lose their temper less often.

3. H_0: $\mu_d \leq 60$, H_a: $\mu_d > 60$
Reject H_0 if $t \geq 2.896$
$\overline{d} = 81.111$, $s_d = 86.955$, $t = 0.728$
Fail to reject H_0; there is not sufficient evidence, at the 0.01 level of significance, to support the claim that the course increases a student's SAT scores by more than 60 points.

5. H_0: $\mu_d = 0$; H_a: $\mu_d \neq 0$
Reject H_0 if $|t| \geq 1.860$
$\overline{d} = 2.889$, $s_d = 5.349$, $t = 1.620$
Fail to reject H_0; there is not sufficient evidence, at the 0.10 level of significance, to support the claim that boys from the same parents do not necessarily have the same weight.

7. H_0: $\mu_d \leq 30$, H_a: $\mu_d > 30$
Reject H_0 if $t \geq 1.415$
$\overline{d} = 31.625$, $s_d = 1.506$, $t = 3.052$
Reject H_0; there is sufficient evidence, at the 0.10 level of significance, to support the teacher's claim that after a year of lessons, students will increase their stamina for standing in the correct position by more than 30 minutes.

9. H_0: $\mu_d \leq 10$, H_a: $\mu_d > 10$
Reject H_0 if $t \geq 1.895$
$\overline{d} = 11.500$, $s_d = 0.756$, $t = 5.612$
Reject H_0; the evidence does support the publisher's claim that the new textbook does increase students' scores by more than 10 points, at the 0.05 level of significance.

Section 11.4 Exercises

1. −0.79

3. −0.86

5. $H_0: p_1 - p_2 \leq 0, H_a: p_1 - p_2 > 0$
$z = 2.34, p = 0.0094$
Reject H_0; there is sufficient evidence at the 0.01 level to support the claim that single women purchase more houses than single men.

7. $H_0: p_1 - p_2 \geq 0, H_a: p_1 - p_2 < 0$
$z = -1.64, p = 0.0505$
Fail to reject H_0; there is not sufficient evidence at the 0.05 level to support the university's claim that the retention rates have improved.

9. $H_0: p_1 - p_2 \leq 0, H_a: p_1 - p_2 > 0$
$z = 0.40, p = 0.3446$
Fail to reject H_0; there is not sufficient evidence at the 0.01 level to support the wives' tale that women who eat chocolate have happier babies.

11. $H_0: p_1 - p_2 = 0; H_a: p_1 - p_2 \neq 0$
$z = 0.50, p = 0.6170$
Fail to reject H_0; there is not evidence at the 0.05 level to say that women and men are unequally likely to get speeding tickets.

Section 11.5 Exercises

1. $H_0: \sigma_1^2 \geq \sigma_2^2$
$H_a: \sigma_1^2 < \sigma_2^2$

3. $H_0: \sigma_1^2 \leq \sigma_2^2$
$H_a: \sigma_1^2 > \sigma_2^2$

5. $H_0: \sigma_1^2 = \sigma_2^2$
$H_a: \sigma_1^2 \neq \sigma_2^2$

7. 1.0379

9. 0.8211

11. $F \leq 0.4632$
Reject H_0

13. $F \leq 0.2062$
Fail to reject H_0

15. $F \geq 1.7675$
Reject H_0

17. $F \leq 0.3629$ or
$F \geq 2.5731$
Fail to reject H_0

19. $F \leq 0.4550$ or
$F \geq 2.2429$
Fail to reject H_0

21. $H_0: \sigma_1^2 \geq \sigma_2^2, H_a: \sigma_1^2 < \sigma_2^2$
Reject H_0 if $F \leq 0.3146$
$F = 0.5531$
Fail to reject H_0; at the 0.05 level of significance, there is not sufficient evidence to say that Titleist golf balls have a smaller variance of distance than the store brand.

23. $H_0: \sigma_1^2 \leq \sigma_2^2, H_a: \sigma_1^2 > \sigma_2^2$
Reject H_0 if $F \geq 3.0145$
$F = 3.1124$
Reject H_0; at the 0.10 level of significance, there is sufficient evidence to say that the variance in cholesterol levels of men is greater than that for women.

25. $H_0: \sigma_1^2 = \sigma_2^2, H_a: \sigma_1^2 \neq \sigma_2^2$
Reject H_0 if $F \leq 0.0432$ or $F \geq 23.1545$
$F = 5.2910$
Fail to reject H_0; at the 0.01 level of significance, there is not sufficient evidence to say that the variance in heart rates of smokers is different from that of non-smokers.

Section 11.6 Exercises

1. 2.7534

3. 2.6241

5. 4.0604

7.

Source	SS	DF	MS	F
Treatment	4	5	0.8	0.4364
Error	11	6	1.833	
Total	15	11		

9.

Source	SS	DF	MS	F
Treatment	50	4	12.5	1.5625
Error	24	3	8	
Total	75	7		

11.

Source	SS	DF	MS	F
Treatment	313	6	52.167	1.2421
Error	336	8	42	
Total	649	14		

13.

Source	SS	DF	MS	*F*	*P*-Value	*F* Critical
Treatment	5.7333	2	2.8667	4.7778	0.0298	2.8068
Error	7.2000	12	0.6000			
Total	12.9333	14				

The *p*-value is less than 0.10, so reject the null hypothesis. Thus there is enough evidence at the 0.10 level of significance to support the claim that there is a difference in the number of defective parts produce between the three workers.

15.

Source	SS	DF	MS	*F*	*P*-Value	*F* Critical
Treatment	10.6667	2	5.333	3.9070	0.0390	6.0129
Error	24.5714	18	1.3651			
Total	35.2381	20				

The *p*-value is greater than 0.01, thus fail to reject the null hypothesis. Therefore, there is not enough evidence, at the 0.01 level of significance, to support the claim that the average number of calls is different between shifts.

Chapter 11 Exercises

1. $H_0\colon \mu_1 - \mu_2 \geq 0, H_a\colon \mu_1 - \mu_2 < 0$
$z = -3.34, p = 0.0004$
Reject H_0; therefore, there is sufficient evidence at the 0.05 level of significance to support psychologists' claim that test anxiety does reduce a student's exam score.

3. $H_0\colon \mu_d \leq 0, H_a\colon \mu_d > 0$
Reject H_0 if $t \geq 1.833$
$t = 2.951$
Reject H_0; at the 0.05 level of significance, there is sufficient evidence to support the claim that husbands gain more weight than wives after marriage.

5. $H_0\colon \mu_1 - \mu_2 \geq 0, H_a\colon \mu_1 - \mu_2 < 0$
Reject H_0 if $t \leq -2.624$
$t = -2.818$
Reject H_0; at the 0.01 level of significance, there is sufficient evidence to say the cuts treated with the new ointment healed faster than those without medication.

7. $H_0\colon \mu_d = 0, H_a\colon \mu_d \neq 0$
Reject H_0 if $|t| \geq 3.250$
$t = 2.377$
Fail to reject H_0; at the 0.01 level of significance, there is not sufficient evidence to support the claim that single-gender classrooms do not have the same number of discipline problems as mixed-gender classrooms.

9. $H_0\colon \mu_1 - \mu_2 \geq 0, H_a\colon \mu_1 - \mu_2 < 0$
$z = -2.71, p = 0.0034$
Reject H_0; therefore, there is sufficient evidence at the 0.05 level of significance to support the psychologists' claim that people who attend church regularly are healthier than those who don't.

Index

✦ A

✦ B

✦ C

✦ D

✦ E

✦ F

G

H

I

✦ L

✦ M

✦ N

✦ O

✦ P

✦ Q

✦ R

✦ S

✦ T

✦ U

✦ V

✦ W